ESSENTIALS OF
GEOLOGY

ELEVENTH EDITION

ESSENTIALS OF
GEOLOGY

FREDERICK K. LUTGENS

EDWARD J. TARBUCK

ILLUSTRATED BY

DENNIS TASA

Prentice Hall

Boston Columbus Indianapolis New York San Francisco Upper Saddle River
Amsterdam Cape Town Dubai London Madrid Milan Munich Paris Montréal Toronto
Delhi Mexico City São Paulo Sydney Hong Kong Seoul Singapore Taipei Tokyo

Library of Congress Cataloging-in-Publication Data

Lutgens, Frederick K.
    Essentials of geology / Frederick K. Lutgens, Edward J. Tarbuck ;
illustrated by Dennis Tasa. — 11th ed.
        p. cm.
    Includes bibliographical references and index.
    ISBN 0-321-71472-5
    1. Geology—Textbooks. I. Tarbuck, Edward J. II. Title.
    QE26.3.L87 2012
    550—dc22
                                                          2010034466

Acquisitions Editor, Geology: *Andrew Dunaway*
Marketing Manager: *Maureen McLaughlin*
Project Editor: *Crissy Dudonis*
Assistant Editor: *Sean Hale*
Editorial Assistant: *Michelle White*
Marketing Assistant: *Nicola Houston*
VP/Executive Director, Development: *Carol Trueheart*
Development Editor: *Melissa Parkin*
Managing Editor, Geosciences and Chemistry: *Gina M. Cheselka*
Project Manager: *Edward Thomas*
Full Service/Composition: *Preparé, Inc.*
Production Editor, Full Service: *Francesca Monaco*
Art Director: *Marilyn Perry*
Cover and Interior Design: *Yin Ling Wong*
Senior Technical Art Specialist: *Connie Long*
Art Studio: *Spatial Graphics*
Photo Research Manager: *Elaine Soares*
Photo Researcher: *Kristin Piljay*
Senior Manufacturing and Operations Manager: *Nick Sklitsis*
Operations Specialist: *Maura Zaldivar*
Senior Media Producer: *Angela Bernhardt*
Media Producer: *Ziki Dekel*
Media Supervisor: *Liz Winer*
Associate Media Project Manager: *David Chavez*
Cover Photo: *Hikers in "The Narrows" of the Virgin River, Zion National Park, Utah. (Photo by Stacey May,*
    http://merrilyrollingalong.com)
Title page photo: *Devil's Tower, Wyoming. (Photo by Michael Collier)*

ISBN-10: 0-321-71472-5 / ISBN-13: 978-0-321-71472-5

Prentice Hall
is an imprint of

www.pearsonhighered.com

To Our Grandchildren
Allison and Lauren
Shannon, Amy, Andy, Ali, and Michael
Each is a bright promise for the future

BRIEF CONTENTS

ESSENTIALS OF GEOLOGY

The newest version of *GEODe: Essentials of Geology* interactive learning aid can be accessed from the book's Premium Website (*www.mygeoscienceplace.com*). This dynamic instructional tool reinforces key concepts using tutorials, animations, and interactive exercises. The *GEODe: Essentials of Geology* icon appears throughout the book wherever a text discussion has a corresponding *GEODe* activity on mygeoscienceplace.com.

GEODe: Essentials of Geology

CONTENTS

● Indicates sections with a corresponding GEODe activity on mygeoscienceplace.com.

4 Volcanoes and Volcanic Hazards 90

5 Weathering and Soils 122

● Indicates sections with a corresponding GEODe activity on mygeoscienceplace.com.

GEODe ESSENTIALS OF GEOLOGY

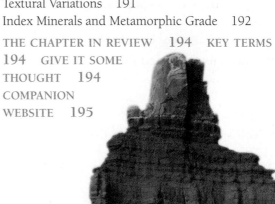

● Indicates sections with a corresponding
GEODe activity on mygeoscienceplace.com.

● Indicates sections with a corresponding GEODe activity on mygeoscienceplace.com.

The Eleventh Edition of *Essentials of Geology*, like its predecessors, is a college-level text for students taking their first and perhaps only course in geology. The book is intended to be a meaningful, nontechnical survey for people with little background in science. Usually students are taking this class to meet a portion of their college's or university's general requirements.

In addition to being informative and up-to-date, a major goal of *Essentials of Geology* is to meet the need of beginning students for a readable and user-friendly text, a book that is a highly usable tool for learning the basic principles and concepts of geology.

New Active Learning Approach

Focus on Concepts

Each chapter begins with "*Focus on Concepts*"—a series of questions that alert readers to important ideas.

FOCUS ON CONCEPTS

To assist you in learning the important concepts in this chapter, focus on the following questions:

⊙ What evidence was used to support the continental drift hypothesis?

⊙ What was one of the main objections to the continental drift hypothesis?

⊙ What is the theory of plate tectonics?

⊙ In what major way does the plate tectonics theory depart from the continental drift hypothesis?

⊙ What are the three types of plate boundaries?

⊙ Where does new lithosphere form?

⊙ How do mountain systems such as the Himalayas form?

⊙ What type of plate motion occurs along a transform fault boundary?

⊙ What evidence is used to support the plate tectonics theory?

⊙ What are the major driving forces for plate tectonics?

⊙ What models have been proposed to explain the driving mechanism for plate motion?

Concept Checks

Within the chapter each major section concludes with *"Concept Checks"* that allow students to monitor their understanding and comprehension of significant facts and ideas.

CONCEPT CHECK 15.2

❶ What was the first line of evidence that led early investigators to suspect the continents were once connected?

❷ Describe the four kinds of evidence that Wegener and his supporters gathered to substantiate the continental drift hypothesis.

❸ Explain why the discovery of the fossil remains of *Mesosaurus* in both South America and Africa, but nowhere else, supports the continental drift hypothesis.

❹ Early in the 20th century, what was the prevailing view of how land animals migrated across vast expanses of open ocean?

❺ How did Wegener account for the existence of glaciers in the southern landmasses at a time when areas in North America, Europe, and Asia supported lush tropical swamps?

Give It Some Thought

Chapters conclude with a new section called *"Give It Some Thought."* It includes six to ten questions and problems that challenge learners by involving them in activities requiring higher-order thinking skills that include the synthesis, analysis, and application of material in the chapter.

GIVE IT SOME THOUGHT

❶ After referring to the section in Chapter 1 entitled "The Nature of Scientific Inquiry" answer the following:

a. What observation led Alfred Wegener to develop his continental drift hypothesis?

b. What evidence did he gather to support his proposal?

c. Why was the continental drift hypothesis rejected by the majority of the scientific community?

d. Do you think Wegener followed the basic principles of scientific inquiry? Support your answer.

❷ Referring to the accompanying diagrams that illustrate the three types of convergent plate boundaries, complete the following:

a. Identify each type of convergent boundary.

b. Volcanic island arcs develop on what type of crust?

c. Why are volcanoes largely absent where two continental blocks collide?

d. Describe two ways that oceanic–oceanic convergent boundaries are different from oceanic–continental boundaries? How are they similar?

A.

B.

C.

Distinguishing Features

Readability

The language of this book is straightforward and *written to be understood*. Clear, readable discussions with a minimum of technical language are the rule. The frequent headings and subheadings help students follow discussions and identify the important ideas presented in each chapter. In the Eleventh Edition, we have improved readability by examining the chapter organization and flow and by writing in a more personal style. Significant portions of several chapters were substantially rewritten in an effort to make the material more understandable.

Maintaining a Focus on Basic Principles and Instructor Flexibility

The main focus of the Eleventh Edition remains the same as in the first ten: to foster student understanding of basic geological principles. As much as possible, we have attempted to provide the reader with a sense of the observational techniques and reasoning processes that constitute the discipline of geology.

The organization of the text remains intentionally traditional. Following the overview of geology in Chapter 1, we turn to a discussion of Earth materials and the related processes of volcanism and weathering. Next, we explore the geological work of gravity, water, wind, and ice in modifying and sculpting landscapes. After this look at external processes, we examine Earth's internal structure and the processes that deform rocks and give rise to mountains. The text concludes with chapters on geologic time, Earth history, and global climate change. This organization accommodates the study of minerals and rocks in the laboratory, which usually comes early in the course.

Realizing that some instructors may prefer to organize their courses somewhat differently, we made each of the chapters self-contained so that they may be taught in a different sequence. Thus, the instructor who wishes to discuss earthquakes, plate tectonics, crustal deformation, and mountain building prior to dealing with erosional processes may do so without difficulty. We also chose to provide a brief introduction to plate tectonics in Chapter 1 so that this important and basic theory could be incorporated in appropriate places throughout the text.

A Strong Visual Component

Geology is highly visual, so art and photographs play a critical role in an introductory textbook. As in the ten previous editions Dennis Tasa, a gifted artist and respected geological illustrator, has worked closely with the authors to plan and produce the diagrams, maps, graphs, and sketches that are so basic to student understanding. Dozens of the figures are new or redrawn. The result is art that is clearer and easier to understand.

Our aim is to get the *maximum effectiveness* from the visual component of the book. Michael Collier aided us greatly in this quest. Dozens of his extraordinary aerial photographs are used in the new edition. Michael is an award-winning geologist-photographer. Among his many awards is the American Geological Institute Award for Outstanding Contribution to the Public Understanding of the Geosciences. We are fortunate to have had Michael's assistance in *Essentials of Geology, Eleventh Edition.*

New to the Eleventh Edition

- Significant updating and revision of content. With the goal of keeping the text current and highly readable for beginning students many discussions, case studies, and examples have been revised. See the section on *Revised and Updated Content* for many of the particulars.

- An all-new chapter on *Global Climate Change.* Chapter 20 looks at many links between geology and the climate system. It examines how climate change is detected and, after an overview of atmospheric "basics," explores several important natural causes of climate change. The chapter concludes with an up-to-date examination of human impact on global climate—a critical global environmental issue. The chapter includes a rich array of graphs, diagrams, maps, and images.

- An even stronger art program. Dozens of figures are new or redrawn. The result is art that is clearer and easier to understand. Numerous diagrams and maps are paired with photographs for greater effectiveness. Many new and revised art pieces also have additional labels that "narrate" the process being illustrated and/or "guide" readers as they examine the figure.

- More than 150 new, high-quality photos and satellite images. New "Geologist's Sketch" illustrations accompany many important photographs and satellite images. Each sketch, which resembles what a geologist might put in a field notebook, helps the student identify important and sometimes subtle aspects of an image. This author–artist collaboration helps make an already strong visual component even stronger and more effective.

- An improved design. Anyone familiar with the previous edition will notice that we have continued the user-friendly "look." The design makes even more effective use of the wider pages. The placement and integration of photographs and art is less traditional and more creative. The scattering of *Did You Know?* items through each chapter adds interest. The authors and editors feel the improved design provides a more effective and engaging visual impact.

Revised and Updated Content

The Eleventh Edition of *Essentials of Geology* represents a thorough revision. Every part of the book was examined carefully with the dual goals of keeping topics current and improving the clarity of text discussions. Those familiar with the previous edition will see much that is new. The list of specifics is long. Examples include the following:

- Much of Chapter 2, "Matter and Minerals," was rewritten in an effort to make a sometimes difficult (chemistry-oriented) topic easier for the beginning student to understand. Discussions of basic atomic structure, bonding, crystals and crystallization, compositional and structural variations, and the silicates received the strongest attention.

- Chapter 3, "Igneous Rocks and Intrusive Activity," was formerly titled "Igneous Rocks." The new chapter title reflects the reorganization that has occurred. The discussions of intrusive activity that were formerly in Chapter 4 are now the concluding portion of Chapter 3. They logically follow the sections on the "Origin of Magma" and "How Magma Evolves." The entire discussion of "Intrusive Igneous Activity" has been revised and rewritten as has the section "Igneous Textures: What Can They Tell Us?"

- In Chapter 4, "Volcanoes and Volcanic Hazards," the discussion of pyroclastic flows is completely revised and rewritten, as is the section on calderas.

- Chapter 7, "Metamorphism and Metamorphic Rocks," includes revised and rewritten sections on "What Drives Metamorphism?" and "Metamorphic Environments."

- There is much that is new in Chapter 8, "Mass Wasting: The Work of Gravity," including a new section on *The Potential for Landslides* and new material on the earthquake-triggered landslides that devastated parts of China in May 2008.

- Among the chapters that focus on external processes Chapter 9, "Running Water," received the most attention. Discussions progress in a clearer, more logical manner for the beginning student. Topics that received special attention include the nature of drainage basins, streamflow characteristics, shaping stream valleys, and bedrock vs. alluvial channels.

- Chapter 10, "Groundwater," begins with a new introduction and has an *all-new* section on *Groundwater—A Basic Resource.*

- Chapter 11, "Glaciers and Glaciation," includes a new section on *Glaciers in Retreat,* and a new discussion devoted to *Greenland's Glacial Budget.*

- Chapter 13, "Shorelines," has new material on rip currents and a completely revised and rewritten section on "Hurricanes—The Ultimate Coastal Hazard" that includes an *all-new* discussion on *Profile of a Hurricane.*

- Chapter 14, "Earthquakes and Earth's Interior," begins with *An Earthquake Disaster in Haiti.* A new section on *Earthquakes and Faults* explores the important relationship between these phenomena. Sections devoted to magnitude scales and earthquake prediction were substantially revised.

- Much is new and updated in Chapter 15, "Plate Tectonics: A Scientific Revolution Unfolds." In addition to new introductory material, several sections were substantially revised including those dealing with fossil evidence for continental drift and forces that drive plate motions.

- The first portion of Chapter 17 on *Crustal Deformation* has been completely reorganized and substantially rewritten. In the second part of the chapter the section on "Collisional Mountain Belts" was revised.

- In Chapter 18, "Geologic Time," the updated section on the geologic time scale reflects the most recent changes to this basic tool, while a new section clarifies some of the terminology associated with the time scale.

The Teaching and Learning Package
For the Instructor

Instructor's Resource Center (IRC) on DVD The IRC puts your lecture resources all in one easy-to-reach place:

- All of the line art, tables, and photos from the text in .jpg files
- Animations of more than 100 key geological processes
- Three PowerPoint® presentations for each chapter
- Images of Earth photo gallery
- Instructor's Manual in Microsoft Word
- Test Bank in Microsoft Word
- TestGen test generation and management software
- Digital Transparency Acetates

Animations The *Geoscience Animation Library* includes more than 100 animations illustrating many difficult-to-visualize topics. Created through a unique collaboration among five of Prentice Hall's leading geoscience authors, these animations represent a significant leap forward in lecture presentation aids.

Available on CD-ROM and from the Instructor Resource Center for download, the animations are provided as Flash files and as links in PowerPoint® slides for both Windows and Mac. The **Fifth Edition** animations now include audio narration and a text transcript, with new controls that enable instructors to turn the text and audio on or off. Student versions of the animations are included on the mygeoscienceplace website, along with associate assessments making them easily assignable and accessible for the first time.

PowerPoints Found on the IRC are three PowerPoint files for each chapter. Cut down on your preparation time, no matter what your lecture needs:

1. **Art and Photos** – All of the line art, tables, and photos from the text, preloaded into PowerPoint slides for easy integration into your presentation
2. **Lecture Outline** – This set averages 35 slides per chapter and includes customizable lecture outlines with supporting art.
3. **Classroom Response System ("Clicker") Questions** – These PowerPoint presentation quizzes are authored for use in conjunction with any classroom response system.

"Images of Earth" Photo Gallery Supplement your personal and text-specific slides with this amazing collection of over 300 geologic photos contributed by Marli Miller (University of Oregon) and other professionals in the field. The photos are grouped by geologic concept and available on the IRC on DVD.

Instructor's Manual with Printed Test Bank The *Instructor's Manual* contains learning objectives, chapter outlines, answers to Concept Check questions, and suggested, short demonstrations to enhance your lecture. The *Printed Test Bank* incorporates art and averages 75 multiple-choice, true/false, short answer, and critical thinking questions per chapter.

TestGen Available on the IRC, use this electronic version of the *Test Bank* to customize and manage your tests. Create multiple versions, add or edit questions, add illustrations – your customization needs are easily addressed by this powerful software.

Course Management Prentice Hall offers instructor and student media for the Eleventh Edition of *Essentials of Geology* in formats compatible with Blackboard and other course management platforms. Contact your local sales representative for more information.

For the Student

The student resources to accompany *Essentials of Geology, Eleventh Edition* have been further refined with the goal of focusing students' efforts and improving their understanding of the concepts of geology.

MyGeosciencePlace The Lutgens *Essentials of Geology* premium website on www.mygeoscienceplace.com has been completely revamped. It now houses over 100 animations from Geoscience Animation Library, GEODe: Essentials of Geology; and a Pearson eText. By migrating these three dynamic media assets online we have made them all easily assignable and allowed student responses to be easily assessed. Chapter quizzes, web links, and flashcards are all included for further study.

A 24/7 Personal Study Portal
mygeoscienceplace is the web portal of the various Companion Websites for Pearson's geoscience textbooks. Offering a variety of resources for students and professors, each Companion Website is unique to the book and the course. Our most popular features include: eText versions of the textbook with linked/integrated multimedia, interactive maps, videos, animations, quizzes and assessments, flashcards, annotated resources for further exploration, RSS Feeds, weblinks, and Class Manager & GradeTracker Gradebook functionality for instructors.

GEODe: Essentials of Geology Now available online! Somewhere between a text and a tutor, this well-received, student-tested product provides broad topic coverage and reinforces key concepts by using animations, tutorials, interactive exercises, and review activities.

Acknowledgments

Writing a college textbook requires the talents and cooperation of many individuals. We value the excellent work of Mark Watry and Teresa Tarbuck of Spring Hill College whose talents helped us improve Chapter 2, "Matter and Minerals." They helped make the chapter more readable, engaging, and up-to-date.

Working with Dennis Tasa, who is responsible for all of the text's outstanding illustrations and much of the developmental work on *GEODe: Essentials of Geology*, is always special for us. We value not only his outstanding artistic talents and imagination but his friendship as well.

Sincere thanks to Michael Collier, whose contributions as an aerial photographer and geologist added greatly to this project. Collaborating with Michael was a special pleasure.

Great thanks also go to those colleagues who prepared in-depth reviews. Their critical comments and thoughtful input helped guide our work and clearly strengthened the text. Special thanks to:

Mara Chen, Salisbury University
Ed Garnero, Arizona State University
Tim Gunderson, College of Southern Idaho
Paul Hudak, University of North Texas
Jaime Marso, Grossmont College and College of Alameda
Steven Petsch, University of Massachusetts, Amherst
Ashish Sinha, California State University, Domingues Hills
Peter Wampler, Grand Valley State University
Chi-yuen Wang, University of California at Berkeley
Ben Wolfe, Metropolitan Community College, Blue River
Jay R. Yett, Orange Coast College

As always, we want to acknowledge the team of professionals at Pearson Prentice Hall. We sincerely appreciate the company's continuing strong support for excellence and innovation. All are committed to producing the best textbooks possible. Special thanks to our geology editor, Andy Dunaway, and to our conscientious project manager, Crissy Dudonis, for a job well done. The production team led by Ed Thomas and Francesca Monaco did an outstanding job. Kristin Piljay's photo research assistance was also a great help. All are true professionals with whom we are very fortunate to be associated.

Fred Lutgens
Ed Tarbuck

1 An Introduction to Geology

THE SPECTACULAR ERUPTION OF A VOLCANO, THE TERROR brought by an earthquake, the magnificent scenery of a mountain valley, and the destruction created by a landslide all are subjects for the geologist.

The study of geology deals with many fascinating and practical questions about our physical environment. Will there soon be another great earthquake in California? What was the Ice Age like? Will there be another? Will oil be found if a well is drilled at this location?

Iceland's Eyjafjallajökull volcano erupting on April 17, 2010. The plume of volcanic ash spread over much of Europe and severely disrupted air traffic. (Photo by Joanna Vestey/Corbis)

To assist you in learning the important concepts in this chapter, focus on the following questions:

- How does physical geology differ from historical geology?
- What is the fundamental difference between uniformitarianism and catastrophism?
- What is relative dating? What are some principles of relative dating?
- How does a scientific hypothesis differ from a scientific theory?
- What are Earth's four major "spheres"?
- Why can Earth be regarded as a system?
- What is the rock cycle? Which geologic interrelationships are illustrated by the cycle?
- How did Earth and the other planets in our solar system originate?
- What criteria were used to establish Earth's layered structure?
- What are the major features of the continents and ocean basins?
- What is the theory of plate tectonics? How do the three types of plate boundaries differ?

The Science of Geology

The subject of this text is **geology**, from the Greek *geo*, "Earth," and *logos*, "discourse." It is the science that pursues an understanding of planet Earth. Geology is traditionally divided into two broad areas—physical and historical. **Physical geology** examines the materials composing Earth and seeks to understand the many processes that operate beneath and upon its surface (**FIGURE 1.1**). The aim of **historical geology**, on the other hand, is to understand the origin of Earth and its development through time. Thus, it strives to establish a chronological arrangement of the multitude of physical and biological changes that have occurred in the geologic past. The study of physical geology logically precedes the study of Earth history because we must first understand how Earth works before we attempt to unravel its past. It should also be pointed out that physical and historical geology are divided into many areas of specialization. **TABLE 1.1** provides a partial list.

TABLE 1.1

Different Areas of Geologic Study*

Archaeological Geology	Ocean Sciences
Biogeosciences	Paleoclimatology
Engineering Geology	Paleontology
Forensic Geology	Petrology
Geochemistry	Planetary Geology
Geomorphology	Sedimentary Geology
Geophysics	Seismology
History of Geology	Structural Geology
Hydrogeology	Tectonics
Medical Geology	Volcanology
Mineralogy	

* Many of these areas of study represent interest sections and specialties of associated societies affiliated with the Geological Society of America (**www.geosociety.org**) and the American Geophysical Union (**www.agu.org**), two professional societies to which many geologists belong.

Every chapter of this book represents one or more areas of specialization in geology.

To understand Earth is challenging because our planet is a dynamic body with many interacting parts and a complex history. Throughout its long existence, Earth has been changing. In fact, it is changing as you read this page and will continue to do so into the foreseeable future. Sometimes the

FIGURE 1.1 Rocks contain information about the processes that produce them. This large exposure of igneous rock in California's Sierra Nevada was once a molten mass found deep within Earth. Geologists study the processes that create and modify mountains such as these. *(Photo by Brian Bailey/Aurora Photos, Inc.)*

changes are rapid and violent, as when landslides or volcanic eruptions occur. Just as often, change takes place so slowly that it goes unnoticed during a lifetime. Scales of size and space also vary greatly among the phenomena that geologists study. Sometimes they must focus on phenomena that are submicroscopic, and at other times they must deal with features that are continental or global in scale.

Geology is perceived as a science that is done in the out of doors, and rightly so. A great deal of geology is based on measurements, observations, and experiments conducted in the field. But geology is also done in the laboratory where, for example, the study of various Earth materials provides insights into many basic processes. Moreover, the development of sophisticated computer models allows for the simulation of many of our planet's complex systems. Frequently, geology requires an understanding and application of knowledge and principles from physics, chemistry, and biology. Geology is a science that seeks to expand our knowledge of the natural world and our place in it.

CONCEPT CHECK 1.1

❶ Name and distinguish between the two broad subdivisions of geology.

Geology, People, and the Environment

The primary focus of this book is to develop an understanding of basic geological principles, but along the way we will explore numerous important relationships between people and the natural environment. Many of the problems and issues addressed by geology are of practical value to people.

Natural hazards are a part of living on Earth. Every day they adversely affect millions of people worldwide and are responsible for staggering damages. Among the hazardous Earth processes studied by geologists are volcanoes, floods, earthquakes, and landslides. Of course, geologic hazards are simply *natural* processes. They become hazards only when people try to live where these processes occur. **FIGURE 1.2** illustrates this point as does the chapter-opening photo.

According to the United Nations, in 2008, for the first time, more people lived in cities than in rural areas. This global trend toward urbanization concentrates millions of people into megacities, many of which are vulnerable to natural hazards. Coastal sites are becoming more vulnerable because development often destroys natural defenses such as wetlands and sand dunes. In addition, there is a growing threat associated with human influences on the Earth system such as sea level rise that is linked to global climate change.* Other megacities are exposed to seismic (earthquake) and volcanic hazards where inappropriate land use and poor construction practices, coupled with rapid population growth, are increasing vulnerability.

Resources represent another important focus of geology that is of great practical value to people. They include water and soil, a great variety of metallic and nonmetallic minerals, and energy (**FIGURE 1.3**). Together they form the very foundation of modern civilization. Geology deals not only with the formation and occurrence of these vital resources but also with maintaining supplies and the environmental impact of their extraction and use.

Complicating all environmental issues is rapid world population growth and everyone's aspiration to a better standard of living. This means a ballooning demand for resources and a growing pressure for people to dwell in environments having significant geologic hazards.

Not only do geologic processes have an impact on people, but we humans can dramatically influence geologic processes as well. For example, river flooding is natural, but the magnitude and frequency of flooding can be changed significantly by human activities such

DID YOU KNOW?
Each year an average American requires huge quantities of Earth materials. Imagine receiving your annual share in a single delivery. A large truck would pull up to your home and unload 12,965 lbs. of stone, 8945 lbs. of sand and gravel, 895 lbs. of cement, 395 lbs. of salt, 361 lbs. of phosphate, and 974 lbs. of other nonmetals. In addition, there would be 709 lbs. of metals, including iron, aluminum, and copper.

*The idea of the Earth system is explored later in the chapter. Global climate change and its effects are the focus of Chapter 20.

FIGURE 1.2 Rescue workers search for victims and survivors after this apartment complex collapsed during a magnitude 8.8 earthquake that struck Chile on February 27, 2010. Geologic hazards are *natural* processes. They only become hazards when people try to live where these processes occur. *(Photo by Jose Luis Saavedra/Reuters/Corbis)*

FIGURE 1.3 The open pit copper mine at Morenci, Arizona, is one of the largest producers of copper in the United States. When demand for copper is strong, the mine operates nonstop, processing 700,000 tons of rock each day and producing about 840 million pounds of copper per year. *(Photo by Michael Collier)*

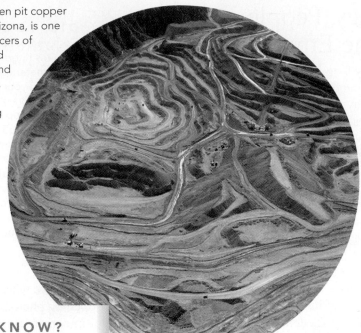

FIGURE 1.4 Record-setting floods along the Red River of the North in late March 2009, in the Fargo, North Dakota/Moorhead, Minnesota area. Natural hazards are part of living on Earth. *(Photo by Scott Olson/Getty Images)*

as clearing forests, building cities, and constructing dams (**FIGURE 1.4**). Unfortunately, natural systems do not always adjust to artificial changes in ways that we can anticipate. Thus, an alteration to the environment that was intended to benefit society sometimes has the opposite effect.

CONCEPT CHECK 1.2

① List at least three different geologic hazards.

Historical Notes about Geology

The nature of our Earth—its materials and its processes—has been a focus of study for centuries. Writings about fossils, gems, earthquakes, and volcanoes date back to the Greeks, more than 2300 years ago. Certainly, the most influential Greek philosopher was Aristotle. Unfortunately, Aristotle's explanations about the natural world were not derived from keen observations and experiments, as is modern science. Instead, they were arbitrary pronouncements based on the limited knowledge of his day. He believed that rocks were created under the "influence" of the stars and that earthquakes occurred when air in the ground was heated by central fires and escaped explosively! When confronted with a fossil fish, he explained that "a great many fishes live in the earth motionless and are found when excavations are made." Although Aristotle's explanations may have been adequate for his day, they unfortunately continued to be expounded for many centuries, thus thwarting the acceptance of more up-to-date ideas.

Catastrophism

In the mid-1600s James Ussher, Anglican Archbishop of Armagh, Primate of all Ireland, published a major work that had immediate and profound influences. A respected scholar of the Bible, Ussher constructed a chronology of human and Earth history in which he determined that Earth was only a few thousand years old, having been created in 4004 BC. Ussher's treatise earned widespread acceptance among Europe's scientific and religious leaders, and his chronology was soon printed in the margins of the Bible itself.

During the seventeenth and eighteenth centuries the doctrine of **catastrophism** strongly influenced people's thinking about Earth. Briefly stated, catastrophists believed that Earth's landscapes had been shaped primarily by great catastrophes. Features such as mountains and canyons, which today we know take great periods of time to form, were explained as having been produced by sudden and often worldwide

FIGURE 1.5 James Hutton (1726–1797) a founder of modern geology. *(Photo courtesy of the Natural History Museum, London)*

events. Hutton carefully cited verifiable observations to support his ideas.

For example, when he argued that mountains are sculpted and ultimately destroyed by weathering and the work of running water, and that their wastes are carried to the oceans by processes that can be observed, Hutton said, "We have a chain of facts which clearly demonstrates . . . that the materials of the wasted mountains have traveled through the rivers"; and further, "There is not one step in all this progress . . . that is not to be actually perceived." He then went on to summarize this thought by asking a question and immediately providing the answer: "What more can we require? Nothing but time."

Geology Today

Today the basic tenets of uniformitarianism are just as viable as in Hutton's day. We realize more strongly than ever that the present gives us insight into the past and that the physical, chemical, and biological laws that govern geological processes remain unchanged through time. However, we also understand that the doctrine should not be taken too literally. To say that geological processes in the past were the same as those occurring today is not to suggest that they always had the same relative importance or operated at precisely the same rate. Moreover, some important geologic processes are not currently observable, but evidence that they occur is well established. For example, we know that Earth has experienced impacts from large meteorites even though we have no

disasters produced by unknown causes that no longer operate. This philosophy was an attempt to fit the rates of Earth processes to the then current ideas on the age of Earth.

The Birth of Modern Geology

Against this backdrop of Aristotle's views and an Earth created in 4004 BC, a Scottish physician and gentleman farmer named James Hutton published *Theory of the Earth* in 1795 (**FIGURE 1.5**). In this work Hutton put forth a fundamental principle that is a pillar of geology today: **uniformitarianism**. It states that the *physical, chemical, and biological laws that operate today also operated in the geologic past.* In other words, the forces and processes that we observe shaping our planet today have been at work for a very long time. Thus, to understand ancient rocks, we must first understand present-day processes and their results. This idea is commonly stated as *the present is the key to the past.*

Prior to Hutton's *Theory of the Earth*, no one had effectively demonstrated that geological processes can continue over extremely long periods of time. Hutton persuasively argued that forces that appear small could, over long spans of time, produce effects just as great as those resulting from sudden catastrophic

This peak is in the Andes of South America. *(Photo by Fotosearch/ photolibrary.com)*

geology

human witnesses. Such events altered Earth's crust, modified its climate, and strongly influenced life on the planet.

The acceptance of uniformitarianism meant the acceptance of a very long history for Earth. Although processes vary in their intensity, they still take a very long time to create or destroy major landscape features (**FIGURE 1.6**). For example, geologists have established that mountains once existed in portions of present-day Minnesota, Wisconsin, Michigan, and Manitoba. Today the region consists of low hills and plains.

Erosion (processes that wear land away) gradually destroyed these peaks. The rock record contains evidence that shows Earth has experienced many such cycles of mountain building and erosion.

In the chapters that follow we will be examining the materials that compose our planet and the processes that modify it. It is important to remember that although many features of our physical landscape may seem to be unchanging in terms of the decades over which we observe them, they are nevertheless changing, but on time scales of hundreds, thousands, or even many millions of years. Concerning the ever changing nature of Earth through great expanses of geologic time, James Hutton made a statement that was to become his most famous. In concluding his classic 1788 paper, published in the *Transactions of the Royal Society of Edinburgh*, he stated, "The result, therefore, of our present enquiry is, that we find no vestige of a beginning, no prospect of an end."

❶ Describe Aristotle's influence on geology.

❷ Contrast catastrophism and uniformitarianism. How did each view the age of Earth?

Geologic Time

Although Hutton and others recognized that geologic time is exceedingly long, they had no methods to accurately determine the age of Earth. However, in 1896 radioactivity was discovered. Using radioactivity for dating was first attempted in 1905 and has been refined ever since. Geologists are now able to assign fairly accurate dates to events in Earth history.* For example, we know that the dinosaurs died out about 65 million years ago. Today the age of Earth is put at about 4.6 billion years.

The Magnitude of Geologic Time

The concept of geologic time is new to many nongeologists. People are accustomed to dealing with increments of time that are measured in hours, days, weeks, and years. Our history books often examine events over spans of centuries, but even a century is difficult to appreciate fully. For most of us, someone or something that is 90 years old is *very old*, and a 1000-year-old artifact is *ancient*.

By contrast, those who study geology must routinely deal with vast time periods—millions or billions (thousands of millions) of years. When viewed in the context of Earth's 4.6-billion-year history, a geologic event that occurred 100 million years ago may be characterized as "recent" by a geologist, and a rock sample that has been dated at 10 million years may be called "young." An appreciation for the magnitude of geologic time is important in the study of geology because many processes are so gradual that vast spans of time are needed before significant changes occur.

FIGURE 1.6 The Grand Canyon of the Colorado River in northern Arizona. The rocks exposed here represent hundreds of millions of years of Earth history. It also took millions of years for weathering and erosion to create the canyon. Geologic processes often act so slowly that changes may not be visible during an entire human lifetime. The relative ages of the rock layers in the canyon can be determined by applying the law of superposition. The youngest rocks are on top, and the oldest are at the bottom. Geologists must routinely deal with ancient materials and events that occurred in the distant geologic past. *(Photo by Jens Hilberger/agefotostock)*

*Chapter 18 is devoted to a much more complete discussion of geologic time, and Chapter 19 presents an overview of Earth history.

A.

B.

FIGURE 1.7 Fossils are important tools for the geologist. In addition to being very important in relative dating, fossils can be useful environmental indicators. **A.** *Archaeopteryx*, a primitive bird that lived during the Jurassic Period (see Geologic Time Scale in Figure 1.8). *(Photo by Michael Collier).* **B.** A fossil fish of Eocene age from the Green River Formation in Wyoming. *(Photo by Francois Gohier/Photo Researchers, Inc.)*

How long is 4.6 billion years? If you were to begin counting at the rate of one number per second and continued 24 hours a day, 7 days a week and never stopped, it would take about two lifetimes (150 years) to reach 4.6 billion!

Relative Dating and the Geologic Time Scale

During the 19th century, long before the discovery of radioactivity, which eventually allowed for the establishment of reliable *numerical* dates, a geologic time scale was developed using principles of relative dating. **Relative dating** means that events are placed in their proper sequence or order without knowing their age in years. This is done by applying principles such as the **law of superposition.** This basic rule applies to materials that were originally deposited at Earth's surface, such as layers of sedimentary rock and volcanic lava flows. The law simply states that the youngest layer is on top, and the oldest layer is on the bottom (assuming that nothing has turned the layers upside down, which sometimes happens). Stated another way, a layer is older than the ones above it and younger than the ones below. Arizona's Grand Canyon provides a fine example where the oldest rocks are located in the inner gorge while the youngest rocks are found on the rim (See Figure 1.6). So the law of superposition establishes the *sequence* of rock layers but not, of course, their numerical ages. Today such a proposal appears to be elementary, but 300 years ago it amounted to a major breakthrough in scientific reasoning by establishing a rational basis for relative time measurements.

Fossils, the remains or traces of prehistoric life, were also essential to the development of a geologic time scale (**FIGURE 1.7**). Fossils are the basis for the **principle of fossil succession,** which states that fossil organisms succeed one another in a definite and determinable order, and therefore any time period can be recognized by its fossil content. This principle was laboriously worked out over decades by collecting fossils from countless rock layers around the world. Once established, it allowed geologists to identify rocks of the same age in widely separated places and to build the *geologic time scale* shown in **FIGURE 1.8**.

FIGURE 1.8 The geologic time scale divides the vast 4.6-billion-year history of Earth into eons, eras, periods, and epochs. We presently live in the Holocene epoch of the Quaternary period. This period is part of the Cenozoic era, which is the latest era of the Phanerozoic eon. Numbers on the time scale represent time in millions of years before the present. These dates were added long after the time scale had been established using relative dating techniques. The Precambrian accounts for more than 88 percent of geologic time. *(Data from International Commission on Stratigraphy and the U.S. Geological Survey)*

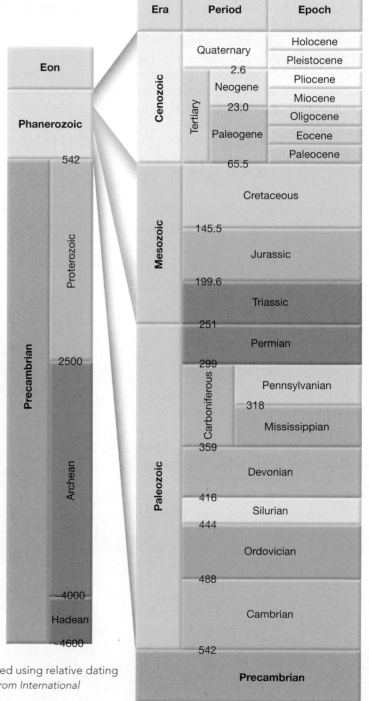

Era	Period	Epoch
	Quaternary	Holocene
		Pleistocene
		2.6
	Neogene	Pliocene
		Miocene
		23.0
	Paleogene	Oligocene
		Eocene
		Paleocene
		65.5

Cenozoic (Tertiary — Neogene/Paleogene)

Era	Period
Mesozoic	Cretaceous
	145.5
	Jurassic
	199.6
	Triassic
	251
Paleozoic	Permian
	299
Carboniferous	Pennsylvanian
	318
	Mississippian
	359
	Devonian
	416
	Silurian
	444
	Ordovician
	488
	Cambrian
	542
	Precambrian

Eon

Phanerozoic	542
Precambrian — Proterozoic	2500
Archean	
	~4000
Hadean	
	~4600

Notice that units having the same designations do not necessarily extend for the same number of years. For example, the Cambrian period lasted about 54 million years, whereas the Silurian period spanned only about 28 million years. As we will emphasize again in Chapter 18, this situation exists because the basis for establishing the time scale was not the regular rhythm of a clock, but rather the changing character of life forms through time. Specific dates were added long after the time scale was established. A glance at Figure 1.8 also reveals that the Phanerozoic eon is divided into many more units than earlier eons even though it encompasses only about 12 percent of Earth history. The meager fossil record for these earlier eons is the primary reason for the lack of detail on this portion of the time scale. Without abundant fossils, geologists lose a very important tool for subdividing geologic time.

CONCEPT CHECK 1.4

❶ How old is Earth?

❷ Describe two principles used to develop the geologic time scale.

tions about what should or should not be expected, given certain facts or circumstances. For example, by knowing how oil deposits form, geologists are able to predict the most favorable sites for exploration and, perhaps as important, to avoid regions having little or no potential.

The development of new scientific knowledge involves some basic logical processes that are universally accepted. To determine what is occurring in the natural world, scientists collect scientific *"facts"* through observation and measurement (**FIGURE 1.9**). Because some error is inevitable, the accuracy of a particular measurement or observation is always open to question. Nevertheless, these data are essential to science and serve as the springboard for the development of scientific theories.

Hypothesis

Once facts have been gathered and principles have been formulated to describe a natural phenomenon, investigators try to explain how or why things happen in the manner observed. They often do this by constructing a tentative (or untested) explanation, which is called a scientific

hypothesis. It is best if an investigator can formulate more than one hypothesis to explain a given set of observations. If an individual scientist is unable to devise multiple hypotheses others in the scientific community will almost always develop alternative explanations. A spirited debate frequently ensues. As a result, extensive research is conducted by proponents of opposing hypotheses, and the results are made available to the wider scientific community in scientific journals.

Before a hypothesis can become an accepted part of scientific knowledge, it must pass objective testing and analysis. (If a hypothesis cannot be tested, it is not scientifically useful, no matter how interesting it might seem.) The verification process requires that *predictions* be made based on the hypothesis being considered and that the predictions be tested by comparing them against objective observations of nature. Put another way, hypotheses must fit observations other than those used to formulate them in the first place. Hypotheses that fail rigorous testing are ultimately discarded. The history of science is littered with discarded hypotheses. One of the best known is the Earth-centered model of the

The Nature of Scientific Inquiry

As members of a modern society, we are constantly reminded of the benefits derived from science. But what exactly is the nature of scientific inquiry? Developing an understanding of how science is done and how scientists work is an important theme that appears throughout this book. You will explore the difficulties in gathering data and some of the ingenious methods that have been developed to overcome these difficulties. You will also see many examples of how hypotheses are formulated and tested, as well as learn about the evolution and development of some major scientific theories.

All science is based on the assumption that the natural world behaves in a consistent and predictable manner that is comprehensible through careful, systematic study. The overall goal of science is to discover the underlying patterns in nature and then to use this knowledge to make predic-

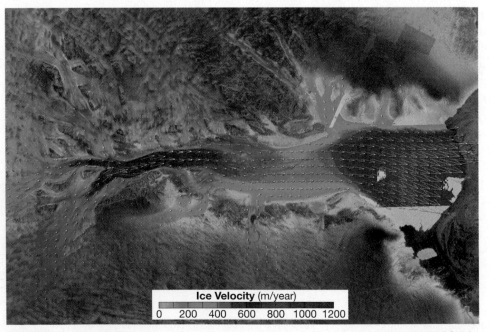

FIGURE 1.9 Scientific facts are gathered in many ways, including laboratory studies and field observations and measurements. Satellite images are another important source of data. This satellite image provides detailed information about the movement of Antarctica's Lambert Glacier. Such information is basic to understanding changes in the behavior of the glacier over time. The ice velocities are determined from pairs of images obtained 24 days apart, using a technique called radar interferometry. *(NASA)*

Ice Velocity (m/year)
0   200   400   600   800   1000   1200

universe—a proposal that was supported by the apparent daily motion of the Sun, Moon, and stars around Earth. As the mathematician Jacob Bronowski so ably stated, "Science is a great many things, but in the end they all return to this: Science is the acceptance of what works and the rejection of what does not."

Theory

When a hypothesis has survived extensive scrutiny and when competing ones have been eliminated, a hypothesis may be elevated to the status of a scientific **theory**. In everyday language we may say "That's only a theory," but a scientific theory is a well-tested and widely accepted view that the scientific community agrees best explains certain observable facts.

Some theories that are extensively documented and extremely well supported are comprehensive in scope. For example, the theory of plate tectonics, which you will learn about shortly, provides the framework for understanding the origin of mountains, earthquakes, and volcanic activity. In addition, plate tectonics explains the evolution of the continents and the ocean basins through time—ideas that are explored in some detail in later chapters.

Scientific Methods

The processes just described, in which scientists gather facts through observations and formulate scientific hypotheses and theories, is called the *scientific method*. Contrary to popular belief, the scientific method is not a standard recipe that scientists apply in a routine manner to unravel the secrets of our natural world. Rather, it is an endeavor that involves creativity and insight. Rutherford and

Ahlgren put it this way: "Inventing hypotheses or theories to imagine how the world works and then figuring out how they can be put to the test of reality is as creative as writing poetry, composing music, or designing skyscrapers."*

There is no fixed path that scientists always follow that leads unerringly to scientific knowledge. Nevertheless, many scientific investigations involve the following steps: (1) collection of scientific facts (data) through observation and measurement (**FIGURE 1.10**); (2) the formulation of questions that relate to the facts and the development of one or more working hypotheses that may answer these questions; (3) development of observations and experiments to test the hypotheses; and (4) the acceptance, modification, or rejection of the hypotheses based on extensive testing.

Other scientific discoveries may result from purely theoretical ideas that stand up to extensive examination. Some researchers use high-speed computers to simulate what is happening in the "real" world. These models are useful when dealing with natural processes that occur on very long time scales or take place in extreme or inaccessible locations. Still other scientific advancements have been made when a totally unexpected happening occurred during an experiment. These serendipitous discoveries are more than pure luck; for as Louis Pasteur stated, "In the field of observation, chance favors only the prepared mind."

Scientific knowledge is acquired through several avenues, so it might be best to describe the nature of scientific inquiry as the *methods* of science rather than *the* scientific method. In addition, it should always be remembered that even the most compelling scientific theories are still simplified explanations of the natural world.

*F. James Rutherford and Andrew Ahlgren, *Science for All Americans* (New York: Oxford University Press, 1990), p. 7.

Arizona's Monument Valley. *(Photo by Ron Chapple Stock/ photolibrary.com)*

FIGURE 1.10 Paleontologists study ancient life. Here scientists are excavating the remains of *Albertasaurus*, a large carnivore similar to *Tyrannosaurus rex*, which lived during the late Cretaceous Period. The excavation site is near Red Deer River, Alberta, Canada. The aim of historical geology is to understand the development of Earth and its life through time. Fossils are essential tools in that quest. *(Photo by Richard T. Nowitz/Science Source/Photo Researchers, Inc.)*

Do Glaciers Move? An Application of the Scientific Method

In Figure 1.9 you learned that specialized instruments aboard satellites allow us to monitor glaciers from space. This, of course, is a recent breakthrough. In the 19th century learning about the behavior of glaciers was far more difficult and laborious. The study of glaciers provides an early application of the scientific method. High in the Alps of Switzerland and France, small glaciers exist in the upper portions of some valleys. In the late 18th and early 19th centuries, people who farmed and herded animals in these valleys suggested that glaciers in the upper reaches of the valleys had previously been much larger and had occupied downvalley areas. They based their explanation on the fact that the valley floors were littered with angular boulders and other rock debris that seemed identical to the materials that they could see in and near the glaciers at the heads of the valleys.

Although the explanation of these observations seemed logical, others did not accept the notion that masses of ice hundreds of meters thick were capable of movement. The disagreement was settled after a simple experiment was designed and carried out to test the hypothesis that glacial ice can move.

Markers were placed in a straight line completely across an alpine glacier. The position of the line was marked on the valley walls so that if the ice moved, the change in position could be detected. After a year or two the results were clear. The markers on the glacier had advanced down the valley, proving that glacial ice indeed moves. In addition, the experiment demonstrated that ice within a glacier does not move at a uniform rate, because the markers in the center advanced farther than did those along the margins. Although most glaciers move too slowly for direct visual detection, the experiment succeeded in demonstrating that movement nevertheless occurs. In the years that followed, this

experiment was repeated many times with greater accuracy using more modern surveying techniques. Each time, the basic relationships established by earlier attempts were verified.

The experiment illustrated in **FIGURE 1.11** was carried out at Switzerland's Rhône Glacier later in the 19th century. It not only traced the movement of markers within the ice but also mapped the position of the glacier's terminus. Notice that even though the ice within the glacier was advancing, the ice front was retreating. As often occurs in science, experiments and observations designed to test one hypothesis yield new information that requires further analysis and explanation.

CONCEPT CHECK 1.5

1 How is a scientific hypothesis different from a scientific theory?

2 List the basic steps in many scientific investigations.

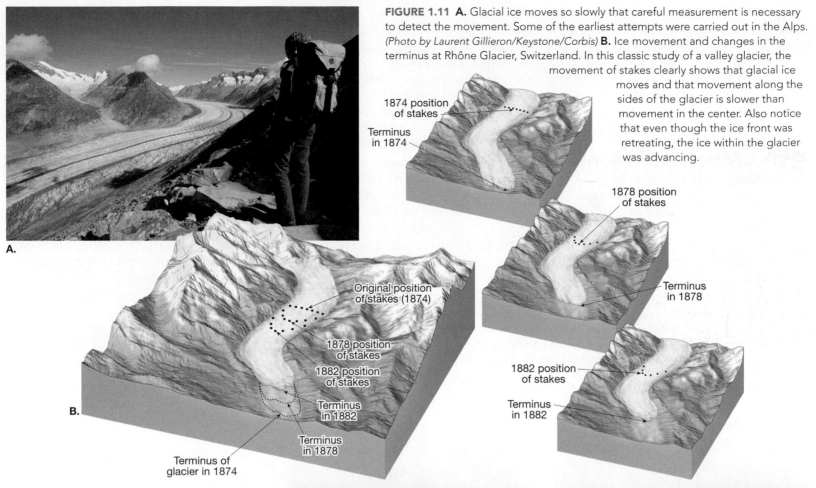

FIGURE 1.11 A. Glacial ice moves so slowly that careful measurement is necessary to detect the movement. Some of the earliest attempts were carried out in the Alps. *(Photo by Laurent Gillieron/Keystone/Corbis)* **B.** Ice movement and changes in the terminus at Rhône Glacier, Switzerland. In this classic study of a valley glacier, the movement of stakes clearly shows that glacial ice moves and that movement along the sides of the glacier is slower than movement in the center. Also notice that even though the ice front was retreating, the ice within the glacier was advancing.

1874 position of stakes

Terminus in 1874

1878 position of stakes

Terminus in 1878

1882 position of stakes

Terminus in 1882

Original position of stakes (1874)

1878 position of stakes

1882 position of stakes

Terminus in 1882

Terminus in 1878

Terminus of glacier in 1874

A.

B.

Earth's Spheres

The images in **FIGURE 1.12** are considered to be classics because they let humanity see Earth differently than even before. Figure 1.12A, known as "Earthrise," was taken when the *Apollo 8* astronauts orbited the Moon for the first time in December 1968. As the spacecraft rounded the Moon, Earth appeared to rise above the lunar surface. Figure 1.12B, referred to as "The Blue Marble," is perhaps the most widely reproduced image of Earth and was taken in December 1972 by the crew of *Apollo 17* during the last lunar mission. These early views profoundly altered our conceptualizations of Earth and remain powerful images decades after they were first viewed. Seen from space, Earth is breathtaking in its beauty and startling in its solitude. The photos remind us that our home is, after all, a planet—small, self-contained, and in some ways even fragile. Bill Anders, the *Apollo 8* astronaut who took the "Earthrise" photo, expressed it this way, "We came all

A.

B.

FIGURE 1.12 A. View, called "Earthrise," that greeted the *Apollo 8* astronauts as their spacecraft emerged from behind the Moon. *(NASA Headquarters)* **B.** Africa and Arabia are prominent in this classic image, called "The Blue Marble" taken from *Apollo 17*. The tan cloud-free zones over the land coincide with major desert regions. The band of clouds across central Africa is associated with a much wetter climate that in places sustains tropical rain forests. The dark blue of the oceans and the swirling cloud patterns remind us of the importance of the oceans and the atmosphere. Antarctica, a continent covered by glacial ice, is visible at the South Pole. *(NASA)*

this way to explore the Moon, and the most important thing is that we discovered Earth."

As we look more closely at our planet from space, it becomes apparent that Earth is much more than rock and soil. In fact, the most conspicuous features in Figure 1.12A are not continents but swirling clouds suspended above the surface and the vast global ocean. These features emphasize the importance of water to our planet.

The closer view of Earth from space shown in Figure 1.12B helps us appreciate why the physical environment is traditionally divided into three major parts: the water portion of our planet, the hydro-sphere; Earth's gaseous envelope, the atmosphere; and, of course, the solid Earth, or geosphere. It needs to be emphasized that our environment is highly integrated and is not dominated by rock, water, or air alone. Rather, it is characterized by continuous interactions as air comes in contact with rock, rock with water, and water with air. Moreover, the biosphere, which is the totality of all plant and animal life on our planet, interacts with each of the three physical realms and is an equally integral part of the planet. Thus, Earth can be thought of as consisting of four major spheres: the hydrosphere, atmosphere, geosphere, and biosphere.

The interactions among Earth's four spheres are incalculable. **FIGURE 1.13** provides us with one easy-to-visualize example. The shoreline is an obvious meeting place for rock, water, and air. In this scene, ocean waves that were created by the drag of air moving across the water are breaking against the rocky shore. The force of the water can be powerful, and the erosional work that is accomplished can be great.

Hydrosphere

Earth is sometimes called the *blue* planet or, as we saw in Figure 1.12B—"The Blue Marble." Water more than anything else

FIGURE 1.13 The shoreline is one obvious example of an *interface*—a common boundary where different parts of a system interact. In this scene, ocean waves (*hydrosphere*) that were created by the force of moving air (*atmosphere*) break against a rocky shore (*geosphere*). The force of the water can be powerful, and the erosional work that is accomplished can be great. *(Photo by Radius Images/photolibrary.com)*

makes Earth unique. The **hydrosphere** is a dynamic mass of water that is continually on the move, evaporating from the oceans to the atmosphere, precipitating back to the land, and running back to the ocean again. The global ocean is certainly the most prominent feature of the hydrosphere, blanketing nearly 71 percent of Earth's surface to an average depth of about 3800 meters (12,500 feet). It accounts for about 97 percent of Earth's water (**FIGURE 1.14**). However, the hydrosphere also includes the freshwater found underground and in streams, lakes, and glaciers. Moreover, water is an important component of all living things.

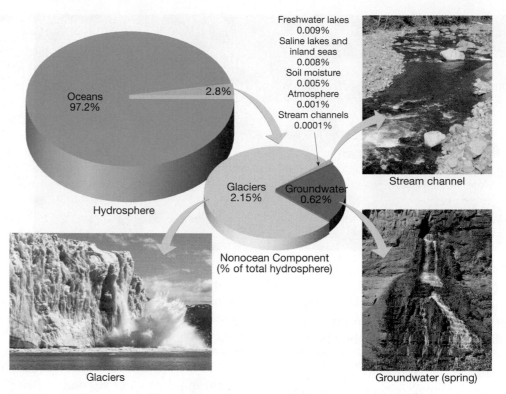

Freshwater lakes 0.009%
Saline lakes and inland seas 0.008%
Soil moisture 0.005%
Atmosphere 0.001%
Stream channels 0.0001%

Oceans 97.2%   2.8%

Hydrosphere

Glaciers 2.15%   Groundwater 0.62%

Nonocean Component (% of total hydrosphere)

Stream channel

Glaciers

Groundwater (spring)

FIGURE 1.14 Distribution of Earth's water. The oceans clearly dominate. When we consider only the nonocean component, ice sheets and glaciers represent nearly 85 percent of Earth's fresh water. Groundwater accounts for just over 14 percent. When only liquid freshwater is considered, the significance of groundwater is obvious. *(Glacier photo by Bernhard Edmaier/Photo Researchers, Inc.; stream photo by E. J. Tarbuck; and groundwater photo by Michael Collier)*

Although these latter sources constitute just a tiny fraction of the total, they are much more important than their meager percentage indicates. Streams, glaciers, and groundwater are responsible for creating many of our planet's varied landforms, as well as the freshwater that is so vital to life on land.

DID YOU KNOW?
The volume of ocean water is so large that if Earth's solid mass were perfectly smooth (level) and spherical, the oceans would cover Earth's entire surface to a uniform depth of more than 2000 m (1.2 mi)!

Atmosphere

Earth is surrounded by a life-giving gaseous envelope called the **atmosphere.** When we watch a high-flying jet plane cross the sky, it seems that the atmosphere extends upward for a great distance. However, when compared to the thickness (radius) of the solid Earth (about 6400 kilometers or 4000 miles), the atmosphere is a very shallow layer. One-half lies below an altitude of 5.6 kilometers (3.5 miles), and 90 percent occurs within just 16 kilometers (10 miles) of Earth's surface (**FIGURE 1.15**). Despite its modest dimensions, this thin blanket of air is an integral part of the planet. It not only provides the air that we breathe but also

acts to protect us from the Sun's intense heat and dangerous ultraviolet radiation. The energy exchanges that continually occur between the atmosphere and the surface and between the atmosphere and space produce the effects we call weather and climate.

If, like the Moon, Earth had no atmosphere, our planet would not only be lifeless but many of the processes and interactions that make the surface such a dynamic place could not operate. Without weathering and erosion, the face of our planet might more closely resemble the lunar surface, which has not changed appreciably in nearly 3 billion years.

FIGURE 1.15 This unique image of Earth's atmosphere merging with the emptiness of space resembles an abstract painting. It was taken over western China in June 2007 by a Space Shuttle crew member. The thin silvery streaks (called *noctilucent clouds*) high in the blue area are at a height of about 80 kilometers (50 miles). The atmosphere at this altitude is *very* thin. Air pressure here is less than a thousandth of that at sea level. The thin reddish zone in the lower portion of the image is the densest part of the atmosphere. It is here, in a layer called the *troposphere*, that practically all weather and cloud formation occur. Ninety percent of Earth's atmosphere occurs within just 16 kilometers (10 miles) of the surface. *(NASA)*

Biosphere

The **biosphere** includes all life on Earth. Ocean life is concentrated in the sunlit surface waters of the sea (**FIGURE 1.16**). Most life on land is also concentrated near the surface, with tree roots and burrowing animals reaching a few meters underground and flying insects and birds reaching a kilometer or so above Earth. A surprising variety of life forms are also adapted to extreme environments. For example, on the ocean floor where pressures are extreme and no light penetrates, there are places where vents spew hot, mineral-rich fluids that support communities of exotic life forms. On land, some bacteria thrive in rocks as deep as 4 kilometers (2.5 miles) and in boiling hot springs. Moreover, air currents can carry microorganisms many kilometers into the atmosphere. But even when we consider these extremes, life still must be thought of as being confined to a narrow band very near Earth's surface.

Plants and animals depend on the physical environment for the basics of life. However, organisms do not just respond to their physical environment. Indeed, the biosphere powerfully influences the other three spheres. Without life, the makeup and nature of the geosphere, hydrosphere, and atmosphere would be very different.

Geosphere

Beneath the atmosphere and the oceans is the solid Earth, or **geosphere**. The geosphere extends from the surface to the center of the planet, a depth of 6400 kilometers, making it by far the largest of Earth's four spheres. Much of our study of the solid Earth focuses on the more accessible surface features. Fortunately, many of these features represent the outward expressions of the dynamic behavior of Earth's interior. By examining the most prominent surface features and their global extent, we can obtain clues to the dynamic processes that have shaped our planet. A first look at the structure of Earth's interior and at the major surface features of the geosphere will come later in the chapter.

Soil, the thin veneer of material at Earth's surface that supports the growth of plants, may be thought of as part of all four spheres. The solid portion is a mixture of weathered rock debris (geosphere) and

FIGURE 1.16 The hydrosphere is home to a significant portion of Earth's biosphere. Modern coral reefs are unique and complex examples and are home to about 25 percent of all marine species. Because of this diversity, they are sometimes referred to as the ocean equivalent of rain forests. *(Photo by Darryl Leniuk/age footstock)*

DID YOU KNOW?
Primitive life first appeared in the oceans about 4 billion years ago and has been spreading and diversifying ever since.

organic matter from decayed plant and animal life (biosphere). The decomposed and disintegrated rock debris is the product of weathering processes that require air (atmosphere) and water (hydrosphere). Air and water also occupy the open spaces between the solid particles and are considered important soil components.

CONCEPT CHECK 1.6

❶ List and briefly describe Earth's four spheres.

Earth as a System

Anyone who studies Earth soon learns that our planet is a dynamic body with many separate but interacting parts or *spheres*. The hydrosphere, atmosphere, biosphere, and geosphere and all of their components can be studied separately. However, the parts are not isolated. Each is related in some way to the others to produce a complex and continuously interacting whole that we call the *Earth system*.

Earth System Science

A simple example of the interactions among different parts of the Earth system occurs every winter as moisture evaporates from the Pacific Ocean and subsequently falls as rain in the hills of southern California, triggering destructive landslides. A case study in Chapter 8 explores such an event (see p. 204). The processes that move water from the hydrosphere to the atmosphere and then to the solid Earth have a profound impact on the plants and animals (including humans) that inhabit the affected regions. **FIGURE 1.17** provides another example.

Scientists have recognized that to more fully understand our planet they must learn how its individual components (land, water, air, and life forms) are interconnected. This endeavor, called **Earth system science** aims to study Earth as a *system* composed of numerous interacting parts, or *subsystems*. Rather than looking through the limited lens of only one of the traditional sciences—geology, atmospheric science, chemistry, biology, and so forth—Earth system science attempts to integrate the knowledge of several academic fields. Using this interdisciplinary approach, we hope to achieve the level of understanding necessary to comprehend and solve many of our global environmental problems.

WHAT IS A SYSTEM? Most of us hear and use the term *system* frequently. We may service our car's cooling *system*, make use of the city's transportation *system*, and participate in the political *system*. A news report might inform us of an approaching weather *system*. Further, we know that Earth is just a small part of a larger system known as the solar *system*, which in turn is a *subsystem* of the even larger system called the Milky Way Galaxy.

Loosely defined, a **system** can be any size group of interacting parts that form a complex whole. Most natural systems are driven by sources of energy that move matter and/or energy from one place to another. A simple analogy is a car's cooling system, which contains liquid (usually water and antifreeze) that is driven from the engine to the radiator and back again. The role of this system is to transfer heat generated by combustion in the engine to the radiator, where moving air removes it from the system. Hence, the term cooling system.

Systems like a car's cooling system are self-contained with regard to matter and are called **closed systems**. Although energy moves freely in and out of a closed system, no matter (liquid in the case of our auto's cooling system) enters or leaves the system. (This assumes you do not get a leak in your radiator.) By contrast, most natural systems

FIGURE 1.17 This image provides an example of interactions among different parts of the Earth system. Aerial view of Caraballeda, Venezuela, covered by material from a massive debris flow (popularly called a mud slide in the press). In December 1999, extraordinary rains triggered this debris flow and thousands of others along this mountainous coastal zone. Caraballeda was located at the mouth of a steep canyon. An estimated 19,000 lives were lost. *(Photo by Kimberly White/Reuters/Corbis/Bettmann)*

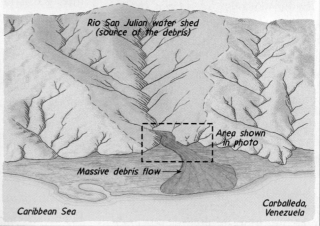

Geologist's Sketch

are **open systems** and are far more complicated than the foregoing example. In an open system both energy and matter flow into and out of the system. In a weather system such as a hurricane factors such as the quantity of water vapor available for cloud formation, the amount of heat released by condensing water vapor, and the flow of air into and out of the storm can fluctuate a great deal. At times the storm may strengthen; at other times it may remain stable or weaken.

FEEDBACK MECHANISMS. Most natural systems have mechanisms that tend to enhance change, as well as other mechanisms that tend to resist change and thus stabilize the system. For example, when we get too hot, we perspire to cool down. This cooling phenomenon works to stabilize our body temperature and is referred to as a **negative feedback mechanism.** Negative feedback mechanisms work to maintain the system as it is or, in other words, to maintain the status quo. By contrast, mechanisms that enhance or drive change are called **positive feedback mechanisms.**

Most of Earth's systems, particularly the climate system, contain a wide variety of negative and positive feedback mechanisms. For example, substantial scientific evidence indicates that Earth has entered a period of global warming. One consequence of global warming is that some of the world's glaciers and ice caps have begun to melt. Highly reflective snow- and ice-covered surfaces are gradually being replaced by brown soils, green trees, or blue oceans, all of which are darker, so they absorb more sunlight. Therefore, as Earth warms and some snow and ice melt, our planet absorbs more sunlight. The result is a positive feedback that contributes to the warming.

On the other hand, an increase in global temperature also causes greater evaporation of water from Earth's land–sea surface. One result of having more water vapor in the air is an increase in cloud cover. Because cloud tops are white and highly reflective, more sunlight is reflected back to space, which diminishes the amount of sunshine reaching Earth's surface and thus reduces global temperatures. Further, warmer temperatures tend to promote the growth of vegetation. Plants in

turn remove carbon dioxide (CO_2) from the air. Since carbon dioxide is one of the atmosphere's *greenhouse gases*, its removal has a negative impact on global warming.*

In addition to natural processes, we must consider the human element. Extensive cutting and clearing of the tropical rain forests and the burning of fossil fuels (oil, natural gas, and coal) result in an increase in atmospheric CO_2. Such activity is contributing to the increase in global temperature that our planet is experiencing. One of the daunting tasks for Earth system scientists is to predict what the climate will be like in the future by taking into account many variables, including technological changes, population trends, and the overall impact of the numerous competing positive and negative feedback mechanisms. Chapter 20 on "Global Climate Change" explores this topic in some detail.

The Earth System

The Earth system has a nearly endless array of subsystems in which matter is recycled over and over again. One example that you will learn about in Chapter 6 traces the movements of carbon among Earth's four spheres. It shows us, for example, that the carbon dioxide in the air and the carbon in living things and in certain sedimentary rocks are all part of a subsystem described by the *carbon cycle*.

CYCLES IN THE EARTH SYSTEM. A more familiar loop or subsystem is the *hydrologic cycle*. It represents the unending circulation of Earth's water among the hydrosphere, atmosphere, biosphere, and geosphere. Water enters the atmosphere by evaporation from Earth's surface and by transpiration from plants. Water vapor condenses in the atmosphere to form clouds, which in turn produce precipitation that falls back to Earth's surface. Some of the rain that falls onto the land sinks in to be taken up by plants or become groundwater, and some flows across the surface toward the ocean.

Viewed over long time spans, the rocks of the geosphere are constantly forming, changing, and reforming. The loop that involves the processes by which one rock changes to another is called the rock cycle and will be discussed at some length in the following section. The cycles of the Earth system, such as the hydrologic and rock cycles, are not independent of one another. To the contrary, there are many places where they have an interface. An **interface** is a common boundary where different parts of a system come in contact and interact. For example weathering at the surface gradually disintegrates and decomposes solid rock. The work of gravity and running water may eventually move this material to another place and deposit it. Later, groundwater percolating through the debris may leave behind mineral matter that cements the grains together into solid rock (a rock that is often very different from the rock we started with). This changing of one rock into another, which is part of the rock cycle, could not have occurred without the movement of water through the hydrologic cycle. There are many places where one cycle or loop in the Earth system has an interface with and is a basic part of another.

ENERGY FOR THE EARTH SYSTEM. The Earth system is powered by energy from two sources. The Sun drives external processes that occur in the atmosphere, hydrosphere, and at Earth's surface. Weather and climate, ocean circulation, and erosional processes such as rivers, glaciers, wind, and waves are driven by energy from the Sun. Earth's interior is the second source of energy. Heat remaining from when our planet formed, and heat that is continuously generated by decay of radioactive elements, power the internal processes that produce volcanoes, earthquakes, and mountains.

*Greenhouse gases absorb heat energy emitted by Earth and thus help keep the atmosphere warm.

FIGURE 1.18 When Mount St. Helens erupted in May 1980 (see inset), the area in the larger image was buried by a volcanic mudflow. Now plants are reestablished and new soil is forming. *(Photo by Jack Dykinga/Getty Images; inset by U.S. Geological Survey)*

As we learn about Earth, it becomes increasingly clear that despite significant separations in distance or time, many processes are connected, and a change in one component can influence the entire system.

Humans are *part of* the Earth system, a system in which the living and nonliving components are entwined and interconnected. Therefore, our actions produce changes in all of the other parts. When we burn gasoline and coal, build breakwaters along the shoreline, dispose of our wastes, and clear the land, we cause other parts of the system to respond, often in unforeseen ways (**FIGURE 1.19**). Throughout this book you will learn about many of Earth's subsystems: the hydrologic system, the tectonic (mountain-building) system, and the rock cycle, to name a few. Remember that these components *and we humans* are all part of the complex interacting whole we call the Earth system.

THE PARTS ARE LINKED. The parts of the Earth system are linked so that a change in one part can produce changes in any or all of the other parts. For example, when a volcano erupts, lava from Earth's interior may flow out at the surface and block a nearby valley. This new obstruction influences the region's drainage system by creating a lake or causing streams to change course. The large quantities of volcanic ash and gases that can be emitted during an eruption might be blown high into the atmosphere and influence the amount of solar energy that can reach Earth's surface. The result could be a drop in air temperatures over the entire hemisphere.

Where the surface is covered by lava flows or a thick layer of volcanic ash, existing soils are buried. This causes the soil-forming processes to begin anew to transform the new surface material into soil (**FIGURE 1.18**). The soil that eventually forms will reflect the interactions among many parts of the Earth system—the volcanic parent material, the type and rate of weathering, and the impact of biological activity. Of course, there will also be significant changes in the biosphere. Some organisms and their habitats will be eliminated by the lava and ash, whereas new settings for life, such as the lake, will be created. The potential climate change can also impact sensitive life forms.

The Earth system is characterized by processes that vary on spatial scales from fractions of millimeters to thousands of kilometers. Time scales for Earth's processes range from milliseconds to billions of years.

CONCEPT CHECK 1.7

❶ How is an open system different from a closed system?

❷ Contrast positive and negative feedback mechanisms.

❸ What are the two sources of energy for the Earth system?

FIGURE 1.19 People are part of the Earth system. This image of Earth's city lights at night shows the geographic distribution of settlements and helps us appreciate the intensity of human presence in many parts of the world. *(NASA)*

The Rock Cycle: One of Earth's Subsystems

An Introduction to Geology:
ESSENTIALS OF GEOLOGY The Rock Cycle

Rock is the most common and abundant material on Earth. To a curious traveler, the variety seems nearly endless. When a rock is examined closely, we find that it consists of smaller crystals or grains called minerals. *Minerals* are chemical compounds (or sometimes single elements), each with its own composition and physical properties. The grains or crystals may be microscopically small or easily seen with the unaided eye.

The nature and appearance of a rock is strongly influenced by the minerals that compose it. In addition, a rock's *texture*—the size, shape, and/or arrangement of its constituent minerals—also has a significant effect on its appearance. A rock's mineral composition and texture, in turn, are a reflection of the geologic processes that created it (**FIGURE 1.20**).

The characteristics of the rocks in **FIGURE 1.21** provided geologists with the clues they needed to determine the processes that formed them. This is true of all rocks. Such analyses are critical to an understanding of our planet. This understanding has many practical applications, as in the search for basic mineral and energy resources and the solution of environmental problems.

Geologists divide rocks into three major groups: igneous, sedimentary, and metamorphic. In Figure 1.21, the lava flow in northern Arizona is classified as igneous, the sandstone in Utah's Zion National Park is sedimentary, and the schist exposed at the bottom of the Grand Canyon is metamorphic.

In the preceding section, you learned that Earth is a system. This means that our planet consists of many interacting parts that form a complex whole. Nowhere is this idea better illustrated than when we examine the rock cycle (**FIGURE 1.22**). The **rock cycle** allows us to view many of the interrelationships among different parts of the Earth system. Knowledge of the rock cycle will help you more clearly understand the idea that each rock group is linked to the others by the processes that act upon and within the planet. You can consider the rock cycle to be a simplified but useful overview of physical geology. What follows is a brief introduction to the rock cycle. Learn the rock cycle well; you will be examining its interrelationships in greater detail throughout this book.

A.

B.

FIGURE 1.20 Texture and mineral composition are important rock properties that allow geologists to determine much about how and where the rock formed. **A.** Gneiss is a metamorphic rock and **B.** Granite porphyry is an igneous rock. *(Photos by E. J. Tarbuck)*

A.

B.

C.

FIGURE 1.21 A. This fine-grained rock, called *basalt*, is part of a lava flow from Sunset Crater in northern Arizona. It formed when molten rock erupted at Earth's surface hundreds of years ago and solidified. *(Photo by Dennis Tasa)* **B.** This rock is exposed in the walls of southern Utah's Zion National Park. This layer, known as the Navajo Sandstone, consists of durable grains of the glassy mineral quartz that once covered this region with mile after mile of drifting sand dunes. *(Photo by Jamie & Judy Wild www.DanitaDelimont.com)* **C.** This rock unit, known as the Vishnu Schist, is exposed in the inner gorge of the Grand Canyon. Its formation is associated with environments far below Earth's surface where temperatures and pressures are high and with ancient mountain-building processes that occurred in Precambrian time. *(Photo by Dennis Tasa)*

ROCK CYCLE

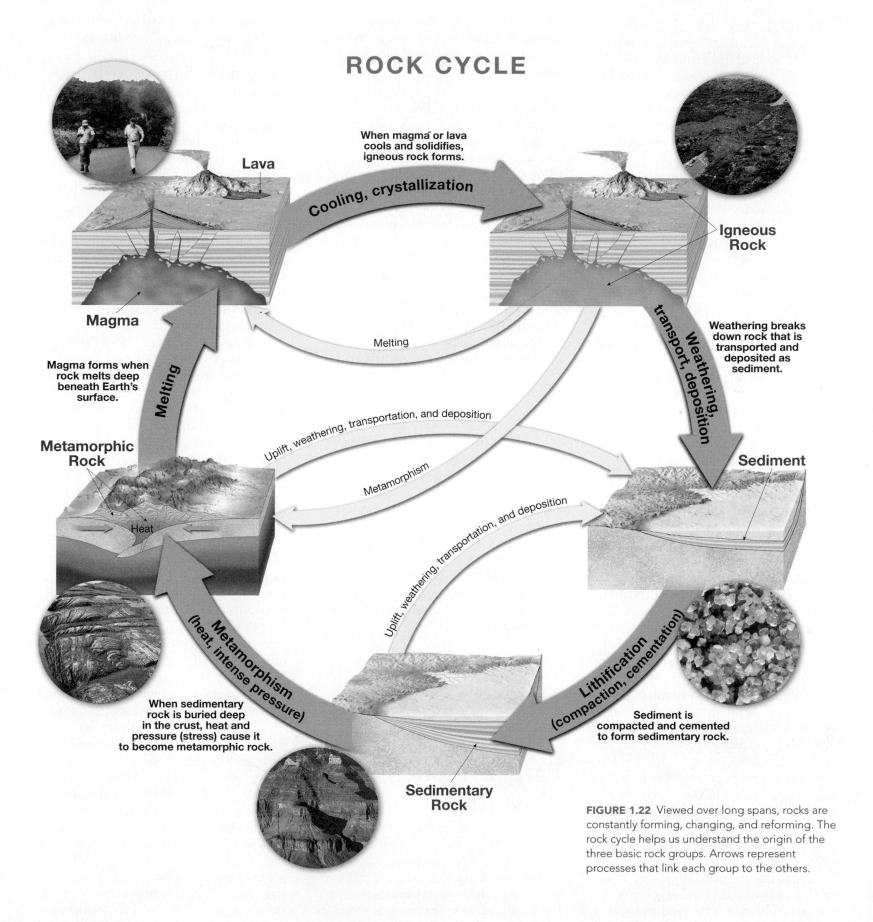

When magma or lava cools and solidifies, igneous rock forms.

Lava

Cooling, crystallization

Igneous Rock

Magma forms when rock melts deep beneath Earth's surface.

Magma

Melting

Melting

Weathering breaks down rock that is transported and deposited as sediment.

Weathering, transport, deposition

Uplift, weathering, transportation, and deposition

Metamorphic Rock

Metamorphism

Sediment

Heat

Uplift, weathering, transportation, and deposition

Metamorphism (heat, intense pressure)

When sedimentary rock is buried deep in the crust, heat and pressure (stress) cause it to become metamorphic rock.

Lithification (compaction, cementation)

Sediment is compacted and cemented to form sedimentary rock.

Sedimentary Rock

FIGURE 1.22 Viewed over long spans, rocks are constantly forming, changing, and reforming. The rock cycle helps us understand the origin of the three basic rock groups. Arrows represent processes that link each group to the others.

The Basic Cycle

We begin at the top of Figure 1.22. **Magma** is molten material that forms inside Earth. Eventually magma cools and solidifies. This process, called *crystallization*, may occur either beneath the surface or, following a volcanic eruption, at the surface. In either situation, the resulting rocks are called **igneous rocks** (*ignis* = fire).

If igneous rocks are exposed at the surface, they will undergo *weathering*, in which the day-in and day-out influences of the atmosphere slowly disintegrate and decompose rocks. The materials that result are often moved downslope by gravity before being picked up and transported by any of a number of erosional agents, such as running water, glaciers, wind, or waves. Eventually these particles and dissolved substances, called **sediment**, are deposited. Although most sediment ultimately comes to rest in the ocean, other sites of deposition include river floodplains, desert basins, swamps, and sand dunes.

Next the sediments undergo *lithification*, a term meaning "conversion into rock." Sediment is usually lithified into **sedimentary rock** when compacted by the weight of overlying layers or when cemented as percolating groundwater fills the pores with mineral matter.

If the resulting sedimentary rock is buried deep within Earth and involved in the dynamics of mountain building or intruded by a mass of magma, it will be subjected to great pressures and/or intense heat. The sedimentary rock will react to the changing environment and turn into the third rock type, **metamorphic rock.** When metamorphic rock is subjected to additional pressure changes or to still higher temperatures, it will melt, creating magma, which will eventually crystallize into igneous rock, starting the cycle all over again.

Although rocks may seem to be unchanging masses, the rock cycle shows that they are not. The changes, however, take time—great amounts of time. The rock cycle is operating all over the world, but in different stages. Today new magma is forming under the island of Hawaii, while the Colorado Rockies are slowly being worn down by weathering and erosion. Some of this weathered debris is eventually carried to the Gulf of Mexico, where it is deposited, adding to the already substantial mass of sediment that has accumulated there.

Alternative Paths

The paths shown in the basic cycle are not the only ones that are possible. To the contrary, other paths are just as likely to be followed as those described in the preceding section. These alternatives are indicated by the blue arrows in Figure 1.22.

Igneous rocks, rather than being exposed to weathering and erosion at Earth's surface, may remain deeply buried. Eventually these masses may be subjected to the strong compressional forces and high temperatures associated with mountain building. When this occurs, they are transformed directly into metamorphic rocks.

Metamorphic and sedimentary rocks, as well as sediment, do not always remain buried. Rather, overlying layers may be stripped away, exposing the once buried rock. When this happens, the material is attacked by weathering processes and turned into new raw materials for sedimentary rocks.

Where does the energy that drives Earth's rock cycle come from? Processes driven by heat from Earth's interior are responsible for forming igneous and metamorphic rocks. Weathering and the movement of weathered material are external processes powered by energy from the Sun. External processes produce sedimentary rocks.

CONCEPT CHECK 1.8

❶ Sketch and label a basic rock cycle. Make sure your sketch includes alternative paths.

Early Evolution of Earth

Recent earthquakes caused by displacements of Earth's crust, along with lavas erupted from active volcanoes, represent only the latest in a long line of events by which our planet has attained its present form and structure. The geologic processes operating in Earth's interior can be best understood when viewed in the context of much earlier events in Earth history.

Origin of Planet Earth

This section describes the most widely accepted views of the origin of our solar system. The theory summarized here represents the most consistent set of ideas available to explain what we know about our solar system today.

Our scenario begins about 13.7 billion years ago with the *Big Bang*, an incomprehensibly large explosion that sent all matter of the universe flying outward at incredible speeds. In time, the debris from this explosion, which was almost entirely hydrogen and helium, began to cool and condense into the first stars and galaxies. It was in one of these galaxies, the Milky Way, that our solar system and planet Earth took form.

Earth is one of eight planets that, along with several dozen moons and numerous smaller bodies, revolve around the Sun. The orderly nature of our solar system leads most researchers to conclude that Earth and the other planets formed at essentially the same time and from the same primordial material as the Sun. The **nebular theory** states that the bodies of our solar system evolved from an enormous rotating cloud called the **solar nebula** (**FIGURE 1.23**). Besides the hydrogen and helium atoms generated during the Big Bang, the solar nebula consisted of microscopic dust grains and the ejected matter of long-dead stars. (Nuclear fusion in stars converts hydrogen and helium into the other elements found in the universe.)

Nearly 5 billion years ago this huge cloud of gases and minute grains of heavier elements began to slowly contract due to the gravitational interactions among its particles (**FIGURE 1.24**). Some external influence, such as a shock wave traveling from a catastrophic explosion (*supernova*), may have triggered the collapse. As this slowly spiraling nebula contracted, it rotated faster and faster for the same reason ice skaters do when they draw their arms toward their bodies. Eventually the inward pull of gravity came into balance with the outward force caused by the rotational motion of the nebula (Figure 1.23). By this time the once vast cloud had assumed a flat disk shape

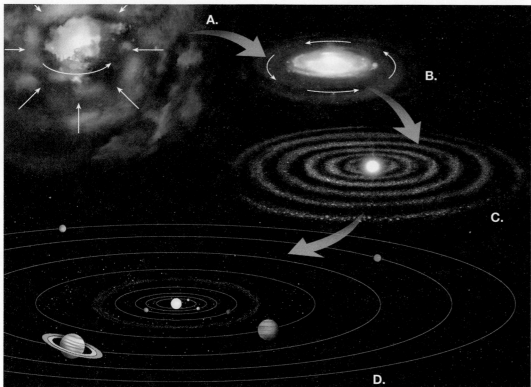

FIGURE 1.23 Formation of the solar system according to the nebular theory. **A.** The birth of our solar system began as dust and gases (nebula) started to gravitationally collapse. **B.** The nebula contracted into a rotating disk that was heated by the conversion of gravitational energy into thermal energy. **C.** Cooling of the nebular cloud caused rocky and metallic material to condense into tiny particles. **D.** Repeated collisions caused the dust-size particles to gradually coalesce into asteroid-size bodies. Within a few million years these bodies accreted into the planets.

FIGURE 1.24 Lagoon Nebula. It is in glowing clouds like these that gases and dust particles become concentrated into stars. *(Courtesy of National Optical Astronomy Observatories)*

with a large concentration of material at its center called the *protosun* (pre-Sun). (Astronomers are fairly confident that the nebular cloud formed a disk because similar structures have been detected around other stars.)

During the collapse, gravitational energy was converted to thermal energy (heat), causing the temperature of the inner portion of the nebula to dramatically rise. At these high temperatures, the dust grains broke up into molecules and extremely energetic atomic particles. However, at distances beyond the orbit of Mars, the temperatures probably remained quite low. At −200 °C, the tiny particles in the outer portion of the nebula were likely covered with a thick layer of ices made of frozen water, carbon dioxide, ammonia, and methane. (Some of this material still resides in the outermost reaches of the solar system in a region called the *Oort cloud*.) The disk-shaped cloud also contained appreciable amounts of the lighter gases hydrogen and helium.

The formation of the Sun marked the end of the period of contraction and thus the end of gravitational heating. Temperatures in the region where the inner planets now reside began to decline. The decrease in temperature caused those substances

with high melting points to condense into tiny particles that began to coalesce (join together). Materials such as iron and nickel and the elements of which the rock-forming minerals are composed—silicon, calcium, sodium, and so forth—formed metallic and rocky clumps that orbited the Sun (Figure 1.23). Repeated collisions caused these masses to coalesce into larger asteroid-size bodies, called *planetesimals*, which in a few tens of millions of years accreted into the four inner planets we call Mercury, Venus, Earth, and Mars (**FIGURE 1.25**). Not all of these clumps of matter were incorporated into the planetesimals. Those rocky and metallic pieces that remained in orbit are called *meteorites* when they survive an impact with Earth.

As more and more material was swept up by these growing planetary bodies, the high-velocity impact of nebular debris caused their temperature to rise. Because of their relatively high temperatures and weak gravitational fields, the inner planets were unable to accumulate much of the lighter components of the nebular cloud. The lightest of these, hydrogen and helium, were eventually whisked from the inner solar system by the solar winds.

At the same time that the inner planets were forming, the larger, outer planets (Jupiter, Saturn, Uranus, and Neptune), along with their extensive satellite systems, were also developing (Figure 1.25). Because of low temperatures far from the Sun, the material from which these planets formed contained a high percentage of ices—water, carbon dioxide, ammonia, and methane—as well as rocky and metallic debris. The accumulation of ices accounts in part for the large size and low density of the outer planets. The two most massive planets, Jupiter and Saturn, had a surface gravity sufficient to attract and hold large quantities of even the lightest elements—hydrogen and helium.

Formation of Earth's Layered Structure

As material accumulated to form Earth (and for a short period afterward), the high-velocity impact of nebular debris and the decay of radioactive elements caused the temperature of our planet to steadily increase. During this time of intense heating, Earth became hot enough

that iron and nickel began to melt. Melting produced liquid blobs of heavy metal that sank toward the center of the planet. This process occurred rapidly on the scale of geologic time and produced Earth's dense iron-rich core.

The early period of heating resulted in another process of chemical differentiation, whereby melting formed buoyant masses of molten rock that rose toward the surface, where they solidified to produce a primitive crust. These rocky materials were enriched in oxygen and "oxygen-seeking" elements, particularly silicon and aluminum, along with lesser amounts of calcium, sodium, potassium, iron, and magnesium. In addition, some heavy metals such as gold, lead, and uranium, which have low melting points or were highly soluble in the ascending molten masses, were scavenged from Earth's interior and concentrated in the developing crust. This early period of chemical segregation established the three basic divisions of Earth's interior—the iron-rich *core*; the thin *primitive crust*; and Earth's largest layer, called the *mantle*, which is located between the core and crust.

An important consequence of this early period of chemical differentiation is that large quantities of gaseous materials were allowed to escape from Earth's interior, as happens today during volcanic eruptions. By this process a primitive atmosphere gradually evolved. It is on this planet, with this atmosphere, that life as we know it came into existence.

Following the events that established Earth's basic structure, the primitive crust was lost to erosion and other geologic processes, so we have no direct record of its makeup. When and exactly how the continental crust—and thus Earth's first landmasses—came into existence is a matter of ongoing research. Nevertheless, there is

FIGURE 1.25 The planets drawn to scale.

general agreement that the continental crust formed gradually over the last 4 billion years. (The oldest rocks yet discovered are isolated fragments found in the Northwest Territories of Canada that have radiometric dates of about 4 billion years.) In addition, as you will see in subsequent chapters, Earth is an evolving planet whose continents (and ocean basins) have continually changed shape and even location during much of this period.

CONCEPT CHECK 1.9

❶ Name and briefly outline the theory that describes the formation of our solar system.

❷ Name the inner planets and the outer planets. Describe basic differences in size and composition.

Earth's Internal Structure

An Introduction to Geology:

ESSENTIALS OF GEOLOGY Earth's Layered Structure

In the preceding section, you learned that the segregation of material that began early in Earth's history resulted in the formation of three layers defined by their chemical composition—the crust, mantle, and core. In addition to these compositionally distinct layers, Earth can be divided into layers based on physical properties. The physical properties used to define such zones include whether the layer is solid or liquid and how weak or strong it is. Knowledge of both types of layered structures is essential to our understanding of basic geologic processes, such as volcanism, earthquakes, and mountain building (**FIGURE 1.26**).

Earth's Crust

The **crust**, Earth's relatively thin, rocky outer skin, is of two different types—continental crust and oceanic crust. Both share the word "crust," but the similarity ends there. The oceanic crust is roughly 7 kilometers (5 miles) thick and composed of the dark igneous rock *basalt*. By contrast, the continental crust averages about 35 kilometers (22 miles) thick

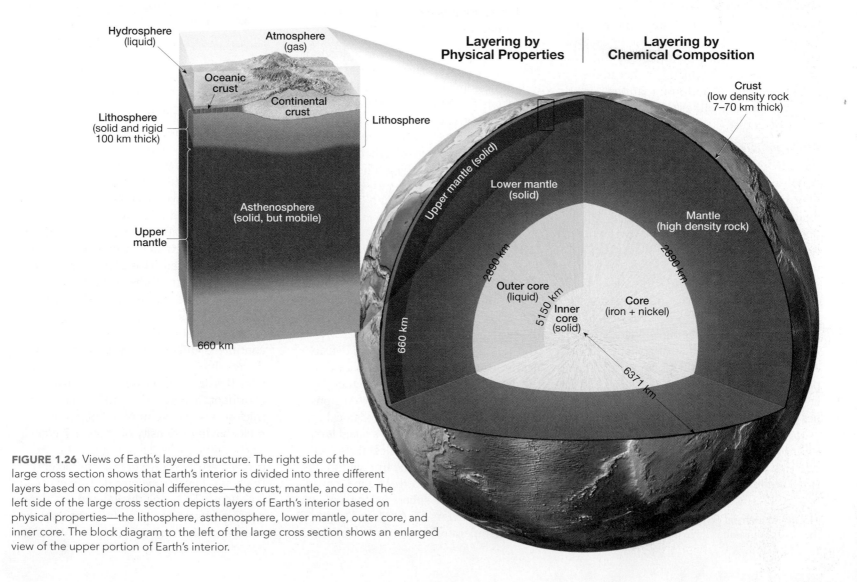

FIGURE 1.26 Views of Earth's layered structure. The right side of the large cross section shows that Earth's interior is divided into three different layers based on compositional differences—the crust, mantle, and core. The left side of the large cross section depicts layers of Earth's interior based on physical properties—the lithosphere, asthenosphere, lower mantle, outer core, and inner core. The block diagram to the left of the large cross section shows an enlarged view of the upper portion of Earth's interior.

but may exceed 70 kilometers (40 miles) in some mountainous regions such as the Rockies and Himalayas. Unlike the oceanic crust, which has a relatively homogeneous chemical composition, the continental crust consists of many rock types. Although the upper crust has an average composition of a *granitic rock* called *granodiorite*, it varies considerably from place to place.

Continental rocks have an average density of about 2.7 g/cm^3, and some have been discovered that are 4 billion years old. The rocks of the oceanic crust are younger (180 million years or less) and denser (about 3.0 g/cm^3) than continental rocks.*

Earth's Mantle

More than 82 percent of Earth's volume is contained in the **mantle**, a solid, rocky shell that extends to a depth of nearly 2900 kilometers (1800 miles). The boundary between the crust and mantle represents a significant change in chemical composition. The dominant rock type in the uppermost mantle is *peridotite*, which is richer in the metals magnesium and iron than the minerals found in either the continental or oceanic crust.

THE UPPER MANTLE. The upper mantle extends from the crust-mantle boundary down to a depth of about 660 kilometers (410 miles). The upper mantle can be divided into two different parts. The top portion of the upper mantle is part of the stiff *lithosphere*, and beneath that is the weaker **asthenosphere.**

The **lithosphere** (sphere of rock) consists of the entire crust and uppermost mantle and forms Earth's relatively cool, rigid outer shell. Averaging about 100 kilometers in thickness, the lithosphere is more than 250 kilometers thick below the oldest portions of the continents. Beneath this stiff layer to a depth of about 350 kilometers lies a soft, comparatively weak layer known as the *asthenosphere* ("weak sphere"). The top portion of the asthenosphere has a temperature/pressure regime that results in a small amount of melting. Within this weak

*Liquid water has a density of 1 g/cm^3; therefore, the density of basalt is three times that of water.

zone the lithosphere is mechanically detached from the layer below. The result is that the lithosphere is able to move independently of the asthenosphere, a fact we will consider later in the chapter and in Chapter 15.

It is important to emphasize that the strength of various Earth materials is a function of both their composition and the temperature and pressure of their environment. You should not get the idea that the entire lithosphere behaves like a brittle solid similar to rocks found on the surface. Rather, the rocks of the lithosphere get progressively hotter and weaker (more easily deformed) with increasing depth. At the depth of the uppermost asthenosphere, the rocks are close enough to their melting temperature (some melting may actually occur) that they are very easily deformed. Thus, the uppermost asthenosphere is weak because it is near its melting point, just as hot wax is weaker than cold wax.

THE LOWER MANTLE. From a depth of 660 kilometers (nearly 410 miles) to the top of the core, at a depth of 2900 kilometers (1800 miles), is the **lower mantle.** Because of an increase in pressure (caused by the weight of the rock above) the mantle gradually strengthens with depth. Despite their strength however, the rocks within the lower mantle are very hot and capable of very gradual flow.

Earth's Core

The composition of the **core** is thought to be an iron-nickel alloy with minor amounts of oxygen, silicon, and sulfur—elements that readily form compounds with iron. At the extreme pressure found in the core, this iron-rich material has an average density of nearly 11 g/cm^3 and approaches 14 times the density of water at Earth's center.

The core is divided into two regions that exhibit very different mechanical strengths. The **outer core** is a *liquid layer* 2270 kilometers (1410 miles) thick. It is the movement of metallic iron within this zone that generates Earth's magnetic field. The **inner core** is a sphere having a radius of 1216 kilometers (754 miles). Despite its higher temperature, the iron in the inner

core is *solid* due to the immense pressures that exist in the center of the planet.

CONCEPT CHECK 1.10

❶ List and briefly describe Earth's compositional layers.
❷ Contrast the lithosphere and the asthenosphere.

The Face of Earth

The two principal divisions of Earth's surface are the continents and the ocean basins (**FIGURE 1.27**). A significant difference between these two areas is their relative levels. The continents are remarkably flat features that have the appearance of plateaus protruding above sea level. With an average elevation of about 0.8 kilometer (0.5 mile), continents lie close to sea level, except for limited areas of mountainous terrain. By contrast, the average depth of the ocean floor is about 3.8 kilometers (2.4 miles) below sea level.

The elevation difference between the continents and ocean basins is primarily the result of differences in their respective densities and thicknesses. Recall that the continents average 35 to 40 kilometers in thickness and are composed of granitic rocks having a density of about 2.7 g/cm^3. The basaltic rocks that comprise the oceanic crust average only 7 kilometers thick and have an average density of about 3.0 g/cm^3. Thus, the thicker and less dense continental crust is more buoyant than the oceanic crust. As a result, continental crust floats on top of the deformable rocks of the mantle at a higher level than oceanic crust for the same reason that a large, empty (less dense) cargo ship rides higher than a small, loaded (more dense) one.

Major Features of the Continents

The largest features of the continents can be grouped into two distinct categories: (1) extensive, flat stable areas that have been eroded nearly to sea level and (2) uplifted regions of deformed rocks that make up present-day mountain belts. Notice in **FIGURE 1.28** that the young mountain belts tend to be long, narrow features at the margins of continents, and the flat, stable areas are typically located in the interior of the continents.

MOUNTAIN BELTS. The most prominent topographic features of the continents are linear mountain belts. Although the distribution of mountains appears to be random, this is not the case. When the youngest mountains are considered (those less than 100 million years old), we find that they are located principally in two major zones. The circum-Pacific belt (the region surrounding the Pacific Ocean) includes the mountains of the western Americas and continues into the western Pacific in the form of volcanic island arcs. Island arcs are active mountainous regions composed largely of volcanic rocks and deformed sedimentary rocks. Examples include the Aleutian Islands, Japan, the Philippines, and New Guinea (Figure 1.27).

The other major mountainous belt extends eastward from the Alps through Iran and the Himalayas and then dips southward into Indonesia. Careful examination of mountainous terrains reveals that most are places where thick sequences of rocks have been squeezed and highly deformed, as if placed in a gigantic vise. Older mountains are also found on the continents. Examples include the Appalachians in the eastern United States and the Urals in Russia. Their once lofty peaks are now worn low, the result of millions of years of erosion.

THE STABLE INTERIOR. Unlike the young mountain belts, which have formed within the last 100 million years, the interiors of the continents have been relatively stable (undisturbed) for the last 600 million years or even longer. Typically, these regions were involved in mountain-building episodes much earlier in Earth's history.

Within the stable interiors are areas known as **shields**, which are expansive, flat regions composed of deformed crystalline rock. Notice in Figure 1.28 that the Canadian Shield is exposed in much of the northeastern part of North America. Age determinations for various shields have shown that they are truly ancient regions. All contain Precambrian-age rocks that are more than 1 billion years old, with some samples approaching 4 billion years in age (see Figure 1.8 to review the geologic time scale). These oldest-known rocks exhibit evidence of enormous forces that have folded and faulted them and altered them with great heat and pressure. Thus, we conclude that these rocks were once part of an ancient mountain system that has since been eroded away to produce these expansive, flat regions.

Other flat areas of the stable interior exist in which highly deformed rocks, like those found in the shields, are covered by a relatively thin veneer of sedimentary rocks. These areas are called **stable platforms.** The sedimentary rocks in stable platforms are nearly horizontal except where they have been warped to form large basins or domes. In North America a major portion of the stable platforms is located between the Canadian Shield and the Rocky Mountains (Figure 1.28).

Major Features of the Ocean Basins

If all water were drained from the ocean basins, a great variety of features would be seen, including linear chains of volcanoes, deep canyons, extensive plateaus, and large expanses of monotonously flat plains. In fact, the scenery would be nearly as diverse as that on the continents (see Figure 1.27).

During the past 60 years, oceanographers using modern depth-sounding equipment have gradually mapped significant portions of the ocean floor. From these studies they have defined three major regions: *continental margins, deep-ocean basins, and oceanic (mid-ocean) ridges.*

CONTINENTAL MARGINS. The **continental margin** is that portion of the seafloor adjacent to major landmasses. It may include the *continental shelf*, the *continental slope*, and the *continental rise.*

Although land and sea meet at the shoreline, this is not the boundary between the continents and the ocean basins. Rather, along most coasts a gently sloping platform, called the **continental shelf,** extends seaward from the shore. Because it is underlain by continental crust, it is considered a flooded extension of the continents. A glance at Figure 1.27 shows that the width of the continental shelf is variable. For example, it is broad along the East and Gulf coasts of the United States but relatively narrow along the Pacific margin of the continent.

The boundary between the continents and the deep-ocean basins lies along the **continental slope,** which is a relatively steep dropoff that extends from the outer edge of the continental shelf to the floor of the deep ocean (Figure 1.27). Using this as the dividing line, we find that about 60 percent of Earth's surface is represented by ocean basins and the remaining 40 percent by continents.

DID YOU KNOW?
Ocean depths are often expressed in *fathoms.* One fathom equals 1.8 m or 6 ft, which is about the distance of a person's outstretched arms. The term is derived from how depth-sounding lines were brought back on board a vessel by hand. As the line was hauled in, a worker counted the number of arm lengths collected. By knowing the length of the person's outstretched arms, the amount of line taken in could be calculated. The length of one fathom was later standardized to 6 ft.

FIGURE 1.27 Major surface features of the geosphere.

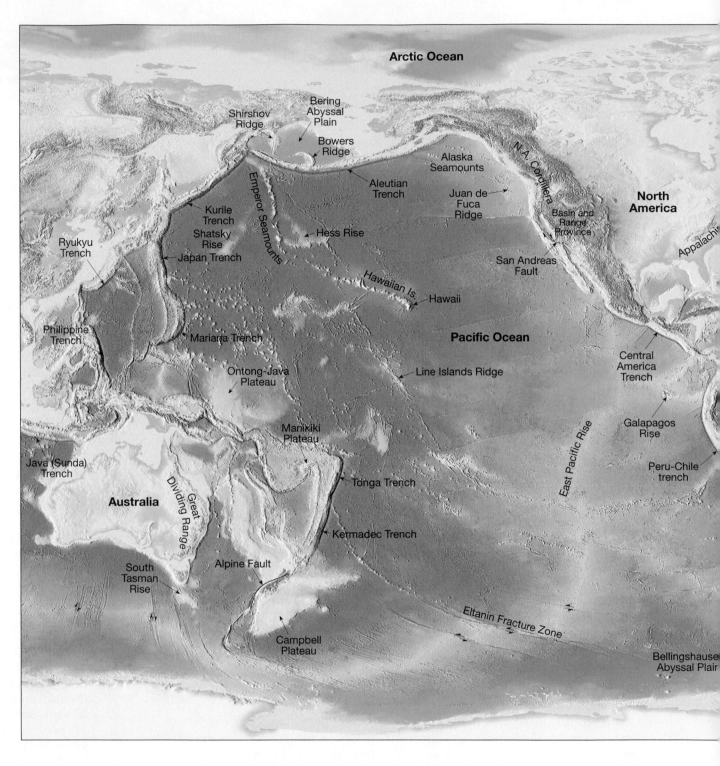

In regions where trenches do not exist, the steep continental slope merges into a more gradual incline known as the **continental rise.** The continental rise consists of a thick accumulation of sediments that moved downslope from the continental shelf to the deep-ocean floor.

DEEP-OCEAN BASINS. Between the continental margins and oceanic ridges lie the **deep-ocean basins.** Parts of these regions consist of incredibly flat features called **abyssal plains.** The ocean floor also contains extremely deep depressions that are occasionally more than 11,000 meters (36,000 feet) deep. Although these **deep-ocean trenches** are relatively narrow and represent only a small fraction of the ocean floor, they are nevertheless very significant features. Some trenches are located adjacent to young mountains that flank the continents. For example, in Figure 1.27 the Peru–Chile trench off the west coast of South America parallels the Andes Mountains. Other trenches parallel linear island chains called *volcanic island arcs.*

Dotting the ocean floor are submerged volcanic structures called **seamounts,** which sometimes form long narrow chains. Volcanic activity has also produced several large *lava plateaus,* such as the Ontong Java Plateau located northeast on New Guinea. In addition, some submerged plateaus are

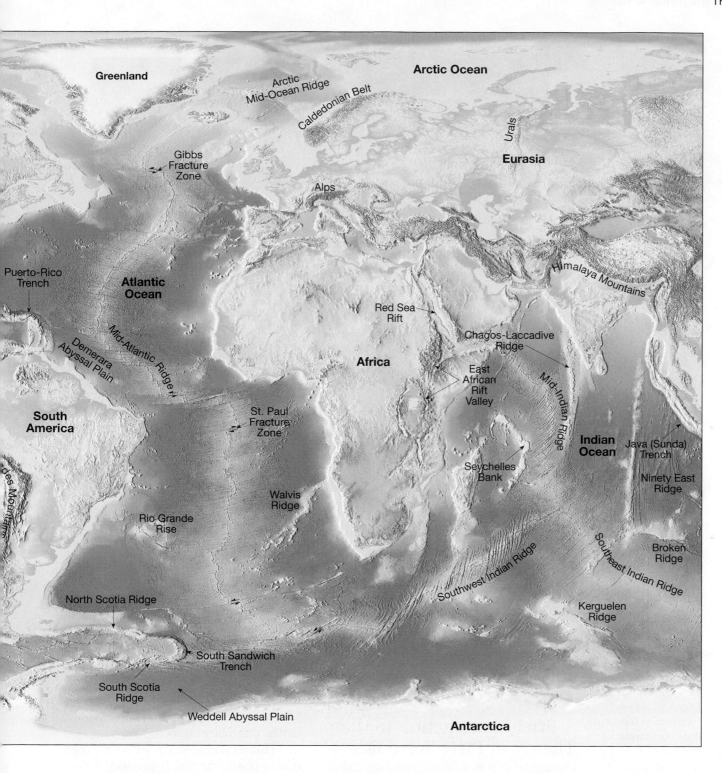

composed of continental-type crust. Examples include the Campbell Plateau southeast of New Zealand and the Seychelles Bank northeast of Madagascar.

OCEANIC RIDGES. The most prominent feature on the ocean floor is the **oceanic** or **mid-ocean ridge.** As shown in Figure 1.27, the Mid-Atlantic Ridge and the East Pacific Rise are parts of this system. This broad elevated feature forms a continuous belt that winds for more than 70,000 kilometers (43,000 miles) around the globe in a manner similar to the seam of a baseball. Rather than consisting of highly deformed rock, like most mountains on the continents, the oceanic ridge system consists of layer upon layer of igneous rock that has been fractured and uplifted.

Understanding the topographic features that comprise the face of Earth is critical to our understanding of the mecha-nisms that have shaped our planet. What is the significance of the enormous ridge system that extends through all the world's oceans? What is the connection, if any, between young, active mountain belts and deep-ocean trenches? What forces crumple rocks to produce majestic mountain ranges? These are questions that will be addressed in some of the coming chapters as we investigate the dynamic processes that shaped our planet in the

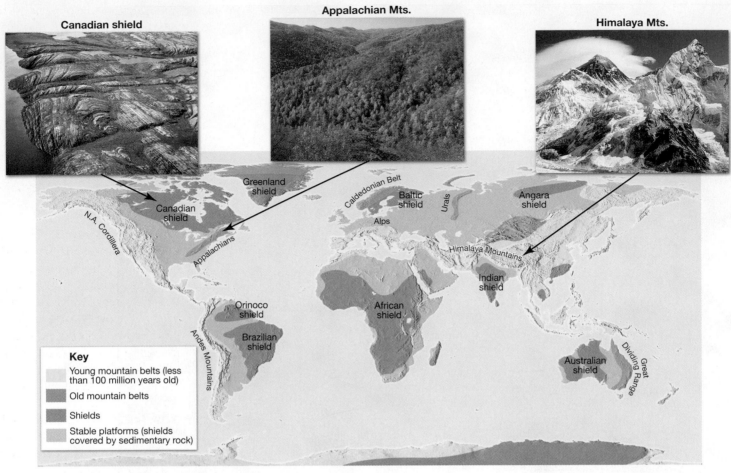

FIGURE 1.28 This map shows the general distribution of Earth's mountain belts, stable platforms, and shields. *[Photo by Robert Hildebrand Photography (left), CORBIS (middle), Images Source Pink/Alamy (right)]*

geologic past and will continue to shape it in the future.

CONCEPT CHECK 1.11

❶ Describe the general distribution of Earth's youngest mountains.

❷ What is the difference between shields and stable platforms?

❸ What are the three major regions of the ocean floor and some features associated with each?

Dynamic Earth

 Plate Tectonics:
ESSENTIALS OF GEOLOGY **Introduction**

Earth is a dynamic planet! If we could go back in time a few hundred million years, we would find the face of our planet dramatically different from what we see today. There would be no Mount St. Helens,

Rocky Mountains, or Gulf of Mexico. Moreover, we would find continents having different sizes and shapes and located in different positions than today's landmasses. By contrast, over the past few billion years the Moon's surface has remained essentially unchanged—only a few craters have been added.

A Brief Introduction to the Theory of Plate Tectonics

During the past several decades a great deal has been learned about the workings of our dynamic planet. This span has seen an unequaled revolution in our understanding of Earth. The revolution had its beginnings in the early part of the 20th century with the radical proposal of *continental drift*—the idea that continents move about the face of the planet. This proposal contradicted the long established view that continents and ocean basins are permanent and stationary features on the face of Earth. For that

reason, the notion of drifting continents was received with great skepticism and even ridicule. More than 50 years passed before enough data were gathered to transform this controversial hypothesis into a sound theory that wove together the basic processes known to operate on Earth. The theory that finally emerged, called the **theory of plate tectonics**, provided geologists with the first comprehensive model of Earth's internal workings.

This section serves as a brief introduction to plate tectonics. It provides background and perspective that will be helpful in some upcoming discussions, especially those dealing with volcanoes and igneous activity (Chapter 4) and metamorphic rocks (Chapter 7). Later Chapter 15, "Plate Tectonics: A Scientific Revolution Unfolds," explores the theory in detail.

According to the plate tectonics model, Earth's rigid outer shell (*lithosphere*) is broken into numerous slabs called **lithospheric plates** or simply **plates**, which are in

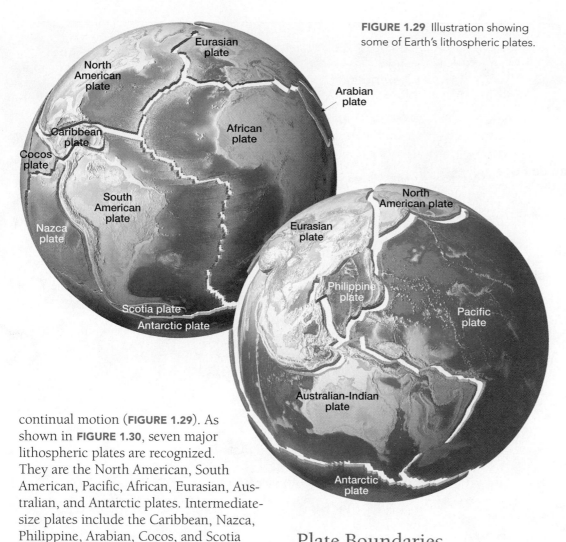

FIGURE 1.29 Illustration showing some of Earth's lithospheric plates.

continual motion (**FIGURE 1.29**). As shown in **FIGURE 1.30**, seven major lithospheric plates are recognized. They are the North American, South American, Pacific, African, Eurasian, Australian, and Antarctic plates. Intermediate-size plates include the Caribbean, Nazca, Philippine, Arabian, Cocos, and Scotia plates. In addition, over a dozen smaller plates have been identified but are not shown in Figure 1.30. Note that several large plates include an entire continent plus a large area of seafloor (for example, the South American plate). However, none of the plates is defined entirely by the margins of a single continent.

The lithospheric plates move relative to each other at a very slow but continuous rate that averages about 5 centimeters (2 inches) a year. This movement is ultimately driven by the unequal distribution of heat within Earth. Hot material found deep in the mantle moves slowly upward and serves as one part of our planet's internal convective system. Concurrently, cooler, denser slabs of lithosphere descend back into the mantle, setting Earth's rigid outer shell in motion. Ultimately, the titanic, grinding movements of Earth's lithospheric plates generate earthquakes, create volcanoes, and deform large masses of rock into mountains.

Plate Boundaries

Lithospheric plates move as coherent units relative to all other plates. Although the interiors of plates may experience some deformation, all major interactions among individual plates (and therefore most deformation) occur along their boundaries. In fact, the first attempts to outline plate boundaries were made using locations of earthquakes. Later work showed that plates are bounded by three distinct types of boundaries, which are differentiated by the type of relative movement they exhibit. These boundaries are depicted at the bottom of Figure 1.30 and are briefly described here:

1. **Divergent boundaries**—where plates move apart, resulting in upwelling of material from the mantle to create new seafloor (Figure 1.30A).

2. **Convergent boundaries**—where plates move together, resulting in the subduction (consumption) of oceanic lithosphere into the mantle (Figure 1.30B).

Convergence can also result in the collision of two continental margins to create a major mountain system.

3. **Transform fault boundaries**—where plates grind past each other without the production or destruction of lithosphere (Figure 1.30C).

If you examine Figure 1.30, you can see that each large plate is bounded by a combination of these boundaries. Movement along one boundary requires that adjustments be made at the others.

DIVERGENT BOUNDARIES. Plate spreading (divergence) occurs mainly along the oceanic ridge. As plates pull apart, the fractures created are immediately filled with molten rock that wells up from the asthenosphere below (**FIGURE 1.31**). This hot material slowly cools to become solid rock, producing new slivers of seafloor. This happens again and again over millions of years, adding thousands of square kilometers of new seafloor.

This mechanism has created the floor of the Atlantic Ocean during the past 160 million years and is appropriately called **seafloor spreading** (Figure 1.31). Because seafloor spreading is the dominant process associated with divergent boundaries, these zones are sometimes referred to as *spreading centers*. The rate of seafloor spreading varies considerably from one spreading center to another. Spreading rates of only 2.5 centimeters (1 inch) per year are typical in the North Atlantic, whereas much faster rates (20 centimeters, 8 inches per year) have been measured along the East Pacific Rise. Even the most rapid rates of spreading are slow on the scale of human history. Nevertheless, the slowest rate of lithosphere production is rapid enough to have created all of Earth's ocean basins over the last 200 million years. In fact, none of the ocean floor that has been dated exceeds 180 million years in age.

Along divergent boundaries where molten rock emerges, the oceanic lithosphere is elevated, because it is hot and occupies more volume than do cooler rocks. Worldwide, this elevated zone (the oceanic ridge) extends for over 70,000 kilometers (43,000 miles) through all major ocean basins (Figure 1.27). As new lithosphere is formed along the oceanic

FIGURE 1.30 Mosaic of rigid plates that constitute Earth's outer shell. *(After W. B. Hamilton, U.S. Geological Survey)*

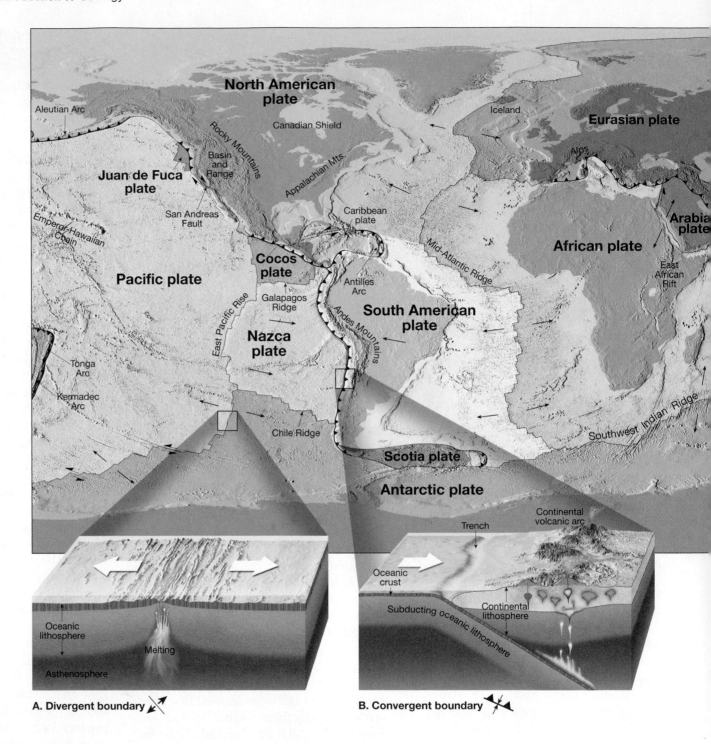

A. Divergent boundary

B. Convergent boundary

ridge, it is slowly yet continually displaced from the zone of upwelling along the ridge axis. Thus, it begins to cool and contract, thereby increasing in density. This thermal contraction accounts for the greater ocean depths that exist away from the ridge. In addition, cooling causes the mantle rocks below the oceanic crust to strengthen, thereby adding to the plate's thickness. Stated another way, the thickness of oceanic lithosphere is age dependent. The older (cooler) it is, the greater its thickness.

CONVERGENT BOUNDARIES. Although new lithosphere is constantly being added at the oceanic ridges, the planet is not growing in size—its total surface area remains constant. To accommodate the newly created lithosphere, older oceanic plates return to the mantle along *convergent boundaries*. As two plates slowly converge, the leading edge of one slab is bent downward, allowing it to slide beneath the other. The surface expression produced by the descending plate is a *deep-ocean trench*,

like the Peru–Chile trench illustrated in Figures 1.27, 1.30, and 1.31.

Plate margins where oceanic crust is being consumed are called **subduction zones.** Here, as the subducted plate moves downward, it enters a high-temperature, high-pressure environment. Some subducted materials, as well as more voluminous amounts of the asthenosphere located above the subducting slab, melt and migrate upward into the overriding plate. Occasionally this molten rock may reach

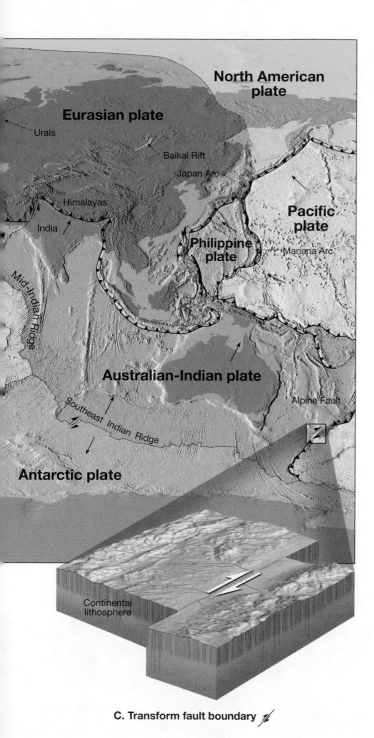

C. Transform fault boundary ⚡

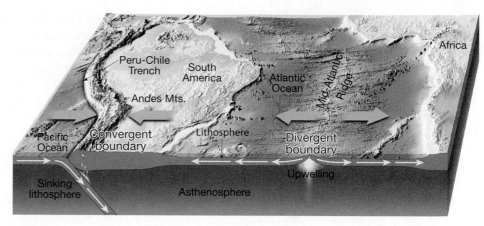

FIGURE 1.31 Convergent boundaries occur where two plates move together, as along the western margin of South America. Divergent boundaries are located where adjacent plates move away from one another. The Mid-Atlantic Ridge is such a boundary.

denser oceanic lithosphere sinks into the asthenosphere (Figure 1.31). The classic convergent boundary of this type occurs along the western margin of South America where the Nazca plate descends beneath the adjacent continental block. Here subduction along the Peru–Chile trench gave rise to the Andes Mountains, a linear chain of deformed rocks capped with numerous volcanoes—a number of which are still active.

The simplest type of convergence occurs where one oceanic plate is thrust beneath another. At such sites subduction results in the production of magma in a manner similar to that in the Andes, except volcanoes grow from the floor of the ocean rather than on a continent. If this activity is sustained, it will eventually build a chain of volcanic structures that emerge from the sea as a *volcanic island arc*. Most volcanic island arcs are found in the Pacific Ocean, as exemplified by the Aleutian, Mariana, and Tonga islands.

As we saw earlier, when an oceanic plate is subducted beneath continental lithosphere, an Andean-type mountain range develops along the margin of the continent. However, if the subducting plate also contains continental lithosphere, continued subduction eventually brings the two continents together (**FIGURE 1.32**). Whereas oceanic lithosphere is relatively dense and sinks into the

the surface, where it gives rise to explosive volcanic eruptions such as Mount St. Helens in 1980. However, much of this molten rock never reaches the surface; rather, it solidifies at depth and acts to thicken the crust.

Whenever slabs of continental lithosphere and oceanic lithosphere converge, the continental plate being less dense remains "floating," while the

FIGURE 1.32 When two plates containing continental lithosphere collide, complex mountains are formed. The formation of the Himalayas represents a relatively recent example.

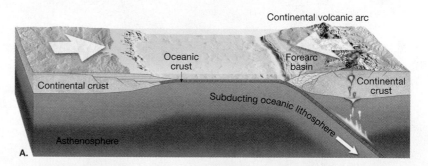

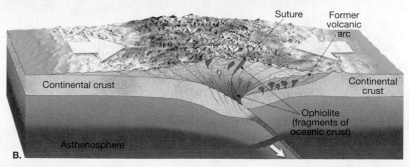

asthenosphere, continental lithosphere is buoyant, which prevents it from being subducted to any great depth. The result is a collision between the two continental blocks (Figure 1.32). Such a collision occurred when the subcontinent of India "rammed" into Asia and produced the Himalayas—the most spectacular mountain range on Earth (**FIGURE 1.33**). During this collision, the continental crust buckled, fractured, and was generally shortened and thickened. In addition to the Himalayas, several other major mountain systems, including the Alps, Appalachians, and Urals, formed during continental collisions along convergent plate boundaries.

TRANSFORM FAULT BOUNDARIES.
Transform fault boundaries are located where plates grind past each other without either generating new lithosphere or consuming old lithosphere. These faults form in the direction of plate movement and were first discovered in association with offsets in oceanic ridges (Figure 1.30C).

Although most transform faults are located along oceanic ridges, some slice through the continents. Two examples are the earthquake-prone San Andreas Fault of California (**FIGURE 1.34**) and the Alpine Fault of New Zealand. Along the San Andreas Fault the Pacific plate is moving toward the northwest, relative to the adja-

cent North American plate (Figure 1.30). The movement along this boundary does not go unnoticed. As these plates pass, strain builds in the rocks on opposite sides of the fault. Occasionally the rocks adjust, releasing energy in the form of a great earthquake of the type that devastated San Francisco in 1906.

CHANGING BOUNDARIES. Although the total surface area of Earth does not change, individual plates may diminish or grow in area depending on the distribution of convergent and divergent boundaries. For example, the Antarctic and African plates are almost entirely bounded by spreading centers and hence are growing larger. By contrast, the Pacific plate is being subducted along much of its perimeter and is therefore diminishing in area. At the current rate, the Pacific would close completely in 300 million years—but this is unlikely because changes in plate boundaries will probably occur before that time.

FIGURE 1.33 Mt. Everest in the Himalayas. This towering mountain chain is associated with plate convergence and began forming about 50 million years ago as India collided with Asia. *(Photo by Stephan Auth/photolibrary.com)*

New plate boundaries are created in response to changes in the forces acting on the lithosphere. For example, a relatively new divergent boundary is located in Africa, in a region known as the East African Rift Valleys. If spreading continues in this region, the African plate will split into two plates, separated by a new ocean basin. At other locations plates carrying continental crust are moving toward each other. Eventually these continents may collide and be sutured together. Thus, the boundary that once separated these plates disappears, and two plates become one.

As long as temperatures within the interior of our planet remain significantly higher than those at the surface, material within Earth will continue to circulate. This internal flow, in turn, will keep the rigid outer shell of Earth in motion. Thus, while Earth's internal heat engine is operating, the positions and shapes of the continents and ocean basins will change and Earth will remain a dynamic planet.

In the remaining chapters we will examine in more detail the workings of our dynamic planet in light of the plate tectonics theory.

DID YOU KNOW?
The average rate at which plates move relative to each other is roughly the same rate at which human fingernails grow.

San Andreas Fault

FIGURE 1.34 California's San Andreas Fault is an example of a transform fault boundary. *(Photo by Michael Collier)*

CONCEPT CHECK 1.12

❶ What are the three basic types of plate boundaries and the relative movement associated with each?

❷ Name the type of plate boundary with which each of the following is associated: deep-ocean trench, seafloor spreading, San Andreas Fault, subduction, oceanic ridge, and Himalaya Mountains.

CHAPTER ONE
An Introduction to Geology in Review

◉ *Geology* means "the study of Earth." The two broad areas of the science of geology are (1) *physical geology*, which examines the materials composing Earth and the processes that operate beneath and upon its surface; and (2) *historical geology*, which seeks to understand the origin of Earth and its development through time.

◉ The relationship between people and the natural environment is an important focus of geology. This includes natural hazards, resources, and human influences on geologic processes.

◉ During the 17th and 18th centuries, *catastrophism* influenced the formulation of explanations about Earth. Catastrophism states that Earth's landscapes developed over short time spans primarily as a result of great catastrophes. By contrast, *uniformitarianism*, one of the fundamental principles of modern geology advanced by *James Hutton* in the late 1700s, states that the physical, chemical, and biological laws that operate today have also operated in the geologic past. The idea is often summarized as "*The present is the key to the past.*" Hutton argued that processes that appear to be slow acting could, over long spans of time, produce effects that are just as great as those resulting from sudden catastrophic events.

◉ Using the principles of *relative dating*, the placing of events in their proper sequence or order without knowing their age in years, scientists developed a geologic time scale during the 19th century. Relative dates can be established by applying such principles as the *law of superposition* and the *principle of fossil succession*.

⊙ All science is based on the assumption that the natural world behaves in a consistent and predictable manner. The process by which scientists gather facts and formulate scientific *hypotheses* and *theories* is called the *scientific method*. To determine what is occurring in the natural world, scientists often (1) collect facts, (2) ask questions and develop hypotheses that may answer these questions, (3) develop observations and experiments to test the hypotheses, and (4) accept, modify, or reject hypotheses on the basis of extensive testing. Other discoveries represent purely theoretical ideas that have stood up to extensive examination. Still other scientific advancements have been made when a totally unexpected happening occurred during an experiment.

⊙ Earth's physical environment is traditionally divided into three major parts: the solid Earth, or *geosphere*; the water portion of our planet, the *hydrosphere*; and Earth's gaseous envelope, the *atmosphere*. In addition, the *biosphere*, the totality of life on Earth, interacts with each of the three physical realms and is an equally integral part of Earth.

⊙ Although each of Earth's four spheres can be studied separately, they are all related in a complex and continuously interacting whole that we call the *Earth system*. *Earth system science* uses an interdisciplinary approach to integrate the knowledge of several academic fields in the study of our planet and its global environmental problems.

⊙ A *system* is a group of interacting parts that form a complex whole. *Closed systems* are those in which energy moves freely in and out, but matter does not enter or leave the system. In an open system, both energy and matter flow into and out of the system.

⊙ Most natural systems have mechanisms that tend to enhance change, called *positive feedback mechanisms*, and other mechanisms, called *negative feedback mechanisms*, that tend to resist change and thus stabilize the system.

⊙ The two sources of energy that power the Earth system are (1) the Sun, which drives the external processes that occur in the atmosphere, hydrosphere, and at Earth's surface, and (2) heat from Earth's interior that powers the internal processes that produce volcanoes, earthquakes, and mountains.

⊙ The *rock cycle* is one of the many cycles or loops of the Earth system in which matter is recycled. The rock cycle is a means of viewing many of the interrelationships of geology. It illustrates the origin of the three basic rock groups and the role of various geologic processes in transforming one rock type into another.

⊙ The *nebular* theory describes the formation of the solar system. The planets and Sun began forming about 5 billion years ago from a large cloud of dust and gases. As the cloud contracted, it began to rotate and assume a disk shape. Material that was gravitationally pulled toward the center became the *protosun*. Within the rotating disk, small centers, called *planetesimals*, swept up more and more of the cloud's debris. Because of the high temperatures near the Sun, the inner planets were unable to accumulate many of the elements that vaporize at low temperatures. Because of the very cold temperatures existing far from the Sun, the large outer planets consist of huge amounts of ices and lighter materials. These substances account for the comparatively large sizes and low densities of the outer planets.

⊙ Earth's internal structure is divided into layers based on differences in chemical composition and on the basis of changes in physical properties. Compositionally, Earth is divided into a thin outer *crust*, a solid rocky *mantle*, and a dense *core*. Other layers, based on physical properties, include the lithosphere, asthenosphere, lower mantle, outer core, and inner core.

⊙ Two principal divisions of Earth's surface are the *continents* and *ocean basins*. A significant difference is their relative levels. The elevation differences between continents and ocean basins is primarily the result of differences in their respective densities and thicknesses.

⊙ The largest features of the continents can be divided into two categories: *mountain belts* and the *stable interior*. The ocean floor is divided into three major topographic units: *continental margins*, *deep-ocean basins*, and *oceanic (mid-ocean) ridges*.

⊙ The *theory of plate tectonics* provides a comprehensive model of Earth's internal workings. It holds that Earth's rigid outer lithosphere consists of several segments called *lithospheric plates* that are slowly and continually in motion relative to one another. Most earthquakes, volcanic activity, and mountain building are associated with the movements of these plates.

⊙ The three distinct types of plate boundaries are (1) *divergent boundaries*, where plates move apart; (2) *convergent boundaries*, where plates move together, causing one to go beneath another, or where plates collide, which occurs when the leading edges are made of continental crust; and (3) *transform fault boundaries*, where plates slide past one another.

Key Terms

abyssal plains (p. 26)
asthenosphere (p. 24)
atmosphere (p. 13)
biosphere (p. 14)
catastrophism (p. 4)
closed systems (p. 15)
continental margin (p. 25)
continental rise (p. 26)

continental shelf (p. 25)
continental slope (p. 25)
convergent boundaries (p. 29)
core (p. 24)
crust (p. 23)
deep-ocean basins (p. 26)
deep-ocean trenches (p. 26)
divergent boundaries (p. 29)

Earth system science (p. 15)
fossils (p. 7)
fossil succession,
   principle of (p. 7)
geology (p. 2)
geosphere (p. 14)
historical geology (p. 2)
hydrosphere (p. 13)

hypothesis (p. 8)
igneous rocks (p. 20)
inner core (p. 24)
interface (p. 16)
lithosphere (p. 24)
lithospheric plates (p. 28)
lower mantle (p. 24)
magma (p. 20)

GIVE IT SOME THOUGHT

❶ Refer to Figure 1.17. Which of the four main components of the Earth System (atmosphere, biosphere, geosphere, hydrosphere) were involved in the natural disaster at Caraballeda, Venezuela? Describe how each of the components you list contributed to the event depicted in the photo.

❷ Earth is 4.6 billion years old. This number does not come from studying Earth rocks since Earth's oldest rocks only date to about 4 billion years. The number comes from studying lunar rocks and meteorites. Based on what you know about the formation of our solar system, how old do you think the other planets are? Why did you assign this age or these ages?

❸ Why do geologists look to meteorites to tell us about the internal structure of Earth?

❹ After entering a dark room, you turn on a wall switch but the light does not come on. Suggest at least three hypotheses that might explain this observation.

❺ Each of the following statements may either be a hypothesis (H), a theory (T), or an observation (O). Use one of these letters to identify each statement. Briefly explain each choice.

  a. A scientist proposes that a recently discovered ring-shaped structure is the remains of an ancient meteorite crater.

  b. The Redwall Formation in the Grand Canyon is composed primarily of limestone.

  c. Earth is composed of several large plates that move and interact with each other.

  d. Since 1885, the terminus of Canada's Athabasca Glacier has receded 1.5 kilometers.

❻ Consider the possible results of the following scenario and describe one positive and one negative feedback. *Earth is getting warmer, consequently evaporation is increasing.*

❼ Refer to Figure 1.22. How does the rock cycle diagram, in particular the process arrows, support the fact that sedimentary rocks are the most abundant rock type on the surface of Earth?

❽ Look at the concept map linking the four spheres of the Earth system. All of the spheres are linked by arrows representing processes by which the spheres interact and influence each other. For each arrow, describe at least one process.

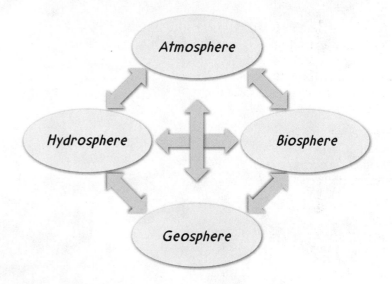

❾ Refer to Figure 1.27 showing major features and topography of Earth's solid surface. Give an example of where you would expect to find (a) very old, deformed rocks, and (b) young undisturbed rocks. Explain your choices.

Companion Website

www.mygeoscienceplace.com

The *Essentials of Geology, 11e* companion Website contains numerous multimedia resources accompanied by assessments to aid in your study of the topics in this chapter. The use of this site's learning tools will help improve your understanding of geology. Utilizing the access code that accompanies this text, visit **www.mygeoscienceplace.com** in order to:

- **Review** key chapter concepts.
- **Read** with links to the Pearson eText and to chapter-specific web resources.
- **Visualize** and comprehend challenging topics using the learning activities in *GEODe: Essentials of Geology* and the *Geoscience Animations Library*.
- **Test** yourself with online quizzes.

Matter and Minerals

EARTH'S CRUST AND OCEANS ARE THE SOURCE OF A WIDE VARIETY OF useful and essential minerals. Most people are familiar with the common uses of many basic metals, including aluminum in beverage cans, copper in electrical wiring, and gold and silver in jewelry. But some people are not aware that pencil lead contains the greasy-feeling mineral graphite and that bath powders and many cosmetics contain the mineral talc. Moreover, many do not know that drill bits impregnated with diamonds are employed by dentists to drill through tooth enamel, or that the common mineral quartz is the source of silicon for computer chips. In fact, practically every manufactured product contains materials obtained from minerals.

Crystals of elbaite, a colorful member of the tournaline group. The white mineral, albite, is a member of the plagioclase feldspar group. (Photo by Jeffrey Scovil)

To assist you in learning the important concepts in this chapter, focus on the following questions:

- ◉ What are minerals, and how are they different from rocks?
- ◉ What are the smallest particles of matter?
- ◉ How do atoms bond?
- ◉ How do isotopes of the same element vary, and why are some isotopes radioactive?
- ◉ What are some of the physical and chemical properties of minerals? How can these properties distinguish one mineral from another?
- ◉ What are the eight elements that make up most of Earth's continental crust?
- ◉ What is the most abundant mineral group? What do all minerals within this group have in common?
- ◉ When is the term *ore* used with reference to a mineral? What are the common ores of iron and lead?

Minerals: Building Blocks of Rocks

Matter and Minerals

ESSENTIALS OF GEOLOGY Introduction

We begin our discussion of Earth materials with an overview of **mineralogy** (*mineral* = mineral, *ology* = the study of), because minerals are the building blocks of rocks. In addition, minerals have been employed by humans for both useful and decorative purposes for thousands of years (**FIGURE 2.1**). The first minerals mined were flint and chert, which people fashioned into weapons and cutting tools. As early as 3700 BC, Egyptians began mining gold, silver, and copper; and by 2200 BC humans discovered how to combine copper with tin to make bronze, a strong, hard alloy. Later, humans developed a process to extract iron from minerals such as hematite—a discovery that marked the decline of the Bronze Age. By about 800 BC, iron-working technology had advanced to the point that weapons and many everyday objects were made of iron rather than copper, bronze, or wood. During the Middle Ages, mining of a variety of minerals was common throughout Europe, and the impetus for the formal study of minerals was in place.

The term *mineral* is used in several different ways. For example, those concerned with health and fitness extol the benefits of vitamins and minerals. The mining industry typically uses the word when referring to anything taken out of the ground, such as coal, iron ore, or sand and gravel. The guessing game known as *Twenty Questions* usually begins with the question, *Is it animal, vegetable, or mineral?* What criteria do geologists use to determine whether something is a mineral?

Geologists define **mineral** as *any naturally occurring inorganic solid that possesses an orderly crystalline structure and can be represented by a chemical formula.* Thus, Earth materials that are classified as minerals exhibit the following characteristics:

1. **Naturally occurring.** Minerals form by natural, geologic processes. Synthetic materials, meaning those produced in a laboratory or by human intervention, are not considered minerals.

2. **Solid substance.** Only solid crystalline substances are considered minerals. Ice (frozen water) fits this criterion and is considered a mineral, whereas liquid water and water vapor do not. The exception is mercury, which is found in its liquid form in nature.

FIGURE 2.1 Collection of well-developed quartz crystals found near Hot Springs, Arkansas. *(Photo by Jeff Scovil)*

3. **Orderly crystalline structure.** Minerals are crystalline substances, which means their atoms are arranged in an orderly, repetitive manner (**FIGURE 2.2**). This orderly packing of atoms is reflected in the regularly shaped objects called crystals. Some naturally occurring solids, such as volcanic glass (obsidian), lack a repetitive atomic structure and are not considered minerals.

4. **Generally inorganic.** Inorganic crystalline solids, such as ordinary table salt (halite), that are found naturally in the ground are considered minerals. (Organic compounds, on the other hand, are generally not. Sugar, a crystalline solid like salt but which comes from sugarcane or sugar beets, is a common example of such an organic compound.) Many marine animals secrete inorganic compounds, such as calcium carbonate (calcite), in the form of shells and coral reefs. If these materials are buried and become part of the rock record, they are considered minerals by geologists.

5. **Can be represented by a chemical formula.** Most minerals are chemical compounds having compositions that can be expressed by a chemical formula. For example, the common mineral quartz has the formula SiO_2, which indicates that quartz consists of silicon (Si) and oxygen (O) atoms in a ratio of one-to-two. This proportion of silicon to oxygen is true for any sample of pure quartz, regardless of its origin. However, the compositions of some minerals vary *within specific, well-defined limits*. This occurs because certain elements can substitute for others of similar size without changing the mineral's internal structure. An example is the mineral olivine in which either the element magnesium (Mg) or the element iron (Fe) may occupy the same site in the crystal structure. Therefore, olivine's formula, $(Mg, Fe)_2SiO_4$, expresses variability in the relative amounts of magnesium and iron. However, the ratio of magnesium plus iron (Mg + Fe) to silicon (Si) and oxygen (O) remains fixed at 2:1:4.

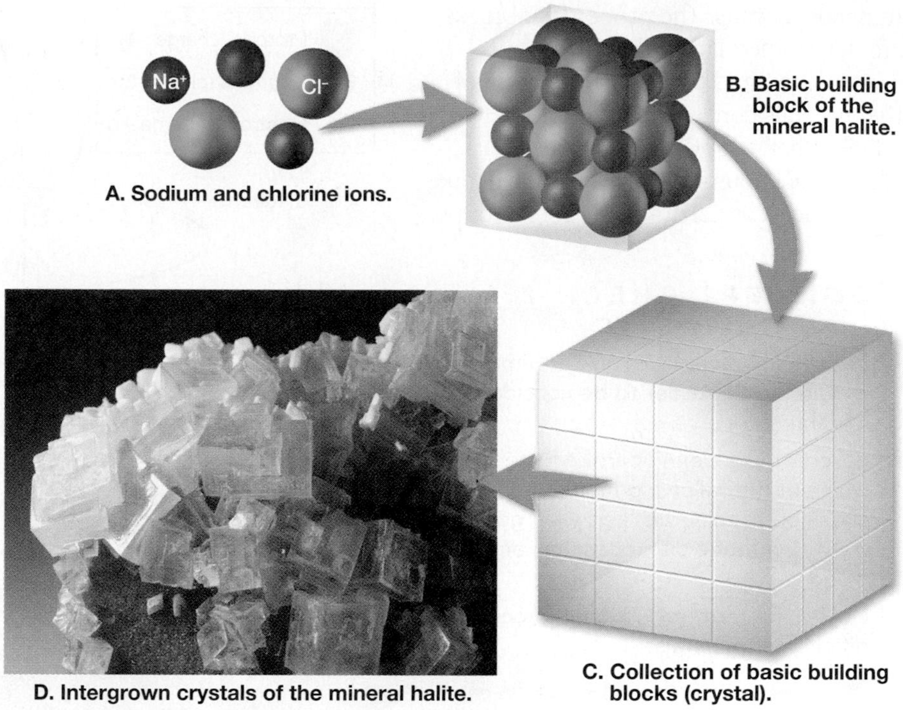

A. Sodium and chlorine ions.

B. Basic building block of the mineral halite.

C. Collection of basic building blocks (crystal).

D. Intergrown crystals of the mineral halite.

FIGURE 2.2 This diagram illustrates the orderly arrangement of sodium and chloride ions in the mineral halite. The arrangement of atoms into basic building blocks having a cubic shape results in regularly shaped cubic crystals. *(Photo by Dennis Tasa)*

Some rocks are composed of nonmineral matter. These include the volcanic rocks *obsidian* and *pumice,* which are noncrystalline glassy substances, and *coal,* which consists of solid organic debris.

Although this chapter deals primarily with the nature of minerals, keep in mind that most rocks are simply aggregates of

In contrast to minerals, rocks are more loosely defined. Simply, a **rock** is any solid mass of mineral, or mineral-like, matter that occurs naturally as part of our planet. Most rocks, like the common rock granite shown in **FIGURE 2.3**, occur as aggregates of several different minerals. The term *aggregate* implies that the minerals are joined in such a way that their individual properties are retained. Note that the mineral constituents of granite can be easily identified. However, some rocks are composed almost entirely of one mineral. A common example is the sedimentary rock *limestone,* which consists of impure masses of the mineral calcite.

Granite (Rock)

FIGURE 2.3 Most rocks are aggregates of two or more minerals. Shown here is a hand sample of the igneous rock granite and three of its major constituent minerals. *(Photos by E. J. Tarbuck)*

Quartz (Mineral)

Hornblende (Mineral)

Feldspar (Mineral)

DID YOU KNOW?
Archaeological finds show that more than 2000 years ago the Romans used lead for pipes to transport water within their buildings. In fact, Roman smelting of lead and copper ores between 500 BC and AD 300 caused a small but significant rise in atmospheric pollution, as recorded in Greenland ice cores.

minerals. Because the properties of rocks are determined largely by the chemical composition and crystalline structure of the minerals contained within them, we will first consider these Earth materials. Then, in Chapters 3, 6, and 7 we will take a closer look at Earth's major rock groups.

CONCEPT CHECK 2.1

❶ List five characteristics an Earth material should have in order to be considered a mineral.

❷ Based on the definition of a mineral, which of the following materials are not classified as minerals, and why: gold; water; synthetic diamonds; ice; and wood.

❸ Define the term *rock*. How do rocks differ from minerals?

Atoms: Building Blocks of Minerals

When minerals are carefully examined, even under optical microscopes, the innumerable tiny particles of their internal structures are not discernable. Nevertheless, all matter, including minerals, is composed of minute building blocks called **atoms**—the smallest particles that cannot be chemically split. Atoms, in turn contain even smaller particles—*protons* and *neutrons* located in a central **nucleus** that is surrounded by *electrons* (**FIGURE 2.4**).

Properties of Protons, Neutrons, and Electrons

Protons and **neutrons** are very dense particles with almost identical masses. By contrast, **electrons** have a negligible mass, about $1/2000^{th}$ that of a proton. For comparison, if a proton or a neutron had the mass of a baseball, an electron would have the mass of a single grain of rice.

Both protons and electrons share a fundamental property called *electrical charge*. Protons have an electrical charge of $+1$, and electrons have a charge of -1. Neutrons, as the name suggests, have no charge. The charge of protons and electrons are equal in magnitude, but opposite in polarity, so when these two particles are paired, the charges cancel each other. Since matter typically contains equal numbers of positively charged protons and negatively charged electrons, most substances are electrically neutral.

In illustrations, electrons are sometimes shown orbiting the nucleus in a manner that resembles the planets of our solar system orbiting the Sun (Figure 2.4A). However, electrons do not actually behave this way. A more realistic depiction shows electrons

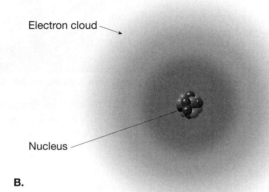

as a cloud of negative charges surrounding the nucleus (Figure 2.4B). Studies of the arrangements of electrons show that they move about the nucleus in regions called *principal shells*, each with an associated energy level. In addition, each shell can hold a specific number of electrons, with the outermost shell containing **valence electrons** that interact with other atoms to form chemical bonds.

Most of the atoms in the universe (except hydrogen and helium) were created inside massive stars by nuclear fusion and released into interstellar space during hot, fiery supernova explosions. As this ejected material cooled, the newly formed nuclei attracted electrons to complete their atomic structure. At the temperatures found at Earth's surface, all free atoms (not bonded to other atoms) have a full complement of electrons—one for each proton in the nucleus.

Elements: Defined by Their Number of Protons

The simplest atoms have only one proton in their nuclei, whereas others have more than 100. The number of protons in the nucleus of an atom, called the **atomic number,** determines its chemical nature. All atoms with the same number of protons have the same chemical and physical properties. Together, a group of the same kind of atoms is called an **element.** There are about 90 naturally occurring elements and 23 that have been synthesized. You are probably familiar with the names of many

FIGURE 2.4 Two models of the atom. **A.** A very simplified view of the atom. The central nucleus consists of protons and neutrons encircled by high-speed electrons. **B.** This model of the atom shows spherically shaped electron clouds (shells) surrounding a central nucleus. The nucleus contains virtually all of the mass of the atom. The remainder of the atom is the space in which the light, negatively charged electrons reside. (The relative sizes of the nuclei shown are greatly exaggerated.)

Protons (charge +1)

Neutrons (charge 0)

Electrons (charge –1)

Electron

Nucleus

A.

Electron cloud

Nucleus

B.

DID YOU KNOW?

The purity of gold is expressed by the number of *karats*, where 24 karats is pure gold. Gold less than 24 karats is an alloy (mixture) of gold and another metal, usually copper or silver. For example, 14-karat gold contains 14 parts gold (by weight) mixed with 10 parts of other metals.

elements including carbon, nitrogen, and oxygen. All carbon atoms have six protons, whereas all nitrogen atoms have seven protons, and all oxygen atoms have eight.

Elements are organized so that those with similar properties line up in columns. This arrangement, called the **periodic table**, is shown in **FIGURE 2.5**. Each element has been assigned a one- or two-letter symbol. The atomic numbers and masses are also included for each element.

Atoms of the naturally occurring elements are the basic building blocks of Earth's minerals. A few minerals, such as native copper, diamonds, and gold, are made entirely of atoms of only one element. However, most elements tend to join with atoms of other elements to form **chemical compounds.** Most minerals are chemical compounds composed of atoms of two or more elements.

CONCEPT CHECK 2.2

❶ List the three main particles of an atom and explain how they differ from one another.

❷ Make a simple sketch of an atom and label its three main particles.

❸ What is the significance of valence electrons?

Why Atoms Bond

Except for a group of elements known as the noble gases, atoms bond to one another under the conditions (temperatures and pressures) that occur on Earth. Some atoms bond to form *ionic compounds*, some form *molecules*, and still others form *metallic substances*. Why does this happen? Experiments show that electrical forces hold atoms together and bond them to each other. These electrical attractions lower the total energy of the bonded atoms, which, in turn, generally makes them more stable. Consequently, atoms that are bonded in compounds tend to be more stable than atoms that are free (not bonded).

As was noted earlier, valence (outer shell) electrons are generally involved in chemical bonding. **FIGURE 2.6** shows a shorthand way of representing the number of valence electrons. Notice that the elements in Group I have one valence electron, those in Group II have two valence electrons, and so on, up to eight valence electrons in Group VIII.

Octet Rule

The noble gases (except helium) have very stable electron arrangements with eight valence electrons and, therefore, tend to

FIGURE 2.5 Periodic table of the elements.

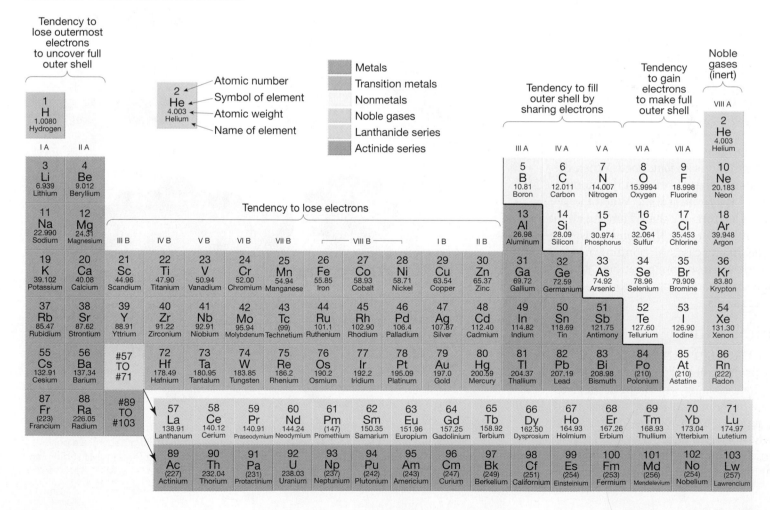

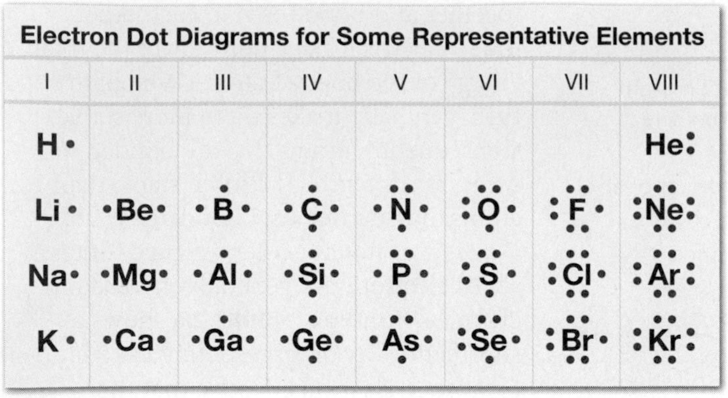

Electron Dot Diagrams for Some Representative Elements							
I	II	III	IV	V	VI	VII	VIII
H •							He:
Li •	•Be•	•B•	•Ċ•	•N̈•	:Ö•	:F̈•	:N̈e:
Na•	•Mg•	•Äl•	•S̈i•	•P̈•	:S̈•	:C̈l•	:Är:
K•	•Ca•	•Ga•	•Ge•	•Äs•	:S̈e•	:B̈r•	:K̈r:

FIGURE 2.6 Dot diagrams for some representative elements. Each dot represents a valence electron found in the outermost principal shell.

lack chemical reactivity. Many other atoms gain, lose, or share electrons during chemical reactions to end up with electron arrangements of the noble gases. This observation led to a chemical guideline known as the **octet rule**: *Atoms tend to gain, lose, or share electrons until they are surrounded by eight valence electrons.* Although there are exceptions to the octet rule, it is a useful *rule of thumb* for understanding chemical bonding.

When an atom's outer shell does not contain eight electrons, it is likely to chemically bond to other atoms to fill its shell. A **chemical bond** is the transfer or sharing of electrons that allows each atom to attain a full valence shell of electrons. Some atoms do this by transferring all of their valence electrons to other atoms so that an inner shell becomes the full valence shell.

When the valence electrons are transferred between the elements to form ions, the bond is an *ionic bond.* When the electrons are shared between the atoms, the bond is a *covalent bond.* When the valence electrons are shared among all the atoms in a substance, the bonding is *metallic.* In any case, the bonding atoms get stable electron configurations, which usually consist of eight electrons in their outmost shells.

Ionic Bonds: Electrons Transferred

Perhaps the easiest type of bond to visualize is the *ionic bond*, in which one atom gives up one or more of its valence electrons to another atom to form **ions**—*positively and negatively charged atoms.* The atom that loses electrons becomes a positive ion, and the atom that gains electrons becomes a

negative ion. Oppositely charged ions are strongly attracted to one another and join to form ionic compounds.

Consider the ionic bonding that occurs between sodium (Na) and chlorine (Cl) to produce sodium chloride, the mineral halite—common table salt. Notice in **FIGURE 2.7A** that sodium gives up its single valence electron to chlorine. As a result,

sodium now has a stable configuration with eight electrons in its outermost shell. By acquiring the electron that sodium loses, chlorine (which has seven valence electrons) gains the eighth electron needed to complete its outermost shell. Thus, through the transfer of a single electron, both the sodium and chlorine atoms have acquired a stable electron configuration.

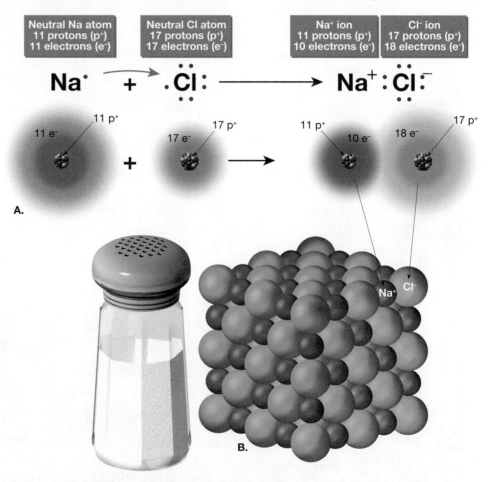

FIGURE 2.7 Chemical bonding of sodium chloride (table salt). **A.** Through the transfer of one electron in the outer shell of a sodium atom to a chlorine atom, sodium becomes a positive ion and chlorine a negative ion. **B.** Diagram illustrating the arrangement (packing) of sodium and chlorine ions in table salt.

After electron transfer takes place, the atoms are no longer electrically neutral. By giving up one electron, a neutral sodium atom becomes positively charged (with 11 protons and 10 electrons). Similarly, by acquiring one electron, a neutral chlorine atom becomes negatively charged (with 17 protons and 18 electrons). We know that ions with like charges repel, and those with unlike charges attract. Thus, an **ionic bond** is the attraction of oppositely charged ions to one another, producing an electrically neutral compound.

FIGURE 2.7B illustrates the arrangement of sodium and chlorine ions in ordinary table salt. Notice that salt consists of alternating sodium and chlorine ions, positioned in such a manner that each positive ion is attracted to and surrounded on all sides by negative ions, and vice versa. This arrangement maximizes the attraction between ions with opposite charges while minimizing the repulsion between ions with identical charges. Thus, ionic compounds consist of an orderly arrangement of oppositely charged ions assembled in a definite ratio that provides overall electrical neutrality.

The properties of a chemical compound are dramatically different from the properties of the various elements comprising it. For example, sodium is a soft silvery metal that is extremely reactive and poisonous. If you were to consume even a small amount of elemental sodium, you would need immediate medical attention. Chlorine, a green poisonous gas, is so toxic that it was used as a chemical weapon during World War I. Together, however, these elements produce sodium chloride, a harmless flavor enhancer that we call table salt. Thus, when elements combine to form compounds their properties change significantly.

Covalent Bonds: Electrons Shared

Sometimes the forces that hold atoms together cannot be understood on the basis of the attraction of oppositely charged ions. One example is the hydrogen molecule (H_2), in which the two hydrogen atoms are held together tightly and no ions are present. The strong attractive force that holds two hydrogen atoms together results from a **covalent bond,** *a chemical bond formed by the sharing of a pair of electrons between atoms.*

Imagine two hydrogen atoms (each with one proton and one electron) approaching one another so that their electron clouds overlap (**FIGURE 2.8**). Once they meet, the electron configuration will change so that both electrons will primarily occupy the space between the atoms. In other words, the two electrons are shared by both hydrogen atoms and attracted simultaneously by the positive charge of the proton in the nucleus of each atom. The attraction between the electrons and both nuclei holds these atoms together. Although ions do not exist in hydrogen molecules, the force that holds these atoms together arises from the attraction of oppositely charged particles—protons in the nuclei and electrons shared by the atoms.

Metallic Bonds: Electrons Free to Move

In **metallic bonds,** the valence electrons are free to move from one atom to another so that all atoms share the available valence electrons. This type of bonding is found in metals such as copper, gold, aluminum, and silver, and in alloys such as brass and bronze. Metallic bonding accounts for the high electrical conductivity of metals, the ease with which metals are shaped, and numerous other special properties.

CONCEPT CHECK 2.3

❶ What is the difference between an atom and an ion?

❷ What occurs in an atom to produce a positive ion? What occurs in an atom to produce a negative ion?

❸ Briefly distinguish between ionic and covalent bonding and the role that electrons play in both.

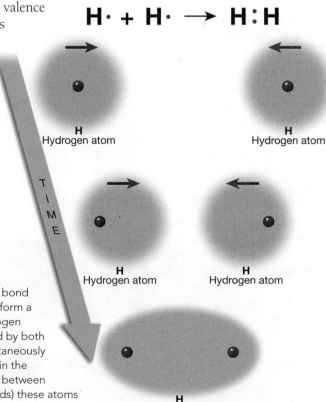

FIGURE 2.8 Formation of a covalent bond between two hydrogen atoms (H) to form a hydrogen molecule (H_2). When hydrogen atoms bond, the electrons are shared by both hydrogen atoms and attracted simultaneously by the positive charge of the proton in the nucleus of each atom. The attraction between electrons and both nuclei holds (bonds) these atoms together.

Isotopes and Radioactive Decay

The **mass number** of an atom is simply the total number of its protons and neutrons. All atoms of a particular element have the same number of protons but they may have varying numbers of neutrons. Atoms with the same number of protons but different numbers of neutrons are **isotopes** of that element. Isotopes of the same element are labeled by placing the mass number after the element's name or symbol. For example, carbon has three well-known isotopes. One has a mass number of 12 (carbon-12), another has a mass number of 13 (carbon-13), and the third, carbon-14, has a mass number of 14. Carbon-12 must also have six neutrons to give it a mass number of 12. Carbon-14, on the other hand, has six protons plus eight neutrons to give it a mass number of 14.

In chemical behavior, all isotopes of the same element are nearly identical. To distinguish among them is like trying to differentiate identical twins, with one weighing slightly more than the other. Because isotopes of the same element exhibit the same chemical behavior, they often become parts of the same mineral. For example, when the mineral calcite ($CaCO_3$) forms, some of its carbon atoms are carbon-12, and some are carbon-14.

The nuclei of most atoms are stable. However, many elements do have isotopes in which the nuclei are unstable—carbon-14 is one example of an unstable isotope. In this context, *unstable* means that the nuclei change through a random process called **radioactive decay.** During radioactive decay, unstable isotopes radiate energy and emit particles. The rates at which unstable isotopes decay are measurable. Therefore, certain radioactive atoms are used to determine the ages of fossils, rocks, and minerals. A discussion of radioactive decay and its applications in dating past geologic events appears in Chapter 18.

CONCEPT CHECK 2.4

❶ What is an isotope?

Physical Properties of Minerals

 Matter and Minerals

ESSENTIALS OF GEOLOGY **Physical Properties of Minerals**

Minerals have definite crystalline structures and chemical compositions that give them unique sets of physical and chemical properties shared by all samples of that mineral. For example, all specimens of halite have the same hardness, the same density, and break in a similar manner. Because a mineral's internal structure and chemical composition are difficult to determine without the aid of sophisticated tests and equipment, the more easily recognized physical properties are frequently used in identification.

Optical Properties

Of the many optical properties of minerals, their luster, their ability to transmit light, their color, and their streak are most frequently used for mineral identification.

LUSTER. The appearance or quality of light reflected from the surface of a mineral is known as **luster.** Minerals that have the appearance of metals, regardless of color, are said to have a *metallic luster* (**FIGURE 2.9**). Some metallic minerals, such as native copper and galena, develop a dull coating or tarnish when exposed to the atmosphere. Because they are not as shiny as samples with freshly broken surfaces, these samples are often said to exhibit a *submetallic luster.*

Most minerals have a *nonmetallic luster* and are described using various adjectives such as *vitreous* or *glassy.* Other nonmetallic minerals are described as having a *dull* or *earthy luster* (a dull appearance like soil) or a *pearly luster* (such as a pearl or the inside of a clamshell). Still others exhibit a *silky luster* (like satin cloth) or a *greasy luster* (as though coated in oil).

THE ABILITY TO TRANSMIT LIGHT. Another optical property used in the identification of minerals is the ability to transmit light. When no light is transmitted, the mineral is described as *opaque;* when light, but not an image, is transmitted through a mineral it is said to be *translucent.* When both light and an image are visible through the sample, the mineral is described as *transparent.*

COLOR. Although **color** is generally the most conspicuous characteristic of any mineral, it is considered a diagnostic property of only a few minerals. Slight impurities in the common mineral quartz, for example, give it a variety of tints including pink, purple, yellow, white, gray, and even black (**FIGURE 2.10**). Other minerals, such as tourmaline, also exhibit a variety of hues, with multiple colors sometimes occurring in the same sample. Thus, the use of color as a means of identification is often ambiguous or even misleading.

FIGURE 2.9 The freshly broken sample of galena (right) displays a metallic luster, while the sample on the left is tarnished and has a submetallic luster. *(Photo courtesy of E. J. Tarbuck)*

FIGURE 2.10 Quartz. Some minerals, such as quartz, occur in a variety of colors. These samples include crystal quartz (colorless), amethyst (purple quartz), citrine (yellow quartz), and smoky quartz (gray to black). *(Photo courtesy of E. J. Tarbuck)*

STREAK. The color of the mineral in powdered form, called **streak,** is often useful in identification. A mineral's streak is obtained by rubbing it across a *streak plate* (a piece of unglazed porcelain) and observing the color of the mark it leaves (**FIGURE 2.11**). Although the color of a mineral may vary from sample to sample, its streak is usually consistent in color.

Streak can also help distinguish between minerals with metallic luster and those with nonmetallic luster. Metallic minerals generally have a dense, dark streak, whereas minerals with nonmetallic luster typically have a light colored streak.

It should be noted that not all minerals produce a streak when rubbed across a streak plate. For example, the mineral quartz is harder than a porcelain streak plate. Therefore, no streak is observed using this method.

FIGURE 2.11 Although the color of a mineral is not always helpful in identification, the streak, which is the color of the powdered mineral, can be very useful. *(Photo by Dennis Tasa)*

Crystal Shape or Habit

Mineralogists use the term **crystal shape** or **habit** to refer to the common or characteristic shape of a crystal or aggregate of crystals. A few minerals exhibit somewhat regular polygons that are helpful in their identification. For example, magnetite crystals sometimes occur as octahedrons, garnets often form dodecahedrons, and halite and fluorite crystals tend to grow as cubes or near cubes. While most minerals have only one common habit, a few have two or more characteristic crystal shapes such as the pyrite sample shown in **FIGURE 2.12**.

Aragonite crystals.
(Photo by Ross Frid)

minerals

Mineral Strength

How easily minerals break or deform under stress is determined by the type and strength of the chemical bonds that hold the crystals together. Mineralogists use terms including *tenacity*, *hardness*, *cleavage*, and *fracture* to describe mineral strength and how minerals break when stress is applied.

TENACITY. The term **tenacity** describes a mineral's toughness, or its resistance to breaking or deforming. Minerals that are ionically bonded, such as fluorite and halite, tend to be *brittle* and shatter into small pieces when struck. By contrast, minerals with metallic bonds, such as native copper, are *malleable*, or easily hammered into different shapes. Minerals, including gypsum and talc, that can be cut into thin shavings are described as *sectile*. Still others, notably the micas, are *elastic* and will bend and snap back to their original shape after the stress is released.

HARDNESS. One of the most useful diagnostic properties is **hardness,** a measure of the resistance of a mineral to abrasion or scratching. This property is determined by rubbing a mineral of unknown hardness against one of known hardness, or vice versa. A numerical value of hardness can by obtained by using the **Mohs scale** of hardness, which consists of 10 minerals arranged in order from 1 (softest) to 10 (hardest), as shown in **FIGURE 2.14A**. It should be noted that the Mohs scale is a relative ranking, and it does not imply that mineral number 2, gypsum, is twice as hard as mineral 1, talc. In fact, gypsum is only slightly harder than talc, as **FIGURE 2.14B** indicates.

In the laboratory, other common objects can be used to determine the hardness of a mineral. These include a human fingernail, which has a hardness of about 2.5, a copper penny (3.5), and a piece of glass (5.5). The mineral gypsum, which has a

FIGURE 2.12 Although most minerals exhibit only one common crystal shape, some, such as pyrite, have two or more characteristic habits. *(Photos by Dennis Tasa)*

By contrast, some minerals rarely develop perfect geometric forms. Many of these, however, develop other characteristic shapes useful for identification. Some minerals tend to grow equally in all three dimensions, whereas others tend to be elongated in one direction, or flattened if growth in one dimension is suppressed. Commonly used terms to describe these and other crystal habits include *equant* (equidimensional), *bladed, fibrous, tabular, prismatic, platy, blocky,* and *botryoidal.* Some of these habits are pictured in **FIGURE 2.13**.

A. Bladed

C. Banded

B. Prismatic

D. Botryoidal

FIGURE 2.13 Some common crystal habits. **A. *Bladed.*** Elongated crystals that are flattened in one direction. **B. *Prismatic.*** Elongated crystals with faces that are parallel to a common direction. **C. *Banded.*** Minerals that have stripes or bands of different color or texture. **D. *Botryoidal.*** Groups of intergrown crystals resembling a bunch of grapes. *(Photos by Dennis Tasa)*

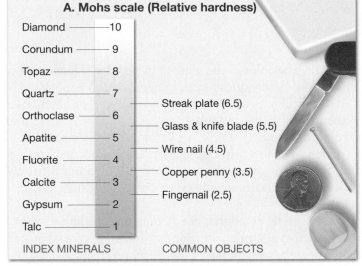

A. Mohs scale (Relative hardness)

INDEX MINERALS		COMMON OBJECTS
Diamond	10	
Corundum	9	
Topaz	8	
Quartz	7	
Orthoclase	6	Streak plate (6.5)
Apatite	5	Glass & knife blade (5.5)
Fluorite	4	Wire nail (4.5)
Calcite	3	Copper penny (3.5)
Gypsum	2	Fingernail (2.5)
Talc	1	

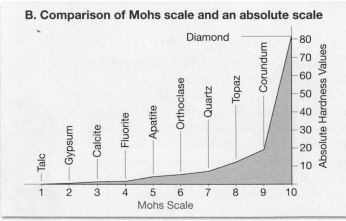

B. Comparison of Mohs scale and an absolute scale

FIGURE 2.14 Hardness scales. **A.** Mohs scale of hardness, with the hardness of some common objects. **B.** Relationship between Mohs relative hardness scale and an absolute hardness scale.

hardness of 2, can be easily scratched with a fingernail. On the other hand, the mineral calcite, which has a hardness of 3, will scratch a fingernail but will not scratch glass. Quartz, one of the hardest common minerals, will easily scratch glass. Diamonds, hardest of all, scratch anything, including other diamonds.

CLEAVAGE. In the crystal structure of many minerals, some atomic bonds are weaker than others. It is along these weak bonds that minerals tend to break when they are stressed. **Cleavage** (*Kleiben* = carve) is the tendency of a mineral to break (cleave) along planes of weak bonding. Not all minerals have cleavage, but those that do can be identified by the relatively smooth, flat surfaces that are produced when the mineral is broken.

The simplest type of cleavage is exhibited by the micas (**FIGURE 2.15**). Because these minerals have very weak bonds in one direction, they cleave to form thin, flat sheets. Some minerals have excellent cleavage in one, two, three, or more directions, whereas others exhibit fair or poor cleavage, and still others have no cleavage at all. When minerals break evenly in more than one direction, cleavage is described by *the number of cleavage directions and the angle(s) at which they meet* (**FIGURE 2.16**).

Each cleavage surface that has a different orientation is counted as a different direction of cleavage. For example, some minerals cleave to form six-sided cubes. Because cubes are defined by three different sets of parallel planes that intersect at 90-degree angles, cleavage is described as *three directions of cleavage that meet at 90 degrees.*

Do not confuse cleavage with crystal shape. When a mineral exhibits cleavage, it will break into pieces that all have the same geometry. By contrast, the smooth-sided quartz crystals shown in Figure 2.1 (p. 38) do not have cleavage. If broken, they fracture into shapes that do not resemble one another or the original crystals.

FRACTURE. Minerals having chemical bonds that are equally, or nearly equally, strong in all directions exhibit a property called **fracture.** When minerals fracture, most produce uneven surfaces and are described as exhibiting *irregular fracture.* However, some minerals, such as

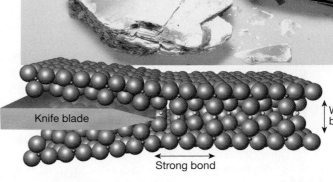

FIGURE 2.15 The thin sheets shown here were produced by splitting a mica (muscovite) crystal parallel to its perfect cleavage. *(Photo by Chip Clark)*

Knife blade

Weak bond

Strong bond

Number of Cleavage Directions	Shape	Sketch	Directions of Cleavage	Sample
1	Flat sheets	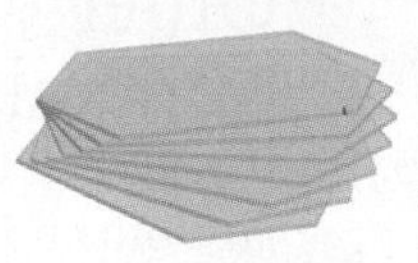		 Muscovite
2 at 90°	Elongated form with rectangle cross section (prism)	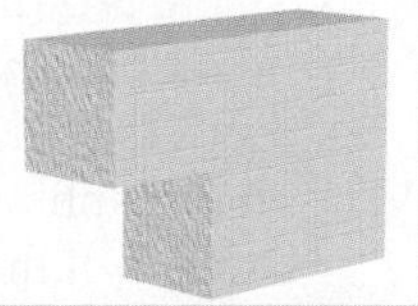	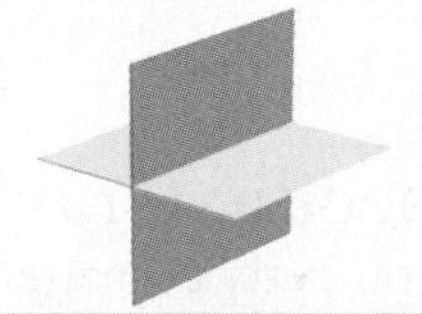	 Feldspar
2 not at 90°	Elongated form with parallelogram cross section (prism)			 Hornblende
3 at 90°	Cube	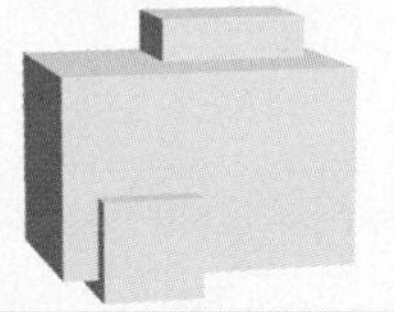	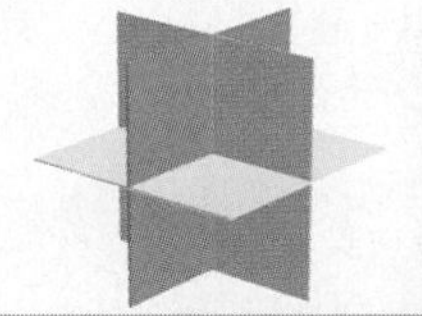	 Halite
3 not at 90°	Rhombohedron	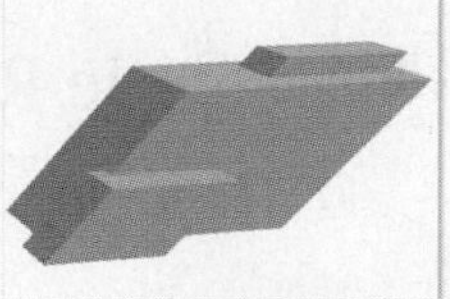		 Calcite
4	Octahedron			 Fluorite

FIGURE 2.16 Common cleavage directions exhibited by minerals. *(Photos by E. J. Tarbuck and Dennis Tasa)*

FIGURE 2.17 Conchoidal fracture. The smooth curved surfaces result when minerals break in a glasslike manner. *(Photo by E. J. Tarbuck)*

quartz, break into smooth, curved surfaces resembling broken glass. Such breaks are called *conchoidal fractures* (**FIGURE 2.17**). Still other minerals exhibit fractures that produce splinters or fibers that are referred to as *splintery* and *fibrous fracture*, respectively.

Density and Specific Gravity

Density, an important property of matter, is defined as mass per unit volume. Mineralogists often use a related measure called **specific gravity** to describe the density of minerals. Specific gravity is a number representing the ratio of a mineral's weight to the weight of an equal volume of water.

Most common rock-forming minerals have a specific gravity of between 2 and 3. For example, quartz has a specific gravity of 2.65. By contrast, some metallic minerals such as pyrite, native copper, and magnetite are more than twice as dense and thus have more than twice the specific gravity as quartz. Galena, an ore of lead, has a specific gravity of roughly 7.5, whereas the specific gravity of 24-karat gold is approximately 20.

With a little practice, you can estimate the specific gravity of a mineral by hefting it in your hand. Ask yourself, does this mineral feel about as "heavy" as similar sized rocks you have handled? If the answer is "yes," the specific gravity of the sample will likely be between 2.5 and 3.

Other Properties of Minerals

In addition to the properties discussed thus far, some minerals can be recognized by other distinctive properties. For example, halite is ordinary salt, so it can be quickly identified through taste. Talc and graphite both have distinctive feels; talc feels soapy, and graphite feels greasy. Further, the streaks of many sulfur-bearing minerals emit odors like rotten eggs. A few minerals, such as magnetite, have a high iron content and can be picked up with a magnet, while some varieties (lodestone) are natural magnets and will pick up small iron-

based objects such as pins and paper clips (see Figure 2.28A, p. 57).

Moreover, some minerals exhibit special optical properties. For example, when a transparent piece of calcite is placed over printed text, the letters appear twice. This optical property is known as *double refraction* (**FIGURE 2.18**).

One very simple chemical test involves placing a drop of dilute hydrochloric acid from a dropper bottle onto a freshly broken mineral surface. Using this technique, certain minerals, called carbonates, will effervesce (fizz) as carbon dioxide gas is released (**FIGURE 2.19**). This test is especially

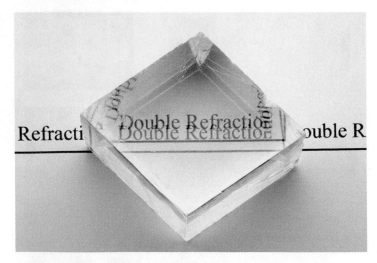

FIGURE 2.18 Double refraction illustrated by the mineral calcite. *(Photo by Chip Clark)*

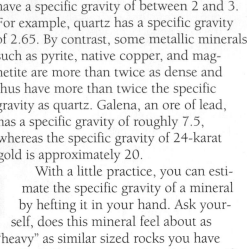

FIGURE 2.19 Calcite reacting with a weak acid. *(Photo by Chip Clark)*

useful in identifying the common carbonate mineral calcite.

Mineral Groups

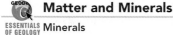

Matter and Minerals

ESSENTIALS OF GEOLOGY Minerals

Over 4000 minerals have been named, and several new ones are identified each year. Fortunately, for students who are beginning to study minerals, no more than a few dozen are abundant! Collectively, these few make up most of the rocks of Earth's crust and, as such, are often referred to as the **rock-forming minerals.**

Although less abundant, many other minerals are used extensively in the manufacture of products and are called *economic minerals.* However, rock-forming minerals and economic minerals are not mutually exclusive groups. When found in large deposits, some rock-forming minerals are economically significant. One example is the mineral calcite, which is the primary component of the sedimentary rock limestone and has many uses including being used in the production of cement.

It is worth noting that *only eight elements* make up the vast majority of the rock-forming minerals and represent more than 98 percent (by weight) of the continental crust (**FIGURE 2.20**). These elements, in order of abundance, are oxygen (O), silicon (Si), aluminum (Al), iron (Fe), calcium (Ca), sodium (Na), potassium (K), and magnesium (Mg). As shown in Figure 2.20, silicon and oxygen are by far the most common elements in Earth's crust. Furthermore, these two elements readily combine to form the basic "building block" for the most common mineral group, the **silicates.**

More than 800 silicate minerals are known, and they account for more than 90 percent of Earth's crust.

Because other mineral groups are far less abundant in Earth's crust than the silicates, they are often grouped together under the heading **nonsilicates.** Although not as common as silicates, some nonsilicate minerals are very important economically. They provide us with iron and aluminum to build our automobiles, gypsum for plaster and drywall for home construction, and copper wire that carries electricity and connects us to the Internet. Some common nonsilicate mineral groups include the carbonates, sulfates, and halides. In addition to their economic importance, these mineral groups include members that are major constituents in sediments and sedimentary rocks.

We first discuss the most common mineral group, the silicates, and then consider some of the prominent nonsilicate mineral groups.

The Silicates

Every silicate mineral contains the two most abundant elements of Earth's crust, oxygen and silicon. Further, most contain one or more of the other common elements. Together, these elements give rise to hundreds of silicate minerals with a wide variety of properties, including hard quartz, soft talc, sheet-like mica, fibrous asbestos, green olivine, and blood-red garnet.

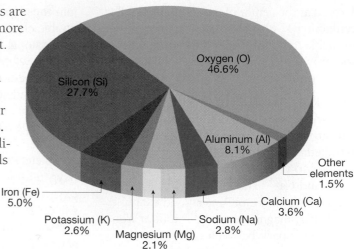

FIGURE 2.20 Relative abundance of the eight most abundant elements in the continental crust.

Silicate Structures

All silicate minerals have the same fundamental building block, the **silicon–oxygen tetrahedron** (SiO_4^{4-}) This structure consists of four oxygen ions (each 2^-) that are covalently bonded to one comparatively small silicon ion (4^+) forming a *tetrahedron*—a pyramid shape with four identical faces (**FIGURE 2.21**). These tetrahedra are not chemical compounds, but rather complex ions (SiO_4^{4-}) having a net charge of -4. To become electrically balanced, these complex ions bond to other positively charged metal ions. Specifically, each O^{2-} has one of its valence electrons bonding with the Si^{4+}

FIGURE 2.21 Two representations of the silicon–oxygen tetrahedron. **A.** The four large spheres represent oxygen ions, and the blue sphere represents a silicon ion. The spheres are drawn in proportion to the radii of the ions. **B.** An expanded view of the tetrahedron that has an oxygen ion at each of the four corners.

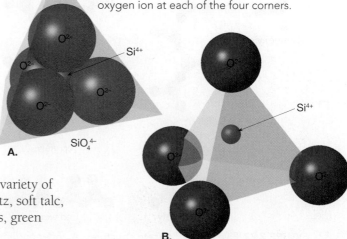

located at the center of the tetrahedron. The remaining 1$^-$ charge on each oxygen is available to bond with another positive ion or with the silicon ion in an adjacent tetrahedron.

MINERALS WITH INDEPENDENT TETRAHEDRA.

One of the simplest silicate structures consists of independent tetrahedra that have their four oxygen ions bonded to positive ions, such as Mg^{2+}, Fe^{2+}, and Ca^{2+}. The mineral olivine, with the formula $MgFe_2SiO_4$ is a good example. In olivine, magnesium (Mg^{2+}) and/or iron (Fe^{2+}) ions pack between comparatively large independent SiO_4 tetrahedra, forming a dense three-dimensional structure. Garnet, another common silicate, is also composed of independent tetrahedra ionically bonded by positive ions. Both olivine and garnet form dense, hard, equidimensional crystals that lack cleavage.

MINERALS WITH CHAIN OR SHEET STRUCTURES.

One reason for the great variety of silicate minerals is the ability of SiO_4 tetrahedra to link to one another in a variety of configurations. This important phenomenon, called **polymerization**, is achieved by the sharing of one, two, three, or all four of the oxygen atoms with adjacent tetrahedra. Vast numbers of tetrahedra join together to form single chains, double chains, sheet structures, or three-dimensional frameworks as shown in **FIGURE 2.22**.

To see how oxygen atoms are shared between adjacent tetrahedra, select one of the silicon ions (small blue spheres) near the middle of the single-chain shown in Figure 2.22B. Notice that this silicon ion is completely surrounded by four larger oxygen ions. Also notice that, of the four oxygen atoms, half are bonded to two silicon atoms, whereas the other two are not shared in this manner. It is the linkage across the shared oxygen ions that join the tetrahedra into a chain structure. Now examine a silicon ion near the middle of the sheet structure (Figure 2.22D) and count the number of shared and unshared oxygen ions surrounding it. As you likely observed, the sheet structure is the result of three of the four oxygen atoms being shared by adjacent tetrahedra.

MINERALS WITH THREE-DIMENSIONAL FRAMEWORKS.

In the most common silicate structure, all four oxygen ions are shared, producing a complex three-dimensional framework (Figure 2.22E). Quartz, a hard, durable mineral, has the simplest structure in which all of the oxygens are shared. Because its structure is electrically neutral, quartz (SiO_2) contains no positive ions—other than silicon.

The ratio of oxygen ions to silicon ions differs in each type of silicate structure. In independent tetrahedra (SiO_4) there are four oxygen ions for every silicon ion. In single chains, the oxygen-to-silicon ratio is 3:1 (SiO_3), and in three-dimensional frameworks as found in quartz the ratio is 2:1 (SiO_2). As more oxygen ions are shared, the percentage of silicon in the structure increases. Silicate minerals are, therefore, described as having a low or high silicon content based on their ratio of oxygen to silicon. Minerals with three-dimensional structures in which all four oxygen ions are shared have the highest silicon content. Minerals composed of independent tetrahedra have the lowest. This difference in silicon content is important, as you will see in Chapter 3.

Joining Silicate Structures

Except for quartz (SiO_2) the basic structure (chains, sheets, or three-dimensional frameworks) of most silicate minerals has a net negative charge. Therefore, metal ions are required to bring the overall charge into balance and to serve as the "mortar" that holds these structures together. The positive ions that most often link silicate structures

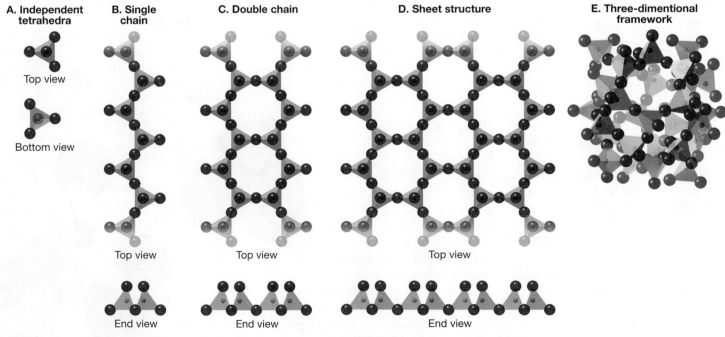

A. Independent tetrahedra — Top view — Bottom view
B. Single chain — Top view — End view
C. Double chain — Top view — End view
D. Sheet structure — Top view — End view
E. Three-dimentional framework

FIGURE 2.22 Five types of silicate structures. **A.** Independent tetrahedra. **B.** Single chains. **C.** Double chains. **D.** Sheet structures. **E.** Three-dimensional framework.

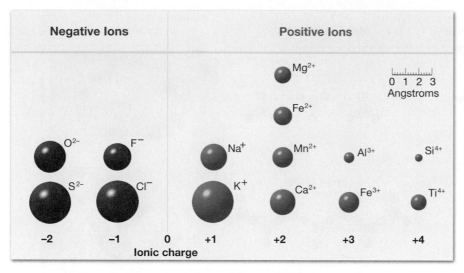

FIGURE 2.23 The relative sizes and charges of ions commonly found in minerals. Ionic radii are usually expressed in angstroms (1 angstrom equals 10^{-8} cm).

are iron (Fe^{2+}), magnesium (Mg^{2+}), potassium (K^{1+}), sodium (Na^{1+}), aluminum (Al^{3+}), and calcium (Ca^{2+}). These positively charged ions bond with the unshared oxygen ions that occupy the corners of the silicate tetrahedra (FIGURE 2.23).

As a general rule, the hybrid covalent bonds between silicon and oxygen are stronger than the ionic bonds that hold one silicate structure to the next. Consequently, properties such as cleavage, and to some extent hardness, are controlled by the nature of the silicate framework. Quartz (SiO_2), which has only silicon–oxygen bonds, has great hardness and lacks cleavage, mainly because of equally strong bonds in all directions. By contrast, the mineral talc (the source of talcum powder), has a sheet structure. Magnesium ions occur between the sheets and weakly join them together. The slippery feel of talcum powder is due to the silicate sheets sliding relative to one another, in much the same way sheets of carbon atoms slide in graphite, giving it its lubricating properties.

Recall that atoms of similar size can substitute freely for one another without altering a mineral's structure. For example, in the mineral olivine, iron (Fe^{2+}) and magnesium (Mg^{2+}) substitute for each other. This also holds true for the third most common element in Earth's crust, aluminum (Al^{3+}), which often substitutes for silicon (Si) in the center of silcon–oxygen tetrahedra.

Because most silicate structures will readily accommodate two or more different positive ions at a given bonding site, individual specimens of a particular mineral may contain varying amounts of certain elements. As a result, many silicate minerals form a *mineral group* that exhibits a range of compositions between two end members. Examples include the olivines, pyroxenes, amphiboles, micas, and feldspars.

CONCEPT CHECK 2.7

❶ Sketch a silicon–oxygen tetrahedron.

❷ Explain the following statement:
Silicate minerals with three-dimensional structures have the highest silicon content, while those composed of independent tetrahedra have the lowest.

Common Silicate Minerals

Matter and Minerals

ESSENTIALS OF GEOLOGY Mineral Groups

The major groups of silicate minerals and common examples are given in FIGURE 2.24. The feldspars are, by far, the most plentiful silicate group, comprising more than 50 percent of Earth's crust. Quartz, the second most abundant mineral in the continental crust, is the only common mineral made completely of silicon and oxygen.

Most silicate minerals form when molten rock cools and crystallizes. Cooling can occur at or near Earth's surface (low temperature and pressure) or at great depths (high temperature and pressure). The environment during crystallization and the chemical composition of the molten rock determine, to a large degree, which

Uncut emerald crystal, a variety of the mineral beryl. *(Photo by Jeffrey Scovill)*

gemstone

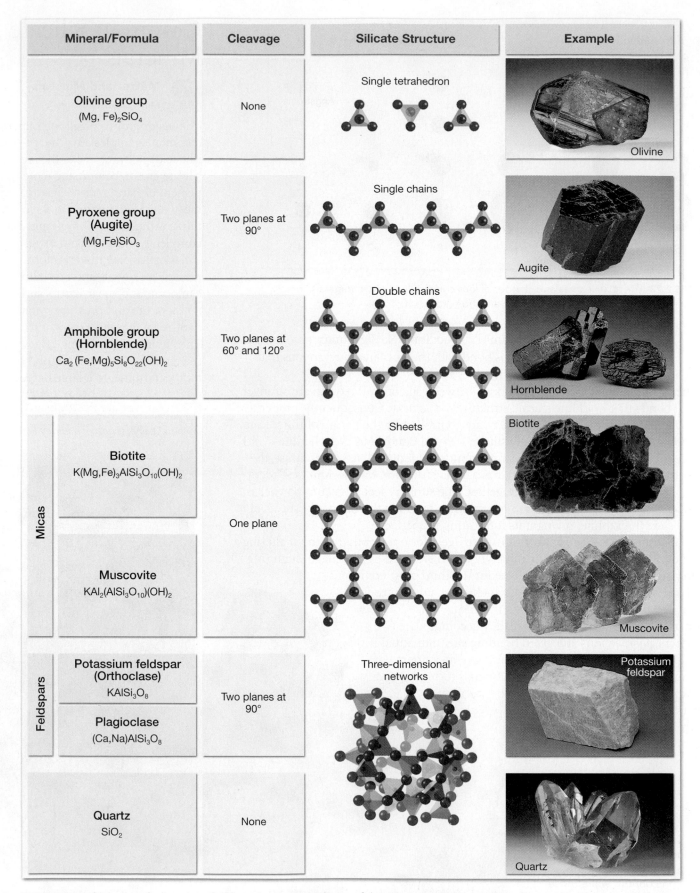

Mineral/Formula	Cleavage	Silicate Structure	Example
Olivine group $(Mg, Fe)_2SiO_4$	None	Single tetrahedron	Olivine
Pyroxene group (Augite) $(Mg,Fe)SiO_3$	Two planes at 90°	Single chains	Augite
Amphibole group (Hornblende) $Ca_2(Fe,Mg)_5Si_8O_{22}(OH)_2$	Two planes at 60° and 120°	Double chains	Hornblende
Micas — **Biotite** $K(Mg,Fe)_3AlSi_3O_{10}(OH)_2$ **Muscovite** $KAl_2(AlSi_3O_{10})(OH)_2$	One plane	Sheets	Biotite Muscovite
Feldspars — **Potassium feldspar (Orthoclase)** $KAlSi_3O_8$ **Plagioclase** $(Ca,Na)AlSi_3O_8$	Two planes at 90°	Three-dimensional networks	Potassium feldspar
Quartz SiO_2	None		Quartz

FIGURE 2.24 Common silicate minerals. Note that the complexity of the silicate structure increases from top to bottom. *(Photos by Dennis Tasa and E. J. Tarbuck)*

minerals are produced. For example, the silicate mineral olivine crystallizes at high temperatures, whereas quartz crystallizes at much lower temperatures.

In addition, some silicate minerals form at Earth's surface from the weathered products of other silicate minerals. Still others are formed under the extreme pressures associated with mountain building. Each silicate mineral, therefore, has a structure and a chemical composition that *indicate the conditions under which it formed.* By carefully examining the mineral constituents of rocks, geologists can usually determine the circumstances under which the rocks formed.

We will now examine some of the most common silicate minerals, which we divide into two major groups on the basis of their chemical makeup.

The Light Silicates

The **light (or nonferromagnesian) silicates** are generally light in color and have a specific gravity of about 2.7, which is considerably less than the dark (ferromagnesian) silicates. These differences are mainly attributable to the presence or absence of iron and magnesium. The light silicates contain varying amounts of aluminum, potassium, calcium, and sodium rather than iron and magnesium.

FELDSPAR GROUP. *Feldspar,* the most common mineral group, can form under a wide range of temperatures and pressures, which partially accounts for its abundance (**FIGURE 2.25**). All feldspars have similar physical properties. They have two planes of cleavage meeting at or near 90-degree angles, are relatively hard (6 on the Mohs scale), and have a luster that ranges from

FIGURE 2.25 Estimated percentages (by volume) of the most common minerals in Earth's crust.

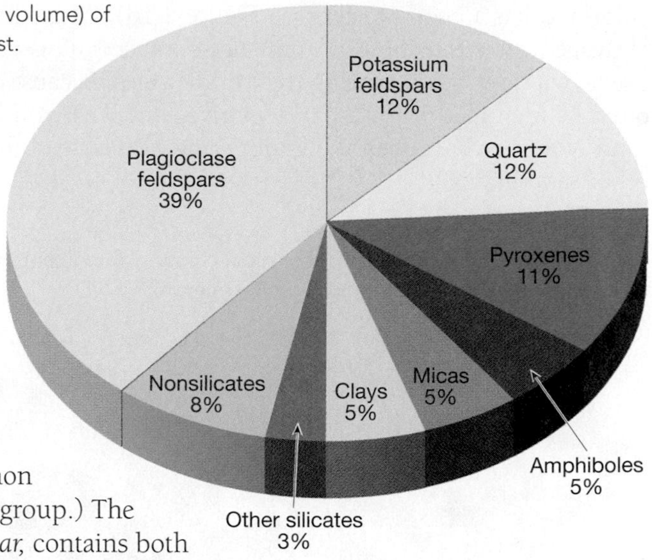

Potassium feldspars 12%

Quartz 12%

Pyroxenes 11%

Plagioclase feldspars 39%

Amphiboles 5%

Other silicates 3%

Nonsilicates 8%

Clays 5%

Micas 5%

glassy to pearly. As one component in a rock, feldspar crystals can be identified by their rectangular shape and rather smooth shiny faces (see Figure 2.24).

Two different feldspar structures exist. One group of feldspar minerals contains potassium ions in its structure and is therefore referred to as *potassium feldspar.* (*Orthoclase* and *microcline* are common members of the potassium feldspar group.) The other group, called *plagioclase feldspar,* contains both sodium and calcium ions that freely substitute for one another depending on the environment during crystalization.

Potassium feldspar is usually light cream, salmon pink, or occasionally bluish-green in color. The plagioclase feldspars, on the other hand, range in color from white to medium gray. However, color should not be used to distinguish these groups. The only way to distinguish the feldspars physically is to look for a multitude of fine parallel lines, called *striations.* Striations are found on some cleavage planes of plagioclase feldspar but are not present on potassium feldspar (**FIGURE 2.26**).

QUARTZ. *Quartz* is the only common silicate mineral consisting entirely of silicon and oxygen. As such, the term *silica* is applied to quartz, which has the chemical formula SiO_2. Because the structure of quartz contains a ratio of two oxygen ions (O^{2-}) for every silicon ion (Si^{4+}), no other positive ions are needed to attain neutrality.

In quartz, a three-dimensional framework is developed through the complete sharing of oxygen by adjacent silicon atoms. Thus, all of the bonds in quartz are of the strong silicon–oxygen type. Consequently, quartz is hard, resistant to weathering, and does not have cleavage. When broken, quartz generally exhibits conchoidal fracture (see Figure 2.17). When pure, quartz is clear and, if allowed to grow without interference, will develop hexagonal crystals that develop pyramid-shaped ends. However, like most other clear minerals, quartz is often colored by inclusions of various ions (impurities) and forms without developing good crystal faces. The most common varieties of quartz are milky (white), smoky (gray), rose (pink), amethyst (purple), and rock crystal (clear) (see Figure 2.10).

MUSCOVITE. *Muscovite* is a common member of the mica family. It is light in

Geologist's Sketch

Striations

FIGURE 2.26 These parallel lines, called striations, are a distinguishing characteristic of plagioclase feldspar. *(Photo by E. J. Tarbuck)*

color and has a pearly luster (see Figure 2.15). Like other micas, muscovite has excellent cleavage in one direction. In thin sheets, muscovite is clear, a property that accounts for its use as window "glass" during the Middle Ages. Because muscovite is very shiny, it can often be identified by the sparkle it gives a rock. If you have ever looked closely at beach sand, you may have seen the glimmering brilliance of the mica flakes scattered among the other sand grains.

Basalt is a common igneous rock composed of a large percentage of dark silicate minerals. *(Photo by Roger Ressmeyer/CORBIS)*

CLAY MINERALS. *Clay* is a term used to describe a category of complex minerals that, like the micas, have a sheet structure. Unlike other common silicates, such as quartz and feldspar, most clay minerals originate as products of the chemical weathering of other silicate minerals. Thus, clay minerals make up a large percentage of the surface material we call soil. Because of the importance of soil in agriculture, and because of its role as a supporting material for buildings, clay minerals are extremely important to humans. In addition, clays account for nearly half the volume of sedimentary rocks. Clay minerals are generally very fine grained, which makes identification difficult, unless studied microscopically. Their layered structure and weak bonding between layers give them a characteristic feel when wet. Clays are common in shales, mudstones, and other sedimentary rocks.

One of the most common clay minerals is *kaolinite,* which is used in the manufacture of fine china and as a coating for *high-gloss paper,* such as that used in this textbook. Further, some clay minerals absorb large amounts of water, which allows them to swell to several times their normal size. These clays have been used commercially in a variety of ingenious ways, including as an additive to thicken milkshakes in fast-food restaurants.

The Dark Silicates

The **dark (or ferromagnesian) silicates** are those minerals containing ions of iron (iron = *ferro*) and/or magnesium in their structure. Because of their iron content, ferromagnesian silicates are dark in color and have a greater specific gravity, between

3.2 and 3.6, than nonferromagnesian silicates. The most common dark silicate minerals are olivine, the pyroxenes, the amphiboles, dark mica (biotite), and garnet.

OLIVINE GROUP. *Olivine* is a family of high-temperature silicate minerals that are black to olive green in color and have a glassy luster and a conchoidal fracture (see Figure 2.24). Transparent olivine is occasionally used as a gemstone called peridot. Rather than developing large crystals, olivine commonly forms small, rounded crystals that give olivine-rich rocks a granular appearance. Olivine and related forms are thought to constitute up to 50 percent of Earth's upper mantle.

PYROXENE GROUP. The *pyroxenes* are a group of complex minerals that are important components in dark colored igneous rocks. The most common member, *augite,* is a black, opaque mineral with two directions of cleavage that meet at nearly a 90-degree angle. Augite is one of the dominant minerals in basalt, a common igneous rock of the oceanic crust and volcanic areas on the continents.

AMPHIBOLE GROUP. *Hornblende* is the most common member of a chemically complex group of minerals called *amphiboles* (see Figure 2.24). Hornblende is usually dark green to black in color, and except for its cleavage angles, which are about 60 degrees and 120 degrees, it is very similar in appearance to augite. In a rock, hornblende often forms elongated crystals. This helps distinguish it from pyroxene, which forms rather blocky crystals. Hornblende is found in igneous rocks, where it often makes up the dark portion of an otherwise light-colored rock.

BIOTITE. *Biotite* is the dark, iron-rich member of the mica family (see Figure 2.24). Like other micas, biotite possesses a sheet structure that gives it excellent cleavage in one direction. Biotite also has a shiny black appearance that helps distinguish it from the other dark ferromagnesian minerals. Like hornblende, biotite is a common constituent of igneous rocks, including the rock granite.

GARNET. *Garnet* is similar to olivine in that its structure is composed of individual tetrahedra linked by metallic ions. Also like olivine, garnet has a glassy luster, lacks cleavage, and exhibits conchoidal fracture. Although the colors of garnet are varied, this mineral is most often brown to deep red. Garnet readily forms equidimensional crystals that are most commonly found in metamorphic rocks. When garnets are transparent, they may be used as gemstones.

CONCEPT CHECK 2.8

1. What do ferromagnesian minerals have in common? List three examples of ferromagnesian minerals.

2. What do muscovite and biotite have in common? How do they differ?

3. Should color be used to distinguish between orthoclase and plagioclase feldspar? What is the most effective means of distinguishing them?

Important Nonsilicate Minerals

 Matter and Minerals
ESSENTIALS OF GEOLOGY **Mineral Groups**

Nonsilicate minerals are typically divided into groups, based on the negatively charged ion or complex ion that the members have in common (**TABLE 2.1**). For example, the *oxides* contain the negative oxygen ion (O^{2-}), which is bonded to one or more kinds of positive ions. Thus, within each mineral group, the basic structure and type of bonding is similar. As a result, the minerals in each group have similar physical properties that are useful in mineral identification.

Although the nonsilicates make up only about 8 percent of Earth's crust, some minerals, such as gypsum, calcite, and halite, occur as constituents in sedimentary rocks in significant amounts. Furthermore, many others are important economically. Table 2.1 lists some of the nonsilicate mineral groups and a few examples of each. A brief discussion of a few of the more common nonsilicate minerals follows.

Some of the most common nonsilicate minerals belong to one of three classes of minerals—the carbonates (CO_3^{2-}), the sulfates (SO_4^{2-}), and the halides (Cl^{1-}, F^{1-}, Br^{1-}). The carbonate minerals are much simpler structurally than the silicates. This mineral group is composed of the carbonate ion (CO_3^{2-}) and one or more kinds of positive ions. The two most common carbonate minerals are *calcite,* $CaCO_3$ (calcium carbonate), and *dolomite,* $CaMg(CO_3)_2$ (calcium/magnesium carbonate). Because these minerals are similar both physically and chemically, they are difficult to distinguish from each other. Both have a vitreous luster, a hardness between 3 and 4, and nearly perfect rhombic cleavage. They can, however, be distinguished by using dilute hydrochloric acid. Calcite reacts vigorously with this acid, whereas dolomite reacts much more slowly. Calcite and dolomite are usually found together as the primary constituents in the sedimentary rocks limestone and dolostone. When calcite is the dominant mineral, the rock is called *limestone,* whereas *dolostone* results from a predominance of dolomite. Limestone has many uses, including as road aggregate, as building stone, and as the main ingredient in Portland cement.

Two other nonsilicate minerals frequently found in sedimentary rocks are *halite* and *gypsum.* Both minerals are commonly found in thick layers that are the last vestiges of ancient seas that have long since evaporated (**FIGURE 2.27**). Like limestone, both are important nonmetallic resources. Halite is the mineral name for common table salt (NaCl). Gypsum ($CaSO_4 \cdot 2 H_2O$), which is calcium sulfate with water bound into the structure, is the mineral of which plaster and other similar building materials are composed.

Most nonsilicate mineral classes contain members that are prized for their economic value. This includes the oxides, whose members hematite and magnetite are important ores of iron (**FIGURE 2.28**). Also significant are the sulfides, which are basically compounds of sulfur (S) and one or more metals. Examples of important sulfide minerals include galena (lead), sphalerite (zinc), and chalcopyrite (copper). In addition, native elements, including gold, silver, and carbon (diamonds), plus a host of other nonsilicate minerals—fluorite (flux in making steel), corundum (gemstone, abrasive), and uraninite (a uranium source)—are important economically.

CONCEPT CHECK 2.9

❶ List six common nonsilicate mineral groups. What key ion(s) or element(s) define each group?

❷ What simple test can be used to distinguish calcite from dolomite?

❸ List eight common nonsilicate minerals and their economic uses.

TABLE 2.1

Common Nonsilicate Mineral Groups

Mineral Groups [key ion(s) or element(s)]	Mineral Name	Chemical Formula	Economic Use
Carbonates (CO_3^{2-})	Calcite	$CaCO_3$	Portland cement, lime
	Dolomite	$CaMg(CO_3)_2$	Portland cement, lime
Halides (Cl^{1-}, F^{1-}, Br^{1-})	Halite	NaCl	Common salt
	Fluorite	CaF_2	Used in steelmaking
	Sylvite	KCl	Fertilizer
Oxides (O^{2-})	Hematite	Fe_2O_3	Ore of iron, pigment
	Magnetite	Fe_3O_4	Ore of iron
	Corundum	Al_2O_3	Gemstone, abrasive
	Ice	H_2O	Solid form of water
Sulfides (S^{2-})	Galena	PbS	Ore of lead
	Sphalerite	ZnS	Ore of zinc
	Pyrite	FeS_2	Sulfuric acid production
	Chalcopyrite	$CuFeS_2$	Ore of copper
	Cinnabar	HgS	Ore of mercury
Sulfates (SO_4^{2-})	Gypsum	$CaSO_4 \cdot 2H_2O$	Plaster
	Anhydrite	$CaSO_4$	Plaster
	Barite	$BaSO_4$	Drilling mud
Native elements (single elements)	Gold	Au	Trade, jewelry
	Copper	Cu	Electrical conductor
	Diamond	C	Gemstone, abrasive
	Sulfur	S	Sulfa drugs, chemicals
	Graphite	C	Pencil lead, dry lubricant
	Silver	Ag	Jewelry, photography
	Platinum	Pt	Catalyst

FIGURE 2.27 Thick bed of halite (salt) at an underground mine in Grand Saline, Texas. Note person for scale. *(Photo by Tom Bochsler)*

Mineral Resources

Earth's crust and oceans are the source of a wide variety of useful and essential substances. In fact, practically every manufactured product contains materials obtained from minerals. Table 2.1 lists important mineral groups in this category.

Mineral resources are the endowment of useful minerals ultimately available commercially. Resources include already identified deposits from which minerals can be extracted profitably, called **reserves**, as well as known deposits that are not yet economically or technologically recoverable. Deposits inferred to exist, but not yet discovered, are also considered mineral resources.

The term **ore** is used to denote those useful metallic minerals that can be mined at a profit (Figure 2.28). In common usage, the term ore is also applied to some non-metallic minerals such as fluorite and sulfur. However, materials used for such purposes as building stone, road aggregate, abrasives, ceramics, and fertilizers are not usually called ores; rather, they are classified as industrial rocks and minerals.

Recall that more than 98 percent of Earth's crust is composed of only eight elements, and except for oxygen and silicon, all other elements make up a relatively small fraction of common crustal rocks (see Figure 2.20). Indeed, the natural concentrations of many elements are exceedingly small. A deposit containing the average percentage of a valuable element such as gold has no economic value, because the cost of extracting it greatly exceeds the value of the gold that could be recovered.

To have economic value, an element must be concentrated above the level of its average crustal abundance. For example, copper makes up about 0.0135 percent of the crust. For a deposit to be considered as copper ore, it must contain a concentration that is about 100 times this amount. Aluminum, on the other hand, represents 8.13 percent of the crust and can be extracted profitably when it is found in concentrations only about four times its average crustal percentage.

It is important to realize that a deposit may become profitable to extract or lose its profitability because of economic changes. If demand for a metal increases and prices rise sufficiently, the status of a previously unprofitable deposit changes, and it

FIGURE 2.28 Magnetite **A.** and hematite **B.** are both oxides and are both important ores of iron. *(Photos by E. J. Tarbuck)*

A.

B.

becomes an ore. The status of unprofitable deposits may also change if a technological advance allows the ore to be extracted at a lower cost than before.

Conversely, changing economic factors can turn a once profitable ore deposit into an unprofitable deposit that can no longer be called an ore. This situation was illustrated at the copper mining operation located at Bingham Canyon, Utah, one of the largest open-pit mines on Earth (**FIGURE 2.29**). Mining was halted there in 1985 because outmoded equipment had driven the cost of extracting the copper beyond the current selling price. The owners responded by replacing an antiquated 1000-car railroad with conveyor belts and pipelines for transporting the ore and waste. These devices achieved a cost reduction of nearly 30 percent and returned this mining operation to profitability.

Over the years, geologists have been keenly interested in learning how natural processes produce localized concentrations of essential minerals. One well-established fact is that occurrences of valuable mineral resources are closely related to the rock cycle. That is, the mechanisms that generate igneous, sedimentary, and metamorphic rocks, including the processes of weathering and erosion, play a major role in producing concentrated accumulations of useful elements.

Moreover, with the development of the theory of plate tectonics, geologists have added another tool for understanding the processes by which one rock is transformed into another. As these rock-forming processes are examined in the following chapters, we will consider their role in producing some of our important mineral resources.

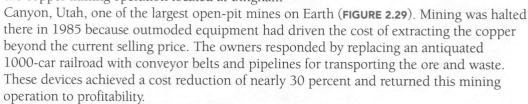

CONCEPT CHECK 2.10

❶ Contrast a mineral *resource* and a mineral *reserve*.

❷ What might cause a mineral deposit that had not been considered an ore to become reclassified as an ore?

FIGURE 2.29 Aerial view of Bingham Canyon copper mine near Salt Lake City, Utah. Although the amount of copper in the rock is less than 1 percent, the huge volume of material removed and processed each day (about 200,000 tons) yields enough metal to be profitable. *(Photo by Michael Collier)*

CHAPTER TWO
Matter and Minerals in Review

⊙ A *mineral* is any naturally occurring inorganic solid that possesses an orderly crystalline structure and can be represented by a chemical formula. Most *rocks* are aggregates composed of two or more minerals.

⊙ All matter, including minerals, is composed of minute, indivisible particles called *atoms*—the building blocks of minerals. Each atom has a *nucleus,* which contains *protons* (particles with positive electrical charges) and *neutrons* (particles with neutral electrical charges). Surrounding the nucleus of an atom in regions called *principal shells,* are *electrons,* which have negative electrical charges. The number of protons in an atom's nucleus determines its *atomic number* and the name of the element. An *element* is a large collection of electrically neutral atoms, all having the same atomic number.

⊙ Atoms combine with each other to form more complex substances called *compounds.* Atoms bond by gaining, losing, or sharing electrons with other atoms. In *ionic bonding,* one or more electrons are transferred from one atom to another, giving the atoms a net positive or negative charge. The resulting electrically charged atoms are called *ions.* Ionic compounds consist of oppositely charged ions assembled in a regular, crystalline structure that allows for the maximum attraction of ions, given their sizes. Another type of bond, the *covalent bond,* is produced when atoms share electrons.

⊙ *Isotopes* are variants of the same element that have a different *mass number* (the total number of neutrons plus protons found in an atom's nucleus). Some isotopes are unstable and disintegrate naturally through a process called *radioactivity.*

⊙ The properties of minerals include *crystal shape (habit), luster, color, streak, tenacity, hardness, cleavage, fracture,* and *density* or *specific gravity.* In addition, a number of special physical and chemical properties (*taste, smell, elasticity, feel, magnetism, double refraction,* and *chemical reaction to hydrochloric acid*) are useful in identifying certain minerals. Each mineral has a unique set of properties that can be used for identification.

⊙ Of the nearly 4000 minerals, no more than a few dozen make up most of the rocks of Earth's crust and, as such, are classified as rock-forming minerals. Eight elements (oxygen, silicon, aluminum, iron, calcium, sodium, potassium, and magnesium) make up the bulk of these minerals and represent over 98 percent (by weight) of Earth's continental crust.

⊙ The most common mineral group is the *silicates.* All silicate minerals have the negatively charged *silicon–oxygen tetrahedron* as their fundamental building block. In some silicate minerals the tetrahedra are joined in chains (the pyroxene and amphibole groups); in others, the tetrahedra are arranged into sheets (the micas—biotite and muscovite) or three-dimensional networks (the feldspars and quartz). The tetrahedra and various silicate structures are often bonded together by the positive ions of iron, magnesium, potassium, sodium, aluminum, and calcium. Each silicate mineral has a structure and a chemical composition that indicates the conditions under which it formed.

⊙ The *nonsilicate* mineral groups, which contain several economically important minerals, include the *oxides* (e.g., the mineral hematite, mined for iron), *sulfides* (e.g., the mineral sphalerite, mined for zinc, and the mineral galena, mined for lead), *sulfates, halides,* and *native elements* (e.g., gold and silver). The more common nonsilicate rock-forming minerals include the *carbonate minerals,* calcite and dolomite. Two other nonsilicate minerals frequently found in sedimentary rocks are halite and gypsum.

⊙ *Mineral resources* are the endowment of useful minerals ultimately available commercially. Resources include already identified deposits from which minerals can be extracted profitably, called *reserves,* as well as known deposits that are not yet economically or technologically recoverable. Deposits inferred to exist, but not yet discovered, are also considered mineral resources. The term *ore* is used to denote those useful metallic minerals that can be mined for a profit, as well as some nonmetallic minerals, such as fluorite and sulfur, that contain useful substances.

Key Terms

atoms (p. 40)
atomic number (p. 40)
chemical bond (p. 42)
chemical compounds (p. 41)
cleavage (p. 47)
color (p. 44)
covalent bond (p. 43)
crystal shape (p. 45)
dark silicates (p. 54)
density (p. 48)
electrons (p. 40)
element (p. 40)

ferromagnesian silicates (p. 54)
fracture (p. 47)
habit (p. 45)
hardness (p. 46)
ions (p. 42)
ionic bond (p. 43)
isotopes (p. 44)
light silicates (p. 53)
luster (p. 44)
mass number (p. 44)
metallic bonds (p. 43)
mineral (p. 38)

mineralogy (p. 38)
mineral resources (p. 56)
Mohs scale (p. 46)
neutrons (p. 40)
nonferromagnesian silicates
  (p. 53)
nonsilicates (p. 49)
nucleus (p. 40)
octet rule (p. 42)
ore (p. 57)
periodic table (p. 41)
polymerization (p. 50)

protons (p. 40)
radioactive decay (p. 44)
reserves (p. 56)
rock (p. 39)
rock-forming minerals (p. 49)
silicates (p. 49)
silicon–oxygen tetrahedron
  (p. 49)
specific gravity (p. 48)
streak (p. 45)
tenacity (p. 46)
valence electrons (p. 40)

GIVE IT SOME THOUGHT

❶ Using the geologic definition of *mineral* as your guide, determine which of the items on the list are minerals and which are not. If not a mineral, explain.

  a. gold nugget

  b. sea water

  c. quartz

  d. cubic zirconia

  e. obsidian

  f. ruby

  g. glacial ice

  h. amber

  Refer to the Periodic Table of the Elements (Figure 2.5) to help you answer questions 2, 3, and 4:

❷ If the number of protons in a neutral atom is 92 and its mass number is 238:

  a. What is the name of that element?

  b. How many electrons does it have?

  c. How many neutrons does it have?

❸ Which of the following elements is more likely to form chemical bonds: Xenon (Xe) or Sodium (Na)? Explain why.

❹ The information below refers to three isotopes of the element potassium. Using this information, determine the appropriate number of protons and neutrons for each isotope. Label each isotope in the manner used in the chapter.

  Atomic Number = 19   Atomic Number = 19   Atomic Number = 19
  Atomic Mass = 39   Atomic Mass = 40   Atomic Mass = 41

❺ Gold has a specific gravity of almost 20. A 5-gallon bucket of water weighs 40 pounds. How much would a 5-gallon bucket of gold weigh?

❻ The continents have an average elevation of about 1 kilometer *above* sea level, whereas the ocean basins have an average elevation of about 4 kilometers *below* sea level. Based on what you know about minerals and their properties, name at least two minerals that are likely abundant in the continental crust, and two minerals that are likely abundant in the oceanic crust. Explain why you chose those minerals.

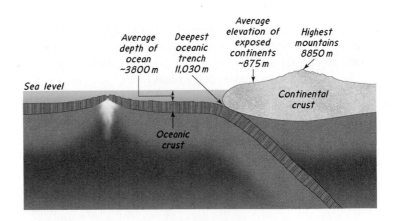

❼ Each of the following statements describes a silicate mineral or mineral group. In each case, provide the appropriate name:

  a. the most common member of the amphibole group

  b. the most common nonferromagnesian member of the mica family

  c. the only common silicate mineral made entirely of silicon and oxygen

  d. a high-temperature silicate with a name that is based on its color

  e. a silicate mineral that is characterized by striations

  f. a silicate mineral that originates as a product of chemical weathering

Companion Website

The *Essentials of Geology, 11e* companion Website contains numerous multimedia resources accompanied by assessments to aid in your study of the topics in this chapter. The use of this site's learning tools will help improve your understanding of geology. Utilizing the access code that accompanies this text, visit **www.mygeoscienceplace.com** in order to:

- **Review** key chapter concepts.
- **Read** with links to the Pearson eText and to chapter-specific web resources.
- **Visualize** and comprehend challenging topics using the learning activities in *GEODe: Essentials of Geology* and the *Geoscience Animations Library.*
- **Test** yourself with online quizzes.

Igneous Rocks and Intrusive Activity

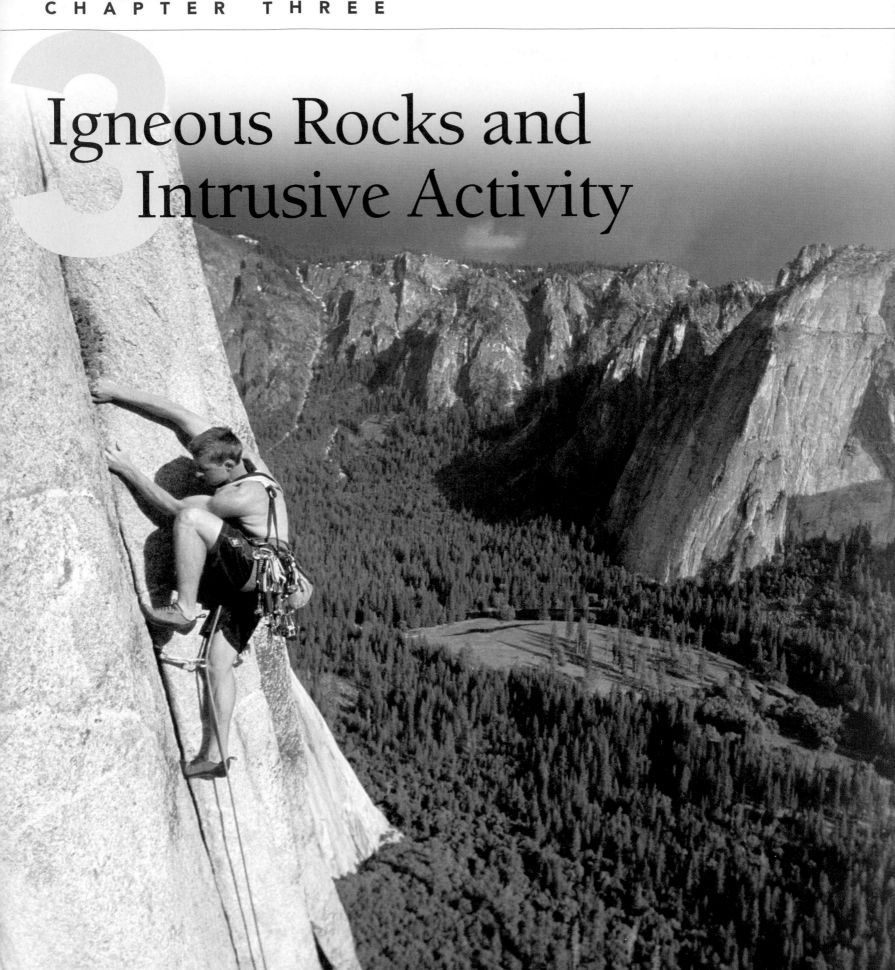

IGNEOUS ROCKS AND METAMORPHIC ROCKS DERIVED FROM IGNEOUS "parents" make up about 95 percent of Earth's crust. Furthermore, the mantle, which accounts for more than 82 percent of Earth's volume, is composed entirely of igneous rock. Thus, Earth can be described as a huge mass of igneous rock covered with a thin veneer of sedimentary rock and having a relatively small iron-rich core. Consequently, a basic knowledge of igneous rocks is essential to our understanding of the structure, composition, and internal workings of our planet.

Climber scaling the vertical face of El Capitan in California's Yosemite National Park. (Photo by Corey Rich/Aurora Photos)

To assist you in learning the important concepts in this chapter, focus on the following questions:

- How are igneous rocks formed?
- How does magma differ from lava?
- What two criteria are used to classify igneous rocks?
- How does the rate of cooling of magma influence the crystal size of minerals in igneous rocks?
- How is the mineral makeup of an igneous rock related to Bowen's reaction series?
- In what ways are granitic rocks different from basaltic rocks?
- What criteria are used to classify intrusive igneous bodies?
- How are economic deposits of gold, silver, and many other metals formed?

Magma: The Parent Material of Igneous Rock

 Igneous Rocks

ESSENTIALS OF GEOLOGY **Introduction**

In our discussion of the rock cycle, it was noted that **igneous rocks** (*ignis* = fire) form as molten rock cools and solidifies. Considerable evidence supports the idea that the parent material for igneous rocks, called **magma**, is formed by melting that occurs at various levels within Earth's crust and upper mantle to depths of perhaps 250 kilometers (about 150 miles).

Once formed, a magma body buoyantly rises toward the surface because it is less dense than the surrounding rocks. (When rock melts it takes up more space and, hence, it becomes less dense than the surrounding solid rock.) Occasionally molten rock reaches Earth's surface where it is called **lava** (**FIGURE 3.1**). Sometimes lava is emitted as fountains that are produced when escaping gases propel it from a magma chamber. On other occasions, magma is explosively ejected, producing dramatic steam and ash eruptions. However, not all eruptions are violent; many volcanoes emit quiet outpourings of very fluid lava.

The Nature of Magma

Magma is completely or partly molten rock, which on cooling solidifies to form an igneous rock composed of silicate minerals. Most magmas consist of three distinct parts—a *liquid component,* a *solid component,* and a *gaseous phase.*

The liquid portion, called **melt**, is composed mainly of mobile ions of the eight most common elements found in Earth's crust—silicon and oxygen, along with lesser amounts of aluminum, potassium, calcium, sodium, iron, and magnesium.

The solid components (if any) in magma are silicate minerals that have already crystallized from the melt. As a magma body cools, the size and number of crystals increase. During the last stage of cooling, a magma body is like a "crystalline mush" with only small amounts of melt.

The gaseous components of magma, called **volatiles**, are materials that will vaporize (form a gas) at surface pressures. The most common volatiles found in magma are water vapor (H_2O), carbon dioxide (CO_2), and sulfur dioxide (SO_2), which are confined by the immense pres-

FIGURE 3.1 Fluid basaltic lava emitted from Hawaii's Kilauea volcano. (*Photo by U.S. Geological Survey*)

sure exerted by the overlying rocks. These gases tend to separate from the melt as it moves toward the surface (low-pressure environment). As the gases build up, they may eventually propel magma from the vent. When deeply buried magma bodies crystallize, the remaining volatiles collect as hot, water-rich fluids that migrate through the surrounding rocks. These hot fluids play an important role in metamorphism and will be considered in Chapter 7.

From Magma to Crystalline Rock

To better understand how magma crystallizes, let us consider how a simple crystalline solid melts. Recall that in any crystalline solid, the ions are arranged in a closely packed regular pattern. However, they are not without some motion—they exhibit a sort of restricted vibration about fixed points. As temperature rises, ions vibrate more rapidly and consequently collide with ever-increasing vigor with their neighbors. Thus, heating causes the ions to occupy more space, which in turn causes the solid to expand. When the ions are vibrating rapidly enough to overcome the force of their chemical bonds, melting occurs. At this stage the ions are able to slide past one another, and the orderly crystalline structure disintegrates. Thus, melting converts a solid consisting of tight, uniformly packed ions into a liquid composed of unordered ions moving randomly about.

In the process called **crystallization**, cooling reverses the events of melting. As the temperature of the liquid drops, ions pack more closely together as their rate of movement slows. When cooled sufficiently, the forces of the chemical bonds will again confine the ions to an orderly crystalline arrangement.

When magma cools, it is generally the silicon and oxygen atoms that link together first to form silicon–oxygen tetrahedra, the basic building blocks of the silicate minerals. As magma continues to lose heat to its surroundings, the tetrahedra join with each other and with other ions to form embryonic crystal nuclei. Slowly each nucleus grows as ions lose their mobility and join the crystalline network.

The earliest formed minerals have space to grow and tend to have better-developed crystal faces than do the later ones that occupy the remaining spaces. Eventually all of the melt is transformed into a solid mass of interlocking silicate minerals that we call an *igneous rock*.

As you will see, the crystallization of magma is much more complex than just described. Whereas a simple compound, such as water, solidifies at a specific temperature, crystallization of magma with its diverse chemistry spans a temperature range of 200 °C, or more. In addition, magmas differ from one another in terms of their chemical composition, the amount of volatiles they contain, and the rate at which they cool. Because all of these factors influence the crystallization process, the appearance and mineral make-up of igneous rocks varies widely.

CONCEPT CHECK 3.1

❶ What is magma? How does magma differ from lava?

❷ List the three major components of magma.

❸ What is melt? What is a volatile?

Igneous Processes

Igneous rocks form in two basic settings. Magma may crystallize at depth or lava may solidify at Earth's surface. When magma loses its mobility before reaching the surface it eventually crystallizes to form **intrusive igneous rocks.** They are also known as **plutonic rocks**—after Pluto, the god of the lower world in classical mythology. Intrusive igneous rocks are

coarse-grained and consist of visible mineral crystals. These rocks are observed at the surface in locations where uplifting and erosion have stripped away the overlying rocks. Exposures of intrusive igneous rocks occur in many places, including Mount Washington, New Hampshire; Stone Mountain, Georgia; the Black Hills of South Dakota; and Yosemite National Park, California (**FIGURE 3.2**).

Igneous rocks that form when molten rock solidifies *at the surface* are classified as **extrusive igneous rocks.** They are also called **volcanic rocks**—after the Roman fire god, Vulcan. Extrusive igneous rocks form when lava solidifies, in which case they tend to be fine-grained, or when volcanic debris falls to Earth's surface. Extrusive igneous rocks are abundant in western portions of the Americas where they make up the volcanic peaks of the Cascade Range and the Andes Mountains. In addition, many oceanic islands, including the Hawaiian chain and Alaska's Aleutian Islands, are composed almost entirely of extrusive igneous rocks.

CONCEPT CHECK 3.2

❶ In what basic settings do intrusive and extrusive igneous rocks originate?

Igneous Compositions

 Igneous Rocks

ESSENTIALS OF GEOLOGY **Igneous Compositions**

Igneous rocks are composed mainly of silicate minerals. Chemical analyses show that silicon and oxygen are by far the most abundant constituents of igneous rocks. These two elements, plus ions of aluminum (Al), calcium (Ca),

FIGURE 3.2 Mount Rushmore National Memorial, located in the Black Hills of South Dakota, is carved from intrusive igneous rocks. *(Photo by Barbara A. Harvey/Shutterstock)*

sodium (Na), potassium (K), magnesium (Mg), and iron (Fe), make up roughly 98 percent, by weight, of most magmas. In addition, magma contains small mounts of many other elements, including titanium and manganese, and trace amounts of much rarer elements such as gold, silver, and uranium.

As magma cools and solidifies, these elements combine to form two major groups of silicate minerals. The *dark (ferromagnesian) silicates* are rich in iron and/or magnesium and comparatively low in silica. *Olivine, pyroxene, amphibole,* and *biotite mica* are the common dark silicate minerals of Earth's crust. By contrast, the *light (nonferromagnesian) silicates* contain greater amounts of potassium, sodium, and calcium rather than iron and magnesium. As a group, nonferromagnesian minerals are richer in silica than the dark silicates. The light silicates include *quartz, muscovite mica,* and the most abundant mineral group, the *feldspars.* Feldspars make up at least 40 percent of most igneous rocks. Thus, in addition to feldspar, igneous rocks contain some combination of the other light and/or dark silicates listed above.

Granitic (Felsic) Versus Basaltic (Mafic) Compositions

Despite their great compositional diversity, igneous rocks (and the magmas from which they form) can be divided into broad groups according to their proportions of light and dark minerals (**FIGURE 3.3**). Near one end of the continuum are rocks composed almost entirely of light-colored silicates—quartz and feldspar. Igneous rocks in which these are the dominant minerals have a **granitic composition.** Geologists also refer to granitic rocks as being **felsic,** a term derived from *feldspar* and *silica* (quartz). In addition to quartz and feldspar, most granitic rocks contain about 10 percent dark silicate minerals, usually biotite mica and amphibole. Granitic rocks are rich in silica (about 70 percent) and are major constituents of the continental crust.

FIGURE 3.3 Mineralogy of common igneous rocks and the magmas from which they form. (After Dietrich, Daily, and Larsen)

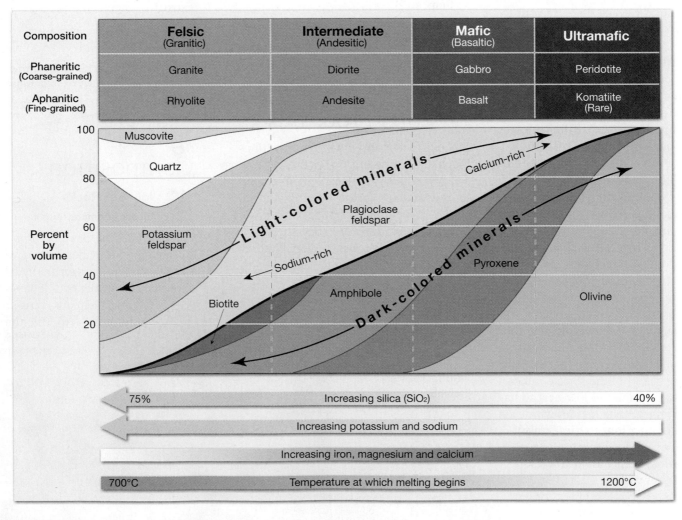

Rocks that contain substantial dark silicate minerals and calcium-rich plagioclase feldspar (but no quartz) are said to have a **basaltic composition** (see Figure 3.3). Basaltic rocks contain a high percentage of ferromagnesian minerals, so geologists also refer to them as **mafic** (from *ma*gnesium and *f*errum, the Latin name for iron). Because of their iron content, mafic rocks are typically darker and denser than granitic rocks. Basaltic rocks make up the ocean floor as well as many of the volcanic islands located within the ocean basins. Basalt also forms extensive lava flows on the continents.

Other Compositional Groups

As you can see in Figure 3.3, rocks with a composition between granitic and basaltic rocks are said to have an **intermediate composition,** or **andesitic composition** after the common volcanic rock *andesite*. Intermediate rocks contain at least 25 percent dark silicate minerals, mainly amphibole, pyroxene, and biotite mica with the other dominant mineral being plagioclase feldspar. This important category of igneous rocks is associated with volcanic activity that is typically confined to the margins of the continents.

Another important igneous rock, *peridotite,* contains mostly olivine and pyroxene and thus falls on the opposite side of the compositional spectrum from granitic rocks (see Figure 3.3). Because peridotite is composed almost entirely of ferromagnesian minerals, its chemical composition is referred to as **ultramafic.** Although ultramafic rocks are rare at Earth's surface, peridotite is the main constituent of the upper mantle.

Silica Content as an Indicator of Composition

An important aspect of the chemical composition of igneous rocks is silica (SiO_2) content. Typically, the silica content of crustal rocks ranges from a low of about 40 percent in ultramafic rocks to a high of more than 70 percent in granitic rocks (see Figure 3.3). The percentage of silica in igneous rocks actually varies in a systematic manner that parallels the abundance of other elements. For example, rocks that are relatively low in silica contain large amounts of iron, magnesium, and calcium. By contrast, rocks high in silica contain very little iron, magnesium, or calcium but are enriched with sodium and potassium. Consequently, the chemical makeup of an igneous rock can be inferred directly from its silica content.

Further, the amount of silica present in magma strongly influences its behavior. Granitic magma, which has a high silica content, is quite viscous ("thick") and may erupt at temperatures as low as 700 °C. On the other hand, basaltic magmas are low in silica and are generally more fluid. Basaltic magmas also erupt at higher temperatures than granitic magmas—usually at temperatures between 1100 and 1250 °C and are completely solid when cooled to 1000 °C.

In summary, igneous rocks can be divided into broad groups according to the proportions of light and dark minerals they contain. *Granitic (felsic) rocks,* which are composed almost entirely of the light-colored minerals quartz and feldspar, are at one end of the compositional spectrum (see Figure 3.3). *Basaltic (mafic) rocks,* which contain abundant dark silicate minerals in addition to plagioclase feldspar, make up the other major igneous rock group of Earth's crust. Between these groups are rocks with an *intermediate (andesitic) composition. Uultramafic rocks,* which lack light-colored minerals, lie at the far end of the compositional spectrum from granitic rocks.

CONCEPT CHECK 3.3

1. Igneous rocks are composed mainly of which group of minerals?
2. Differentiate between felsic and mafic igneous rocks.
3. In what way are the dark (ferromagnesian) silicate minerals different from the light (nonferromagnesian) silicate minerals?

DID YOU KNOW?

People have been making glass in roughly the same way for at least 2000 years. The process involves melting certain Earth materials and cooling the liquid quickly before the atoms have time to form an orderly crystalline structure. This is the same way that natural glass, called obsidian, is generated from lava. It is possible to produce glass from a variety of materials, but most commercial glass is produced from quartz sand and lesser amounts of carbonate minerals.

Igneous Textures: What Can They Tell Us?

Igneous Rocks

ESSENTIALS OF GEOLOGY Igneous Textures

The term **texture** is used to describe the overall appearance of a rock based on the size, shape, and arrangement of its mineral grains (**FIGURE 3.4**). Texture is an important property because it reveals a great deal about the environment in which the rock formed. This fact allows geologists to make inferences about a rock's origin based on careful observations of grain size and other characteristics of the rock.

Factors Affecting Crystal Size

Three factors influence the textures of igneous rocks: (1) *the rate at which molten rock cools;* (2) *the amount of silica present;* and (3) *the amount of dissolved gases in the magma.* Among these, the rate of cooling tends to be the dominant factor.

A very large magma body located many kilometers beneath Earth's surface will cool over a period of perhaps tens to hundreds of thousands of years. Initially, relatively few crystal nuclei form. Slow cooling permits ions to migrate freely until they eventually join one of the existing crystalline structures. Consequently, slow cooling promotes the growth of fewer but larger crystals.

On the other hand, when cooling occurs rapidly—for example, in a thin lava flow—the ions quickly lose their mobility

FIGURE 3.4 Igneous rock textures. **A.** During a volcanic eruption in which silica-rich lava is ejected into the atmosphere, a frothy glass called pumice may form. **B.** Rocks that exhibit a *pyroclastic texture* are a result of the consolidation of rock fragments that were ejected during a violent volcanic eruption. **C.** Igneous rocks that crystallize at or near Earth's surface cool quickly and often exhibit a *fine-grained (aphanitic) texture.* **D.** A *porphyritic texture* results when magma that already contains some large crystals migrates to a new location where the rate of cooling increases. The resulting rock consists of larger crystals *(phenocrysts)* embedded within a matrix of smaller crystals *(groundmass).* **E.** *Coarse-grained (phaneritic)* igneous rocks form when magma slowly crystallizes at depth. *(Photos by E. J. Tarbuck)*

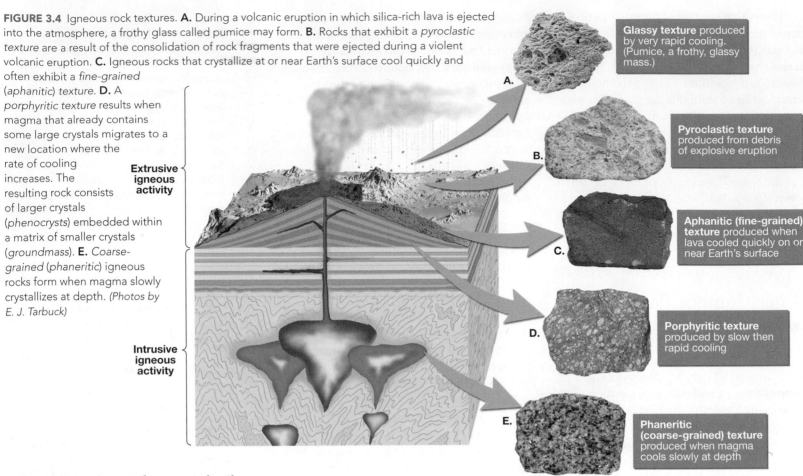

Glassy texture produced by very rapid cooling. (Pumice, a frothy, glassy mass.)

Pyroclastic texture produced from debris of explosive eruption

Aphanitic (fine-grained) texture produced when lava cooled quickly on or near Earth's surface

Porphyritic texture produced by slow then rapid cooling

Phaneritic (coarse-grained) texture produced when magma cools slowly at depth

Extrusive igneous activity

Intrusive igneous activity

and readily combine to form crystals. This results in the development of numerous embryonic nuclei, all of which compete for the available ions. The result is a solid mass of tiny intergrown crystals.

When molten material is quenched quickly, there may not be sufficient time for the ions to arrange into an ordered crystalline network. Rocks that consist of unordered ions that are "frozen" randomly in place are referred to as **glass.**

Types of Igneous Textures

As you saw, the effect of cooling on rock textures is fairly straightforward. Slow cooling promotes the growth of large crystals, whereas rapid cooling tends to generate small crystals.

APHANITIC (FINE-GRAINED) TEXTURE. Igneous rocks that form at the surface, or as small intrusive masses within the upper crust where cooling is relatively rapid, exhibit a **fine-grained texture,** also termed an **aphanitic texture** (*a* = not, *phaner* = visible). The crystals that make up aphanitic rocks are so small that individual minerals can only be distinguished with

FIGURE 3.5 Comparison of aphanitic (fine-grained) and phaneritic (coarse-grained) igneous rock textures. The smaller images were generated using a polarizing microscope that produces photomicrographs of thin, transparent rock slices. **A.** Basalt is a fine-grained igneous rock composed mainly of plagioclase feldspar and pyroxene. **B.** Granite is a coarse-grained igneous rock composed mainly of quartz, potassium feldspar, and lesser amounts of mica and dark minerals. *(Photos courtesy of E. J. Tarbuck)*

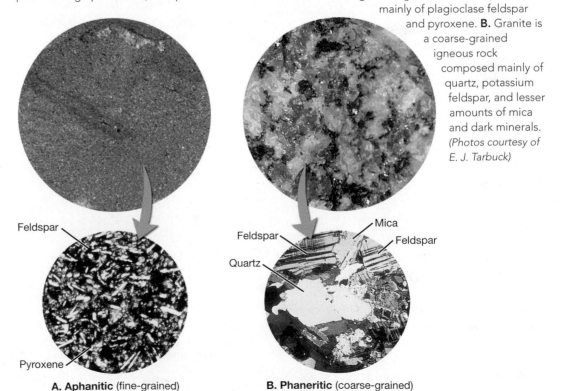

Feldspar

Pyroxene

A. Aphanitic (fine-grained)

Feldspar

Quartz

Mica

Feldspar

B. Phaneritic (coarse-grained)

the aid of a polarizing microscope or other sophisticated techniques (**FIGURE 3.5**). Therefore, we commonly characterize fine-grained rocks as being light, intermediate, or dark in color. Using this system of grouping, light-colored aphanitic rocks are those containing primarily light-colored nonferromagnesian silicate minerals and so forth.

Common features of many extrusive rocks are the voids left by gas bubbles that escape as lava solidifies. These nearly spherical openings are called *vesicles,* and the rocks that contain them are said to have a **vesicular texture**. Rocks that exhibit a vesicular texture usually form in the upper zone of a lava flow, where cooling occurs rapidly enough to preserve the openings produced by the expanding gas bubbles (**FIGURE 3.6**).

PHANERITIC (COARSE-GRAINED) TEXTURE. When large masses of magma slowly crystallize at great depth, they form igneous rocks that exhibit a **coarse-grained texture** also referred to as a **phaneritic texture** (*phaner* = visible). Coarse-grained rocks consist of a mass of intergrown crystals that are roughly equal in size and large enough so that the individual minerals can be identified without the aid of a microscope (Figure 3.5B). Geologists often use a small magnifying lens to aid in identifying minerals in a phaneritic rock.

PORPHYRITIC TEXTURE. A large mass of magma may require tens to hundreds of thousands of years to solidify. Because different minerals crystallize under different environmental conditions (temperatures and pressure), it is possible for crystals of one mineral to become quite large before others even begin to form. Should molten rock containing some large crystals move to a different environment—for example, by erupting at the surface—the remaining liquid portion of the lava would cool more quickly. The resulting rock, which has large crystals embedded in a matrix of smaller crystals, is said to have a **porphyritic texture** (see Figure 3.4D). The large crystals in such a rock are referred to as **phenocrysts** (*pheno* = show, *cryst* = crystal), whereas the matrix of smaller crystals is called **groundmass**. A rock with a *porphyritic* texture is termed a **porphyry**.

GLASSY TEXTURE. During some volcanic eruptions, molten rock is ejected into the atmosphere, where it is quenched quickly. Rapid cooling of this type may generate rocks having a **glassy texture**. Glass results when unordered ions are "frozen in place" before they are able to unite into an orderly crystalline structure.

Obsidian, a common type of natural glass, is similar in appearance to a dark chunk of manufactured glass (**FIGURE 3.7**). Because of its excellent conchodial fracture and ability to hold a sharp, hard edge, obsidian was a prized material from which Native Americans chipped arrowheads and cutting tools (see Figure 3.7 inset).

FIGURE 3.6 The larger image shows a lava flow erupting from Hawaii's Kilauea volcano. The rock sample photo is a close-up showing the vesicular texture of the rock. Vesicles are small holes left by escaping gas bubbles. *(Photo by J. D. Griggs, U.S. Geological Survey/rock sample photo courtesy of E. J. Tarbuck)*

As this lava cools, it is transformed into a solid mass of interlocking minerals we call an igneous rock. *(Photo by Griggs/USGS)*

FIGURE 3.7 Obsidian, a natural glass, was used by Native Americans for making arrowheads and cutting tools. *(Photo by E. J. Tarbuck; Inset photo by Jeffrey Scovil)*

lava

FIGURE 3.8 This obsidian flow was extruded from a vent along the south wall of Newberry Caldera, Oregon. Note the road for scale. *(Photo by Marli Miller)*

Geologist's Sketch

Cinder cones

South wall of Newberry Caldera

Lava erupted from here

Obsidian flow

Road

FIGURE 3.9 Rocks that exhibit a pyroclastic texture are the result of the consolidation of rock fragments ejected during a violent volcanic eruption. *(Photo by Steve Kaufman/Peter Arnold, Inc.)*

Obsidian flows, a few hundred feet thick, are evidence that rapid cooling is not the only mechanism that produces a glassy texture (**FIGURE 3.8**). Magmas with a high silica content tend to form long, chainlike structures (polymerization) before crystallization is complete. These structures, in turn, impede ionic transport and increase the magma's viscosity. (*Viscosity* is a measure of a fluid's resistance to flow.)

Granitic magma, which is rich in silica, may be extruded as an extremely viscous mass that eventually solidifies to form obsidian. By contrast, basaltic magma, which is low in silica, forms very fluid lavas that, upon cooling, usually generate fine-grained crystalline rocks. However, the surface of basaltic lava may be quenched rapidly enough to form a thin, glassy skin. Moreover, Hawaiian volcanoes sometimes generate lava fountains, which spray basaltic lava tens of meters into the air. Such activity can produce strands and tiny blobs of volcanic glass called *Pele's hair* and *Pele's tears* respectively, after the Hawaiian goddess of volcanoes.

PYROCLASTIC (FRAGMENTAL) TEXTURE. Another group of igneous rocks is formed from the consolidation of individual rock fragments that are ejected during a violent volcanic eruption. The ejected particles might be very fine ash, molten blobs, or large angular blocks torn from the walls of the vent during the eruption (**FIGURE 3.9**). Igneous rocks composed of these rock fragments are said to have a **pyroclastic texture** or a **fragmental texture** (see Figure 3.4B).

A common type of pyroclastic rock, called *welded tuff,* is composed of fine fragments of glass that remained hot enough during their flight to fuse together upon impact. Other pyroclastic rocks are composed of fragments that solidified before impact and became cemented together at some later time. Because pyroclastic rocks are made of individual particles or fragments rather than interlocking crystals, their textures often appear to be more similar to those exhibited by sedimentary rocks than to those associated with igneous rocks.

PEGMATITIC TEXTURE. Under special conditions, exceptionally coarse-grained igneous rocks, called **pegmatites,** may form. These rocks, which are composed of interlocking crystals all larger than a centimeter in diameter, are said to have a **pegmatitic texture.** Most pegmatites occur as small masses or thin veins situated around the margins of large intrusive bodies.

Pegmatites form late in the crystallization of a magma, when water and other volatiles, such as carbon dioxide,

DID YOU KNOW?

Enormous crystals have been found in a type of igneous rock called *pegmatite*. For example, single crystals of the mineral feldspar the size of a house have been unearthed in North Carolina and Norway, while huge muscovite (mica) plates weighing 85 tons have been mined in India. In addition, crystals as large as telephone poles (over 40 feet in length) of the lithium-bearing mineral spodumene have been found in the Black Hills of South Dakota.

chlorine, and fluorine, make up an unusually high percentage of the melt. Because ion migration is enhanced in these fluid-rich environments, the crystals that form are abnormally large. Thus, the large crystals in pegmatites are not the result of inordinately long cooling histories; rather, they are the consequence of a fluid-rich environment that enhances crystallization.

The composition of most pegmatites is similar to that of granite. Thus, pegmatites contain large crystals of quartz, feldspar, and muscovite. However, some contain significant quantities of relatively rare and hence valuable elements—gold, tungsten, beryllium, and the rare Earth elements that are used in modern technological devices, including superconductors and hybrid autos.

CONCEPT CHECK 3.4

1. How does the rate of cooling influence crystal size? What other factors influence the texture of igneous rocks?

2. The statements that follow relate to terms used to describe igneous rock textures. For each statement, identify the appropriate term.
   a. openings produced by escaping gases
   b. obsidian exhibits this texture
   c. a matrix of fine crystals surrounding phenocrysts
   d. crystals are too small to be seen without a microscope
   e. a texture characterized by two distinctly different crystal sizes
   f. coarse-grained, with crystals of roughly equal size
   g. exceptionally large crystals exceeding 1 centimeter in diameter

3. Why are the crystals in pegmatites so large?

4. What does a porphyritic texture indicate about an igneous rock?

Naming Igneous Rocks

 Igneous Rocks

ESSENTIALS OF GEOLOGY   **Naming Igneous Rocks**

Igneous rocks are most often classified, or grouped, on the basis of their texture and mineral composition (**FIGURE 3.10**). The various igneous textures result mainly from different cooling histories, whereas the minerology of an igneous rock is the consequence of the chemical makeup of its parent magma. Because igneous rocks are classified on the basis of their mineral composition and texture, two rocks may have similar mineral constituents but have different textures and hence different names.

Felsic (Granitic) Igneous Rocks

GRANITE. *Granite* is perhaps the best known of all igneous rocks (**FIGURE 3.11**). This is partly because of its natural beauty, which is enhanced when it is polished, and partly because of its abundance in the

FIGURE 3.10 Classification of major igneous rocks based on mineral composition and texture. Coarse-grained rocks are plutonic, solidifying deep underground. Fine-grained rocks are volcanic, or solidify as shallow, thin plutons. Ultramafic rocks are dark, dense rocks, composed almost entirely of minerals containing iron and magnesium. Although relatively rare at or near Earth's surface, these rocks are major constituents of the upper mantle.

Chemical Composition		Felsic (Granitic)	Intermediate (Andesitic)	Mafic (Basaltic)	Ultramafic
Dominant Minerals		Quartz Potassium feldspar Sodium-rich plagioclase feldspar	Amphibole Sodium- and calcium-rich plagioclase feldspar	Pyroxene Calcium-rich plagioclase feldspar	Olivine Pyroxene
Accessory Minerals		Amphibole Muscovite Biotite	Pyroxene Biotite	Amphibole Olivine	Calcium-rich plagioclase feldspar
T E X T U R E	Phaneritic (coarse-grained)	**Granite**	**Diorite**	**Gabbro**	**Peridotite**
	Aphanitic (fine-grained)	**Rhyolite**	**Andesite**	**Basalt**	**Komatiite** (rare)
	Porphyritic	 "Porphyritic" precedes any of the above names whenever there are appreciable phenocrysts			Uncommon
	Glassy	 **Obsidian** (compact glass) **Pumice** (frothy glass)			
	Pyroclastic (fragmental)	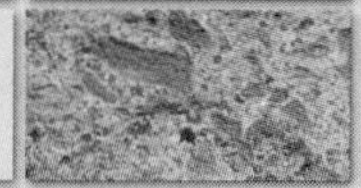 **Tuff** (fragments less than 2 mm) **Volcanic Breccia** (fragments greater than 2 mm)			
Rock Color (based on % of dark minerals)		0% to 25%	25% to 45%	45% to 85%	85% to 100%

Texture	Composition		
	Felsic (Granitic)	**Intermediate** (Andesitic)	**Mafic** (Basaltic)
Phaneritic (course-grained)	Granite	Diorite	Gabbro
Aphanitic (fine-grained)	Rhyolite	Andesite	Basalt
Porphyritic	Granite porphyry	Andesite porphyry	Basalt porphyry

FIGURE 3.11 Common igneous rocks. *(Photos by E. J. Tarbuck)*

continental crust (**FIGURE 3.12**). Slabs of polished granite are commonly used for tombstones and monuments and as building stones. Well-known areas in the United States where granite is quarried include Barre, Vermont; Mount Airy, North Carolina; and St. Cloud, Minnesota.

Granite is a coarse-grained rock composed of about 25 percent quartz and roughly 65 percent feldspar, mostly potassium-and sodium-rich varieties. Quartz crystals, which are roughly spherical in shape, are often glassy and clear to light gray in color. By contrast, feldspar crystals are generally white to gray or salmon pink in color and exhibit a rectan-

gular rather than spherical shape. Other minor constituents of granite include muscovite and some dark silicates, particularly biotite and amphibole. Although the dark components generally make up less than 10 percent of most granites, dark minerals appear to be more prominent than their percentage would indicate.

When potassium feldspar is dominant and dark pink in color, granite appears reddish (see Figure 3.11). This variety of granite is popular for monuments and as a building stone. However, the feldspar grains are more often white to gray, so when they are mixed with lesser amounts of dark silicates, granite appears light gray

in color. In addition, some granites have a porphyritic texture. These specimens contain elongated feldspar crystals a few centimeters in length that are scattered among smaller crystals of quartz and amphibole.

Granite is a very abundant rock. However, it has become common practice among geologists to apply the term *granite* to any coarse-grained intrusive rock composed predominantly of light silicate minerals. We will follow this practice for the sake of simplicity. You should keep in mind that this use of the term *granite* covers rocks having a wide range of mineral compositions.

RHYOLITE. Rhyolite is the extrusive equivalent of granite and, like granite, is composed essentially of the light-colored silicates (see Figure 3.11). This fact accounts for its color, which is usually buff to pink or occasionally very light gray. Rhyolite is fine-grained and frequently contains glass fragments and voids, indicating rapid cooling in a surface environment. When rhyolite contains phenocrysts they are small and composed of either quartz or potassium feldspar. In contrast to granite, which is widely distributed as large plutonic masses, rhyolite deposits are less common and generally less voluminous. Yellowstone Park is one well-known exception. Here, rhyolite lava flows and thick ash deposits of similar composition are extensive.

OBSIDIAN. *Obsidian* is a dark-colored glassy rock that usually forms when silica-rich lava is quenched quickly (**FIGURE 3.13**). In contrast to the orderly arrangement of ions characteristic of minerals, *the ions in glass are unordered.* Consequently, glassy rocks such as obsidian are not composed of minerals in the same sense as most other rocks.

Although usually black or reddish-brown in color, obsidian has a composition that is more akin to light-colored igneous rocks such as granite, rather than to dark rocks such as

DID YOU KNOW?
Although Earth's upper mantle is often depicted in diagrams as a reddish-colored layer to show its high temperature, its color is actually a dark green, but how do we know it is green? On occasion, unmelted fragments of mantle rocks are brought to Earth's surface by rapidly rising magma and are ejected from volcanoes. These mantle rocks, called *peridotite*, are composed mainly of olivine, which, as the name suggests, is olive green, and pyroxene, which is blackish green.

FIGURE 3.12 Rocks contain information about the processes that produce them. This mass of granitic rock in Yosemite National Park, California, was once a molten mass deep within Earth's crust. *(Photo by Enrique R. Aguirre/agefotostock)*

A. Obsidan flow.

FIGURE 3.13 Obsidian is a dark-colored, glassy rock formed from silica-rich lava. The scene in part A shows the leading edge of an obsidian lava flow located south of Mono Lake, California. The large banded block in the lower portion of the image contains white layers of porous pumice-like material. *(Photos by E. J. Tarbuck)*

B. Hand sample of obsidan.

are quite noticeable, whereas in others the pumice resembles fine shards of intertwined glass. Because of the large percentage of voids, many samples of pumice will float when placed in water. Oftentimes, flow lines are visible in pumice, indicating that some movement occurred before solidification was complete. Moreover, pumice and obsidian can often be found in the same rock mass, where they exist in alternating layers.

Intermediate (Andesitic) Igneous Rocks

ANDESITE. Andesite is a medium-gray, fine-grained rock of volcanic origin. Its name comes from South America's Andes Mountains, where numerous volcanoes are composed of this rock type. In addition to the volcanoes of the Andes and the Cascade Range, many of the volcanic structures occupying the continental margins that surround the Pacific Ocean are of andesitic composition. Andesite commonly exhibits a porphyritic texture (see Figure 3.11). When this is the case, the phenocrysts are often light, rectangular crystals of plagioclase feldspar or black, elongated amphibole crystals. Andesite often resembles rhyolite, so their identification usually requires microscopic examination to verify mineral make-up.

DIORITE. *Diorite* is the plutonic equivalent of andesite. It is a phaneritic rock that looks somewhat similar to gray granite. However, it can be distinguished from granite by the absence of visible quartz crystals and because it contains a higher percentage of dark silicate minerals. The mineral makeup of diorite is primarily sodium-rich plagioclase feldspar and amphibole, with lesser amounts of biotite. Because the light-colored feldspar grains and dark amphibole crystals appear to be roughly equal in abundance, diorite has a salt-and-pepper appearance (see Figure 3.11).

basalt. Obsidian's dark color results from small amounts of metallic ions in an otherwise relatively clear, glassy substance. If you examine a thin edge, obsidian will appear nearly transparent (see Figure 3.7).

PUMICE. *Pumice* is a volcanic rock with a glassy texture that forms when large amounts of gas escape through silica-rich lava to generate a gray, frothy mass (**FIGURE 3.14**). In some samples, the voids

← 2 cm →

FIGURE 3.14 Pumice, a glassy rock, is very lightweight because it contains numerous vesicles. *(Inset photo by Chip Clark)*

Mafic (Basaltic) Igneous Rocks

BASALT. *Basalt* is a very dark green to black, aphanitic rock composed primarily of pyroxene and calcium-rich plagioclase feldspar, with lesser amounts of olivine and amphibole (see Figure 3.11). When porphyritic, basalt commonly contains small light-colored

FIGURE 3.15 The Columbia River basalts. At some locations these once-fluid lava flows are more than 3 kilometers thick. Although basalt is usually black, it often weathers to a reddish-brown color. *(Photo by Willianborg)*

Today, we know that silica-rich lava is too viscous (thick) to flow more than a few miles from a vent.

Pyroclastic rocks composed mainly of particles larger than ash are called *volcanic breccia*. The particles in volcanic breccia can consist of streamlined fragments that solidified in air, blocks broken from the walls of the vent, crystals, and glass fragments.

Unlike most igneous rock names, such as granite and basalt, the terms *tuff* and *volcanic breccia* do not imply mineral composition. Thus, they are frequently used with a modifier, as, for example, rhyolite tuff.

CONCEPT CHECK 3.5

❶ The classification of igneous rocks is based largely on two criteria. Name these criteria.

❷ How are granite and rhyolite different? In what way are they similar?

❸ Compare and contrast each of the following pairs of rocks:
  a. granite and diorite
  b. basalt and gabbro
  c. andesite and rhyolite

❹ How do tuff and volcanic breccia differ from other igneous rocks such as granite and basalt?

feldspar phenocrysts or green, glassy-appearing olivine phenocrysts embedded in a dark groundmass.

Basalt is the most common extrusive igneous rock. Many volcanic islands, such as the Hawaiian Islands and Iceland, are composed mainly of basalt. Further, the upper layers of the oceanic crust consist of basalt. In the United States, large portions of central Oregon and Washington were the sites of extensive basaltic outpourings (**FIGURE 3.15**). At some locations, these once-fluid basaltic flows have accumulated to thicknesses approaching 3 kilometers.

GABBRO. Gabbro is the intrusive equivalent of basalt (see Figure 3.11). Like basalt, it tends to be dark green to black in color and composed primarily of pyroxene and calcium-rich plagioclase feldspar. Although gabbro is uncommon in the continental crust, it makes up a significant percentage of oceanic crust.

Pyroclastic Rocks

Pyroclastic rocks are composed of fragments ejected during a volcanic eruption. One of the most common pyroclastic rocks, called *tuff*, is composed mainly of tiny, ash-size fragments that were later cemented together (see Figure 3.4B). In situations where the ash particles remained hot enough to fuse, the rock is called *welded tuff*. Although welded tuff consists mostly of tiny glass shards, it may contain walnut-size pieces of pumice and other rock fragments.

Welded tuffs blanket vast portions of formerly active volcanic areas in the western United States (**FIGURE 3.16**). Some of these tuff deposits are hundreds of feet thick and extend for tens of miles from their source. Most formed millions of years ago as volcanic ash spewed from large volcanic structures (calderas) in an avalanche style, spreading laterally at speeds approaching 100 kilometers per hour. Early investigators of these deposits incorrectly classified them as rhyolite lava flows.

FIGURE 3.16 Outcrop of welded tuff from Valles Caldera near Los Alamos, New Mexico. Tuff is composed mainly of ash-sized particles and may contain larger fragments of pumice or other volcanic rocks. *(Photo by Marli Miller)*

Origin of Magma

Most magma originates in the uppermost mantle. The greatest quantities are produced at divergent plate boundaries in association with seafloor spreading. Lesser amounts form at subduction zones, where oceanic lithosphere descends into the mantle. In addition, magma can originate far from plate boundaries.

Generating Magma from Solid Rock

Based on evidence from the study of earthquake waves, *Earth's crust and mantle are composed primarily of solid, not molten, rock.* Although the outer core is fluid, this iron-rich material is very dense and remains deep within Earth. So, where does magma come from?

INCREASE IN TEMPERATURE. Most magma originates when essentially solid rock, located in the crust and upper mantle, melts. The most obvious way to generate magma from solid rock is to raise the temperature above the rock's melting point.

Workers in underground mines know that temperatures get higher as they go deeper. Although the rate of temperature change varies considerably from place to place, it *averages* about 25 °C per kilometer in the *upper* crust. This increase in temperature with depth, known as the **geothermal gradient,** is somewhat higher beneath the oceans than beneath the continents. As shown in **FIGURE 3.17**, when a typical geothermal gradient is compared to the melting point curve for the mantle rock peridotite, the temperature at which peridotite melts is everywhere higher than the geothermal gradient. Thus, under normal conditions, the mantle is solid. As you will see, tectonic processes exist that can increase the geothermal gradient sufficiently to trigger melting. In addition, other mechanisms exist that trigger melting by reducing the temperature at which peridotite begins to melt.

DECREASE IN PRESSURE: DECOMPRESSION MELTING. If temperature were the only factor that determined whether or not rock melts, our planet would be a molten ball covered with a thin, solid outer shell. This, of course, is not the case. The reason is that pressure also increases with depth.

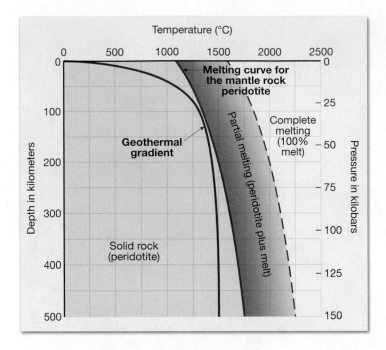

FIGURE 3.17 A schematic diagram illustrating a typical geothermal gradient (increase in temperature with depth) for the crust and upper mantle. Also illustrated is an idealized curve that depicts the melting point temperatures for the mantle rock peridotite. Notice that when the geothermal gradient is compared to the melting point curve for peridotite, the temperature at which peridotite melts is everywhere higher than the geothermal gradient. Thus, under normal conditions the mantle is solid. Special circumstances are required to generate magma.

Melting, which is accompanied by an increase in volume, *occurs at higher temperatures at depth* because of greater confining pressure. Consequently, an increase in confining pressure causes an increase in the rock's melting temperature. Conversely, reducing confining pressure lowers a rock's melting temperature. When confining pressure drops sufficiently, **decompression melting** is triggered.

Decompression melting occurs where hot, solid mantle rock ascends in zones of convective upwelling, thereby moving into regions of lower pressure. This process is responsible for generating magma along divergent plate boundaries (oceanic ridges) where plates are rifting apart (**FIGURE 3.18**).

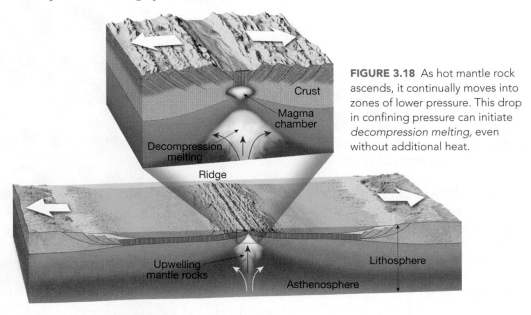

FIGURE 3.18 As hot mantle rock ascends, it continually moves into zones of lower pressure. This drop in confining pressure can initiate *decompression melting,* even without additional heat.

Below the ridge crest, hot mantle rock rises and melts replacing the material that shifted horizontally away from the ridge axis. Decompression melting also occurs within ascending mantle plumes.

ADDITION OF VOLATILES. Another important factor affecting the melting temperature of rock is its water content. Water and other volatiles act as salt does to melt ice. That is, volatiles cause rock to melt at lower temperatures. Further, the effect of volatiles is magnified by increased pressure. Deeply buried "wet" rock has a much lower melting temperature than "dry" rock of the same composition (see Figure 3.17). Therefore, in addition to a rock's composition, its temperature, depth (confining pressure), and water content determine whether it exists as a solid or liquid.

Volatiles play an important role in generating magma at convergent plate boundaries where cool slabs of oceanic lithosphere descend into the mantle (**FIGURE 3.19**). As an oceanic plate sinks, both heat and pressure drive water from the subducting crustal rocks. These fluids, which are very mobile, migrate into the wedge of hot mantle that lies directly above. The addition of water lowers the melting temperature of peridotite sufficiently to generate some melt. Laboratory studies have shown that the temperature at which peridotite begins to melt can be lowered by as much as 100 °C by the addition of only 0.1 percent water.

Melting of peridotite generates basaltic magma having a temperature of 1200 °C or higher. When enough mantle-derived basaltic magma forms, it will buoyantly rise toward the surface. In a continental setting, basaltic magma may "pond" beneath crustal rocks, which have a lower density and are already near their melting temperature. This may result in some melting of the crust and the formation of a secondary, silica-rich magma.

In summary, magma can be generated three ways: (1) when an *increase in temperature* causes a rock to exceed its melting point; (2) in zones of upwelling *a decrease in pressure* (without the addition of heat) can result in *decompression melting;* and (3) the *introduction of volatiles*

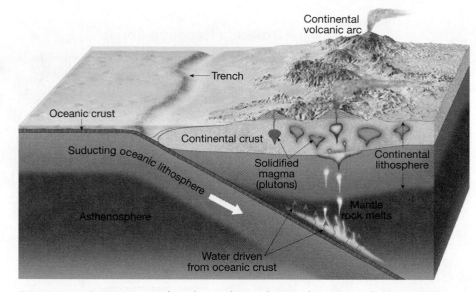

FIGURE 3.19 As an oceanic plate descends into the mantle, water and other volatiles are driven from the subducting crustal rocks into the mantle above. These volatiles lower the melting temperature of hot mantle rock sufficiently to trigger melting.

(principally water) can lower the melting temperature of hot mantle rock sufficiently to generate magma.

CONCEPT CHECK 3.6

❶ What is the geothermal gradient?

❷ Describe decompression melting.

❸ How does the introduction of volatiles trigger melting?

❹ In which plate tectonic settings would you expect magma to form?

How Magmas Evolve

Because a large variety of igneous rocks exists, it is logical to assume that a wide variety of magmas must also exist. However, geologists have observed that, over time, a volcano may extrude lavas exhibiting quite different compositions. Data of this type led them to examine the possibility that magma might change (evolve) and thus become the parent to a variety of igneous rocks. To explore this idea, a pioneering investigation into the crystallization of magma was carried out by N. L. Bowen in the first quarter of the 20th century.

Granite is an intrusive igneous rock that is especially abundant in Earth's continental crust. *(Photo by E. J. Tarbuck)*

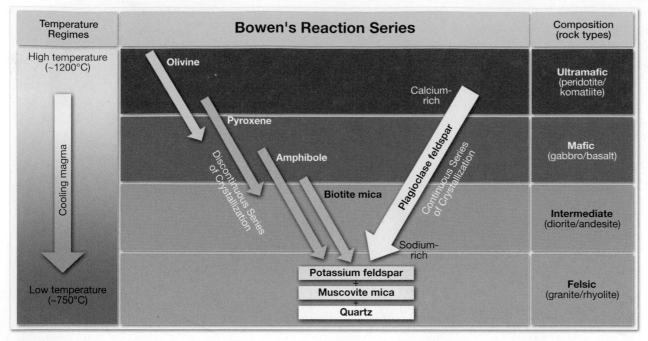

Temperature Regimes	Bowen's Reaction Series	Composition (rock types)

FIGURE 3.20 Bowen's reaction series shows the sequence in which minerals crystallize from a magma. Compare this figure to the mineral composition of the rock groups in Figure 3.10. Note that each rock group consists of minerals that crystallize in the same temperature range.

Bowen's Reaction Series and the Composition of Igneous Rocks

Recall that ice freezes at a single temperature, whereas basaltic magma crystallizes over a range of at least 200 °C of cooling. In a laboratory setting Bowen and his coworkers demonstrated that as a basaltic magma cools, minerals tend to crystallize in a systematic fashion based on their melting points. As shown in **FIGURE 3.20**, the first mineral to crystallize is the ferromagnesian mineral olivine. Further cooling generates calcium-rich plagioclase feldspar as well as pyroxene, and so forth down the diagram.

During the crystallization process, the composition of the remaining liquid portion of the magma also continually changes. For example, at the stage when about a third of the magma has solidified, the melt will be nearly depleted of iron, magnesium, and calcium, because these elements are major constituents of the earliest-formed minerals. The removal of these elements causes the melt to become enriched in sodium and potassium. Further, because the original basaltic magma contained about 50 percent silica (SiO_2), the crystallization of the earliest-formed mineral olivine, which is only about 40 percent silica, leaves the remaining melt richer in SiO_2. Thus, the silica component of the melt becomes enriched as the magma evolves.

Bowen also demonstrated that if the solid components of a magma remain in contact with the remaining melt, they will chemically react and change mineralogy as shown in Figure 3.20. For this reason, this arrangement of minerals became known as **Bowen's reaction series.** As you will see, in nature the earliest-formed minerals can be separated from the melt, thus halting any further chemical reaction.

The diagram of Bowen's reaction series in Figure 3.20 depicts the sequence in which minerals crystallize from a magma of basaltic composition under laboratory conditions. Evidence that this highly idealized crystallization model approximates what can happen in nature comes from the analysis of igneous rocks. In particular, we find that minerals that form in the same general temperature regime depicted on Bowen's reaction series are found together in the same igneous rocks. For example, notice in Figure 3.20 that the minerals quartz, potassium feldspar, and muscovite, which are located in the same region of Bowen's diagram, are typically found together as major constituents of the plutonic igneous rock *granite.*

MAGMATIC DIFFERENTI-ATION. Bowen demonstrated that minerals crystallize from magma in a systematic fashion. But how do Bowen's findings account for the great diversity of igneous rocks? It has been shown that, at one or more stages during the crystallization of magma, a separation of various components can occur. One mechanism that allows this to happen is called **crystal settling.** This process occurs when the earlier-formed minerals are more dense (heavier) than the liquid portion and sink toward the bottom of the magma chamber, as shown in **FIGURE 3.21**. (Crystallization can also occur along the cool margins of a magma body.) When the remaining melt solidifies—either in place or in another location if it migrates into fractures in the surrounding rocks—it will form a rock with a mineralogy much different from the parent magma (Figure 3.21). The formation of one or more secondary magmas from a single parent magma is called **magmatic differentiation.**

One consequence of magnetic differentiation of basaltic (mafic) magma is that it

DID YOU KNOW?
Quartz watches actually contain quartz crystals that allow them to keep time. Before quartz watches, timepieces used some sort of oscillating mass or tuning fork. Cogs and wheels converted this mechanical movement to the movement of the hands. It turns out that if voltage is applied to a quartz crystal, it will oscillate with a consistency that is hundreds of times better for timing than a tuning fork. Because of this property, and modern integrated-circuit technology, quartz watches are now built so cheaply that they are sometimes given away in cereal boxes. Modern watches that employ mechanical movements are very expensive indeed.

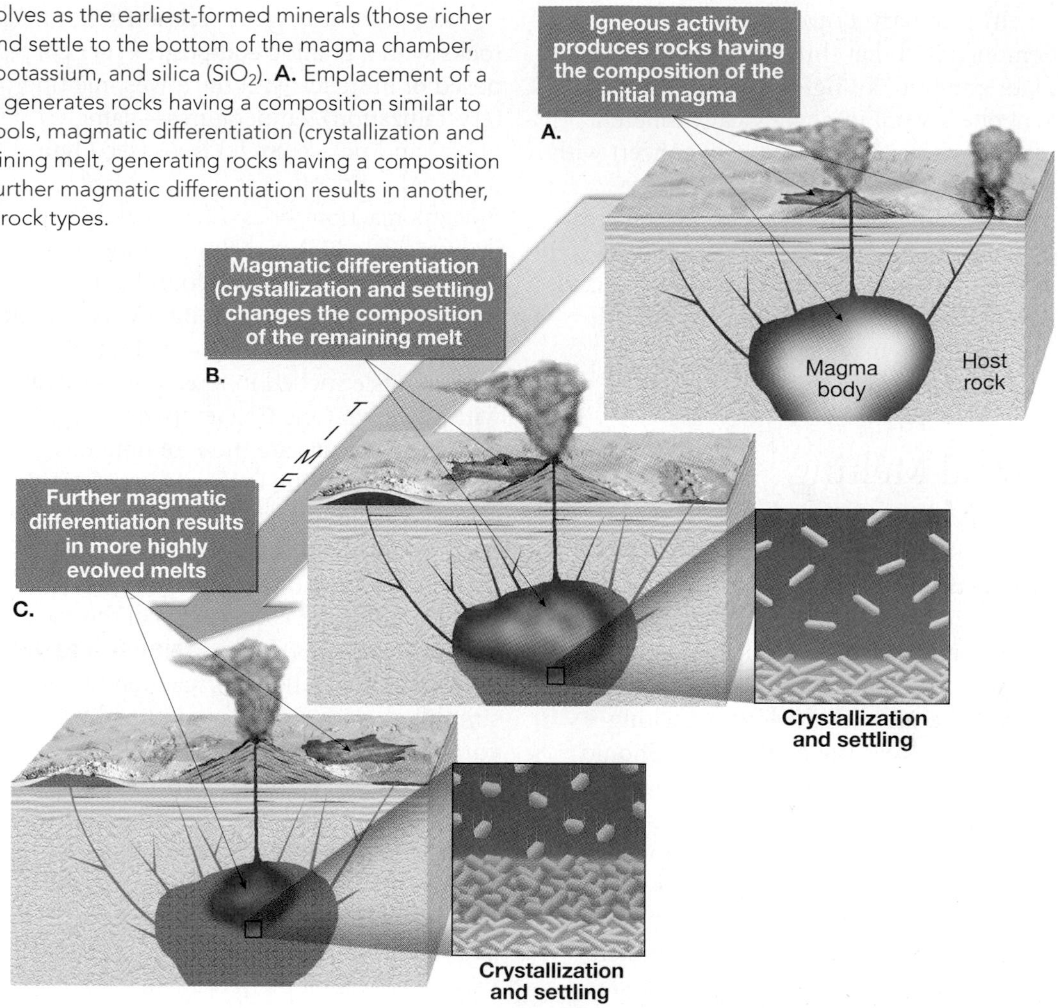

FIGURE 3.21 Illustration of how a magma evolves as the earliest-formed minerals (those richer in iron, magnesium, and calcium) crystallize and settle to the bottom of the magma chamber, leaving the remaining melt richer in sodium, potassium, and silica (SiO_2). **A.** Emplacement of a magma body and associated igneous activity generates rocks having a composition similar to that of the initial magma. **B.** As the magma cools, magmatic differentiation (crystallization and settling) change the composition of the remaining melt, generating rocks having a composition quite different than the original magma. **C.** Further magmatic differentiation results in another, more highly evolved, melt with its associated rock types.

tends to generate evolved silica-rich felsic melts of low density that can rise toward the top of the magma chamber. The result is a compositionally zoned magma reservoir with dense, basaltic melt stratified below less dense, silica-rich melt.

Assimilation and Magma Mixing

Bowen successfully demonstrated that through magmatic differentiation a parent magma can generate several mineralogically different igneous rocks. However, more recent work indicates that magmatic differentiation cannot, by itself, account for the entire compositional spectrum of igneous rocks.

Once a magma body forms its composition can also change through the incorporation of foreign material. For example, as magma migrates through the crust, it may incorporate some of the surrounding host rock, a process called **assimilation** (**FIGURE 3.22A**). In a near-surface environment where rocks are brittle, magma causes numerous cracks in

the overlying rock as it pushes upward. The force of the injected magma is often sufficient to dislodge blocks of "foreign" rock, which melt and are incorporated into the magma body.

Another means by which the composition of magma can be altered is **magma mixing**. This process occurs when one magma body intrudes another that has a different composition (Figure 3.22C). Once combined, convective flow may stir the two magmas and generate a mass having a composition that is a blend of the two. Magma mixing may occur during the ascent of two chemically distinct magma bodies as the more buoyant mass overtakes the more slowly moving mass.

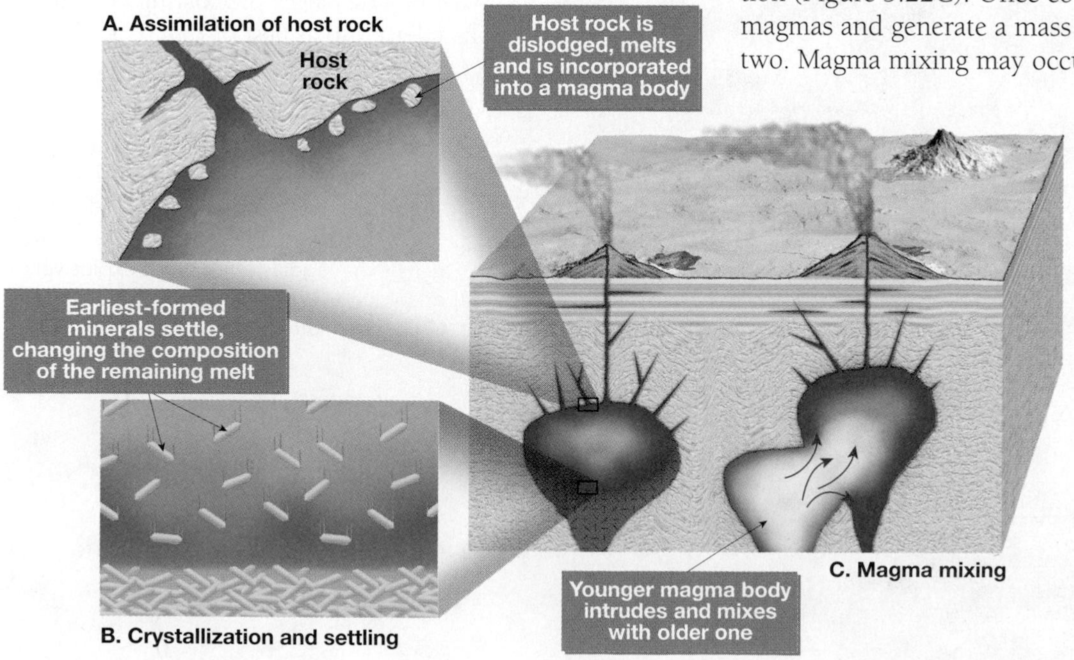

Igneous activity produces rocks having the composition of the initial magma

A.

Magmatic differentiation (crystallization and settling) changes the composition of the remaining melt

B.

Magma body   Host rock

Further magmatic differentiation results in more highly evolved melts

C.

Crystallization and settling

Crystallization and settling

DID YOU KNOW?
Although the United States is the largest consumer of industrial diamonds it has no commercial source of this mineral. It does, however, manufacture large quantities of synthetic diamonds for industrial use.

A. Assimilation of host rock

Host rock

Host rock is dislodged, melts and is incorporated into a magma body

Earliest-formed minerals settle, changing the composition of the remaining melt

B. Crystallization and settling

Younger magma body intrudes and mixes with older one

C. Magma mixing

FIGURE 3.22 This illustration shows three ways that the composition of a magma body may be altered: **A.** Assimilation of host rock; **B.** Crystallization and settling (magmatic differentiation); and **C.** Magma mixing.

In summary, Bowen successfully demonstrated that through magmatic differentiation, a single parent magma can generate several mineralogically different igneous rocks. This process, in concert with magma mixing and contamination by crustal rocks, accounts in part for the great diversity of magmas and igneous rocks. We will next look at another important process, partial melting, which is also responsible for generating magmas having varying compositions.

Partial Melting and Magma Composition

Recall that the crystallization of basaltic magma occurs over a temperature range of at least 200 °C. As you might expect, melting, the reverse process, spans a similar temperature range. As rock begins to melt, those minerals with the lowest melting temperatures are the first to melt. Should melting continue, minerals with higher melting points begin to melt and the composition of the magma steadily approaches the overall composition of the rock from which it was derived. Most often, however, melting is not complete. The incomplete melting of rocks is known as **partial melting,** a process that produces most magma.

Recall from Bowen's reaction series that rocks with a granitic composition are composed of minerals with the lowest melting (crystallization) temperatures—namely, quartz and potassium feldspar (see Figure 3.20). Also note that as we move up Bowen's reaction series, the minerals have progressively higher melting temperatures and that olivine, which is found at the top, has the highest melting point. When a rock undergoes partial melting it will form a melt that is enriched in ions from minerals with the lowest melting temperatures. The unmelted crystals are those of minerals with higher melting temperatures. Separation of these two fractions would yield a melt with a chemical composition that is richer in silica and nearer to the granitic (felsic) end of the spectrum than the rock from which it was derived (**FIGURE 3.23**).

In summary, Bowen's reaction series is a simplified guide to understanding how partial melting produces rocks of various compositions. As rock begins to melt, those minerals with the lowest melting temperatures are the first to melt. As a result, partial melting produces magmas enriched in low-melting temperature components and richer in silica than the parent rock. Partial melting of the ultramafic rock peridotite produces a magma of basaltic (mafic) composition, while partial melting of the mafic rock basalt generates magma of intermediate (andesitic) composition. Granitic (felsic) magmas evolve from basaltic magmas and/or the melting and assimilation of silica-rich crustal rocks.

CONCEPT CHECK 3.7

❶ What is magmatic differentiation? How might this process lead to the formation of several different igneous rocks from a single magma?

❷ Relate the classification of igneous rocks to Bowen's reaction series.

❸ What is partial melting?

❹ How does the composition of a melt produced by partial melting compare with the composition of the parent rock?

Intrusive Igneous Activity

 Volcanoes

ESSENTIALS OF GEOLOGY **Intrusive Igneous Activity**

Although volcanic eruptions can be violent and spectacular events, most magma is emplaced and crystallizes at depth, without fanfare. Therefore, understanding the igneous processes that occur deep underground is as important to geologists as the study of volcanic events.

When magma rises through the crust, it forcefully displaces preexisting crustal rocks referred to as *host* or *country rock.*

FIGURE 3.23 1996 eruption of Mount Ruapehu, Tongariro National Park, New Zealand. Volcanoes that border the Pacific Ocean are fed largely by magmas that have intermediate or felsic compositions. These silica-rich magmas often erupt explosively, generating large plumes of volcanic dust and ash. *(Photo by ImageState/Alamy)*

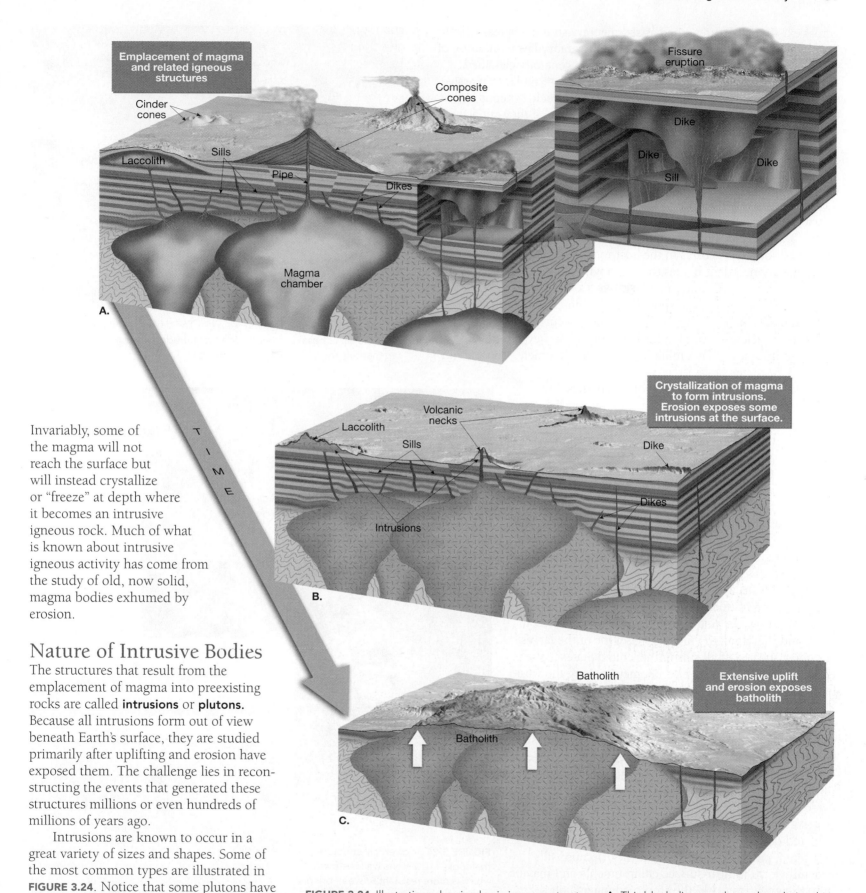

Invariably, some of the magma will not reach the surface but will instead crystallize or "freeze" at depth where it becomes an intrusive igneous rock. Much of what is known about intrusive igneous activity has come from the study of old, now solid, magma bodies exhumed by erosion.

Nature of Intrusive Bodies

The structures that result from the emplacement of magma into preexisting rocks are called **intrusions** or **plutons.** Because all intrusions form out of view beneath Earth's surface, they are studied primarily after uplifting and erosion have exposed them. The challenge lies in reconstructing the events that generated these structures millions or even hundreds of millions of years ago.

Intrusions are known to occur in a great variety of sizes and shapes. Some of the most common types are illustrated in **FIGURE 3.24**. Notice that some plutons have a tabular (tabletop) shape, whereas others are best described as massive. Also, observe that some of these bodies cut

FIGURE 3.24 Illustrations showing basic igneous structures. **A.** This block diagram shows the relationship between volcanism and intrusive igneous activity. **B.** This view illustrates the basic intrusive igneous structures, some of which have been exposed by erosion long after their formation. **C.** After millions of years of uplifting and erosion, a batholith is exposed at the surface.

across existing structures, such as sedimentary strata; whereas others form when magma is injected between sedimentary layers. Because of these differences, intrusive igneous bodies are generally classified according to their shape as either **tabular** (*tabula* = table) or **massive** and by their orientation with respect to the host rock. Igneous bodies are said to be **discordant** (*discordare* = to disagree) if they cut across existing structures and **concordant** (*concordare* = to agree) if they form parallel to features such as sedimentary strata.

Tablular Intrusive Bodies: Dikes and Sills

Tabular intrusive bodies are produced when magma is forcibly injected into a fracture or zone of weakness, such as a bedding surface (see Figure 3.24). **Dikes** are discordant bodies that cut across bedding surfaces or other structures in the host rock. By contrast, **sills** are nearly horizontal, concordant bodies that form when magma exploits weaknesses between sedimentary beds (**FIGURE 3.25**). In general, dikes serve as tabular conduits that transport magma, whereas sills store magma. Dikes and sills are typically shallow features, occurring where the host rocks are sufficiently brittle to fracture. Although they can range in thickness from less than a millimeter to over a kilometer, most are in the 1- to 20-meter range.

Dikes and sills can occur as solitary bodies, but dikes in particular tend to form in roughly parallel groups called *dike swarms*. These multiple structures reflect the tendency for fractures to form in sets when tensional forces stretch brittle country rock. Dikes can also occur radiating from an eroded volcanic neck, like spokes on a wheel. In these situations the active ascent of magma generated fissures in the volcanic cone out of which lava flowed.

Dikes frequently weather more slowly than the surrounding rock. Consequently, when exposed by erosion, dikes tend to have a wall-like appearance, as shown in **FIGURE 3.26**. Because dikes and sills are relatively uniform in thickness and can extend for many kilometers they are assumed to be the product of very fluid, and therefore, mobile magmas. One of the largest and most studied of all sills in the United States is the Palisades Sill. Exposed for 80 kilometers along the west bank of the Hudson River in southeastern New York and northeastern New Jersey, this sill is about 300 meters thick. Because it is resistant to erosion, the Palisades Sill forms an imposing cliff that can be easily seen from the opposite side of the Hudson.

In many respects, sills closely resemble buried lava flows. Both are tabular and can have a wide aerial extent, and both may exhibit columnar jointing (**FIGURE 3.27**). **Columnar joints** form as igneous rocks cool and develop shrinkage fractures that produce elongated, pillar-like columns. Further, because sills generally form in near-surface environments and may be only a few meters thick, the emplaced magma often cools quickly enough to generate a fine-grained texture. (Recall that most intrusive igneous bodies have a coarse-grained texture.)

FIGURE 3.25 Salt River Canyon, Arizona. The dark, essentially horizontal band is a sill of basaltic composition that intruded horizontal layers of sedimentary rock. *(Photo by E. J. Tarbuck)*

FIGURE 3.26 The vertical structure that looks like a stone wall is a dike, which is more resistant to weathering than the surrounding rock. This dike is located west of Granby, Colorado, near Arapaho National Forest. *(Photo by R. Jay Fleisher)*

Rock surrounding dike is more easily eroded

Dike made of resistant igneous rock

Weathered debris

Geologist's Sketch

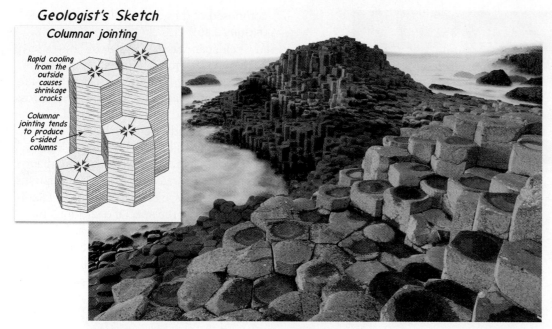

Geologist's Sketch

Columnar jointing

Rapid cooling from the outside causes shrinkage cracks

Columnar jointing tends to produce 6-sided columns

FIGURE 3.27 Columnar jointing in basalt, Giants Causeway National Park, Northern Ireland. These five-to seven-sided columns are produced by contraction and fracturing that result as a lava flow or sill gradually cools. *(Photo by John Lawrence/Getty Images)*

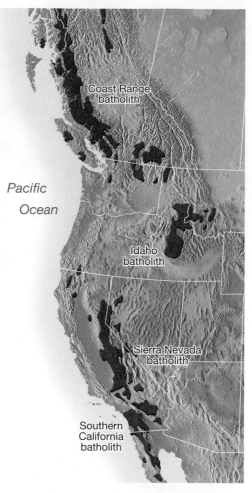

Coast Range batholith

Pacific Ocean

Idaho batholith

Sierra Nevada batholith

Southern California batholith

FIGURE 3.28 Granitic batholiths that occur along the western margin of North America. These gigantic, elongated bodies consist of numerous plutons that were emplaced during the last 150 million years of Earth's history.

Massive Intrusive Bodies: Batholiths, Stocks, and Laccoliths

BATHOLITHS. By far the largest intrusive igneous bodies are **batholiths** (*bathos* = depth, *lithos* = stone). Batholiths occur as mammoth linear structures several hundreds of kilometers long and up to 100 kilometers wide (**FIGURE 3.28**). The Sierra Nevada batholith, for example, is a continuous granitic structure that forms much of the Sierra Nevada in California. An even larger batholith extends for over 1800 kilometers (1100 miles) along the Coast Mountains of western Canada and into southern Alaska. Although batholiths can cover a large area, recent gravitational studies indicate that most are less than 10 kilometers (6 miles) thick. Some are even thinner. The Coastal batholith of Peru, for example, is essentially a flat slab with an average thickness of only 2 to 3 kilometers (1 to 2 miles).

Batholiths are almost always made up of felsic and intermediate rock types and are often referred to as "granite batholiths." Large granite batholiths consist of hundreds of plutons that intimately crowd against or penetrate one another. These bulbous masses were emplaced over spans of millions of years. The intrusive activity that created the Sierra Nevada batholith, for example, occurred nearly continuously over a 130-million-year period that ended about 80 million years ago (see Figure 3.28).

STOCKS. By definition, a plutonic body must have a surface exposure greater than 100 square kilometers (40 square miles) to be considered a batholith. Smaller plutons of this type are termed **stocks**. However, many stocks appear to be portions of much larger intrusive bodies that would be called batholiths if they were fully exposed.

LACCOLITHS. A 19th century study by G. K. Gilbert of the U.S. Geological Survey in the Henry Mountains of Utah produced the first clear evidence that igneous intrusions can lift the sedimentary strata they penetrate. Gilbert named the igneous intrusions he observed **laccoliths,** which he envisioned as igneous rock forcibly injected between sedimentary strata, so as to arch the beds above, while leaving those below relatively flat. It is now known that the five major peaks of the Henry Mountains are not laccoliths, but stocks. However, these central magma bodies are the source material for branching offshoots that are true laccoliths, as Gilbert defined them (**FIGURE 3.29**).

Numerous other granitic laccoliths have since been identified in Utah. The largest is a part of the Pine Valley Mountains located north of St. George, Utah. Others are found in the La Sal Mountains near Arches National Park and in the Abajo Mountains directly to the south.

CONCEPT CHECK 3.8

❶ Describe each of the four basic intrusive features (dikes, sills, laccoliths, and batholiths).

❷ What is the largest of all intrusive igneous bodies? Is it tabular or massive? Concordant or discordant?

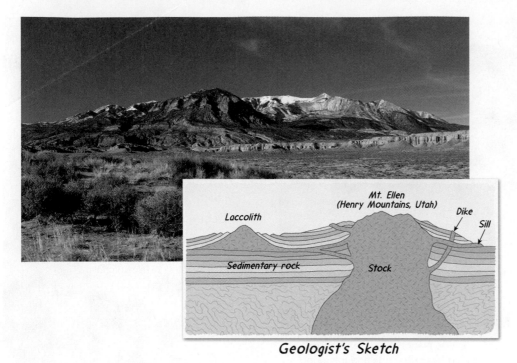

FIGURE 3.29 Mount Ellen, the northernmost of five peaks that make up Utah's Henry Mountains. Although the main intrusions in the Henry Mountains are stocks, numerous laccoliths formed as offshoots of these structures. *(Photo by Michael DeFreitas North America/Alamy)*

Mineral Resources and Igneous Processes

Given the growth of the middle class in countries such as China, India, and Brazil, the demand for metallic natural resources has increased exponentially in recent years. Some of the most important accumulations of metals, such as gold, silver, copper, mercury, lead, platinum, and nickel, are produced by igneous processes (**TABLE 3.1**). These mineral resources result from processes that concentrate desirable materials to the extent that they can be profitably extracted. Therefore, knowledge of how and where these important materials are likely to be concentrated is vital to our well-being.

The igneous processes that generate some of these metal deposits are quite straightforward. For example, as a large basaltic magma body cools, the heavy minerals that crystallize early tend to settle to the lower portion of the magma chamber. This type of magmatic segregation serves to concentrate selected metals producing major deposits of chromite (ore of chromium), magnetite, and platinum. Layers of chromite, interbedded with other heavy minerals, are mined at Montana's

Stillwater Complex, whereas the Bushveld Complex in South Africa contains over 70 percent of the world's known reserves of platinum.

Magmatic segregation is also important in the late stages of the magmatic process. This is particularly true of granitic magmas in which the residual melt can become enriched in rare elements and some heavy metals. Further, because water and other volatile substances do not crystallize along with the bulk of the magma body, these fluids make up a high percentage of the melt during the final phase of solidification. Crystallization in a fluid-rich environment, where ion migration is enhanced, results in the formation of crystals several centimeters, or even a few meters, in length. The resulting rocks, called *pegmatites*, are

FIGURE 3.30 This pegmatite in the Black Hills of South Dakota was mined for its large crystals of spodumene, an important source of lithium. Arrows are pointing to impressions left by crystals. Note person in upper center of photo for scale. *(Photo by James G. Kirchner)*

composed of these unusually large crystals (**FIGURE 3.30**).

Feldspar masses the size of houses have been quarried from a pegmatite located in North Carolina. Gigantic hexagonal crystals of muscovite measuring a few meters across have been found in Ontario, Canada. In the Black Hills, spodumene crystals as thick as telephone poles have been mined (Figure 3.30). The largest of these was more than 12 meters (40 feet) long. Not all pegmatites contain such large crystals, but these examples emphasize the special conditions that must exist during their formation.

Most pegmatites are granitic in composition and consist of unusually large crystals of quartz, feldspar, and muscovite. Feldspar is used in the production of ceramics, and muscovite is used for electrical insulation and glitter. Further, pegmatites often contain some of the least abundant elements. Minerals containing the elements lithium, cesium, uranium, and the rare earths are occasionally found. Moreover, some pegmatites contain semiprecious gems such as beryl, topaz, and tourmaline. Most pegmatites are located within large igneous masses or as dikes or veins that cut into the host rock that surrounds the magma chamber (**FIGURE 3.31**).

Not all late-stage magmas produce pegmatites, nor do all have a granitic composition. Rather, some magmas become enriched in iron or occasionally copper. For example, at Kirava, Sweden, magma composed of over 60 percent magnetite

TABLE 3.1

Occurrences of Metallic Minerals

Metal	Principal Ores	Geological Occurrences
Aluminum	Bauxite	Residual product of weathering
Chromium	Chromite	Magmatic segregation
Copper	Chalcopyrite Bornite Chalcocite	Hydrothermal deposits; contact metamorphism; enrichment by weathering processes
Gold	Native gold	Hydrothermal deposits; placers
Iron	Hematite Magnetite Limonite	Banded sedimentary formations; magmatic segregation
Lead	Galena	Hydrothermal deposits
Magnesium	Magnesite Dolomite	Hydrothermal deposits
Manganese	Pyrolusite	Residual product of weathering
Mercury	Cinnabar	Hydrothermal deposits
Molybdenum	Molybdenite	Hydrothermal deposits
Nickel	Pentlandite	Magmatic segregation
Platinum	Native platinum	Magmatic segregation; placers
Silver	Native silver Argentite	Hydrothermal deposits; enrichment by weathering processes
Tin	Cassiterite	Hydrothermal deposits; placers
Titanium	Ilmenite	Magmatic segregation; placers
Tungsten	Wolframite	Pegmatites; contact metamorphic deposits; placers
Uranium	Uraninite (pitchblende)	Pegmatites; sedimentary deposits
Zinc	Sphalerite	Hydrothermal deposits

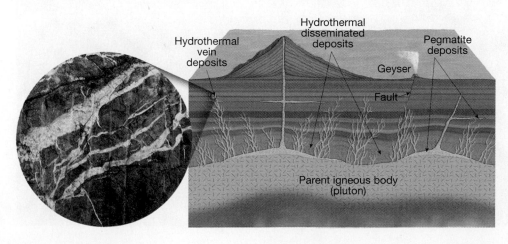

FIGURE 3.31 Illustration of the relationship between a parent igneous body and the associated pegmatite and hydrothermal deposits. Inset photo shows light-colored vein deposits emplaced along a series of fractures in dark-colored metamorphic rock. *(Photo by James E. Patterson)*

solidified to produce one of the largest iron deposits in the world.

Among the best known and most important ore deposits are those generated from hot, ion-rich fluids called **hydrothermal (hot-water) solutions.*** Included in this group are the gold deposits of the Homestake mine in South Dakota; the lead, zinc, and silver ores near Coeur d'Alene, Idaho; the silver deposits of the Comstock Lode in Nevada; and the copper ores of Michigan's Keweenaw Peninsula.

The majority of hydrothermal deposits originate from hot, metal-rich fluids that are remnants of late-stage magmatic processes. During solidification, liquids plus various metallic ions accumulate near the top of the magma chamber. Because of their mobility, these ion-rich solutions can migrate great distances through the surrounding rock before they are eventually

---

*Because these hot, ion-rich fluids tend to chemically alter the host rock, this process is called hydrothermal metamorphism and is discussed in Chapter 7.

---

deposited, usually as sulfides of various metals (Figure 3.31). Some of these fluids move along openings such as fractures or bedding planes, where they cool and precipitate metallic ions to produce *vein deposits*. Many of the most productive deposits of gold, silver, and mercury occur as hydrothermal vein deposits (**FIGURE 3.32**).

Another important type of accumulation generated by hydrothermal activity is called a *disseminated deposit*. Rather than being concentrated in narrow veins and dikes, these ores are distributed as minute masses throughout the entire rock mass. Much of the world's copper is extracted from disseminated deposits, including those at Chuquicamata, Chile, and the huge Bingham Canyon copper mine in Utah. Because these accumulations contain only 0.4 to 0.8 percent copper, between 125 and 250 kilograms of ore must be mined for every kilogram of metal recovered. The environmental impact of these large excavations, including the problems of waste disposal, is significant.

Another economically important mineral with an igneous origin is diamond. Although best known as gems, diamonds are used extensively as abrasives. Diamonds are thought to originate at depths of nearly 200 kilometers, where the confining pressure is great enough to generate this high-pressure form of carbon. Once crystallized, the diamonds are carried upward through pipe-shaped conduits that increase in diameter toward the surface. In diamond-bearing pipes, nearly the entire pipe contains diamond crystals that are disseminated throughout an ultramafic rock called *kimberlite*. The most productive kimberlite pipes are those in South Africa. The only equivalent source of diamonds in the United States is located near Murfreesboro, Arkansas, but this deposit is exhausted and serves today merely as a tourist attraction.

CONCEPT CHECK 3.9

❶ List two types of hydrothermal deposits.

FIGURE 3.32 High grade gold ore from a mine in Ghana, West Africa. *(Photo by Greenshoots Communications/Alamy)*

CHAPTER THREE
Igneous Rocks and Intrusive Activity in Review

⊙ *Igneous rocks* form when *magma cools* and solidifies. *Extrusive,* or *volcanic,* igneous rocks result when *lava* cools at the surface. Magma that solidifies at depth produces *intrusive,* or *plutonic,* igneous rocks.

⊙ As magma cools, the ions that compose it arrange themselves into orderly patterns—a process called *crystallization.* Slow cooling results in the formation of relatively large crystals. Conversely, when cooling occurs rapidly, the outcome is a solid mass consisting of tiny intergrown crystals. When molten material is quenched instantly, a mass of unordered atoms, referred to as *glass,* forms.

⊙ The mineral composition of an igneous rock is the consequence of the chemical make-up of the parent magma and the environment of crystallization. Igneous rocks are divided into broad compositional groups based on the percentage of dark and light silicate minerals they contain. *Felsic rocks* (e.g., granite and rhyolite) are composed mostly of the light-colored silicate minerals potassium feldspar and quartz. Rocks of *intermediate* composition, (e.g., andesite and diorite) are rich in plagioclase feldspar and amphibole. *Mafic rocks* (e.g., basalt and gabbro) contain abundant olivine, pyroxene, and calcium-rich plagioclase feldspar. They are high in iron, magnesium, and calcium, low in silica, and dark gray to black in color.

⊙ The texture of an igneous rock refers to the overall appearance of the rock based on the size, shape, and arrangement of its mineral grains. The most important factor influencing the texture of igneous rocks is the rate at which magma cools. Common igneous rock textures include *aphanitic (fine-grained),* with grains too small to be distinguished without the aid of a microscope; *phaneritic (coarse-grained),* with intergrown crystals that are roughly equal in size and large enough to be identified without the aid of a microscope; *porphyritic,* which has larger crystals (*phenocrysts*) embedded in a matrix of smaller crystals (*groundmass*); and *glassy.*

⊙ The mineral make-up of an igneous rock is ultimately determined by the chemical composition of the magma from which it crystallizes. N. L. Bowen discovered that as magma cools in the laboratory, those minerals with higher melting points crystallize before minerals with lower melting points. *Bowen's reaction series* illustrates the sequence of mineral formation within magma.

⊙ During the crystallization of magma, if the earlier-formed minerals are more dense than the liquid portion, they will settle to the bottom of the magma chamber during a process called *crystal settling.* Owing to the fact that crystal settling removes the earlier-formed minerals, the remaining melt will form a rock with a chemical composition that is different from the parent magma.

The process of developing more than one magma type from a common magma is called *magmatic differentiation.*

⊙ Once a magma body forms, its composition can change through the incorporation of foreign material, a process termed *assimilation,* or by *magma mixing.*

⊙ Magma originates from essentially solid rock of the crust and mantle. In addition to a rock's composition, its temperature, depth (confining pressure), and water content determine whether it exists as a solid or liquid. Thus, magma can be generated by *raising a rock's temperature,* as occurs when a hot mantle plume "ponds" beneath crustal rocks. A *decrease in pressure* can cause *decompression melting.* Further, the *introduction of volatiles* (water) can lower a rock's melting point sufficiently to generate magma. A process called *partial melting* produces a melt made of the low-melting-temperature minerals, which are higher in silica than the original rock. Thus, magmas generated by partial melting are nearer to the felsic end of the compositional spectrum than are the rocks from which they formed.

⊙ Intrusive igneous bodies are classified according to their *shape* and by their *orientation with respect to the country or host rock,* generally sedimentary or metamorphic rock. The two general shapes are *tabular* (sheet-like) and *massive.* Intrusive igneous bodies that cut across existing sedimentary beds are said to be *discordant;* those that form parallel to existing sedimentary beds are *concordant.*

⊙ *Dikes* are tabular, discordant igneous bodies produced when magma is injected into fractures that cut across rock layers. Nearly horizontal, tabular, concordant bodies, called *sills,* form when magma is injected along the bedding surfaces of sedimentary rocks. In many respects, sills closely resemble buried lava flows. *Batholiths,* the largest intrusive igneous bodies, sometimes make up large linear mountains, as exemplified by the Sierra Nevada. *Laccoliths* are similar to sills but form from less fluid magma that collects as a lens-shaped mass that arches overlying strata upward.

⊙ Some of the most important accumulations of metals, such as gold, silver, lead, and copper, are produced by igneous processes. The best-known and most important ore deposits are generated from *hydrothermal* (hot-water) *solutions.* Hydrothermal deposits are thought to originate from hot, metal-rich fluids that are remnants of late-stage magmatic processes. These ion-rich solutions move along fractures or bedding planes, cool, and precipitate the metallic ions to produce *vein deposits.* In a *disseminated deposit* (e.g., much of the world's copper deposits), the ores from hydrothermal solutions are distributed as minute masses throughout the entire rock mass.

Key Terms

andesitic composition (p. 67)
aphanitic texture (p. 68)
assimilation (p. 79)
basaltic composition (p. 67)
batholiths (p. 83)
Bowen's reaction series (p. 78)
coarse-grained texture (p. 69)
columnar joints (p. 82)
concordant (p. 82)
crystallization (p. 65)
crystal settling (p. 78)
decompression melting (p. 76)
dikes (p. 82)
discordant (p. 82)

extrusive igneous rocks (p. 65)
felsic (p. 66)
fine-grained texture (p. 68)
fragmental texture (p. 70)
geothermal gradient (p. 76)
glass (p. 68)
glassy texture (p. 69)
groundmass (p. 69)
granitic composition (p. 66)
hydrothermal solutions (p. 86)
igneous rocks (p. 64)
intermediate composition
   (p. 67)
intrusions (p. 81)

intrusive igneous rocks (p. 65)
laccoliths (p. 83)
lava (p. 64)
mafic (p. 67)
magma (p. 64)
magma mixing (p. 79)
magmatic differentiation (p. 78)
massive (p. 82)
melt (p. 64)
partial melting (p. 80)
pegmatites (p. 70)
pegmatitic texture (p. 71)
phaneritic texture (p. 69)
phenocrysts (p. 69)

plutonic rocks (p. 65)
plutons (p. 81)
porphyritic texture (p. 69)
porphyry (p. 69)
pyroclastic texture (p. 70)
sills (p. 82)
stocks (p. 83)
tabular (p. 82)
texture (p. 67)
ultramafic (p. 67)
vesicular texture (p. 69)
volatiles (p. 64)
volcanic rocks (p. 65)

GIVE IT SOME THOUGHT

❶ Would you expect all of the crystals in an intrusive igneous rock to be the same size? Explain why or why not.

❷ Apply your understanding of igneous rock textures to describe the cooling history of each of the igneous rocks pictured here.

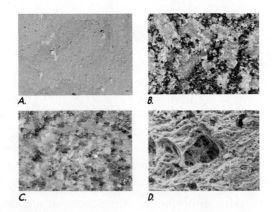

A.   B.

C.   D.

❸ Is it possible for two igneous rocks to have the same mineral composition but be different rocks? Use an example to explain your answer.

❹ Using Figure 3.3 and Figure 3.10, classify the following igneous rocks:

  a. An aphanitic rock containing about 30 percent calcium-rich plagioclase feldspar, 55 percent pyroxene, and 15 percent olivine.

  b. A phaneritic rock containing about 20 percent quartz, 40 percent potassium feldspar, 20 percent sodium-rich plagioclase feldspar, a few percent muscovite, and the remainder dark-colored silicate.

  c. An aphanitic rock containing about 50 percent plagioclase feldspar, 35 percent amphibole, 10 percent pyroxene, and minor amounts of other light-colored silicates.

  d. A phaneritic rock made mainly of olivine and pyroxene, with lesser amounts of calcium-rich plagioclase feldspar.

❺ During a hike you pick up an interesting looking igneous rock. It is dark in color and has a fine, granular appearance with small, fairly round crystals of a glassy, deep green mineral scattered throughout. Answer the following questions about this rock:

  a. What mineral is the small, rounded, glassy green crystals likely to be?

  b. Did the magma from which this rock formed likely originate in the mantle or in the crust? Explain.

  c. Was the magma likely a very high temperature magma or a fairly low temperature magma? Why?

  d. What other minerals might make up the rest of this rock?

❻ A common misconception about Earth's mantle is that it is a thick layer of molten rock. Explain, in your own words, why Earth's mantle is actually solid under normal conditions.

❼ Describe two situations in which mantle rock can melt without an increase in temperature. How do these magma-generating mechanisms relate to plate tectonics?

8. Use your understanding of Bowen's reaction series (Figure 3.20) to explain how partial melting can generate magmas having different compositions.

9. Each statement describes how an intrusive feature appears when exposed at Earth's surface due to erosion. Name the feature.

   a. A dome-shaped mountainous structure flanked by upturned layers of sedimentary rocks.

   b. A vertical wall-like feature a few meters wide and hundreds of meters long.

   c. A huge expanse of granitic rock forming a mountainous terrain tens of kilometers wide.

   d. A relatively thin layer of basalt sandwiched between layers of sedimentary rocks exposed on the side of a canyon.

10. During a field trip with your geology class you visit an exposure of rock layers similar to the one sketched here. A fellow student suggests that the layer of basalt is a sill. You, however, disagree. Why do you think the other student is incorrect? What is a more likely explanation for the basalt layer?

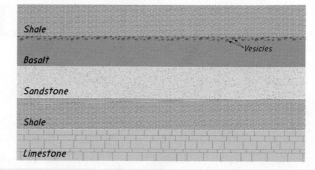

Companion Website

www.mygeoscienceplace.com

The *Essentials of Geology, 11e* companion Website contains numerous multimedia resources accompanied by assessments to aid in your study of the topics in this chapter. The use of this site's learning tools will help improve your understanding of geology. Utilizing the access code that accompanies this text, visit **www.mygeoscienceplace.com** in order to:

- **Review** key chapter concepts.
- **Read** with links to the Pearson eText and to chapter-specific web resources.
- **Visualize** and comprehend challenging topics using the learning activities in *GEODe: Essentials of Geology* and the *Geoscience Animations Library*.
- **Test** yourself with online quizzes.

Volcanoes and Volcanic Hazards

WHY DO VOLCANOES LIKE MOUNT ST. HELENS ERUPT

explosively, whereas others like the Hawaiian volcano Kilauea are relatively quiet? Why do volcanoes occur in chains like the Aleutian Islands or the Cascade Range?

Why do some volcanoes form on the ocean floor, while others occur on the continents?

This chapter will deal with these and similar questions as we explore the nature and movement of magma and lava.

A recent eruption of Italy's Mount Etna. *(Photo by Marco Fulle)*

To assist you in learning the important concepts in this chapter, focus on the following questions:

- ⦿ What primary factors determine the nature of volcanic eruptions? How do these factors affect a magma's viscosity?
- ⦿ What materials are associated with a volcanic eruption?
- ⦿ What are the eruptive patterns and basic characteristics of the three types of volcanoes generally recognized by volcanologists?
- ⦿ What destructive forces are associated with composite volcanoes?
- ⦿ How do calderas form?
- ⦿ What is the source of magma for flood basalts?
- ⦿ What is the relation between volcanic activity and plate tectonics?
- ⦿ What changes in a volcanic landscape can be monitored to detect the movement of magma?

Mount St. Helens Versus Kilauea

On Sunday, May 18, 1980, the largest volcanic eruption to occur in North America in historic times transformed a picturesque volcano into a decapitated remnant (**FIGURE 4.1**). On this date in southwestern Washington State, Mount St. Helens erupted with tremendous force. The blast blew out the entire north flank of the volcano, leaving a gaping hole. In one brief moment, a prominent volcano whose summit had been more than 2900 meters (9500 feet) above sea level was lowered by more than 400 meters (1350 feet).

The event devastated a wide swath of timber-rich land on the north side of the mountain (**FIGURE 4.2**). Trees within a 400-square-kilometer area lay intertwined and flattened, stripped of their branches and appearing from the air like toothpicks strewn about. The accompanying mudflows carried ash, trees, and water-saturated rock debris 29 kilometers (18 miles) down the Toutle River. The eruption claimed 59 lives, some dying from the intense heat and the suffocating cloud of ash and gases, others from being hurled by the blast, and still others from entrapment in the mudflows.

The eruption ejected nearly a cubic kilometer of ash and rock debris. Following the devastating explosion, Mount St. Helens continued to emit great quantities of hot gases and ash. The force of the blast was so strong that some ash was propelled more than 18,000 meters (over 11 miles) into the stratosphere. During the next few days, this very fine-grained material was carried around Earth by strong upper-air winds. Measurable deposits were reported in Oklahoma and Minnesota, with crop damage into central Montana. Meanwhile, ash fallout in the immediate vicinity exceeded 2 meters in depth. The air over Yakima, Washington (130 kilometers to the east), was so filled with ash that residents experienced midnightlike darkness at noon.

Not all volcanic eruptions are as violent as the 1980 Mount St. Helens event. Some volcanoes, such as Hawaii's Kilauea volcano, generate relatively quiet outpourings of fluid lavas. These "gentle" eruptions are not without some fiery displays; occasionally fountains of incandescent lava spray hundreds of meters into the air. Nevertheless, during Kilauea's most recent active phase, which began in 1983, more than 180 homes and a national park visitor center were destroyed.

Testimony to the "quiet nature" of Kilauea's eruptions is the fact that the Hawaiian Volcanoes Observatory has operated on its summit since 1912. This despite the fact that Kilauea has had more than 50 eruptive phases since record keeping began in 1823.

FIGURE 4.1 Before-and-after photographs show the transformation of Mount St. Helens caused by the May 18, 1980, eruption. *(Photo courtesy of U.S. Geological Survey)*

Spirit Lake

CONCEPT CHECK 4.1

❶ Briefly compare the May 18, 1980, eruption of Mount St. Helens to a typical eruption of Hawaii's Kilauea volcano.

FIGURE 4.2 Douglas fir trees were snapped off or uprooted by the lateral blast of Mount St. Helens on May 18, 1980. *(Large photo by Lyn Topinka/AP Photo/U.S. Geological Survey; Inset photo by John M. Burnley/Photo Researchers, Inc.)*

The Nature of Volcanic Eruptions

 **Volcanoes**

 **The Nature of Volcanic Eruptions**

Volcanic activity is commonly perceived as a process that produces a picturesque, cone-shaped structure that periodically erupts in a violent manner, like Mount St. Helens. Although some eruptions may be very explosive, many are not. What determines whether a volcano extrudes magma violently or "gently"? The primary factors include the magma's *composition,* its *temperature,* and the amount of *dissolved gases* it contains. To varying degrees, these factors affect the magma's mobility, or **viscosity** (*viscos* = stickly). The more viscous the material, the greater its resistance to flow. For example, compare syrup to water—syrup is more viscous, and thus more resistant to flow, than water. Magma associated with an explosive eruption may be five times more viscous than magma that is extruded in a quiescent manner.

Factors Affecting Viscosity

The effect of temperature on viscosity is easily seen. Just as heating syrup makes it more fluid (less viscous), the mobility of lava is strongly influenced by temperature. As lava cools and begins to congeal, its mobility decreases and eventually the flow halts.

A more significant factor influencing volcanic behavior is the chemical composition of the magma. Recall that a major difference among various igneous rocks is their silica (SiO_2) content (**TABLE 4.1**). Magmas that produce mafic rocks such as basalt contain about 50 percent silica, whereas magmas that produce felsic rocks (granite and its extrusive equivalent, rhyolite) contain more than 70 percent silica. Intermediate rock types—andesite and diorite—contain about 60 percent silica.

A magma's viscosity is directly related to its silica content—*the more silica in magma, the greater its viscosity.* Silica impedes the flow of magma because silicate structures start to link together into long chains early in the crystallization process. Consequently, rhyolitic (felsic) lavas are very viscous and tend to form comparatively short, thick flows. By contrast, basaltic lavas, which contain less silica, are relatively fluid and have been known to travel 150 kilometers (90 miles) or more before congealing.

The amount of **volatiles** (the gaseous components of magma, mainly water) contained in magma also affects its mobility. Other factors being equal, water dissolved in the magma tends to increase fluidity because it reduces polymerization (formation of long silicate chains) by breaking silicon–oxygen bonds. It follows, therefore, that the loss of gases renders magma (lava) more viscous.

TABLE 4.1

Magmas Have Different Compositions, Which Cause Their Properties to Vary			
Composition	Silica Content	Viscosity	Gas Content
Mafic (Basaltic) magma	Least (~50%)	Least	Least (1–2%)
Intermediate (Andesitic) magma	Intermediate (~60%)	Intermediate	Intermediate (3–4%)
Felsic (Rhyolitic) magma	Most (~70%)	Greatest	Most (4–6%)

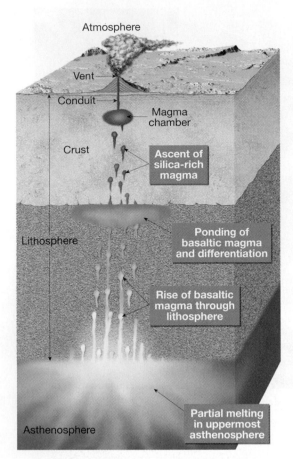

FIGURE 4.3 Schematic drawing showing the movement of magma from its source in the upper asthenosphere through the continental crust. During its ascent, mantle-derived basaltic magmas evolve through the process of magmatic differentiation and by melting and incorporating continental crust. Magmas that feed volcanoes in a continental setting tend to be silica-rich (viscous) and have a high gas content.

Why Do Volcanoes Erupt?

You learned in Chapter 3 that most magma is generated by the partial melting of the rock peridotite in the upper mantle to form magma with a basaltic composition. Once formed, the buoyant molten rock will rise toward the surface (**FIGURE 4.3**). Because the density of crustal rocks tends to decrease the closer they are to the surface, ascending basaltic magma may reach a level where the rocks above are less dense. Should this occur, the molten material begins to collect or pond, forming a magma chamber. As the magma body cools, minerals having high melting temperatures crystallize first, leaving the remaining melt enriched in silica and other less dense components. Some of this highly evolved material may ascend to

the surface to produce a volcanic eruption. In most, but not all, tectonic settings, only a fraction of magma generated at depth ever reaches the surface.

TRIGGERING HAWAIIAN-TYPE ERUPTIONS. Eruptions that involve very fluid basaltic magmas are often triggered by the arrival of a new batch of melt into a near-surface magma reservoir. This can be detected because the summit of the volcano begins to inflate months, or even years, before an eruption begins. The injection of a fresh supply of melt causes the magma chamber to swell and fracture the rock above. This, in turn, mobilizes the magma, which quickly moves upward along the newly formed openings, often generating outpourings of lava for weeks, months, or even years.

THE ROLE OF VOLATILES IN EXPLOSIVE ERUPTIONS. All magmas contain some water and other volatiles that are held in solution by the immense pressure of the overlying rock. Volatiles tend to be most abundant near the tops of magma reservoirs containing highly evolved, silica-rich melts. When magma rises (or the rocks confining the magma fail) a reduction in pressure occurs and the dissolved gases begin to separate from the melt, forming tiny bubbles. This is analogous to opening a warm soda and allowing the carbon dioxide bubbles to escape.

When fluid basaltic magmas erupt, the pressurized gases escape with relative ease. At temperatures of 1000 °C and low near-surface pressures, these gases can quickly expand to occupy hundreds of times their original volumes. On some occasions, these expanding gases propel incandescent lava hundreds of meters into the

FIGURE 4.4 Fluid basaltic lava erupting from Kilauea volcano, Hawaii. *(Photo by Douglas Peebles/Photolibrary)*

air, producing lava fountains (**FIGURE 4.4**). Although spectacular, these fountains are mostly harmless and not generally associated with major explosive events that cause great loss of life and property.

At the other extreme, highly viscous, rhyolitic magmas may produce explosive clouds of hot ash and gases that evolve into buoyant plumes called **eruption columns** that extend thousands of meters into the atmosphere (**FIGURE 4.5**). Because of the high viscosity of silica-rich magma, a significant portion of the volatiles remain dissolved until the magma reaches a shallow depth, where tiny bubbles begin to form and grow. Bubbles grow by two processes, continued separation of gases from the melt and expansion of bubbles as the confining pressure drops. Should the pressure of the expanding magma body exceed the strength of the overlying rock, fracturing occurs. As magma moves up the fractures, a further drop in confining pressure causes more gas bubbles to form and grow. This chain-reaction may generate an explosive event in which magma is literally blown

into fragments (ash and pumice) that are carried to great heights by the hot gases. (As exemplified by the 1980 eruption of Mount St. Helens, the collapse of a volcano's flank can also trigger an energetic explosive eruption.)

When magma in the uppermost portion of the magma chamber is forcefully ejected by the escaping gases, the confining pressure on the molten rock directly below drops suddenly. Thus, rather than a single "bang," volcanic eruptions are really a series of explosions. This process might logically continue until the entire magma chamber is emptied, much like a geyser empties itself of water (see Chapter 10). However, this is generally not the case. It is typically only the magma in the upper part of a magma chamber that has a sufficiently high gas content to trigger a steam-and-ash explosion.

To summarize, the viscosity of magma plus the quantity of dissolved gases and the ease with which they can escape largely determine the nature of a volcanic eruption. In general, hot basaltic magmas contain a smaller gaseous component and permit these gases to escape with relative ease as compared to more highly evolved andesitic and rhyolitic magmas. This explains the contrast between "gentle" outflows of fluid basaltic lavas in Hawaii and the explosive and sometimes catastrophic eruptions of viscous lavas from volcanoes such as Mount St. Helens (1980), Mount Pinatubo in the Philippines (1991), and Soufriere Hills on the island of Montserrat (1995).

CONCEPT CHECK 4.2

❶ List three factors that determine the nature of a volcanic eruption. What role does each play?

❷ Generally, what triggers a Hawaiian-type eruption?

❸ Why is a volcano fed by highly viscous magma likely to be a greater threat to life and property than a volcano supplied with very fluid magma?

Materials Extruded during an Eruption

 Volcanoes

ESSENTIALS OF GEOLOGY **Materials Extruded during an Eruption**

Volcanoes extrude lava, large volumes of gas, and pyroclastic materials (broken rock, lava "bombs," fine ash, and dust). In this section we will examine each of these materials.

Lava Flows

The vast majority of lava on Earth, more than 90 percent of the total volume, is estimated to be basaltic in composition. Andesites and other lavas of intermediate composition account for most of the rest, while rhyolitic (felsic) flows make up as little as 1 percent of the total.

Hot basaltic lavas, which are usually very fluid, generally flow in thin, broad sheets or streamlike ribbons. On the island of Hawaii, these lavas have been clocked at 30 kilometers (19 miles) per hour down steep slopes. However, flow rates of 10 to 300 meters (30 to 1000 feet) per hour are more common. By contrast, the movement of silica-rich, rhyolitic lava may be too slow to perceive. Furthermore, most rhyolitic lavas seldom travel more than a few kilometers from their vents. As you might expect, andesitic lavas, which are intermediate in composition, exhibit characteristics that are between the extremes.

FIGURE 4.5 Steam and ash eruption column from Mount Augustine, Cook Inlet, Alaska. *(Photo by Steve Kaufman/Peter Arnold, Inc.)*

FIGURE 4.6 Lava flows. **A.** A typical slow-moving, basaltic, aa flow. (*Photo by J. D. Griggs, U.S. Geological Survey*) **B.** A typical fluid pahoehoe (ropy) lava. Both of these lava flows erupted from a rift on the flank of Hawaii's Kilauea volcano. (*Photo by Philip Rosenberg/Photolibrary*)

A.

B.

AA AND PAHOEHOE FLOWS. Two types of lava flows are known by their Hawaiian names. The most common of these, **aa** (pronounced ah-ah) **flows**, have surfaces of rough jagged blocks with dangerously sharp edges and spiny projections (**FIGURE 4.6A**). Crossing an aa flow can be a trying and miserable experience. By contrast, **pahoehoe** (pronounced pah-hoy-hoy) **flows** exhibit smooth surfaces that often resemble the twisted braids of ropes (**FIGURE 4.6B**). Pahoehoe means "on which one can walk."

Aa and pahoehoe lavas can erupt from the same vent. However, pahoehoe lavas form at higher temperatures and are more fluid than aa flows. In addition, pahoehoe lavas can change into aa lavas flow, although the reverse (aa to pahoehoe) does not occur.

One factor that facilitates the change from pahoehoe to aa is cooling that occurs as the flow moves away from the vent. Cooling increases viscosity and promotes bubble formation. Escaping gas bubbles produce numerous voids and sharp spines in the surface of the congealing lava. As the molten interior advances, the outer crust is broken further, transforming a relatively smooth surface into an advancing mass of rough, clinkery rubble.

LAVA TUBES. Hardened basaltic flows commonly contain cave-like tunnels called **lava tubes** that were once conduits carrying lava from the volcanic vent to the flow's leading edge (**FIGURE 4.7**). These conduits develop in the interior of a flow where temperatures remain high long after the surface hardens. Lava tubes are important

features because they serve as insulated pathways that facilitate the advance of lava great distances from its source.

Lava tubes are associated with volcanoes that emit fluid basaltic lava and are found in most parts of the world. Even the massive volcanoes on Mars have flows that contain numerous lava tubes. Some lava tubes exhibit extraordinary dimensions—one such structure, Kazumura Cave located on the southeast slope of Hawaii's Mauna Loa volcano, extends for more than 60 kilometers (40 miles).

BLOCK LAVAS. In contrast to fluid basaltic magmas, which can travel many kilometers, andesitic and rhyolitic magmas tend to generate relatively short prominent flows, a few hundred meters to a few kilometers long. Their upper surface consists largely of vesicle-free, detached blocks, hence the name **block lava**. Although similar to aa flows, these lavas consist of blocks with slightly curved, smooth surfaces, rather than the rough, clinkery surfaces.

PILLOW LAVAS. Recall that much of Earth's volcanic output occurs along oceanic ridges (divergent plate boundaries). When outpourings of lava occur on the ocean floor, the flow's outer skin quickly congeals. However, the lava is usually able to move forward by breaking through the hardened surface. This process occurs over and over, as molten basalt is extruded—like toothpaste from a tightly

squeezed tube. The result is a lava flow composed of numerous tube-like structures called **pillow lavas**, stacked one atop the other. Pillow lavas are useful in the reconstruction of geologic history because whenever they are observed, they indicate that the lava flow formed in an underwater environment.

Gases

Magmas contain varying amounts of dissolved gases (*volatiles*) held in the molten rock by confining pressure, just as carbon dioxide is held in cans and bottles of soft drinks. As with soft drinks, as soon as the pressure is reduced, the gases begin to escape. Obtaining gas samples from an erupting volcano is difficult and dangerous, so geologists usually must estimate the amount of gas originally contained within the magma.

The gaseous portion of most magmas makes up from 1 to 6 percent of the total weight, with most of this in the form of water vapor. Although the percentage may be small, the actual quantity of emitted gas can exceed thousands of tons per day.

fractures the rock above. Then, hot blasts of high-pressure gases expand the cracks and develop a passageway to the surface. Once the passageway is completed, the hot gases, armed with rock fragments, erode its walls, producing a larger conduit. Because these erosive forces are concentrated on any protrusion along the pathway, the volcanic pipes that are produced tend to develop a circular shape. As the conduit enlarges, magma moves upward to produce surface activity.

Following an eruptive phase, the volcanic pipe often becomes choked with a mixture of congealed magma and debris that was not thrown clear of the vent. Before the next eruption, a new surge of explosive gases may again clear the conduit.

A.

B.

FIGURE 4.7 Lava flows often develop a solid crust while the molten lava below continues to advance in conduits called *lava tubes*. **A.** View of an active lava tube as seen through the collapsed roof. *(Photo by G. Brad Lewis/SPL/Photo Researchers, Inc.)* **B.** Thurston Lava Tube, Hawaii Volcanoes National Park. *(Photo by Philip Rosenberg/Photolibrary)*

Occasionally, eruptions emit colossal amounts of volcanic gases that rise high into the atmosphere, where they may reside for several years. Some of these eruptions may have an impact on Earth's climate, a topic we will consider in Chapter 20.

The composition of volcanic gases is important because they contribute significantly to our planet's atmosphere. Analyses of samples taken during Hawaiian eruptions indicate that the gas component is about 70 percent water vapor, 15 percent carbon dioxide, 5 percent nitrogen, and 5 percent sulfur dioxide, with lesser amounts of chlorine, hydrogen, and argon. (The relative proportion of each gas varies significantly from one volcanic region to another.) Sulfur compounds are easily recognized by their pungent odor. Volcanoes are also natural sources of air pollution—some emit large quantities of sulfur dioxide, which readily combines with atmospheric gases to form sulfuric acid and other sulfate compounds.

In addition to propelling magma from a volcano, gases play an important role in creating the narrow conduit that connects the magma chamber to the surface. First, swelling of the magma body

Mexico's Colima volcano.
(Photo by Michael Collier)

Pyroclastic Materials

When volcanoes erupt energetically they eject pulverized rock, lava, and glass fragments from the vent. The particles produced are referred to as **pyroclastic materials** (*pyro* = fire, *clast* = fragment). These fragments range in size from very fine dust and sand-sized volcanic ash (less than 2 millimeters) to pieces that weigh several tons.

Ash and *dust* particles are produced when gas-rich viscous magma erupts explosively (see Figure 4.5). As magma moves up in the vent, the gases rapidly expand, generating a melt that resembles the froth that flows from a bottle of champagne. As the hot gases expand explosively, the froth is blown into very fine glassy fragments (**FIGURE 4.8**). When the hot ash falls, the glassy shards often fuse to form a rock called *welded tuff*. Sheets of this material, as well as ash deposits that later consolidate, cover vast portions of the western United States.

Somewhat larger pyroclasts that range in size from small beads to walnuts are known as *lapilli* ("little stones"). These ejecta are commonly called *cinders* (2–64 millimeters). Particles larger than 64 millimeters (2.5 inches) in diameter are called *blocks* when they are made of hardened lava and *bombs* when they are ejected as incandescent lava (see Figure 4.8B and C). Because bombs are semimolten upon ejection, they often take on a streamlined shape as they hurtle through the air (**FIGURE 4.9**). Because of their size, bombs and blocks usually fall near the vent; however, they are occasionally propelled great distances.

A.

B.

C.

FIGURE 4.8 Pyroclastic materials. **A.** *Volcanic ash and small pumice fragments (lapilli) that erupted from Mount St. Helens in 1980. Inset photo is an image obtained using a scanning electron microscope (SEM). This vesicular ash particle exhibits a glassy texture and is roughly the diameter of a human hair.* **B.** *Volcanic block. Volcanic blocks are solid fragments that were ejected from a volcano during an explosive eruption.* **C.** *These basaltic bombs were erupted by Hawaii's Mauna Kea volcano. Volcanic bombs are blobs of lava that are ejected while still molten and often acquire rounded, aerodynamic shapes as they travel through the air. (Photos courtesy of U.S. Geological Survey)*

FIGURE 4.9 Volcanic bombs forming during an eruption of Hawaii's Kilauea volcano. Ejected lava masses take on a streamlined shape as they sail through the air. The bomb in the insert is about 10 centimeters long. *(Photo by Arthur Roy/National Audubon Society; inset photo by E. J. Tarbuck)*

A. Scoria

B. Pumice

FIGURE 4.10 Scoria and pumice are volcanic rocks that exhibit a vesicular texture. Vesicles are small holes left by escaping gas bubbles. **A.** Scoria is usually a product of mafic (basaltic) magma. **B.** Pumice forms during explosive eruptions of viscous magmas having an intermediate (andesitic) or felsic (rhyolitic) composition. *(Photos by E. J. Tarbuck)*

So far we have distinguished various pyroclastic materials based largely on the size of the fragments. Some materials are also identified by their texture and composition. In particular, **scoria** is the name applied to vesicular ejecta that is a product of basaltic magma (**FIGURE 4.10A**). These black to reddish-brown fragments are generally found in the size range of lapilli and resemble cinders and clinkers produced by furnaces used to smelt iron. When magmas with intermediate (andesitic) or felsic (rhyolitic) compositions erupt explosively, they emit ash and the vesicular rock **pumice** (**FIGURE 4.10B**). Pumice is usually lighter in color and less dense than scoria, and many pumice fragments have so many vesicles that they are light enough to float.

CONCEPT CHECK 4.3

❶ Describe pahoehoe and aa lava flows.

❷ List the main gases released during a volcanic eruption. What role do gases have in eruptions?

❸ How do volcanic bombs differ from blocks of pyroclastic debris?

❹ What is scoria? How is scoria different from pumice?

Volcanic Structures and Eruptive Styles

 Volcanoes

ESSENTIALS OF GEOLOGY Volcanic Structures and Eruptive Styles

The popular image of a volcano is that of a solitary, graceful, snow-capped cone, such as Mount Hood in Oregon or Japan's Fujiyama. These picturesque, conical mountains are produced by volcanic activity that occurred intermittently over thousands, or even hundreds of thousands, of years. However, many volcanoes do not fit this image. Cinder cones are quite small and form during a single eruptive phase that lasts a few days to a few years. Other volcanic landforms are not volcanoes at all. For example, Alaska's Valley of Ten Thousand Smokes is a flat-topped deposit consisting of 15cubic kilometers of ash that erupted in less than 60 hours and blanketed a section of river valley to a depth of 200 meters (600 feet).

Volcanic landforms come in a wide variety of shapes and sizes, and each structure has a unique eruptive history. Nevertheless, volcanologists have been able to classify volcanic landforms and determine their eruptive patterns. In this section we will consider the general anatomy of a volcano and look at three major volcanic types: shield volcanoes, cinder cones, and composite cones.

Anatomy of a Volcano

Volcanic activity frequently begins when a fissure (crack) develops in the crust as magma moves forcefully toward the surface. As the gas-rich magma moves up through a fissure, its path is usually localized into a circular **conduit**, or **pipe**, that terminates at a surface opening called a **vent** (**FIGURE 4.11**). Successive eruptions of lava, pyroclastic material, or frequently a combination of both, often separated by long periods of inactivity, eventually build the cone-shaped structure we call a **volcano.**

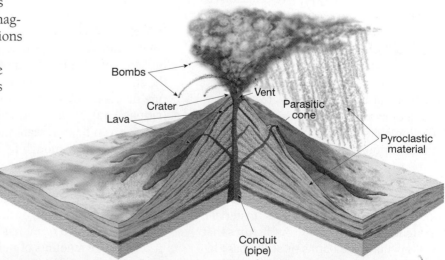

Bombs

Crater

Lava

Vent

Parasitic cone

Pyroclastic material

Conduit (pipe)

FIGURE 4.11 Anatomy of a "typical" composite cone (see also Figures 4.13 and 4.16 for a comparison with a shield and cinder cone, respectively).

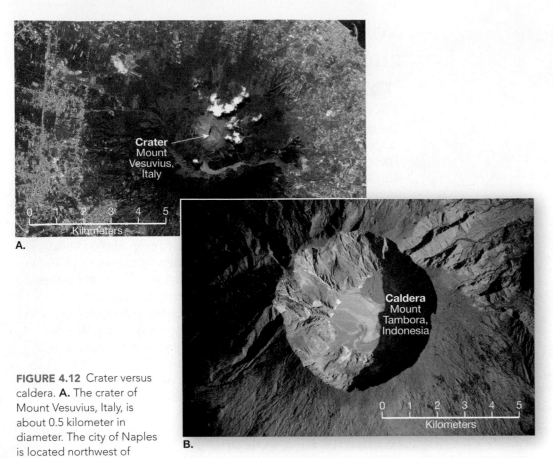

FIGURE 4.12 Crater versus caldera. **A.** The crater of Mount Vesuvius, Italy, is about 0.5 kilometer in diameter. The city of Naples is located northwest of Vesuvius, whereas Pompeii, the Roman town that was buried by an eruption in AD 79, is located southeast of the volcano. **B.** The huge caldera—6 kilometers in diameter—formed when Tambora's peak was removed during an explosive eruption in 1815. *(Photos courtesy of NASA)*

Located at the summit of most volcanoes is a somewhat funnel-shaped depression, called a **crater** (*crater* = a bowl). Volcanoes that are built primarily of pyroclastic materials typically have craters that form by gradual accumulation of volcanic debris on the surrounding rim. Other craters form during explosive eruptions as the rapidly ejected particles erode the crater walls. Craters also form when the summit area of a volcano collapses following an eruption (**FIGURE 4.12**). Some volcanoes have very large circular depressions called *calderas* that have diameters greater than one kilometer and in rare cases can exceed 50 kilometers. We will consider the formation of various types of calderas later in this chapter.

During early stages of growth most volcanic discharges come from a central summit vent. As a volcano matures, material also tends to be emitted from fissures that develop along the flanks or at the base of the volcano. Continued activity from a flank eruption may produce a small

parasitic cone (*parasitus* = one who eats at the table of another). Italy's Mount Etna, for example, has more than 200 secondary vents, some of which have built parasitic cones. Many of these vents, however, emit only gases and are appropriately called **fumaroles** (*fumus* = smoke).

Shield Volcanoes

Shield volcanoes are produced by the accumulation of fluid basaltic lavas and exhibit the shape of a broad, slightly domed structure that resembles a warrior's shield (**FIGURE 4.13**). Most shield volcanoes begin on the ocean floor as seamounts, a few of which grow large enough to form volcanic islands. In fact, with the exception of the volcanic islands that form above subduction zones, most other oceanic islands are either single shield volcanoes, or more often the coalescence of two or more shields built upon massive amounts of pillow lavas. Examples include the Canary Islands, the Hawaiian Islands, the Galapagos, and Easter

Island. In addition, some shield volcanoes form on continental crust. Included in this group are several volcanic structures located in East Africa.

MAUNA LOA: A CLASSIC SHIELD VOLCANO. Extensive study of the Hawaiian Islands confirms that they are constructed of a myriad of thin basaltic lava flows averaging a few meters thick intermixed with relatively minor amounts of pyroclastic ejecta. Mauna Loa is one of five overlapping shield volcanoes that together comprise the Big Island of Hawaii (see Figure 4.13). From its base on the floor of the Pacific Ocean to its summit, Mauna Loa is over 9 kilometers (6 miles) high, exceeding the height of Mount Everest. This massive pile of basaltic rock has a volume of 80,000 cubic kilometers that was extruded over a span of about 1 million years. The volume of material composing Mauna Loa is roughly 200 times greater than the amount composing a large composite cone such as Mount Rainier (**FIGURE 4.14**). Although the shield volcanoes that comprise islands are often quite large, some are more modest in size. In addition, an estimated 1 million basaltic submarine volcanoes (seamounts) of various sizes dot the ocean floor.

The flanks of Mauna Loa have gentle slopes of only a few degrees. The low angle results because very hot, fluid lava travels "fast and far" from the vent. In addition, most of the lava (perhaps 80 percent) flows through a well-developed system of lava tubes (see Figure 4.7). This greatly increases the distance lava can travel before it solidifies. Thus, lava emitted near the summit often reaches the sea, thereby adding to the width of the cone at the expense of its height.

Another feature common to many active shield volcanoes is a large, steepwalled caldera that occupies the summit (see Figure 4.26). Calderas on large shield volcanoes form when the roof above the magma chamber collapses. This usually occurs as the magma reservoir empties following a large eruption or as magma migrates to the flank of a volcano to feed a fissure eruption.

In the final stage of growth, shield volcanoes are more sporadic, and pyroclastic

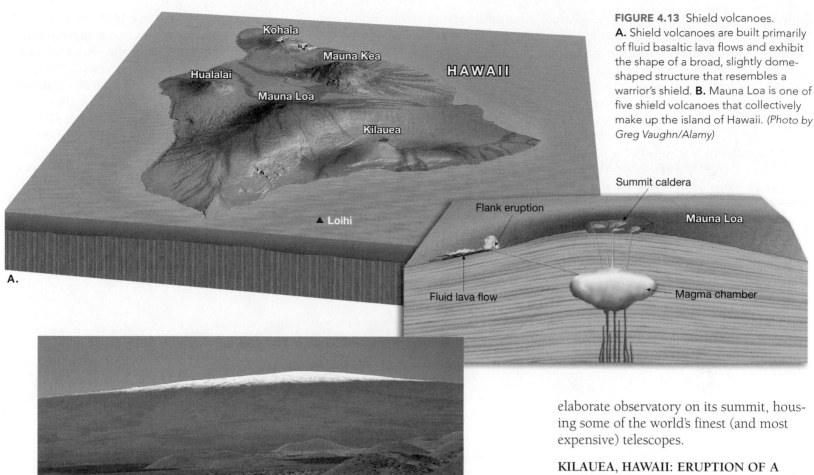

FIGURE 4.13 Shield volcanoes.
A. Shield volcanoes are built primarily of fluid basaltic lava flows and exhibit the shape of a broad, slightly dome-shaped structure that resembles a warrior's shield. **B.** Mauna Loa is one of five shield volcanoes that collectively make up the island of Hawaii. *(Photo by Greg Vaughn/Alamy)*

elaborate observatory on its summit, housing some of the world's finest (and most expensive) telescopes.

KILAUEA, HAWAII: ERUPTION OF A SHIELD VOLCANO. Kilauea, the most active and intensely studied shield volcano in the world, is located on the island of Hawaii in the shadow of Mauna Loa. More than 50 eruptions have been witnessed here since record keeping began in 1823. Several months before each eruptive phase, Kilauea inflates as magma gradually migrates upward and accumulates in a central reservoir located a few kilometers below the summit. For up to 24 hours in advance of an eruption, swarms of small earthquakes warn of the impending activity.

Most of the recent activity on Kilauea has occurred along the flanks of the volcano in a region called the East Rift Zone. A rift eruption here in 1960 engulfed the coastal village of Kapoho, located nearly 30 kilometers (20 miles) from the source. The longest and largest rift eruption ever recorded on Kilauea began in 1983 and continues to this day, with no signs of abating.

The first discharge began along a 6-kilometer (4-mile) fissure where a 100-meter (300-foot) high "curtain of fire"

ejections are more common. Further, lavas increase in viscosity, resulting in thicker, shorter flows. These eruptions tend to steepen the slope of the summit area, which often becomes capped with clusters of cinder cones. This may explain why

Mauna Kea, which is a more mature volcano that has not erupted in historic times, has a steeper summit than Mauna Loa, which erupted as recently as 1984. Astronomers are so certain that Mauna Kea is "over the hill" that they have built an

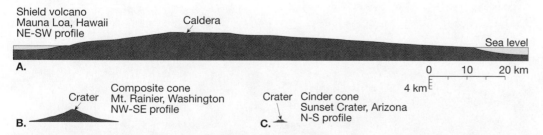

FIGURE 4.14 Profiles comparing scales of different volcanoes. **A.** Profile of Mauna Loa, Hawaii, the largest shield volcano in the Hawaiian chain. Note size comparison with Mount Rainier, Washington, a large composite cone. **B.** Profile of Mount Rainier, Washington. Note how it dwarfs a typical cinder cone. **C.** Profile of Sunset Crater, Arizona, a typical steep-sided cinder cone.

formed as red-hot lava was ejected skyward (**FIGURE 4.15**). When the activity became localized, a cinder and spatter cone, given the Hawaiian name *Puu Oo,* was built. Over the next 3 years the general eruptive pattern consisted of short periods (hours to days) when fountains of gas-rich lava sprayed skyward. Each event was followed by nearly a month of inactivity.

By the summer of 1986 a new vent opened 3 kilometers downrift. Here smooth-surfaced pahoehoe lava formed a lava lake. Occasionally the lake overflowed, but more often lava escaped through tunnels to feed flows that moved down the

southeastern flank of the volcano toward the sea. These flows destroyed nearly a hundred rural homes, covered a major roadway, and eventually reached the sea. Lava has been intermittently pouring into the ocean ever since, adding new land to the island of Hawaii.

Cinder Cones

As the name suggests, **cinder cones** (also called **scoria cones**) are built from ejected lava fragments that take on the appearance of cinders or clinkers as they begin to harden in flight (**FIGURE 4.16**). These pyro-

clastic fragments range in size from fine ash to bombs that may exceed a meter in diameter. However, most of the volume of a cinder cone consists of pea- to walnut-sized lapilli that are markedly vesicular and have a black to reddish-brown color. (Recall that these vesicular rock fragments are called *scoria.*) Although cinder cones are composed mostly of loose pyroclastic material, they sometimes extrude lava. On such occasions the discharges most often come from vents located at or near the base rather than from the summit crater.

Cinder cones have very simple, distinctive shapes determined by the slope that

FIGURE 4.15 Lava "curtain" extruded along the East Rift Zone, Kilauea, Hawaii. *(Photo by Greg Vaughn/Alamy)*

DID YOU KNOW?
According to legend Pele, the Hawaiian goddess of volcanoes, makes her home at the summit of Kilauea volcano. Evidence for her existence is "Pele's hair," thin, delicate strands of glass, which are soft and flexible and have a golden-brown color. This threadlike volcanic glass forms when blobs of hot lava are spattered and shredded by escaping gases.

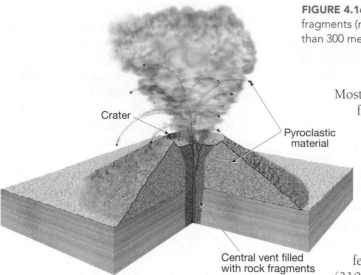

FIGURE 4.16 Cinder cones are built from ejected lava fragments (mostly cinders and bombs) and are usually less than 300 meters (1000 feet) in height.

Most cinder cones are produced by a single, short-lived eruptive event. One study found that half of all cinder cones examined were constructed in less than one month and that 95 percent formed in less than 1 year. However, in some cases, they remain active for several years. Parícutin, shown in **FIGURE 4.18**, had an eruptive cycle that spanned 9 years. Once the event ceases, the magma in the "plumbing" connecting the vent to the magma source solidifies, and the volcano usually does not erupt again. (One exception is Cerro Negro, a cinder cone in Nicaragua, which has erupted more than 20 times since it formed in 1850.) As a consequence of this short life span, cinder cones are small, usually between 30 meters (100 feet) and 300 meters (1000 feet). A few rare examples exceed 700 meters (2100 feet) in height.

Cinder cones number in the thousands around the globe. Some occur in volcanic fields such as the one near Flagstaff, Arizona, which consists of about 600 cones. Others are parasitic cones that are found on the flanks of larger volcanoes.

loose pyroclasic material maintains as it comes to rest (**FIGURE 4.17**). Because cinders have a high angle of repose (the steepest angle at which material remains stable), cinder cones are steep-sided, having slopes between 30 and 40 degrees. In addition, cinder cones have large, deep craters in relation to the overall size of the structure. Although relatively symmetrical, many cinder cones are elongated and higher on the side that was downwind during the eruptions.

PARÍCUTIN: LIFE OF A GARDEN-VARIETY CINDER CONE. One of the very few volcanoes studied by geologists from its very beginning is the cinder cone called Parícutin, located about 320 kilometers (200 miles) west of Mexico City. In 1943 its eruptive phase began in a cornfield owned by Dionisio Pulido, who witnessed the event as he prepared the field for planting.

For two weeks prior to the first eruption, numerous Earth tremors caused apprehension in the nearby village of Parícutin. Then, on February 20,

DID YOU KNOW?
A short distance off the south coast of Hawaii is a submarine volcano called Loihi. Although very active, it has another 3000 feet to go before it breaks the surface and becomes another island in the Hawaiian chain.

FIGURE 4.17 SP Crater, a cinder cone located in the San Francisco Peaks volcanic field north of Flagstaff, Arizona. The lava flow in the upper part of the image originated from the base of the cinder cone. Inset photo shows a large bomb on the slopes of SP Crater. *(Photos by Michael Collier)*

FIGURE 4.18 The village of San Juan Parangaricutiro was engulfed by aa lava from Parícutin. Only the church towers remain. *(Photo by Michael Collier)*

sulfurous gases began billowing from a small depression that had been in the cornfield for as long as people could remember. During the night hot, glowing rock fragments were ejected from the vent, producing a spectacular fireworks display. Explosive discharges continued, throwing hot fragments and ash occasionally as high as 6000 meters (20,000 feet) into the air. Larger fragments fell near the crater, some remaining incandescent as they rolled down the slope. These built an aesthetically pleasing cone, while finer ash fell over a much larger area, burning and eventually covering the village of Parícutin. In the first day the cone grew to 40 meters (130 feet), and by the fifth day it was more than 100 meters (330 feet) high.

The first lava flow came from a fissure that opened just north of the cone, but after a few months flows began to emerge from the base of the cone itself. In June 1944, a clinkery aa flow 10 meters (30 feet) thick moved over much of the village of San Juan Parangaricutiro, leaving only the church steeple exposed (see Figure 4.18). After 9 years of intermittent pyroclastic explosions and nearly continuous discharge of lava from vents at its base, the activity ceased almost as quickly as it had begun. Today, Parícutin is just another one of the scores of cinder cones dotting the landscape in this region of Mexico. Like the others, it will not erupt again.

Composite Cones

Earth's most picturesque yet potentially dangerous volcanoes are **composite cones** or **stratovolcanoes.** Most are located in a relatively narrow zone that rims the Pacific Ocean, appropriately called the *Ring of Fire*

(see Figure 4.32). This active zone consists of a chain of continental volcanoes that are distributed along the west coast of the Americas, including the large cones of the Andes in South America and the Cascade Range of the western United States and Canada. The latter group includes Mount St. Helens, Mount Shasta, and Mount Garibaldi. The most active regions in the Ring of Fire are located along curved belts of volcanic cones situated adjacent to the deep-ocean trenches of the northern and western Pacific. This nearly continuous chain of volcanoes stretches from the Aleutian Islands to Japan and the Philippines and to the North Island of New Zealand. These impressive volcanic structures are manifestations of processes that occur in the mantle in association with *subduction zones*.

The classic composite cone is a large, nearly symmetrical structure consisting of alternating layers of explosively erupted cinders and ash interbedded with lava flows. A few composite cones, notably Italy's Etna and Stromboli, display very persistent eruption

FIGURE 4.19 Japan's Fujiyama exhibits the classic form of a composite cone—steep summit and gently sloping flanks. *(Photo by Koji Nakano/Getty Images/Sebun)*

activity, and molten lava has been observed in their summit craters for decades. Stromboli is so well known for eruptions that eject incandescent blobs of lava that it has been referred to as the "Lighthouse of the Mediterranean." Mount Etna, on the other hand, has erupted, on average, once every 2 years since 1979.

Just as shield volcanoes owe their shape to fluid basaltic lavas, composite cones reflect the viscous nature of the material from which they are made. In general, composite cones are the product of gas-rich magma having an andesitic composition. However, many composite cones also emit various amounts of fluid basaltic lava and occasionally pyroclastic material having rhyolitic composition. Relative to shields, the silica-rich magmas typical of composite cones generate thick viscous lavas that travel less than a few kilometers. In addition, composite cones are noted for generating explosive eruptions that eject huge quantities of pyroclastic material.

A conical shape, with a steep summit area and more gradually sloping flanks, is typical of many large composite cones. This classic profile, which adorns calendars and postcards, is partially a consequence of the way viscous lavas and pyroclastic ejecta contribute to the growth of the cone. Coarse fragments ejected from the summit crater tend to accumulate near their source. Because of their high

angle of repose, coarse materials contribute to the steep slopes of the summit area. Finer ejecta, on the other hand, are deposited as a thin layer over a large area. This acts to flatten the flank of the cone. In addition, during the early stages of growth, lavas tend to be more abundant and flow greater distances from the vent than lavas do later in the volcano's history. This contributes to the cone's broad base. As the volcano matures, the shorter flows that come from the central vent serve to armor and strengthen the summit area. Consequently, steep slopes exceeding 40 degrees are sometimes possible. Two of the most perfect cones—Mount Mayon in the Philippines and Fujiyama in Japan—exhibit the classic form we expect of a composite cone, with its steep summit and gently sloping flanks (**FIGURE 4.19**).

Despite the symmetrical forms of many composite cones, most have complex histories. Huge mounds of volcanic debris surrounding these structures provide evidence that large sections of these volcanoes slid downslope as massive landslides. Others develop horseshoe-shaped depressions at their summits as a result of explosive lateral eruptions—as occurred during the 1980 eruption of Mount St. Helens. Often, so much rebuilding has occurred since these eruptions that no trace of the amphitheater-shaped scars remain.

Many composite cones have numerous small, parasitic cones on their flanks, while others, such as Crater Lake, have been truncated by the collapse of their summit (see **FIGURE 4.25**). Still others have a lake in their crater that may be hot and muddy. Such lakes are often highly acidic because of the influx of sulfur and chlorine gases that react with water to produce sulfuric (H_2SO_4) and hydrochloric acid (HCl).

CONCEPT CHECK 4.4

❶ Compare a volcanic crater to a caldera.

❷ Compare and contrast the three main types of volcanoes (consider size, composition, shape, and eruptive style).

❸ Name a prominent volcano for each of the three types of volcanoes.

❹ Briefly compare the eruptions of Kilauea and Parícutin.

Living in the Shadow of a Composite Cone

More than 50 volcanoes have erupted in the United States in the past 200 years (**FIGURE 4.20**). Fortunately, the most explosive of these eruptions occurred in sparsely inhabited regions of Alaska. On a global scale many destructive eruptions have occurred during the past few thousand years, a few of which may have influenced the course of human civilization.

Nuée Ardente: A Deadly Pyroclastic Flow

One of the most destructive forces of nature is the **pyroclastic flow**, which consists of hot gases infused with incandescent ash and larger lava fragments. Also referred to as **nuée ardentes** (*glowing avalanches*), these

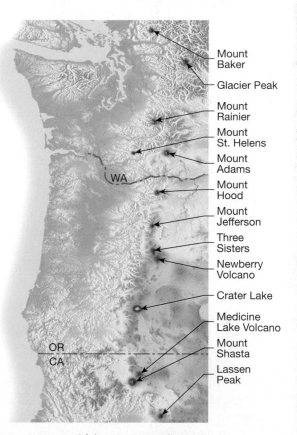

FIGURE 4.20 Of the 13 potentially active volcanoes in the Cascade Range, 11 have erupted in the past 4000 years and 7 in just the past 200 years. More than 100 eruptions, most of which were explosive, have occurred in the past 4000 years. Mount St. Helens is the most active volcano in the Cascades. Its eruptions have ranged from relatively quiet outflows of lava to explosive events much larger than that of May 18, 1980.

fiery flows are capable of racing down steep volcanic slopes at speeds that can exceed 200 kilometers (125 miles) per hour (**FIGURE 4.21**). Nuée ardentes are composed of two parts—a low-density cloud of hot expanding gases containing fine ash particles and a ground-hugging portion that contains most of the material in the flow.

Driven by gravity, pyroclastic flows tend to move in a manner similar to snow avalanches. They are mobilized by volcanic gases released from the lava fragments and by the expansion of heated air that is over-taken and trapped in the moving front. These gases reduce friction between the fragments and the ground. Strong turbulent flow is another important mechanism that aids in the transport of ash and pumice fragments downslope in a nearly friction-less environment (Figure 4.21). This helps explain why some nuée ardente deposits are found more than 100 kilometers (60 miles) from their source.

Sometimes, powerful hot blasts that carry small amounts of ash separate from the main body of a pyroclastic flow. These low-density clouds, called *surges,* can be deadly but seldom have sufficient force to destroy buildings in their paths. Neverthe-less, on June 3, 1991, a hot ash cloud from Japan's Unzen volcano engulfed and burned hundreds of homes and moved cars as much as 80 meters (250 feet).

Pyroclastic flows may originate in a variety of volcanic settings. Some occur when a powerful eruption blasts pyroclastic material out of the side of a volcano—the lateral eruption of Mount St. Helens in 1980, for example. More frequently, how-ever, nuée ardentes are generated by the collapse of tall eruption columns during an explosive event. When gravity eventually overcomes the initial upward thrust pro-vided by the escaping gases, the ejecta begin to fall, sending massive amounts of incandescent blocks, ash, and pumice cascading downslope.

In summary, pyroclastic flows are a mixture of hot gases and pyroclastic materi-als moving along the ground, driven prima-rily by gravity. In general, flows that are fast and highly turbulent can transport fine particles for distances of 100 kilometers or more.

THE DESTRUCTION OF ST. PIERRE. In 1902, an infamous nuée ardente and associated surge from Mount Pelée, a small volcano on the Caribbean island of Martinique, destroyed the port town of St. Pierre. Although the main pyroclastic flow was largely confined to the valley of Riviere Blanche, the fiery surge spread south of the river and quickly engulfed the entire city. The destruction happened in moments and was so devastating that almost all of St. Pierre's 28,000 inhabitants were killed. Only one person on the outskirts of town—a prisoner protected in a dungeon—and a few people on ships in the harbor were spared (**FIGURE 4.22**).

Within days of this calamitous eruption, scientists arrived on the scene. Although St. Pierre was mantled by only a thin layer of volcanic debris, they discovered that masonry walls nearly a meter thick were knocked over like dominoes, large trees were uprooted, and cannons were torn from their mounts. A further reminder of the destructive force of this nuée ardente is preserved in the ruins of the mental hospital. One of the immense steel chairs that had been used to confine alcoholic patients can be seen today, contorted, as though it were made of plastic.

THE DESTRUCTION OF POMPEII. One well documented event of historic proportions was the AD 79 eruption of the Italian volcano we now call Vesuvius. Prior to this eruption, Vesuvius had been dormant for centuries and had vineyards adorning its sunny slopes. On August 24, however, the tranquility ended, and in less than 24 hours the city of Pompeii (near Naples) and more than 2000 of its 20,000 residents perished. Some were entombed beneath a layer of pumice nearly 3 meters (10 feet) thick, while others were encased within a layer of ash (**FIGURE 4.23B**). They remained this way for nearly 17 centuries, until the city was excavated, giving archaeologists a superbly detailed picture of ancient Roman life (**FIGURE 4.23A**).

By reconciling historical records with detailed scientific studies of the region, volcanol-ogists have pieced together the chronology of the destruction of Pompeii. The eruption most likely began as steam discharges on the morning of August 24. By early afternoon fine ash and pumice fragments formed a tall eruptive cloud. Shortly thereafter, debris from this

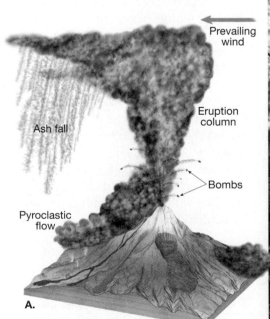

FIGURE 4.21 Pyroclastic flows. **A.** Illustration of a fiery ash and pumice flow racing down the slope of a volcano. **B.** Pyroclastic flow moving rapidly down the forested slopes of Mt. Unzen toward a Japanese village. *(Photo by Yomiuri/AP Photo)*

FIGURE 4.22 The photo on the left shows St. Pierre as it appeared shortly after the eruption of Mount Pelée, 1902. *(Reproduced from the collection of the Library of Congress)* The photo on the right shows St. Pierre before the eruption. Many vessels are anchored offshore, as was the case on the day of the eruption. *(Photo courtesy of The Granger Collection, New York)*

cloud began to shower Pompeii, which was located 9 kilometers (6 miles) downwind of the volcano. Many people fled during this early phase of the eruption. For the next several hours pumice fragments as large as 5 centimeters (2 inches) fell on Pompeii.

FIGURE 4.23 The Roman city of Pompeii was destroyed in AD 79 during an eruption of Mount Vesuvius. **A.** Ruins of Pompeii. Excavation began in the 18th century and continues today. *(Photo by Roger Ressmeyer/ CORBIS)* **B.** Plaster casts of several victims of the AD 79 eruption of Mount Vesuvius. *(Photo by Leonard von Matt/Photo Researchers, Inc.)*

A.

B.

One historical record of the eruption states that some people tied pillows to their heads in order to fend off the flying fragments.

The rain of pumice continued for several hours, accumulating at the rate of 12 to 15 centimeters (5 to 6 inches) per hour. Most of the roofs in Pompeii eventually gave way. Despite the accumulation of more than 2 meters of pumice, many of the people that had not evacuated Pompeii were probably still alive the next morning. Then, suddenly and unexpectedly, a surge of searing hot ash and gas swept rapidly down the flanks of Vesuvius. This deadly pyroclastic flow killed an estimated 2000 people who had somehow managed to survive the pumice fall. Most died instantly as a result of inhaling the hot, ash-laden gases. Their remains were quickly buried by the falling ash. Rain then caused the ash to become rock hard before their bodies had time to decay. The subsequent decomposition of the bodies produced cavities in the hardened ash that replicated their forms and, in some cases, even preserved facial expressions. Nineteenth-century excavators found these cavities and created casts of the corpses by pouring plaster of Paris into the voids (Figure 4.23B).

Today, Vesuvius towers over the Naples skyline. Such an image should prompt us to consider how volcanic crises might be managed in the future.

Lahars: Mudflows on Active and Inactive Cones

In addition to violent eruptions, large composite cones may generate a type of very fluid mudflow referred to by its Indonesian name **lahar**. These destructive flows occur when volcanic debris becomes saturated with water and rapidly moves down steep volcanic slopes, generally following gullies and stream valleys. Some lahars may be triggered when magma is emplaced near the surface, causing large volumes of ice and snow to melt. Others are generated when heavy rains saturate weathered volcanic deposits. Thus, lahars may occur even when a volcano is *not* erupting.

When Mount St. Helens erupted in 1980, several lahars were generated. These flows and accompanying flood waters raced down nearby river valleys at speeds exceeding 30 kilometers per hour. These raging rivers of mud destroyed or severely damaged nearly all the homes and bridges along their paths. Fortunately, the area was not densely populated (**FIGURE 4.24**).

In 1985, deadly lahars were produced during a small eruption of Nevado del Ruiz, a 5300-meter (17,400-foot) volcano in the Andes Mountains of Colombia. Hot pyroclastic material melted ice and snow that capped the mountain (*nevado* means *snow* in Spanish) and sent torrents of ash and debris down three major river valleys that flank the volcano. Reaching speeds of 100 kilometers (60 miles) per hour, these mudflows tragically took 25,000 lives.

Mount Rainier, Washington, is considered by many to be America's most dangerous volcano because, like Nevado del Ruiz, it has a thick, year-round mantle of snow and glacial ice. Adding to the risk is the fact that more than 100,000 people live in the valleys around Rainier, and many homes are built on deposits left by lahars that flowed down the volcano hundreds or thousands of years ago. A future eruption, or perhaps just a period of extraordinary rainfall, may produce lahars that could take similar paths.

CONCEPT CHECK 4.5

1. Describe the nature of a pyroclastic flow, also referred to as a nuée ardente.

2. Contrast the destruction of Pompeii with the destruction of St. Pierre (time frame, volcanic material, and nature of destruction).

3. Briefly describe a lahar.

FIGURE 4.24 Lahars are mud flows that originate on volcanic slopes. This lahar raced down the Muddy River, located southeast of Mount St. Helens, following the May 18, 1980, eruption. Notice the former height of this fluid mud flow as recorded by the mud flow line on the tree trunks. Note person (circled) for scale. *(Photo by Lyn Topinka/U.S. Geological Survey)*

Other Volcanic Landforms

The most obvious volcanic structure is a cone, but other distinctive and important landforms are also associated with volcanic activity.

Calderas

Calderas (*caldaria* = a cooking pot) are large depressions with diameters that exceed 1 kilometer and have a somewhat circular form. (Those less than a kilometer across are called *collapse pits* or *craters.*) Most calderas are formed by one of the following processes: (1) the collapse of the summit of a large composite volcano following an explosive eruption of silica-rich pumice and ash fragments (*Crater Lake-type calderas*); (2) the collapse of the top of a shield volcano caused by subterranean drainage from a central magma chamber (*Hawaiian-type calderas*); and (3) the collapse of a large area, caused by the discharge of colossal volumes of silica-rich pumice and ash along ring fractures (*Yellowstone-type calderas*).

CRATER LAKE-TYPE CALDERAS. Crater Lake, Oregon, is situated in a caldera that has a maximum diameter of 10 kilometers (6 miles) and is 1175 meters (more than 3800 feet) deep. This caldera formed about 7000 years ago when a composite cone, later named Mount Mazama, violently extruded 50 to 70 cubic kilometers of pyroclastic material (**FIGURE 4.25**). With the loss of support, 1500 meters (nearly a mile) of the summit of this once-prominent cone collapsed. After the collapse, rainwater filled the caldera. Later volcanic activity built a small cinder cone in the caldera. Today this cone, called Wizard Island, provides a mute reminder of past activity.

HAWAIIAN-TYPE CALDERAS. Although some calderas are produced by a *collapse following an explosive eruption,* many are not. For example, Hawaii's active shield volcanoes, Mauna Loa and Kilauea, both have large calderas at their summits. Kilauea's measures 3.3 by 4.4 kilometers (about 2 by 3 miles) and is 150 meters

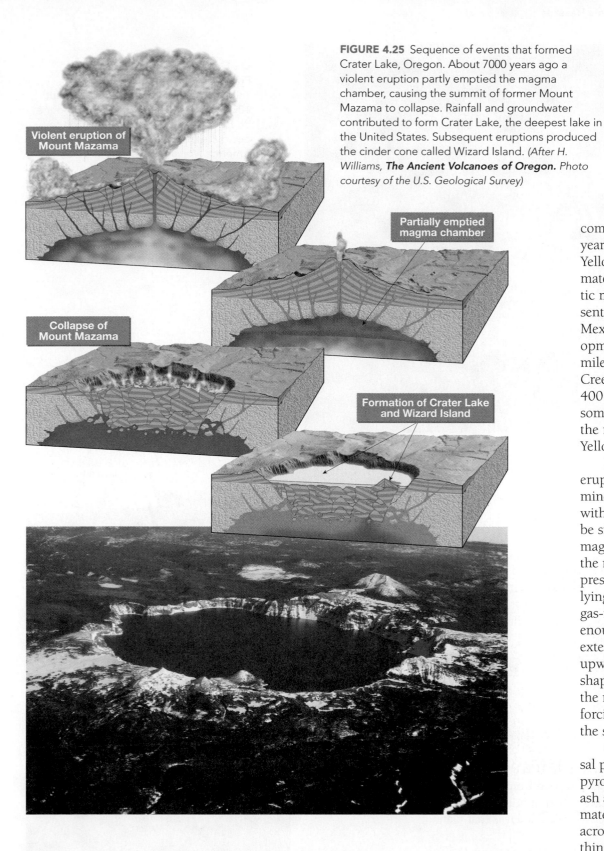

FIGURE 4.25 Sequence of events that formed Crater Lake, Oregon. About 7000 years ago a violent eruption partly emptied the magma chamber, causing the summit of former Mount Mazama to collapse. Rainfall and groundwater contributed to form Crater Lake, the deepest lake in the United States. Subsequent eruptions produced the cinder cone called Wizard Island. *(After H. Williams,* **The Ancient Volcanoes of Oregon.** *Photo courtesy of the U.S. Geological Survey)*

Violent eruption of Mount Mazama

Partially emptied magma chamber

Collapse of Mount Mazama

Formation of Crater Lake and Wizard Island

comparison to what happened 630,000 years ago in the region now occupied by Yellowstone National Park, when approximately 1000 cubic kilometers of pyroclastic material erupted. This supereruption sent showers of ash as far as the Gulf of Mexico and resulted in the eventual development of a caldera 70 kilometers (43 miles) across. It also gave rise to the Lava Creek Tuff, a hardened ash deposit that is 400 meters (more than 1200 feet) thick in some places. Vestiges of this event are the many hot springs and geysers in the Yellowstone region.

Based on the extraordinary volume of erupted material, researchers have determined that the magma chambers associated with Yellowstone-type calderas must also be similarly monstrous. As more and more magma accumulates, the pressure within the magma chamber begins to exceed the pressure exerted by the weight of the overlying rocks. An eruption occurs when the gas-rich magma raises the overlying strata enough to create vertical fractures that extend to the surface. Magma surges upward along these cracks forming a ring-shaped eruption. With a loss of support, the roof of the magma chamber collapses forcing even more gas-rich magma toward the surface.

Caldera-forming eruptions are of colossal proportions, ejecting huge volumes of pyroclastic materials, mainly in the form of ash and pumice fragments. Typically, these materials form pyroclastic flows that sweep across the landscape, destroying most living things in their paths. Upon coming to rest, the hot fragments of ash and pumice fuse together, forming a welded tuff that closely resembles a solidified lava flow. Despite the immense size of these calderas, their eruptions are brief, lasting hours to perhaps a few days.

(500 feet) deep (**FIGURE 4.26**). The walls of this caldera are almost vertical, and as a result it looks like a vast, nearly flat-botomed pit. Kilauea's caldera formed by gradual subsidence as magma slowly drained laterally from the underlying magma chamber to the East Rift Zone, leaving the summit unsupported.

YELLOWSTONE-TYPE CALDERAS.
Historic and destructive eruptions such as Mount St. Helens and Vesuvius pale in

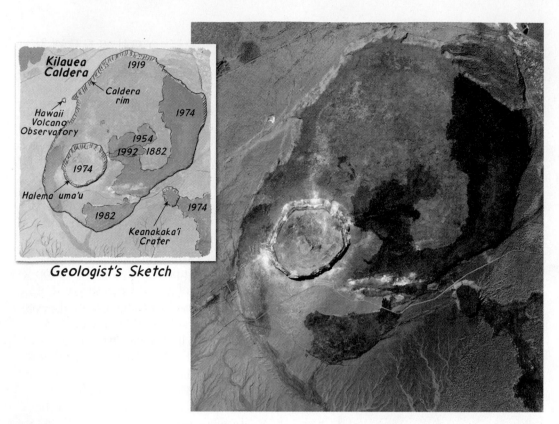

Geologist's Sketch

FIGURE 4.26 Kilauea's summit caldera measures 3.3 by 4.4 kilometers (about 2 by 3 miles) and is 150 meters (500 feet) deep. More recent flows are dark, while the older flows are pale because the iron in the lava oxidizes into "rust." Halema´uma´u, the crater in the southwest part of the caldera, was filled with a molten lava lake as recently as 1924. *(Photo courtesy of NASA)*

Unlike calderas associated with shield volcanoes or composite cones, these depressions are so large and poorly defined that many remained undetected until high-quality aerial and satellite images became available. Other examples of large calderas located in the United States are California's Long Valley caldera, LaGarita caldera, located in the San Juan Mountains of southern Colorado, and the Valles caldera west of Los Alamos, New Mexico. These and similar calderas found around the globe are the largest volcanic structures on Earth. Volcanologists have compared their destructive force with that of the impact of a small asteroid. Fortunately, no eruption of this type has occurred in historic times.

Fissure Eruptions and Basalt Plateaus

The greatest volume of volcanic material is extruded from fractures in the crust called **fissures** (*fissura* = to split). Rather than building a cone, these long, narrow cracks tend to emit low-viscosity basaltic lavas that blanket a wide area (**FIGURE 4.27**).

The Columbia Plateau in the northwestern United States is the product of this type of activity (**FIGURE 4.28**). Numerous **fissure eruptions** have buried the landscape creating a lava plateau nearly a mile thick. Some of the lava remained molten long enough to flow 150 kilometers (90 miles) from its source. The term **flood basalts** appropriately describes these deposits.

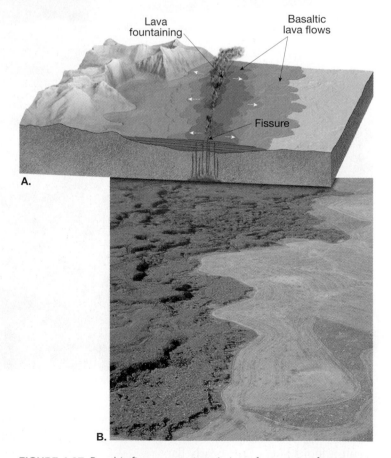

FIGURE 4.27 Basaltic fissure eruption. **A.** Lava fountaining from a fissure and formation of fluid lava flows called *flood basalts*. **B.** Photo of basalt flows near Idaho Falls. *(Photo by John S. Shelton)*

A.

B.

FIGURE 4.28 Volcanic areas that comprise the Columbia Plateau in the Pacific Northwest. **A.** The Columbia River basalts cover an area of nearly 200,000 square kilometers (80,000 square miles). Activity here began about 17 million years ago as lava began to pour out of large fissures, eventually producing a basalt plateau with an average thickness of more than 1 kilometer. **B.** Basalt flows exposed along Dry Falls in eastern Washington State. *(Photo by Wolfgang Kaehler/Alamy)*

Massive accumulations of basaltic lava, similar to those of the Columbia Plateau, occur elsewhere in the world. One of the largest examples is the Deccan Traps, a thick sequence of flat-lying basalt flows covering nearly 500,000 square kilometers (195,000 square miles) of west central India. When the Deccan Traps formed about 66 million years ago, nearly 2 million cubic kilometers of lava were extruded in less than 1 million years. Several other huge deposits of flood basalts, including the Ontong Java Plateau, are found on the floor of the ocean.

Lava Domes

In contrast to mafic lavas, silica-rich felsic lavas are so viscous they hardly flow at all. As the thick lava is "squeezed" out of the vent, it often produces a dome-shaped mass called a **lava dome** (**FIGURE 4.29**). Most lava domes are only a few tens of meters high, but some are more than 1 kilometer high.

Lava domes come in a variety of shapes that range from pancake-like flows to steep-sided plugs that were pushed upward like pistons. Most develop over a period of several years following an explosive eruption of gas-rich magma. A recent example is the dome that continues to grow in the crater of Mount St. Helens (Figure 4.29). A second dome building event began in October 2004. Although these eruptive phases produced some ash plumes, they were benign compared to the May 18, 1980, eruption.

Geologist's Sketch

Steam and ash

Lava dome

FIGURE 4.29 This lava dome began to develop following the May, 1980, eruption of Mount St. Helens. *(Photo by Lyn Topinka/U.S. Geological Survey)*

FIGURE 4.30 These lava domes are part of a 17-kilometer-long (10-mile-long) chain of rhyolitic domes that form Mono Craters, California. The blocky nature of the domes is hidden beneath a cover of pyroclastic debris. The last eruption here occurred about 600 years ago. *(Photo by E. J. Tarbuck)*

Although lava domes often form on the summit of a composite cone they can also develop on the flanks of volcanoes. In addition, some domes occur as isolated features, whereas others form linear chains. One example is the line of rhyolitic and obsidian domes at Mono Craters, California (**FIGURE 4.30**).

Volcanic Pipes and Necks

Most volcanoes are fed magma through short conduits, called *pipes,* that connect a magma chamber to the surface. One rare type of pipe, called a *diatreme*, extends to depths that exceed 200 kilometers (125 miles). Magmas that migrate upward through diatremes travel rapidly enough that they undergo very little alteration during their ascent. Geologists consider these unusually deep pipes to be "windows" into Earth that allow us to view rock normally found only at great depths.

The best-known volcanic pipes are the diamond-bearing structures of South Africa. The rocks filling these pipes originated at depths of at least 150 kilometers (90 miles), where pressure is high enough to generate diamonds and other high-pressure minerals. The process of transporting essentially unaltered magma

(along with diamond inclusions) through 150 kilometers of solid rock is exceptional. This fact accounts for the scarcity of natural diamonds.

Volcanoes on land are continually being lowered by weathering and erosion. Cinder cones are easily eroded because they are composed of unconsolidated materials. However, all volcanoes will eventually succumb to erosion. As erosion progresses, the rock occupying a volcanic pipe is often more resistant and may remain standing above the surrounding terrain long after most of the cone has vanished. Shiprock, New Mexico, is a classic example of this structure, which geologists call a **volcanic neck** (**FIGURE 4.31**). Higher

Dike

Shiprock (volcanic neck)

Original volcanic cone

Dike

Geologist's Sketch

FIGURE 4.31 Shiprock, New Mexico, is a volcanic neck. This structure, which stands over 420 meters (1380 feet) high, consists of igneous rock that crystallized in the vent of a volcano that has long since been eroded away. *(Photo by Dennis Tasa)*

than many skyscrapers, Shiprock is but one of many such landforms that protrude conspicuously from the red desert landscapes of the American Southwest.

CONCEPT CHECK 4.6

❶ Describe the formation of Crater Lake. Compare it to the caldera found on shield volcanoes, such as Kilauea.

❷ Extensive pyroclastic flow deposits are associated with which volcanic structure?

❸ How do the eruptions that created the Columbia Plateau differ from eruptions that create large composite cones?

❹ What is Shiprock, New Mexico, and how did it form?

Plate Tectonics and Volcanic Activity

Geologists have known for decades that the global distribution of volcanism is not random. Most active volcanoes are located along the margins of the ocean basins—notably within the circum-Pacific belt known as the *Ring of Fire* (**FIGURE 4.32**). These volcanoes consist mainly of composite cones that emit volatile-rich magma having an intermediate (andesitic) composition

and that occasionally produce awe-inspiring eruptions.

A second group includes the basaltic shields that emit very fluid lavas. These volcanic structures comprise most of the islands of the deep ocean basins, including the Hawaiian Islands, the Galapagos Islands, and Easter Island. In addition, this group includes many active submarine volcanoes that dot the ocean floor; particularly notable are the innumerable small seamounts that occur along the axis of the midocean ridge. At these depths, the pressures are so great that the gases that are emitted quickly dissolve in the seawater and never reach the surface. Thus, first-hand knowledge of these eruptions is limited, coming mainly from deep-diving submersibles.

A third group includes volcanic structures that appear to be somewhat randomly distributed in the interiors of the continents. None are found in Australia nor in the eastern two-thirds of North and South America. Africa is notable because it has many potentially active volcanoes including Mount Kilimanjaro, the highest point on the continent (5895 meters, 19,454 feet). When compared to volcanism in ocean basins, volcanism on continents is more diverse, ranging from eruptions of very fluid basaltic lavas, like those that generated the Columbia Plateau, to explosive

eruptions of silica-rich rhyolitic magma as occurred in Yellowstone.

Until the late 1960s, geologists had no explanation for the apparently haphazard distribution of continental volcanoes, nor were they able to account for the almost continuous chain of volcanoes that circles the margin of the Pacific basin. With the development of the theory of plate tectonics, the picture was greatly clarified. Recall that most primary (unaltered) magma originates in the upper mantle and that the mantle is essentially solid, *not molten*, rock. The basic connection between plate tectonics and volcanism is that *plate motions provide the mechanisms by which mantle rocks melt to generate magma.*

We will examine three zones of igneous activity and their relationship to plate boundaries. These active areas are located (1) along convergent plate boundaries where plates move toward each other and one sinks beneath the other; (2) along divergent plate boundaries, where plates move away from each other and new seafloor is created; and (3) areas within the plates proper that are not associated with any plate boundary. These three volcanic settings are depicted in **FIGURE 4.33**. (If you are unclear as to how magma is generated, study the section entitled "Origin of Magma" in Chapter 3 before proceeding.)

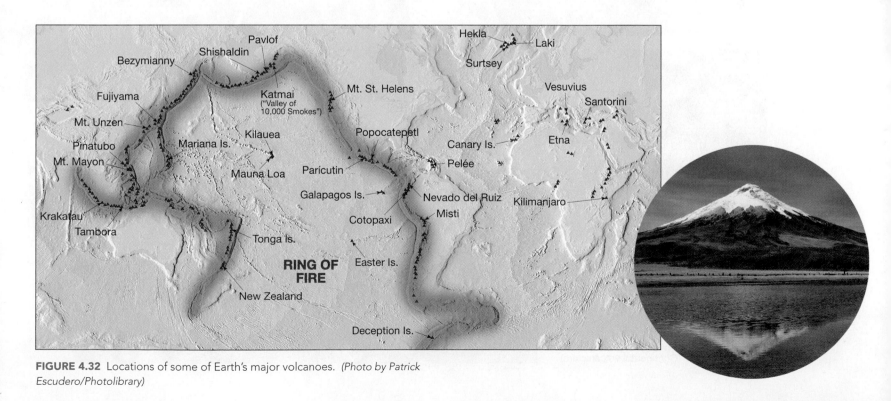

FIGURE 4.32 Locations of some of Earth's major volcanoes. *(Photo by Patrick Escudero/Photolibrary)*

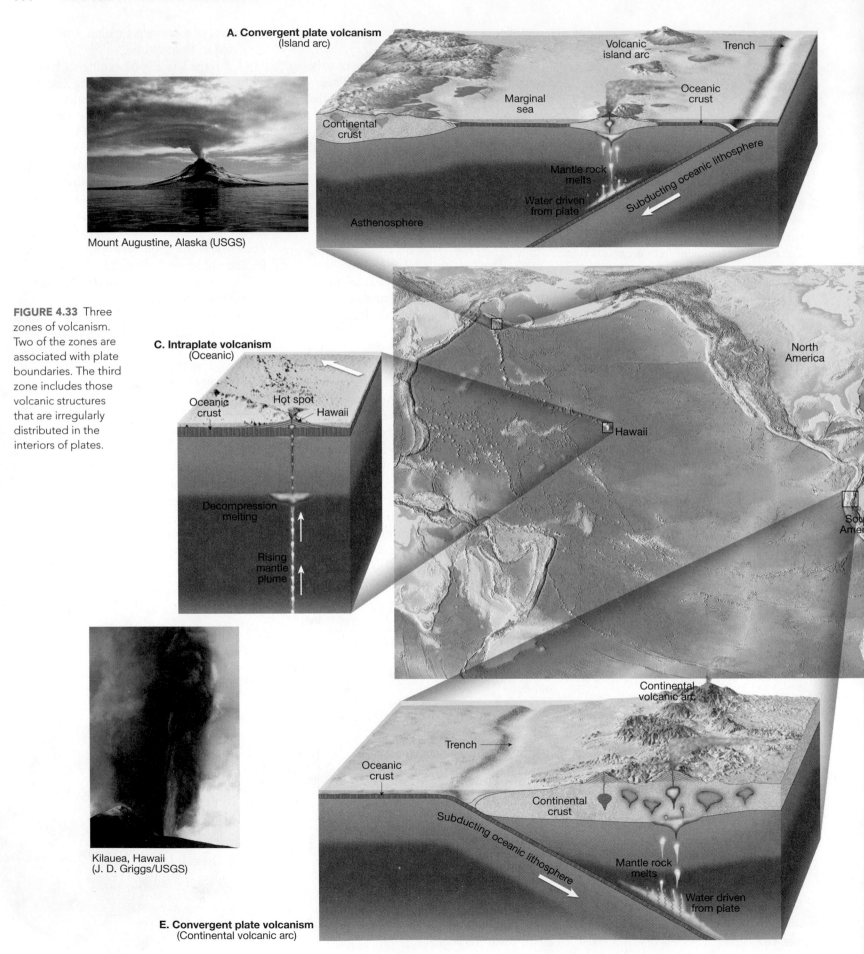

A. Convergent plate volcanism
(Island arc)

Volcanic island arc

Trench

Marginal sea

Oceanic crust

Continental crust

Mantle rock melts

Subducting oceanic lithosphere

Water driven from plate

Asthenosphere

Mount Augustine, Alaska (USGS)

FIGURE 4.33 Three zones of volcanism. Two of the zones are associated with plate boundaries. The third zone includes those volcanic structures that are irregularly distributed in the interiors of plates.

C. Intraplate volcanism
(Oceanic)

Hot spot

Oceanic crust

Hawaii

Decompression melting

Rising mantle plume

North America

Hawaii

South Amer

Kilauea, Hawaii
(J. D. Griggs/USGS)

Continental volcanic arc

Trench

Oceanic crust

Continental crust

Subducting oceanic lithosphere

Mantle rock melts

Water driven from plate

E. Convergent plate volcanism
(Continental volcanic arc)

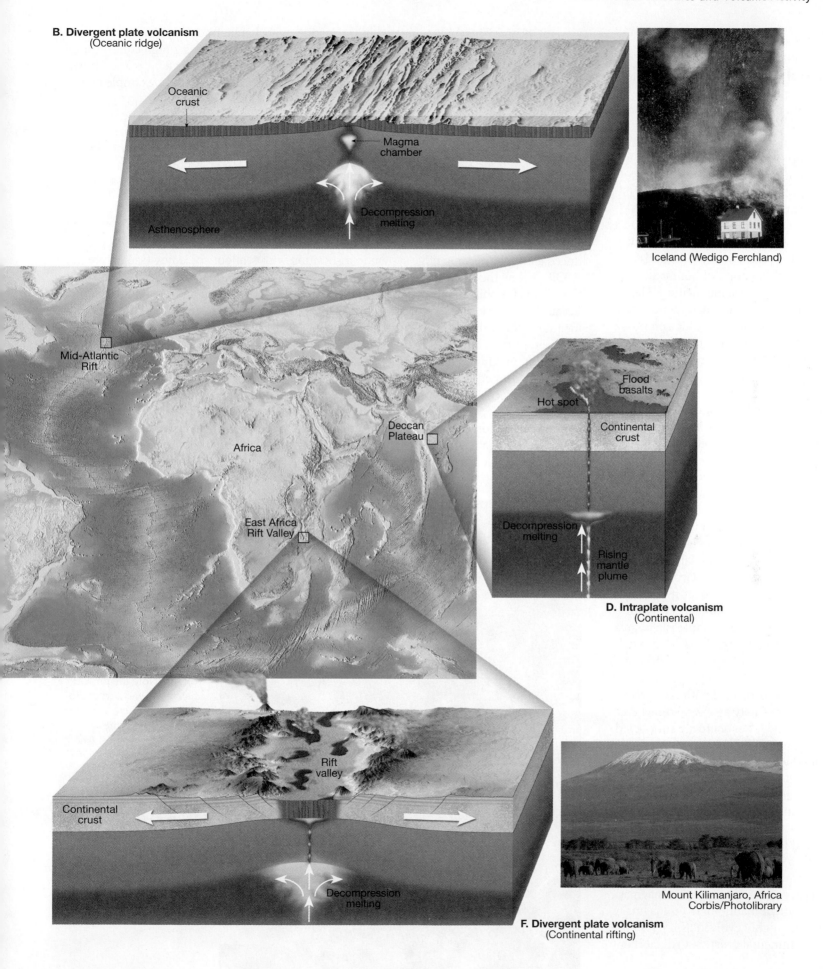

B. Divergent plate volcanism
(Oceanic ridge)

Oceanic crust

Magma chamber

Asthenosphere

Decompression melting

Iceland (Wedigo Ferchland)

Mid-Atlantic Rift

Africa

Deccan Plateau

East Africa Rift Valley

Hot spot

Flood basalts

Continental crust

Decompression melting

Rising mantle plume

D. Intraplate volcanism
(Continental)

Rift valley

Continental crust

Decompression melting

Mount Kilimanjaro, Africa
Corbis/Photolibrary

F. Divergent plate volcanism
(Continental rifting)

Volcanism at Convergent Plate Boundaries

Recall that at convergent plate boundaries slabs of oceanic crust are bent as they descend into the mantle, generating a deep-ocean trench. As a slab sinks deeper into the mantle, the increase in temperature and pressure drives volatiles (mostly water) from the oceanic crust. These mobile fluids migrate upward into the wedge-shaped piece of mantle located between the subducting slab and the overriding plate. Once the sinking slab reaches a depth of about 100 kilometers, these water-rich fluids reduce the melting point of hot mantle rock sufficiently to trigger some melting. The partial melting of mantle rock (peridotite) generates magma with a basaltic composition. After a sufficient quantity of magma has accumulated, it slowly migrates upward.

Volcanism at a convergent plate margin results in the development of a slightly curved chain of volcanoes called a *volcanic arc*. These volcanic chains develop roughly parallel to the associated trench—at distances of 200 to 300 kilometers (100 to 200 miles). Volcanic arcs can be constructed on oceanic, or continental, lithosphere. Those that develop within the ocean and grow large enough for their tops to rise above the surface are labeled *island archipelagos* in most atlases. Geologists prefer the more descriptive term **volcanic island arcs**, or simply **island arcs** (Figure 4.33A). Several young volcanic island arcs border the western Pacific basin, including the Aleutians, the Tongas, and the Marianas.

Volcanism associated with convergent plate boundaries may also develop where slabs of oceanic lithosphere are subducted under continental lithosphere to produce a **continental volcanic arc** (Figure 4.33E). The mechanisms that generate these mantle-derived magmas are essentially the same as those operating at island arcs. The major difference is that continental crust is much thicker and is composed of rocks having a higher silica content than oceanic crust. Hence, through the assimilation of silica-rich crustal rocks, plus extensive magmatic differentiation, a mantle-derived magma may become highly evolved as it rises through continental crust. Stated another way, the primary magmas generated in the mantle may change from a comparatively dry, fluid basaltic magma to a viscous andesitic or rhyolitic magma having a high concentration of volatiles as it moves up through the continental crust. The volcanic chain of the Andes Mountains along the western margin of South America is perhaps the best example of a mature continental volcanic arc.

Since the Pacific basin is essentially bordered by convergent plate boundaries and associated subduction zones, it is easy to see why the irregular belt of explosive volcanoes we call the Ring of Fire formed in this region (**FIGURE 4.34**). The volcanoes of the Cascade Range in the northwestern United States, including Mount Hood, Mount Rainier, and Mount Shasta, are included in this group.

Volcanism at Divergent Plate Boundaries

The greatest volume of magma (perhaps 60 percent of Earth's total yearly output) is produced along the oceanic ridge system in association with seafloor spreading (see Figure 4.33B). Below the ridge axis where lithospheric plates are continually being pulled apart, the solid yet mobile mantle responds to the decrease in overburden and rises to fill the rift. Recall that as rock rises, it experiences a decrease in confining pressure and undergoes melting without the addition of heat. This process, called *decompression melting,* is the most common process by which mantle rocks melt.

Partial melting of mantle rock at spreading centers produces basaltic magma. Because this newly formed magma is less dense than the mantle rock from which it was derived, it rises and collects in reservoirs located just beneath the ridge crest. About 10 percent of this melt eventually migrates upward along fissures to erupt on the ocean floor. This activity continuously adds new basaltic rock to plate margins, temporarily welding them together, only to break again as spreading continues. Along some ridges, outpourings of bulbous pillow lavas build numerous small seamounts.

FIGURE 4.34 Mount Unzen, Japan. This composite cone produced several fiery pyroclastic flows between 1900 and 1995. The most destructive eruption occurred in 1991 when a searing surge killed 43 people and burned over 500 homes and one school. Pumice- and ash-flow deposits (light color) can be seen in the valley leading down the volcano. Note the protective channels (lower part of the image) constructed to divert flows away from the surrounding villages. *(Photo by Michael S. Yamashita/CORBIS)*

DID YOU KNOW?
At 14,411 feet, Washington's Mount Rainier is the tallest of the 15 great volcanoes that make up the backbone of the Cascade Range. Although considered active, its summit is covered by more than 25 alpine glaciers.

Although most spreading centers are located along the axis of an oceanic ridge, some are not. In particular, the East African Rift is a site where continental lithosphere is being pulled apart (see Figure 4.33F). In this setting, magma is generated by decompression melting in the same manner as along the oceanic ridge system. Vast outpourings of fluid lavas as well as basaltic shield volcanoes are common in this region.

Intraplate Volcanism

We know why igneous activity is initiated along plate boundaries, but why do eruptions occur in the interiors of plates? Hawaii's Kilauea is considered the world's most active volcano, yet it is situated thousands of kilometers from the nearest plate boundary in the middle of the vast Pacific plate (Figure 4.33C). Other sites of **intraplate volcanism** (meaning "within the plate") include the Canary Islands, Yellowstone, and several volcanic centers that you may be surprised to learn are located in the Sahara Desert of Africa.

Geologists now recognize that most intraplate volcanism occurs where a mass of hotter than normal mantle material called a **mantle plume** ascends toward the surface (Figure 4.33C). Although the depth at which (at least some) mantle plumes originate is still hotly debated, some appear to form deep within Earth at the core–mantle boundary. These plumes of solid yet mobile mantle rock rise toward the surface in a manner similar to the blobs that form within a lava lamp. (These are the lamps that contain two immiscible liquids in a glass container. As the base of the lamp is heated, the denser liquid at the bottom becomes buoyant and forms blobs that rise to the top.) Like the blobs in a lava lamp, a mantle plume has a bulbous head that draws out a narrow stalk beneath it as it rises. Once the plume head nears the top of the mantle, decompression melting generates basaltic magma that may eventually trigger volcanism at the surface.

The result is a localized volcanic region a few hundred kilometers across called a **hot spot** (Figure 4.33C). More than 40 hot spots have been identified, and most have persisted for millions of years. The land surface surrounding a hot spot is often elevated because it is buoyed up by the rising plume of warm low-density material. Furthermore, by measuring the heat flow in these regions, geologists have determined that the mantle beneath hot spots must be 100 to 150 °C hotter than normal mantle material.

Mantle plumes are responsible for the vast outpourings of basaltic lava that created the large basalt plateaus including the Siberian Traps in Russia, India's Deccan Plateau, and the Ontong Java Plateau in the western Pacific. The most widely accepted explanation for these eruptions, which emit extremely large volumes of basaltic lava over relatively short time intervals, involves a plume with a monsterous head and a long, narrow tail (**FIGURE 4.35A**). Upon reaching the base of the lithosphere, these unusually hot, massive heads begin to melt. Melting progresses rapidly, causing the burst of volcanism that emits voluminous outpourings of lava to form a huge basalt plateau in a matter of a million or so years (**FIGURE 4.35B**). The comparatively short initial eruptive phase is followed by tens of millions of years of less voluminous activity, as the plume tail slowly rises to the surface. Extending away from most large flood basalt provinces is a chain of volcanic structures, similar to the Hawaiian chain, that terminates over an active hot spot marking the current position of the remaining tail of the plume (**FIGURE 4.35C**).

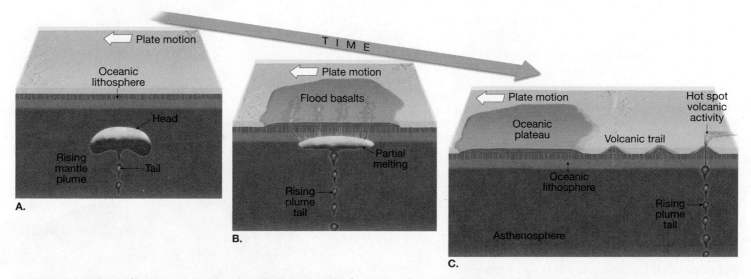

FIGURE 4.35 Model of hot-spot volcanism thought to explain the formation of oceanic plateaus and the volcanic islands associated with these features. **A.** A rising mantle plume with large bulbous head and narrow tail. **B.** Rapid decompression melting of the head of a mantle plume produces vast outpourings of basalt to generate the oceanic plateau. Large basaltic plateaus can also form on continental crust—examples include the Columbia Plateau in the northwestern United States and India's Deccan Plateau. **C.** Later, less voluminous activity caused by the rising plume tail produces a linear volcanic chain on the seafloor.

CONCEPT CHECK 4.7

❶ Are volcanoes in the Ring of Fire generally described as relatively quiet or violent? Name a volcano that would support your answer.

❷ How is magma generated along convergent plate boundaries?

❸ Volcanism at divergent plate boundaries is associated with which rock type? What causes rocks to melt in these regions?

❹ What is the source of magma for intraplate volcanism?

❺ At which type of plate boundary is the greatest quantity of magma generated?

FIGURE 4.36 Seattle, Washington, with Mount Rainier in the background. *(Photo by Ken Straiton/Corbis)*

Living with Volcanoes

About 10 percent of Earth's population lives in the vicinity of an active volcano. In fact, several major cities including Seattle, Washington; Mexico City, Mexico; Tokyo, Japan; Naples, Italy; and Quito, Ecuador are located on or near a volcano (**FIGURE 4.36**).

Until recently, the dominant view of western societies was that humans possess the wherewithal to subdue volcanoes and other types of catastrophic natural hazards. Today it is apparent that volcanoes are not only very destructive but unpredictable as well. With this awareness, a new attitude is developing—"How do we live with volcanoes?"

Volcanic Hazards

Volcanoes produce a wide variety of potential hazards that can kill people and wildlife, as well as destroy property (**FIGURE 4.37**). Perhaps the greatest threats to life are pyroclastic flows. These hot mixtures of gas, ash, and pumice that sometimes exceed 800 °C, race down the flanks of volcanoes, giving people little chance to escape.

Lahars, which can occur even when a volcano is quiet, are perhaps the next most dangerous volcanic hazard. These mixtures of volcanic debris and water can flow for tens of kilometers down steep volcanic slopes at speeds that may exceed 100 kilometers (60 miles) per hour. Lahars pose a potential threat to many communities downstream from glacier-clad volcanoes such as Mount Rainier. Other potentially destructive mass-wasting events include the rapid collapse of the volcano's summit or flank.

Other obvious hazards include explosive eruptions that can endanger people and property hundreds of miles from a volcano (**FIGURE 4.38**). During the past 15 years, at least 80 commercial jets have been damaged by inadvertently flying into clouds of volcanic ash. One of these was

a near crash that occurred in 1989 when a Boeing 747, with more than 300 passengers aboard, encountered an ash cloud from Alaska's Redoubt volcano. All four engines stalled after they became clogged with ash. Fortunately, the engines were restarted at the last minute and the aircraft managed to land safely in Anchorage.

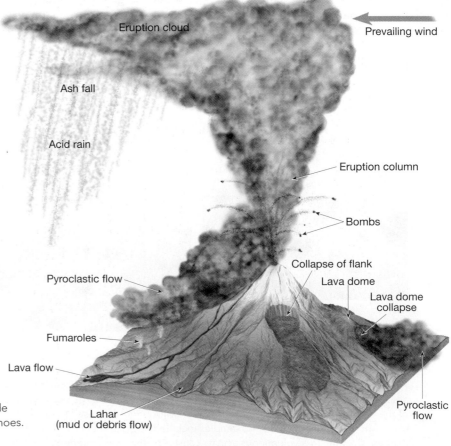

Eruption cloud

Prevailing wind

Ash fall

Acid rain

Eruption column

Bombs

Collapse of flank

Lava dome

Lava dome collapse

Pyroclastic flow

Fumaroles

Lava flow

Lahar (mud or debris flow)

Pyroclastic flow

FIGURE 4.37 Simplified drawing showing a wide variety of natural hazards associated with volcanoes. *(After U.S. Geological Survey)*

B.

A.

FIGURE 4.38 Volcanic eruptions can endanger people and property far from a volcano. **A.** The eruption of Iceland's Eyjafjallajökull volcano sent ash high into the atmosphere on April 16, 2010. The thick plume of ash drifted over Europe, causing airlines to cancel thousands of flights, leaving hundreds of thousands of travelers stranded. *(AP Photo by Brynjar Gauti)* **B.** Satellite image of the ash plume from Eyjafjallajökull volcano. *(NASA)*

Monitoring Volcanic Activity

Today a number of volcano monitoring techniques are employed, with most of them aimed at detecting the movement of magma from a subterranean reservoir (typically several kilometers deep) toward the surface. The four most noticeable changes in a volcanic landscape caused by the migration of magma are: (1) changes in the pattern of volcanic earthquakes; (2) expansion of a near-surface magma chamber, which leads to inflation of the volcano; (3) changes in the amount and/or composition of the gases that are released from a volcano; and (4) an increase in ground temperature caused by the implacement of new magma.

Almost a third of all volcanoes that have erupted in historic times are now monitored using seismographs, instruments that detect earthquake tremors. In general,

a sharp increase in seismic unrest followed by a period of relative quiet has been shown to be a precursor for many volcanic eruptions. However, some large volcanic structures have exhibited lengthy periods of seismic unrest. For example, Rabaul caldera in New Guinea recorded a strong increase in seismicity in 1981. This activity lasted 13 years and finally culminated with an eruption in 1994. Occasionally, a large earthquake triggers a volcanic eruption, or at least disturbs the volcano's plumbing. Kilauea, for example, began to erupt after the Kalapana earthquake of 1975.

The roof of a volcano may rise as new magma accumulates in its interior—a phenomenon that precedes many volcanic eruptions. Because the accessibility of many volcanoes is limited, remote sensing devices, including lasers, Doppler radar, and Earth orbiting satellites, are often used to determine whether or not a volcano is swelling. The recent discovery of ground

doming at Three Sisters volcanoes in Oregon was first detected using radar images obtained from satellites.

Volcanologists also frequently monitor the gases that are released from volcanoes in an effort to detect even minor changes in their amount and/or composition. Some volcanoes show an increase in sulfur dioxide (SO_2) emissions months or years prior to an eruption. On the other hand, a few days prior to the 1991 eruption of Mount Pinatubo, emissions of carbon dioxide (CO_2) dropped dramatically.

The development of remote sensing devices has greatly increased our ability to monitor volcanoes. These instruments and techniques are particularly useful for monitoring eruptions in progress. Photographic images and infrared (heat) sensors can detect lava flows and volcanic columns rising from a volcano. Furthermore, satellites can detect ground deformation as well as monitor SO_2 emissions.

The overriding goal of all monitoring is to discover precursors that may warn of an imminent eruption. This is accomplished by first diagnosing the current condition of a volcano and then using this baseline data to predict its future behavior. Stated another way, a volcano must be observed over an extended period to recognize significant changes from its "resting state."

CHAPTER FOUR
Volcanoes and Volcanic Hazards in Review

⊙ The primary factors that determine the nature of volcanic eruptions include the magma's *composition,* its *temperature,* and the *amount of dissolved gases* it contains. As lava cools, it begins to congeal and, as *viscosity* increases, its mobility decreases. The *viscosity of magma is also directly related to its silica content.* Rhyolitic (felsic) lava, with its high silica content (over 70 percent), is very viscous and forms short, thick flows. Basaltic (mafic) lava, with a lower silica content (about 50 percent), is more fluid and may travel a long distance before congealing. Dissolved gases tend to make magma more fluid and, as they expand, provide the force that propels molten rock from the volcano.

⊙ The materials associated with a volcanic eruption include (1) *lava flows* (*pahoehoe* flows, which resemble twisted braids; and *aa* flows, consisting of rough, jagged blocks; both form from basaltic lavas); (2) *gases* (primarily *water vapor*); and (3) *pyroclastic material* (pulverized rock and lava fragments blown from the volcano's vent, which include *ash, pumice, lapilli, cinders, blocks,* and *bombs*).

⊙ Successive eruptions of lava from a central vent result in a mountainous accumulation of material known as a *volcano.* Located at the summit of many volcanoes is a steep-walled depression called a *crater. Shield cones* are broad, slightly domed volcanoes built primarily of fluid, basaltic lava. *Cinder cones* have steep slopes composed of pyroclastic material. *Composite cones,* or *stratovolcanoes,* are large, nearly symmetrical structures built of interbedded lavas and pyroclastic deposits. Composite cones produce some of the most violent volcanic activity. Often associated with a violent eruption is a *nuée ardente,* a fiery cloud of hot gases infused with incandescent ash that races down steep volcanic slopes. Large composite cones may also generate a type of mudflow known as a *lahar.*

⊙ Most volcanoes are fed by *conduits* or *pipes.* As erosion progresses, the rock occupying the pipe, which is often more resistant, may remain standing above the surrounding terrain as a *volcanic neck.* The summits of some volcanoes have large, nearly circular depressions called *calderas* that result from collapse. Calderas also form on shield volcanoes by subterranean drainage from a central magma chamber, and the largest calderas form by the discharge of colossal volumes of silica-rich pumice along ring fractures. Although volcanic eruptions from a central vent are the most familiar, by far the largest amounts of volcanic material are extruded from cracks in the crust called *fissures.* The term *flood basalts* describes the fluid basaltic lava flows that cover an extensive region in the northwestern United States known as the Columbia Plateau. When silica-rich magma is extruded, *pyroclastic flows,* consisting largely of ash and pumice fragments, usually result.

⊙ *Most active volcanoes are associated with plate boundaries.* Active areas of volcanism are found along mid-ocean ridges where seafloor spreading is occurring (*divergent plate boundaries*), in the vicinity of ocean trenches where one plate is being subducted beneath another (*convergent plate boundaries*), and in the interiors of plates themselves (*intraplate volcanism*). Rising plumes of hot mantle rock are the source of most intraplate volcanism.

Key Terms

aa flows (p. 96)
block lava (p. 96)
calderas (p. 108)
cinder cones (p. 102)
composite cones (p. 104)
conduit (p. 99)
continental volcanic arc (p. 116)
crater (p. 100)
eruption columns (p. 94)
fissures (p. 110)

fissure eruptions (p. 110)
flood basalts (p. 110)
fumaroles (p. 100)
hot spot (p. 117)
intraplate volcanism (p. 117)
island arcs (p. 116)
lahar (p. 108)
lava dome (p. 111)
lava tubes (p. 96)
mantle plume (p. 117)

nuée ardente (p. 105)
pahoehoe flows (p. 96)
parasitic cone (p. 100)
pillow lavas (p. 96)
pipe (p. 99)
pumice (p. 99)
pyroclastic flow (p. 105)
pyroclastic materials (p. 98)
scoria (p. 99)
scoria cones (p. 102)

shield volcanoes (p. 100)
stratovolcanoes (p. 104)
vent (p. 99)
viscosity (p. 93)
volcanic island arcs (p. 116)
volcanic neck (p. 112)
volcano (p. 99)
volatiles (p. 93)

GIVE IT SOME THOUGHT

1 Match each of these volcanic regions with one of the three zones of volcanism (convergent plate boundaries, divergent plate boundaries, or intraplate volcanism):

a. Crater Lake
b. Hawaii's Kilauea
c. Mount St. Helens
d. East African Rift
e. Yellowstone
f. Vesuvius
g. Deccan Plateau
h. Mount Etna

2 Examine the accompanying photo and answer the following questions:

a. What type of volcano is it? What features helped you make a decision?

b. What is the eruptive style of such volcanoes? Describe the likely composition and viscosity of the magma.

c. Which one of the three zones of volcanism is the likely setting for this volcano?

d. Name a city that is vulnerable to the effects of a volcano of this type.

(Photo by Simon Phipps)

3 Divergent boundaries, such as the Mid-Atlantic ridge, are characterized by outpourings of basaltic lava. Answer the following questions about divergent boundaries and their associated lavas:

a. What is the source of these lavas?

b. What causes the source rocks to melt?

c. Describe a divergent boundary that would be associated with lava other than basalt. Why did you choose it, and what type of lava would you expect to erupt there?

4 Explain why volcanic activity occurs in places other than plate boundaries.

5 For each of the accompanying four sketches, identify the geologic setting (zone of volcanism). Which of these settings will most likely generate explosive eruptions? Which will produce outpouring of fluid basaltic lavas?

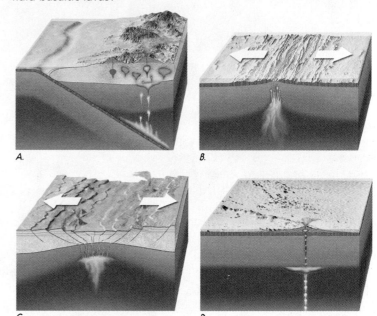

6 Assume you want to monitor a volcano that has erupted several times in the recent past but appears to be quiet now. How might you determine if magma were actually moving through the crust beneath the volcano? Suggest at least two phenomena you would observe or measure.

7 Imagine you are a geologist charged with the task of choosing three sites where state-of-the-art volcano monitoring systems will be deployed. The sites can be anywhere in the world, but the budget and number of experts you can employ to oversee the operations are limited. What criteria would you use to select these sites? List some potential choices and your reasons for considering them.

8 Explain why an eruption of Mount Rainier, similar to the one that occurred at Mount St. Helens in 1980, would be considerably more destructive.

Companion Website

www.mygeoscienceplace.com

The *Essentials of Geology, 11e* companion Website contains numerous multimedia resources accompanied by assessments to aid in your study of the topics in this chapter. The use of this site's learning tools will help improve your understanding of geology. Utilizing the access code that accompanies this text, visit **www.mygeoscienceplace.com** in order to:

- **Review** key chapter concepts.
- **Read** with links to the Pearson eText and to chapter-specific web resources.
- **Visualize** and comprehend challenging topics using the learning activities in *GEODe: Essentials of Geology* and the *Geoscience Animations Library*.
- **Test** yourself with online quizzes.

Weathering and Soils

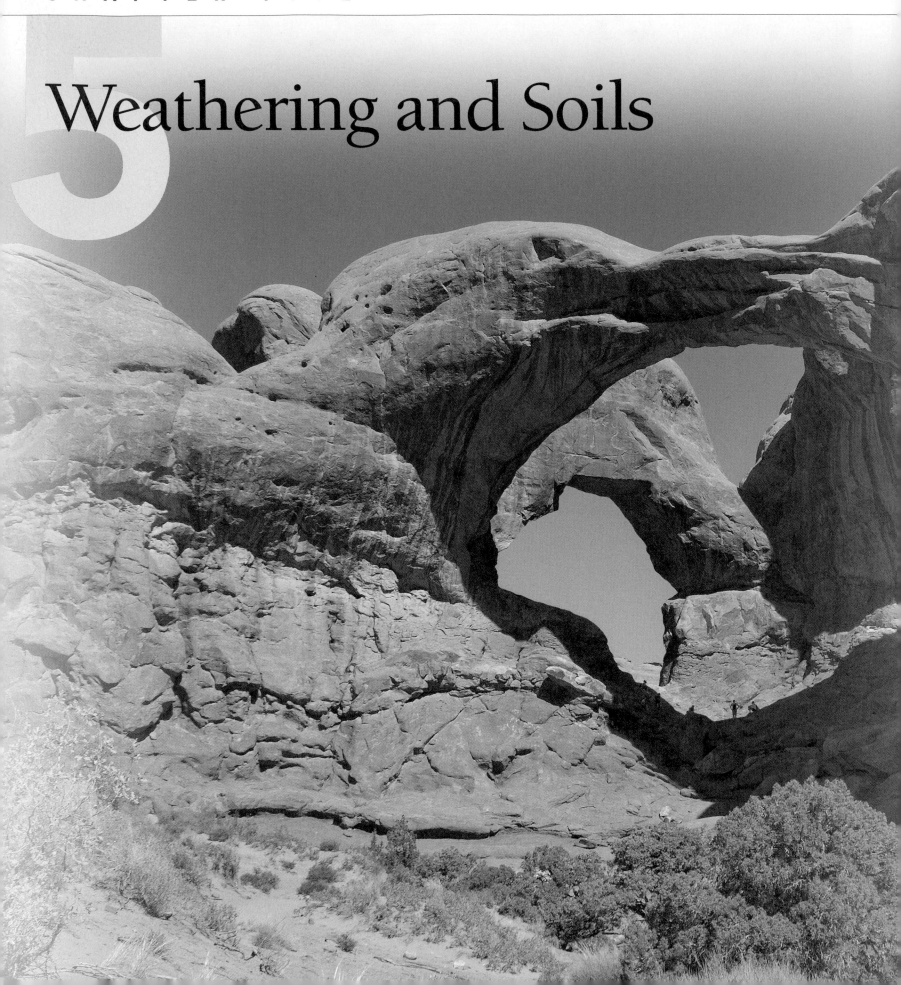

E ARTH'S SURFACE IS CONSTANTLY CHANGING. ROCK IS DISINTEGRATED and decomposed, moved to lower elevations by gravity, and carried away by water, wind, or ice. In this manner Earth's physical landscape is sculpted. This chapter focuses on the first step of this never-ending process—weathering.

Mechanical and chemical weathering contributed greatly to the formation of the arches and other rock formations in Utah's Arches National Park. *(Photo by Dennis Tasa)*

To assist you in learning the important concepts in this chapter, focus on the following questions:

- What are Earth's external processes, and what roles do they play in the rock cycle?

- What are the two main categories of weathering? In what ways are they different?

- What factors determine the rate at which rock weathers?

- What is soil? What are the factors that control soil formation?

- What factors influence natural rates of soil erosion? What impact have humans had?

- How is weathering related to the formation of ore deposits?

FOCUS ON CONCEPTS

Earth's External Processes

 Weathering and Soil

ESSENTIALS OF GEOLOGY **External versus Internal Processes**

Weathering, mass wasting, and erosion are called **external processes** because they occur at or near Earth's surface and are powered by energy from the Sun. External processes are a basic part of the rock cycle because they are responsible for transforming solid rock into sediment.

To the casual observer, the face of Earth may appear to be without change, unaffected by time. In fact, 200 years ago, most people believed that mountains, lakes, and deserts were permanent features of an Earth that was thought to be no more than a few thousand years old. Today we know that Earth is 4.6 billion years old and that mountains eventually succumb to weathering and erosion, lakes fill with sediment or are drained by streams, and deserts come and go with changes in climate.

Earth is a dynamic body. Some parts of Earth's surface are gradually elevated by mountain building and volcanic activity. These **internal processes** derive their energy from Earth's interior. Meanwhile, opposing external processes are continually breaking rock apart and moving the debris to lower elevations. The latter processes include:

1. **Weathering**—the physical breakdown (disintegration) and chemical alteration (decomposition) of rocks at or near Earth's surface.

2. **Mass wasting**—the transfer of rock and soil downslope under the influence of gravity.

3. **Erosion**—the physical removal of material by mobile agents such as water, wind, or ice.

In this chapter we will focus on rock weathering and the products generated by this activity. However, weathering cannot be easily separated from mass wasting and erosion, because as weathering breaks rocks apart, mass wasting and erosion remove the rock debris. This transport of material by mass wasting and erosion further disintegrates and decomposes the rock.

FIGURE 5.1 Arizona's Monument Valley. When weathering accentuates differences in rocks, spectacular landforms are sometimes created. As the rock gradually disintegrates and decomposes, mass wasting and erosion remove the products of weathering. *(Photo by Michael Collier)*

CONCEPT CHECK 5.1

❶ List examples of external and internal processes.
❷ From where do these processes derive their energy?

Weathering

 Weathering and Soil

ESSENTIALS OF GEOLOGY **Types of Weathering**

Weathering goes on all around us, but it seems like such a slow and subtle process that it is easy to underestimate its importance. It is worth remembering that weathering is a basic part of the rock cycle and thus a key process in the Earth system.

Weathering is also important to humans—even to those of us who are not studying geology. For example, many of the life-sustaining minerals and elements found in soil, and ultimately in the food we eat, were freed from solid rock by weathering processes. As the chapter-opening photo, **FIGURE 5.1**, and many other images in this book illustrate, weathering also contributes to the formation of some of Earth's most spectacular scenery. Of course, these same processes are also responsible for causing the deterioration of many of the structures we build (**FIGURE 5.2**).

All materials are susceptible to weathering. Consider, for example, the fabricated product concrete, which closely resembles the sedimentary rock called conglomerate. A newly poured concrete sidewalk has a smooth, fresh, unweathered look. However, not many years later, the same sidewalk will appear chipped, cracked, and rough, with pebbles exposed at the surface. If a tree is nearby, its roots may heave and buckle the concrete as well. The same natural processes that eventually break apart a concrete sidewalk act to disintegrate rock.

Weathering occurs when rock is mechanically fragmented (disintegrated) and/or chemically altered (decomposed). **Mechanical weathering** is accomplished by physical forces that break rock into smaller and smaller pieces without changing the rock's mineral composition. **Chemical weathering** involves a chemical transformation of rock into one or more new compounds. These two concepts can be illustrated with a piece of paper. The paper can be disintegrated by tearing it into smaller and smaller pieces, whereas decomposition occurs when the paper is set afire and burned.

Why does rock weather? Simply, weathering is the response of Earth materials to a changing environment. For instance, after millions of years of uplift and erosion, the rocks overlying a large intrusive igneous body may be removed, exposing it at the surface. The mass of crystalline rock, which formed deep below ground where temperatures and pressures are much greater than at the surface, is now subjected to a very different and comparatively hostile surface environment. In response, this rock mass will gradually change. This transformation of rock is what we call *weathering*.

In the following sections, we will discuss the various modes of mechanical and chemical weathering. Although we will consider these two categories separately, keep in mind that mechanical and chemical weathering processes usually work simultaneously in nature and reinforce each other.

FIGURE 5.2 Even the most "solid" monuments that people erect eventually yield to the day-in and day-out attack of weathering processes. Temple of Olympian Zeus, Athens, Greece. *(Photo by CORBIS)*

CONCEPT CHECK 5.2

❶ What are the two basic categories of weathering?
❷ How do the products of each category differ?

Mechanical Weathering

 Weathering and Soil

ESSENTIALS OF GEOLOGY **Mechanical Weathering**

When a rock undergoes *mechanical weathering*, it is broken into smaller and smaller pieces, each retaining the characteristics of the original material. The end result is many small pieces from a single large one. **FIGURE 5.3** shows that breaking a rock into smaller pieces increases the surface area available for chemical attack. An analogous situation occurs when sugar is added to a liquid. In this situation, a cube of sugar will dissolve much more slowly than an equal volume of sugar granules because the cube has much less surface area available for dissolution. Hence, by breaking rocks into smaller pieces, mechanical weathering increases the amount of surface area available for chemical weathering.

In nature, four physical processes are especially important in breaking rocks into smaller fragments: frost wedging, salt crystal growth, expansion resulting from unloading, and biological activity.

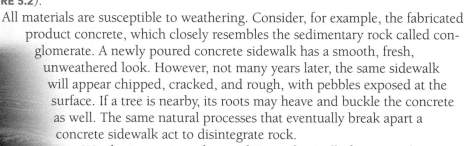

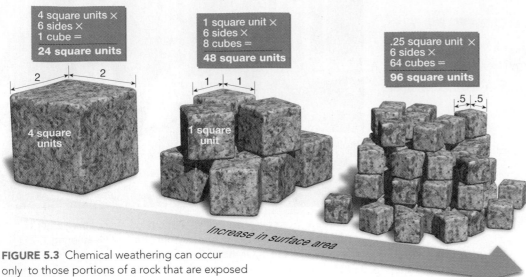

FIGURE 5.3 Chemical weathering can occur only to those portions of a rock that are exposed to the elements. Mechanical weathering breaks rock into smaller and smaller pieces, thereby increasing the surface area available for chemical attack.

In addition, although the work of erosional agents such as wind, glacial ice, rivers, and waves is usually considered separately from mechanical weathering, it is nevertheless important to point out that as these mobile agents move rock debris, they relentlessly disintegrate these materials.

Frost Wedging

If you leave a glass bottle of water in the freezer a bit too long, you will find the bottle fractured. The bottle breaks because water has the unique property of expanding about 9 percent upon freezing. This is also the reason that poorly insulated or exposed water pipes rupture during frigid weather. You might expect this same process to fracture rocks in nature. This is, in fact, the basis for the traditional explanation of **frost wedging**. After water works its way into the cracks in rock, the freezing water

FIGURE 5.4 Frost wedging. As water freezes, it expands, exerting a force great enough to break rock. When frost wedging occurs in a setting such as this, the broken rock fragments fall to the base of the cliff and create a cone-shaped accumulation known as a talus slope. (Photo by Tom Bean/Corbis)

enlarges the cracks and angular fragments are eventually produced. (**FIGURE 5.4**)

For many years, the conventional wisdom was that most frost wedging occurred in this way. Recently, however, research has shown that frost wedging can also occur in a different way.[*] It has long been known that when moist soils freeze,

[*]Bernard Hallet, "Why Do Freezing Rocks Break?," *Science*, Vol. 314, November 2006, pp. 1092–93.

they expand or *frost heave* due to the growth of ice lenses. These masses of ice grow larger because they are supplied with water migrating from unfrozen areas as thin liquid films. As more water accumulates and freezes, the soil is heaved upward. A similar process occurs within the cracks and pore spaces of rocks. Lenses of ice grow larger as they attract liquid water from surrounding pores. The growth of these ice masses gradually weakens the rock, causing it to fracture.

Salt Crystal Growth

Another expansive force that can split rocks is created by the growth of salt crystals. Rocky shorelines and arid regions are common settings for this process. It begins when sea spray from breaking waves or salty groundwater penetrates crevices and pore spaces in rock. As this water

DID YOU KNOW?
The intense heat from a brush or forest fire can cause flakes of rock to spall from boulders or bedrock. As the rock surface becomes overheated, a thin layer expands and shatters.

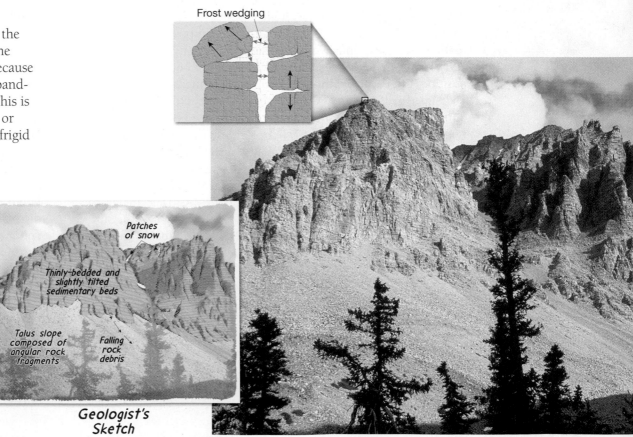

Frost wedging

Geologist's Sketch

evaporates, salt crystals form. As these crystals gradually grow larger, they weaken the rock by pushing apart the surrounding grains or enlarging tiny cracks.

This same process can contribute to crumbling roadways where salt is spread to melt snow and ice in winter. The salt dissolves in water and seeps into cracks that quite likely originated from frost action. When the water evaporates, the growth of salt crystals further breaks the pavement.

Sheeting

When large masses of igneous rock, particularly granite, are exposed by erosion, concentric slabs begin to break loose. The process generating these onionlike layers is called **sheeting.** It is thought that this occurs, at least in part, because of the great reduction in pressure when the overlying rock is eroded away, a process called *unloading.* Accompanying this unloading, the outer layers expand more than the rock below and thus separate from the rock body (**FIGURE 5.5**). Continued weathering eventually causes the slabs to separate and spall off, creating **exfoliation domes.** Excellent examples of exfoliation domes include Stone Mountain, Georgia, and Half Dome (Figure 5.5C), and Liberty Cap in Yosemite National Park.

Deep underground mining provides us with another example of how rocks behave once the confining pressure is removed. Large rock slabs sometimes explode off the walls of newly cut mine tunnels because of the abruptly reduced pressure. Evidence of this type, plus the fact that fracturing occurs parallel to the floor of a quarry when large blocks of rock are removed, strongly supports the process of unloading as the cause of sheeting.

Although many fractures are created by expansion, others are produced by contraction as igneous materials cool (see Figure 4.31, p. 112), and still others by tectonic forces during mountain building. Fractures produced by these activities often form a definite pattern and are called *joints* (**FIGURE 5.6**). Joints are important rock structures that allow water to penetrate deeply and start the process of weathering long before the rock is exposed.

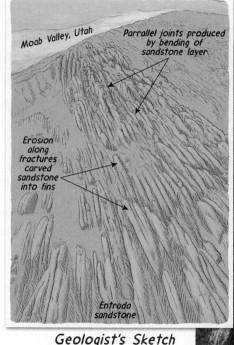

Geologist's Sketch

FIGURE 5.6 Aerial view of nearly parallel joints near Moab, Utah. *(Photo by Michael Collier)*

FIGURE 5.5 Sheeting is caused by the expansion of crystalline rock as erosion removes the overlying material. When the deeply buried pluton in **A.** is exposed at the surface following uplift and erosion in **B.**, the igneous mass fractures into thin slabs. The photo in **C.** is of the summit of Half Dome in Yosemite National Park, California. It is an exfoliation dome and illustrates the onionlike layers created by sheeting. *(Photo by Gary Moon/agefotostock)*

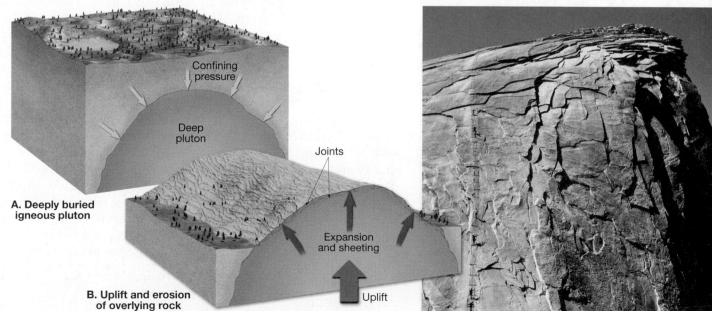

A. Deeply buried igneous pluton

B. Uplift and erosion of overlying rock

C. Exfoliation dome

Biological Activity

Both mechanical and chemical weathering are accomplished by the activities of organisms. Plant roots in search of minerals and water grow into fractures, and as the roots grow, they wedge the rock apart (**FIGURE 5.7**). Burrowing animals further break down the rock by moving fresh material to the surface, where physical and chemical processes can more effectively attack it. Of course, where rock has been blasted in search of minerals or for construction, the impact of humans is particularly noticeable.

There are numerous ways that organisms play a role in chemical weathering. For example, plant roots, fungi, and lichens that occupy fractures or may encrust a rock produce acids that promote decomposition. Moreover, some bacteria are capable of extracting compounds from minerals and

FIGURE 5.7 Root wedging widens fractures in rocks and aids the process of mechanical weathering. Mt. Sanitas, Boulder, Colorado. (Photo by Kristin Piljay)

using the energy from the compound's chemical bonds to supply their life needs. These primitive "mineral-eating" life forms can live at depths as great as a few kilometers.

CONCEPT CHECK 5.3

❶ When a rock is mechanically weathered, how does its surface area change? How does this influence chemical weathering?

❷ Explain how water can cause mechanical weathering.

❸ How does an exfoliation dome form?

❹ How do joints promote weathering?

❺ How does biological activity contribute to weathering?

FIGURE 5.8 Iron reacts with oxygen to form iron oxide as seen on these rusted barrels. (Photo by Steven Robertson/iStockphoto)

Chemical Weathering

 Weathering and Soil

ESSENTIALS OF GEOLOGY **Chemical Weathering**

In the preceding discussion of mechanical weathering you learned that breaking rock into smaller pieces aids chemical weathering by increasing the surface area available for chemical attack. It should also be pointed out that chemical weathering contributes to mechanical weathering. It does so by weakening the outer portions of some rocks which, in turn, makes them more susceptible to being broken by mechanical weathering processes.

Chemical weathering involves the complex processes that break down rock components and internal structures of minerals. Such processes convert the constituents to new minerals or release them to the surrounding environment. During this transformation, the original rock decomposes into substances that are stable in the surface environment. Consequently, the products of chemical weathering will remain essentially unchanged as long as they remain in an environment similar to the one in which they formed.

Water and Carbonic Acid

Water is by far the most important agent of chemical weathering. Although pure water is nonreactive, a small amount of dissolved material is generally all that is needed to activate it. Oxygen dissolved in water will oxidize some materials. For example, when an iron nail is found in moist soil, it will have a coating of rust (iron oxide), and if the time of exposure has been long, the nail will be so weak that it can be broken as easily as a toothpick. When rocks containing iron-rich minerals oxidize, a yellow to reddish-brown rust will appear on the surface (**FIGURE 5.8**).

Carbon dioxide (CO_2) dissolved in water (H_2O) forms carbonic acid (H_2CO_3), the same weak acid produced when soft drinks are carbonated. Rain dissolves some carbon dioxide as it falls through the atmosphere, and additional amounts released by decaying organic matter are acquired as the water percolates through the soil. Carbonic acid ionizes to form the very reactive hydrogen ion (H^+) and the bicarbonate ion (HCO_3^-)

Acids such as carbonic acid readily decompose many rocks and produce certain products that are water soluble. For example, the mineral calcite ($CaCO_3$), which composes the common building stones marble and limestone, is easily attacked by even a weakly acidic solution. The overall reaction by which calcite dissolves in water containing carbon dioxide is:

$$CaCO_3 + (H^+ + HCO_3^-) \longrightarrow Ca^{2+} + 2HCO_3^-$$

calcite   carbonic acid   calcium ion   bicarbonate
                                                        ion

FIGURE 5.9 Missouri's Meramec Caverns. The dissolving power of carbonic acid plays an important part in forming limestone caverns. *(Photo by Michael Szönyi/Photolibrary)*

During this process, the insoluble calcium carbonate is transformed into soluble products. In nature, over periods of thousands of years, large quantities of limestone are dissolved and carried away by underground water. This activity is clearly evidenced by the large number of subsurface caverns found in every one of the contiguous 48 states (**FIGURE 5.9**).

How Granite Weathers

To illustrate how rock chemically weathers when attacked by carbonic acid, we will consider the weathering of granite, the most abundant continental rock. Recall that granite consists mainly of quartz and potassium feldspar. The weathering of the potassium feldspar component of granite takes place as follows:

$$2\ KAlSi_3O_8 \ + \ 2(H^+ + HCO_3^-) \ + \ H_2O \ \longrightarrow \ Al_2Si_2O_5(OH)_4 \ + \ 2K^+ \ + \ 2HCO_3^- \ + \ 4SiO_2$$

potassium feldspar	carbonic acid	water	clay mineral	potassium ion	bicarbonate ion	silica ion

in solution

In this reaction, the hydrogen ions (H^+) attack and replace potassium ions (K^+) in the feldspar structure, thereby disrupting the crystalline network. Once removed, the potassium is available as a nutrient for plants or becomes the soluble salt potassium bicarbonate ($KHCO_3$) which may be incorporated into other minerals or carried to the ocean in dissolved form by streams.

The most abundant products of the chemical breakdown of feldspar are residual clay minerals. Clay minerals are the end products of weathering and are very stable under surface conditions. Consequently, clay minerals make up a high percentage of the inorganic material in soils. Moreover, the most abundant sedimentary rock, shale, contains a high proportion of clay minerals.

In addition to the formation of clay minerals during the weathering of feldspar, some silica is removed from the feldspar structure and is carried away by groundwater (water beneath Earth's surface). This dissolved silica will eventually precipitate to produce nodules of chert or flint, or it will fill in the pore spaces between sediment grains, or it will be carried to the ocean, where microscopic animals will remove it from the water to build hard silica shells.

To summarize, the weathering of potassium feldspar generates a residual clay mineral, a soluble salt (potassium bicarbonate), and some silica, which enters into solution.

TABLE 5.1
Products of Weathering

Mineral	Residual Products	Material in Solution
Quartz	Quartz grains	Silica
Feldspars	Clay minerals	Silica, K^+, Na^+, Ca^{2+}
Amphibole (hornblende)	Clay minerals	Silica, Ca^{2+}, Mg^{2+}
	Limonite	
	Hematite	
Olivine	Limonite	Silica, Mg^{2+}
	Hematite	

Quartz, the other main component of granite, is *very resistant* to chemical weathering; it remains substantially unaltered when attacked by weak acidic solutions. As a result, when granite weathers, the feldspar crystals dull and slowly turn to clay, releasing the once-interlocked quartz grains, which still retain their fresh, glassy appearance. Although some quartz remains in the soil, much is eventually transported to the sea or to other sites of deposition, where it becomes the main constituent of such features as sandy beaches and sand dunes. In time these quartz grains may become lithified to form the sedimentary rock *sandstone*.

DID YOU KNOW?
The only common mineral that is very resistant to both mechanical and chemical weathering is quartz.

Weathering of Silicate Minerals

TABLE 5.1 lists the weathered products of some of the most common silicate minerals. Remember that silicate minerals make up most of Earth's crust and that these minerals are composed essentially of only eight elements. When chemically weathered, silicate minerals yield sodium, calcium, potassium, and magnesium ions that form soluble products, which may be removed by groundwater. The element iron combines with oxygen, producing relatively insoluble iron oxides, which give soil a reddish-brown or yellowish color. Under most conditions the three remaining elements—aluminum, silicon, and oxygen—join with water to produce residual clay minerals. However, even the highly insoluble clay minerals are very slowly removed by subsurface water.

Spheroidal Weathering

In addition to altering the internal structure of minerals, chemical weathering causes physical changes. For instance, when angular rock masses chemically weather as water enters along joints, they tend to take on a spherical shape. Gradually the corners and edges of the angular blocks become more rounded. The corners are attacked most readily because of their greater surface area, as compared to the edges and faces. This

Bryce Canyon. *(Photo by Joe Cornish/Photolibrary)*

A.

FIGURE 5.10 A. Spheroidal weathering is evident in this exposure of granite in California's Joshua Tree National Park. Because the rocks are attacked more vigorously on the corners and edges, they take on a spherical shape. The lines visible in the rock are called *joints*. Joints are important rock structures that allow water to penetrate and start the weathering process long before the rock is exposed. *(Photo by E. J. Tarbuck)* **B.** Sometimes successive shells are loosened as the weathering process continues to penetrate ever deeper into the rock. *(Photo by Martin Schmidt, Jr.)*

process, called **spheroidal weathering,** gives the weathered rock a more rounded or spherical shape (**FIGURE 5.10A**).

Sometimes during the formation of spheroidal boulders, successive shells separate from the rock's main body (**FIGURE 5.10B**). Eventually the outer shells break off, allowing the chemical-weathering activity to penetrate deeper into the boulder. This spherical scaling results because the minerals in the rock increase in size through the addition of water to their structure as they weather to clay. This increased bulk exerts an outward force that causes concentric layers of rock to break loose and fall off. Hence, chemical weathering does produce forces great enough to cause mechanical weathering.

This type of spheroidal weathering, in which shells spall off, should not be confused with the phenomenon of sheeting discussed earlier. In sheeting, the fracturing occurs as a result of unloading, and the rock layers that separate from the main body are largely unaltered at the time of separation.

CONCEPT CHECK 5.4

① How is carbonic acid formed in nature?

② What occurs when carbonic acid reacts with limestone?

③ What products result when carbonic acid reacts with potassium feldspar?

④ Explain how the rounded boulders in Figure 5.10A formed.

B.

Rates of Weathering

 Weathering and Soil

ESSENTIALS OF GEOLOGY **Rates of Weathering**

Several factors influence the type and rate of rock weathering. We have already seen how mechanical weathering affects the rate of weathering. By breaking rock into smaller pieces, the amount of surface area exposed to chemical weathering is increased. Other important factors examined here include rock characteristics and climate.

Rock Characteristics

Rock characteristics encompass all of the chemical traits of rocks, including mineral composition and solubility. In addition, any physical features such as joints (cracks) can be important because they allow water to penetrate rock and start the process of weathering long before the rock is exposed.

The variations in weathering rates due to the mineral constituents can be demonstrated by comparing old headstones made

from different rock types. Head-stones of granite, which are composed of silicate minerals, are relatively resistant to chemical weathering. We can see this by examining the inscriptions on the headstones shown in **FIGURE 5.11**. In contrast to the granite, the marble headstone shows signs of extensive chemical alteration over a relatively short period of time. Marble is composed of calcite (calcium carbonate), which readily dissolves even in a weakly acidic solution.

The silicates, the most abundant mineral group, weather in essentially the same order as their order of crystallization. By examining Bowen's reaction series (see Figure 3.20, p. 78), you can see that olivine crystallizes first and is therefore the least resistant to chemical weathering, whereas quartz, which crystallizes last, is the most resistant.

DID YOU KNOW?

The moon has no atmosphere, no water, and no biological activity. Therefore, the weathering processes we are familiar with on Earth are lacking on the Moon. However, all lunar terrains are covered with a soil-like layer of gray debris, called *lunar regolith*, derived from a few billion years of bombardment by meteorites. The rate of change at the lunar surface is so slow that the footprints left by *Apollo* astronauts will likely remain fresh-looking for millions of years.

FIGURE 5.11 An examination of headstones reveals the rate of chemical weathering on diverse rock types. The granite headstone (left) was erected four years before the marble headstone (right). The inscription date of 1872 on the marble monument is nearly illegible. *(Photos by E. J. Tarbuck)*

Climate

Climatic factors, particularly temperature and moisture, are crucial to the rate of rock weathering. One important example from mechanical weathering is that the frequency of freeze–thaw cycles greatly affects the amount of frost wedging. Temperature and moisture also exert a strong influence on rates of chemical weathering and on the kind and amount of vegetation present. Regions with lush vegetation generally have a thick mantle of soil rich in decayed organic matter from which chemically active fluids such as carbonic and humic acids are derived.

The optimal environment for chemical weathering is a combination of warm temperatures and abundant moisture. In polar regions chemical weathering is ineffective because frigid temperatures keep the available moisture locked up as ice, whereas in arid regions there is insufficient moisture to foster rapid chemical weathering.

Human activities can influence the composition of the atmosphere, which in turn can impact the rate of chemical weathering. One well-known example is acid rain (**FIGURE 5.12**).

FIGURE 5.12 Acid rain accelerates the chemical weathering of stone monuments and structures, including this building facade in Leipzig, Germany. As a consequence of burning large quantities of coal and petroleum, more than 27 million tons of sulfur and nitrogen oxides are released into the atmosphere each year in the United States. Through a series of complex chemical reactions, some of these pollutants are converted into acids that then fall to Earth's surface as rain or snow. *(Photo by Doug Plummer/Photo Researchers, Inc.)*

Differential Weathering

Masses of rock do not weather uniformly. Take a moment to look back at the photo of Shiprock, New Mexico, in Figure 4.27 (p. 110). The durable volcanic neck protrudes high above the surrounding terrain. A glance at the chapter-opening photo shows an additional example of this phenomenon, called **differential weathering.** The results vary in scale from the rough, uneven surface of the marble headstone in Figure 5.11 to the boldly sculpted exposures in Bryce Canyon (**FIGURE 5.13**).

Many factors influence the rate of rock weathering. Among the most important are variations in the composition of the rock. More resistant rock protrudes as ridges or pinnacles, or as steeper cliffs on an irregular hillside (see Figure 6.4, p. 153). The number and spacing of joints can also be a significant factor (see Figure 5.10A). Differential weathering and subsequent erosion are responsible for creating many unusual and sometimes spectacular rock formations and landforms.

Soil

Soil covers most land surfaces. Along with air and water, it is one of our most indispensable resources. Also, like air and water, soil is taken for granted by many of us. The following quote helps put this vital layer in perspective.

Science, in recent years, has focused more and more on the Earth as a planet, one that for all we know is unique—where a thin blanket of air, a thinner film of water, and the thinnest veneer of soil combine to support a web of life of wondrous diversity in continuous change.[*]

Soil has accurately been called "the bridge between life and the inanimate world." All life—the entire biosphere—owes its existence to a dozen or so elements that must ultimately come from Earth's crust. Once weathering and other processes create soil, plants carry out the intermediary role of assimilating the necessary elements and making them available to animals, including humans.

An Interface in the Earth System

When Earth is viewed as a system, soil is referred to as an interface—a common boundary where different parts of a system

DID YOU KNOW?
Sometimes soils become buried and preserved. Later, if these ancient soils, called *paleosols,* are uncovered, they can provide useful clues to climates and the nature of landscapes thousands or millions of years ago.

[*]Jack Eddy. "A Fragile Seam of Dark Blue Light," in *Proceedings of the Global Change Research Forum,* U.S. Geological Survey Circular 1086, (1993), p. 15.

interact. This is an appropriate designation because soil forms where the geosphere, the atmosphere, the hydrosphere, and the biosphere meet. Soil is a material that develops in response to complex environmental interactions among different parts of the Earth system. Over time, soil gradually evolves to a state of equilibrium or balance with the environment. Soil is dynamic and sensitive to almost every aspect of its surroundings. Thus, when environmental changes occur, in climate, vegetative cover, or animal (including human) activity, the soil responds. Any such change produces a gradual alteration of soil characteristics until a new balance is reached. Although thinly distributed over the land surface, soil functions as a fundamental interface, providing an excellent example of the integration among many parts of the Earth system.

What Is Soil?

With few exceptions, Earth's land surface is covered by **regolith,** the layer of rock and mineral fragments produced by weathering. Some would call this material soil, but soil is more than an accumulation of weathered debris. **Soil** is a combination of mineral and organic matter, water, and air—that portion of the regolith that supports the growth of plants. Although the proportions of the major components in soil vary, the same four components always are present to some extent (**FIGURE 5.14**). About one-half of the total volume of a good quality surface soil is a mixture of disintegrated and decomposed rock (mineral matter) and *humus,* the decayed remains of animal and plant life (organic matter). The remaining half consists of pore spaces among the solid particles where air and water circulate.

Although the mineral portion of the soil is usually much greater than the organic portion, humus is an essential component. In addition to being an important source of plant nutrients, humus enhances the soil's ability to retain water. Because plants require air and water to live and grow, the portion of the soil consisting of pore spaces that allow for the circulation of these fluids is as vital as the solid soil constituents.

Soil water is far from "pure" water; instead, it is a complex solution containing many soluble nutrients. Soil water not only provides the necessary moisture for the chemical reactions that sustain life, it also supplies plants with nutrients in a form they can use. The pore spaces that are not filled with water contain air. This air is the source of necessary oxygen and carbon dioxide for most microorganisms and plants that live in the soil.

Controls of Soil Formation

Soil is the product of the complex interplay of several factors. The most important of these are parent material, time, climate, plants and animals, and topography. Although all of these factors are interdependent, their roles will be examined separately.

Parent Material

The source of the weathered mineral matter from which soils develop is called the **parent material** and is a major factor influencing a newly forming soil. Gradually it undergoes physical and chemical changes as the processes of soil formation progress. Parent material might be the underlying bedrock, or it can be a layer of unconsolidated deposits, as in a stream valley. When the parent material is bedrock, the soils are termed *residual soils.* By contrast, those developed on unconsolidated sediment are called *transported soils* (**FIGURE 5.15**). Note that transported soils form *in place* on parent materials that have been carried from elsewhere and deposited by gravity, water, wind, or ice.

The nature of the parent material influences soils in two ways. First, the type of parent material affects the rate of weathering and thus the rate of soil formation. (Consider the weathering rates of granite

FIGURE 5.14 Soil is an essential resource that we often take for granted. Soil is not a living entity, but it contains a great deal of life. Moreover, this complex medium supports nearly all plant life, which in turn supports animal life. The pie chart shows the composition (by volume) of a soil in good condition for plant growth. Although percentages vary, each soil is composed of mineral and organic matter, water, and air. *(Photo by Colin Molyneux/Getty Images)*

25% air

25% water

45% mineral matter

5% organic matter

versus marble.) Also, because unconsolidated deposits are already partly weathered and provide more surface area for chemical weathering, soil development on such material usually progresses more rapidly. Second, the chemical make-up of the parent material affects the soil's fertility. This influences the character of the natural vegetation the soil can support.

At one time the parent material was thought to be the primary factor causing differences among soils. However, soil scientists came to understand that other factors, especially climate, are more important. In fact, it was found that similar soils often develop from different parent materials and that dissimilar soils develop from the same parent material. Such discoveries reinforce the importance of the other soil-forming factors.

FIGURE 5.13 Differential weathering is illustrated by these sculpted rock pinnacles in Bryce National Park, Utah. *(Photo by Angelo Cavalli/Photolibrary)*

Time

Time is an important component of *every* geological process, and soil formation is no exception. The nature of soil is strongly influenced by the length of time that processes have been operating. If weathering has been going on for a comparatively short time, the parent material strongly influences the characteristics of the soil. As weathering processes continue, the influence of parent material on soil is overshadowed by the other soil-forming factors, especially climate. The amount of time required for various soils to evolve cannot be specified, because the soil-forming processes act at varying rates under different circumstances. However, as a rule, the longer a soil has been forming, the thicker it becomes and the less it resembles the parent material.

No soil development because of very steep slope

Transported soil developed on stream-deposited sediment

Residual soil is developed on bedrock

Thick soil

Bedrock

Thinner soil on steep slope because of erosion

Unconsolidated deposits

FIGURE 5.15 The parent material for residual soils is the underlying bedrock, whereas transported soils form on unconsolidated deposits. Also note that as slopes become steeper, soil becomes thinner. *(Left and center photos by E. J. Tarbuck; right photo by Grilly Bernard/Getty Images, Inc./Stone Allstock)*

Climate

Climate is the most influential control of soil formation. Just as temperature and precipitation are the climatic elements that influence people the most, so too are they the elements that exert the strongest impact on soil formation. Variations in temperature and precipitation determine whether chemical or mechanical weathering predominates. They also greatly influence the rate and depth of weathering. For instance, a hot, wet climate may produce a thick layer of chemically weathered soil in the same amount of time that a cold, dry climate produces a thin mantle of mechanically weathered debris. Also, the amount of precipitation influences the degree to which various materials are removed (leached) from the soil, thereby affecting soil fertility. Finally, climatic conditions are important factors controlling the type of plant and animal life present.

Plants and Animals

The biosphere plays a vital role in soil formation. The types and abundance of organisms present have a strong influence on the physical and chemical properties of a soil (**FIGURE 5.16**). In fact, for well-developed soils in many regions, the significance of natural vegetation in influencing soil type is frequently implied in the description used by soil scientists. Such phrases as *prairie soil, forest soil*, and *tundra soil* are common.

Plants and animals furnish organic matter to the soil. Certain bog soils are composed almost entirely of organic matter, whereas desert soils may contain only a tiny percentage. Although the quantity of organic matter varies substantially among soils, it is the rare soil that completely lacks it.

The primary source of organic matter is plants, although animals and the uncountable microorganisms also contribute. When organic matter decomposes, important nutrients are supplied to plants, as well as to animals and microorganisms living in the soil. Consequently, soil fertility depends in part on the amount of organic matter present. Furthermore, the decay of plant and animal remains causes the formation of various organic acids. These complex acids hasten the weathering process. Organic matter also has a high water-holding ability and thus aids water retention in a soil.

Microorganisms, including fungi, bacteria, and single-celled protozoa, play an active role in the decay of plant and animal remains. The end product is humus, a material that no longer resembles the plants and animals from which it was formed. In addition, certain microorganisms aid soil fertility because they have the ability to convert atmospheric nitrogen into soil nitrogen.

Earthworms and other burrowing animals act to mix the mineral and

FIGURE 5.16 The nature of the vegetation in an area can have a significant influence on soil formation. **A.** The northern coniferous forest. The organic litter received by the soil from the conifers is high in acid resins, which contributes to an accumulation of acid in the soil. As a result, intensive acid leaching is an important soil-forming process here. *(Photo by Bill Brooks/Alamy)* **B.** The relatively meager vegetation in Arizona's Sonoran Desert is very different in character from the northern coniferous forest. Desert soils are typically thin and lack much organic matter. *(Photo by Russ Bishop/age fotostock)*

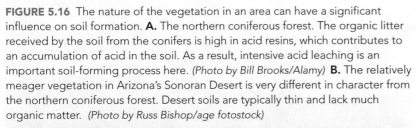

organic portions of a soil. Earthworms, for example, feed on organic matter and thoroughly mix soils in which they live, often moving and enriching many tons per acre each year. Burrows and holes also aid the passage of water and air through the soil.

Topography

The lay of the land can vary greatly over short distances. Such variations in topography can lead to the development of a variety of localized soil types. Many of the differences exist because the length and steepness of slopes have a significant impact on the amount of erosion and the water content of soil.

On steep slopes, soils are often poorly developed. In such situations little water can soak in; as a result, soil moisture may be insufficient for vigorous plant growth. Further, because of accelerated erosion on steep slopes, the soils are thin or nonexistent (see Figure 5.15).

In contrast, waterlogged soils in poorly drained bottomlands have a much different character. Such soils are usually thick and dark. The dark color results from the large quantity of organic matter that accumulates because saturated conditions retard the decay of vegetation. The optimum terrain for soil development is a flat-to-undulating upland surface. Here we find good drainage, minimum erosion, and sufficient infiltration of water into the soil.

Slope orientation, or the direction a slope is facing, also is significant. In the midlatitudes of the Northern Hemisphere, a south-facing slope receives a great deal more sunlight than a north-facing slope. In fact, a steep north-facing slope may receive no direct sunlight at all. The difference in the amount of solar radiation received causes substantial differences in soil temperature and moisture, which in turn influence the nature of the vegetation and the character of the soil.

Although we have dealt separately with each of the soil-forming factors, remember that all of them work together to form soil. No single factor is responsible for a soil's character. Rather, it is the combined influence of parent material, time, climate, plants and animals, and topography that determines this character.

CONCEPT CHECK 5.7

❶ List the five basic controls of soil formation.
❷ Which factor is most influential in soil formation?
❸ How might the direction a slope is facing influence soil formation?

The Soil Profile

Because soil-forming processes operate from the surface downward, variations in composition, texture, structure, and color gradually evolve at varying depths. These vertical differences, which usually become more pronounced as time passes, divide the soil into zones or layers known as **horizons.** If you were to dig a trench in soil, you would see that its walls are layered. Such a vertical section through all of the soil horizons constitutes the **soil profile** (**FIGURE 5.17**).

FIGURE 5.18 presents an idealized view of a well-developed soil profile in which five horizons are identified. From the surface downward, they are designated as *O, A, E, B,* and *C,* respectively. These five horizons are common to soils in temperate regions. The characteristics and extent of development of horizons vary in different environments. Thus, different localities exhibit soil profiles that can contrast greatly with one another.

The *O* horizon consists largely of organic material. This is in contrast to the layers beneath it that consist mainly of mineral matter. The upper portion of the *O* horizon is primarily plant litter such as loose leaves and other organic debris that are still recognizable. By contrast, the lower portion of the *O* horizon is made up of partly decomposed organic matter (humus) in which plant structures can no longer be identified. In addition to plants, the *O* horizon is teeming with microscopic life, including bacteria, fungi, algae, and insects. All of these organisms contribute oxygen, carbon dioxide, and organic acids to the developing soil.

Underlying the organic-rich *O* horizon is the *A* horizon. This zone is largely mineral matter, yet biological activity is high and humus is generally present at up to 30 percent in some instances. Together the *O* and *A* horizons make up what is commonly called *topsoil.* Below the *A* horizon, the *E* horizon is a light-colored layer that contains little organic material. As water percolates downward through this zone, finer particles are carried away. This washing out of the fine soil components is termed **eluviation.** Water percolating downward also dissolves soluble inorganic soil components and carries them to deeper zones. This depletion of soluble materials from the upper soil is termed **leaching.**

Spheroidal weathering. *(Photo by Bill Hatcher/National Geographic)*

weathering

Immediately below the *E* horizon is the *B* horizon, or *subsoil*. Much of the material removed from the *E* horizon by eluviation is deposited in the *B* horizon, which is often referred to as the *zone of accumulation*.

FIGURE 5.17 A soil profile is a vertical cross section from the surface through all of the soil's horizons and into the parent material. **A.** The profile shows a well-developed soil in southeastern South Dakota. *(Photo by E. J. Tarbuck)* **B.** The boundaries between horizons in this soil in Puerto Rico are indistinct, giving it a relatively uniform appearance. *(Photo courtesy of Soil Science Society of America)*

FIGURE 5.18 Well-developed soils show distinct layers called *horizons*. This is an idealized soil profile from a humid climate in the middle latitudes.

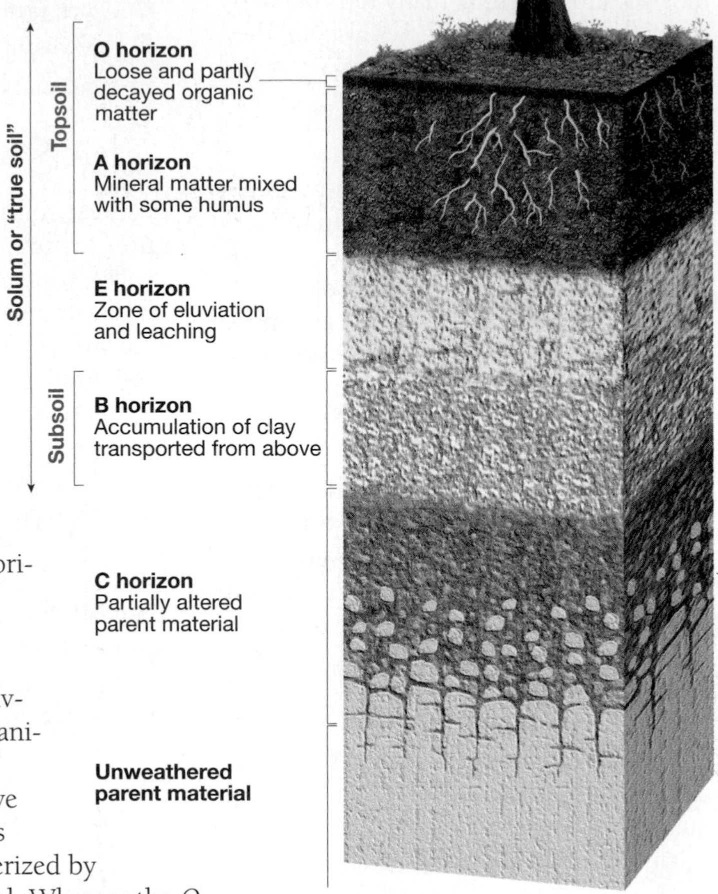

O horizon
Loose and partly decayed organic matter

A horizon
Mineral matter mixed with some humus

E horizon
Zone of eluviation and leaching

B horizon
Accumulation of clay transported from above

C horizon
Partially altered parent material

Unweathered parent material

Solum or "true soil"
Topsoil
Subsoil

The accumulation of the fine clay particles enhances water retention in the subsoil. However, in extreme cases clay accumulation can form a very compact and impermeable layer called *hardpan*. The *O, A, E,* and *B* horizons together constitute the **solum,** or true soil. It is in the solum that the soil-forming processes are active and that living roots and other plant and animal life are largely confined.

Below the solum and above the unaltered parentmaterial is the *C* horizon, a layer characterized by partially altered parent material. Whereas the *O, A, E,* and *B* horizons bear little resemblance to the parent material, it is easily identifiable in the *C* horizon. Although this material is undergoing changes that will eventually transform it into soil, it has not yet crossed the threshold that separates regolith from soil.

The characteristics and extent of development can vary greatly among soils in different environments. The boundaries between soil horizons may be very distinct, or the horizons may blend gradually from one to another. A well-developed soil profile indicates that environmental conditions have been relatively stable over an extended time span and that the soil is mature. By contrast, some soils lack horizons altogether. Such soils are called *immature*, because soil building has been going on for only a short time. Immature soils are also characteristic of steep slopes where erosion continually strips away the soil, preventing full development.

CONCEPT CHECK 5.8

❶ Sketch and label the main soil horizons in a well-developed soil profile.

❷ Describe the following features or processes: eluviation, leaching, zone of accumulation, hardpan.

Classifying Soils

There are many variations from place to place and from time to time among the factors that control soil formation. These differences lead to a bewildering variety of soil types. To cope with such variety, it is essential to devise some means of classifying the vast array of data to be studied. By establishing groups consisting of items that have certain important characteristics in common, order and simplicity are introduced. Bringing order to large quantities of information not only aids comprehension and understanding but also facilitates analysis and explanation.

In the United States, soil scientists have devised a system for classifying soils known as the **Soil Taxonomy**. It emphasizes the physical and chemical properties of the soil profile and is organized on the basis of observable soil characteristics. There are six hierarchical categories of classification, ranging from *order*, the broadest category,

to *series*, the most specific category. The system recognizes 12 soil orders and more than 19,000 soil series.

The names of the classification units are combinations of syllables, most of which are derived from Latin or Greek. The names are descriptive. For example, soils of the order Aridosol (from the Latin *aridus*, dry, and *solum*, soil) are characteristically dry soils in arid regions. Soils in the order Inceptisols (from Latin *inceptum*, beginning, and *solum*, soil) are soils with only the beginning or inception of profile development.

Brief descriptions of the 12 basic soil orders are provided in **TABLE 5.2**.

FIGURE 5.19 shows the complex worldwide distribution pattern of the Soil Taxonomy's 12 soil orders. Like many classification systems, the Soil Taxonomy is not suitable for every purpose. It is especially useful for agricultural and related land-use purposes, but it is not a useful system for engineers who are preparing evaluations of potential construction sites.

CONCEPT CHECK 5.9

❶ Why are soils classified?

TABLE 5.2
Basic Soil Orders

Alfisols	Moderately weathered soils that form under boreal forests or broadleaf deciduous forests, rich in iron and aluminum. Clay particles accumulate in a subsurface layer in response to leaching in moist environments. Fertile, productive soils, because they are neither too wet nor too dry.
Andisols	Young soils in which the parent material is volcanic ash and cinders deposited by recent volcanic activity.
Aridosols	Soils that develop in dry places; insufficient water to remove soluble minerals; may have an accumulation of calcium carbonate, gypsum, or salt in subsoil; low organic content. (see Figure 5.16B)
Entisols	Young soils having limited development and exhibiting properties of the parent material. Productivity ranges from very high for some formed on recent river deposits to very low for those forming on shifting sand or rocky slopes.
Gelisols	Young soils with little profile development that occur in regions with permafrost. Low temperatures and frozen conditions for much of the year slow soil-forming processes.
Histosols	Organic soils with little or no climatic implications. Can be found in any climate where organic debris can accumulate to form a bog soil. Dark, partially decomposed organic material commonly referred to as *peat*.
Inceptisols	Weakly developed young soils in which the beginning (inception) of profile development is evident. Most common in humid climates, they exist from the Arctic to the tropics. Native vegetation is most often forest.
Mollisols	Dark, soft soils that have developed under grass vegetation, generally found in prairie areas. Humus-rich surface horizon that is rich in calcium and magnesium. Soil fertility is excellent. Also found in hardwood forests with significant earthworm activity. Climatic range is boreal or alpine to tropical. Dry seasons are normal. (see Figure 5.17A)
Oxisols	Soils that occur on old land surfaces unless parent materials were strongly weathered before they were deposited. Generally found in the tropics and subtropical regions. Rich in iron and aluminum oxides, oxisols are heavily leached, hence poor soils for agricultural activity. (see Figure 5.17B)
Spodosols	Soils found only in humid regions on sandy material. Common in northern coniferous forests (see Figure 5.16A) and cool humid forests. Beneath the dark upper horizon of weathered organic material lies a light-colored horizon of leached material, the distinctive property of this soil.
Ultisols	Soils that represent the products of long periods of weathering. Water percolating through the soil concentrates clay particles in the lower horizons (argillic horizons). Restricted to humid climates in the temperate regions and the tropics, where the growing season is long. Abundant water and a long frost-free period contribute to extensive leaching, hence poorer soil quality.
Vertisols	Soils containing large amounts of clay, which shrink upon drying and swell with the addition of water. Found in subhumid to arid climates, provided that adequate supplies of water are available to saturate the soil after periods of drought. Soil expansion and contraction exert stresses on human structures.

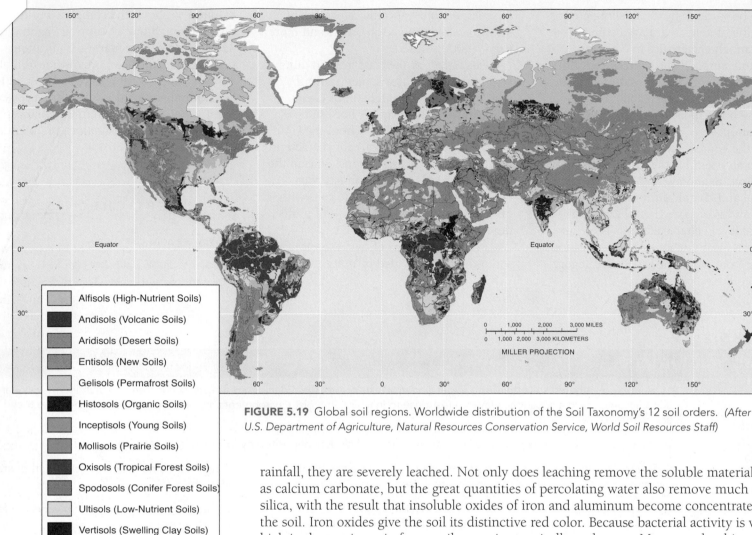

Legend:

- Alfisols (High-Nutrient Soils)
- Andisols (Volcanic Soils)
- Aridisols (Desert Soils)
- Entisols (New Soils)
- Gelisols (Permafrost Soils)
- Histosols (Organic Soils)
- Inceptisols (Young Soils)
- Mollisols (Prairie Soils)
- Oxisols (Tropical Forest Soils)
- Spodosols (Conifer Forest Soils)
- Ultisols (Low-Nutrient Soils)
- Vertisols (Swelling Clay Soils)
- Rock Land
- Shifting Sands
- Ice/Glacier

FIGURE 5.19 Global soil regions. Worldwide distribution of the Soil Taxonomy's 12 soil orders. *(After U.S. Department of Agriculture, Natural Resources Conservation Service, World Soil Resources Staff)*

Clearing the Tropical Rain Forest—A Case Study of Human Impact on Soil

Thick red soils are common in the wet tropics and subtropics. They are the end product of extreme chemical weathering. Because lush tropical rain forests are associated with these soils, we might assume they are fertile and have great potential for agriculture. However, just the opposite is true—they are among the poorest soils for farming. How can this be?

Because rain forest soils develop under conditions of high temperature and heavy rainfall, they are severely leached. Not only does leaching remove the soluble materials such as calcium carbonate, but the great quantities of percolating water also remove much of the silica, with the result that insoluble oxides of iron and aluminum become concentrated in the soil. Iron oxides give the soil its distinctive red color. Because bacterial activity is very high in the tropics, rain forest soils contain practically no humus. Moreover, leaching destroys fertility because most plant nutrients are removed by the large volume of downward-percolating water. Therefore, even though the vegetation may be dense and luxuriant, the soil itself contains few available nutrients.

Most nutrients that support the rain forest are locked up in the trees themselves. As vegetation dies and decomposes, the roots of the rain forest trees quickly absorb the nutrients before they are leached from the soil. The nutrients are continuously recycled as trees die and decompose.

Therefore, when forests are cleared to provide land for farming or to harvest the timber, most of the nutrients are removed as well (**FIGURE 5.20**). What remains is a soil that contains little to nourish planted crops.

The clearing of rain forests not only removes plant nutrients but also accelerates erosion. When vegetation is present, its roots anchor the soil, and its leaves and branches provide a canopy that protects the ground by deflecting the full force of the frequent heavy rains.

The removal of vegetation also exposes the ground to strong direct sunlight. When baked by the Sun, these tropical soils can harden to a bricklike consistency and become practically impenetrable to water and crop roots. In only a few years, soils in a freshly cleared area may no longer be cultivable.

The term *laterite*, which is often applied to these soils, is derived from the Latin word *latere*, meaning "brick," and was first applied to the use of this material for brick-making in India and Cambodia. Laborers simply excavated the soil, shaped it, and allowed it to harden in the Sun. Ancient but still well-preserved structures built of laterite remain standing today in the wet tropics (**FIGURE 5.21**). Such structures have withstood centuries

FIGURE 5.20 Clearing the Amazon rain forest in Surinam. The thick lateric soil is highly leached. *(Photo by Wesley Bocxe/Photo Researchers)*

Soil Erosion

Soils are just a tiny fraction of all Earth materials, yet they are a vital resource. Because soils are necessary for the growth of rooted plants, they are the very foundation of the human life-support system. Just as human ingenuity can increase the agricultural productivity of soils through fertilization and irrigation, soils can be damaged or destroyed by careless activities. Despite their basic role in providing food, fiber, and other basic materials, soils are among our most abused resources.

Perhaps this neglect and indifference has occurred because a substantial amount of soil seems to remain even where soil erosion is serious. Nevertheless, although the loss of fertile topsoil may not be obvious to the untrained eye, it is a growing problem as human activities expand and disturb more and more of Earth's surface.

of weathering because all of the original soluble materials were already removed from the soil by chemical weathering. Laterites are therefore virtually insoluble and very stable.

In summary, we have seen that some rain forest soils are highly leached products of extreme chemical weathering in the warm, wet tropics. Although they may be associated with lush tropical rain forests, these soils are unproductive when vegetation is removed. Moreover, when cleared of plants, these soils are subject to accelerated erosion and can be baked to bricklike hardness by the Sun.

CONCEPT CHECK 5.10

❶ Why are soils in tropical rainforests not well suited for farming?

❷ What are some detrimental effects of tropical deforestation?

FIGURE 5.21 This ancient temple at Angkor Wat, Cambodia was built in the 12th century of bricks made of laterite. *(Photo by Nestor Noci/Shutterstock)*

How Soil Is Eroded

Soil erosion is a natural process. It is part of the constant recycling of Earth materials that we call the *rock cycle*. Once soil forms, erosional forces, especially water and wind, move soil components from one place to another. Every time it rains, raindrops strike the land with surprising force (**FIGURE 5.22**). Each drop acts like a tiny bomb, blasting movable soil particles out of their positions in the soil mass. Then, water flowing across the surface carries away the dislodged soil particles. Because the soil is moved by thin sheets of water, this process is termed *sheet erosion*.

After a thin, unconfined sheet has flowed for a relatively short distance,

DID YOU KNOW?

It has been estimated that between 3 and 5 million acres of prime U.S. farmland are lost each year through mismanagement (including soil erosion) and conversion to nonagricultural uses. According to the United Nations, since 1950 more than one-third of the world's farmable land has been lost to soil erosion.

FIGURE 5.22 When it is raining, millions of water drops are falling at velocities approaching 10 meters per second (35 kilometers per hour). When water drops strike an exposed surface, soil particles may splash as high as 1 meter into the air and land more than a meter away from the point of raindrop impact. Soil dislodged by splash erosion is more easily moved by sheet erosion. *(Photo courtesy of U.S. Department of Agriculture)*

threads of current typically develop, and tiny channels called *rills* begin to form. Still deeper cuts in the soil, known as *gullies*, are created as rills enlarge (**FIGURE 5.23**). When normal farm cultivation cannot eliminate the channels, we know the rills have grown large enough to be called gullies. Although most dislodged soil particles move only a short distance during each rainfall, substantial quantities eventually leave the fields and make their way downslope to a stream. Once in the stream channel, these soil particles, which can now be called *sediment*, are transported downstream and eventually deposited elsewhere.

Rates of Erosion

We know that soil erosion is the ultimate fate of practically all soils. In the past, erosion occurred at slower rates than it does today because more of the land surface was covered and protected by trees, shrubs, grasses, and other plants. However, human activities such as farming, logging, and construction, which remove or disrupt the natural vegetation, have greatly accelerated the rate of soil erosion. Without the stabilizing effect of plants, the soil is more easily swept away by the wind or carried downslope by sheet wash.

Natural rates of soil erosion vary greatly from one place to another and depend on soil characteristics as well as

such factors as climate, topography, and type of vegetation. Over a broad area, erosion caused by surface runoff may be estimated by determining the sediment loads of the streams that drain the region. When studies of this kind were made on a global scale, they indicated that prior to the appearance of humans, sediment transport by rivers to the ocean amounted to just over 9 billion metric tons per year (1 metric ton = 1000 kilograms). By contrast, the amount of material currently transported to the sea by rivers is about 24 billion metric tons per year, or more than two and a half times the earlier rate.

A.

FIGURE 5.23 A. Soil erosion from this field in northeastern Wisconsin is obvious. *(Photo by D. P. Burnside/Photo Researchers, Inc.)* **B.** Gully erosion is severe in this poorly protected soil in southern Colombia. *(Photo by Carl Purcell/ Photo Researchers, Inc.)*

It is more difficult to measure the loss of soil due to wind erosion. However, the removal of soil by wind is generally much less significant than erosion by flowing water except during periods of prolonged drought. When dry conditions prevail, strong winds can remove large quantities of soil from unprotected fields (**FIGURE 5.24**). Such was the case in the 1930s. During that period, large dust storms plagued the Great Plains. Because of the size and severity of these storms, the region came to be called the Dust Bowl, and the time period the Dirty Thirties. The heart of the Dust Bowl was nearly 100 million acres in the panhandles of Texas and Oklahoma and adjacent parts of Colorado, New Mexico, and Kansas (**FIGURE 5.25**). At times, dust storms were so severe that they were called "black blizzards" and "black rollers" because visibility was sometimes reduced to only a few feet.

B.

What caused the Dust Bowl? Clearly the fact that portions of the Great Plains experience some of North America's strongest winds is important. However, it was the expansion of agriculture during an unusually wet period that set the stage for the disastrous period of soil erosion. Mechanization allowed the rapid transformation of the grass-covered prairies of this semiarid region into farms. As long as precipitation was adequate, the soil remained in place. However, when a prolonged drought struck in the 1930s, the unprotected soils were vulnerable to the wind. The result was severe soil loss, crop failure, and economic hardship.*

In many regions the rate of soil erosion is significantly greater than the rate of soil formation. This means that a renewable resource has become nonrenewable in these places. At present, it is estimated that topsoil is eroding faster than it forms on more than one-third of the world's croplands.

*For a readable and engaging account of this time you can read *The Worst Hard Times: The Untold Story of those Who Survived the Great American Dust Bowl*, by Timothy Egan, Houghton Mifflin Co., 2006.

FIGURE 5.25 A dust storm blackens the Colorado sky in this historic image from the Dust Bowl of the 1930s. *(Photo courtesy of the U.S.D.A./Natural Resources Conservation Service)*

The result is lower productivity, poorer crop quality, reduced agricultural income, and an ominous future.

Sedimentation and Chemical Pollution

Another problem related to excessive soil erosion involves the deposition of sediment. Each year in the United States hundreds of millions of tons of eroded soil are deposited in lakes, reservoirs, and streams. The detrimental impact of this process can be significant. For example, as more and more sediment is deposited in a reservoir, the capacity of the reservoir is reduced, limiting its usefulness for flood control, water supply, and/or hydroelectric power generation. In addition, sedimentation in streams and other waterways can restrict navigation and lead to the need for costly dredging operations.

In some cases soil particles are contaminated with pesticides used in farming. When these chemicals are introduced into a lake or reservoir, the quality of the water supply is threatened, and aquatic organisms may be endangered. In addition to pesticides, nutrients found naturally in soils as well as those added by agricultural fertilizers make their way into streams and lakes, where they stimulate the growth of plants. Over a period of time, excessive nutrients accelerate the process by which plant growth leads to the depletion of oxygen and an early death of the lake.

The availability of good soils is critical if the world's rapidly growing population is to be fed. On every continent, unnecessary

A.

B.

FIGURE 5.26 These images illustrate some of the ways that soil erosion can be reduced. **A.** In this scene from a farm in northeastern Iowa, corn (tan) and hay (green) have been planted in strips that follow the contours of the hillsides. This pattern of crop planting can reduce soil loss due to water erosion by significantly slowing the rate at which water runs off. **B.** Windbreaks protecting wheat fields in North Dakota. These flat expanses are susceptible to wind erosion especially when the fields are bare. The rows of trees slow the wind and deflect it upward, which decreases the loss of fine soil particles. *(Photos courtesy of U.S.D.A./Natural Resources Conservation Service)*

soil loss is occurring because appropriate conservation measures are not being used. Although it is a recognized fact that soil erosion can never be completely eliminated, soil conservation programs can substantially reduce the loss of this basic resource (**FIGURE 5.26**). Windbreaks (rows of trees), terracing, and plowing along the contours of hills are effective measures, as are special tillage practices and crop rotation.

CONCEPT CHECK 5.11

❶ How have human activities affected the rate of soil erosion?

❷ What are two detrimental effects of soil erosion aside from the loss of topsoil?

Weathering and Ore Deposits

Weathering creates many important mineral deposits by concentrating minor amounts of metals that are scattered through unweathered rock into economically valuable concentrations. Such a transformation is often termed **secondary enrichment** and takes place in one of two ways. In one situation, chemical weathering coupled with downward-percolating water removes undesired materials from decomposing rock, leaving the desired elements enriched in the upper zones of the soil. The second way is basically the reverse of the first. That is, the desirable elements that are found in low concentrations near the surface are removed and carried to lower zones, where they are redeposited and become more concentrated.

Bauxite

The formation of *bauxite*, the principal ore of aluminum, is one important example of an ore created as a result of enrichment by weathering processes (**FIGURE 5.27**). Although aluminum is the third most abundant element in Earth's crust, economically valuable concentrations of this important metal are not common because most aluminum is tied up in silicate minerals, from which it is extremely difficult to extract.

FIGURE 5.27 Bauxite is the principal ore of aluminum and forms as a result of weathering processes under tropical conditions. Its color varies from red or brown to nearly white. *(Photo by E. J. Tarbuck)*

Bauxite forms in rainy tropical climates. When aluminum-rich source rocks are subjected to the intense and prolonged chemical weathering of the tropics, most of the common elements, including calcium, sodium, and potassium, are removed by leaching. Because aluminum is extremely insoluble, it becomes concentrated in the soil (as bauxite, a hydrated aluminum oxide). Thus, the formation of bauxite depends on climatic conditions in which chemical weathering and leaching are pronounced, plus, of course, the presence of aluminum-rich source rock. In a similar manner, important deposits of nickel and cobalt develop from igneous rocks rich in silicate minerals such as olivine.

There is significant concern regarding the mining of bauxite and other residual deposits because they tend to occur in environmentally sensitive areas of the tropics. Mining is preceded by the removal of tropical vegetation, thus destroying rainforest ecosystems. Moreover, the thin moisture-retaining layer of organic matter is also disturbed. When the soil dries out in the hot sun, as has been mentioned, it becomes bricklike and loses its moisture-retaining qualities. Such soil cannot be productively farmed nor can it support significant forest growth. The long-term consequences of bauxite mining are clearly of concern for developing countries in the tropics, where this important ore is mined.

Other Deposits

Many copper and silver deposits result when weathering processes concentrate metals that are dispersed through a low-grade primary ore. Usually such enrichment occurs in deposits containing pyrite (FeS_2), the most common and widespread sulfide mineral. Pyrite is important because when it chemically weathers, sulfuric acid forms, which enables percolating waters to dissolve the ore metals. Once dissolved, the metals gradually migrate downward through the primary ore body until they are precipitated. Deposition takes place because of changes that occur in the chemistry of the solution when it reaches the groundwater zone (the zone beneath the surface where all pore spaces are filled with water). In this manner, the small percentage of dispersed metal can be removed from a large volume of rock and redeposited as a higher-grade ore in a smaller volume of rock.

CONCEPT CHECK 5.12

❶ How can weathering create an ore deposit? What important ore is an example?

CHAPTER FIVE
Weathering and Soils in Review

⊙ External processes include (1) *weathering*—the disintegration and decomposition of rock at or near Earth's surface; (2) *mass wasting*—the transfer of rock material downslope under the influence of gravity; and (3) *erosion*—the removal of material by a mobile agent, usually water, wind, or ice. They are called *external processes* because they occur at or near Earth's surface and are powered by energy from the Sun. By contrast, *internal processes*, such as volcanism and mountain building, derive their energy from Earth's interior.

⊙ *Mechanical weathering* is the physical breaking up of rock into smaller pieces. Rocks can be broken into smaller fragments by *frost wedging* (where water works its way into cracks or voids in rock and upon freezing expands and enlarges the openings), *salt crystal growth, unloading* (expansion and breaking due to a great reduction in pressure when the overlying rock is eroded away), and *biological activity* (by humans, burrowing animals, plant roots, etc.).

⊙ *Chemical weathering* alters a rock's chemistry, changing it into different substances. Water is by far the most important agent of chemical weathering. Oxygen dissolved in water will *oxidize* iron-rich minerals, while carbon dioxide (CO_2) dissolved in water forms carbonic acid, which attacks and alters rock. The chemical weathering of silicate minerals frequently produces (1) soluble products containing sodium, calcium, potassium, and magnesium ions, as well as silica in solution; (2) insoluble iron oxides, including limonite and hematite; and (3) clay minerals.

⊙ The rate at which rock weathers depends on such factors as (1) *particle size*—small pieces generally weather faster than large pieces; (2) *mineral make-up*—calcite readily dissolves in mildly acidic solutions, and silicate minerals that form first from magma are least resistant to chemical weathering; and (3) *climatic factors*, particularly temperature and moisture. Frequently, rocks exposed at Earth's surface do not weather at the same rate. This *differential weathering* of rocks is influenced by such factors as mineral make-up and degree of jointing.

⊙ *Soil*—that portion of the *regolith* (the layer of rock and mineral fragments produced by weathering) that supports the growth of plants—is a combination of mineral and organic matter, water, and air. About half of the total volume of a good-quality soil is a mixture of disintegrated and decomposed rock (mineral matter) and *humus* (the decayed remains of animal and plant life); the remaining half consists of pore spaces, where air and water circulate. The most important factors that control soil formation are *parent material, time, climate, plants and animals,* and *topography*.

⊙ Soil-forming processes operate from the surface downward and produce zones or layers in the soil called *horizons*. From the surface downward, the soil horizons are respectively designated as *O* (largely organic matter), *A* (largely mineral matter), *E* (where the fine soil components and soluble materials have been removed by eluviation and leaching), *B* (*subsoil*, often referred to as the *zone of accumulation*), and *C* (partially altered parent material). Together the *O* and *A* horizons make up what is commonly called the *topsoil*.

⊙ In the United States, soils are classified using a system known as the *Soil Taxonomy*. It is based on physical and chemical properties of the soil profile and includes six hierarchical categories. The system is especially useful for agricultural and related land-use purposes.

⊙ Soil erosion is a natural process. It is part of the constant recycling of Earth materials that we call the *rock cycle*. Once in a stream channel, soil particles, which can now be called *sediment*, are transported downstream and eventually deposited. *Rates of soil erosion* vary from one place to another and depend on the soil's characteristics as well as on such factors as climate, slope, and type of vegetation. Human activities have greatly accelerated the rate of soil erosion in many areas.

⊙ Weathering creates ore deposits by concentrating minor amounts of metals into economically valuable deposits. The process, often called *secondary enrichment*, is accomplished by either (1) removing undesirable materials and leaving the desired elements enriched in the upper zones of the soil or (2) removing and carrying the desirable elements to lower soil zones where they are redeposited and become more concentrated. *Bauxite*, the principal ore of aluminum, is one important ore created as a result of enrichment by weathering processes. In addition, many copper and silver deposits result when weathering processes concentrate metals that were formerly dispersed through low-grade primary ore.

Key Terms

chemical weathering (p. 125)
differential weathering (p. 132)
eluviation (p. 137)
erosion (p. 124)
exfoliation domes (p. 127)
external processes (p. 124)

frost wedging (p. 126)
horizons (p. 137)
internal processes (p. 124)
leaching (p. 137)
mass wasting (p. 124)
mechanical weathering (p. 125)

parent material (p. 133)
regolith (p. 133)
secondary enrichment (p. 144)
sheeting (p. 127)
soil (p. 133)
soil profile (p. 137)

Soil Taxonomy (p. 139)
solum (p. 138)
spheroidal weathering (p. 131)
weathering (p. 124)

GIVE IT SOME THOUGHT

❶ Imagine you are a geologist who was hired to study the geology of a small mountainous island in the tropics. When you arrive at the shore of the island to begin your study, you find that transportation to the island's interior will not be available until the next day. Rather than lose a day of investigation, how might you begin your study of the island's geology having access only to the beach?

❷ Describe how plants promote mechanical and chemical weathering but inhibit erosion.

❸ Granite and basalt are exposed at Earth's surface in a hot, wet region. Will mechanical weathering or chemical weathering predominate? Which rock will weather most rapidly? Why?

❹ The level of carbon dioxide (CO_2) in the atmosphere has been increasing for more than a century. Do you think this increase tends to accelerate or slow down the rate of chemical weathering of Earth's surface rocks? Explain your conclusion.

❺ When you studied Chapter 3, you learned that feldspars are very common minerals in igneous rocks. When you learn about the common minerals that compose sedimentary rocks in Chapter 6, you will find that feldspars are relatively rare. Applying what you have learned about chemical weathering, explain why this is true. Based on this explanation, what mineral might you expect to be common in sedimentary rocks that is not found in igneous rocks?

❻ If a historic monument or building were particularly valuable to your community, how would you determine how rapidly the structure might deteriorate and whether it needed protection?

❼ The accompanying photo shows a footprint on the Moon left by an Apollo astronaut in material popularly called *lunar soil*. Does this material meet the definition we use for soil on Earth? Explain why or why not. You may want to refer to Figure 5.14.

❽ Using the map of global soil regions in Figure 5.19, identify the main soil order in the region adjacent to South America's Amazon River and the predominant soil order in the American Southwest. Briefly contrast these soils. Do they have anything in common? Referring to Table 5.2, Figure 5.16B, and Figure 5.20 might be helpful.

❾ Explain this apparent contradiction: Lush tropical rainforests typically grow in infertile soils.

Companion Website   PEARSON mygeoscience™ place

www.mygeoscienceplace.com

The *Essentials of Geology, 11e* companion Website contains numerous multimedia resources accompanied by assessments to aid in your study of the topics in this chapter. The use of this site's learning tools will help improve your understanding of geology. Utilizing the access code that accompanies this text, visit **www.mygeoscienceplace.com** in order to:

- **Review** key chapter concepts.
- **Read** with links to the Pearson eText and to chapter-specific web resources.
- **Visualize** and comprehend challenging topics using the learning activities in *GEODe: Essentials of Geology* and the *Geoscience Animations Library.*
- **Test** yourself with online quizzes.

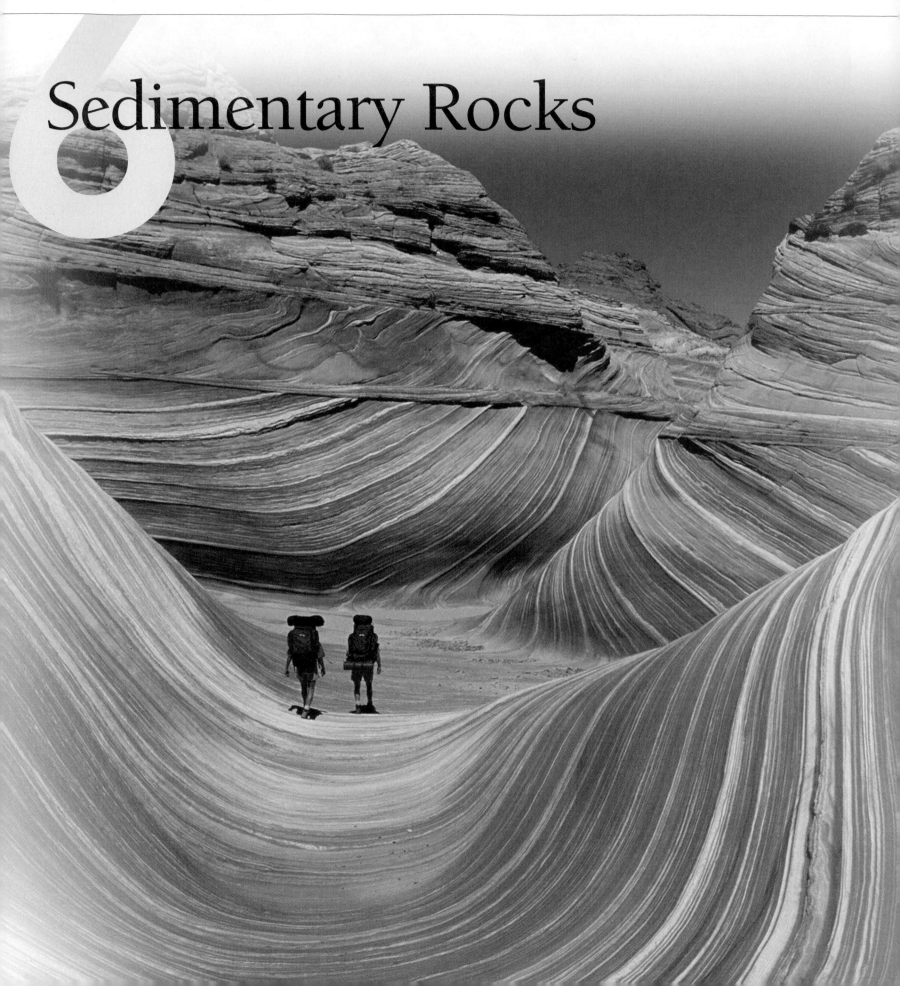

6 Sedimentary Rocks

THE PRECEDING CHAPTER PROVIDED THE BACKGROUND YOU needed to understand the origin of sedimentary rocks.

Recall that weathering of existing rocks begins the process. Next, gravity and erosional agents remove the products of weathering and carry them to a new location where they are deposited. Usually the particles are broken down further during this transport phase. Following deposition, this material, which is now called sediment, becomes lithified (turned to rock).

Sandstone in Arizona's Vermillion Cliffs Wilderness Area. *(Photo by James Kay/DanitaDelimont.com)*

To assist you in learning the important concepts in this chapter, focus on the following questions:

- What is a sedimentary rock? About what percentage of Earth's crust is sedimentary?
- How is sediment turned into sedimentary rock?
- What are the two general categories of sedimentary rocks, and how does each type form?
- What is the primary basis for distinguishing among various types of detrital sedimentary rocks? What criterion is used to subdivide the chemical rocks?
- What are sedimentary structures? Why are these structures useful to geologists?
- What are the two broad groups of nonmetallic mineral resources?
- Which energy resources are associated with sedimentary rocks?

The Importance of Sedimentary Rocks

Most of solid Earth consists of igneous and metamorphic rocks. Geologists estimate these two categories represent 90 to 95 percent of the outer 16 kilometers (10 miles) of the crust. Nevertheless, most of Earth's solid surface consists of either sediment or sedimentary rock! Across the ocean floor, which represents about 70 percent of Earth's solid surface, virtually everything is covered by sediment. Igneous rocks are exposed only at the crest of mid-ocean ridges and at some volcanic areas. Thus, while sediment and sedimentary rocks make up only a small percentage of Earth's crust, they are concentrated at or near the surface—the interface among the geosphere, hydrosphere, atmosphere, and biosphere. Because of this unique position, sediments and the rock layers that they eventually form contain evidence of past conditions and events at the surface (**FIGURE 6.1**). Furthermore, it is sedimentary rocks that contain fossils, which are vital tools in the study of the geologic past. This group of rocks provides geologists with much of the basic information they need to reconstruct the details of Earth history.

Such study is not only of interest for its own sake but has practical value as well. Coal, which provides a significant portion of our electrical energy, is classified as a sedimentary rock. Moreover, other major energy sources—oil, natural gas, and uranium—are derived from sedimentary rocks. So are major sources of iron, aluminum, manganese, and phosphate fertilizer, plus numerous materials essential to the construction industry, such as cement and aggregate. Sediments and sedimentary rocks are also the primary reservoir of groundwater. Thus, an understanding of this group of rocks and the processes that form and modify them is basic to locating additional supplies of many important resources.

FIGURE 6.1 Sedimentary rocks are exposed at Earth's surface more than igneous and metamorphic rocks. Because they contain fossils and other clues about the geologic past, sedimentary rocks are important in the study of Earth history. *(Photo by Scott T. Smith/ DanitaDelimont.com)*

CONCEPT CHECK 6.1

❶ How does the volume of sedimentary rocks in Earth's crust compare to igneous and metamorphic rocks?

❷ Why are sedimentary rocks important?

Origins of Sedimentary Rock

Sedimentary Rocks

ESSENTIALS OF GEOLOGY Introduction

FIGURE 6.2 illustrates the portion of the rock cycle that occurs near Earth's surface—the part that pertains to sediments and sedimentary rocks. A brief overview of these processes provides a useful perspective.

• Weathering begins the process. It involves the physical disintegration and chemical decomposition of preexisting igneous, metamorphic, and sedimentary rocks. Weathering generates a variety of products, including various solid particles and ions in

DID YOU KNOW?
Coal is responsible for slightly more than half the electricity generated in the United States.

solution. These are the raw materials for sedimentary rocks.

• Soluble constituents are carried away by runoff and groundwater. Solid particles are frequently moved downslope by gravity, a process termed mass wasting, before running water, groundwater, wind, and glacial ice remove them. Transportation moves these materials from the sites where they originated to locations where they accumulate. The transport of sediment is usually intermittent. For example, during a flood, a rapidly moving river moves large quantities of sand and gravel. As the flood waters recede, particles are temporarily deposited, only to be moved again by a subsequent flood.

• Deposition of solid particles occurs when wind and water currents slow down and as glacial ice melts. The word *sedimentary* actually refers to this process. It is derived from the Latin *sedimentum*,

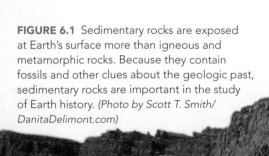

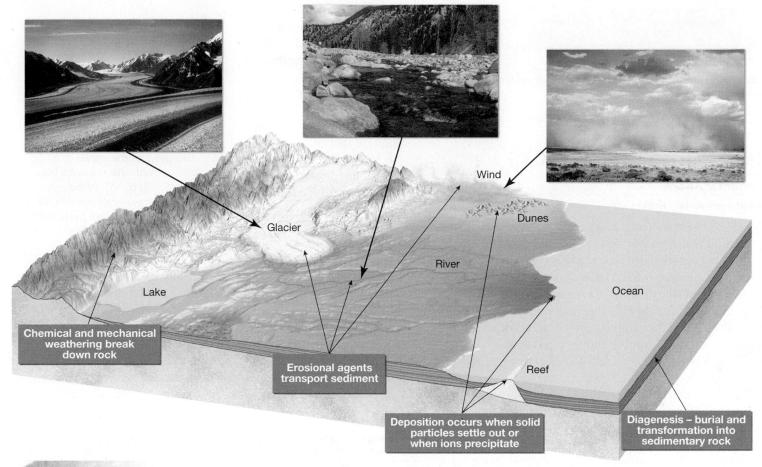

FIGURE 6.2 This diagram outlines the portion of the rock cycle that pertains to the formation of sedimentary rocks. Weathering, transportation, deposition, and diagenesis represent the basic processes involved. *(Left and center photos by E. J. Tarbuck; right photo by Jenny Elia Pfeiffer/Corbis)*

which means "to settle," a reference to solid material settling out of a fluid (water or air). The mud on the floor of a lake, a delta at the mouth of a river, a gravel bar in a stream bed, the particles in a desert sand dune, and even household dust are examples.

• The deposition of material dissolved in water is not related to the strength of wind or water currents. Rather, ions in solution are removed when chemical or temperature changes cause material to crystallize and precipitate or when organisms remove dissolved material to build shells.

• As deposition continues, older sediments are buried beneath younger layers and gradually converted to sedimentary rock (lithified) by compaction and cementation. This and other changes are referred to as *diagenesis* (*dia* = change ; *genesis* = origin), a collective term for all of the changes (short of metamorphism) that take place in texture, composition, and other physical properties after sediments are deposited.

Because there are a variety of ways that the products of weathering are transported, deposited, and transformed into solid rock, three categories of sedimentary rocks are recognized. As the overview reminded us, sediment has two principal sources. First, it may be an accumulation of material that

originates and is transported as solid particles derived from both mechanical and chemical weathering. Deposits of this type are termed *detrital*, and the sedimentary rocks that they form are called **detrital sedimentary rocks**.

The second major source of sediment is soluble material produced largely by chemical weathering. When these ions in solution are precipitated by either inorganic or biologic processes, the material is known as *chemical sediment*, and the rocks formed from it are called **chemical sedimentary rocks**.

The third category is **organic sedimentary rocks**. The primary example is coal. This black combustible rock consists of organic carbon from the remains of plants that died and accumulated on the floor of a swamp. The bits and pieces of undecayed plant material that constitute the "sediments" in coal are quite unlike the weathering products that make up detrital and chemical sedimentary rocks.

CONCEPT CHECK 6.2

❶ Outline the steps that would transform an exposure of granite in the mountains into various sedimentary rocks.

❷ List and briefly distinguish among the three basic sedimentary rock categories.

Detrital Sedimentary Rocks

Sedimentary Rocks

Types of Sedimentary Rocks

Though a wide variety of minerals and rock fragments (clasts) may be found in detrital rocks, clay minerals and quartz are the chief constituents of most sedimentary rocks in this category. Recall from Chapter 5 that clay minerals are the most abundant product of the chemical weathering of silicate minerals, especially the feldspars. Clays are fine-grained minerals with sheetlike crystalline structures similar to the micas. The other common mineral, quartz, is abundant because it is extremely durable and very resistant to chemical weathering. Thus, when igneous rocks such as granite are attacked by weathering processes, individual quartz grains are freed.

Other common minerals in detrital rocks are feldspars and micas. Because chemical weathering rapidly transforms these minerals into new substances, their presence in sedimentary rocks indicates that erosion and deposition were fast enough to preserve some of the primary minerals from the source rock before they could be decomposed.

Particle size is the primary basis for distinguishing among various detrital sedimentary rocks. **FIGURE 6.3** presents the size categories for particles making up detrital rocks. Particle size is not only a convenient method of dividing detrital rocks, but the sizes of the component grains also provide useful information about environments of deposition. Currents of water or air sort the particles by size—the stronger the current, the larger the particle size carried. Gravels, for example, are moved by swiftly flowing rivers as well as by landslides and glaciers. Less energy is required to transport sand; thus, it is common to such features as windblown dunes and some river deposits and beaches. Very little energy is needed to transport clay, so it settles very slowly. Accumulations of these tiny particles are generally associated with the quiet waters of a lake, lagoon, swamp, or certain marine environments.

Common detrital sedimentary rocks, in order of increasing particle size, are shale, sandstone, and conglomerate or breccia. We will now look at each type and how it forms.

Shale

Shale is a sedimentary rock consisting of silt- and clay-size particles (see Figure 6.3). These fine-grained detrital rocks account for well over half of all sedimentary rocks. The particles in these rocks are so small that they cannot be readily identified without great

magnification and for this reason make shale more difficult to study and analyze than most other sedimentary rocks.

Much of what can be learned is based on particle size. The tiny grains in shale indicate that deposition occurs as the result of gradual settling from relatively quiet, nonturbulent currents. Such environments include lakes, river floodplains, lagoons, and portions of the deep-ocean basins. Even in these quiet environments, there is usually enough turbulence to keep clay-size particles suspended almost indefinitely. Consequently, much of the clay is deposited only after the individual particles coalesce to form larger aggregates.

Sometimes the chemical composition of the rock provides additional information. One example is black shale, which is black because it contains abundant organic matter (carbon). When such a rock is found, it strongly implies that deposition occurred in an

DID YOU KNOW?
The term *clay* may be confusing because it has more than one meaning. In the context of detrital particle size, the term *clay* refers to those grains less than 1/256 millimeter, thus being microscopic (see Figure 6.3). However, the term *clay* is also used as the name of a group of silicate minerals (see Chapter 2, p. 54). Although most clay minerals are of clay size, not all clay-size sediment consists of clay minerals!

FIGURE 6.3 Particle size classification for detrital sedimentary rocks. Particle size is the primary basis for distinguishing among various detrital sedimentary rocks. *(Photos by E. J. Tarbuck)*

Size Range (millimeters)	Particle Name	Common Name	Detrital Rock
>256	Boulder	Gravel	Conglomerate 
64–256	Cobble		
4–64	Pebble		or
2–4	Granule		Breccia 
1/16–2	Sand	Sand	Sandstone 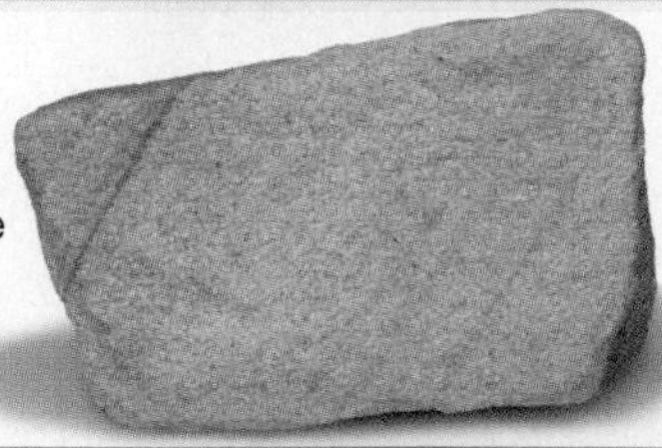
1/256–1/16	Silt	Mud	Shale, Mudstone or Siltstone 
<1/256	Clay		

0 10 20 30 40 50 60 70 mm

oxygen-poor environment such as a swamp, where organic materials do not readily oxidize and decay.

As silt and clay accumulate, they tend to form thin layers commonly referred to as *laminae*. Initially the particles in the laminae are oriented randomly. This disordered arrangement leaves a high percentage of open space (called *pore space*) that is filled with water. However, this situation usually changes with time as additional layers of sediment pile up and compact the sediment below.

During this phase the clay and silt particles take on a more parallel alignment and become tightly packed. This rearrangement of grains reduces the size of the pore spaces and forces out much of the water. Once the grains are pressed closely together, the tiny spaces between particles do not readily permit solutions containing cementing material to circulate. Therefore, shales are often described as being weak because they are poorly cemented and therefore not well lithified. The inability of water to penetrate its microscopic pore spaces explains why shale often forms barriers to the subsurface movement of water and petroleum. Indeed, rock layers that contain groundwater are commonly underlain by shale beds that block further downward movement.[*] The opposite is true for underground reservoirs of petroleum. They are often capped by shale beds that effectively prevent oil and gas from escaping to the surface.

*The relationship between impermeable beds and the occurrence and movement of groundwater is examined in Chapter 10.

It is common to apply the term *shale* to all fine-grained sedimentary rocks, especially in a nontechnical context. However, be aware that there is a more restricted use of the term. In this narrower usage, shale must exhibit the ability to split into thin layers along well-developed, closely spaced planes. This property is termed *fissility*. If the rock breaks into chunks or blocks, the name *mudstone* is applied. Another fine-grained sedimentary rock that, like mudstone, is often grouped with shale but lacks fissility is *siltstone*. As its name implies, siltstone is composed largely of silt-size particles and contains less clay-size material than shale and mudstone.

Although shale is far more common than other sedimentary rocks it does not usually attract as much notice as other, less abundant members of this group. The reason is that shale does not form prominent outcrops as sandstone and limestone often do. Rather, shale crumbles easily and usually forms a cover of soil that hides the unweathered rock below. This is illustrated nicely in the Grand Canyon, where the gentler slopes of weathered shale are quite inconspicuous and overgrown with vegetation, in sharp contrast to the bold cliffs produced by more durable rocks (**FIGURE 6.4**).

Although shal striking cliffs and some deposits hav Certain shales are material for potter Moreover, when m shale is used to ma In the future, one type of shale, called oil shale, may become a valuable energy resource.

Sandstone

Sandstone is the name given rocks in which sand-sized grains predominate (see Figure 6.3). After shale, sandstone is the most abundant sedimentary rock, accounting for approximately 20 percent of the entire group. Sandstones form in a variety of environments and often contain significant clues about their origin, including sorting, particle shape, and composition.

SORTING AND PARTICLE SHAPE.
Sorting is the degree of similarity in particle size in a sedimentary rock. For example, if all the grains in a sample of sandstone are about the same size, the sand is considered *well sorted*. Conversely, if the rock contains mixed large and small particles, the sand is said to be *poorly sorted* (**FIGURE 6.5A**). By studying the degree

FIGURE 6.4 Sedimentary rock layers exposed in the walls of the Grand Canyon, Arizona. Beds of resistant sandstone and limestone produce bold cliffs. By contrast, weaker, poorly cemented shale crumbles and produces a gentler slope of weathered debris in which some vegetation is growing. *(Photo by LH Images/ Alamy)*

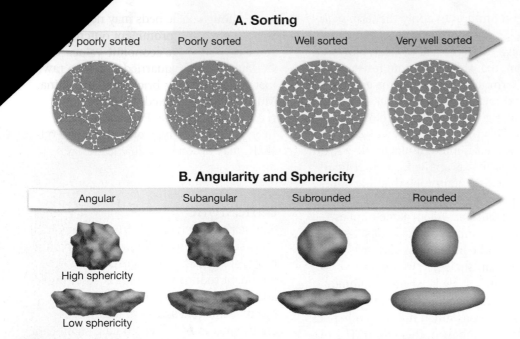

A. Sorting

poorly sorted   Poorly sorted   Well sorted   Very well sorted

B. Angularity and Sphericity

Angular   Subangular   Subrounded   Rounded

High sphericity

Low sphericity

FIGURE 6.5 A. Detrital rocks commonly have a variety of different size clasts. *Sorting* refers to the range of sizes present. Rocks with clasts that are nearly all the same size are considered "well sorted." When sediments are "very poorly sorted," there is a wide range of different sizes. When a rock contains larger clasts surrounded by much smaller ones, the mass of smaller clasts is often referred to as the *matrix*. **B.** Geologists describe a particle's shape in terms of its *angularity* (degree to which the clast's edges and corners are rounded) and *sphericity* (how close the shape of the clast is to a sphere). Transportation reduces the size and angularity of clasts but does not change their general shape.

and corners and become more rounded as they collide with other particles during transport. Thus, rounded grains likely have been airborne or water-borne. Further, the degree of rounding indicates the distance or time involved in the transportation of sediment by currents of air or water. Highly rounded grains indicate that a great deal of abrasion and hence a great deal of transport has occurred.

Very angular grains, on the other hand, imply two things: that the materials were transported only a short distance before they were deposited and that some other medium may have transported them. For example, when glaciers move sediment, the particles are usually made more irregular by the crushing and grinding action of the ice.

In addition to affecting the degree of rounding and the amount of sorting that particles undergo, the length of transport by turbulent air and water currents also influences the mineral composition of a sedimentary deposit. Substantial weathering and long transport lead to the gradual destruction of weaker and less stable minerals, including the feldspars and ferromagnesians. Because quartz is very durable, it is usually the mineral that survives the long trip in a turbulent environment.

The preceding discussion has shown that the origin and history of a sandstone can often be deduced by examining the sorting, roundness, and mineral composition of its constituent grains. Knowing this information allows us to infer that a well-sorted, quartz-rich sandstone consisting of highly rounded grains must be the result of a great deal of transport. Such a rock, in fact, may represent several cycles of weathering, transport, and deposition. We may also conclude that a sandstone containing a significant amount of feldspar and angular grains of ferromagnesian minerals underwent little chemical weathering and transport and was probably deposited close to the source area of the particles.

of sorting, we can learn much about the depositing current. Deposits of windblown sand are usually better sorted than deposits sorted by wave activity. (**FIGURE 6.6**). Particles washed by waves are commonly better sorted than materials deposited by streams. Sediment accumulations that exhibit poor sorting usually result when particles are transported for only a relatively short time and then rapidly deposited. For example, when a turbulent stream reaches the gentler slopes at the base of a steep mountain, its velocity is quickly reduced, and poorly sorted sands and gravels are deposited.

The shapes of sand grains can also help decipher the history of a sandstone (Figure 6.5B). When streams, winds, or waves move sand and other larger sedimentary particles, the grains lose their sharp edges

COMPOSITION. Owing to its durability, quartz is the predominant mineral in most sandstones. When this is the case, the rock may simply be called *quartz sandstone*. When a sandstone contains appreciable quantities of feldspar (25 percent or more), the rock is called *arkose*. In addition to feldspar, arkose usually contains quartz and sparkling bits of mica. The mineral composition of arkose indicates that the grains were derived from granitic source rocks. The particles are generally poorly sorted and angular, which suggests short-distance transport, minimal chemical weathering in a relatively dry climate, and rapid deposition and burial.

A third variety of sandstone is known as *graywacke*. Along with quartz and feldspar, this dark-colored rock contains abundant rock fragments and matrix. *Matrix* refers to the silt- and clay-size particles found in spaces between larger sand grains. More than 15 percent of graywacke's volume is matrix. The poor sorting and angular grains characteristic of graywacke suggest that the particles were transported only a relatively short distance from their source area and then rapidly deposited. Before the sediment could be reworked and sorted further, it was most likely buried by additional layers of material. Graywacke is frequently associated with submarine deposits made by dense sediment-choked torrents called turbidity currents.

DID YOU KNOW?
The Navajo Sandstone pictured in Figure 6.6A represents a vast area of sand dunes that once covered up to 400,000 sq km (156,000 sq mi), an area the size of California.

DID YOU KNOW?
The most important and common material used for making glass is silica, which is usually obtained from the quartz in "clean," well-sorted sandstones.

FIGURE 6.6 A. The orange and yellow cliffs of Utah's Zion National Park expose thousands of feet of Jurassic-age Navajo Sandstone. *(Photo by Dennis Tasa)* **B.** The quartz grains composing the Navajo Sandstone were deposited by wind as dunes similar to these in Colorado's Great Sand Dunes National Park. These sand grains are considered to be well-sorted because they are all practically the same size. *(Photo by Ric Ergenbright / DanitaDelimont.com)*

Conglomerate and Breccia

Conglomerate consists largely of gravels. As Figure 6.3 indicates, these particles may range in size from large boulders to particles as small as garden peas (**FIGURE 6.7**). The particles are commonly large enough to be identified as distinctive rock types; thus, they can be valuable in identifying the source areas of sediments. More often than not, conglomerates are poorly sorted because the openings between the large gravel particles contain sand or mud.

Gravels accumulate in a variety of environments and usually indicate the existence of steep slopes or very turbulent currents. The coarse particles in a conglomerate may reflect the action of energetic mountain streams or result from strong wave activity along a rapidly eroding coast. Some glacial and landslide deposits also contain plentiful gravel.

If the large particles are angular rather than rounded, the rock is called *breccia* (Figure 6.3). Because large particles abrade and become rounded very rapidly during transport, the pebbles and cobbles in a breccia indicate that they did not travel far from their source area before they were deposited. Thus, as with many other sedimentary rocks, conglomerates and breccias contain clues to their history. Their particle sizes reveal the strength of the currents that transported them, whereas the degree of rounding indicates how far the particles traveled. The fragments within a sample identify the source rocks that supplied them.

CONCEPT CHECK 6.3

❶ What minerals are most abundant in detrital sedimentary rocks? In which rocks do these sediments predominate?

❷ What is the primary basis for distinguishing among detrital rocks?

❸ Describe how sediments become sorted.

❹ Distinguish between conglomerate and breccia.

FIGURE 6.7 Gravel deposits along Carbon Creek in Grand Canyon National Park, Arizona. If these poorly sorted sediments were lithified, the rock would be conglomerate. *(Photo by Michael Collier)*

Chemical Sedimentary Rocks

Sedimentary Rocks

ESSENTIALS OF GEOLOGY Types of Sedimentary Rocks

In contrast to detrital rocks, which form from the solid products of weathering, chemical sediments derive from ions that are carried *in solution* to lakes and seas. This material does not remain dissolved in the water indefinitely, however. Some of it precipitates to form chemical sediments. These become rocks such as limestone, chert, and rock salt.

This precipitation of material occurs in two ways. *Inorganic* processes such as evaporation and chemical activity can produce chemical sediments. *Organic* (life) processes of water-dwelling organisms also form chemical sediments, said to be of **biochemical origin**.

One example of a deposit resulting from inorganic chemical processes is the dripstone that decorates many caves (**FIGURE 6.8**). Another is the salt left behind as a body of seawater evaporates. In contrast, many water-dwelling animals and plants extract dissolved mineral matter to form shells and other hard parts. After the organisms die, their skeletons collect by the millions on the floor of a lake or ocean as biochemical sediment (**FIGURE 6.9**).

Layers of sedimentary rock exposed in southern Utah. (Photo by Scott T. Smith/DanitaDelimont.com)

Limestone

Representing about 10 percent of the total volume of all sedimentary rocks, *limestone* is the most abundant chemical sedimentary rock. It is composed chiefly of the mineral calcite ($CaCO_3$) and forms either by inorganic means or as the result of biochemical processes. Regardless of its origin, the mineral composition of all limestone is similar, yet many different types exist. This is true because limestones are produced under a variety of conditions. Those forms having a marine biochemical origin are by far the most common.

CARBONATE REEFS. Corals are one important example of organisms that are capable of creating large quantities of marine limestone. These relatively simple invertebrate animals secrete a calcareous (calcium carbonate) external skeleton. Although they are small, corals are capable of creating massive structures called *reefs*. Reefs consist of coral colonies made up of great numbers of individuals that live side by side on a calcite structure secreted by the animals. In addition, calcium carbonate-secreting algae live with the corals and help cement the entire structure into a solid mass. A wide variety of other organisms also live in and near reefs.

Certainly the best-known modern reef is Australia's Great Barrier Reef, 2000 kilometers (1240 miles) long, but many lesser reefs also exist (**FIGURE 6.10A**). They develop in the shallow, warm waters of the tropics and subtropics equatorward of about 30° latitude. Striking examples exist in the Bahamas and Florida Keys.

Modern corals were not the first reef builders. Earth's first reef-building organisms were photosynthesizing bacteria living during

DID YOU KNOW?

In the western and southwestern United States, sedimentary rocks often exhibit a brilliant array of colors. (For example, see Chapter 5 opening photo, Figure 5.1, Figure 5.13, Figure 6.1, and Figure 6.4). In the walls of Arizona's Grand Canyon we can see layers that are red, orange, purple, gray, brown, and buff. Some of the sedimentary rocks in Utah's Bryce Canyon are a delicate pink color. Sedimentary rocks in more humid places are also colorful, but they are usually covered by soil and vegetation.

The most important "pigments" are iron oxides, and only very small amounts are needed to color a rock. Hematite tints rocks red or pink, whereas limonite produces shades of yellow and brown. When sedimentary rocks contain organic matter, it often colors them black or gray (see Figure 6.3).

strata

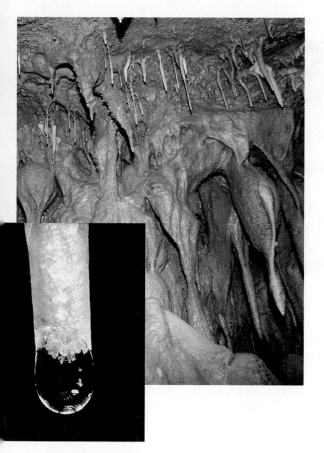

FIGURE 6.9 **A.** This Florida beach consists primarily of shells and shell fragments—biochemical sediment. *(Photo by Donald R. Frazier Photolibrary, Inc./Plany)* **B.** This rock, a variety of limestone called coquina, consists of shell fragments. *(Photo by E.J. Tarbuck)*

A.

B.

Close up

FIGURE 6.8 Because many cave deposits are created by the seemingly endless dripping of water over long time spans, they are commonly called *dripstone*. The material being deposited is calcium carbonate ($CaCO_3$) and the rock is a form of limestone called *travertine*. The calcium carbonate is precipitated when some dissolved carbon dioxide escapes from a water drop. *(Photo by Guillen Photography; Inset photo by Clifford Stroud, National Park Service)*

A.

B.

FIGURE 6.10 **A.** Australia's Great Barrier Reef is the largest reef system in the world. Located off the coast of Queensland, it extends for 2600 kilometers (1600 miles) and consists of more than 2900 individual reefs. *(Photo by Ingo Schulz/Photolibrary)* **B.** This image in Guadalupe Mountains National Park, Texas, shows just a small portion of a huge Permian-age reef complex that once formed a 600-kilometer (nearly 375-mile) loop around the margin of the Delaware Basin. The reef consisted of a rich community of sponges, bryozoans, crinoids, gastropods, calcareous algae, and rare corals. *(Photo by Michael Collier)*

Precambrian time more than 2 billion years ago. From fossil remains, it is known that a variety of organisms have constructed reefs, including bivalves (clams and oysters), bryozoans (coral-like animals), and sponges. Corals have been found in fossil reefs as old as 500 million years, but corals similar to the modern colonial varieties have constructed reefs only during the last 60 million years.

In the United States, reefs of Silurian age (416 to 444 million years ago) are prominent features in Wisconsin, Illinois, and Indiana. In west Texas and adjacent southeastern New Mexico, a massive reef complex formed during the Permian period (251 to 299 million years ago) is strikingly exposed in Guadalupe Mountains National Park (Figure 6.10B).

COQUINA AND CHALK. Although much limestone is the product of biological processes, this origin is not always evident, because shells and skeletons may undergo considerable change before becoming lithified into rock. However, one easily identified biochemical limestone is *coquina*, a coarse rock composed of poorly cemented shells and shell fragments (see Figure 6.9B). Another less obvious but nevertheless familiar example is *chalk*, a soft, porous rock made up almost entirely of the hard parts of microscopic marine organisms smaller than the head of a pin. Among the most famous chalk deposits are those exposed along the southeast coast of England (**FIGURE 6.11**).

INORGANIC LIMESTONES. Limestones having an inorganic origin form when chemical changes or high water temperatures increase the concentration of calcium carbonate to the point that it precipitates. *Travertine*, the type of limestone commonly seen in caves, is an example (Figure 6.8). When travertine is deposited in caves, groundwater is the source of the calcium

FIGURE 6.11 The White Chalk Cliffs, Sussex, England. (Photo by Charles Bowman/Photolibrary)

carbonate. As water droplets become exposed to the air in a cavern, some of the carbon dioxide dissolved in the water escapes, causing calcium carbonate to precipitate.

Another variety of inorganic limestone is *oolitic limestone*. It is a rock composed of small spherical grains called *ooids*. Ooids form in shallow marine waters as tiny seed particles (commonly small shell fragments) and are moved back and forth by currents. As the grains are rolled about in the warm water, which is supersaturated with calcium carbonate, they become coated with layer upon layer of the chemical precipitate (**FIGURE 6.12**).

Dolostone

Closely related to limestone is *dolostone*, a rock composed of the calcium-magnesium carbonate mineral dolomite [$CaMg(CO_3)_2$]. Although dolostone and limestone sometimes closely resemble one another, they can be easily distinguished by observing their reaction to dilute hydrochloric acid. When a drop of acid is placed on limestone, the reaction (fizzing) is obvious. However, unless dolostone is powdered, it will not visibly react to the acid.

The origin of dolostone is not altogether clear and remains a subject of discussion among geologists. No marine organisms produce hard parts of dolomite, and the chemical precipitation of dolomite from seawater occurs only under conditions of unusual water chemistry in certain near-shore sites. Yet, dolostone is abundant in many ancient sedimentary rock successions.

It appears that significant quantities of dolostone are produced when magnesium-rich waters circulate through limestone and convert calcite to dolomite by the replacement of some calcium ions with magnesium ions (a process called *dolomitization*). However, other dolostones lack evidence that they formed by such a process, and their origin remains uncertain.

Chert

Chert is a name used for a number of very compact and hard rocks made of microcrystalline quartz (SiO_2). One well-known form is *flint*, whose dark color results from the

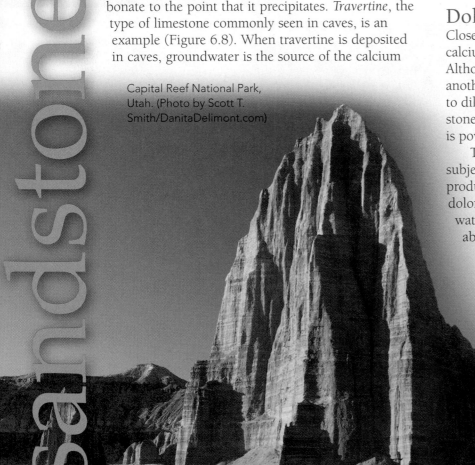

Capital Reef National Park, Utah. (Photo by Scott T. Smith/DanitaDelimont.com)

sandstone

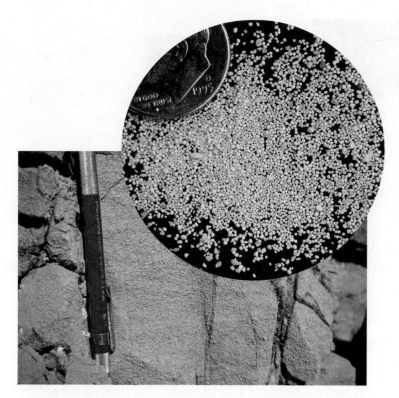

FIGURE 6.12 Oolitic limestone consists of *ooids*, which are small spherical grains formed by the chemical precipitation of calcium carbonate around a tiny nucleus. The carbonate is added in concentric layers as the spheres are rolled back and forth by currents in a warm shallow marine setting. *(Photos by Marli Miller)*

organic matter it contains. *Jasper*, a red variety, gets its bright color from iron oxide. The banded form is usually referred to as *agate* (**FIGURE 6.13**).

Like glass, most chert has a conchoidal fracture. Its hardness, ease of chipping, and ability to hold a sharp edge made chert a favorite of Native Americans for fashioning points for spears and arrows. Because of chert's durability and extensive use, "arrowheads" are found in many parts of North America.

Chert deposits are commonly found in one of two situations: as layered deposits referred to as *bedded cherts* and as *nodules*, somewhat spherical masses varying in diameter from a few millimeters (pea size) to a few centimeters.

Most water-dwelling organisms that produce hard parts make them of calcium carbonate. But some, such as diatoms and radiolarians, produce glasslike silica skeletons. These tiny organisms are able to extract silica even though seawater contains only tiny quantities of the dissolved material. It is from their remains that most bedded cherts are believed to originate. Some bedded cherts occur in association with lava flows and layers of volcanic ash. For these occurrences it is probable that the silica was derived from the decomposition of the volcanic ash and not from biochemical sources. Chert nodules are sometimes referred to as *secondary* or *replacement cherts* and most often occur within beds of limestone. They form when silica originally deposited in one place dissolves, migrates, and then chemically precipitates elsewhere, replacing older material.

Evaporites

Very often, evaporation is the mechanism triggering deposition of chemical precipitates. Minerals commonly precipitated in this fashion include halite (sodium chloride, NaCl), the chief component of *rock salt*, and gypsum (hydrous calcium sulfate, $CaSO_4 \cdot 2H_2O$), the main ingredient of *rock gypsum*. Both have significant commercial importance. Halite is familiar to everyone as the common salt used in cooking and for seasoning foods. Of course, it has many other uses—from melting ice on roads to making hydrochloric acid—and has been considered important enough that people have sought, traded, and fought over it for much of human history. Gypsum is the basic ingredient of plaster of Paris. This material is used most extensively in the construction industry for wallboard and interior plaster.

In the geologic past, many areas that are now dry land were basins, submerged under shallow arms of a sea that had only narrow connections to the open ocean. Under these conditions, seawater continually moved into the bay to replace water lost by evaporation. Eventually the waters of the bay became saturated and salt deposition began. Such deposits are called **evaporites**.

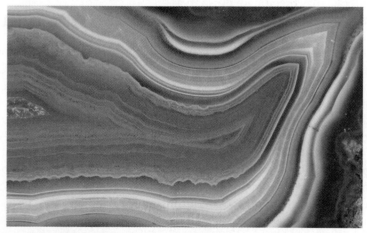

A. Agate

B. Flint   **C.** Jasper   **D.** Arrowhead

FIGURE 6.13 *Chert* is a name used for a number of dense, hard rocks made of microcrystalline quartz. Three examples are shown here. **A.** *Agate* is the banded variety. *(Photo by Jeffrey A. Scovil)* **B.** The dark color of *flint* results from organic matter. **C.** The red variety, called *jasper*, gets its color from iron oxide. *(Photo by E. J. Tarbuck)* **D.** Native Americans frequently made arrowheads and sharp tools from chert. *(Photo by LA VENTA/CORBIS/SYGMA)*

FIGURE 6.14 The Bonneville salt flats in Utah are a well-known example of evaporite deposits. *(Photo by Creatas/Photolibrary)*

When a body of seawater evaporates, the minerals that precipitate do so in a sequence determined by their solubility. Less soluble minerals precipitate first, and more soluble minerals precipitate later as salinity increases. For example, gypsum precipitates when about 80 percent of the seawater has evaporated, and halite settles out when 90 percent of the water has been removed. During the last stages of this process, potassium and magnesium salts precipitate. One of these last-formed salts, the mineral *sylvite*, is mined as a significant source of potassium ("potash") for fertilizer.

On a smaller scale, evaporite deposits can be seen in many desert basins in the western United States. Here, following rains or periods of snowmelt in the mountains, streams flow from the surrounding mountains into an enclosed basin. As the water evaporates, **salt flats** form when dissolved materials are precipitated as a white crust on the ground (**FIGURE 6.14**).

CONCEPT CHECK 6.4

❶ What are the two categories of chemical sedimentary rock? Give an example of a rock that belongs to each category.

❷ Distinguish among limestone, dolostone, and chert.

❸ How do evaporites form? Give an example.

Coal—An Organic Sedimentary Rock

Coal is quite different from other sedimentary rocks. Unlike limestone and chert, which are calcite or silica rich, coal is made mostly of organic matter. Close examination of a piece of coal under a microscope or magnifying glass often reveals plant structures such as leaves, bark, and wood that have been chemically altered but are still identifiable. This supports the conclusion that coal is the end product of large amounts of plant material buried for millions of years (**FIGURE 6.15**).

The initial stage in coal formation is the accumulation of large quantities of plant remains. However, special conditions are required for such accumulations because dead plants normally decompose when exposed to the atmosphere or other oxygen-rich environments. One important environment that allows for the buildup of plant material is a swamp.

Stagnant swamp water is oxygen-deficient, so complete decay (oxidation) of the plant material is not possible. Instead, the plants are attacked by certain bacteria that partly decompose the organic material and liberate oxygen and hydrogen. As these elements escape, the percentage of carbon gradually increases. The bacteria are not able to finish the job of decomposition because they are destroyed by acids liberated from the plants.

The partial decomposition of plants in an oxygen-poor swamp creates a layer of *peat*, a soft brown material in which plant structures are still easily recognized. With shallow burial, peat slowly changes to *lignite*, a soft brown coal. Burial increases the temperature of sediments as well as the pressure on them.

The higher temperatures bring about chemical reactions within the plant materials and yield water and organic gases (volatiles). As the load increases from more sediment on top of the developing coal, the water and volatiles are pressed out and the proportion of *fixed carbon* (the remaining solid combustible material) increases. The greater the carbon content, the greater the coal's energy ranking as a fuel. During burial the coal also becomes increasingly compact. For example, deeper burial transforms lignite into a harder, more compacted black rock called *bituminous* coal. Compared to the peat from which it formed, a bed of bituminous coal may be only 1/10 as thick.

Lignite and bituminous coals are sedimentary rocks. However, when sedimentary layers are subjected to the folding and deformation associated with mountain building, the heat and pressure cause a

FIGURE 6.15 Successive stages in the formation of coal.

SWAMP ENVIRONMENT

Burial

PEAT
(Partially altered plant material)

Compaction

LIGNITE
(Soft, brown coal)

Greater burial

Compaction

BITUMINOUS
(Soft, black coal)

METAMORPHISM

Stress

ANTHRACITE
(Hard, black coal)

further loss of volatiles and water, thus increasing the concentration of fixed carbon. This metamorphoses bituminous coal into *anthracite*, a very hard, shiny, black *metamorphic* rock. Although anthracite is a clean-burning fuel, only a relatively small amount is mined. Anthracite is not widespread and is more difficult and expensive to extract than the relatively flat-lying layers of bituminous coal.

Coal is a major energy resource. Its role as a fuel and some of the problems associated with burning coal are discussed later in this chapter.

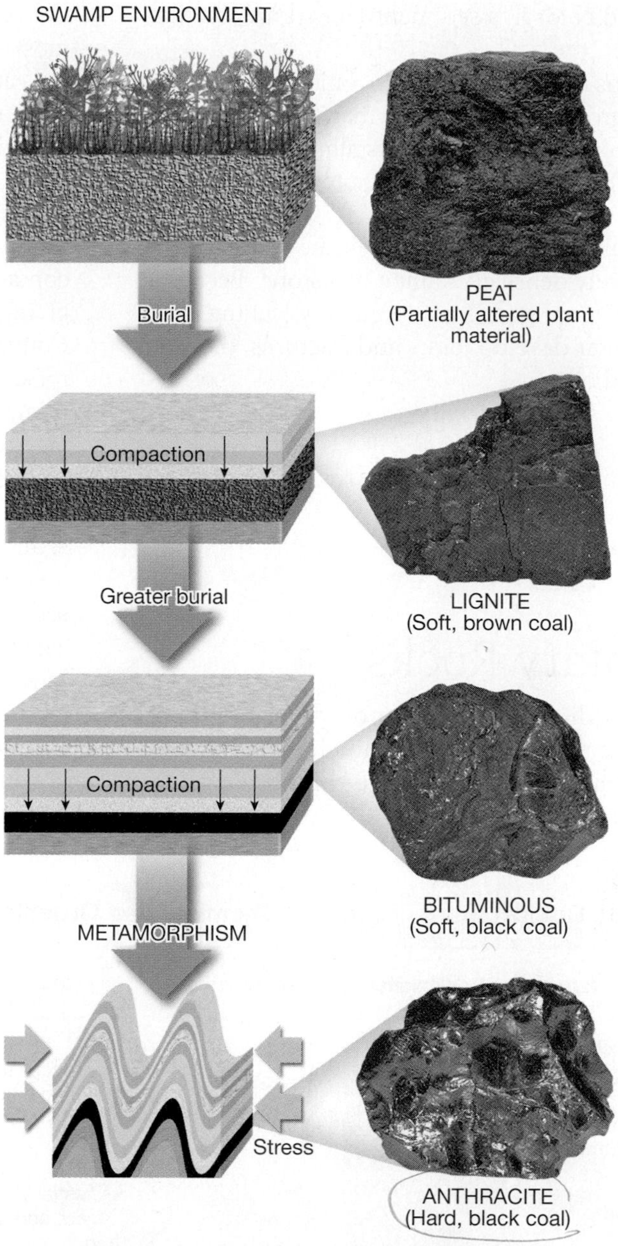

CONCEPT CHECK 6.5

❶ Outline the successive stages in the formation of coal.

Turning Sediment into Sedimentary Rock: Diagenesis and Lithification

A great deal of change can occur to sediment from the time it is deposited until it becomes a sedimentary rock and is subsequently subjected to the temperatures and pressures that convert it to metamorphic rock. The term **diagenesis** (*dia* = change, *genesis* = origin) is a collective term for all of the chemical, physical, and biological changes that take place after sediments are deposited and during and after lithification.

Burial promotes diagenesis because as sediments are buried, they are subjected to increasingly higher temperatures and pressures. Diagenesis occurs within the upper few kilometers of Earth's crust at temperatures that are generally less than 150° to 200 °C. Beyond this somewhat arbitrary threshold, metamorphism is said to occur.

One example of diagenetic change is *recrystallization*, the development of more stable minerals from less stable ones. It is illustrated by the mineral aragonite, the less stable form of calcium carbonate ($CaCO_3$). Aragonite is secreted by many marine organisms to form shells and other hard parts, such as the skeletal structures produced by corals. In some environments, large quantities of these solid materials accumulate as sediment. As burial

takes place, aragonite recrystallizes to the more stable form of calcium carbonate, calcite, the main constituent in the sedimentary rock limestone.

Another example of diagenesis was provided in the preceding discussion of coal. It involved the chemical alteration of organic matter in an oxygen-poor environment. Instead of completely decaying, as would occur in the presence of oxygen, the organic matter is slowly transformed to solid carbon.

Diagenesis includes **lithification** (*lithos* = stone, *fic* = making), the processes by which unconsolidated sediments are transformed into solid sedimentary rocks. Basic lithification processes include compaction and cementation.

The most common physical diagenetic change is **compaction**. As sediment accumulates, the weight of overlying material compresses the deeper sediments. The deeper a sediment is buried, the more it is compacted and the firmer it becomes. As the grains are pressed closer and closer, there is considerable reduction in pore space (the open space between particles). For example, when clays are buried beneath several thousand meters of material, the volume of clay may be reduced by as much as 40 percent. As pore space decreases, much of the water that was trapped in the sediments is driven out. Because sands and other coarse sediments are less compressible, compaction is most significant as a lithification process in fine-grained sedimentary rocks.

Cementation is the most important process by which sediments are converted to sedimentary rock. It is a diagenetic change that involves the crystallization of minerals among the individual sediment grains. Groundwater carries ions in solution. Gradually, the crystallization of new minerals from these ions takes place in the pore spaces, cementing the clasts together. Just as the amount of pore space is reduced during compaction, the addition of cement into a sedimentary deposit reduces its porosity as well.

Calcite, silica, and iron oxide are the most common cements. It is often a relatively simple matter to identify the cementing material. Calcite cement will effervesce with dilute hydrochloric acid. Silica is the hardest cement and thus produces the hardest sedimentary rocks. An orange or dark red color in a sedimentary rock means that iron oxide is present.

Most sedimentary rocks are lithified by means of compaction and cementation. However, some initially form as solid masses of intergrown crystals rather than beginning as accumulations of separate particles that later become solid. Other crystalline sedimentary rocks do not begin that way but rather are transformed into masses of interlocking crystals sometime after the sediment is deposited.

For example, with time and burial, loose sediment consisting of delicate calcareous skeletal debris may be recrystallized into a relatively dense crystalline limestone. Because crystals grow until they fill all the available space, pore spaces are frequently lacking in crystalline sedimentary rocks. Unless the rocks later develop joints and fractures, they will be relatively impermeable to fluids like water and oil.

CONCEPT CHECK 6.6

❶ What is diagenesis?

❷ List three common cements. How might each be identified?

Classification of Sedimentary Rocks

The classification scheme in **FIGURE 6.16** divides sedimentary rocks into two major groups: detrital and chemical/organic. Further, we can see that the main criterion for subdividing the detrital rocks is particle size, whereas the primary basis for distinguishing among different rocks in the chemical group is their mineral composition.

As is the case with many (perhaps most) classifications of natural phenomena, the categories presented in Figure 6.16 are more rigid than the actual state of nature. In reality, many of the sedimentary rocks classified into the chemical group also contain at least small quantities of detrital sediment. Many limestones, for example, contain varying amounts of mud or sand, giving them a "sandy" or "shaly" quality. Conversely, because practically all detrital rocks are cemented with material that was originally dissolved in water, they too are far from being "pure."

As was the case with the igneous rocks examined in Chapter 3, *texture* is a part of sedimentary rock classification. Two major textures are used in the classification of sedimentary rocks: clastic and nonclastic. The term **clastic** is taken from a Greek word meaning "broken." Rocks that display a clastic texture consist of discrete fragments and particles that are cemented and compacted together. Although cement is present in the spaces between particles,

Detrital Sedimentary Rocks				**Chemical and Organic Sedimentary Rocks**			
ClasticTexture (particle size)		**Sediment Name**	**Rock Name**	**Composition**	**Texture**	**Rock Name**	
Coarse (over 2 mm)		Gravel (Rounded particles)	**Conglomerate**	Calcite, CaCO₃	Nonclastic: Fine to coarse crystalline	**Crystalline Limestone**	
		Gravel (Angular particles)	**Breccia**			**Travertine**	
Medium (1/16 to 2 mm)		Sand (If abundant feldspar is present the rock is called **Arkose**)	**Sandstone**		Clastic: Visible shells and shell fragments loosely cemented	**Coquina**	**B i o c h e m i c a l L i m e s t o n e**
					Clastic: Various size shells and shell fragments cemented with calcite cement	**Fossiliferous Limestone**	
Fine (1/16 to 1/256 mm)		Mud	**Siltstone**		Clastic: Microscopic shells and clay	**Chalk**	
Very fine (less than 1/256 mm)		Mud	**Shale or Mudstone**	Quartz, SiO₂	Nonclastic: Very fine crystalline	**Chert (light colored) Flint (dark colored)**	
				Gypsum CaSO₄•2H₂O	Nonclastic: Fine to coarse crystalline	**Rock Gypsum**	
				Halite, NaCl	Nonclastic: Fine to coarse crystalline	**Rock Salt**	
				Altered plant fragments	Nonclastic: Fine-grained organic matter	**Bituminous Coal**	

FIGURE 6.16 Identification of sedimentary rocks. Sedimentary rocks are divided into three groups—detrital, chemical, and organic. The main criterion for naming detrital sedimentary rocks is particle size, whereas the primary basis for distinguishing among chemical sedimentary rocks is their mineral composition.

FIGURE 6.17 Like other evaporites, this sample of rock salt is said to have a nonclastic (crystalline) texture because it is composed of intergrown crystals. *(Photos by E. J. Tarbuck)*

Close up

these openings are rarely filled completely. All detrital rocks have a clastic texture. In addition, some chemical sedimentary rocks exhibit this texture. For example, coquina, the limestone composed of shells and shell fragments, is obviously as clastic as conglomerate or sandstone. The same applies for some varieties of oolitic limestone.

Some chemical sedimentary rocks have a **nonclastic** or **crystalline texture** in which the minerals form a pattern of interlocking crystals. The crystals may be microscopically small or large enough to be visible without magnification. Common examples of rocks with nonclastic textures are evaporites (**FIGURE 6.17**). The materials that make up many other nonclastic rocks may actually have originated as detrital deposits. In these instances, the particles probably consisted of shell fragments and other hard parts rich in calcium carbonate or silica. The clastic nature of the grains was subsequently obliterated or obscured because the particles recrystallized when they were consolidated into limestone or chert.

Nonclastic rocks consist of intergrown crystals, and some may resemble igneous rocks, which are also crystalline. The two rock types are usually easy to distinguish because the minerals that make up nonclastic sedimentary rocks are different from those found in most igneous rocks. For example, rock salt, rock gypsum, and some

forms of limestone consist of intergrown crystals, but the minerals contained within these rocks (halite, gypsum, and calcite) are seldom associated with igneous rocks.

CONCEPT CHECK 6.7

 What is the primary basis for distinguishing (naming) different chemical sedimentary rocks?

❷ Distinguish between clastic and nonclastic. Which texture is associated with all detrital rocks?

Sedimentary Rocks Represent Past Environments

Sedimentary Rocks

ESSENTIALS OF GEOLOGY   **Sedimentary Environments**

Sedimentary rocks are important in the interpretation of Earth history. By understanding the conditions under which sedimentary rocks form, geologists can often deduce the history of a rock, including information about the origin of its component particles, the method of sediment transport, and the nature of the place where the grains eventually came to rest; that is, the environment of deposition. An **environment of deposition** or **sedimentary**

environment is simply a geographic setting where sediment is accumulating. Each site is characterized by a particular combination of geologic processes and environmental conditions. Thus, when a series of sedimentary layers is studied, we can see the successive changes in environmental conditions that occurred at a particular place with the passage of time.

At any given time, the geographic setting and environmental conditions of a sedimentary environment determine the nature of the sediments that accumulate. Therefore, geologists carefully study the sediments in present-day depositional environments because the features they find can also be observed in ancient sedimentary rocks. By applying a thorough knowledge of present-day conditions, geologists attempt to reconstruct the ancient environments and geographical relationships of an area at the time a particular set of sedimentary layers were deposited. Such analyses often lead to the creation of maps depicting the geographic distribution of land and sea, mountains and river valleys, deserts and glaciers, and other environments of deposition. The foregoing description is an excellent example of the application of a fundamental principle of modern geology, namely that "the present is the key to the past."[*] Sedimentary environments are commonly placed into one of three categories: *continental*, *marine*, or *transitional* (shoreline). Each category includes many specific subenvironments. **FIGURE 6.18** is an idealized diagram illustrating a number of important sedimentary environments associated with each category. Later, Chapters 9 through 13 will describe these environments in detail. Each is an area where sediment accumulates and where organisms live and die. Each produces a characteristic sedimentary rock or assemblage that reflects prevailing conditions.

CONCEPT CHECK 6.8

❶ What are the three categories of sedimentary environments?

DID YOU KNOW?
These terms are often used when describing the mineral make-up of sedimentary rocks: *siliceous rocks* contain abundant quartz, *argillaceous* rocks are clay rich, and *carbonate* rocks contain calcite or dolomite.

FIGURE 6.18 Sedimentary environments are those places where sediment accumulates. Each is characterized by certain physical, chemical, and biological conditions. Because each sediment contains clues about the environment in which it was deposited, sedimentary rocks are important in the interpretation of Earth history. A number of important continental, transitional, and marine sedimentary environments are represented in this idealized diagram. (Lake photo by JonArnoldImages/www. DanitaDelimont; Alluvial Fan photo by Marli Miller, Swamp photo by Raymond Gehman/National Geographic; all others by E. J. Tarbuck)

Beach

Sand dunes

Lake

Spit

Lake

Estuary

Deep-sea fans

Turbidity current

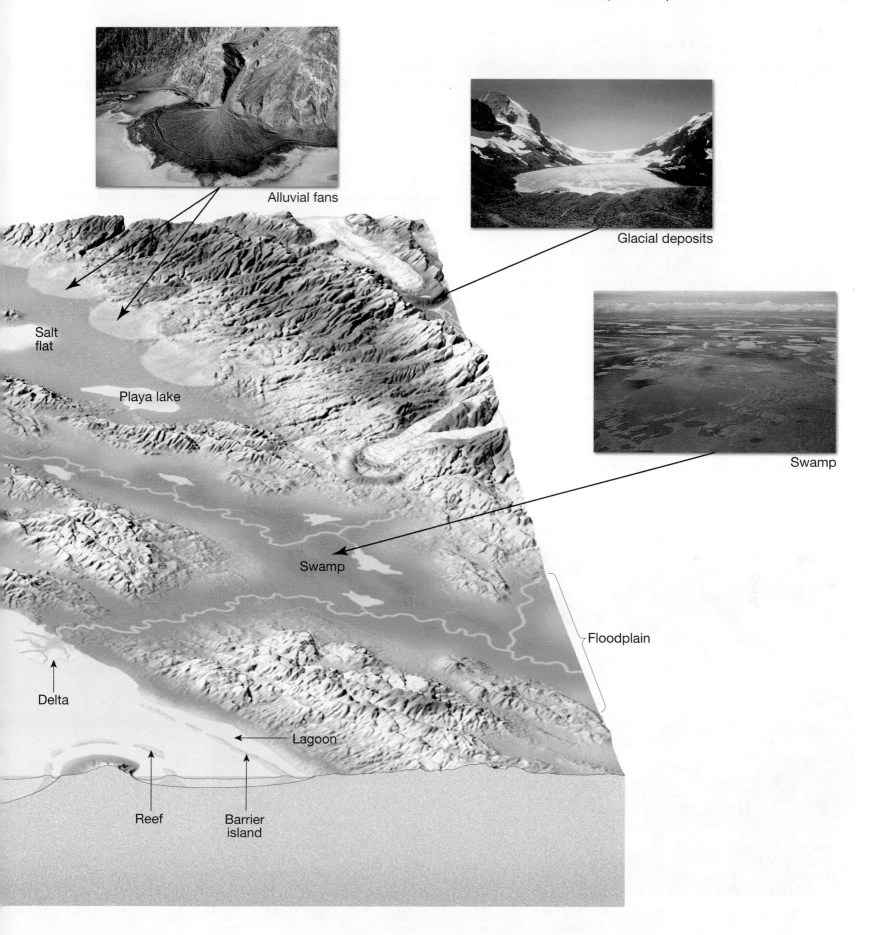

Alluvial fans

Glacial deposits

Salt
flat

Playa lake

Swamp

Swamp

Floodplain

Delta

Lagoon

Reef

Barrier
island

Sedimentary Structures

In addition to variations in grain size, mineral composition, and texture, sediments exhibit a variety of structures. Some, such as graded beds, are created when sediments are accumulating and are a reflection of the transporting medium. Others, such as *mud cracks*, form after the materials have been deposited and result from processes occurring in the environment. When present, sedimentary structures provide additional information that can be useful in the interpretation of Earth history.

Sedimentary rocks form as layer upon layer of sediment accumulates in various depositional environments. These layers, called **strata** or **beds**, are probably *the single most common and characteristic feature of sedimentary rocks*. Each stratum is unique. It may be a coarse sandstone, a fossil-rich limestone, a black shale, and so on. When you look at **FIGURE 6.19**, or look back through this chapter at Figures 6.1 and 6.4, you will see many such layers, each different from the others. The variations in texture, composition, and thickness reflect the different conditions under which each layer was deposited.

The thickness of beds ranges from microscopically thin to tens of meters thick. Separating the strata are **bedding planes,** flat surfaces along which rocks tend to separate or break. Changes in the grain size or in the composition of the sediment being deposited can create bedding planes. Pauses in deposition can also lead to layering because chances are slight that newly deposited material will be exactly the same as previously deposited sediment. Generally, each bedding plane marks the end of one episode of sedimentation and the beginning of another.

Because sediments usually accumulate as particles that settle from a fluid, most strata are originally deposited as horizontal layers. There are circumstances, however, when sediments do not accumulate in horizontal beds. Sometimes when a bed of sedimentary rock is examined, we see layers within it that are inclined to the horizontal. When this occurs, it is called **cross-bedding** and is most characteristic of sand dunes, river deltas, and certain

stream channel deposits (see Figure 6.6A and **FIGURE 6.20**).

Graded beds represent another special type of bedding. In this case the particles within a single sedimentary layer gradually change from coarse at the bottom to fine at the top. Graded beds are most characteristic of rapid deposition from water containing sediment of varying sizes. When a current experiences a rapid energy loss, the largest particles settle first, followed by successively smaller grains. The deposition of a graded bed is most often associated with a turbidity current, a mass of sediment-choked water that is denser than clear water and that moves downslope along the bottom of a lake or ocean (**FIGURE 6.21**).

As geologists examine sedimentary rocks, much can be deduced. A conglomerate, for example, may indicate a high-energy environment, such as a surf zone or rushing stream, where only coarse materials settle out and finer particles are kept suspended. If the rock is arkose, it may signify a dry climate where little chemical alteration of feldspar is possible. Carbonaceous shale is a sign of a low-energy, organic-rich environment, such as a swamp or lagoon.

Other features found in some sedimentary rocks give clues to past environments. Ripple marks are such a feature. **Ripple marks** are small waves of sand that develop on the surface of a sediment layer by the action of moving water or air (**FIGURE 6.22A**). The ridges form at right angles to the direction of motion. If the ripple marks were formed by air or water moving in essentially one direction, their form will be asymmetrical. These *current ripple marks*

FIGURE 6.19 This outcrop of sedimentary strata illustrates the characteristic layering of this group of rocks. Minnewaska State Park, New York. *(Photo by Carr Clifton/National Geographic Stock)*

Limestone outcrop
Minnewaska State Park, NY

Thicker limestone strata

Bedding planes

Thinner limestone beds

Geologist's Sketch

A.

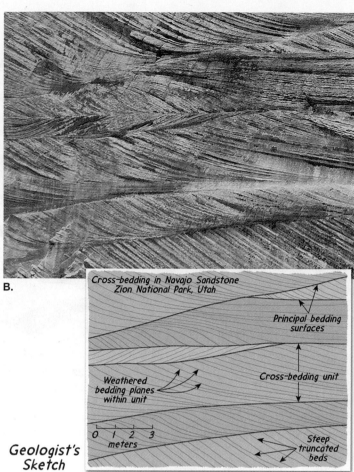

B.

FIGURE 6.20 A. The cutaway section of this sand dune shows cross-bedding. *(Photo by John S. Shelton)* **B.** The cross-bedding of this sandstone indicates it was once a sand dune. *(Photo by Dennis Tasa)*

Geologist's Sketch

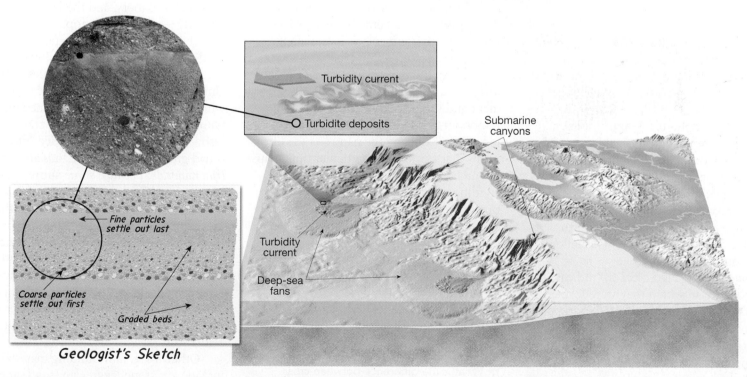

Geologist's Sketch

FIGURE 6.21 Turbidity currents are downslope movements of dense, sediment-laden water. They are created when sand and mud on the continental shelf and slope are dislodged and thrown into suspension. Because such mud-choked water is denser than normal seawater, it flows downslope, eroding and accumulating more sediment. Beds deposited by these currents are called *turbidites*. Each event produces a single bed characterized by a decrease in sediment size from bottom to top, a feature known as a *graded bed*. *(Photo by Marli Miller)*

B.

FIGURE 6.22 A. Ripple marks are produced by currents of water or wind. *(Photo by Ken Hamblin)* **B.** Mud cracks form when wet mud or clay dries out and shrinks. *(Photo by Gary Yeowell/Getty Images, Inc-Stone Allstock)*

A.

will have steeper sides in the downcurrent direction and more gradual slopes on the upcurrent side. Ripple marks produced by a stream flowing across a sandy channel or by wind blowing over a sand dune are two common examples of current ripples. When present in solid rock, they may be used to determine the direction of movement of ancient wind or water currents. Other ripple marks have a symmetrical form. These features, called *oscillation ripple marks*, result from the back-and-forth movement of surface waves in a shallow nearshore environment.

Mud cracks (Figure 6.22B) indicate that the sediment in which they were formed was alternately wet and dry. When exposed to air, wet mud dries out and shrinks, producing cracks. Mud cracks are associated with such environments as tidal flats, shallow lakes, and desert basins.

Fossils, the remains or traces of prehistoric life, are important inclusions in sediment and sedimentary rock. They are important tools for interpreting the geologic past. Knowing the nature of the life forms that existed at a particular time helps researchers decipher past environmental conditions. Further, fossils are important time indicators and play a key role in correlating rocks that are of similar ages but are from different places. Fossils will be examined in more detail in Chapter 18.

CONCEPT CHECK 6.9

❶ What is the single most characteristic feature of sedimentary rocks?

❷ What is the difference between cross-bedding and graded bedding?

❸ How might mud cracks or ripple marks be useful clues about the geologic past?

Nonmetallic Mineral Resources from Sedimentary Rocks

Earth materials that are not used as fuels or processed for the metals they contain are referred to as **nonmetallic mineral resources**. Realize that use of the word "mineral" is very broad in this economic context and is quite different from the geologist's strict definition of mineral found in Chapter 2. Nonmetallic mineral resources are extracted and processed either for the nonmetallic elements they contain or for the physical and chemical properties they possess (**TABLE 6.1**). Although these resources have diverse origins, many are sediments or sedimentary rocks.

People often do not realize the importance of nonmetallic minerals, because they see only the products that resulted from their use and not the minerals themselves. That is, many nonmetallics are used up in the process of creating other products. Examples include the fluorite and limestone that are part of the steelmaking process, the abrasives required to make a piece of machinery, and the fertilizers needed to grow a food crop.

The quantities of nonmetallic minerals used each year are enormous. Per capita consumption of nonfuel resources in the United States is nearly 11 metric tons, of which over 95 percent are nonmetallics (**FIGURE 6.23**). Nonmetallic mineral resources are commonly divided into two broad groups: *building materials* and *industrial minerals*. Because some substances have many different uses, they are found in both categories. Limestone, perhaps the most versatile and widely used rock of all, is the best example. As a building material, it is used not only as crushed rock and building stone but also in making cement. Moreover, as an industrial mineral, limestone is an ingredient in the manufacture of steel and is used in agriculture to neutralize acidic soils.

Other important building materials include cut stone, aggregate (sand, gravel, and crushed rock), gypsum for plaster and wallboard, clay for tile and bricks, and cement, which is made from limestone and shale. Cement and aggregate go into the

TABLE 6.1
Uses of Nonmetallic Minerals

Mineral	Uses
Apatite	Phosphorus fertilizers
Asbestos (chrysotile)	Incombustible fibers
Calcite	Aggregate; steelmaking; soil conditioning; chemicals; cement; building stone
Clay minerals (kaolinite)	Ceramics; china
Corundum	Gemstones; abrasives
Diamond	Gemstones; abrasives
Fluorite	Steelmaking; aluminum refining; glass; chemicals
Garnet	Abrasives; gemstones
Graphite	Pencil lead; lubricant; refractories
Gypsum	Plaster of Paris
Halite	Table salt; chemicals; ice control
Muscovite	Insulator in electrical applications
Quartz	Primary ingredient in glass
Sulfur	Chemicals; fertilizer manufacture
Sylvite	Potassium fertilizers
Talc	Powder used in paints, cosmetics, etc.

CONCEPT CHECK 6.10

❶ List the two groups of nonmetallic resources and some examples of each.

Energy Resources from Sedimentary Rocks

Coal, petroleum, and natural gas are the primary fuels of our modern industrial economy. About 84 percent of the energy consumed in the United States today comes from these basic fossil fuels (**FIGURE 6.24**). Although major shortages of oil and gas will not occur for many years, proven reserves are declining. Despite new exploration, even in very remote regions and severe environments, new sources of oil are not keeping pace with consumption.

Unless large, new petroleum reserves are discovered (which is possible but not likely), a greater share of our future needs

making of concrete, a material that is essential to practically all construction.

A wide variety of resources are classified as industrial minerals. In some instances these materials are important because they are sources of specific chemical elements or compounds. Such minerals are used in the manufacture of chemicals and the production of fertilizers. In other cases their importance is related to the physical properties they exhibit. Examples include minerals such as corundum and garnet, which are used as abrasives. Although supplies are generally plentiful, most industrial minerals are not nearly as abundant as building materials.

Moreover, deposits are far more restricted in distribution and extent. As a result, many of these nonmetallic resources must be transported considerable distances, which of course adds to their cost. Unlike most building materials, which need a minimum of processing before they are ready to use, many industrial minerals require considerable processing to extract the desired substance at the proper degree of purity for its ultimate use.

DID YOU KNOW?

Just 2 kilometers of four-lane highway require more than 85 tons of aggregate. The United States produces about 2 billion tons of aggregate per year. This represents about one half of the entire nonenergy mining volume in the country.

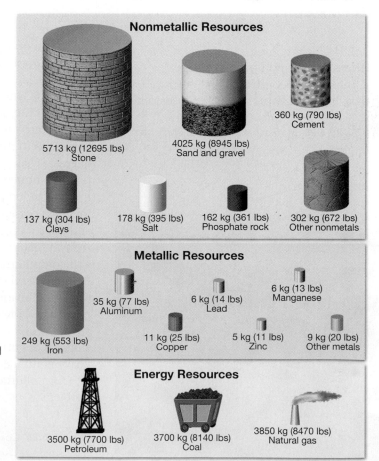

FIGURE 6.23 The annual per capita consumption of nonmetallic and metallic mineral resources for the United States. About 97 percent of the materials used are nonmetallic. The per capita use of oil, coal, and natural gas exceeds 11,000 kilograms. *(After U.S. Geological Survey)*

Nonmetallic Resources
5713 kg (12695 lbs) Stone
4025 kg (8945 lbs) Sand and gravel
360 kg (790 lbs) Cement
137 kg (304 lbs) Clays
178 kg (395 lbs) Salt
162 kg (361 lbs) Phosphate rock
302 kg (672 lbs) Other nonmetals

Metallic Resources
249 kg (553 lbs) Iron
35 kg (77 lbs) Aluminum
11 kg (25 lbs) Copper
6 kg (14 lbs) Lead
5 kg (11 lbs) Zinc
6 kg (13 lbs) Manganese
9 kg (20 lbs) Other metals

Energy Resources
3500 kg (7700 lbs) Petroleum
3700 kg (8140 lbs) Coal
3850 kg (8470 lbs) Natural gas

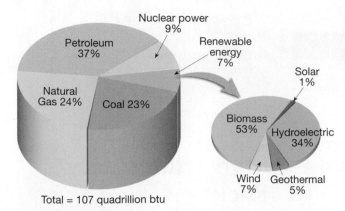

FIGURE 6.24 U.S. energy consumption, 2008. The total was nearly 107 quadrillion Btu. A quadrillion, by the way, is 10 raised to the 12th power, or a million million—a quadrillion Btu is a convenient unit for referring to U.S. energy use as a whole. (Source: U.S. Department of Energy, Energy Information Administration)

will eventually have to come from coal and/or from alternative energy sources such as nuclear, solar, wind, tidal, and hydroelectric power. Two fossil-fuel alternatives—oil sands and oil shale—are sometimes mentioned as promising new sources of liquid fuels. In the following sections, we will briefly examine the fuels that have traditionally supplied our energy needs.

Coal

Along with oil and natural gas, coal is commonly called a **fossil fuel**. Such a designation is appropriate, because each

FIGURE 6.25 Strip mining of coal at Black Mesa, Arizona. This type of mining is common when coal seams are near the surface. (Photo by Richard W. Brooks/Photo Researchers, Inc.)

time we burn coal we are using energy from the Sun that was stored by plants many millions of years ago. We are indeed burning a "fossil."

Coal has been an important fuel for centuries. In the 19th and early 20th centuries, cheap and plentiful coal powered the Industrial Revolution. By 1900, coal was providing 90 percent of the energy used in the United States. Although still important, coal currently provides only about 23 percent of the energy needs of the United States (**FIGURE 6.25**).

More than 80 percent of present-day coal usage is for the generation of electricity. As oil reserves gradually diminish in the years to come, the use of coal may actually increase. Expanded coal production is possible because the world has enormous reserves and the technology to mine coal efficiently. In the United States, coal fields are widespread and contain supplies that should last for hundreds of years (**FIGURE 6.26**).

Although coal is plentiful, its recovery and its use present a number of problems. Surface mining can turn the countryside into a scarred wasteland if careful (and costly) reclamation is not carried out to restore the land. (Today all U.S. surface mines must reclaim the land.) Although underground mining does not scar the landscape to the same degree, it has been costly in terms of human life and health. Underground mining long ago ceased to be a pick-and-shovel operation and is today a highly mechanized and computerized process. Strong federal safety regulations have made U.S. mining quite safe. However, the hazards of roof falls, gas explosions, and working with heavy equipment remain.

Air pollution is a major problem associated with the

burning of coal. Much coal contains significant quantities of sulfur. Despite efforts to remove sulfur before the coal is burned, some remains. When the coal is burned, the sulfur is converted into noxious sulfur-oxide gases. Through a series of complex chemical reactions in the atmosphere, the sulfur oxides are converted to sulfuric acid, which then falls to Earth's surface as rain or snow. This *acid precipitation* can have detrimental ecological effects over widespread areas. (See Figure 5.12, p. 132)

Because none of the problems just mentioned are likely to prevent the increased use of this important and abundant fuel, stronger efforts must be made to correct the problems associated with the mining and use of coal.

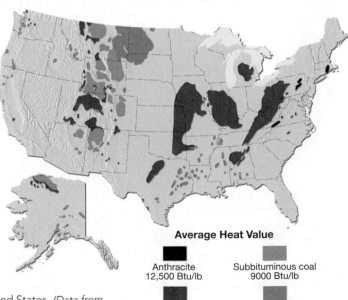

FIGURE 6.26 Coal fields of the United States. (Data from U.S. Geological Survey)

Oil and Natural Gas

Petroleum and natural gas obviously are not rocks, but we class them as mineral resources because they *come from* sedimentary rocks. They are found in similar environments and typically occur together. Both consist of various hydrocarbon compounds (compounds consisting of hydrogen and carbon) mixed together. They may also contain small quantities of other elements, such as sulfur, nitrogen, and oxygen. Like coal, petroleum and natural gas are biological products derived from the remains of organisms. However, the environments in which they form are very different, as are the organisms. Coal is formed mostly from plant material that accumulated in a swampy environment above sea level. Oil and gas are derived from the remains of both plants and animals having a marine origin.

Petroleum formation is complex and not completely understood. Nevertheless, we know that it begins with the accumulation of sediment in ocean areas that are rich in plant and animal remains. These accumulations must occur where biological activity is high, such as nearshore areas. However, most marine environments are oxygen rich, which leads to the decay of organic remains before they can be buried by other sediments. Therefore, accumulations of oil and gas are not as widespread as the marine environments that support abundant biological activity. This limiting factor notwithstanding, large quantities of organic matter are buried and protected from oxidation in many offshore sedimentary basins. With increasing burial over millions of years, chemical reactions gradually transform some of the original organic matter into the liquid and gaseous hydrocarbons we call petroleum and natural gas.

Unlike the solid organic matter from which they formed, the newly created petroleum and natural gas are *mobile*. These fluids are gradually squeezed from the compacting, mud-rich layers where they originate into adjacent permeable beds such as sandstone, where openings between sediment grains are larger. Because all of this occurs underwater, the rock layers containing the oil and gas are already saturated with water. But oil and gas are less dense than water, so they migrate upward through the water-filled pore spaces of the enclosing rocks. Unless something acts to halt this upward migration, the fluids will eventually reach the surface. There the volatile components will evaporate.

Sometimes the upward migration is halted. A geologic environment that allows for economically significant amounts of oil and gas to accumulate underground is termed an **oil trap**. Several geologic structures can act as oil traps. All have two basic conditions in common: a porous, permeable **reservoir rock** that will yield petroleum and natural gas in sufficient quantities to make drilling worthwhile; and a **cap rock**, such as shale, that is virtually impermeable to oil and gas. The cap rock keeps the upwardly mobile oil and gas from escaping at the surface.

CONCEPT CHECK 6.11

❶ Coal has the advantage of being plentiful. What are some of coal's disadvantages?

❷ What is an oil trap? List two conditions common to all traps.

The Carbon Cycle and Sedimentary Rocks

To illustrate the movement of material and energy in the Earth system, let us take a brief look at the *carbon cycle* (**FIGURE 6.27**). Pure carbon is relatively rare in nature. It is found predominantly in two minerals: diamond and graphite. Most carbon is bonded chemically to other elements to form compounds such as carbon dioxide, calcium carbonate, and the hydrocarbons found in coal and petroleum. Carbon is also the basic building block of life because it readily combines with hydrogen and oxygen to form the fundamental organic compounds that compose living things.

In the atmosphere, carbon is found mainly as carbon dioxide (CO_2). Atmospheric carbon dioxide is significant because it is a greenhouse gas, which means it is an efficient absorber of energy emitted by Earth and thus influences the heating of the atmosphere. Because many of the processes that operate on Earth involve carbon dioxide, this gas is constantly moving into and out of the atmosphere (**FIGURE 6.28**). For example, through the process of photosynthesis, plants absorb carbon dioxide from the atmosphere to produce the essential organic compounds needed for growth. Animals that consume these plants (or consume other animals that eat plants) use these organic

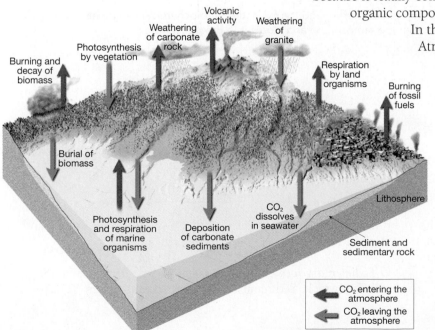

FIGURE 6.27 Simplified diagram of the carbon cycle, with emphasis on the flow of carbon between the atmosphere and the hydrosphere, geosphere, and biosphere. The colored arrows show whether the flow of carbon is into or out of the atmosphere.

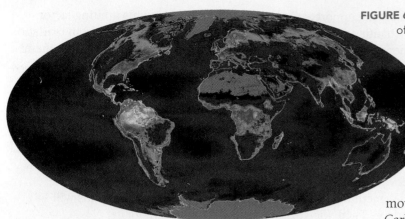

FIGURE 6.28 This map was created using space-based measurements of a range of plant properties and shows the net productivity of vegetation on land and in the oceans in 2002. It is calculated by determining how much CO_2 is taken up by vegetation during photosynthesis minus how much is given off during respiration. Scientists expect this global measure of biological activity to yield new insights into Earth's complex carbon cycle. (*NASA image*)

Net Primary Productivity (kgC/m²/year)

0   1   2   3

compounds as a source of energy and, through the process of respiration, return carbon dioxide to the atmosphere. (Plants also return some CO_2 to the atmosphere via respiration.) Further, when plants die and decay or are burned, this biomass is oxidized, and carbon dioxide is returned to the atmosphere.

Not all dead plant material decays immediately back to carbon dioxide. A small percentage is deposited as sediment. Over long spans of geologic time, considerable biomass is buried with sediment. Under the right conditions, some of these carbon-rich deposits are converted to fossil fuels—coal, petroleum, or natural gas. Eventually some of the fuels are recovered (mined or pumped from a well) and burned to run factories and to fuel our transportation system. One result of fossil-fuel combustion is the release of huge quantities of CO_2 into the atmosphere.

Certainly, one of the most active parts of the carbon cycle is the movement of CO_2 from the atmosphere to the biosphere and back again. Carbon also moves from the geosphere and hydrosphere to the atmosphere and back again. For example, volcanic activity early in Earth's history is thought to be the source of much of the carbon dioxide found in the atmosphere. One way that carbon dioxide makes its way back to the hydrosphere and then to the solid Earth is by first combining with water to form carbonic acid (H_2CO_3), which then attacks the rocks that compose the geosphere. One product of this chemical weathering of solid rock is the soluble bicarbonate ion ($2HCO_3^-$), which is carried by groundwater and streams to the ocean. Here water-dwelling organisms extract this dissolved material to produce hard parts of calcium carbonate ($CaCO_3$). When the organisms die, these skeletal remains settle to the ocean floor as biochemical sediment and become sedimentary rock. In fact, the geosphere is by far Earth's largest depository of carbon, where it is a constituent of a variety of rocks, the most abundant being limestone. Eventually the limestone may be exposed at Earth's surface, where chemical weathering will cause the carbon stored in the rock to be released to the atmosphere as CO_2.

In summary, carbon moves among all four of Earth's major spheres. It is essential to every living thing in the biosphere. In the atmosphere, carbon dioxide is an important greenhouse gas. In the hydrosphere, carbon dioxide is dissolved in lakes, rivers, and the ocean. In the geosphere, carbon is contained in carbonate sediments and sedimentary rocks and is stored as organic matter dispersed through sedimentary rocks and deposits of coal and petroleum.

CONCEPT CHECK 6.12

❶ Describe how chemical weathering and the formation of biochemical sediment removes carbon from the atmosphere and stores it in the geosphere.

CHAPTER SIX
Sedimentary Rocks in Review

⊙ Sedimentary rocks account for about 5 to 10 percent of Earth's outer 16 kilometers (10 miles). Because they are concentrated at Earth's surface, the importance of this group is much greater than this percentage implies. Sedimentary rocks contain much of the basic information needed to reconstruct Earth history. In addi-

tion, this group is associated with many important energy and mineral resources.

⊙ *Sedimentary rock* consists of *sediment* that has been *lithified* into solid rock. Sediment has two principal sources: (1) as *detrital material*, which originates and is transported as solid particles from both mechanical and chemical weathering, which, when lithified, forms detrital sedimentary rocks; and (2) from soluble material produced largely by chemical weathering, which, when precipitated, forms *chemical sedimentary rocks*. Coal is the primary example of a third group called *organic sedimentary rocks*, which consist of organic carbon from the remains of partially altered plant material.

⦿ *Particle size* is the primary basis for distinguishing among various detrital sedimentary rocks. The size of the particles in a detrital rock indicates the energy of the medium that transported them. For example, gravels are moved by swiftly flowing rivers, whereas less energy is required to transport sand. Common detrital sedimentary rocks include *shale* (silt- and clay-size particles), *sandstone*, and *conglomerate* (rounded gravel-size particles) or *breccia* (angular gravel-size particles).

⦿ Precipitation of chemical sediments occurs in two ways: (1) by *inorganic processes*, such as evaporation and chemical activity; or (2) by *organic processes* of water-dwelling organisms that produce sediments of *biochemical origin*. *Limestone*, the most abundant chemical sedimentary rock, consists of the mineral calcite ($CaCO_3$) and forms either by inorganic means or as the result of biochemical processes. Inorganic limestones include *travertine*, which is commonly seen in caves, and *oolitic limestone*, consisting of small spherical grains of calcium carbonate. Other common chemical sedimentary rocks include *dolostone* (composed of the calcium-magnesium carbonate mineral dolomite), *chert* (made of microcrystalline quartz), and *evaporites* (such as rock salt and rock gypsum).

⦿ *Diagenesis* refers to all of the physical, chemical, and biological changes that occur after sediments are deposited and during and after the time they are turned into sedimentary rock. Burial promotes diagenesis. Diagenesis includes lithification.

⦿ *Lithification* refers to the processes by which unconsolidated sediments are transformed into solid sedimentary rock. Most sedimentary rocks are lithified by means of *compaction* and/or *cementation*. Compaction occurs when the weight of overlying materials compresses the deeper sediments. Cementation, the most important process by which sediments are converted to sedimentary rock, occurs when soluble cementing materials, such as *calcite, silica*, and *iron oxide*, are precipitated onto sediment grains, fill open spaces, and join the particles. Although most sedimentary rocks are lithified by compaction or cementation, certain chemical rocks, such as the evaporites, initially form as solid masses of intergrown crystals.

⦿ Sedimentary rocks are divided into three groups: *detrital, chemical*, and *organic*. All detrital rocks have a *clastic texture*, which consists of discrete fragments and particles that are cemented and compacted together. The main criterion for subdividing the detrital rocks is particle size. Common detrital rocks include *conglomerate, sandstone*, and *shale*. The primary basis for distinguishing among different rocks in the chemical group is their mineral composition. Some chemical rocks, such as those deposited when seawater evaporates, have a *nonclastic (crystalline) texture* in which the minerals form a pattern of interlocking crystals. However, in reality, many of the sedimentary rocks classified into the chemical group also contain at least small quantities of detrital sediment. Common chemical rocks include *limestone, chert*, and *rock gypsum*. Coal is the primary example of an organic sedimentary rock.

⦿ Sedimentary environments are those places where sediment accumulates. They are grouped into continental, marine, and transitional (shoreline) environments. Each is characterized by certain physical, chemical, and biological conditions. Because sediment contains clues about the environment in which it was deposited, sedimentary rocks are important in the interpretation of Earth's history.

⦿ Layers, called *strata* or *beds*, are probably the single most characteristic feature of sedimentary rocks. Other features found in some sedimentary rocks, such as *ripple marks, mud cracks, cross-bedding, graded bedding*, and *fossils*, also give clues to past environments.

⦿ Earth materials that are not used as fuels or processed for the metals they contain are referred to as *nonmetallic resources*. Many are sediments or sedimentary rocks. The two broad groups of nonmetallic resources are *building materials* and *industrial minerals*. Limestone, perhaps the most versatile and widely used rock of all, is found in both groups.

⦿ *Coal, petroleum*, and *natural gas*, the *fossil fuels* of our modern economy, are all associated with sedimentary rocks. Coal originates from large quantities of plant remains that accumulate in an oxygen-deficient environment, such as a swamp. More than 80 percent of present-day coal usage is for the generation of electricity. Air pollution from the sulfur-oxide gases that form from burning most types of coal is a significant environmental problem.

⦿ Oil and natural gas, which commonly occur together in the pore spaces of some sedimentary rocks, consist of various *hydrocarbon compounds* (compounds made of hydrogen and carbon) mixed together. Petroleum formation is associated with the accumulation of sediment in ocean areas rich in plant and animal remains that have become buried and isolated in an oxygen-deficient environment. As the mobile petroleum and natural gas form, they migrate and accumulate in adjacent permeable beds such as sandstone. If the upward migration is halted by an impermeable rock layer, referred to as a *cap rock*, a geologic environment develops that allows for economically significant amounts of oil and gas to accumulate underground in what is termed an *oil trap*.

Key Terms

bedding planes (p. 166)
beds (strata) (p. 166)
biochemical origin (p. 156)
cap rock (p. 171)
cementation (p. 161)
chemical sedimentary
   rocks (p. 151)
clastic (p. 162)
compaction (p. 161)
cross-bedding (p. 166)

crystalline texture (p. 163)
detrital sedimentary
   rocks (p. 151)
diagenesis (p. 161)
environment of deposition
   (p. 163)
evaporites (p. 159)
fossil fuel (p. 170)
fossils (p. 168)
graded beds (p. 166)

lithification (p. 161)
mud cracks (p. 168)
nonclastic (p. 163)
nonmetallic mineral
   resources (p. 168)
oil trap (p. 171)
organic sedimentary rocks
   (p. 151)
reservoir rock (p. 171)
ripple marks (p. 166)

salt flats (p. 160)
sedimentary environment
   (p. 163)
sorting (p. 153)
strata (beds) (p. 166)

GIVE IT SOME THOUGHT

❶ Dust collecting on furniture is an everyday example of a sedimentary process. Provide another example of a sedimentary process that might be observed in or around where you live.

❷ Discuss two reasons sedimentary rocks are more likely to contain fossils than igneous rocks.

❸ Imagine you are studying sedimentary deposits in an abandoned river channel. You see layers of well-sorted sand alternating with coarser-grained, poorly sorted material consisting of sand, gravel, and rock fragments. How would you interpret the history of this river channel based on these layers of sediment? What might the layers indicate about possible changes in climate conditions and the flow of water in the river in the past?

❹ If you hiked to a mountain peak and found limestone at the top, what would that indicate about the likely geologic history of the rock atop the mountain?

❺ In which of the environments illustrated in Figure 6.18 would you expect to find:

   a. A deposit consisting of several beds in which sediment sizes in each bed decrease from bottom to top.

   b. An evaporite deposit.

   c. A well-sorted sand deposit.

   d. A deposit that includes a high percentage of partially decomposed plant material.

❻ While on a field trip with your geology class, you stop at an outcrop of sandstone. An examination with a hand lens shows that the sandstone is poorly sorted and rich in feldspar and quartz. Your instructor tells you that the sediment was derived from one of two sites in the area. Select the most likely site and explain your choice. What name is given to this type of sandstone?

Site #1 A nearby exposure of weathered basaltic lava flows.

Site #2 An outcrop of granite at the previous field trip stop up the road.

❼ Every year about 20,000 pounds of stone, sand and gravel are mined for each person in the United States.

   a. Calculate how many pounds of stone, sand, and gravel will be needed for an individual during an 80-year lifespan.

   b. If one cubic yard of rock weighs roughly 1700 pounds, how big a hole (cubic yards) must be dug to supply an individual with 80 years' worth of stone, sand, and gravel?

   c. A typical pickup truck can carry about a half cubic yard of rock. How many pickup truck loads would be necessary during the 80-year span?

Companion Website

The *Essentials of Geology, 11e* companion Website contains numerous multimedia resources accompanied by assessments to aid in your study of the topics in this chapter. The use of this site's learning tools will help improve your understanding of geology. Utilizing the access code that accompanies this text, visit **www.mygeoscienceplace.com** in order to:

- **Review** key chapter concepts.
- **Read** with links to the Pearson eText and to chapter-specific web resources.
- **Visualize** and comprehend challenging topics using the learning activities in *GEODe: Essentials of Geology* and the *Geoscience Animations Library*.
- **Test** yourself with online quizzes.

7 Metamorphism and Metamorphic Rocks

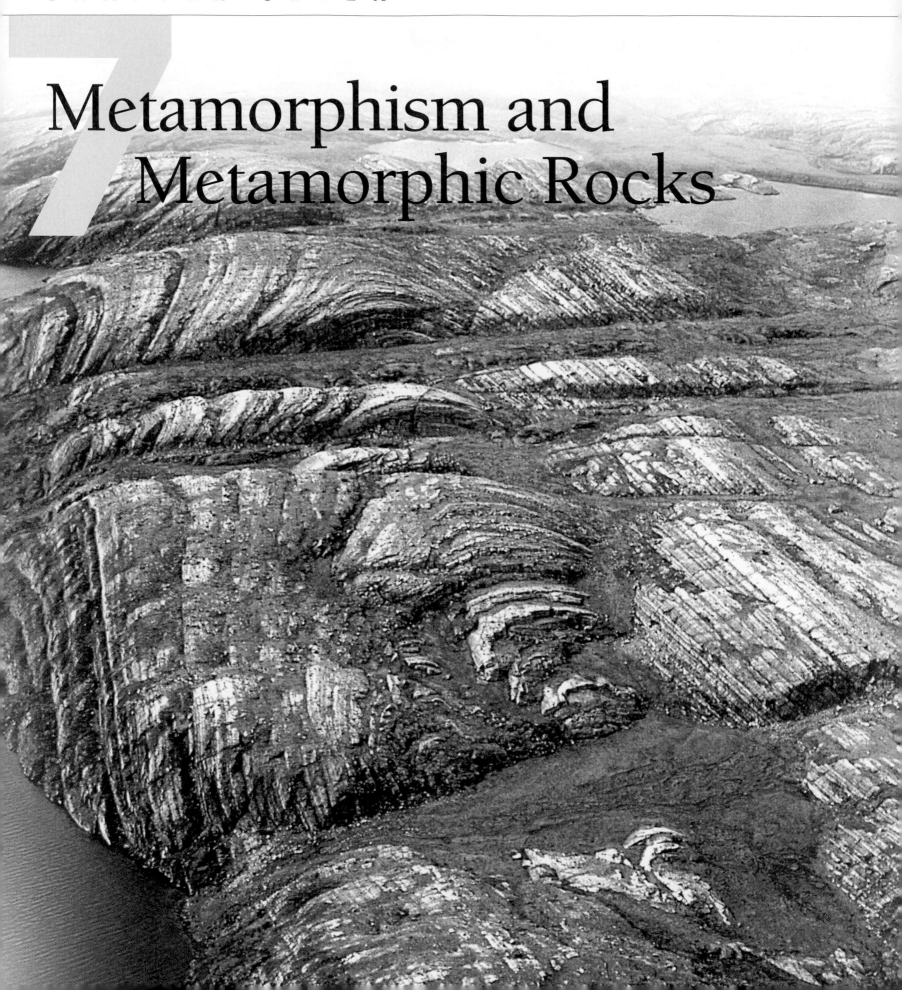

T HE FOLDED AND METAMORPHOSED ROCKS SHOWN IN THE PHOTO to the left were once flat-lying sedimentary strata. Compressional forces of unimaginable magnitude and temperatures hundreds of degrees above surface conditions prevailed for perhaps thousands or millions of years to produce the deformation displayed by these rocks. Under such extreme conditions, solid rock responds by folding, fracturing, and often flowing. This chapter looks at the tectonic forces that forge metamorphic rocks and how these rocks change in appearance, mineral content, and sometimes even in overall chemical composition.

To assist you in learning the important concepts in this chapter, focus on the following questions:

⊙ What are metamorphic rocks, and how do they form?

⊙ In which three geologic settings is metamorphism most likely to occur?

⊙ What are the agents of metamorphism?

⊙ What are the two textural divisions of metamorphic rocks and the conditions associated with each?

⊙ What are the names, textures, and compositions of the common metamorphic rocks?

⊙ How is the intensity, or degree, of metamorphism reflected in the texture and mineral content of metamorphic rocks?

Aerial view of metamorphic rocks exposed in the Canadian Shield.
(Photo by Robert Hildebrand)

FOCUS ON CONCEPTS

What Is Metamorphism?

Metamorphic Rocks

ESSENTIALS OF GEOLOGY **Introduction**

Recall from the discussion of the rock cycle that metamorphism is the transformation of one rock type into another. Metamorphic rocks are produced from preexisting igneous, sedimentary, or even other metamorphic rocks. Thus, every metamorphic rock has a **parent rock**—the rock from which it was formed.

Metamorphism, which means to "change form," is a process that leads to changes in the mineral content, texture, and sometimes the chemical composition of rocks. Metamorphism takes place where preexisting rock is subjected to new conditions, usually elevated temperatures and pressures, that are significantly different from those in which it initially formed. In response to these new conditions, the rock gradually changes until a state of equilibrium with the new environment is achieved.

The intensity of metamorphism can vary substantially from one environment to another. For example, in low-grade metamorphic environments, the common sedimentary rock *shale* becomes the more compact metamorphic rock *slate*. Hand samples of these rocks are sometimes difficult to distinguish, illustrating that the transition from sedimentary to metamorphic is often gradual and the changes can be subtle.

In more extreme environments, metamorphism causes a transformation so complete that the identity of the parent rock cannot be determined. In high-grade metamorphism, such features as bedding planes, fossils, and vesicles that existed in the parent rock are obliterated. Further, when rocks deep in the crust (where temperatures are high) are subjected to directed pressure, the entire mass may deform, producing large-scale structures, mainly folds (**FIGURE 7.1**).

In the most extreme metamorphic environments, the temperatures approach

FIGURE 7.1 Deformed and folded gneiss, Anza Borrego Desert State Park, California. *(Photo by A. P. Trujillo/APT Photos)*

those at which rocks melt. However, *during metamorphism the rock remains essentially solid—* when complete melting occurs, we have entered the realm of igneous activity.

CONCEPT CHECK 7.1

❶ Metamorphism means to "change form." Describe how a rock may change during metamorphism.

❷ Briefly describe what is meant by the statement, "every metamorphic rock has a *parent rock*?"

What Drives Metamorphism?

Metamorphic Rocks

ESSENTIALS OF GEOLOGY **Agents of Metamorphism**

The agents of metamorphism include *heat, pressure (stress), and chemically active fluids.* During metamorphism, rocks are usually subjected to all three metamorphic agents simultaneously. However, the degree of metamorphism and the contribution of each agent vary greatly from one environment to another.

Heat as a Metamorphic Agent

The most important factor driving metamorphism is *heat* because it provides the energy needed to drive the chemical reactions that result in the recrystallization of existing minerals and/or the formation of new minerals. Recall from the discussion of igneous rocks that an increase in temperature causes the ions within a mineral to vibrate more rapidly. Even in a crystalline solid, where ions are strongly bonded, this elevated level of activity allows individual atoms to migrate more freely between sites in the crystalline structure.

CHANGES CAUSED BY HEAT. When Earth materials are heated, especially those that form in low-temperature environments, they are affected in two ways. First, heating promotes recrystallization of mineral grains. This is particularly true of sedimentary and volcanic

rocks that are composed of fine-grained clay and silt sized particles. Higher temperatures promote crystal growth in which fine particles join together to form larger grains with the same mineral composition.

Second, when rocks are heated, they eventually reach a temperature at which one or more minerals become chemically unstable. When this occurs, the constituent atoms begin to arrange themselves into crystalline structures that are more stable in the new high-temperature environment. These chemical reactions create new minerals with stable configurations that have an overall composition roughly equivalent to that of the original rock. (In some environments ions may actually migrate into or out of a rock, thereby changing its overall chemical composition.)

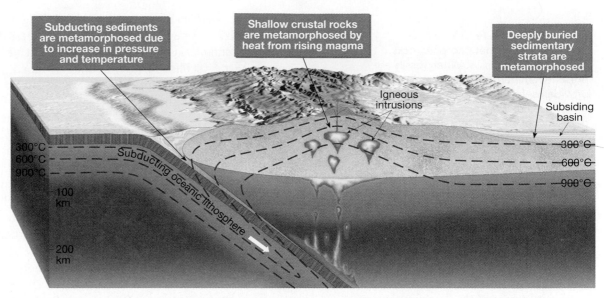

FIGURE 7.2 The geothermal gradient and its role in metamorphism. Notice how the geothermal gradient is lowered by the subduction of relatively cool oceanic lithosphere. By contrast, thermal heating is evident where magma intrudes the upper crust.

To summarize, imagine being a rock collector who is crossing a region where metamorphic rocks have been uplifted and then exposed by erosional processes. If you are traveling from an area where metamorphism was less intense to one where it had been more intense, you would expect to observe two changes that were largely attributable to temperature. The average size of the crystals in the rock samples you collected would increase, and your analysis of the rocks would show a change in mineral content.

WHAT IS THE SOURCE OF HEAT? Earth's internal heat comes mainly from energy that is continually being released by radioactive decay and thermal energy that remains from the time when our planet was forming. Recall that temperatures increase with depth at a rate known as the *geothermal gradient*. In the upper crust, this increase in temperature averages about 25 °C per kilometer (**FIGURE 7.2**). Thus, rocks that formed at Earth's surface will experience a gradual increase in temperature if they are transported to greater depths. When buried to a depth of about 8 kilometers (5 miles), where temperatures are about 200 °C, clay minerals tend to become unstable and begin to recrystallize into new minerals, such as chlorite and muscovite, that are stable in this environment. Chlorite is a micalike mineral formed by the metamorphism of dark (iron and magnesium rich) silicate minerals. However, many silicate minerals, particularly those found in crystalline igneous rocks—quartz and feldspar for example—remain stable at these temperatures. Thus, metamorphic changes in these minerals generally occur at much greater depths.

Environments where rocks may be carried to great depths and heated include convergent plate boundaries where slabs of sediment-laden oceanic crust are being subducted. Rocks may also become deeply buried in large basins where gradual subsidence results in very thick accumulations of sediment (see Figure 7.2). These basins, exemplified by the Gulf of Mexico, are known to develop low-grade metamorphic conditions near the base of the pile. In addition, continental collisions, which result in crustal thickening by folding and faulting, cause some rocks to be uplifted while others are thrust downward where elevated temperatures may cause metamorphism.

Heat may also be transported from the mantle into even the shallowest layers of the crust by igneous intrusions. Rising mantle plumes, upwelling at mid-ocean ridges, and magma generated by partial melting of mantle rock at subduction zones are three examples. Whenever magma forms and buoyantly rises toward the surface, metamorphism occurs. When magma intrudes relatively cool rocks at shallow depths, the host rock is "baked." This process, called *contact metamorphism,* will be considered later in the chapter.

Confining Pressure and Differential Stress

Pressure, like temperature, also increases with depth as the thickness of the overlying rock increases. Buried rocks are subjected to **confining pressure**, which is analogous to water pressure, in which the forces are applied equally in all directions (**FIGURE 7.3A**). The deeper you go in the ocean, the greater the confining pressure. The same is true for buried rock. Confining pressure causes the spaces between mineral grains to close, producing a more compact rock having a greater density. Furthermore,

as confining pressure increases some minerals recrystallize into new minerals that have the same chemical composition but a more compact crystalline form. Confining pressure does *not,* however, fold and deform rocks like those shown in Figure 7.1.

In addition to confining pressure, rocks may be subjected to directed pressure. This occurs, for example, at convergent plate boundaries where slabs of lithosphere collide. Here the forces that deform rock are unequal in different directions and are referred to as **differential stress** (Figure 7.3B). (A more in-depth discussion of *differential stress* is provided in Chapter 17).

Unlike confining pressure, which "squeezes" rock equally in all directions, differential stresses are greater in one direction than in others. As shown in

Figure 7.3B, rocks subjected to differential stress are shortened in the direction of greatest stress and elongated, or lengthened, in the direction perpendicular to that stress. As a result, the rocks involved are often *folded* or *flattened* (similar to when you step on a rubber ball). Along convergent plate boundaries the greatest differential stress is directed roughly horizontal in the direction of plate motion, and the least pressure is in the vertical direction. Consequently, in these settings the crust is greatly shortened (horizontally) and thickened (vertically). Although, differential stresses are generally small when compared to confining pressure, they are important in creating the various large scale structures and textures exhibited by metamorphic rocks.

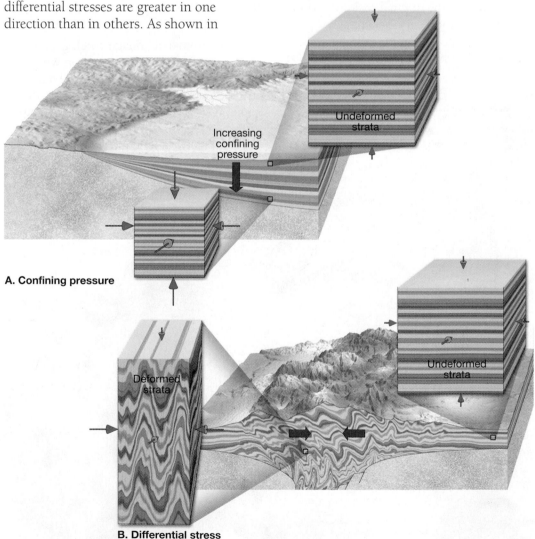

A. Confining pressure

B. Differential stress

FIGURE 7.3 Confining pressure and differential stress as metamorphic agents. **A.** In a depositional environment, as confining pressure increases, rocks deform by decreasing in volume. **B.** During mountain building, rocks subjected to differential stress are shortened in the direction that pressure is applied and lengthened in the direction perpendicular to that force.

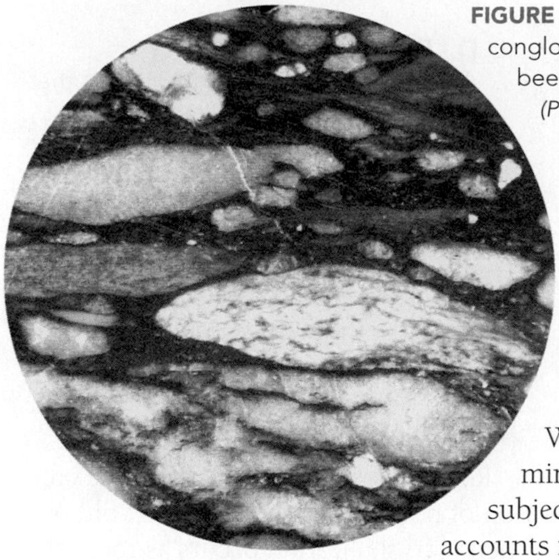

FIGURE 7.4 Metaconglomerate, also called stretched pebble conglomerate. These once nearly spherical pebbles have been heated and flattened into elongated structures. *(Photo by E. J. Tarbuck)*

In surface environments where temperatures are comparatively low, rocks are *brittle* and tend to fracture when subjected to differential stress. Continued deformation grinds and pulverizes the mineral grains into small fragments. By contrast, in high-temperature environments rocks are *ductile*. When rocks exhibit ductile behavior, their mineral grains tend to flatten and elongate when subjected to differential stress (**FIGURE 7.4**). This accounts for their ability to deform by flowing (rather than fracturing) to generate intricate folds.

Chemically Active Fluids

Many minerals, including clays, micas, and amphiboles, are hydrated—meaning they contain water in their crystalline structures. Elevated temperatures and pressures cause the dehydration of these minerals. Once expelled, these hot fluids promote recrystallization by enhancing the migration of mineral matter.

As discussed earlier, the metamorphism of shale to slate involves clay minerals that recrystallize to form mica and chlorite minerals. Hot fluids enhance this process by dissolving and transporting ions from one site in the crystal structure to another. In increasingly hotter environments these fluids become correspondingly more reactive.

In some metamorphic environments, hot fluids transport mineral matter over considerable distances. This occurs, for example, when hot, mineral-rich fluids are expelled from a magma body as it cools and solidifies. If the rocks that surround the pluton differ markedly in composition from the invading fluids, there may be an exchange of ions between the fluids and host rocks. When this occurs, the overall chemical composition of the surrounding rock changes. When substantial chemical change accompanies metamorphism the process is called **metasomatism.**

The Importance of Parent Rock

Most metamorphic rocks have the same overall chemical composition as the parent rock from which they formed, except for the possible loss or acquisition of volatiles such as water (H_2O) and carbon dioxide (CO_2). Therefore, when trying to establish the parent material from which metamorphic rocks were derived, the most important clue comes from their chemical composition.

Consider the large exposures of the metamorphic rock marble found high in the Alps of southern Europe. Because marble and the common sedimentary rock limestone have the same mineral content (calcite, $CaCO_3$), it seems reasonable to conclude that limestone is the parent rock of marble. Furthermore, because limestone usually forms in warm, shallow marine environments we can surmise that considerable deformation must have occurred to convert limy deposits in a shallow sea into marble crags in the lofty Alps.

The mineral makeup of the parent rock also largely determines the degree to which each metamorphic agent will cause change. For example, when magma forces its way into surrounding rock, high temperatures and hot fluids may alter the host rock. If the host rock is composed of minerals that are comparatively unreactive, such as quartz grains in sandstone, any alterations that may occur will be confined to a narrow zone next to the pluton. However, when the host rock is limestone, which is highly reactive, the zone of metamorphism may extend far from the intrusion.

DID YOU KNOW?
Glacial ice is a metamorphic rock that exhibits ductile flow much like hot rocks buried deep within Earth's crust. Although we think of glacial ice as being cold, it is in fact "hot," relative to its melting temperature. Therefore, we should not be surprised that the ice within a glacier gradually flows downslope in response to the force of gravity.

CONCEPT CHECK 7.2

❶ What drives metamorphism?

❷ Why is heat considered the most important agent of metamorphism?

❸ How is confining pressure different from differential stress?

❹ What role do chemically active fluids play in metamorphism?

❺ In what two ways can the parent rock affect the metamorphic process?

Metamorphic Textures

Metamorphic Rocks

ESSENTIALS OF GEOLOGY **Textural and Mineralogical Changes**

Recall that the term **texture** is used to describe the size, shape, and arrangement of grains within a rock. Most igneous

Earth's interior is the source of heat that drives metamorphism. *(Photo by Hubert Stadler/ Corbis)*

heat

and many sedimentary rocks consist of mineral grains that have a random orientation and thus appear the same when viewed from any direction. By contrast, deformed metamorphic rocks that contain platy minerals (micas) and/or elongated minerals (amphiboles), typically display some kind of *preferred orientation* in which the mineral grains exhibit a parallel to subparallel alignment. Like a fistful of pencils, rocks containing elongated minerals that are oriented parallel to each other will appear different when viewed from the side than when they are viewed head-on. A rock that exhibits a preferred orientation of its minerals is said to possess *foliation*.

Foliation

The term **foliation** refers to any planar (nearly flat) arrangement of mineral grains or structural features within a rock. Although foliation may occur in some sedimentary and even a few types of igneous rocks, it is a fundamental characteristic of regionally metamorphosed rocks—that is, rock units that have been strongly deformed, mainly by folding. In metamorphic environments, foliation is ultimately driven by compressional stresses that shorten rock units, causing mineral grains in preexisting rocks to develop parallel, or nearly parallel, alignments. Examples of foliation include the parallel alignment of platy minerals; the parallel alignment of flattened pebbles; compositional banding in which the separation of dark and light minerals generates a layered appearance; and rock cleavage where rocks can be easily

split into tabular slabs. These diverse types of foliation can form in many different ways, including:

1. Rotation of platy and/or elongated mineral grains into a parallel or nearly parallel orientation.
2. Recrystallization that produces new minerals with grains that exhibit a preferred orientation.
3. Mechanisms that change spherically shaped grains into elongated shapes that are aligned in a preferred orientation.

The rotation of existing mineral grains is the easiest of these mechanisms to envision. **FIGURE 7.5** illustrates the mechanics by which platy or elongated minerals are rotated. Note that the new alignment is roughly perpendicular to the direction of maximum shortening. Although physical rotation of platy minerals contributes to the development of foliation in low-grade metamorphism, other mechanisms dominate in more extreme environments.

Recall that recrystallization is the creation of new mineral grains out of old ones. When recrystallization occurs as rock is being subjected to differential stresses, any elongated and platy minerals that form tend to recrystallize perpendicular to the direction of maximum stress. Thus, the newly formed mineral grains will possess a parallel alignment and the metamorphic rock containing them will exhibit foliation.

Mechanisms that change the shapes of existing grains are especially important for

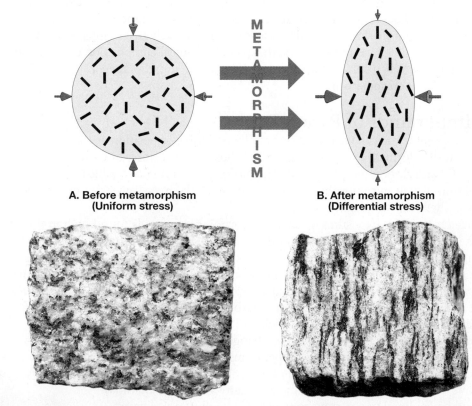

A. Before metamorphism (Uniform stress)

B. After metamorphism (Differential stress)

FIGURE 7.5 Mechanical rotation of platy or elongated mineral grains. **A.** Existing mineral grains keep their random orientation if force is uniformly applied. **B.** As differential stress causes rocks to flatten, mineral grains rotate toward the plane of flattening. *(Photos by E. J. Tarbuck)*

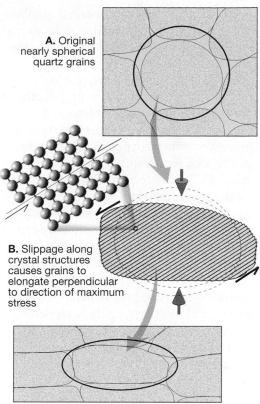

A. Original nearly spherical quartz grains

B. Slippage along crystal structures causes grains to elongate perpendicular to direction of maximum stress

D. Flattened rock containing elongated quartz grains

FIGURE 7.6 Development of preferred orientations of minerals that have roughly spherical crystals, such as quartz. This mechanism for changing the shape of mineral grains occurs when units of the mineral's crystalline structure slide relative to one another.

the development of preferred orientations in rocks that contain minerals such as quartz, calcite, and olivine—minerals that normally develop roughly spherical crystals.

These processes operate in metamorphic environments where differential stresses exist. A change in grain shape can occur as units of a mineral's crystalline structure slide relative to one another along discrete planes, thereby distorting the grain as shown in **FIGURE 7.6**. This type of gradual solid-state flow involves slippage that disrupts the crystal lattice as atoms shift positions. This process involves the breaking of existing chemical bonds and the formation of new ones.

The shape of a mineral may also change as ions move from a highly stressed location along the margin of the grain to a less-stressed position on the same grain. Mineral matter dissolves where grains are in contact with each other (areas of high

stress) and precipitates in pore spaces (areas of low stress). As a result, the mineral grains tend to become elongated in the direction of maximum stress. This mechanism is aided by hot, chemically active fluids.

Foliated Textures

Various types of foliation exist, depending largely upon the grade of metamorphism and the mineral content of the parent rock. We will look at three: *rock* or *slaty cleavage, schistosity,* and *gneissic texture.*

ROCK OR SLATY CLEAVAGE. Rock cleavage refers to closely spaced, flat surfaces along which rocks split into thin slabs when hit with a hammer. Rock cleavage develops in various metamorphic rocks but is best displayed in slates, which exhibit an excellent splitting property called **slaty cleavage.**

Depending on the metamorphic environment and the composition of the parent rock, rock cleavage develops in a number of ways. In a low-grade metamorphic environment, rock cleavage is known to

develop where beds of shale (and related sedimentary rocks) are strongly folded and metamorphosed to form slate. The process begins as platy grains are kinked and bent—generating microscopic folds having limbs (sides) that are roughly aligned (**FIGURE 7.7**). With further deformation, this new alignment is enhanced as old grains break down and recrystallize preferentially in the direction of the newly developed orientation. In this manner the rock develops narrow parallel zones where mica flakes are concentrated. These features alternate with zones containing quartz and other mineral grains that do not exhibit a pronounced linear orientation. It is along these very thin zones of platy mineral that slate splits (**FIGURE 7.8**).

Because slate typically forms during the low-grade metamorphism of shale, evidence of the original sedimentary bedding planes is often preserved. However, as Figure 7.7D illustrates, the orientation of slate's cleavage usually develops at an oblique angle to the original sedimentary layers. Thus, unlike shale, which splits along bedding planes, slate often splits

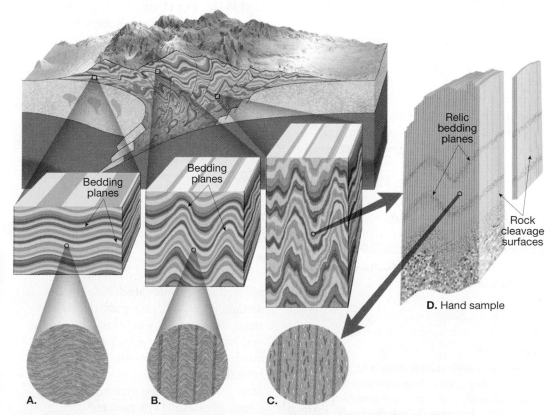

FIGURE 7.7 Development of rock cleavage. As shale is strongly folded (**A., B.**) and metamorphosed to form slate, the developing mica flakes are bent into microfolds. **C.** Further metamorphism results in the recrystallization of mica grains along the limbs of these folds to enhance the foliation. **D.** This hand sample of slate illustrates rock cleavage and its orientation to relic bedding surfaces.

FIGURE 7.8 Excellent slaty cleavage is exhibited by the rock in this slate quarry. Because slate breaks into flat slabs, it has many uses. *(Photo by Fred Bruemmer/Photolibrary)* The inset photo shows the use of slate for the roof of this house in Switzerland. *(Photo by E. J. Tarbuck)*

across them. Other metamorphic rocks, such as schists and gneisses, may also split along planar surfaces and exhibit rock cleavage.

SCHISTOSITY. Under higher temperature-pressure regimes, the minute mica and chlorite crystals in slate begin to grow. When these platy minerals are large enough to be discernible with the unaided eye and exhibit a planar or layered structure, the rock is said to exhibit a type of foliation called **schistosity.** Rocks having this texture are referred to as *schists.* In addition to platy minerals, schists often contain deformed quartz and feldspar crystals that appear flat or lens-shaped and are hidden among the mica grains.

GNEISSIC TEXTURE. During high-grade metamorphism, ion migration can result in the segregation of minerals as shown in **FIGURE 7.9.** Notice that the dark biotite crystals and light silicate minerals (quartz and feldspar) have separated, giving the rock a banded appearance called **gneissic texture.** A metamorphic rock with this texture is called *gneiss* (pronounced "nice"). Although foliated, gneisses will not usually split as easily as slates and some schists.

Other Metamorphic Textures

Not all metamorphic rocks exhibit a foliated texture. Those that *do not* are referred to as **nonfoliated.** Nonfoliated metamorphic rocks typically develop in environments where deformation is minimal and the parent rocks are composed of minerals that exhibit equidimensional crystals, such as quartz or calcite. For example, when a fine-grained limestone (made of calcite) is metamorphosed by the intrusion of a hot magma body, the small calcite grains recrystallize to form larger interlocking crystals. The resulting rock, *marble,* exhibits large, equidimensional

grains that are randomly oriented, similar to those in a coarse-grained igneous rock.

Another texture common to metamorphic rocks consists of unusually large grains, called *porphyroblasts,* that are surrounded by a fine-grained matrix of other minerals. **Porphyroblastic textures** develop in a wide range of rock types and metamorphic environments when minerals in the parent rock recrystallize to form new minerals. During recrystallization certain metamorphic minerals, including garnet, staurolite, and andalusite, often develop *a small number of very large crystals.* By contrast, minerals such as muscovite, biotite, and quartz typically form *a large number of very small grains.* As a result, when metamorphism generates the minerals garnet, biotite, and muscovite in the same setting, the rock will contain large crystals (porphyroblasts) of garnet embedded in a finer-grained matrix of biotite and muscovite (**FIGURE 7.10**).

CONCEPT CHECK 7.3

❶ What is foliation? Distinguish between *slaty cleavage, schistosity,* and *gneissic textures.*

❷ Briefly describe the three mechanisms by which minerals develop a preferred orientation.

❸ List some changes that might occur to a rock in response to metamorphic processes.

FIGURE 7.9 This rock displays a gneissic texture. Notice that the dark biotite flakes and light silicate minerals are segregated, giving the rock a banded or layered appearance. *(Photo by E. J. Tarbuck)*

Common Metamorphic Rocks

Metamorphic Rocks

ESSENTIALS OF GEOLOGY **Common Metamorphic Rocks**

Recall that metamorphism causes many changes in rocks, including increased density, change in grain size, reorientation of mineral grains into a planar arrangement known as foliation, and the transformation of low-temperature minerals into high-temperature minerals. Moreover, the introduction of ions may generate new minerals, some of which are economically important.

The major characteristics of some common metamorphic rocks are summarized in **FIGURE 7.11**. Notice that metamorphic rocks can be broadly classified by the type of foliation exhibited and to a lesser extent on the chemical composition of the parent rock.

Close up of porphyroblast

FIGURE 7.10 Garnet-mica schist. The dark red garnet crystals (porphyroblasts) are embedded in a matrix of fine-grained micas. *(Photo by E. J. Tarbuck)*

Rock Name		Texture		Grain Size	Comments	Original Parent Rock
Slate	Increasing Metamorphism	Foliated		Very fine	Excellent rock cleavage, smooth dull surfaces	Shale, mudstone, or siltstone
Phyllite				Fine	Breaks along wavy surfaces, glossy sheen	Shale, mudstone, or siltstone
Schist				Medium to Coarse	Micaceous minerals dominate, scaly foliation	Shale, mudstone, or siltstone
Gneiss				Medium to Coarse	Compositional banding due to segregation of minerals	Shale, granite, or volcanic rocks
Migmatite				Medium to Coarse	Banded rock with zones of light-colored crystalline minerals	Shale, granite, or volcanic rocks
Mylonite		Weakly Foliated		Fine	When very fine-grained, resembles chert, often breaks into slabs	Any rock type
Metaconglomerate				Coarse-grained	Stretched pebbles with preferred orientation	Quartz-rich conglomerate
Marble		Nonfoliated		Medium to coarse	Interlocking calcite or dolomite grains	Limestone, dolostone
Quartzite				Medium to coarse	Fused quartz grains, massive, very hard	Quartz sandstone
Hornfels				Fine	Usually, dark massive rock with dull luster	Any rock type
Anthracite				Fine	Shiny black rock that may exhibit conchoidal fracture	Bituminous coal
Fault breccia				Medium to very coarse	Broken fragments in a haphazard arrangement	Any rock type

FIGURE 7.11 Classification of common metamorphic rocks.

Foliated metamorphic rocks

Slate

Phyllite

Schist

Gneiss

Nonfoliated metamorphic rocks

Marble

Quartzite

FIGURE 7.12 Common metamorphic rocks. *(Photos by E. J. Tarbuck)*

Foliated Rocks

SLATE. Slate is a very fine-grained (less than 0.5 millimeter) foliated rock composed of minute mica flakes that are too small to be visible (**FIGURE 7.12**). Thus, slate generally appears dull and closely resembles shale. A noteworthy characteristic of slate is its excellent rock cleavage, or tendency to break into flat slabs (see Figure 7.8).

Slate is most often generated by the low-grade metamorphism of shale, mudstone, or siltstone. Less frequently it is produced when volcanic ash is metamorphosed. Slate's color depends on its mineral constituents. Black (carbonaceous) slate contains organic material; red slate gets its color from iron oxide; and green slate usually contains chlorite.

PHYLLITE. Phyllite represents a gradation in the degree of metamorphism between slate and schist. Its constituent platy minerals are larger than those in slate but not yet large enough to be readily identifiable with the unaided eye. Although phyllite appears similar to slate, it can be easily distinguished from slate by its glossy sheen and its sometimes wavy surface (Figure 7.12). Phyllite usually exhibits rock cleavage and is composed mainly of very fine crystals of either muscovite or chlorite, or both.

SCHIST. Schists are medium- to coarse-grained metamorphic rocks in which platy minerals predominate. These flat components commonly include the micas (muscovite and biotite), which display a planar alignment that gives the rock its foliated texture. In addition, schists contain smaller amounts of other minerals, often quartz and feldspar. Schists composed mostly of dark minerals (amphiboles) are known. Like slate, the parent rock of many schists is shale, which has undergone medium- to high-grade metamorphism during major mountain-building episodes.

The term *schist* describes the texture of a rock and as such it is used to describe rocks having a wide variety of chemical compositions. To indicate the composition, mineral names are used. For example, schists composed primarily of muscovite and biotite are called *mica schist* (Figure 7.12). Depending upon the degree of metamorphism and composition of the parent rock, mica schists often contain *accessory minerals,* some of which are unique to metamorphic rocks. Some common accessory minerals that occur as porphyroblasts include *garnet, staurolite,* and *sillimanite,* in which case the rock is called *garnet-mica schist, staurolite-mica schist,* and so forth (see Figure 7.10).

In addition, schists may be composed largely of the minerals chlorite or talc, in

FIGURE 7.13 Marble, because of its workability, is a widely used building stone. The exterior of the Taj Mahal is constructed primarily of the metamorphic rock marble. *(Photo by Steve Vider/Superstock)*

which case they are called *chlorite schist* and *talc schist,* respectively. Both chlorite and talc schists can form when rocks with a basaltic composition undergo metamorphism. Others contain the mineral *graphite,* which is used as pencil "lead," graphite fibers (used in fishing rods), and lubricant (commonly for locks).

GNEISS. *Gneiss* is the term applied to medium- to coarse-grained banded metamorphic rocks in which granular and elongated (as opposed to platy) minerals predominate. The most common minerals in gneiss are quartz, potassium feldspar, and sodium-rich plagioclase feldspar. Most gneisses also contain lesser amounts of biotite, muscovite, and amphibole that develop a preferred orientation. Some gneisses will split along the layers of platy minerals, but most break in an irregular fashion. Recall that during high-grade metamorphism the light and dark components separate, giving gneisses their characteristic banded or layered appearance. Thus, most gneisses consist of alternating bands of white or reddish feldspar-rich zones and layers of dark ferromagnesian minerals (Figure 7.12).

These banded gneisses often exhibit evidence of deformation, including folds and sometimes faults (see Figure 7.1).

Most gneisses have a felsic composition and are often derived from granite or its fine-grained equivalent, rhyolite. However, many form from the high-grade metamorphism of shale. In this instance, gneiss represents the last rock in the sequence of shale, slate, phyllite, schist, and gneiss. Like schists, gneisses may also include large crystals of accessory minerals such as garnet and staurolite. Gneisses made up primarily of dark minerals such as those that compose basalt also occur. For example, an amphibole-rich rock that exhibits a gneissic texture is called *amphibolite*.

Nonfoliated Rocks

MARBLE. Marble is a coarse, crystalline metamorphic rock whose parent was limestone or dolostone (Figure 7.12). Pure marble is white and composed essentially of the mineral calcite. Because of its relative softness (hardness of 3), marble is easy to cut and shape. White marble is particularly prized as a stone from which to create monuments and statues, such as the Taj Mahal in India (**FIGURE 7.13**). Unfortunately, marble's composition of calcium carbonate causes it to weather when exposed to acid rain.

The parent rocks from which most marbles form contain impurities that color the stone. Thus, marble can be pink, gray, green, or even black and may contain a variety of accessory minerals (chlorite, mica, garnet, and wollastonite). When marble forms from limestone interbedded with shales, it will appear banded and exhibit visible foliation. When deformed, these banded marbles may develop highly contorted mica-rich folds that give the rock a rather artistic design. Hence, these decorative marbles have been used as a building stone since prehistoric times.

QUARTZITE. Quartzite is a very hard metamorphic rock formed from quartz sandstone (Figure 7.12). Under moderate- to high-grade metamorphism, the quartz grains in sandstone fuse together. The recrystallization is often so complete that when broken, quartzite will split through the quartz grains rather than along their boundaries. In some instances, sedimentary features such as cross-bedding are preserved and give the rock a banded appearance. Pure quartzite is white, but iron oxide may produce reddish or pinkish stains, while dark mineral grains may impart a gray color.

CONCEPT CHECK 7.4

❶ Slate and phyllite resemble each other. How might you distinguish one from the other?

❷ In the rock mica schist, what does *mica* indicate and what does *schist* indicate?

❸ How are marble and quartzite similar? How are they different?

Metamorphic Environments

There are many environments in which metamorphism occurs. Most are in the vicinity of plate margins, and several are associated with igneous activity. We will consider the following types of metamorphism: (1) *contact* or *thermal metamorphism*; (2) *hydrothermal metamorphism*; (3) *burial and subduction zone metamorphism*; (4) *regional metamorphism*; (5) *metamorphism along faults*; (6) *impact metamorphism* and with the exception of impact metamorphism, there is considerable overlap among the types.

Contact or Thermal Metamorphism

Contact or **thermal metamorphism** occurs when rocks immediately surrounding a molten igneous body are "baked" and therefore altered from their original state. The altered rocks occur in a zone called a metamorphic **aureole** (**FIGURE 7.14**). The emplacement of small intrusions such as dikes and sills typically form aureoles only a few centimeters thick, while large igneous plutons that generate batholiths can produce aureoles extending outward for several kilometers.

In addition to the size of the magma body, the mineral composition of the host rock and the availability of water greatly affect the size of the aureole produced. In chemically active rock such as limestone, the zone of alteration can be 10 kilometers (6 miles) thick. These large aureoles often consist of distinct *zones of metamorphism*. Near the magma body, high-temperature minerals such as garnet may form, whereas farther away low-grade minerals such as chlorite are produced.

Although contact metamorphism is not entirely restricted to shallow crustal depths, it is most easily recognized when it occurs in this setting. Here, the temperature contrast between the molten body and the surrounding host rock is large. Because contact metamorphism does not involve directed pressure, rocks found within a metamorphic aureole are usually not foliated.

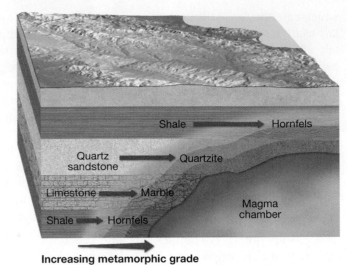

Increasing metamorphic grade

FIGURE 7.15 Contact metamorphism of shale yields hornfels, while contact metamorphism of quartz sandstone and limestone produces quartzite and marble, respectively.

DID YOU KNOW?
Because marble can be carved readily, it has been used for centuries for buildings and memorials. Examples of important structures whose exteriors are clad in marble include the Parthenon in Greece, the Taj Mahal in India, and the Washington Monument in the United States.

During contact metamorphism of mudstones and shales, the clay minerals are baked as if placed in a kiln. The result is a very hard, fine-grained metamorphic rock called *hornfels* (**FIGURE 7.15**). Hornfels can form from a variety of materials

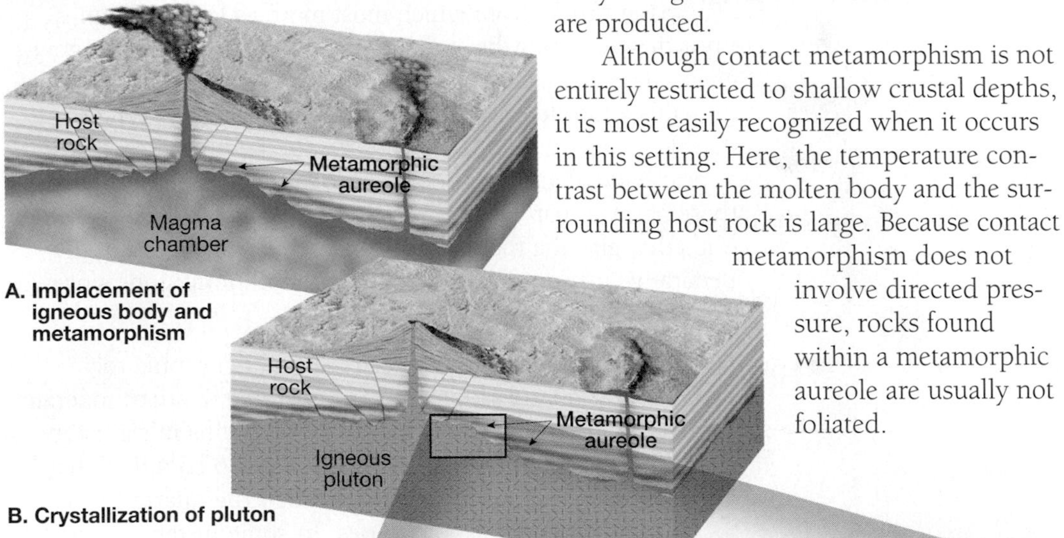

FIGURE 7.16 Hydrothermal metamorphism can occur at shallow crustal depths in regions where hot springs and geysers are active. (*Photo by Philippe Clement/Nature Picture Library*)

A. Implacement of igneous body and metamorphism

B. Crystallization of pluton

FIGURE 7.14 Contact metamorphism produces a zone of alteration called an *aureole* around an intrusive igneous body. In the photo, the dark layer, called a *roof pendant*, consists of metamorphosed host rock adjacent to the upper part of the light-colored igneous pluton. The term *roof pendant* implies that the rock was once the roof of a magma chamber. Sierra Nevada, near Bishop, California. (*Photo by John S. Shelton*)

C. Uplift and erosion expose pluton and metamorphic cap rock

including volcanic ash and basalt. In some cases, large grains of metamorphic minerals, such as garnet and staurolite, may form, giving the hornfels a porphyroblastic texture (see Figure 7.10).

Hydrothermal Metamorphism

When hot, ion-rich fluids circulate through fissures and cracks in rock, a chemical alteration called **hydrothermal metamorphism** occurs (**FIGURE 7.16**). This type of metamorphism is often closely associated with the emplacement of magma. As large magma bodies cool and solidify, silica-rich fluids (mainly water) are driven into the host rocks. When the host rock is highly fractured, mineral matter contained in these **hydrothermal solutions** may precipitate to form a variety of minerals, some of which are economically important. If the host rocks are permeable and highly reactive, such as the carbonate rock limestone, silicate-rich hydrothermal solutions react to produce a variety of calcium-rich silicate minerals. Recall that a metamorphic process that alters the overall chemical composition of a rock unit is called *metasomatism*.

As our understanding of plate tectonics grew, it became clear that the most widespread occurrence of

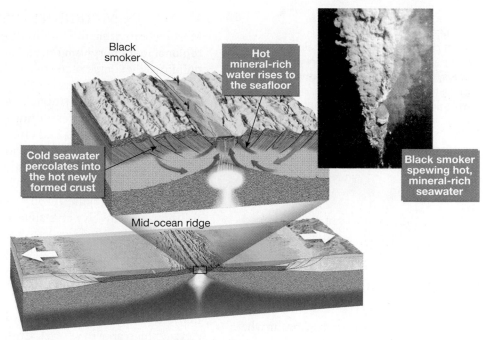

FIGURE 7.17 Hydrothermal metamorphism along a mid-ocean ridge. *(Photo by R. Ballard/Woods Hole)*

hydrothermal metamorphism is along the axis of the mid-ocean ridge system. As plates move apart, upwelling magma from the mantle generates new seafloor. As seawater percolates through the young, hot oceanic crust, it is heated and chemically reacts with the newly formed basaltic rocks (**FIGURE 7.17**). The result is the conversion of ferromagnesian minerals, such as olivine and pyroxene, into hydrated silicates, such as serpentine, chlorite, and talc. In addition, calcium-rich plagioclase feldspars in basalt become increasingly sodium enriched as the salt (NaCl) in seawater exchanges Na ions for Ca ions.

Hydrothermal solutions circulating through the seafloor also remove large amounts of metals, such as iron, cobalt, nickel, silver, gold, and copper, from the newly formed crust. These hot, metal-rich fluids eventually rise along fractures and gush from the seafloor at temperatures of about 350 °C, generating particle-filled clouds called *black smokers*. Upon mixing with the cold seawater, sulfides and carbonate minerals containing these heavy metals precipitate to form metallic deposits, some of which are economically valuable. This process is believed to be the origin of the copper ores mined today on the Mediterranean island of Cyprus.

Burial and Subduction Zone Metamorphism

Burial metamorphism tends to occur where massive amounts of sedimentary or volcanic material accumulates in a subsiding basin (see Figure 7.2). Here, low-grade metamorphic conditions may be attained within the deepest layers. Confining pressure and geothermal heat drive the recrystallization of the constituent minerals—changing the texture and/or mineral content of the rock without appreciable deformation.

The depth required for burial metamorphism varies from one location to another, depending mainly on the prevailing geothermal gradient. Metamorphism typically begins at depths of about 8 kilometers (5 miles), where temperatures are about 200 °C. However, in areas that exhibit large geothermal gradients and where molten rock has been emplaced near the surface, such as near the Salton Sea in California and in northern New Zealand, drilling operations have collected metamorphic minerals from depths of only a few kilometers.

Rocks and sediments can also be carried to great depths along convergent boundaries where oceanic lithosphere is being subducted. This phenomenon, called **subduction zone metamorphism**, differs from burial metamorphism in that differential stresses play a major role in deforming rock as it is metamorphosed. Furthermore, metamorphic rocks that form along subduction zones are often further metamorphosed by the collision of two continental blocks.

Regional Metamorphism

Most metamorphic rock is produced by **regional metamorphism** during mountain building when large segments of Earth's crust are intensely deformed along convergent plate boundaries (**FIGURE 7.18**). This activity occurs most often during continental collisions. Sediments and crustal rocks that form the margins of the colliding continental blocks are folded and faulted, causing them to shorten and thicken like a rumpled carpet (Figure 7.18B). Continental collisions also involve crystalline continental basement rocks, as well as slices of oceanic crust that once floored the intervening ocean basin.

The general thickening of the crust that occurs during mountain building results in buoyant lifting, in which deformed rocks are elevated high above sea level. Crustal thickening also results in the deep burial of large quantities of rock as crustal blocks are thrust one beneath another. Deep in the roots of mountains, elevated temperatures caused by deep burial are responsible for the most productive and intense metamorphic activity within a mountain belt. Often, these deeply buried rocks become heated to their melting point. As a result, magma collects until it forms bodies large enough to rise buoyantly

and intrude the overlying metamorphic and sedimentary rocks (Figure 7.18B). Consequently, the cores of many mountain ranges consist of folded and faulted metamorphic rocks, often intertwined with igneous bodies. Over time, these deformed rock masses are uplifted, and erosion removes the overlying material to expose the igneous and metamorphic rocks that comprise the central core of the mountain range.

Other Metamorphic Environments

Other types of metamorphism, that generate relatively small amounts of metamorphic rock, tend to be localized.

METAMORPHISM ALONG FAULT ZONES. Near the surface, rock behaves like a brittle solid. Consequently, movement along a

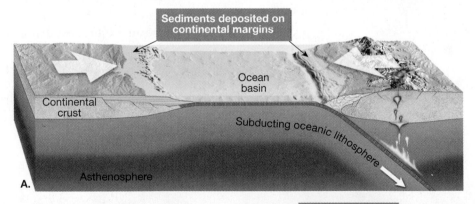

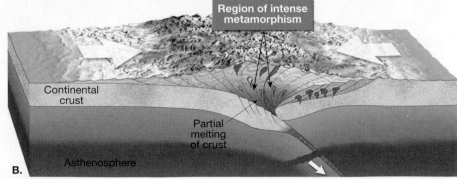

FIGURE 7.18 Regional metamorphism occurs where rocks are squeezed between two converging lithospheric plates during mountain building.

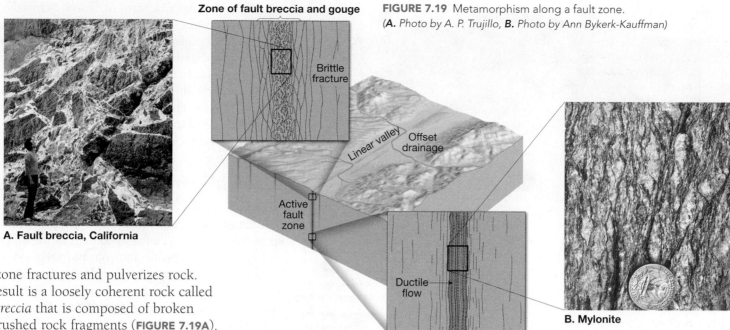

Zone of fault breccia and gouge

FIGURE 7.19 Metamorphism along a fault zone. (**A.** Photo by A. P. Trujillo, **B.** Photo by Ann Bykerk-Kauffman)

Brittle fracture

Linear valley   Offset drainage

Active fault zone

Ductile flow

A. Fault breccia, California

Zone of mylonite

B. Mylonite

fault zone fractures and pulverizes rock. The result is a loosely coherent rock called *fault breccia* that is composed of broken and crushed rock fragments (**FIGURE 7.19A**). Displacements along California's San Andreas Fault have created a zone of fault breccia and related rock types more than 1000 kilometers long and up to 3 kilometers wide.

In some shallow fault zones a soft, uncemented claylike material called *fault gouge* is also produced. Fault gouge is formed by the crushing and grinding of rock material during fault movement. The resulting crushed material is further altered by groundwater that infiltrates the porous fault zone.

Much of the deformation associated with fault zones occurs at great depth and thus at high temperatures. In this environment preexisting minerals deform by ductile flow. As large slabs of rock move in opposite directions, the minerals in the fault zone between them tend to form elongated grains that give the rock a foliated or lineated appearance (Figure 7.19B). Rocks formed in these zones of intense ductile deformation are termed *mylonites* (*mylo* = a mill, *ite* = a stone).

DID YOU KNOW?
Although most of central North America has a rather subdued topography, the underlying bedrock is similar to rocks found in the cores of metamorphosed mountain belts. This strongly supports the view that these rocks once formed the roots of ancient mountain chains that may have risen as high as the present-day Himalayas.

IMPACT METAMORPHISM. Impact (or **shock**) **metamorphism** occurs when high-speed projectiles called *meteorites* (fragments of comets or asteroids) strike Earth's surface. Upon impact the energy of the once rapidly moving meteorite is transformed into heat energy and shock waves that pass through the surrounding rocks. The result is pulverized, shattered, and sometimes melted rock.

The products of these impacts, called *impactiles,* include mixtures of fused fragmented rock plus glass-rich ejecta that resemble volcanic bombs. In some cases, a very dense form of quartz (*coesite*) and minute *diamonds* are found. These high-pressure minerals provide convincing evidence that pressures and temperatures as great as those existing in the upper mantle must have been attained for at least a brief moment.

CONCEPT CHECK 7.5

❶ Distinguish between contact metamorphism and regional metamorphism. Which creates the greater quantity of metamorphic rock?

❷ Where does most hydrothermal metamorphism occur?

❸ Describe burial metamorphism.

❹ With which type of plate boundary is regional metamorphism associated?

❺ Why do metamorphic rocks often comprise the interiors of Earth's major mountain belts?

Metamorphic Zones

In areas affected by metamorphism, there are usually systematic variations in the mineral content and texture of the rocks that can be observed as we traverse the region. These differences are clearly related to variations in the degree of metamorphism experienced in each metamorphic zone.

Textural Variations

When we begin with a clay-rich sedimentary rock such as shale or mudstone, a gradual increase in metamorphic intensity is accompanied by a general coarsening of the grain size. Thus, we observe shale changing to a fine-grained slate, which then forms phyllite and,

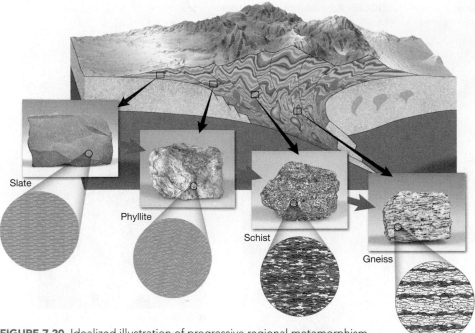

Slate

Phyllite

Schist

Gneiss

FIGURE 7.20 Idealized illustration of progressive regional metamorphism. From left to right, we progress from low-grade metamorphism (slate) to high-grade metamorphism (gneiss). *(Photos by E. J. Tarbuck)*

broadly folded Appalachians of central Pennsylvania, the rocks that once formed flat-lying beds are folded and display a preferred orientation of platy mineral grains as exhibited by well-developed slaty cleavage. As we move farther eastward into the intensely deformed crystalline Appalachians, we find large outcrops of schists. The most intense zones of metamorphism are found in Vermont and New Hampshire, where gneissic rocks outcrop.

Index Minerals and Metamorphic Grade

In addition to textural changes, we encounter corresponding changes in mineral content as we shift from regions of low-grade metamorphism to regions of high-grade metamorphism. An idealized transition in mineralogy that results from the regional metamorphism of shale is shown in **FIGURE 7.21**. The first new mineral to form as shale changes to slate is chlorite. At higher temperatures flakes of muscovite and biotite begin to dominate. Under more extreme conditions, metamorphic rocks may contain garnet and staurolite crystals. At temperatures approaching the melting point of rock, sillimanite forms. Sillimanite is a high-temperature metamorphic mineral used to make refractory porcelains such as those used in spark plugs.

through continued recrystallization, generates a coarse-grained schist (**FIGURE 7.20**). Under more intense conditions a gneissic texture that exhibits layers of dark and light minerals may develop. This systematic transition in metamorphic textures can be observed as we approach the Appalachian Mountains from the west. Beds of shale, which once extended over large areas of the eastern United States, still occur as nearly flat-lying strata in Ohio. However, in the

Through the study of metamorphic rocks in their natural settings (called *field studies*) and through experimental studies, researchers have learned that certain minerals, such as those in Figure 7.21, are good indicators of the metamorphic environment in which they formed. Using these **index minerals,** geologists distinguish among different zones of regional metamorphism. For example, the mineral chlorite begins to form when temperatures are relatively low, less than 200 °C (**FIGURE 7.22**). Thus, rocks that contain chlorite (usually slates) are referred to as *low-grade*. By contrast, the mineral sillimanite only forms in extreme environments where temperatures exceed 500 °C, and rocks containing it are

FIGURE 7.21 The typical transition in mineral content that results from the progressive metamorphism of shale.

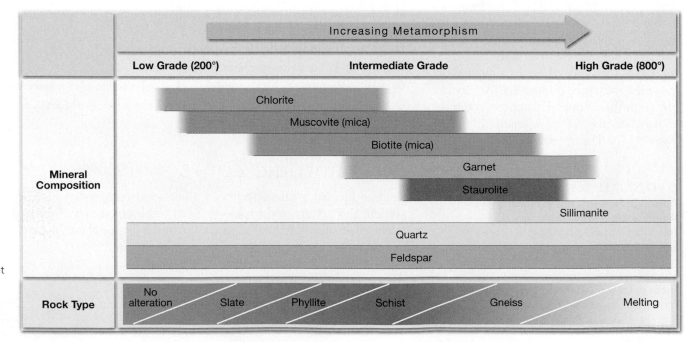

Increasing Metamorphism

	Low Grade (200°)	Intermediate Grade	High Grade (800°)
Mineral Composition	Chlorite	Muscovite (mica) · Biotite (mica) · Garnet · Staurolite	Sillimanite
	Quartz		
	Feldspar		
Rock Type	No alteration · Slate	Phyllite · Schist	Gneiss · Melting

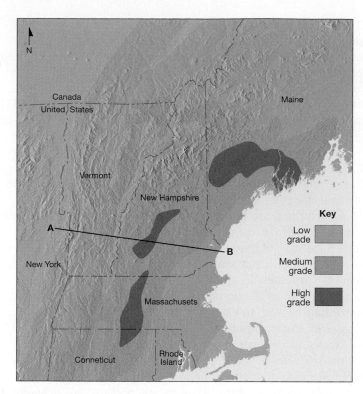

FIGURE 7.22 Generalized map showing zones of metamorphic intensities across New England.

FIGURE 7.23 Migmatite. The lightest-colored layers are igneous rock composed of quartz and feldspar, whereas the darker layers have a metamorphic origin.
(Photo by Harlan H. Roepke)

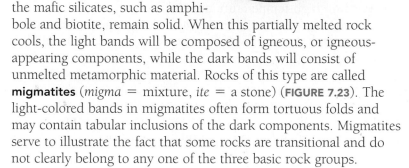

considered *high-grade.* By mapping the occurrences of index minerals, geologists are in effect mapping zones of varying metamorphic grade. *Grade* is a term used in a relative sense to refer to the conditions of temperature (or sometimes pressure) to which a rock has been subjected.

MIGMATITES. In the most extreme environments, even the highest-grade metamorphic rocks undergo change. For example, gneissic rocks may be heated sufficiently to cause melting to begin. However, recall from our discussion of igneous rocks that different

minerals melt at different temperatures. The light-colored silicates, usually quartz and potassium feldspar, have the lowest melting temperatures and begin to melt first, whereas the mafic silicates, such as amphibole and biotite, remain solid. When this partially melted rock cools, the light bands will be composed of igneous, or igneous-appearing components, while the dark bands will consist of unmelted metamorphic material. Rocks of this type are called **migmatites** (*migma* = mixture, *ite* = a stone) (**FIGURE 7.23**). The light-colored bands in migmatites often form tortuous folds and may contain tabular inclusions of the dark components. Migmatites serve to illustrate the fact that some rocks are transitional and do not clearly belong to any one of the three basic rock groups.

CONCEPT CHECK 7.6

❶ Briefly describe the textural changes that occur in the transformation of slate to phyllite, phyllite to schist, and then schist to gneiss.

❷ How do geologists use index minerals?

❸ How are gneisses and migmatites related?

CHAPTER 7
Metamorphism and Metamorphic Rocks in Review

◉ *Metamorphism* is the transformation of one rock type into another. *Metamorphic rocks* form from preexisting rocks (either igneous, sedimentary, or other metamorphic rocks) that have been altered by the agents of metamorphism, which include *heat, pressure (stress),* and *chemically active fluids.* During metamorphism the material essentially remains solid. The changes that occur in metamorphosed rocks are textural as well as mineralogical.

◉ The mineral makeup of the parent rock determines, to a large extent, the degree to which each metamorphic agent will cause change. Heat is the most important agent because it provides the energy to drive chemical reactions that result in the recrystalliza-

tion of minerals. Pressure, like temperature, also increases with depth. When subjected to *confining pressure,* minerals may recrystallize into more compact forms. During mountain building, rocks are subjected to *differential stress,* which tends to shorten them in the direction pressure is applied and lengthen them in the direction perpendicular to that force. At depth, rocks are warm and *ductile,* which accounts for their ability to deform by flowing when subjected to differential stresses. Chemically active fluids, most commonly water containing ions in solution, also enhance the metamorphic process by dissolving minerals and aiding the migration and precipitation of this material at other sites.

⊙ *The grade of metamorphism is reflected in the texture and mineral content of metamorphic rocks.* During regional metamorphism, rocks typically display a *preferred orientation* called *foliation.* Foliation develops as platy or elongated minerals are rotated into parallel alignment, recrystallize to form new grains that exhibit a preferred orientation, or are plastically deformed into flattened grains that exhibit a planar alignment. *Rock cleavage* is a type of foliation in which rocks split cleanly into thin slabs along surfaces where platy minerals are aligned. *Schistosity* is a type of foliation defined by the parallel alignment of medium- to coarse-grained platy minerals. During high-grade metamorphism, ion migrations can cause minerals to segregate into distinct layers or bands. Metamorphic rocks with a banded texture are called *gneiss.* Metamorphic rocks composed of only one mineral forming equidimensional crystals often appear *nonfoliated.* *Marble* (metamorphosed limestone) is often nonfoliated. Further, metamorphism can cause the transformation of low-temperature minerals into high-temperature minerals and, through the introduction of ions from *hydrothermal solutions,* generate new minerals, some of which form economically important metallic ore deposits.

⊙ Common foliated metamorphic rocks include *slate, phyllite,* various types of *schists* (e.g., garnet-mica schist), and *gneiss.* Nonfoliated rocks include *marble* (parent rock—limestone) and *quartzite* (most often formed from quartz sandstone).

⊙ The four geologic environments in which metamorphism commonly occurs are (1) *contact* or *thermal metamorphism,* (2) *hydrothermal metamorphism,* (3) *burial and subduction zone metamorphism,* and (4) *regional metamorphism.* Contact metamorphism occurs when rocks are in contact with an igneous body, resulting in the formation of zones of alteration around the magma called *aureoles.* Most contact metamorphic rocks are fine-grained, dense, tough rocks of various chemical compositions. Because directional pressure is not a major factor, these rocks are not generally foliated. Hydrothermal metamorphism occurs where hot, ion-rich fluids circulate through rock and cause chemical alteration of the constituent minerals. Most hydrothermal alteration occurs along the mid-ocean ridge system, where seawater migrates through hot oceanic crust and chemically alters newly formed basaltic rocks. Metallic ions that are removed from the crust are eventually carried to the floor of the ocean, where they precipitate from black smokers to form metallic deposits, some of which may be economically important. Regional metamorphism takes place at considerable depths over an extensive area and is associated with the process of mountain building. A gradation in the degree of change usually exists in association with regional metamorphism, in which the intensity of metamorphism (low- to high-grade) is reflected in the texture and mineral content of the rocks. In the most extreme metamorphic environments, rocks called *migmatites* fall into a transition zone *somewhere between* "true" igneous rocks and "true" metamorphic rocks.

Key Terms

aureole (p. 188)
burial metamorphism (p. 190)
confining pressure (p. 180)
contact metamorphism (p. 188)
differential stress (p. 180)
foliation (p. 182)
gneissic texture (p. 184)

hydrothermal metamorphism (p. 189)
hydrothermal solutions (p. 189)
impact metamorphism (p. 191)
index minerals (p. 192)
metamorphism (p. 178)
metasomatism (p. 181)

migmatites (p. 193)
nonfoliated (p. 184)
parent rock (p. 178)
porphyroblastic textures (p. 184)
regional metamorphism (p. 190)
rock cleavage (p. 183)
schistosity (p. 184)

shock metamorphism (p. 191)
slaty cleavage (p. 183)
subduction zone metamorphism (p. 190)
texture (p. 181)
thermal metamorphism (p. 188)

GIVE IT SOME THOUGHT

❶ Each of the following statements describes one or more characteristics of a particular metamorphic rock. For each statement, identify the metamorphic rock that is being described:
  a. calcite-rich and often nonfoliated
  b. loosely coherent and composed of broken fragments that formed along a fault zone
  c. represents a grade of metamorphism between slate and schist
  d. very fine-grained and foliated; excellent rock cleavage
  e. foliated and composed predominately of platy materials
  f. composed of alternating bands of light and dark silicate minerals
  g. hard and nonfoliated; resulting from contact metamorphism

❷ Refer to Figure 7.5 showing the formation of foliation in response to differential stress. Select a location on Earth where rocks may be experiencing or may have experienced the metamorphic conditions necessary to form foliation. Describe the location, and explain what is happening or has happened that would result in foliated rocks.

❸ Refer to Figure 7.11, the classification chart for common metamorphic rocks, and answer the following:
  a. Identify a case in which the metamorphic rock may be stronger than its parent rock, and explain why.
  b. Identify a case in which the metamorphic rock may be weaker than its parent rock, and explain why.
  c. In general, do you think metamorphic rocks are stronger or weaker than igneous rocks? How about sedimentary rocks? Explain your answers.

❹ One of the rock outcrops in the accompanying photos consists mainly of metamorphic rock. Which do you think it is? Explain why you ruled out the other rock bodies.

A.   B.   C.

❺ Examine the accompanying photos that show the geology of the Grand Canyon. Notice that most of the canyon consists of layers of sedimentary rocks, but if you were to hike into the inner gorge you would encounter the Vishnu schist, a metamorphic rock.

   a. What process might have been responsible for the formation of the Vishnu schist? How does this process differ from the processes that formed the sedimentary rocks that are atop the Vishnu schist?

   b. What does the Vishnu schist tell you about the history of the Grand Canyon prior to the formation of the canyon itself?

   c. Why is the Vishnu schist visible at Earth's surface?

   d. Is it likely that rocks similar to the Vishnu schist exist elsewhere but are not exposed at Earth's surface? Explain.

A. Grand Canyon

B. Close up of Vishnu Schist (dark color)

❻ Refer to the accompanying diagram and match each labeled area with the appropriate environment listed below:

   a. contact metamorphism
   b. subduction metamorphism
   c. regional metamorphism
   d. burial metamorphism
   e. hydrothermal metamorphism

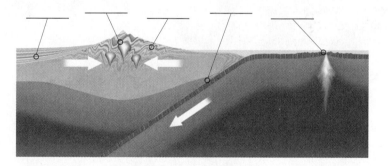

❼ Based on the information provided in Figure 7.22, complete the following:

   a. Describe how the metamorphic grade changes from west to east across New England along line A-B.

   b. How might these metamorphic rocks have formed?

   c. Are these zones of metamorphism consistent with the current tectonic setting of New England?

Companion Website

mygeoscience place

www.mygeoscienceplace.com

The *Essentials of Geology, 11e* companion Website contains numerous multimedia resources accompanied by assessments to aid in your study of the topics in this chapter. The use of this site's learning tools will help improve your understanding of geology. Utilizing the access code that accompanies this text, visit **www.mygeoscienceplace.com** in order to:

- **Review** key chapter concepts.
- **Read** with links to the Pearson eText and to chapter-specific web resources.
- **Visualize** and comprehend challenging topics using the learning activities in *GEODe: Essentials of Geology* and the *Geoscience Animations Library*.
- **Test** yourself with online quizzes.

8 Mass Wasting: The Work of Gravity

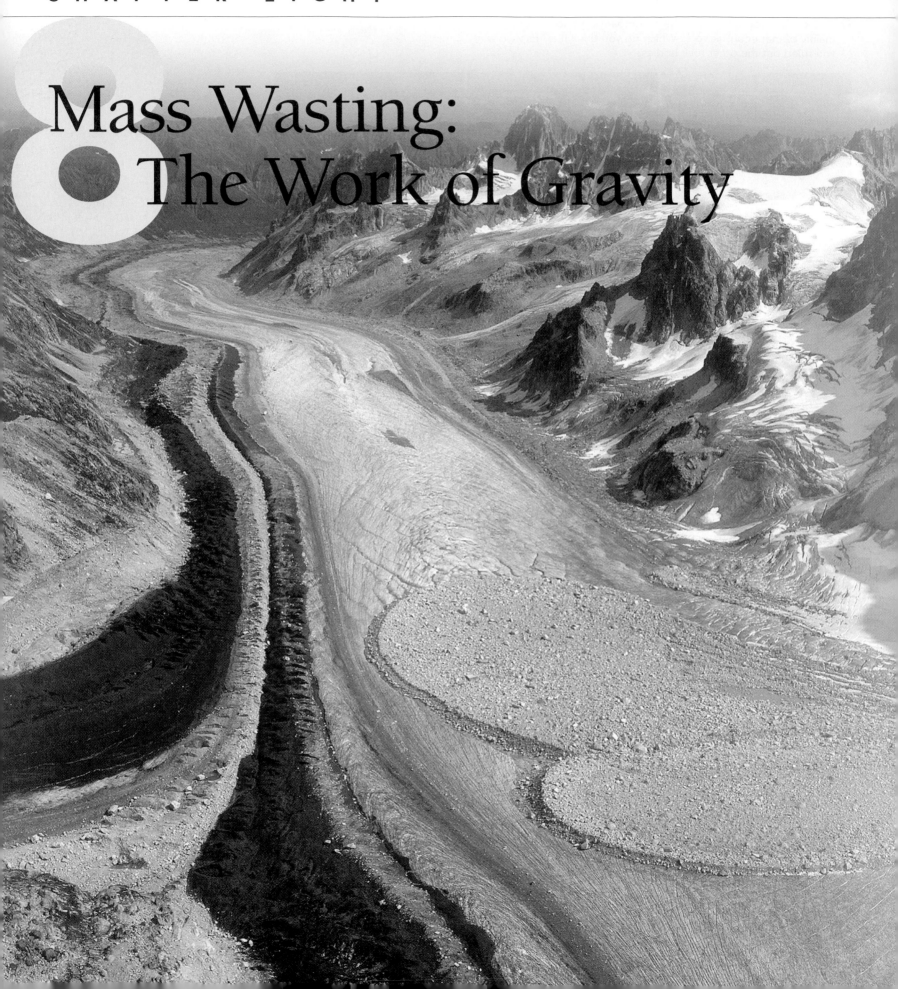

ALTHOUGH MOST SLOPES MAY APPEAR TO BE STABLE AND UNCHANGING THE force of gravity causes material to move. At one extreme, the movement may be gradual and practically imperceptible. At the other extreme, it may consist of a roaring debris flow or a thundering rock avalanche. Landslides are a worldwide natural hazard. When these hazardous processes lead to loss of life and property, they become natural disasters

Landslide debris atop Buckskin Glacier in Alaska's Denali National Park. In the evolution of many landforms, gravity-driven mass-wasting processes transfer the products of weathering downslope where an erosional agent, in this case a valley glacier, carries the debris away. (Photo by Michael Collier)

To assist you in learning the important concepts in this chapter, focus on the following questions:

⊙ What is the process of mass wasting?

⊙ What role does mass wasting play in the development of valleys?

⊙ What are the controls and triggers of mass wasting?

⊙ What criteria are used to divide and describe the various types of mass wasting?

⊙ What are the general characteristics of slump, rockslide, debris flow, earthflow, and creep?

Landslides as Natural Disasters

Even in areas with steep slopes, catastrophic landslides are relatively rare occurrences. As a result people living in susceptible areas often do not appreciate the risk of living where they do. However, media reports remind us that such events occur with some regularity around the world. The three examples described here occurred during a span of just four months.

On October 5, 2005, torrential rains from Hurricane Stan triggered mudflows in Guatemala. A slurry of mud 1 kilometer wide and up to 12 meters (40 feet) deep buried the village of Panabaj. Death toll estimates for the area approached 1400 people. Flows such as this can travel at speeds of 50 kilometers (30 miles) per hour down rugged mountain slopes.

Just three days later, on October 8, 2005, a magnitude 7.6 earthquake struck the Kashmir region between India and Pakistan. Compounding the tragic effects caused directly by the severe ground shaking were hundreds of landslides triggered by the quake and its many aftershocks. Rockfalls and debris slides thundered down steep mountain slopes and were focused into narrow valleys where many people made their homes. The landslides also blocked roads and trails, delaying attempts to reach those in need (**FIGURE 8.1A**).

On February 17, 2006, only a few months after the tragedies in Kashmir

and Central America, a lethal mudflow triggered by extraordinary rains buried a small town on the Philippine island of Leyte (Figure 8.1B). A mass of mud engulfed this remote coastal area to depths as great as 10 meters (30 feet). Although an accurate count of fatalities was difficult, as many as 1800 people perished. This region is prone to such events due in part to the fact that deforestation has denuded the nearby mountain slopes. In the pages that follow you will take a closer look at rapid mass-wasting events in an attempt to better understand their causes and effects.

CONCEPT CHECK 8.1

❶ What events triggered the landslides shown in Figure 8.1?

FIGURE 8.1 A. On October 8, 2005, a major earthquake in Kashmir triggered hundreds of landslides including the one shown here. *(Photo by AP Photo/Burhan Ozblici)* **B.** In February 2006, heavy rains triggered this mudflow that buried a small town on the Philippine island of Leyte. *(Photo by AP Photo/Pat Roque)*

A.

B.

rockfall

Rockfall blocking a Montana highway. *(Photo by AP/Wide World Photo)*

ROAD CLOSED

This home in Pacific Palisades, California, was destroyed by a landslide triggered by the January 1994 Northridge earthquake. *(Photo by Chromo Sohm/Corbis/The Stock Market)*

Mass Wasting and Landform Development

Landslides are spectacular examples of a common geologic process called mass wasting. **Mass wasting** refers to the downslope movement of rock, regolith, and soil under the direct influence of gravity. It is distinct from the erosional processes that are examined in subsequent chapters because mass wasting does not require a transporting medium such as water, wind, or glacial ice.

The Role of Mass Wasting

Earth's surface is never perfectly flat but instead consists of slopes. Some are steep and precipitous; others are moderate or gentle. Some are long and gradual; others are short and abrupt. Slopes can be mantled with soil and covered by vegetation or consist of barren rock and rubble. Taken together, slopes are the most common elements in our physical landscape.

In the evolution of most landforms, mass wasting is the step that follows weathering. By itself weathering does not produce significant landforms. Rather, landforms develop as the products of weathering are removed from the places where they originate. Once weathering weakens rock and breaks it apart, mass wasting transfers the debris downslope, where a stream, acting as a conveyor belt, usually carries it away. Although there may be many intermediate stops along the way, the sediment is eventually transported to its ultimate destination: the sea.

The combined effects of mass wasting and running water produce stream valleys, which are the most common and conspicuous of Earth's landforms. If streams alone were responsible for creating the valleys in which they flow, valleys would be very narrow features. However, the fact that most river valleys are much wider than they are deep is a strong indication of the significance of mass-wasting processes in supplying material to streams. This is

DID YOU KNOW?
Although many people, including geologists, frequently use the word *landslide*, the term has no specific definition in geology. Rather, it is a popular nontechnical term used to describe all relatively rapid forms of mass wasting.

FIGURE 8.2 The walls of the Grand Canyon extend far from the channel of the Colorado River. This results primarily from the transfer of weathered debris downslope to the river and its tributaries by mass-wasting processes. *(Photo by Bryan Brazil/Shutterstock)*

Colorado River

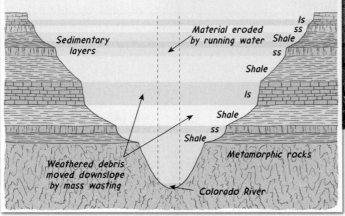

Geologist's Sketch

illustrated by the Grand Canyon (**FIGURE 8.2**). The walls of the canyon extend far from the Colorado River, owing to the transfer of weathered debris downslope to the river and its tributaries by mass-wasting processes. In this manner, streams and mass wasting combine to modify and sculpt the surface. Of course, glaciers, groundwater, waves, and wind are also important agents in shaping landforms and developing landscapes.

Slopes Change through Time

It is clear that if mass wasting is to occur, there must be slopes down which rock, soil, and regolith can move. It is Earth's mountain-building and volcanic processes that produce these slopes through sporadic changes in the elevations of landmasses and the ocean floor. If dynamic internal processes did not continually produce regions having higher elevations, the system that moves debris to lower elevations would gradually slow and eventually cease.

Most rapid and spectacular mass-wasting events occur in areas of rugged, geologically young mountains. Newly formed mountains are rapidly eroded by rivers and glaciers into regions characterized by steep and unstable slopes. It is in such settings that massive destructive landslides, such as those described at the beginning of the chapter, occur. As mountain building subsides, mass wasting and erosional processes lower the land. Through time, steep and rugged mountain slopes give way to gentler, more subdued terrain. Thus, as a landscape ages, massive and rapid mass-wasting processes give way to smaller, less dramatic downslope movements.

CONCEPT CHECK 8.2

❶ What is the controlling force of mass wasting?

❷ Sketch or describe how mass wasting contributes to the development of a valley.

Controls and Triggers of Mass Wasting

Mass Wasting

ESSENTIALS OF GEOLOGY Controls and Triggers of Mass Wasting

Gravity is the controlling force of mass wasting, but several factors play an important role in overcoming inertia and creating downslope movements. Long before a landslide occurs, various processes work to weaken slope material, gradually making it more and more susceptible to the pull of gravity. During this span, the slope remains stable but gets closer and closer to being unstable. Eventually, the strength of the slope is weakened to the point that something causes it to cross the threshold from stability to instability. Such an event that

FIGURE 8.3 Many large debris flows and floods were triggered by the torrential rains that accompanied Hurricane Mitch when it struck Honduras in October 1998. It was that country's worst natural disaster in 200 years. Pictured here is the El Berrinche landslide, one of two that destroyed portions of the city of Tegucigalpa, killing more than 1000 people and damming the Rio Choluteca. The mass of debris flow material was estimated to be 6 million cubic meters, enough to fill nearly 300 thousand average dump trucks! *(Photos by Michael Collier)*

initiates downslope movement is called a *trigger*. Remember that the trigger is not the sole cause of the mass-wasting event, but just the last of many causes. Among the common factors that trigger mass-wasting processes are saturation of material with water, oversteepening of slopes, removal of anchoring vegetation, and ground vibrations from earthquakes.

The Role of Water

Mass wasting is sometimes triggered when heavy rains or periods of snowmelt saturate surface materials. This was the case in October 1998 when torrential rains associated with Hurricane Mitch triggered devastating debris flows in Central America (**FIGURE 8.3**). A case study of another rain-triggered mass-wasting event that occurred at La Conchita, California, in January 2005 is found at the end of this section.

When the pores in sediment become filled with water, the cohesion among particles is destroyed, allowing them to slide past one another with relative ease. For example, when sand is slightly moist, it sticks together quite well. However, if enough water is added to fill the openings between the grains, the sand will ooze out in all directions (**FIGURE 8.4**). Thus,

saturation reduces the internal resistance of materials, which are then easily set in motion by the force of gravity. When clay is wetted, it becomes very slick—another example of the lubricating effect of water. Water also adds considerable weight to a mass of material. The added weight in itself may be enough to cause the material to slide or flow downslope.

Oversteepened Slopes

Oversteepening of slopes is another trigger of many mass movements. There are many situations in nature where oversteepening takes place. A stream undercutting a valley wall and waves pounding against the base of a cliff are but two familiar examples. Furthermore, through their activities, people often create oversteepened and

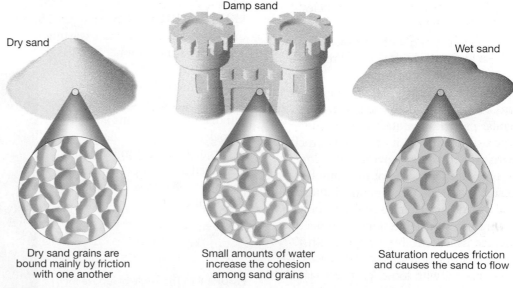

Dry sand

Damp sand

Wet sand

Dry sand grains are bound mainly by friction with one another

Small amounts of water increase the cohesion among sand grains

Saturation reduces friction and causes the sand to flow

FIGURE 8.4 The effect of water on mass wasting can be great. When little or no water is present, friction among the closely packed soil particles on the slope holds them in place. When the soil is saturated, the grains are forced apart and friction is reduced, allowing the soil to move downslope.

Removal of Vegetation

Plants protect against erosion and contribute to the stability of slopes because their root systems bind soil and regolith together. In addition, plants shield the soil surface from the erosional effects of raindrop impact (see Figure 5.22, p. 142). Where plants are lacking, mass wasting is enhanced, especially if slopes are steep and water is plentiful. When anchoring vegetation is removed by forest fires or by people (for timber, farming, or development), surface materials frequently move downslope.

In July 1994 a severe wildfire swept Storm King Mountain west of Glenwood Springs, Colorado, denuding the slopes of vegetation. Two months later heavy rains resulted in numerous debris flows, one of which blocked Interstate 70 and threatened to dam the Colorado River. A 5-kilometer (3-mile) length of the highway was inundated with tons of rock, mud, and burned trees. The closure of Interstate 70 imposed costly delays on this major highway.

Wildfires are inevitable in the western United States and fast-moving, highly destructive debris flows triggered by intense rainfall are one of the most dangerous post-fire hazards (**FIGURE 8.7**). Such events are particularly dangerous because they tend to occur with little warning. Their mass and speed make them especially destructive. Post-fire debris flows are most common in the two years after a fire. Some of the largest debris-flow events have been triggered by the very first intense rain event following the wildfire. It takes much less rain to trigger debris flows in burned areas than in unburned areas. In southern California, as little as 7 millimeters (0.3 inch) of rain in 30 minutes has triggered debris flows.

How large can these flows be? According to the U.S. Geological Survey, documented debris flows from burned areas in southern California and other western states have ranged in volume from as small as 600 cubic meters to as large as 300,000 cubic meters. This larger volume is enough material to cover a football field with mud and rocks to a depth of about 65 meters (almost 215 feet)!

In addition to eliminating plants that anchor the soil, fire can promote mass wasting in other ways. Following a wildfire, the upper part of the soil may become dry and loose. As a result, even in dry weather, the soil tends to move down steep slopes. Moreover, fire can also "bake" the ground, creating a water-repellant layer at a shallow depth. This nearly impermeable barrier prevents or slows the infiltration of water, resulting in increased surface runoff during rains. The consequence can be dangerous torrents of viscous mud and rock debris.

Earthquakes as Triggers

Conditions that favor mass wasting may exist in an area for a long time without movement occurring. An additional factor is sometimes necessary to trigger the movement. Among the more important and dramatic triggers are earthquakes. An earthquake and its aftershocks can dislodge enormous volumes of rock and unconsolidated material. The event in Kashmir described at the beginning of this chapter is one tragic example (see Figure 8.1A).

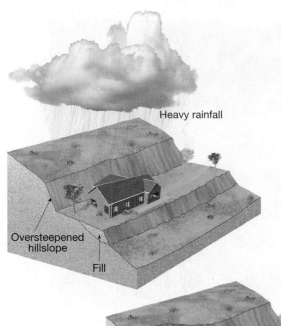

FIGURE 8.5
When slopes are oversteepened and made unstable, they are prime sites for mass wasting. Natural processes such as stream and wave erosion can oversteepen slopes. Changing the slope to accommodate a new house or road can also lead to instability and a destructive mass wasting event.

unstable slopes that become prime sites for mass wasting (**FIGURE 8.5**).

Unconsolidated, granular particles (sand-sized or coarser) assume a stable slope called the **angle of repose.** This is the steepest angle at which material remains stable (**FIGURE 8.6**). Depending on the size and shape of the particles, the angle varies from 25 to 40 degrees. The larger, more angular particles maintain the steepest slopes. If the angle is increased, the rock debris will adjust by moving downslope.

Oversteepening is not just important because it triggers movements of unconsolidated granular materials. Oversteepening also produces unstable slopes and mass movements in cohesive soils, regolith, and bedrock. The response will not be immediate, as with loose, granular material, but sooner or later one or more mass-wasting processes will eliminate the oversteepening and restore stability to the slope.

FIGURE 8.6 The angle of repose for this granular material is about 30° *(Photo by G. Leavens/Photo Researchers, Inc.)*

FIGURE 8.7 During summer, wildfires are common occurrences in many parts of the West. Millions of acres are burned each year. The loss of anchoring vegetation sets the stage for accelerated mass wasting. *(Photo by Raymond Gehman/National Geographic Stock)*

EXAMPLES FROM CALIFORNIA AND CHINA. A memorable United States example occurred in January 1994 when a quake struck the Los Angeles region of southern California. Named for its epicenter in the town of Northridge, the 6.7-magnitude event produced estimated losses of $20 billion. Some of the losses were the result of more than 11,000 landslides in an area of about 10,000 square kilometers (3900 square miles) that were set in motion by the quake (see the photo on pp.198–199). Most were shallow rock falls and slides, but some were much larger and filled canyon bottoms with jumbles of soil, rock, and plant debris. The debris in canyon bottoms created a secondary threat because of its ability to mobilize during rainstorms, producing debris flows. Such flows are common and often disastrous in southern California.

On May 12, 2008, a magnitude 7.9 earthquake struck near the city of Chengdu in China's Sichuan Province. Compounding the tragic effects caused directly by the severe ground shaking were hundreds of landslides triggered by the quake and its many aftershocks (**FIGURE 8.8**). Rock avalanches and debris slides thundered down steep mountain slopes burying buildings and blocking

roads and rail lines. The landslides also dammed rivers creating more than two dozen lakes. Earthquake-created lakes present a dual danger. Apart from the upstream floods that occur as the lake builds behind the natural dam, the piles of rubble that form the dam may be unstable. Another quake or simply the pressure of water could burst the dam, sending a wall of water downstream. Such floods may also occur when water begins to cascade over the top of the dam. The largest of the lakes created by the May 12 earthquake, Tangji-ashan Lake, threatened roughly 1.3 million people. In this instance, a disaster was avoided when Chinese engineers successfully breached the landslide dam and safely drained the lake.

LIQUEFACTION. Intense ground shaking during earthquakes can cause water-saturated surface materials to lose their strength and behave as fluidlike masses that flow. This process, called *liquefaction*, was a major cause of property damage in Anchorage, Alaska, during the massive 1964 Good Friday earthquake described in Chapter 14.

FIGURE 8.8 When a magnitude 7.9 earthquake struck west northwest of Chengdu in China's Sichuan Province on May 12, 2008, it triggered hundreds of landslides that destroyed roads and bridges. Buildings and rail lines were buried by landslide debris. Several rivers were blocked by debris dams creating lakes which threatened the safely of thousands of people living downstream. This aerial view shows the huge volume of landslide debris that dammed a river and created a lake. Note the heavy equipment for scale. *(Photo by MARK/epa/Corbis)*

Case Study: Landslide Hazards at La Conchita, California

Southern California lies astride a major plate boundary defined by the San Andreas Fault and numerous other related faults that are spread across the region. It is a dynamic environment characterized by rugged mountains and steep-walled canyons. Unfortunately this scenic landscape presents serious geologic hazards. Just as tectonic forces are steadily pushing the landscape upward, gravity is relentlessly pulling it downward. When gravity prevails, landslides occur.

As you might expect, some of the region's landslides are triggered by earthquakes. Many others, however, are related to periods of prolonged and intense rainfall. A tragic example of the latter situation occurred on January 10, 2005, when a massive debris flow (popularly called a *mudslide*) swept through La Conchita, California, a small town located about 80 kilometers (50 miles) northwest of Los Angeles (**FIGURE 8.9**).

Although the rapid torrent of mud took many of the town's inhabitants by surprise, such an event should not have been unexpected. Let's briefly examine the factors that contributed to the deadly debris flow at La Conchita.

The town is situated on a narrow coastal strip about 250 meters (800 feet) wide between the shoreline and a steep (600 foot) bluff. The bluff consists of poorly sorted marine sediments and weakly cemented layers of shale, siltstone, and sandstone.

The deadly 2005 debris flow involved little or no newly failed material, but rather consisted of the remobilization of a portion of a large landslide that destroyed homes in 1995. In fact, historical accounts dating back to 1865 indicate that landslides in the immediate area have been a regular occurrence. Further, geologic evidence shows that landsliding of a variety of types and scales has probably been occurring at La Conchita for thousands of years.

The most significant contributing factor to the tragic 2005 debris flow was prolonged and intense rain. The event occurred at the end of a span that produced near record amounts of rainfall in Southern California. Wintertime rainfall at nearby Ventura totaled 49.3 centimeters (19.4 inches) as compared to an average value of just 12.2 centimeters (4.8 inches). As **FIGURE 8.10** indicates, much of that total fell during the two weeks immediately preceding the debris flow.

This was not the first destructive landslide to strike La Conchita, nor is it likely to be the last. The town's geologic setting and history of rapid mass-wasting events clearly support this notion. When the amount and intensity of rainfall is sufficient, debris flows are to be expected.

Landslides without Triggers?

Do rapid mass-wasting events always require some sort of trigger such as heavy rains or earthquakes? The answer is no. Such events sometimes occur without being triggered. For example, on the afternoon of May 9, 1999, a landslide killed 10 hikers and injured many others at Sacred Falls State Park near Hauula on the north shore of Oahu, Hawaii. The tragic event occurred when a mass of rock from a canyon wall plunged 150 meters (500 feet) down a nearly vertical slope to the valley floor. Because of safety concerns, the park was closed so that landslide specialists from the U.S. Geological Survey could investigate the site. Their study concluded that the landslide occurred *without being triggered* by any discernible external conditions.

Many rapid mass-wasting events occur without a discernible trigger. Slope materials gradually weaken over time under the influence of long-term weathering, infiltration of water, and other physical processes. Eventually, if the strength falls below what is necessary to maintain slope stability, a landslide will occur. The timing of such events is random, and thus accurate prediction is impossible.

Scarp created by 1995 landslide

2005 debris flow

Area outlined at bottom of image on left

FIGURE 8.9 Views of the La Conchita debris flow taken shortly after it occurred in January 2005. The light-colored exposed rock in the upper part of the photo on the left is the main scarp of a slide that occurred 10 years earlier in 1995. The January 2005 event was a remobilization of the 1995 event. The flow was quite thick (viscous) and moved houses in its path rather than flowing around them. (*AP Wide World Photos*)

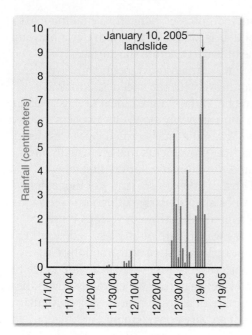

FIGURE 8.10 Daily rainfall at the nearby town of Ventura during the weeks leading up to the January 2005 La Conchita event. Each line on the bar graph shows rainfall for a particular day. The 2005 debris flow occurred at the culmination of the heaviest rainfall of the season. About 80 percent of the season's exceptional total fell in this short span. *(After National Weather Service)*

The Potential for Landslides

FIGURE 8.11 shows the landslide potential for the contiguous United States. All states experience some damage from rapid mass-wasting processes, but it is obvious that not all areas of the country have the same landslide hazard potential. As you might expect, there are greater landslide risks in mountain areas. In the East, landslides are most common in the Appalachian Mountains. In the mountainous parts of the Pacific Northwest, water from heavy rains and melting snow often trigger rapid forms of mass wasting. Coastal California's steep slopes have a high landslide potential. Here mass-wasting events may be triggered by winter storms or the ground shaking associated with earthquakes. Landslides are also triggered when strong wave activity undercuts and oversteepens coastal cliffs.

A glance at the map shows that Florida and the adjacent Atlantic and Gulf coastal plains have some of the lowest landslide potentials because steep slopes are largely absent. In the center of the country, the plains states are relatively flat, so landslide

potential is mostly low-to-moderate. High-potential areas occur along the steep bluffs that flank river valleys.

Classification of Mass-Wasting Processes

There is a broad array of different processes that geologists call mass wasting. Generally, the different types are classified based on the type of material involved, the kind of motion displayed, and the velocity of the movement.

Type of Material

The classification of mass-wasting processes on the basis of the material involved in the movement depends upon whether the descending mass began as unconsolidated material or as bedrock. If soil and regolith dominate, terms such as *debris, mud,* or *earth* are used in the description. In contrast, when a mass of bedrock breaks loose and moves downslope, the term *rock* may be part of the description.

Type of Motion

In addition to characterizing the type of material involved in a mass-wasting event, the way in which the material moves may also be important. Generally, the kind of motion is described as either a fall, a slide, or a flow.

FALL. When the movement involves the freefall of detached individual pieces of any size, it is termed a **fall.** Fall is a common form of movement on slopes that are so steep that loose material cannot remain on the surface. The rock may fall directly to the base of the slope or move in a series of leaps and bounds over other rocks along the way. Many falls result when freeze and thaw cycles and/or the action of plant roots

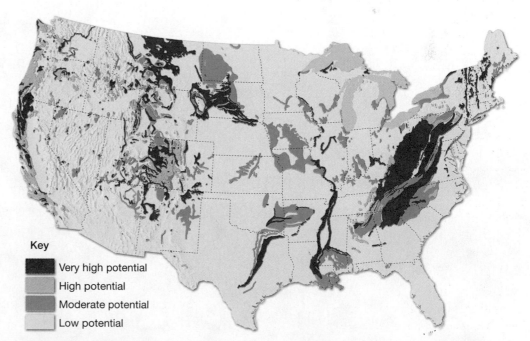

FIGURE 8.11 Landslide potential for the contiguous United States. Risks are clearly greatest in mountain areas where slopes are steepest. Rapid forms of mass wasting occur and can cause damage in all 50 states. Heavy rains, earthquakes, volcanic activity, coastal wave attack, and wildfires can all contribute to widespread slope instability. *(After U.S. Geological Survey)*

Key

■ Very high potential
■ High potential
■ Moderate potential
□ Low potential

loosen rock to the point that gravity takes over. Although signs along bedrock cuts on highways warn of falling rock, few of us have actually witnessed such an event. However, as **FIGURE 8.12** illustrates, they do indeed occur. In fact, this is the primary way in which **talus slopes** are built and maintained (see Figure 5.4, p. 126). Sometimes falls may trigger other forms of downslope movement.

SLIDE. Many mass-wasting processes are described as **slides.** The term refers to mass movements in which there is a distinct zone of weakness separating the slide material from the more stable underlying material. Two basic types of slides are recognized. *Rotational slides* are those in which the surface of rupture is a concave-upward curve that resembles the shape of a spoon and the descending material exhibits a downward and outward rotation. The process termed *slump* is an example. By contrast, a *translational slide* is one in which a mass of material moves along a relatively flat surface such as a joint, fault, or bedding plane. Such slides, called

rockslides, exhibit little rotation or backward tilting.

FLOW. The third type of movement common to mass-wasting processes is termed **flow.** Flow occurs when material moves downslope as a viscous fluid. Most flows are saturated with water and typically move as lobes or tongues.

Rate of Movement

When mass-wasting events make the news, a large quantity of material has in all likelihood moved rapidly down a steep slope and has had a disastrous effect upon people and property. Indeed, during events called **rock avalanches,** rock and debris can hurtle downslope at speeds exceeding 200 kilometers (125 miles) per hour. Researchers now understand that rock avalanches, such as the one that produced the scene in **FIGURE 8.13**, must literally float on air as they move downslope. That is, high velocities result when air becomes trapped

and compressed beneath the falling mass of debris, allowing it to move as a buoyant, flexible sheet across the surface.

Most mass movements, however, do not move with the speed of a rock avalanche. In fact, a great deal of mass wasting is imperceptibly slow. One process that we will examine later, termed *creep*, results in particle movements that are usually measured in millimeters or centimeters per year. Thus, as you can see, rates of movement can be spectacularly sudden or exceptionally gradual. Although various types of mass wasting are often classified as either rapid or slow, such a distinction is highly subjective because there is a wide range of rates between the two extremes. Even the velocity of a single process at a particular site can vary considerably.

CONCEPT CHECK 8.4

❶ What terms describe the way material moves during mass wasting?

❷ Why can rock avalanches move at such great speeds?

FIGURE 8.12 In January 1997, this rockfall blocked Highway 140 near the Arch Rock entrance to Yosemite National Park, California. *(Photo by Roger J. Wyan/AP/ Wide World Photos)*

DID YOU KNOW?
The U.S. Geological Survey estimates that between 25 and 50 people are killed by landslides annually in the United States. The worldwide death toll is much higher.

FIGURE 8.13 Aerial view of the Blackhawk landslide, a prehistoric event that occurred on the north slope of California's San Bernardino Mountains. It is considered to be one of the largest known landslides in North America. This 8-kilometer-long tongue of rubble is about 3 kilometers wide and 9 to 30 meters thick. Research has shown that the mass of debris raced downslope to its resting place on a cushion of compressed air. (Photo by Michael Collier)

Slump

Mass Wasting

ESSENTIALS OF GEOLOGY Types of Mass Wasting

Slump refers to the downward sliding of a mass of rock or unconsolidated material moving as a unit along a curved surface (**FIGURE 8.14**). Usually the slumped material does not travel spectacularly fast or very far. This is a common form of mass wasting, especially in thick accumulations of cohesive materials such as clay. The rupture surface is characteristically spoon-shaped and concave upward or outward. As the movement occurs, a crescent-shaped scarp is created at the head, and the block's upper surface is sometimes tilted backward. Although slump may involve a single mass, it often consists of multiple blocks. Sometimes water collects

between the base of the scarp and the top of the tilted block. As this water percolates downward along the surface of rupture, it may promote further instability and additional movement.

Slump commonly occurs because a slope has been oversteepened. The material on the upper portion of a slope is held in place by the material at the bottom of the slope. As this anchoring material at the base is removed, the material above is made unstable and reacts to the pull of gravity. One relatively common example is a valley wall that becomes oversteepened by a meandering river. The photo in **FIGURE 8.15** provides another example in which a coastal cliff has been undercut by wave action at its base.

Slumping may also occur when a slope is overloaded, causing internal stress on the material below. This type of slump often occurs where weak, clay-rich material underlies layers of stronger, more resistant rock such as sandstone. The seepage of water through the

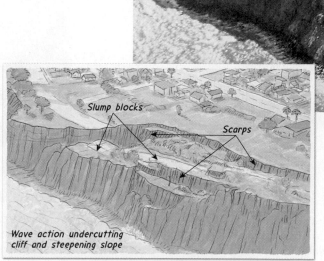

FIGURE 8.15 Slump at Point Fermin, California. Slump is often triggered when slopes become oversteepened by erosional processes such as wave action. (Photo by John S. Shelton)

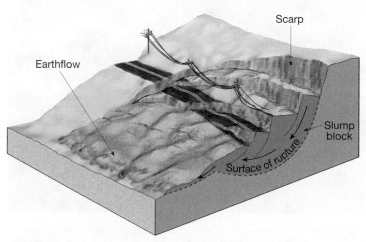

FIGURE 8.14 Slump occurs when material slips downslope en masse along a curved surface of rupture. It is an example of a rotational slide. Earthflows frequently form at the base of the slump.

Geologist's Sketch

upper layers reduces the strength of the clay below, causing slope failure.

Rockslide

Mass Wasting

ESSENTIALS OF GEOLOGY **Types of Mass Wasting**

Rockslides occur when blocks of bedrock break loose and slide down a slope (**FIGURE 8.16**). If the material is largely unconsolidated, the term **debris slide** is used instead. Such events are among the fastest and most destructive mass movements. Usually rockslides take place in a geologic setting where the rock strata are inclined or where joints and fractures exist parallel to the slope. When such a rock unit is undercut at the base of the slope, it loses support and the rock eventually gives way. Sometimes the rockslide is triggered when rain or melting snow lubricates the underlying surface to the point that friction is no longer sufficient to hold the rock unit in place. As a result, rockslides tend to be more common during the spring, when heavy rains and melting snow are most prevalent.

As was mentioned earlier, earthquakes can trigger rockslides and other mass movements. There are many notable examples. The 1811 earthquake at New Madrid, Missouri, caused slides in an area of more than 13,000 square kilometers (5000 square miles) along the Mississippi River valley. On August 17, 1959, a severe earthquake west of Yellowstone National Park triggered a massive slide in the canyon of the Madison River in southwestern Montana. In a matter of moments an estimated 27 million cubic meters of rock, soil, and trees slid into the canyon. The debris dammed the river and buried a campground and highway. More than 20 unsuspecting campers perished.

Not far from the site of the Madison Canyon slide, the classic Gros Ventre rockslide occurred 34 years earlier. The Gros Ventre River flows west from the northernmost part of the Wind River Range in northwestern Wyoming through the Grand Teton National Park and eventually empties into the Snake River. On June 23, 1925, a massive rockslide took place in its valley, just east of the small town of Kelly. In the span of just a few minutes a great mass of sandstone, shale, and soil crashed down the south side of the valley, carrying with it a dense pine forest. The volume of debris, estimated at 38 million cubic meters (50 million cubic yards), created a 70-meter-high dam on the Gros Ventre River. Because the river was completely blocked, a lake was created. It filled so quickly that a house that had been 18 meters (60 feet) above the river was floated off its foundation 18 hours after the slide. In 1927, the lake overflowed the dam, partially draining the lake and resulting in a devastating flood downstream.

Why did the Gros Ventre rockslide take place? **FIGURE 8.17** shows a diagrammatic

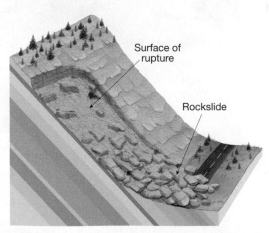

FIGURE 8.16 Rockslides and debris slides are rapid movements that are classified as translational slides in which the material moves along a relatively flat surface with little or no rotation or backward tilting.

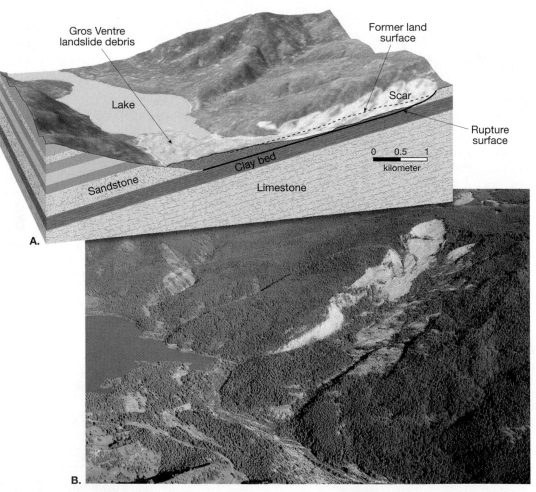

FIGURE 8.17 Part **A.** shows a cross-sectional view of the Gros Ventre rockslide. The slide occurred when the tilted and undercut sandstone bed could no longer maintain its position atop the saturated bed of clay. As the photo in part **B.** illustrates, even though the Gros Ventre rockslide occurred in 1925, the scar left on the side of Sheep Mountain is still a prominent feature. *(Part **A.** after W. C. Alden, "Landslide and Flood at Gros Ventre, Wyoming," Transactions (AIME) 76 (1928); 348. Part **B.** photo by Michael Collier)*

cross-sectional view of the geology of the valley. Notice the following points: (1) The sedimentary strata in this area dip (tilt) 15 to 21 degrees; (2) underlying the bed of sandstone is a relatively thin layer of clay; and (3) at the bottom of the valley, the river had cut through much of the sandstone layer. During the spring of 1925, water from heavy rains and melting snow seeped through the sandstone, saturating the clay below. Because much of the sandstone layer had been cut through by the Gros Ventre River, the layer had virtually no support at the bottom of the slope. Eventually the sandstone could no longer hold its position on the wetted clay, and gravity pulled the mass down the side of the valley. The circumstances at this location were such that the event was inevitable.

CONCEPT CHECK 8.6

❶ Both slump and rockslide move by sliding. How do these processes differ?

❷ What factors led to the massive rockslide at Gros Ventre, Wyoming?

Debris Flow

 Mass Wasting

ESSENTIALS OF GEOLOGY Types of Mass Wasting

Debris flow is a relatively rapid type of mass wasting that involves a flow of soil and regolith containing a large amount of water. Several debris flows are pictured in this chapter—see Figure 8.1B, Figure 8.3, and Figure 8.9. Debris flows are sometimes called **mudflows** when the material is primarily fine-grained. Although they can occur in many different climatic settings, they probably occur more frequently in semiarid mountainous regions. Debris flows, called *lahars*, are also common on the steep slopes of some volcanoes. Because of their fluid properties, debris flows frequently follow canyons and stream channels (**FIGURE 8.18**). In populated areas, debris flows can pose a significant hazard to life and property.

Debris Flows in Semiarid Regions

When a cloudburst or rapidly melting mountain snows create a sudden flood in

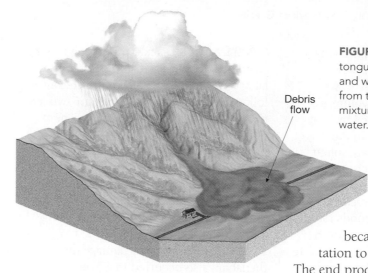

Debris flow

FIGURE 8.18 Debris flow is a moving tongue of well-mixed mud, soil, rock, and water. Its consistency may range from that of wet concrete to a soupy mixture not much thicker than muddy water.

a semiarid region, large quantities of soil and regolith are washed into nearby stream channels because there is usually little vegetation to anchor the surface material. The end product is a flowing tongue of well-mixed mud, soil, rock, and water. Its consistency can range from that of wet concrete to a soupy mixture not much thicker than muddy water. The rate of flow therefore depends not only on the slope but also on the water content. When dense, debris flows are capable of carrying or pushing large boulders, trees, and even houses with relative ease.

Debris flows pose a serious hazard to development in relatively dry mountainous areas such as southern California. Here construction of homes on canyon hillsides and the removal of native vegetation by brush fires and other means have increased the frequency of these destructive events. Moreover, when a debris flow reaches the end of a steep, narrow canyon, it spreads out, covering the area beyond the mouth of the canyon with a mixture of wet debris. This material contributes to the buildup of fanlike deposits, called *alluvial fans*, at canyon mouths.* The fans are relatively easy to build on, often have nice views, and are close to the mountains; in fact, like the nearby canyons, many have become preferred sites for development. Because debris flows occur only sporadically, the public is often unaware of the potential hazard of such sites.

A small rockfall in progress. *(Photo by Herb Dunn/ DunnRight Photography)*

Lahars

Debris flows composed mostly of volcanic materials on the flanks of volcanoes are called **lahars**. The word originated in Indonesia, a volcanic region that has experienced many of these often destructive events. Historically, lahars have been some of the deadliest volcano hazards. They can occur either during an eruption or when a volcano is quiet. They take place when highly unstable layers of ash and debris become saturated with water and flow down

*Alluvial fans are discussed in greater detail in Chapters 9 and 12.

steep volcanic slopes, generally following existing stream channels (**FIGURE 8.19**). Heavy rainfalls often trigger these flows. Others are initiated when large volumes of ice and snow are suddenly melted by heat flowing to the surface from within the volcano or by the hot gases and near-molten debris emitted during a violent eruption.

When Mount St. Helens erupted in May 1980, several lahars were created. The flows and accompanying floods raced down the valleys of the north and south forks of the Toutle River at speeds that were often in excess of 30 kilometers (20 miles) per hour. Fortunately, the affected area was not densely settled. Nevertheless, more than 200 homes were destroyed or severely damaged. Most bridges met a similar fate.

In November 1985, lahars were produced during the eruption of Nevado del Ruiz, a 5300-meter (17,400-foot) volcano in the Andes Mountains of Colombia. The eruption melted much of the snow and ice that capped the uppermost 600 meters

(2000 feet) of the peak, producing torrents of hot, thick mud, ash, and debris. The lahars moved outward from the volcano, following the valleys of three rain-swollen rivers that radiate from the peak. The flow that moved down the valley of the Lagunilla River was the most destructive. It devastated the town of Armero, 48 kilometers (30 miles) from the mountain. Most of the more than 25,000 deaths caused by the event occurred in this once thriving agricultural community.

Death and property damage due to the lahars also occurred in 13 other villages within the 180-square-kilometer (70-square-mile) disaster area. Although a great deal of pyroclastic material was explosively ejected from Nevado del Ruiz, it was the lahars triggered by this eruption that made this such a devastating natural disaster. In fact, it was the worst volcanic disaster since 28,000 people died following the 1902 eruption of Mount Pelée on the Caribbean island of Martinique.*

*A discussion of the Mount Pelée eruption, as well as additional material on lahars, can be found in Chapter 4.

FIGURE 8.19 The eruption of Mount Redoubt, a volcano southwest of Anchorage on Alaska's Kenai Peninsula, sent large debris flows (lahars) down the Drift River valley in April 2009. Channels form a branching pattern just west of Cook Inlet. The dark color of the lahars contrasts sharply with the surrounding snow-covered landscape. *(NASA)*

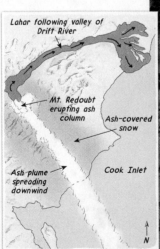

Lahar following valley of Drift River

Mt. Redoubt erupting ash column

Ash-covered snow

Cook Inlet

Ash-plume spreading downwind

N

Geologist's Sketch

CONCEPT CHECK 8.7

❶ Explain why building a home on an alluvial fan may not be a good idea.

❷ How is a lahar different from a debris flow that might occur in southern California?

Earthflow

 Mass Wasting

ESSENTIALS OF GEOLOGY Types of Mass Wasting

We have seen that debris flows are frequently confined to channels in semiarid regions. In contrast, **earthflows** most often form on hillsides in humid areas during times of heavy precipitation or snowmelt. When water saturates the soil and regolith on a hillside, the material may break away, leaving a scar on the slope and forming a tongue- or teardrop-shaped mass that flows downslope (**FIGURE 8.20**).

The materials most commonly involved are rich in clay and silt and contain only small proportions of sand and coarser particles. Earthflows range in size from bodies a few meters long, a few meters wide, and less than a meter deep to masses more than a kilometer long, several hundred meters wide, and more than 10 meters deep.

Because earthflows are quite viscous, they generally move at slower rates than the more fluid debris flows described in a preceding section. They are characterized by a slow and persistent movement and may remain active for periods ranging from days to years. Depending on the steepness of the slope and the material's consistency, measured velocities range from less than 1 millimeter per day up to several meters per day. Over the time span that earthflows are active, movement is typically faster during wet periods than during drier times. In addition to occurring as isolated hillside phenomena, earthflows commonly take place in association with large slumps. In this situation, they may be seen as tongue-like flows at the base of the slump block (look at Figure 8.14).

CONCEPT CHECK 8.8

❶ Contrast earthflows with debris flows.

Slow Movements

 Mass Wasting

ESSENTIALS OF GEOLOGY **Types of Mass Wasting**

Movements such as rockslides, rock avalanches, and lahars are certainly the most spectacular and catastrophic forms of mass wasting. These dangerous events deserve intensive study to enable more effective prediction, timely warnings, and better controls to save lives. However, because of their large size and spectacular nature, they give us a false impression of their importance as a mass-wasting process. Indeed, sudden movements are responsible for moving less material than the slower and far more subtle action of creep. Whereas rapid types of mass wasting are characteristic of mountains and steep hillsides, creep takes place on both steep and gentle slopes and is thus much more widespread.

Creep

Creep is a type of mass wasting that involves the gradual downhill movement of soil and regolith. One factor that contributes to creep is the alternate expansion and contraction of surface material caused by freezing and thawing or wetting and drying. As shown in **FIGURE 8.21**, freezing or wetting lifts particles at right angles to the slope, and thawing or drying allows the particles to fall back to a slightly lower level. Each cycle therefore moves the material a tiny distance downslope. Creep is aided by anything that disturbs the soil. For example, raindrop impact and disturbance by plant roots and burrowing animals may contribute.

Creep is also promoted when the ground becomes saturated with water. Following a heavy rain or snowmelt, waterlogged soil may lose its internal cohesion, allowing gravity to pull the material downslope. Because creep is imperceptibly slow, the process cannot be observed in action. What can be observed, however, are the effects of creep. Creep causes fences and utility poles to tilt and retaining walls to be displaced.

Solifluction

When soil is saturated with water, the soggy mass may flow downslope at a rate of

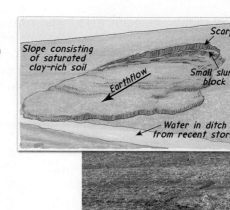

FIGURE 8.20 This small, tongue-shaped earthflow occurred on a newly formed slope along a recently constructed highway. It formed in clay-rich material following a period of heavy rain. Notice the small slump at the head of the earthflow. *(Photo by E. J. Tarbuck)*

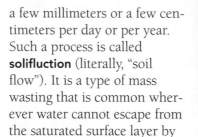

a few millimeters or a few centimeters per day or per year. Such a process is called **solifluction** (literally, "soil flow"). It is a type of mass wasting that is common wherever water cannot escape from the saturated surface layer by infiltrating to deeper levels. A dense clay hardpan in soil or an impermeable bedrock layer can promote solifluction.

Solifluction is a form of mass wasting that is common in regions underlain by **permafrost.** Permafrost refers to the permanently frozen ground that occurs in association with Earth's harsh tundra and ice-cap climates. Solifluction occurs in a zone above the permafrost called the *active layer*, which thaws to a depth of about a meter during the brief high-latitude summer and then refreezes in winter. During the summer season, water is unable to percolate into the impervious permafrost layer below. As a result, the active layer becomes saturated and slowly flows. The process can occur on slopes as gentle as 2 or 3 degrees. Where there is a well-developed mat of vegetation, a solifluction sheet may move in a series of well-defined lobes or as a series of partially overriding folds (**FIGURE 8.22**).

The Sensitive Permafrost Landscape

Many of the mass-wasting events described in this chapter had sudden and disastrous impacts on people. When the activities of people cause ice contained in permanently frozen ground to melt, the impact is more gradual and less deadly. Nevertheless, because permafrost regions are sensitive and fragile landscapes, the scars resulting from poorly planned actions can remain for generations.

Permafrost occurs where summers are too cool to melt more than a shallow layer. Deeper ground remains frozen year-round. When people disturb the surface by removing the insulating vegetation mat or by constructing roads and buildings, the delicate thermal balance is disturbed, and the permafrost can thaw

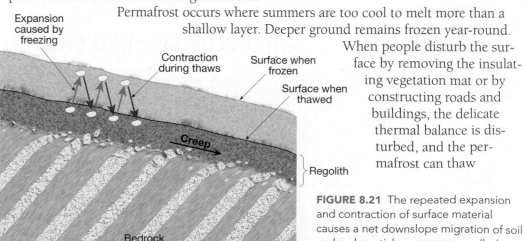

FIGURE 8.21 The repeated expansion and contraction of surface material causes a net downslope migration of soil and rock particles—a process called *creep.*

(**FIGURE 8.23A**). Thawing produces unstable ground that may slide, slump, subside, and undergo severe frost heaving. When a heated structure is built directly on permafrost that contains a high proportion of ice, thawing creates soggy material into which a building can sink. One solution is to place buildings and other structures on piles, like stilts. Such piles allow subfreezing air to circulate between the floor of the building and the soil and thereby keep the ground frozen.

When oil was discovered on Alaska's North Slope, many people were concerned about the building of a pipeline linking the oil fields of Prudhoe Bay to the ice-free port of Valdez 1300 kilometers (800 miles) to the south. There was serious concern that such a massive project might damage the sensitive permafrost environment. Many also worried about possible oil spills.

Because oil must be heated to about 60 °C (140 °F) to flow properly, special engineering procedures had to be developed to isolate this heat from the permafrost. Methods included insulating the pipe, elevating portions of the pipeline above ground level, and even placing cooling devices in the ground to keep it frozen (**FIGURE 8.23B**). The Alaska pipeline is clearly one of the most complex and costly projects ever built in the Arctic tundra. Detailed studies and careful engineering helped minimize adverse effects resulting from the disturbance of frozen ground.

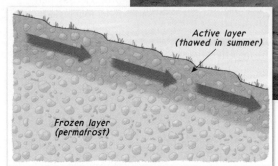

Geologist's Sketch

Active layer (thawed in summer)

Frozen layer (permafrost)

FIGURE 8.22 Solifluction lobes northeast of Fairbanks, Alaska. Solifluction occurs in permafrost regions when the active layer thaws in summer. *(Photo by James E. Patterson)*

DID YOU KNOW?
Permafrost underlies nearly one quarter of Earth's land area. In North America, more than 80 percent of Alaska and about 50 percent of Canada are underlain by permafrost.

CONCEPT CHECK 8.9

❶ Describe the basic mechanisms that contribute to creep.

❷ During what season does solifluction occur? Explain why it occurs only during that season.

FIGURE 8.23 **A.** When a rail line was built across this permafrost landscape in Alaska, the ground subsided. *(Photo by Lynn. A. Yehle, U.S. Geological Survey)* **B.** In places in Alaska, a pipeline is suspended above ground to prevent melting of delicate permafrost. *(Photo by Tom & Pat Leeson/Photo Researchers, Inc.)*

A.

B.

CHAPTER EIGHT
Mass Wasting: The Work of Gravity in Review

⊙ *Mass wasting* refers to the downslope movement of rock, regolith, and soil under the direct influence of gravity. In the evolution of most landforms, mass wasting is the step that follows weathering. The combined effects of mass wasting and erosion by running water produce stream valleys.

⊙ Gravity is the controlling force of mass wasting. Other factors that influence or trigger downslope movements are saturation of the material with water, oversteepening of slopes beyond the *angle of repose*, removal of vegetation, and ground shaking by earthquakes.

⊙ The various processes included under the name of mass wasting are divided and described on the basis of (1) the type of material involved (debris, mud, earth, or rock); (2) the type of motion (fall, slide, or flow); and (3) the rate of movement (rapid or slow).

⊙ The more rapid forms of mass wasting include *slump*, the downward sliding of a mass of rock or unconsolidated material moving as a unit along a curved surface; *rockslide*, blocks of bedrock breaking loose and sliding downslope; *debris flow*, a relatively rapid flow of soil and regolith containing a large amount of water; and *earthflow*, an unconfined flow of saturated, clay-rich soil that most often occurs on a hillside in a humid area following heavy precipitation or snowmelt.

⊙ The slowest forms of mass wasting include *creep*, the gradual downhill movement of soil and regolith, and *solifluction*, the gradual flow of a saturated surface layer that is underlain by an impermeable zone. Common sites for solifluction include regions underlain by *permafrost* (permanently frozen ground associated with tundra and ice-cap climates).

⊙ *Permafrost* covers large portions of North America and Siberia. Thawing produces unstable ground that may slide, slump, subside, and undergo severe frost heaving.

Key Terms

angle of repose (p. 202)
creep (p. 211)
debris flow (p. 209)
debris slide (p. 208)
earthflows (p. 210)

fall (p. 205)
flow (p. 206)
lahars (p. 209)
mass wasting (p. 199)

mudflow (p. 209)
permafrosts (p. 211)
rock avalanches (p. 206)
rockslides (p. 208)

slides (p. 206)
slump (p. 207)
solifluction (p. 211)
talus slopes (p. 206)

GIVE IT SOME THOUGHT

❶ Describe a type of mass wasting that might occur in your home area. Remember to consider characteristics such as climate, surface materials, and steepness of slopes. Does your example have a trigger?

❷ Rivers, groundwater, glaciers, wind, and waves can all move and deposit sediment. Geologists refer to these phenomena as *agents of erosion*. Mass wasting also involves the movement and deposition of sediment, yet it is *not* classified as an agent of erosion. How is mass wasting different?

❸ The concept of external and internal processes was introduced at the beginning of Chapter 5 (p.124).
In the accompanying photo, external processes are clearly active. Describe the role mass wasting is playing. Identify one feature that is the result of mass wasting.

❹ Describe at least one situation in which an internal process might cause or contribute to a mass-wasting event.

❺ Do you think it is likely that landslides frequently occur on the Moon? Explain why or why not.

❻ Mass wasting is influenced by many processes associated with all four spheres of the Earth system. Select three items from the list below. For each, outline a series of events that relate the item to various spheres and to a mass-wasting process. Here is an example that assumes "frost wedging" is an item on the list: *Frost wedging involves rock (geosphere) being broken when water (hydrosphere) freezes. Freeze–thaw cycles (atmosphere) promote frost wedging. When frost wedging loosens a rock on a cliff, the fragment tumbles to the base of the cliff. This event, rockfall, is an example of mass wasting.* Now you give it a try. Use your imagination.

a. Wildfire
b. Spring thaw/melting snow
c. Highway road cut
d. Crashing waves
e. Cavern formation (see Figure 10.25B p. 257)

Companion Website

Running Water

CONSIDER THE BITTERSWEET RELATIONSHIP WE HAVE with rivers. They are vital economic tools—used as highways to move goods, sources of water for irrigation and energy—as well as prime locations for sport and recreation. When considered as part of the Earth system, rivers and streams represent a fundamental link in the constant cycling of our planet's water. Running water is the dominant agent of landscape alteration, eroding more terrain and transporting more sediment than any other process. However, because human populations gravitate toward rivers, flooding in these otherwise valuable valleys represents enormous potential for destruction.

The Godafoss (*waterfall of the gods*) is one of the most spectacular waterfalls in Iceland. (*Photo by Jorgen Larsson/agefotostock*)

To assist you in learning the important concepts in this chapter, focus on the following questions:

- ⊙ What is the hydrologic cycle? What is the source of energy that powers the cycle?
- ⊙ What are the three main zones of a river system?
- ⊙ What is stream discharge, and how is it measured?
- ⊙ What are the factors that determine the velocity of water in a stream?
- ⊙ The work of a stream includes what three processes?
- ⊙ What are the two general types of stream valleys and some features associated with each?
- ⊙ In what way is base level related to a stream's ability to erode?
- ⊙ How are deltas, natural levees, and alluvial fans similar? How are they different?
- ⊙ What are the common drainage patterns produced by streams?
- ⊙ Why do floods occur? What are some basic flood-control strategies?

Earth as a System: The Hydrologic Cycle

Running Water

ESSENTIALS OF GEOLOGY **Hydrologic Cycle**

We live on a planet that is unique in the solar system—it is in just the right location and is just the right size (see Chapter 19). If Earth were appreciably closer to the Sun, water would exist only as a vapor. Conversely, water would be forever frozen if our planet were much farther away. Moreover, Earth is large enough to have a hot mantle that supports conductive flow, which carries water to the surface through volcanism. Water that rose from Earth's interior through mantle convection generated our planet's oceans and atmosphere. Thus, by coincidence of favorable size and location, Earth is the only planet in the solar system with a global ocean and a hydrologic cycle.

Water is found almost everywhere on Earth—in the oceans, glaciers, rivers, lakes, air, soil, and in living tissue (**FIGURE 9.1**). All of these "reservoirs" constitute Earth's hydrosphere, which contains about 1.36 billion cubic kilometers (326 million cubic miles) of water. The vast majority of it, about 97 percent, is stored in the global ocean (**FIGURE 9.2**). Ice sheets and glaciers account for slightly more than 2 percent, leaving less than 1 percent to be divided among lakes, streams, subsurface water, and the atmosphere.

All the rivers run into the sea; yet the sea is not full; unto the place from whence the rivers come, thither they return again.

(Ecclesiastes 1:7)

As the perceptive writer of Ecclesiastes implied, water is constantly moving among Earth's different spheres—the *hydrosphere,* the *atmosphere,* the *geosphere,* and the *biosphere.* This unending circulation of water, called the **hydrologic cycle,** describes what happens as water evaporates from the ocean, plants, and soil, moves through the atmosphere, and eventually falls as precipitation (**FIGURE 9.3**). Precipitation that falls onto the ocean has completed its cycle and is ready to begin another.

When precipitation falls on land, it either soaks into the ground, a process called **infiltration,** flows over the surface as **runoff,** or immediately evaporates. Much of the water that infiltrates or runs off eventually finds its way back to the atmosphere via evaporation from soil, lakes, and streams. In addition, some of the water that soaks into the ground is absorbed by plants, which later release it into the atmosphere. This process is called **transpiration** (*trans* = across, *spiro* = to breathe). Because both evaporation and transpiration involve the transfer of water from the surface directly to the atmosphere, they are often considered together as the combined process of **evapotranspiration.**

More water falls on land as precipitation than is lost by evapotranspiration. The excess is carried back to the ocean mainly by streams—less than 1 percent returns as groundwater. However, much of the water that flows in rivers is not transmitted directly into river channels after falling as precipitation. Instead, a large percentage first soaks into the soil and then gradually flows as groundwater to river channels. In this manner, groundwater provides a form of storage that sustains the flow of streams between storms and during periods of drought.

When precipitation falls in very cold areas—at high elevations or high latitudes—the water may not immediately soak in, run off, or evaporate. Instead, it may become part of a snowfield or a glacier.

FIGURE 9.1 Cumberland Falls State Park near Corbin, Kentucky. *(Photo by Chuck Haney/DanitaDelimont.com)*

In this way, glaciers store large quantities of water. If present-day glaciers were to melt and release their stored water, sea level would rise by several tens of meters worldwide and submerge many heavily populated coastal areas. As you will see in Chapter 11, over the past 2 million years, huge ice sheets have formed and melted on several occasions, each time changing the balance of the hydrologic cycle.

Figure 9.3 also shows that Earth's hydrologic cycle is balanced. Each year, solar energy evaporates about 320,000 cubic kilometers of water from the oceans, but only 284,000 cubic kilometers return to the oceans as precipitation. A balance is achieved by the 36,000 cubic kilometers that are carried to the ocean as runoff. Although runoff makes up a small percentage of the total, running water is, nevertheless, *the single most important erosional agent sculpting Earth's land surface.*

To summarize, the hydrologic cycle represents the continuous movement of water from the oceans to the atmosphere, from the atmosphere to the land, and from the land back to

Freshwater lakes 0.009%
Saline lakes and inland seas 0.008%
Soil moisture 0.005%
Atmosphere 0.001%
Stream channels 0.0001%

Oceans 97.2%   2.8%

Hydrosphere

Glaciers 2.15%   Groundwater 0.62%

Nonocean Component (% of total hydrosphere)

FIGURE 9.2 Distribution of Earth's water.

the sea. The wearing down of Earth's land surface is largely attributable to the last of these steps and is the primary focus of the remainder of this chapter.

CONCEPT CHECK 9.1

❶ Describe or sketch the movement of water through the hydrologic cycle. Once precipitation has fallen on land, what paths might the water take?

❷ What is meant by the term *evapotranspiration*?

❸ Over the oceans, evaporation exceeds precipitation, yet sea level does not drop. Explain why.

DID YOU KNOW?

Each year a field of crops may transpire the equivalent of a water layer 60 cm (2 feet) deep over the entire field. The same area of trees may pump twice this amount into the atmosphere.

FIGURE 9.3 Earth's hydrologic cycle. The primary movement of water through the hydrologic cycle is shown by the large arrows. Each year, solar energy evaporates about 320,000 cubic kilometers of water from the oceans, while evaporation from the land (including lakes and streams) contributes 60,000 cubic kilometers of water. Of this total of 380,000 cubic kilometers of water, about 284,000 cubic kilometers fall back to the ocean, and the remaining 96,000 cubic kilometers fall on the land surface. Of that 96,000 cubic kilometers, only 60,000 cubic kilometers of water return to the atmosphere by evaporation and transpiration, leaving 36,000 cubic kilometers of water to erode the land during the journey back to the oceans.

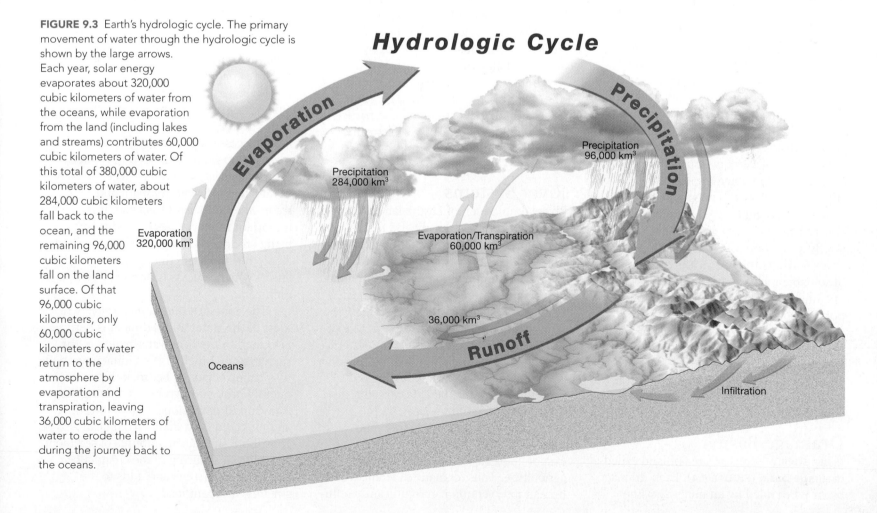

Hydrologic Cycle

Evaporation   Precipitation

Precipitation 284,000 km³

Precipitation 96,000 km³

Evaporation 320,000 km³

Evaporation/Transpiration 60,000 km³

Oceans

36,000 km³

Runoff

Infiltration

Running Water

Running Water

ESSENTIALS OF GEOLOGY **Stream Characteristics**

Recall that most of the precipitation that falls on land either enters the soil (*infiltration*) or remains at the surface, moving downslope as runoff. The amount of water that runs off rather than soaking into the ground depends on several factors: (1) intensity and duration of rainfall; (2) amount of water already in the soil; (3) nature of the surface material; (4) slope of the land, and; (5) the extent and type of vegetation. When the surface material is highly impermeable, or when it becomes saturated, runoff is the dominant process. Runoff is also high in urban areas because large areas are covered by impermeable buildings, roads, and parking lots.

Runoff initially flows in broad, thin sheets across hillslopes by a process called **sheet flow.** This thin, unconfined flow eventually develops threads of current that form tiny channels called **rills.** Rills meet to form gullies, which join to form brooks, creeks, or streams—then, when they reach an undefined size, they are called rivers. Although the terms *river* and *stream* are often used interchangeably, geologists define **stream** as water that flows in a channel, regardless of size. **River,** on the other hand, is a general term for streams that carry substantial amounts of water and have numerous tributaries.

In humid regions, the water to support stream flow comes from two sources, overland flow that sporadically enters the stream and groundwater that slowly, yet continuously, enters the channel. In areas where the bedrock is composed of soluable rocks such as limestone, large openings may exist that facilitate the transport of groundwater to streams. In arid regions, however, the *water table* may be below the level of the stream channel, in which case the stream loses water to the groundwater system by outflow through the streambed.

Drainage Basins

Every stream drains an area of land called a **drainage basin** (**FIGURE 9.4**). Each drainage basin is bounded by an imaginary line called a **divide,** something that is clearly visible as a sharp ridge in some mountainous areas but can be rather difficult to discern in more subdued topographies. The outlet, where the stream exits the drainage basin, is at a lower elevation than the rest of the basin.

Drainage divides range in scale from a small ridge separating two gullies on a hillside to a *continental divide* that splits an entire continent into enormous drainage basins. The Mississippi River has the largest drainage basin in North America, collecting and carrying 40 percent of the flow in the United States (**FIGURE 9.5**).

By looking at the drainage basin in Figure 9.4, it should be obvious that hillslopes cover most of the area. Water erosion on the hillslopes is aided by the impact of falling rain and surface flow, which moves downslope as sheets or in rills toward a stream channel. Hillslope erosion is the main source of fine particles (clays and fine sand) carried in stream channels.

River Systems

Rivers drain much of Earth's land area, with the exception of extremely arid regions and polar areas that are permanently frozen. To a large extent, the variety of rivers that exist is a reflection of the different environments in which they are found. For example, Uruguay's La Plata river drains roughly the same size area as the Nile in Egypt, yet La Plata carries nearly 10 times more water to the ocean. Because its drainage basin is in a tropical climate, La Plata has a huge runoff. By contrast, the Nile, which also originates in a humid region, flows through an expansive arid landscape where significant amounts of water evaporate or are withdrawn to sustain agriculture. Thus, climatic differences and human intervention can significantly influ-

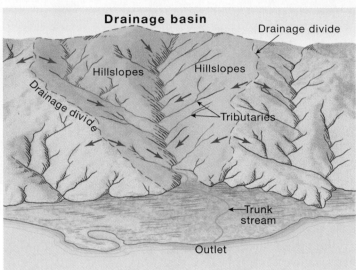

FIGURE 9.4 The drainage basin is the area of land drained by a river and its tributaries.

ence the character of a river. Later, we will examine other factors that contribute to stream variability.

River systems involve not only a network of stream channels, but also the entire drainage basin. Based on the dominant processes operating within them, river systems can be divided into three zones; *sediment production*—where erosion dominates, *sediment transport,* and *sediment deposition* (**FIGURE 9.6**). It is important to recognize that sediment is being eroded, transported, and deposited along the entire length of a stream, regardless of which process is dominant within each zone.

The zone of *sediment production,* where most of the water and sediment is derived, is located in the headwater region of the river system. Much of the sediment carried by streams begins as bedrock that is subsequently broken down by weathering, then transported downslope by mass wasting and overland flow. Bank erosion can also contribute significant amounts of sediment. In addition, scouring of the channel bed deepens the channel and adds to the stream's sediment load.

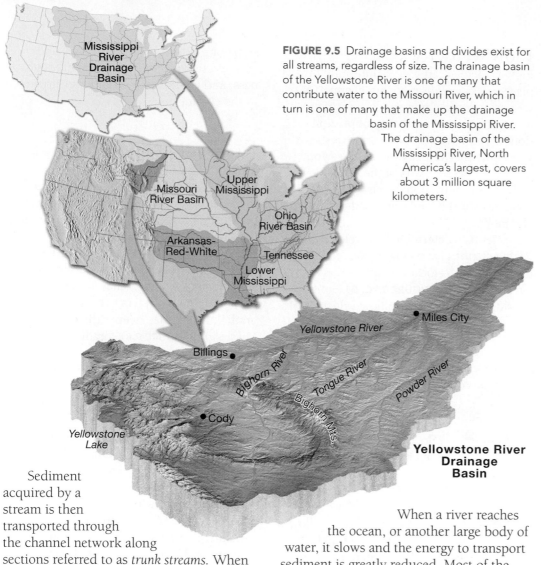

Mississippi River Drainage Basin

Missouri River Basin

Upper Mississippi

Ohio River Basin

Arkansas-Red-White

Tennessee

Lower Mississippi

Miles City

Yellowstone River

Billings

Bighorn River

Tongue River

Powder River

Cody

Bighorn Mts.

Yellowstone Lake

Yellowstone River Drainage Basin

FIGURE 9.5 Drainage basins and divides exist for all streams, regardless of size. The drainage basin of the Yellowstone River is one of many that contribute water to the Missouri River, which in turn is one of many that make up the drainage basin of the Mississippi River. The drainage basin of the Mississippi River, North America's largest, covers about 3 million square kilometers.

Sediment acquired by a stream is then transported through the channel network along sections referred to as *trunk streams*. When trunk streams are in balance, the amount of sediment eroded from their banks equals the amount deposited elsewhere in the channel. Although trunk streams rework their channels over time, they are not a source of sediment nor do they accumulate or store it.

When a river reaches the ocean, or another large body of water, it slows and the energy to transport sediment is greatly reduced. Most of the sediments accumulate at the mouth of the river to form a delta, are reconfigured by wave action to form a variety of coastal features, or are moved far offshore by ocean currents. Because coarse sediments tend to be deposited upstream, it is primarily the fine sediments (clay, silt, and fine sand) that eventually reach the ocean. Taken together, erosion, transportation, and deposition are the processes by which rivers move Earth's surface materials and sculpt landscapes (see Figure 9.6).

FIGURE 9.6
A river system can be divided into three zones based on the dominant processes operating within each zone. These are the zones of sediment production (erosion), sediment transport, and sediment deposition. (*Adapted from Shumm*)

Zone of sediment production (erosion)

Zone of transportation

Trunk stream

Zone of deposition

CONCEPT CHECK 9.2

❶ List several factors that influence infiltration.

❷ Draw a simple sketch of a drainage basin and a divide and label each.

❸ What are the three main parts (zones) of a river system?

Streamflow

Running Water

ESSENTIALS OF GEOLOGY **Stream Characteristics**

The water in river channels travels downslope under the influence of gravity. In very slow flowing streams, water moves in roughly straight-line paths parallel to the stream channel, called **laminar flow.** However, streams typically exhibit **turbulent flow,** where the water moves in an erratic fashion characterized by a series of horizontal and vertical swirling motions. Strong, turbulent behavior occurs in whirlpools and eddies, as well as in rolling whitewater rapids. Even streams that appear smooth on the surface often

Flash flooding in Las Vegas in August 2003. (*Photo by John Locher*)

exhibit turbulent flow near the bottom and sides of the channel where flow resistance is greatest. Turbulence contributes to the stream's ability to erode its channel because it acts to lift sediment from the streambed.

Flow Velocity

Flow velocities can vary significantly from place to place along a stream channel, as well as over time, in response to variations in the amount and intensity of precipitation. If you have ever waded into a stream, you know that velocity increases as you move into deeper parts of the channel. This is the result of frictional resistance, which is greatest near the banks and beds of stream channels.

Scientists determine flow velocities at gaging stations by averaging measurements taken at various locations across the stream's channel (**FIGURE 9.7C, D**). Some sluggish streams have flow velocities of less than 1 kilometer per hour, whereas stretches of some fast-flowing rivers may exceed 30 kilometers per hour.

The ability of a stream to erode and transport material is directly related to its flow velocity. Even slight variations in flow rate can lead to significant changes in the sediment load transported by a stream. Several factors influence flow velocities and, therefore, control a stream's potential to do "work." These factors include: (1) channel slope or gradient; (2) channel size and cross-sectional shape; (3) channel roughness; and (4) the amount of water flowing in the channel.

Gradient and Channel Characteristics

The slope of a stream channel, expressed as the vertical drop of a stream over a specified distance, is called **gradient.** Portions of the lower Mississippi River have very low gradients, about 10 centimeters or less per kilometer. By contrast, some mountain streams have channels that drop at a rate of more than 40 meters per kilometer—a gradient 400 times steeper than the lower Mississippi (**FIGURE 9.8**). Gradient also varies along the length of a particular channel. When the gradient is steeper, more gravitational energy is available to drive channel flow.

As water in a stream channel moves downslope it encounters a significant amount of frictional resistance. The *cross-sectional shape* (a slice taken across the channel) determines, to a large extent, the amount of flow in contact with the banks and bed of the channel. This measure is referred to as the **wetted perimeter.** The most efficient channel is one with the least wetted perimeter for its cross-sectional area. Figure 9.7 compares two channels that differ only in shape—channel A is wide and shallow, channel B is narrow and deep. Although the cross-sectional area of both is identical, channel B has less water in contact with the channel, and therefore, less frictional drag. As a

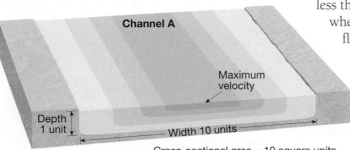

Channel A

Maximum velocity

Depth 1 unit

Width 10 units

Cross-sectional area = 10 square units
Wetted perimeter = 12 units
Ratio = $\frac{10}{12}$ = 0.83

A. Wide, shallow channel

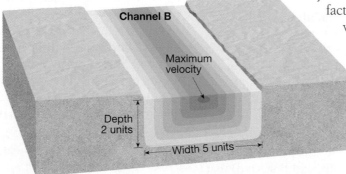

Channel B

Maximum velocity

Depth 2 units

Width 5 units

Cross-sectional area = 10 square units
Wetted perimeter = 9 units
Ratio = $\frac{10}{9}$ = 1.11

B. Narrow, deep channel

Conical cups

Sounding weight

D.

C. Gaging station

FIGURE 9.7 Influence of channel shape on velocity. **A.** Stream A has a wide, shallow channel and a large, wetted perimeter. **B.** The cross-sectional area of channel B is the same as stream A, but has less water in contact with its channel and therefore less frictional drag. Thus, water will flow more rapidly in channel B, all other factors being equal. **C.** Continuous records of stage and discharge are collected by the U.S. Geological Survey at more than 7000 gaging stations in the United States. Average velocities are determined by using measurements from several spots across the stream. This station is on the Rio Grande south of Taos, New Mexico. *(Photo by E. J. Tarbuck)* **D.** Current meter used to measure stream velocity at a gaging station.

TABLE 9.1
World's Largest Rivers Ranked by Discharge

Rank	River	Country	Drainage Area		Average Discharge	
			Square kilometers	Square miles	Cubic meters per second	Cubic feet per second
1	Amazon	Brazil	5,778,000	2,231,000	212,400	7,500,000
2	Congo	Rep. of Congo	4,014,500	1,550,000	39,650	1,400,000
3	Yangtze	China	1,942,500	750,000	21,800	770,000
4	Brahmaputra	Bangladesh	935,000	361,000	19,800	700,000
5	Ganges	India	1,059,300	409,000	18,700	660,000
6	Yenisei	Russia	2,590,000	1,000,000	17,400	614,000
7	Mississippi	United States	3,222,000	1,244,000	17,300	611,000
8	Orinoco	Venezuela	880,600	340,000	17,000	600,000
9	Lena	Russia	2,424,000	936,000	15,500	547,000
10	Parana	Argentina	2,305,000	890,000	14,900	526,000

result, if all other factors are equal, the water will flow more efficiently and at a higher velocity in channel B than in channel A.

Water depth also affects the frictional resistance the channel exerts on flow. Maximum flow velocity occurs when a stream is *bankfull,* before water starts to inundate the floodplain. At this stage, the channel's ratio of the cross-sectional area to wetted perimeter is highest and stream flow is most efficient. Similarly, an increase in channel size increases the ratio of cross-sectional area to wetted perimeter and therefore increases channel efficiency. All other factors being equal, flow velocities are higher in large channels than in small channels.

Most streams have channels that can be described as *rough.* Elements such as boulders, irregularities in the channel bed, and woody debris create turbulence that significantly impedes flow.

Discharge
Streams vary in size from small headwater creeks less than a meter wide to large rivers with widths of several kilometers. The size of a stream channel is largely determined by the amount of water supplied from the drainage basin. The measure most often used to compare the size of streams is **discharge**—the volume of water flowing past a certain point in a given unit of

time. Discharge, usually measured in cubic meters per second or cubic feet per second, is determined by multiplying a stream's cross-sectional area by its velocity.

TABLE 9.1 lists the world's largest rivers in terms of discharge. The largest river in North America, the Mississippi, has a discharge that averages 17,300 cubic meters per second. Nevertheless, that amount is dwarfed by South America's mighty Amazon, which discharges 12 times more water than the Mississippi. In fact, it has been estimated that the flow of the Amazon accounts for about 15 percent of all the fresh water transported to the ocean by all of the world's rivers. Just one day's discharge would supply the water needs of New York City for nine years!

The discharge of a river system changes over time because of variations in the amount of precipitation received by the drainage basin. Studies show that when discharge increases, the width, depth, and flow velocity of the channel all increase predictably. As we saw earlier, when the size of the channel increases, proportionally less water is in contact with the bed and banks of the channel. Thus, friction is reduced, resulting in an increase in the rate of flow.

Changes Downstream
One useful way of studying a stream is to examine its **longitudinal profile.** Such a profile is simply a cross-sectional view of a stream from its source area (called the **head** or **headwaters**) to its **mouth,** the point

FIGURE 9.8 Rapids are common in mountain streams because gradients are steep and channels are rough and irregular. *(Photo by Fogstock Llc/Photolibrary)*

downstream where it empties into another water body—a river, lake, or ocean. By examining **FIGURE 9.9** you will see that the most obvious feature of a typical longitudinal profile is its concave shape—a result of the decrease in slope that occurs from the headwaters to the mouth. In addition, local irregularities exist in the profiles of most streams—the flatter sections may be associated with lakes or reservoirs, and the steeper sections are sites of rapids or waterfalls.

The change in slope observed on most stream profiles is usually accompanied by an increase in discharge, channel size, and a reduction in sediment particle size (**FIGURE 9.10**). For example, data from successive gaging stations show that in humid regions discharge increases toward the mouth. This should come as no surprise because, as we move downstream, more and more tributaries contribute water to the main channel. In the case of the Amazon, for example, about 1000 tributaries join the main river along its 6500-kilometer course across South America.

In order to accommodate the growing volume of water, channel size typically increases downstream as well. Recall that flow velocities are higher in large channels compared to in small channels. Furthermore, observations show a general decline in sediment size downstream, making the channel smoother and more efficient.

Although the channel slope decreases toward a stream's mouth, the flow velocity generally increases. This fact contradicts our intuitive assumptions of swift, narrow headwater streams and wide, placid rivers flowing across more subtle topography. Increases in channel size and discharge and decreases in channel roughness that occur downstream compensate for the decrease in slope—thereby making the stream more efficient (see Figure 9.10). Thus, the average flow velocity is typically lower in headwater streams than in wide, placid rivers just "rollin' along."

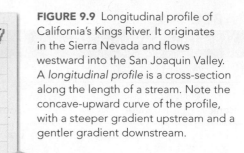

FIGURE 9.9 Longitudinal profile of California's Kings River. It originates in the Sierra Nevada and flows westward into the San Joaquin Valley. A *longitudinal profile* is a cross-section along the length of a stream. Note the concave-upward curve of the profile, with a steeper gradient upstream and a gentler gradient downstream.

CONCEPT CHECK 9.3

1. Compare and contrast laminar and turbulent flow.
2. Define stream *discharge*. How is discharge measured?
3. What typically happens to channel width, channel depth, flow velocity, and discharge between the point where a stream begins and the point where it ends? Briefly explain why these changes occur.

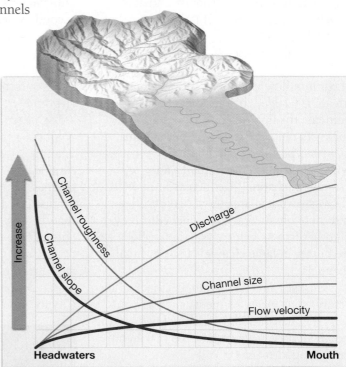

FIGURE 9.10 This graph shows how various properties of a stream channel change from its headwaters to its mouth. Although the gradient decreases toward the mouth, increases in channel size and discharge and decreases in roughness more than offset the decrease in slope. Consequently, stream flow velocity usually increases toward the mouth.

The Work of Running Water

Streams are Earth's most important erosional agents. Not only do they have the ability to downcut and widen their channels, but streams also have the capacity to transport enormous quantities of sediment delivered to them by overland flow, mass wasting, and groundwater. Eventually, much of this material is deposited to create a variety of landforms.

Stream Erosion

A stream's ability to accumulate and transport soil and weathered rock is aided by the work of raindrops, which knock sediment particles loose (see Figure 5.22 on p. 142). When the

ground is saturated, rainwater cannot infiltrate, so it flows downslope, transporting some of the material it dislodges. On barren slopes the flow of muddy water (sheet flow) often produces small channels (rills),

FIGURE 9.11 Potholes in the bed of a stream. The rotational motion of swirling pebbles acts like a drill to create potholes. *(Photo by Elmari Joubert/Alamy)*

into the flow. In addition, the individual sediment grains are also abraded by their many impacts with the channel and with one another. Thus, by scraping, rubbing, and bumping, abrasion erodes a bedrock channel and simultaneously smoothes and rounds the abrading particles. That is why smooth, rounded rocks and pebbles are found in streams. Abrasion also results in a reduction in the size of the sediments transported by streams.

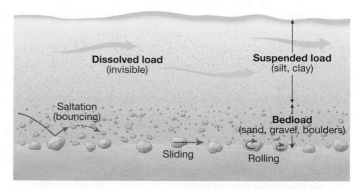

FIGURE 9.12 Streams transport their load of sediment in three ways. The dissolved and suspended loads are carried in the general flow. The bed load includes coarse sand, gravel, and boulders that move by rolling, sliding, and saltation.

Features common to some bedrock channels are circular depressions known as **potholes** created by the abrasive action of particles swirling in fast-moving eddies (**FIGURE 9.11**). The rotational motion of sand and pebbles acts like a drill to bore the holes. As the particles wear down to nothing, they are replaced by new ones that continue to drill the stream bed. Eventually, smooth depressions several meters across and deep may result.

which in time can evolve into larger gullies (see Figure 5.23 on p. 142).

Once flow is confined in a channel, the erosional power of a stream is related to its slope and discharge. The rate of erosion, however, is also dependent on the relative resistance of the bank and bed material. In general, channels composed of unconsolidated materials are more easily eroded than channels cut into bedrock.

When the channel is sandy, the particles are easily dislodged from the bed and banks and lifted into the moving water. Moreover, banks consisting of sandy material are often undercut, dumping even more loose debris into the water to be carried downstream. Banks that consist of coarse gravels or cohesive clay and silt particles tend to be relatively resistant to erosion. Thus, channels with cohesive silty banks are generally more narrow than comparable ones with sandy banks.

Streams cut channels into bedrock through *quarrying* and *abrasion*. **Quarrying** involves the removal of blocks from the bed of the channel. This process is aided by fracturing and weathering that loosen the blocks sufficiently so they are moveable during times of high flow rates. Quarrying is mainly the result of the impact forces exerted by flowing water.

Abrasion is the process by which the bed and banks of a bedrock channel are ceaselessly bombarded by particles carried

Transport of Sediment by Streams

All streams, regardless of size, transport some rock material (**FIGURE 9.12**). Streams also sort the solid sediment they transport because finer, lighter material is carried more readily than larger, heavier particles. Streams transport their load of sediment in three ways: (1) in solution (**dissolved load**), (2) in suspension (**suspended load**), and (3) sliding or rolling along the bottom (**bed load**).

DISSOLVED LOAD. Most of the dissolved load is brought to a stream by groundwater and is dispersed throughout the flow. When water percolates through the ground, it acquires soluble soil compounds. Then it seeps through cracks and pores in bedrock, dissolving additional mineral matter. Eventually much of this mineral-rich water finds its way into streams.

The velocity of streamflow has essentially no effect on a stream's ability to carry its dissolved load—material in the solution goes wherever the stream goes. Precipitation of the dissolved mineral matter occurs when the chemistry of the water changes or when the water enters an inland "sea," located in an arid climate where the rate of evaporation is high.

SUSPENDED LOAD. Most streams carry the largest part of their load in *suspension*. Indeed, the muddy appearance created by suspended sediment is the most obvious portion of a stream's load (**FIGURE 9.13**). Usually only very fine sand, silt, and clay particles are carried this way, but during flood stage larger particles can also be transported in suspension. During

Arizona's Lake Powell is a major reservoir. *(Photo by CORBIS)*

flood stage the total quantity of material carried in suspension also increases dramatically, verifiable by people whose homes have become sites for the deposition of this material. For example, during its flood stage the Yellow River (Hwang Ho) of China is reported to carry an amount of sediment equal in weight to the water that carries it. Rivers such as this are appropriately described as "too thick to drink but too thin to cultivate."

The type and amount of material carried in suspension are controlled by two factors: the flow velocity and the settling velocity of each sediment grain. **Settling velocity** is defined as the speed at which a particle falls through a still fluid. The larger the particle, the more rapidly it settles toward the stream bed. In addition to size, the shape and specific gravity of particles also influence settling velocity. Flat grains sink through water more slowly than spherical grains, and dense particles fall toward the bottom more rapidly than less dense particles. The slower the settling velocity and higher the flow velocity, the longer a sediment particle will stay in suspension and the farther it will be carried downstream.

BED LOAD. Coarse material, including coarse sands, gravels, and even boulders typically move along the bed of the channel as bed load. The particles that make up the bed load move by rolling, sliding, and saltation. Sediment moving by **saltation** appears to jump or skip along the stream bed (see Figure 9.12). This occurs as particles are propelled upward by collisions or lifted by the current and then carried downstream a short distance until gravity pulls them back to the bed of the stream. Particles that are too large or heavy to move by saltation either roll or slide along the bottom, depending on their shapes.

Compared with suspended load, the movement of bed load through a stream network tends to be less rapid and more localized. A study conducted on a glacially fed river in Norway determined that suspended sediments took only a day to exit the drainage basin, while the bed load required several decades to travel the same distance. Depending on the discharge and slope of the channel, coarse gravels may only be moved during times of high flow, while boulders move only during exceptional floods. Once set in motion, large particles are usually carried short distances. Along some stretches of a stream, bed load cannot be carried at all until it is broken into smaller particles.

CAPACITY AND COMPETENCE. A stream's ability to carry solid particles is described using two criteria—*capacity* and *competence*. **Capacity** is the maximum load of solid particles a stream can transport per unit time. The greater the discharge, the greater the stream's capacity for hauling sediment. Consequently, large rivers with high flow velocities have large capacities.

Competence is a measure of a stream's ability to transport particles based on size rather than quantity. Flow velocity is key—swift streams have greater competencies than slow streams, regardless of channel size. A stream's competence increases proportionately to the square of its velocity. Thus, if the velocity of a stream doubles, the impact force of the water increases four times; if the velocity triples, the force increases nine times, and so forth. Hence, large boulders that are often visible during low water and seem immovable can, in fact, be transported during exceptional floods because of the stream's increased competence.

By now it should be clear why the greatest erosion and transportation of sediment occur during periods of high water associated with floods. The increase in discharge results in greater capacity; the increased velocity produces greater competency. Rising velocity makes the water more turbulent, and larger particles are set in motion. In the course of a few days, or perhaps a few hours, a stream at flood stage can erode and transport more sediment than it does during several months of normal flow.

Deposition of Sediment by Streams

Deposition occurs whenever a stream slows, causing a reduction in competence. As its velocity decreases, sediment begins to settle, largest particles first. Thus, stream transport provides a mechanism by which solid particles of various sizes are separated. This process, called **sorting**, explains why particles of similar size are deposited together.

FIGURE 9.13 Colorado River, Grand Canyon National Park. The muddy appearance is a result of suspended sediment. *(Photo by Michael Collier)*

The general term for sediment deposited by streams is **alluvium.** Many different depositional features are composed of alluvium. Some occur within stream channels, some occur on the valley floor adjacent to the channel, and some are found at the mouth of the stream. We will consider the nature of these features later in the chapter.

CONCEPT CHECK 9.4

❶ Describe two processes by which streams cut channels in bedrock.

❷ In what three ways does a stream transport its load?

❸ If you collect a jar of water from a stream, what part of its load will settle to the bottom of the jar? What portion will remain in the water indefinitely? What part of the stream's load would probably not be present in your sample?

❹ Explain the difference between capacity and competency.

❺ What is settling velocity? What factors influence settling velocity?

Stream Channels

A basic characteristic that distinguishes streamflow from overland flow is that it is confined in a channel. A stream channel can be thought of as an open conduit consisting of the streambed and banks that act to confine flow except, of course, during floods.

Although somewhat oversimplified, we can divide stream channels into two basic types. *Bedrock channels* are those in which the streams are actively cutting into solid rock. In contrast, when the bed and banks are composed mainly of unconsolidated sediment or alluvium, the channel is called an *alluvial channel*.

Bedrock Channels

As the name suggests, **bedrock channels** are cut into the underlying strata and typically form in the headwaters of river systems where streams have steep slopes. The energetic flow tends to transport coarse particles that actively abrade the bedrock channel. Potholes are often visible evidence of the erosional forces at work.

Steep bedrock channels often develop a sequence of *pools* and *steps*, relatively flat segments (pools) where alluvium tends to accumulate, and steep segments (steps) where bedrock is exposed. The steep areas contain rapids or, occasionally, waterfalls.

The channel pattern exhibited by streams cutting into bedrock is controlled by the underlying geologic structure. Even when flowing over rather uniform bedrock, streams tend to exhibit winding or irregular patterns rather than flowing in straight channels. Anyone who has gone white-water rafting has observed the steep, winding nature of a stream flowing in a bedrock channel.

Alluvial Channels

Alluvial channels form in sediment that was previously deposited in the valley. When the valley floor reaches sufficient width, material deposited by the stream can form a *floodplain* that borders the channel. Because the banks and beds of alluvial channels are composed of unconsolidated sediment (alluvium) they can undergo major changes in shape as material is continually being eroded, transported, and redeposited. The major factors affecting the shapes of these channels are the average size of the sediment being transported, the channel's gradient, and discharge.

Alluvial channel patterns reflect a stream's ability to transport its load at a uniform rate while expending the least amount of energy. Thus, the size and type of sediment being carried help determine the nature of the stream channel. Two common types of alluvial channels are *meandering channels* and *braided channels*.

MEANDERING CHANNELS. Streams that transport much of their load in suspension generally move in sweeping bends called **meanders** (**FIGURE 9.14**). These streams flow in relatively deep, smooth channels and primarily transport mud (silt and clay), sand, and occasionally fine gravel. The lower Mississippi River exhibits this type of channel.

Meandering channels evolve over time as individual bends migrate across the floodplain. Most of the erosion is focused at the outside of the meander, where velocity and turbulence are greatest. In time,

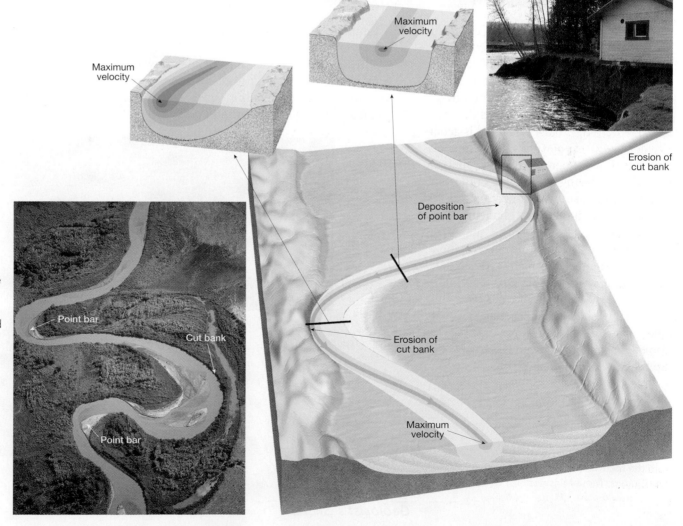

FIGURE 9.14 When a stream meanders, its zone of maximum speed shifts toward the outer bank. The point bars shown here are on the White River near Vernal, Utah. The black and white photo shows erosion of a cut bank along the Newaukum River in Washington State. By eroding its outer bank and depositing material on the inside of the bend, a stream is able to shift its channel. *(Point bar photo by Michael Collier; cut bank photo by P. A. Glancy, U.S. Geological Survey)*

FIGURE 9.15 Oxbow lakes occupy abandoned meanders. As they fill with sediment, oxbow lakes gradually become swampy meander scars. Aerial view of an oxbow lake created by the meandering Green River near Bronx, Wyoming. *(Photo by Michael Collier)*

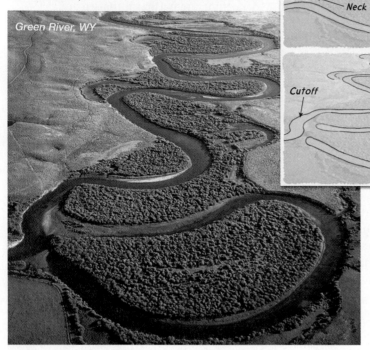

Green River, WY

Geologist's Sketch

material. This allows the next meander upstream to gradually erode the material between the two meanders as shown in **FIGURE 9.15**. Eventually, the river may erode through the narrow neck of land forming a new, shorter channel segment called a **cutoff**. Because of its shape, the abandoned bend is called an **oxbow lake**.

BRAIDED CHANNELS. Some streams consist of a complex network of converging and diverging channels that thread their way among numerous islands or gravel bars (**FIGURE 9.16**). Because these channels have an interwoven appearance, they are said to be **braided**. Braided channels form where a large portion of a stream's load consists of coarse material (sand and gravel) and the stream has a highly variable discharge. Because the bank material is readily erodable, braided channels are wide and shallow.

One setting in which braided streams form is at the end of glaciers where there is a large seasonal variation in discharge. During the summer, large amounts of ice-eroded sediment are dumped into the meltwater streams flowing away from the glacier. However, when flow is sluggish, the stream deposits the coarsest material as elongated structures called *bars*. This process causes the flow to split into several paths around the bars. During the next period of high flow, the laterally shifting channels erode and redeposit much of this coarse sediment, thereby transforming the entire

the outside bank is undermined, especially during periods of high water. Because the outside of a meander is a zone of active erosion, it is often referred to as the **cut bank** (Figure 9.14). Debris acquired by the stream at the cut bank moves downstream where the coarser material is generally deposited as **point bars** on the insides of bends. In this manner, meanders migrate laterally by eroding the outside of the bends and depositing sediment on the inside without appreciably changing their shape.

In addition to migrating laterally, the bends in a channel also migrate down the valley. This occurs because erosion is more effective on the downstream (downslope) side of the meander. Sometimes the downstream migration of a meander is slowed when it reaches a more resistant bank

streambed. In some braided streams, the bars have built semipermanent islands anchored by vegetation.

In summary, meandering channels develop where the load consists largely of fine-grained unconsolidated particles that are transported as suspended load in deep, relatively smooth channels. By contrast, wide, shallow braided channels develop where coarse-grained alluvium is transported mainly as bed load.

FIGURE 9.16 The Knik River, a classic braided stream with multiple channels separated by migrating gravel bars. The Knik is choked with sediment from the meltwater of four glaciers in the Chugach Mountains north of Anchorage, Alaska. *(Photo by Michael Collier)*

Geologist's Sketch

Base Level and Graded Streams

In 1875 John Wesley Powell, the pioneering geologist who first explored the Grand Canyon and later headed the U.S. Geological Survey, introduced the concept of a downward limit to stream erosion, which he called **base level**. A fundamental concept in the study of stream activity, base level is defined as the lowest elevation to which a stream can erode its channel. Essentially this is the level at which the mouth of a stream enters the ocean, a lake, or a trunk stream. Powell determined that two types of base level exist: "We may consider the level of the sea to be a grand base level, below which the dry lands cannot be eroded; but we may also have, for local and temporary purposes, other base levels of erosion."*

Sea level, which Powell called "grand base level," is now referred to as **ultimate base level. Local** or **temporary base levels** include lakes, resistant layers of rock, and rivers that act as base levels for their tributaries. All limit a stream's ability to downcut its channel.

Changes in base level cause corresponding adjustments in the "work" that streams perform. When a dam is built along a stream course, the reservoir that forms behind it raises the base level of the stream (**FIGURE 9.17**). Upstream from the reservoir the stream gradient is reduced, lowering its velocity and hence its sediment-transporting ability. As a result, the stream deposits material, thereby building up its channel. This process continues until the stream again has a gradient sufficient to carry its load. The profile of the new channel will be similar to the old, but somewhat higher.

*Exploration of the Colorado River of the West (Washington, DC: Smithsonian Institution, 1875), p. 203.

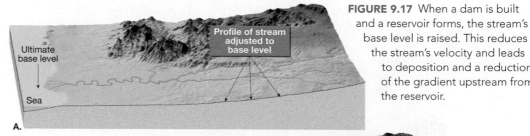

FIGURE 9.17 When a dam is built and a reservoir forms, the stream's base level is raised. This reduces the stream's velocity and leads to deposition and a reduction of the gradient upstream from the reservoir.

If, on the other hand, base level is lowered by a drop in sea level, the stream will have excess energy and downcut its channel to establish a balance with its new base level. Erosion first occurs near the mouth then progresses upstream creating a new stream profile.

Observing streams that adjust their profiles to changes in base level led to the concept of a graded stream. A **graded stream** has the necessary slope and other channel characteristics to maintain the minimum velocity required to transport the material supplied to it. On average, a graded system is neither eroding nor depositing material but is simply transporting it. When a stream reaches equilibrium, it becomes a self-regulating system in which a change in one characteristic causes a change in the others to counteract the effect.

Consider what would happen if displacement along a fault raised a layer of resistant rock along the course of a graded stream. As shown in **FIGURE 9.18**, the resistant rock forms a waterfall and serves as a

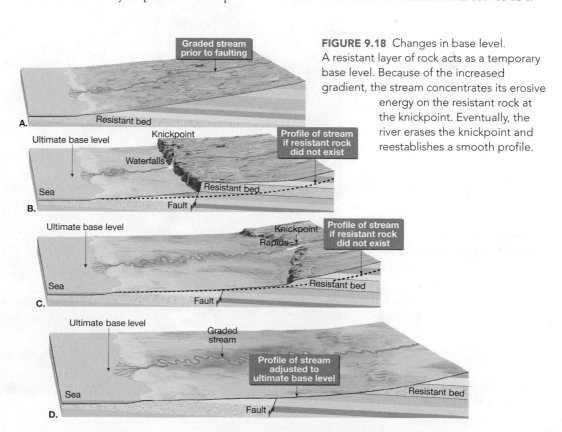

FIGURE 9.18 Changes in base level. A resistant layer of rock acts as a temporary base level. Because of the increased gradient, the stream concentrates its erosive energy on the resistant rock at the knickpoint. Eventually, the river erases the knickpoint and reestablishes a smooth profile.

temporary base level for the stream. Because of the increased gradient, the stream concentrates its erosive energy on the resistant rock along an area called a *knickpoint*. Eventually, the river erases the knickpoint from its path and reestablishes a smooth profile.

CONCEPT CHECK 9.6

❶ Define *base level*.

❷ Describe two ways base level can change.

Shaping Stream Valleys

 Running Water

ESSENTIALS OF GEOLOGY Reviewing Valleys and Stream-related Features

A **stream valley** consists of a channel and the surrounding terrain that directs water to the stream. It includes the *valley floor,* which is the lower, flatter area that is partially or totally occupied by the stream channel, and the sloping *valley walls* that rise above the valley floor on both sides. Alluvial channels often flow in valleys that have wide valley floors consisting of sand and gravel deposited in the channel, and clay and silt deposited by floods. Bedrock channels, on the other hand, tend to be located in narrow V-shaped valleys. In some arid regions, where weathering is slow and rock is particularly resistant, narrow valleys having nearly vertical walls are also found. Stream valleys exist on a continuum from narrow, steep-sided valleys to those that are so flat and wide that the valley walls are not discernable.

Streams, with the aid of weathering and mass wasting, shape the landscape through which they flow. As a result, streams continuously modify the valleys they occupy.

Valley Deepening

When a stream's gradient is steep and the channel is well above base level, downcutting is the dominant activity. Abrasion caused by bed load sliding and rolling along the bottom and the hydraulic power of fast-moving water slowly lower the streambed. The result is usually a V-shaped valley with steep sides. A classic example of a V-shaped valley is the section of the Yellowstone River shown in **FIGURE 9.19**.

The most prominent features of V-shaped valleys are *rapids* and *waterfalls*. Both occur where the stream's gradient increases significantly, a situation usually caused by variations in the erodability of the bedrock into which a stream channel is cutting. Resistant beds create rapids by acting as a temporary base level upstream while allowing downcutting to continue downstream. Recall that, in time, erosion usually eliminates the resistant rock. Waterfalls are places where streams make vertical drops.

Valley Widening

As a stream approaches a graded condition, downcutting becomes less dominant. At this point the stream's channel takes on a meandering pattern, and more of its energy is directed from side to side. As a result, the valley widens as the river cuts away at one bank and then the other. The continuous lateral erosion caused by shifting meanders gradually produces a broad, flat valley floor covered with alluvium (**FIGURE 9.20**). This feature, called a **floodplain**, is appropriately named because when a river overflows its banks during flood stage, it inundates the floodplain. Over time the floodplain will widen to a point where the stream is only actively eroding the valley walls in a few places. In the case of the lower Mississippi River, for example, the distance from one valley wall to another sometimes exceeds 160 kilometers (100 miles).

When a river erodes laterally and creates a floodplain as described, it is called an *erosional floodplain*. Floodplains can be

FIGURE 9.19 V-shaped valley of the Yellowstone River. The rapids and waterfalls indicate that the river is vigorously downcutting. *(Photo by Jorgen Larsson/ agefotostock)*

Geologist's Sketch

depositional as well. *Depositional floodplains* are produced by major fluctuations in conditions, such as changes in base level or climate. The floodplain in California's Yosemite Valley is one such feature; it was produced when a glacier gouged the valley floor about 300 meters (1000 feet) deeper than its former level. After the glacial ice melted, running water refilled the valley with alluvium. The Merced River currently winds across a relatively flat floodplain that forms much of the floor of Yosemite Valley.

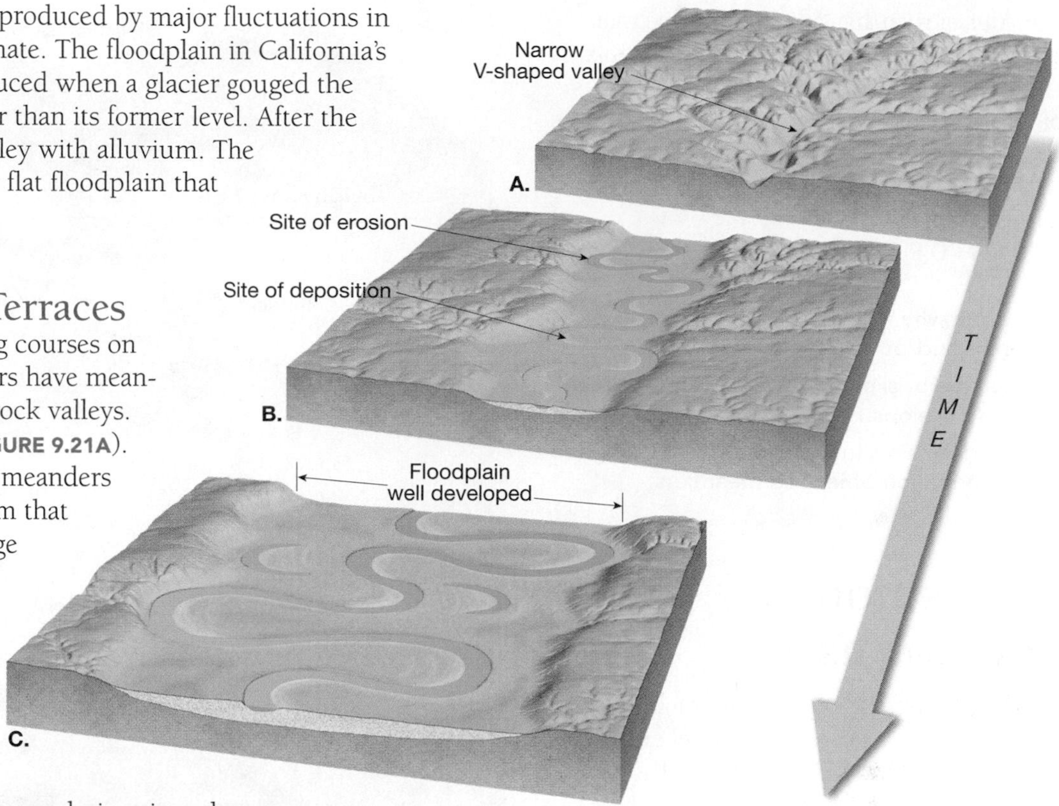

Incised Meanders and Stream Terraces

We usually find streams with highly meandering courses on floodplains in wide valleys. However, some rivers have meandering channels that flow in steep, narrow, bedrock valleys. Such meanders are called **incised meanders** (**FIGURE 9.21A**).

How do these features form? Originally the meanders probably developed on the floodplain of a stream that was in balance with its base level. Then, a change in base level caused the stream to begin downcutting. One of two events likely occurred—either base level dropped or the land upon which the river was flowing was uplifted. For example, regional uplifting of the Colorado Plateau in the southwestern United States generated incised meanders on several rivers. As the plateau gradually rose, meandering rivers began downcutting because of their steepening gradient (**FIGURE 9.21B,C**).

FIGURE 9.20 Development of an erosional floodplain.

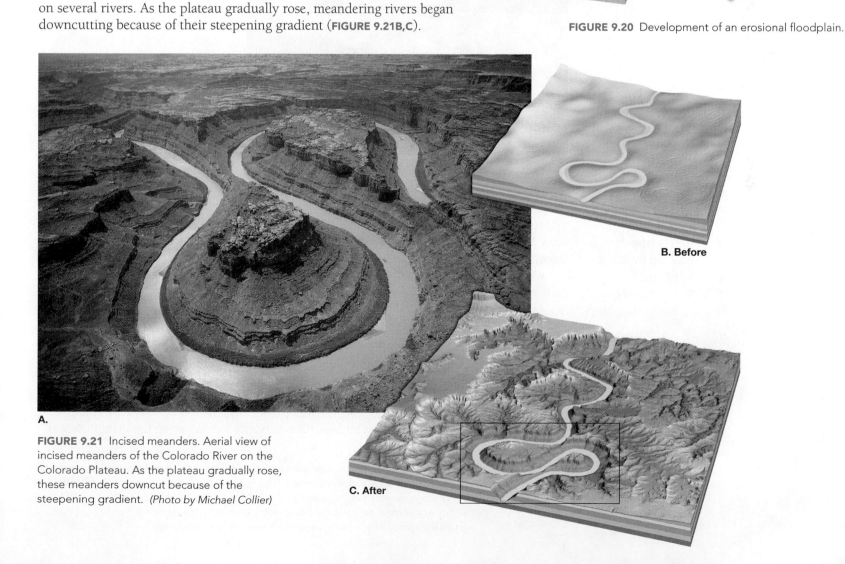

FIGURE 9.21 Incised meanders. Aerial view of incised meanders of the Colorado River on the Colorado Plateau. As the plateau gradually rose, these meanders downcut because of the steepening gradient. *(Photo by Michael Collier)*

After a river has adjusted in this manner, it may once again produce a floodplain at a level below the old one. The remnants of a former floodplain are sometimes present as relatively flat surfaces called **terraces** (**FIGURE 9.22**).

Depositional Landforms

Recall that streams continually pick up sediment in one part of their channel and deposit it downstream. These small-scale channel deposits are called **bars**. Such features, however, are only temporary, as the material will be picked up again and eventually carried to the ocean. In addition to sand and gravel bars, streams also create other depositional features that have somewhat longer life spans. These include *deltas, natural levees,* and *alluvial fans.*

Deltas

Deltas form where sediment-charged streams enter the relatively still waters of a lake, an inland sea, or the ocean (**FIGURE 9.23A**). As the stream's forward motion is slowed, sediment is deposited by the dying current. As a delta grows outward from the shoreline, the stream's gradient continually decreases. This circumstance eventually

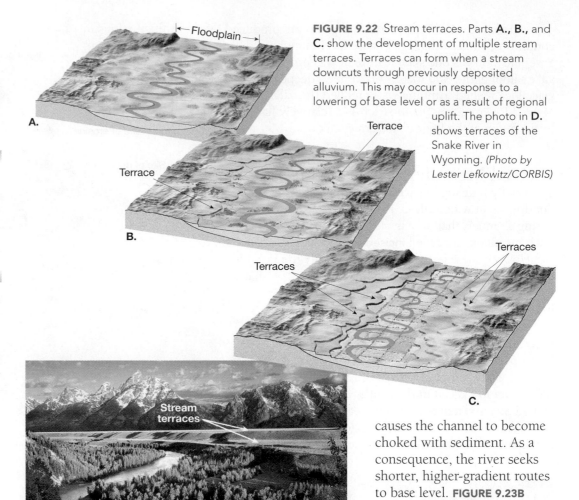

FIGURE 9.22 Stream terraces. Parts **A.**, **B.**, and **C.** show the development of multiple stream terraces. Terraces can form when a stream downcuts through previously deposited alluvium. This may occur in response to a lowering of base level or as a result of regional uplift. The photo in **D.** shows terraces of the Snake River in Wyoming. *(Photo by Lester Lefkowitz/CORBIS)*

causes the channel to become choked with sediment. As a consequence, the river seeks shorter, higher-gradient routes to base level. **FIGURE 9.23B** shows the main channel dividing into several smaller ones, called **distributaries** that carry water away from the main channel in varying paths to base level. After numerous shifts in the main flow from one distributary to the next, a delta may grow into the triangular shape of the Greek letter delta (Δ), although several other shapes exist.

In the ocean, deltas form where the supply of sediment exceeds the rate of marine erosion. Differences in the configurations of shorelines and variations in the nature and strength of wave activity are responsible for the shape and structure of each delta. Many of the world's great rivers have created massive deltas, each with its own peculiarities and typically more complex than the one illustrated in Figure 9.23A.

The Mississippi Delta

The Mississippi River delta resulted from the accumulation of huge quantities of sediment derived from the vast region drained by the river and its tributaries. New Orleans currently rests where there was once ocean. **FIGURE 9.24** shows the portion of the Mississippi delta that has accumulated over the past 6000 years. As the figure illustrates, the delta is actually a series of seven coalescing subdeltas. Each subdelta formed when the main flow was

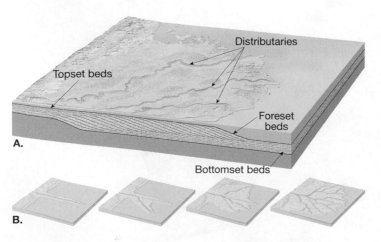

FIGURE 9.23 Formation of a simple delta. **A.** Structure of a simple delta that forms in relatively quiet waters. **B.** Growth of a simple delta. As a stream extends its channel, the gradient is reduced. Frequently, during flood stage the river is diverted to a higher-gradient route, forming a new distributary. Old, abandoned distributaries are gradually invaded by aquatic vegetation and filled with sediment.

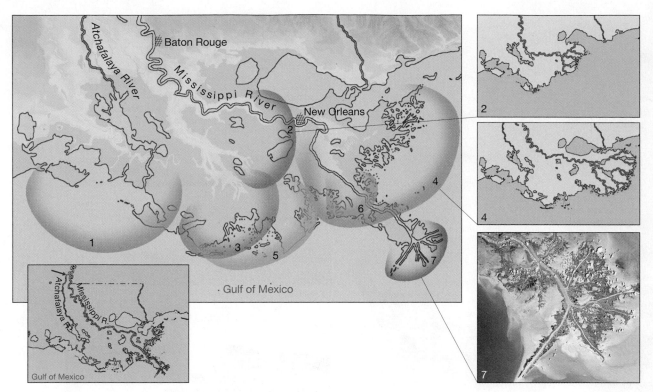

FIGURE 9.24 During the past 6000 years, the Mississippi River has built a series of seven coalescing subdeltas. The numbers indicate the order in which the subdeltas were deposited. The present bird-foot delta (number 7) represents the activity of the past 500 years. *(Image courtesy of JPL/Cal Tech/NASA)* Without ongoing human efforts, the present course will shift and follow the path of the Atchafalaya River. The left inset shows the point where the Mississippi may someday break through (arrow) and the shorter path it would take to the Gulf of Mexico. *(After C. R. Kolb and J. R. Van Lopik)*

diverted from one channel to a shorter, more direct path to the Gulf of Mexico. The individual subdeltas intertwine and partially cover each other to produce a complex structure. It is also apparent from Figure 9.24 that after each channel was abandoned, coastal erosion modified the newly formed subdelta. The present subdelta (number 7 in Figure 9.24), called a *bird-foot* delta because of the configuration of its distributaries, has been built by the Mississippi over the past 500 years.

At present, this active bird-foot delta has extended seaward almost as far as natural forces will allow. In fact, for many years the river has been "struggling" to cut through a narrow neck of land

and shift its course to the Atchafalaya River (see inset in Figure 9.24). If this were to happen, the Mississippi would abandon the lowermost 500-kilometers of its channel in favor of the Atchafalaya's much shorter 225-kilometer route to the Gulf.

In a concerted effort to keep the Mississippi on its present course, a damlike structure was erected at the site where the channel was trying to break through. A flood in 1973 weakened the control structure, and the river again threatened to shift. This event caused the U.S. Corps of Engineers to construct a massive auxiliary dam that was completed in the mid-1980s. For the time being, at least, the inevitable has been avoided, and the Mississippi River continues to flow past Baton Rouge and New Orleans on its way to the Gulf of Mexico.

Natural Levees

Meandering rivers that occupy valleys with broad floodplains tend to build **natural levees** that parallel their channels on both

banks. Natural levees are built by years of successive floods. When a stream overflows onto the floodplain, the water flows over the surface as a broad sheet. Because the flow velocity drops significantly, the coarser portion of the suspended load is immediately deposited adjacent to the channel. As the water spreads across the floodplain, a thin layer of fine sediment is laid down over the valley floor. This uneven distribution of material produces the gentle, almost imperceptible, slope of the natural levee (**FIGURE 9.25**).

The natural levees of the lower Mississippi River rise 6 meters (20 feet) above the adjacent valley floor. The area behind the levee is characteristically poorly drained for the obvious reason that water cannot flow over the levee and into the river. Marshes called **back swamps** often result. When a tributary stream enters a river valley having a substantial natural levee, it often flows for many kilometers through the back swamp before finding an opening where it enters the main river. Such streams are called **yazoo tributaries**, after the Yazoo River, which parallels the lower Mississippi River for more than 300 kilometers.

Alluvial Fans

Alluvial fans are fan-shaped deposits that accumulate along steep mountain fronts (see Figure 12.10 p. 295). When mountain streams emerge onto a relatively flat lowland their gradient drops and they deposit a large portion of their sediment load. Although alluvial fans are more prevalent in arid climates, they are occasionally found in humid regions.

Mountain streams, because of their steep gradients, carry much of their sediment load as coarse sand and gravel. Because alluvial fans are composed of these same coarse materials, water that flows across them readily soaks in. When a stream emerges from its valley onto an alluvial fan, its flow divides itself into several distributary channels. The fan shape is produced because the main flow swings

DID YOU KNOW?
The world's highest uninterrupted waterfall is Venezuela's Angel Falls. Named for American aviator Jimmie Angel, who first sighted the falls from the air in 1933, the water plunges 979 meters (3212 feet).

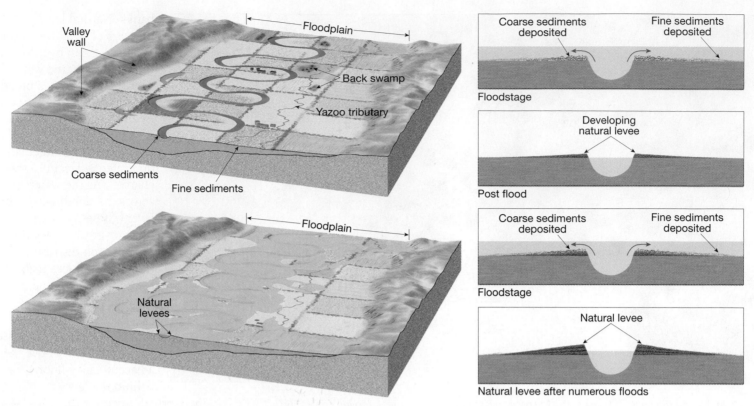

FIGURE 9.25 Natural levees are gently sloping structures that are created by repeated floods. The diagrams on the right show the sequence of development. Because the ground next to the stream channel is higher than the adjacent floodplain, back swamps and yazoo tributaries may develop.

back and forth between the distributaries from a fixed point where the stream exits the mountain.

Between rainy periods in deserts, little or no water flows across an alluvial fan, evident by the many dry channels that cross its surface. Thus, fans in dry regions grow intermittently, receiving considerable water and sediment only during wet periods. As you learned in Chapter 8, steep canyons in dry regions are prime locations for debris flows. Therefore, as would be expected, many alluvial fans have debris-flow deposits interbedded with the coarse alluvium.

CONCEPT CHECK 9.8

❶ What happens when a stream enters the relatively still waters of a lake, an inland sea, or the ocean?

❷ Briefly describe the formation of a natural levee. How is this feature related to back swamps and yazoo tributaries?

❸ Describe the formation of an alluvial fan. How does an alluvial fan differ from a delta?

Drainage Patterns

Drainage systems are interconnected networks of streams that form a variety of patterns. Variances occur primarily in response to the type of material on which the streams developed and/or the structural pattern of fractures, faults, and folds.

The most common, the **dendritic drainage pattern**, is characterized by irregular branching of tributary streams that resemble the branching pattern of a deciduous tree (**FIGURE 9.26A**). In fact, the word *dendritic* means "treelike." This drainage pattern develops whenever underlying bedrock is relatively uniform, such as flat-lying sedimentary strata or massive igneous rocks. Because the underlying material is essentially uniform in its resistance to erosion, it does not control the pattern of streamflow. Rather, the pattern is primarily determined by the slope of land.

When streams diverge from a central area, like spokes from the hub of a wheel, it is called a **radial drainage pattern** (**FIGURE 9.26B**). This pattern typically develops on isolated volcanic cones and domal uplifts.

FIGURE 9.26C illustrates a **rectangular drainage pattern**, with many right-angle bends. This pattern typically develops where the bedrock is crisscrossed by a series of joints. Because fractured rock tends to weather and erode more easily than unbroken rock, the geometric pattern of joints guides the direction of streams as they carve their valleys.

Illustrated in **FIGURE 9.26D** is a **trellis drainage pattern**, a rectangular pattern in which tributary streams are nearly parallel to one another and have the appearance of a garden trellis. This pattern forms in areas underlain by alternating bands of resistant and less resistant rock and is particularly well displayed in the folded Appalachian Mountains, where both weak and strong strata outcrop in nearly parallel belts.

CONCEPT CHECK 9.9

❶ Make a simple sketch of each of the four types of drainage patterns.

❷ How does geology influence each of these drainage patterns?

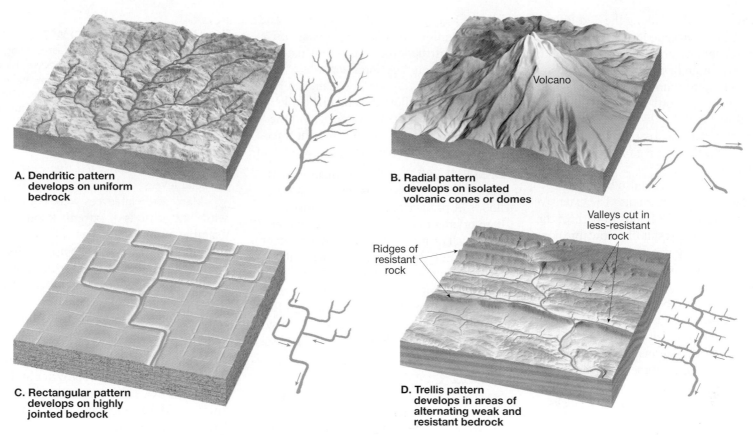

A. Dendritic pattern develops on uniform bedrock

Volcano

B. Radial pattern develops on isolated volcanic cones or domes

C. Rectangular pattern develops on highly jointed bedrock

Ridges of resistant rock

Valleys cut in less-resistant rock

D. Trellis pattern develops in areas of alternating weak and resistant bedrock

FIGURE 9.26 Drainage patterns. **A.** Dendritic. **B.** Radial. **C.** Rectangular. **D.** Trellis.

Floods and Flood Control

Floods occur when the flow of a stream becomes so great that it exceeds the capacity of its channel and overflows its banks (**FIGURE 9.27**). Among the most deadly and most destructive of all geologic hazards, floods are, nevertheless, simply part of the *natural* behavior of streams.

Most floods are caused by atmospheric processes that can vary greatly in both time and space. An hour or less of intense thunderstorm rainfall can trigger floods in small valleys. By contrast, major floods in large river valleys are often the result of an extraordinary series of precipitation events over a broad region for many days or weeks.

Land use planning in river basins requires an understanding of the frequency and magnitude of floods. For any particular river, a relationship exists between the size of a flood and the frequency with which it occurs. The larger the flood, the less often it is expected to occur. You have probably heard of a *100-year flood*. This describes the *recurrence interval* or *return period*, which is

an estimate of how often a flood of a given size can be expected to occur. A 25-year event would be much smaller, but would be four times more likely to occur than a 100-year flood. The relationship between the frequency and magnitude of floods differs from one region to another. For example, in arid climates, large, 100-year floods are typically much more extreme than those in humid areas.

Types of Floods

Floods can be the result of several naturally occurring and human-induced factors. Common flood types include *regional floods*, *flash floods*, *ice-jam floods*, and *dam-failure floods*.

REGIONAL FLOODS. Most regional floods are seasonal. Rapid melting of snow in spring and/or heavy spring rains often overwhelm rivers. For example, the

FIGURE 9.27 The flooding Cedar River covers a large portion of Cedar Rapids, Iowa, on June 14, 2008. *(Photo by AP Photo/Jeff Roberson)*

extensive 1997 flood along the Red River of the North was preceded by an especially snowy winter and an early spring blizzard. Early April brought rapidly rising temperatures, melting the snow in a matter of days, causing a record-breaking 500-year flood. Roughly 4.5 million acres were underwater, and the losses in the Grand Forks, North Dakota, region exceeded $3.5 billion.*

Regional floods can also be the result of numerous heavy rain events. The extensive and costly June 2008 floods in parts of the Midwest were the result of record-breaking rainfall on already waterlogged soils. Indiana experienced its costliest weather disaster in history, but Iowa suffered even greater losses with 83 of its 99 counties declared disaster areas. Nine of the state's rivers were at or above previous record flood levels and millions of acres of productive farmland were submerged. Thousands of people were evacuated, mostly in Cedar Rapids where more than 400 city blocks were under water (see Figure 9.27).

FLASH FLOODS. Flash floods occur with little warning and are potentially deadly because they produce rapid rises in water levels and can have devastating flow velocities. Rainfall intensity and duration, surface conditions, and topography are among the factors that influence flash flooding. Mountainous areas are especially susceptible because steep slopes can funnel runoff into narrow canyons with disastrous consequences, such as the Big Thompson River flood of July 31, 1976, in Colorado. During a four-hour period, more than 30 centimeters (12 inches) of rain fell, overwhelming its small drainage basin. The flash flood in this narrow canyon lasted only a few hours but claimed 139 lives and caused tens of millions of dollars in damages (**FIGURE 9.28**).

Urban areas are also susceptible to flash floods because a high percentage of the surface area is composed of impervious roofs, streets, and parking lots, where infiltration is minimal and runoff is rapid.

ICE-JAM FLOODS. Frozen rivers are especially susceptible to ice-jam floods. As the

level of a stream rises, it breaks up ice and creates ice flows that can accumulate on channel obstructions. Jams of this nature create temporary ice dams across the channel. Water trapped upstream can rise rapidly and overflow the channel banks. When an ice dam fails, water behind the dam is often released with sufficient force to inflict considerable damage downstream.

DAM-FAILURE FLOODS. Human interference with stream systems can also cause floods. A prime example is the failure of a dam or an artificial levee designed to contain small or moderate floods. When larger floods occur, the dam or levee may fail, resulting in the water behind it being released as a flash flood. The bursting of a dam in 1889 on the Little Conemaugh River caused the devastating Johnstown, Pennsylvania, flood that took more than 2200 lives.

Flood Control

Several strategies have been devised to eliminate or lessen the catastrophic impact of floods on our lives and environment. Engineering efforts include the construction of artificial levees, the building of flood-control dams, and river channelization.

ARTIFICIAL LEVEES. *Artificial levees* are earthen mounds built on river banks to increase the volume of water the channel can hold. Levees, used since ancient times, are the most common stream-containment structures. In some locations, concrete

floodwalls are constructed that function as artificial levees.

Many artificial levees were not built to withstand periods of extreme flooding. For example, numerous levees failed during the summer of 1993, when the upper Mississippi and many of its tributaries experienced record flooding (**FIGURE 9.29**). During that event, floodwalls at St. Louis, Missouri, created a bottleneck for the river that led to increased flooding upstream of the city.

FLOOD-CONTROL DAMS. *Flood-control dams* are built to store floodwater and then release it slowly, in a controlled manner. Since the 1920s, thousands of dams have been built on nearly every major river in the United States. Many dams have significant nonflood-related functions such as providing water for irrigated agriculture and for hydroelectric power generation. Many reservoirs are also major regional recreational facilities.

Although dams are effective in reducing flooding and provide other benefits, their construction and maintenance also have significant costs and consequences. For exam-

FIGURE 9.28 The disastrous nature of flash floods is illustrated by the Big Thompson River flood of July 31, 1976, in Colorado. During a four-hour span more than 30 centimeters (12 inches) of rain fell on portions of the river's small drainage basin. This amounted to nearly three-quarters of the average yearly total. The flash flood in the narrow canyon lasted only a few hours but cost 139 people their lives. Damages were estimated at $39 million. *(Phot by U.S. Geological Survey, Denver)*

*Ice jams also contribute to floods on the Red River of the North. See the section on "Ice-Jam Floods" at right.

ple, reservoirs created by dams may cover valuable farmland, forests, historic sites, and scenic valleys. Large dams can also cause significant damage to river ecosystems that have developed over thousands of years.

Furthermore, building dams is not a permanent solution to flooding. Sedimentation behind a dam gradually diminishes the volume of its reservoir, reducing the long-term effectiveness of this flood-control measure.

CHANNELIZATION. *Channelization* involves altering a stream channel in order to make the flow more efficient. This may simply involve clearing a channel of obstructions or dredging a channel to make it wider and deeper. Another alteration involves straightening and thus shortening a channel by creating *artificial cutoffs*. Shortening the stream increases the channel's flow velocity.

Between 1929 and 1942, the Army Corps of Engineers removed 16 meander bends on the lower Mississippi for the purpose of increasing the slope of the channel and thereby reducing the threat of flooding. This river, a vital transportation corridor, was shortened about 240 kilometers (150 miles). These efforts have been somewhat successful in reducing the maximum height of the river during flood stage. However, channel shortening led to increased gradients and accelerated erosion of bank material, both of which necessitated further intervention. Following the creation of artificial cutoffs, massive bank protection was installed along several stretches of the lower Mississippi.

A similar case in which artificial cutoffs accelerated bank erosion occurred on the Black-water River in Missouri, whose meandering course was shortened in 1910. Among the many effects of this project was a significant increase in the channel's width due to increased velocity. One particular bridge over the river collapsed in 1930 because of bank erosion. During the following 17 years the bridge was replaced three times, each requiring a longer span.

A NONSTRUCTURAL APPROACH. All of the flood-control measures described so far employ structural solutions to "control" rivers. These solutions are typically expensive and often give those who reside on the floodplain a false sense of security.

Currently, many scientists and engineers advocate a nonstructural approach to flood control. They suggest that sound floodplain management is an alternative to artificial levees, dams, and channelization. By identifying high-risk flood areas, appropriate zoning regulations can be implemented to minimize development and promote safer, more appropriate land use.

FIGURE 9.29 Water rushes through a break in an artificial levee in Monroe County, Illinois. During the record-breaking 1993 Midwest floods, many artificial levees could not withstand the force of the floodwaters. Sections of many weakened structures were overtopped or simply collapsed. *(Photo by James A. Finley/AP/Wide World Photos)*

CONCEPT CHECK 9.10

❶ Contrast regional floods and flash floods. Which type is more deadly?

❷ List and briefly describe three basic flood-control strategies. What are some drawbacks of each?

❸ Describe what is meant by a *nonstructural* approach to flood control.

CHAPTER NINE
Running Water in Review

⊙ The *hydrologic cycle* describes the continuous interchange of water among the oceans, atmosphere, and continents. Powered by energy from the Sun, it is a global system in which the atmosphere provides the link between the oceans and continents. The processes involved in the hydrologic cycle include *precipitation, evaporation, infiltration* (the movement of water into rocks or soil through cracks and pore spaces), *runoff* (water that flows over the land), and *transpiration* (the release of water vapor to the atmosphere by plants). *Running water is the single most important agent sculpting Earth's land surface.*

⊙ Initially, runoff flows as broad, thin sheets across the ground, called *sheet flow*. After a short distance, threads of current typically develop, and tiny channels called *rills* form.

⊙ The land area that contributes water to a stream is its *drainage basin*. Drainage basins are separated by imaginary lines called *divides*.

⊙ River systems consist of three main parts: the zones of *sediment production, sediment transport,* and *sediment deposition*.

⊙ The factors that determine a stream's *flow velocity* are *gradient* (slope of the stream channel), *shape, size,* and *roughness* of the channel, and the stream's *discharge* (amount of water passing a given point per unit of time). Most often, the gradient and roughness of a stream decrease downstream, while width, depth, discharge, and velocity increase.

⊙ Streams transport their load of sediment in solution (*dissolved load*), in suspension (*suspended load*), and along the bottom of the channel (*bed load*). Much of the dissolved load is contributed by groundwater. Most streams carry the greatest part of their load in suspension.

⊙ A stream's ability to transport solid particles is described using two criteria: *capacity* (the maximum load of solid particles a stream can carry) and *competence* (the maximum particle size a stream can transport). Competence increases as the square of stream velocity, so if velocity doubles, water's force increases fourfold.

⊙ Streams deposit sediment when velocity slows and competence is reduced. Stream deposits are called *alluvium* and may occur as channel deposits called *bars;* as floodplain deposits, such as *natural levees;* and as *deltas* or *alluvial fans* at the mouths of streams.

⊙ Stream channels are of two basic types: *bedrock channels* and *alluvial channels*. Bedrock channels are most commonly found in headwaters or regions where gradients are steep. Rapids and waterfalls are common features. Two types of alluvial channels are *meandering channels* and *braided channels*.

⊙ The two general types of *base level* (the lowest point to which a stream may erode its channel) are (1) *ultimate base level* and (2) *temporary* or *local base level*. Any change in base level will cause a stream to adjust and establish a new balance. Lowering base level will cause a stream to downcut, whereas raising base level results in deposition of material.

⊙ Although many gradations exist, the two general types of stream valleys are (1) *narrow V-shaped valleys* and (2) *wide valleys with flat floors*. Because the dominant activity is downcutting toward base level, narrow valleys often contain *waterfalls* and *rapids*.

⊙ When a stream has cut its channel closer to base level, its energy is directed from side to side, and erosion produces a flat valley floor, or *floodplain*. Streams that flow upon floodplains often move in sweeping bends called *meanders*. Widespread meandering may result in shorter channel segments, called *cutoffs,* and abandoned bends, called *oxbow lakes.*

⊙ Common *drainage patterns* (the form of a network of streams) produced by a main channel and its tributaries include (1) *dendritic*, (2) *radial*, (3) *rectangular*, and (4) *trellis*.

⊙ *Floods* are triggered by heavy rains and/or rapid snowmelt. Sometimes human interference can worsen or even cause floods. Flood-control measures include the building of *artificial levees* and *dams*, as well as *channelization,* which can involve creating *artificial cutoffs*. Many scientists and engineers advocate a non-structural approach to flood control that involves more appropriate land use.

Key Terms

abrasion (p. 223)
alluvial fans (p. 231)
alluvial channels (p. 225)
alluvium (p. 224)
back swamps (p. 231)
bars (p. 230)
base level (p. 227)
bed load (p. 223)
bedrock channels (p. 225)
braided (p. 226)
capacity (p. 224)
competence (p. 224)
cut bank (p. 226)
cutoff (p. 226)
deltas (p. 230)
dendritic drainage pattern (p. 232)

discharge (p. 221)
dissolved load (p. 223)
distributaries (p. 230)
divide (p. 218)
drainage basin (p. 218)
evapotranspiration
 (p. 216)
floods (p. 233)
floodplain (p. 228)
graded stream (p. 227)
gradient (p. 220)
head (headwaters) (p. 221)
hydrologic cycle (p. 216)
incised meanders (p. 229)
infiltration (p. 216)
laminar flow (p. 219)

local (temporary) base level
 (p. 227)
longitudinal profile (p. 221)
meanders (p. 225)
mouth (p. 221)
natural levees (p. 231)
oxbow lake (p. 226)
point bars (p. 226)
potholes (p. 223)
quarrying (p. 223)
radial drainage pattern (p. 232)
rectangular drainage pattern
 (p. 232)
rills (p. 218)
river (p. 218)
runoff (p. 216)

saltation (p. 224)
settling velocity (p. 224)
sheet flow (p. 218)
sorting (p. 224)
stream (p. 218)
stream valley (p. 228)
suspended load (p. 223)
temporary (local) base level
 (p. 227)
terraces (p. 230)
transpiration (p. 216)
trellis drainage pattern (p. 232)
turbulent flow (p. 219)
ultimate base level (p. 227)
wetted perimeter (p. 220)
yazoo tributaries (p. 231)

GIVE IT SOME THOUGHT

❶ A river system consists of three zones based on the dominant process operating in each zone. On the accompanying illustration, match each process with one of the three zones:
 a. sediment production (erosion)
 b. sediment deposition
 c. sediment transportation

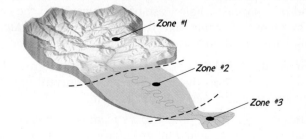

❷ If a stream originates at 6000 meters above sea level and travels 2500 kilometers to the ocean, what is its average gradient in meters per kilometer?

❸ Suppose the stream mentioned in the previous question developed extensive meanders that lengthened its course to 3000 kilometers. Calculate the new gradient. How does meandering affect gradient?

❹ Streamflow is affected by several variables, including discharge, gradient, and channel roughness, size, and shape. Develop a scenario in which a mass wasting event influences a stream's flow. Explain what led up to, or triggered, the event and describe how the mass wasting process influenced the stream's flow.

❺ Describe and explain what happens to a stream's ability to erode during periods of (a) drought and (b) heavy precipitation. Suggest other factors that could change a stream's ability to erode other than a change in precipitation.

❻ Match the description that best describes each of the accompanying photos.

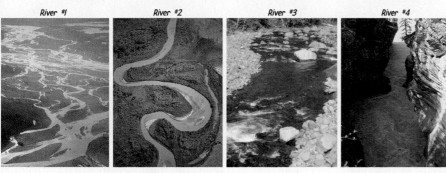

River #1   River #2   River #3   River #4

   a. bedrock channel
   b. alluvial channel (meandering)
   c. alluvial channel (non-meandering)
   d. alluvial channel (braided)

❼ Use the photos referred to in the previous question to answer the following:

   a. Which of these rivers is actively downcutting its channel?
   b. Which is/are mainly transporting coarse sediments?
   c. Which is/are mainly transporting silt and clay-size particles?
   d. Which do you think is/are being fed by water from a melting glacier?

❽ Look at Figure 9.24. Based on the locations of the seven subdeltas of the Mississippi River, describe the history of flow at the mouth of the Mississippi River. Why would people be concerned with the possibility of the Mississippi River shifting from its current path and following the path of the Atchafalaya River?

❾ Identify the main river in your area. For what streams does it act as base level? What is base level for the Mississippi River? The Missouri River?

❿ The accompanying diagram shows the time relationship between a rainstorm and the occurrence of flooding. Based on this graph, when does the flooding occur in relation to the rainstorm?

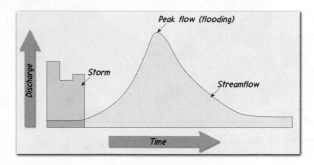

⓫ The accompanying graphs shows lag times between rainfall and peak flow (flooding) for an urban area and a rural area. Which graph (A or B) represents a rural area? Explain your choice.

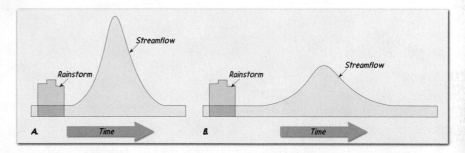

Companion Website

![PEARSON mygeoscience place]

www.mygeoscienceplace.com

The *Essentials of Geology, 11e* companion Website contains numerous multimedia resources accompanied by assessments to aid in your study of the topics in this chapter. The use of this site's learning tools will help improve your understanding of geology. Utilizing the access code that accompanies this text, visit **www.mygeoscienceplace.com** in order to:

- **Review** key chapter concepts.
- **Read** with links to the Pearson eText and to chapter-specific web resources.
- **Visualize** and comprehend challenging topics using the learning activities in *GEODe: Essentials of Geology* and the *Geoscience Animations Library*.
- **Test** yourself with online quizzes.

Groundwater

WORLDWIDE, WELLS AND SPRINGS PROVIDE WATER

for cities, crops, livestock, and industry. In some areas, however, overuse of this basic resource has resulted in water shortages, streamflow depletion, land subsidence, contamination by saltwater, increased pumping costs, and groundwater pollution.

Sunset along the Li River in China's Guilin District. This part of southern China exhibits tower karst development in which groundwater has dissolved large volumes of limestone leaving only these residual towers. *(Photo by Moodboard/Corbis)*

To assist you in learning the important concepts in this chapter, focus on the following questions:

◉ What is the importance of groundwater as a resource and as a geological agent?

◉ What is groundwater, and what factors affect its movement?

◉ How do springs, geysers, wells, and artesian systems form?

◉ What are some environmental problems associated with groundwater?

◉ What features are produced by the geologic work of groundwater?

Importance of Groundwater

Groundwater

ESSENTIALS
OF GEOLOGY **Importance and Distribution of Groundwater**

Although it is hidden from view, vast quantities of water exist in the cracks, crevices, and pore spaces of rock and soil. It occurs almost everywhere beneath Earth's surface and is a major source of water worldwide. Groundwater is a valuable natural resource that provides about half of our drinking water and is essential to the vitality of agriculture and industry (**FIGURE 10.1**). In addition to human uses, groundwater plays a crucial role in sustaining streamflow, especially during protracted dry periods. Many ecosystems depend on groundwater discharge into streams, lakes and wetlands.

Groundwater is one of our most important and widely available resources, yet people's perceptions of the subsurface environment from which it comes are often unclear and incorrect. The reason is that the groundwater environment is largely hidden from view except in caves and mines, and the impressions people gain from these subsurface openings are misleading. Observations on the land surface give an impression that Earth is solid. This view remains when we enter a cave and see water flowing in a channel that appears to have been cut into solid rock.

Because of such observations, many people believe that groundwater occurs only in underground rivers. In reality, most of the subsurface environment is not solid at all. It includes countless tiny *pore spaces* between grains of soil and sediment, plus narrow joints and fractures in bedrock. Together, these spaces add up to an

immense volume. It is in these small openings that groundwater collects and moves.

Considering the entire hydrosphere, or all of Earth's water, only about six-tenths of 1 percent occurs underground. Nevertheless, this small percentage, stored in the rocks and sediments beneath Earth's surface, is a vast quantity. When the oceans are excluded and only sources of fresh water are considered, the significance of groundwater becomes more apparent.

TABLE 10.1 contains estimates of the distribution of fresh water in the hydrosphere. Clearly, the largest volume occurs as glacial ice. Second in

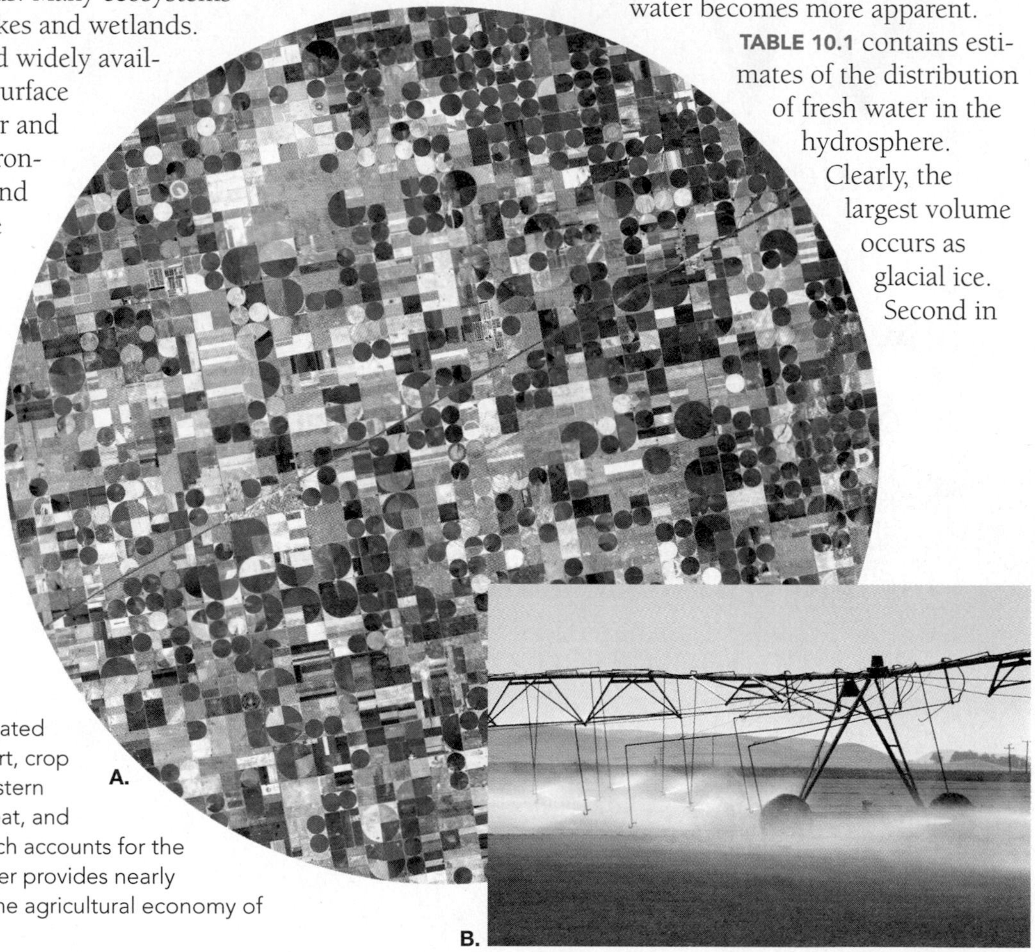

FIGURE 10.1 A. Satellite image of circular crop fields irrigated by center pivot irrigation. Resembling a work of modern art, crop circles cover what was once short grass prairie in southwestern Kansas. Three different crops are being grown—corn, wheat, and sorghum. Each is at a different point of development, which accounts for the varying shades of green and yellow. *(NASA)* **B.** Groundwater provides nearly 216 billion liters (57 billion gallons) per day in support of the agricultural economy of the United States. *(Photo by Clark Dunbar/Corbis)*

TABLE 10.1

Fresh Water of the Hydrosphere

Parts of the Hydrosphere	Volume of Fresh Water (km³)	Share of Total Volume of Fresh Water (percent)	Share of Total Volume of Liquid Fresh Water (percent)
Ice sheets and glaciers	24,000,000	84.945	0
Groundwater	4,000,000	14.158	94.05
Lakes and reservoirs	155,000	0.549	3.64
Soil moisture	83,000	0.294	1.95
Water vapor in the atmosphere	14,000	0.049	0.33
River water	1,200	0.004	0.03
Total	28,253,200	100.00	100.00

Source: U.S. Geological Survey Water Supply Paper 2220, 1987.

rank is groundwater, with slightly more than 14 percent of the total. However, when ice is excluded and just liquid water is considered, more than 94 percent of all fresh water is groundwater. Without question, *groundwater represents the largest reservoir of fresh water that is readily available to humans.* Its value in terms of economics and human well-being is incalculable.

Geologically, groundwater is important as an erosional agent. The dissolving action of groundwater slowly removes rock, allowing surface depressions known as sinkholes to form as well as creating subterranean caverns (**FIGURE 10.2**). Groundwater is also an equalizer of streamflow. Much of the water that flows in rivers is not direct runoff from rain and snowmelt. Rather, a large percentage of precipitation soaks in and then moves slowly underground to stream channels. Groundwater is thus a form of storage that sustains streams during periods when rain does not fall. When we see water flowing in a river during a dry period, it is water from rain that fell at some earlier time and was stored underground.

CONCEPT CHECK 10.1

❶ What percentage of fresh water is groundwater? How does this change if glacial ice is excluded?

❷ What are two geological roles for groundwater?

FIGURE 10.2 A view of the interior of a cavern. The dissolving action of acidic groundwater created the caverns. Later, groundwater deposited the limestone decorations. Carlsbad Caverns National Park, New Mexico. *(Photo by Jules Cowan/Photolibrary)*

Groundwater—A Basic Resource

Water is basic to life. It has been called the "bloodstream" of both the biosphere and society. Each day in the United States we use about 349 billion gallons of fresh water.* About 77 percent comes from surface sources. Groundwater provides the remaining 23 percent (**FIGURE 10.3**). One of the advantages of groundwater is that it exists almost everywhere across the country and, thus, is often available in places that lack reliable surface sources such as lakes and rivers. Water in a groundwater system is stored in subsurface pore spaces and fractures. As water is withdrawn from a well, the connected pore spaces and fractures act as a "pipeline" that allows water to gradually move from one part of the hydrologic system to where it is being withdrawn.

Although people have been digging and drilling wells for thousands of years, extensive use of groundwater is a relatively recent phenomenon—growing rapidly with the development of rural electrification and more effective pumping technologies during the past 80 years. The 79 billion gallons of groundwater withdrawn each day in the United States represents only about 8 percent of the estimated 1 trillion gallons per day of natural replenishment (termed *recharge*). Therefore, our groundwater resource appears to be ample. However, this is misleading because groundwater availability varies widely. For example, in the dry western states, groundwater demand is great, but rainfall to replenish the supply is often scarce.

What are the primary ways that we use groundwater? The U.S. Geological Survey identifies several categories that are shown in Figure 10.3. More groundwater is used for irrigation than for all other uses combined. There are nearly 60 million acres (nearly 243,000 square kilometers or about 93,700 square miles) of irrigated land in the United States. That is an area nearly the size of the state of Wyoming. The vast majority (75 percent) of the irrigated land is in the 17 conterminous western states where annual precipitation is typically less than 20 inches. About 42 percent of the water used for irrigation is groundwater.

Public and domestic uses include water for indoor and outdoor household purposes as well as water used for commercial purposes. Common indoor uses include drinking, cooking, bathing, washing clothes and dishes, and flushing toilets. If you are curious about how much water an average American uses each day for indoor domestic purposes, look at **FIGURE 10.4**. Major outdoor uses are watering lawns and gardens. Water for domestic use may come from a public supply or be self-supplied.** For those who are self-

supplied, practically all (98 percent) rely on groundwater.

Another category, aquaculture, involves water used for fish hatcheries, fish farms, and shellfish farms. Many mining operations require significant quantities of water, as do industrial processes that include petroleum refining and the manufacture of chemicals, plastics, paper, steel, and concrete.

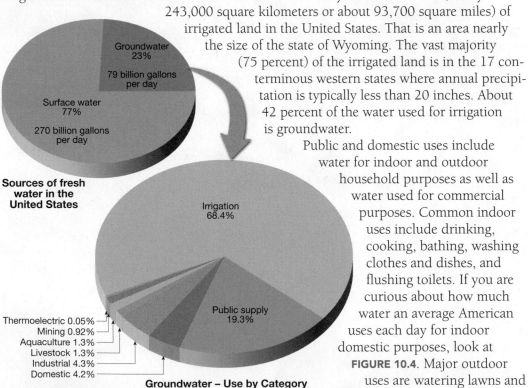

Sources of fresh water in the United States

Groundwater 23%
79 billion gallons per day

Surface water 77%
270 billion gallons per day

Groundwater – Use by Category

Irrigation 68.4%
Public supply 19.3%
Thermoelectric 0.05%
Mining 0.92%
Aquaculture 1.3%
Livestock 1.3%
Industrial 4.3%
Domestic 4.2%

FIGURE 10.3 Each day in the United States we use 345 billion gallons of fresh water. Groundwater is the source of nearly one quarter of the total. More groundwater is used for irrigation than for all other uses combined. *(Data from U.S. Geological Survey)*

DID YOU KNOW?
The term "acre-foot" is often used to express large quantities of water such as the volume of a reservoir or amounts allocated for irrigation. As the term implies, an acre-foot is the volume of water needed to cover one acre of land to a depth of 1 foot. An acre-foot is equal to 1.2 million liters or 326,000 gallons.

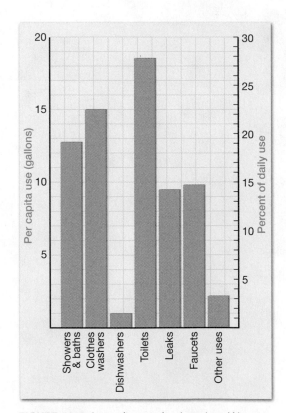

FIGURE 10.4 According to the American Water Works Association, daily indoor per capita use in an average single family home is 69.3 gallons. By installing more efficient water fixtures and regularly checking for leaks, households could reduce this figure by about one third.

❶ What share of U.S. fresh water is provided by groundwater? What is most groundwater used for?

❷ Did the amount of water in any category in Figure 10.4 surprise you?

Distribution of Groundwater

 Groundwater

ESSENTIALS OF GEOLOGY Importance and Distribution of Groundwater

When rain falls, some of the water runs off, some returns to the atmosphere by evaporation and transpiration, and the remainder soaks into the ground. This last path is the primary source of practically all subsurface water. The amount of water that takes each of these paths, however, varies greatly both in time and space. Influential factors include steepness of slope, nature of surface material, intensity of rainfall, and type and amount of vegetation. Heavy rains falling on steep slopes underlain by impervious materials will obviously result in a high percentage of the water running off. Conversely, if rain falls steadily and gently on more gradual slopes composed of materials that are easily penetrated by the water, a much larger percentage of water soaks into the ground.

Some of the water that soaks in does not travel far, because it is held by molecular attraction as a surface film on soil particles. This near-surface zone is called the *zone of soil moisture*. It is crisscrossed by roots, voids left by decayed roots, and animal and worm burrows that enhance the infiltration of rainwater into the soil. Soil water is used by plants in life functions and transpiration. Some water also evaporates directly back into the atmosphere.

Water that is not held as soil moisture will percolate downward until it reaches a zone where all of the open spaces in sediment and rock are completely filled with water. This is the **zone of saturation**. Water within it is called **groundwater**. The upper limit of this zone is known as the **water table**. The area above the water table where the soil, sediment, and rock are not saturated is called the **unsaturated zone** (**FIGURE 10.5**). The pore spaces in this zone contain both air and water. Although a considerable amount of water can be present in the unsaturated zone, this water cannot be pumped by wells because it clings too tightly to rock and soil particles. By contrast, below the water table, the water pressure is great enough to allow water to enter wells, thus permitting groundwater to be withdrawn for use. We will examine wells more closely later in the chapter.

❶ When it rains, what factors influence the amount of water that soaks in?

❷ Define groundwater and relate it to the water table.

The Water Table

 Groundwater

ESSENTIALS OF GEOLOGY Importance and Distribution of Groundwater

The water table, the upper limit of the zone of saturation, is a very significant feature of the groundwater system. The water-table level is important in predicting the productivity of wells, explaining the changes in the flow of springs and streams, and accounting for fluctuations in the levels of lakes.

Variations in the Water Table

The depth of the water table is highly variable and can range from zero, when it is at the surface, to hundreds of meters in some places. An important characteristic of the water table is that its configuration varies seasonally and from year to year because the addition of water to the groundwater system is closely related to the quantity, distribution, and timing of precipitation. Except where the water table is at the surface, we cannot observe it directly. Nevertheless, its elevation can be mapped and studied in detail where wells are numerous, because the water level in wells coincides with the water table (**FIGURE 10.6**). Such maps reveal that the water table is rarely level, as we might expect a table to be. Instead, its shape is usually a subdued replica of the surface topography, reaching its highest elevations beneath hills and then descending toward valleys. Where a wetland (swamp) is encountered, the water table is right at the surface. Lakes and streams generally occupy areas low enough that the water table is above the land surface.

Several factors contribute to the irregular surface of the water table. One

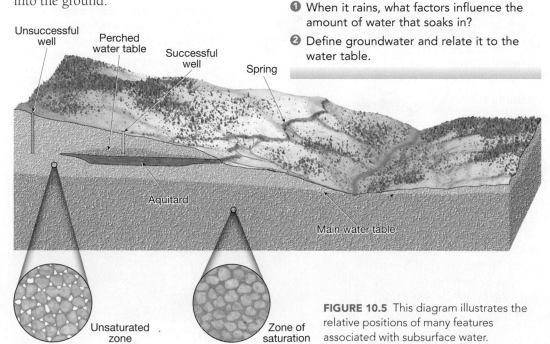

Unsuccessful well

Perched water table

Successful well

Spring

Aquitard

Main water table

Unsaturated zone

Zone of saturation

FIGURE 10.5 This diagram illustrates the relative positions of many features associated with subsurface water.

important influence is the fact that groundwater moves very slowly and at varying rates under different conditions. Because of this, water tends to "pile up" beneath high areas between stream valleys. If rainfall were to cease completely, these water table "hills" would slowly subside and gradually approach the level of the valleys. However, new supplies of rainwater are usually added frequently enough to prevent this. Nevertheless, in times of extended drought, the water table may drop enough to dry up shallow wells. Other causes for the uneven water table are variations in rainfall and permeability from place to place.

FIGURE 10.6 Preparing a map of the water table. The water level in wells coincides with the water table. **A.** First, the locations of wells and the elevation of the water table above sea level are plotted on a map. **B.** These data points are used to guide the drawing of water-table contour lines at regular intervals. On this sample map the interval is 10 feet. Groundwater flow lines can be added to show water movement in the upper portion of the zone of saturation. Groundwater tends to move approximately perpendicular to the contours and down the slope of the water table. (*After U.S Geological Survey*)

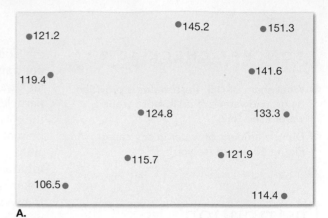

A.

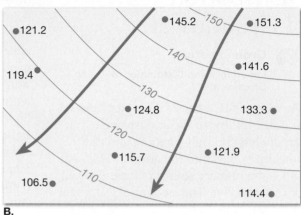

B.

EXPLANATION

● Location of well and elevation of water table above sea level, in feet

⌐120⌐ Water table contour shows elevation of water table, contour interval 10 feet

⟵ Groundwater flow line

Interaction between Groundwater and Streams

The interaction between the groundwater system and streams is a basic link in the hydrologic cycle. It can take place in one of three ways. Streams may gain water from the inflow of groundwater through the streambed. Such streams are called **gaining streams** (**FIGURE 10.7A**). For this to occur, the elevation of the water table must be higher than the level of the surface of the stream. Streams may lose water to the groundwater system by outflow through the streambed. The term **losing stream** is applied to this situation (**FIGURE 10.7B**). When this happens, the elevation of the water table is lower than the surface of the stream. The third possibility is a combination of the first two—a stream gains in some sections and loses in others.

Losing streams can be connected to the groundwater system by a continuous saturated zone, or they can be disconnected from the groundwater system by an unsaturated zone. Compare parts B and C in Figure 10.7. When the stream is disconnected, the water table may have a discernible bulge beneath the stream if the rate of water movement through the streambed and zone of aeration is greater than the rate of groundwater movement away from the bulge.

In some settings, a stream might always be a gaining stream or always be a losing stream. However, in many situations flow direction can vary a great deal along a stream; some sections receive groundwater and other sections lose water to the groundwater system. Moreover, the direction of flow can change over a short time span as the result of storms adding water near the streambank or when temporary flood peaks move down the channel.

Groundwater contributes to streams in most geologic and climatic settings. Even where streams are primarily losing water to the groundwater system, certain sections may receive groundwater inflow during some seasons. In one study of 54 streams in all parts of the United States, the analysis indicated that 52 percent of the streamflow was contributed by groundwater. The groundwater contribution ranged from a low of 14 percent to a maximum of 90 percent. Groundwater is also a major source of water for lakes and wetlands.

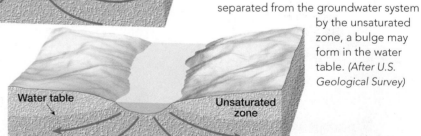

A. Gaining stream

FIGURE 10.7 Interaction between the groundwater system and streams. **A.** Gaining streams receive water from the groundwater system. **B.** Losing streams lose water to the groundwater system. **C.** When losing streams are separated from the groundwater system by the unsaturated zone, a bulge may form in the water table. (*After U.S. Geological Survey*)

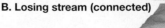

B. Losing stream (connected)

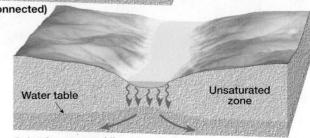

C. Losing stream (disconnected)

CONCEPT CHECK 10.4

❶ A kitchen table is flat. Is this usually the case for water tables? Why?

❷ Contrast a gaining stream and a losing stream.

Factors Influencing the Storage and Movement of Groundwater

The nature of subsurface materials strongly influences the rate of groundwater movement and the amount of groundwater that can be stored. Two factors are especially important—porosity and permeability.

Porosity

Water soaks into the ground because bedrock, sediment, and soil contain countless voids, or openings. These openings are similar to those of a sponge and are often called *pore spaces*. The quantity of groundwater that can be stored depends on the **porosity** of the material, which is the percentage of the total volume of rock or sediment that consists of pore spaces. Voids most often are spaces between sedimentary particles, but also common are joints, faults, cavities formed by the dissolving of soluble rocks such as limestone, and vesicles (voids left by gases escaping from lava).

Variations in porosity can be great. Sediment is commonly quite porous, and open spaces may occupy 10 to 50 percent of the sediment's total volume. Pore space depends on the size and shape of the grains, how they are packed together, the degree of sorting, and in sedimentary rocks,

DID YOU KNOW?
Because of its high porosity, excellent permeability, and great size, the High Plains Acquifer, the largest aquifer in the United States, accumulated huge amounts of groundwater—enough fresh water to fill Lake Huron.

the amount of cementing material. For example, clay may have a porosity as high as 50 percent, whereas some gravels may have only 20 percent voids.

Where sediments are poorly sorted, the porosity is reduced because the finer particles tend to fill the openings among the larger grains (see Figure 6.5, p. 154). Most igneous and metamorphic rocks, as well as some sedimentary rocks, are composed of tightly interlocking crystals, so the voids between the grains may be negligible. In these rocks, fractures must provide the voids.

Permeability, Aquitards, and Aquifers

Porosity alone cannot measure a material's capacity to yield groundwater. Rock or sediment might be very porous yet still not allow water to move through it. The pores must be *connected* to allow water flow, and they must be *large enough* to allow flow. Thus, the **permeability** of a material, its ability to *transmit* a fluid, is also very important.

Groundwater moves by twisting and turning through interconnected small openings. The smaller the pore spaces, the slower the water moves. For example, clay's ability to store water can be great, owing to its high porosity, but its pore spaces are so small that water is unable to move through it. Thus, clay's porosity is high but its permeability is poor.

Impermeable layers that hinder or prevent water movement are termed **aquitards.** Clay is a good example. In contrast, larger particles, such as sand or gravel, have larger pore spaces. Therefore, the water moves with relative ease. Permeable rock strata or sediments that transmit groundwater freely are called **aquifers.** Sands and gravels are common examples.

In summary, you have seen that porosity is not always a reliable guide to the amount of groundwater that can be produced, and permeability is significant in determining the rate of groundwater movement and the quantity of water that might be pumped from a well.

How Groundwater Moves

The movement of water in the atmosphere and on the land surface is relatively easy to visualize, but the movement of groundwater is not. Near the beginning of the chapter we mentioned the common misconception that groundwater occurs in underground rivers that resemble surface streams. Although subsurface streams do exist, they are *not* common. Rather, as you learned in the preceding sections, groundwater exists in the pore spaces and fractures in rock and sediment. Thus, contrary to any impressions of rapid flow that an underground river might evoke, the movement of most groundwater is exceedingly slow, from pore to pore. By exceedingly slow, we mean anywhere from millimeters per year to perhaps a kilometer per year, depending on conditions.

DID YOU KNOW?
The rate of groundwater movement is highly variable. One method of measuring this movement involves introducing dye into a well. The time is measured until the coloring agent appears in another well at a known distance from the first. A typical rate for many aquifers is about 15 m per year (about 4 cm per day).

Spring in Arizona's Marble Canyon.
(Photo by Michael Collier)

A Simple Groundwater Flow System

FIGURE 10.8 depicts a simple example of a groundwater flow system—a three-dimensional body of Earth material saturated with moving groundwater. It shows groundwater moving along flow paths from areas of recharge to a zone of discharge along a stream. Discharge also occurs at springs, lakes, or wetlands, as well as in coastal areas as seeps into bays or the ocean. Transpiration by plants whose roots extend to near the water table represents another form of groundwater discharge.

The energy that makes groundwater move is provided by the force of gravity. In response to gravity, water moves from areas where the water table is high to zones where the water table is lower. Although some water takes the most direct path down the slope of the water table, much of the water follows long, curving paths toward the zone of discharge.

Figure 10.8 shows water percolating into a stream from all possible directions. Some paths clearly turn upward, apparently against the force of gravity, and enter through the bottom of the channel. This is easily explained: The deeper you go into the zone of saturation, the greater the water pressure. Thus, the looping curves followed by water in the saturated zone may be thought of as a compromise between the downward pull of gravity and the tendency of water to move toward areas of reduced pressure. As a result, water at any given height is under greater pressure beneath a hill than beneath a stream channel, and the water tends to migrate toward points of lower pressure.

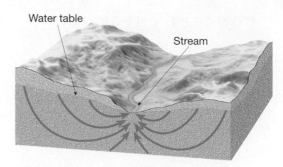

FIGURE 10.8 Arrows indicate groundwater movement through uniformly permeable material. The looping curves may be thought of as a compromise between the downward pull of gravity and the tendency of water to move toward areas of reduced pressure.

Measuring Groundwater Movement

The foundations of our modern understanding of groundwater movement began in the mid-19th century with the work of the French scientist-engineer Henri Darcy. Among the experiments carried out by Darcy was one that showed that the velocity of groundwater flow is proportional to the slope of the water table—the steeper the slope, the faster the water moves (because the steeper the slope, the greater the pressure difference between two points). The water-table slope is known as the **hydraulic gradient** and can be expressed as follows:

$$\text{hydraulic gradient} = \frac{h_1 - h_2}{d}$$

where h_1 is the elevation of one point on the water table, h_2 is the elevation of a second point, and d is the horizontal distance between the two points (**FIGURE 10.9**).

The speleothems found in caves include stalactites, stalagmites, and columns. *(Photo by Rufus Rufus/Photolibrary)*

speleothem

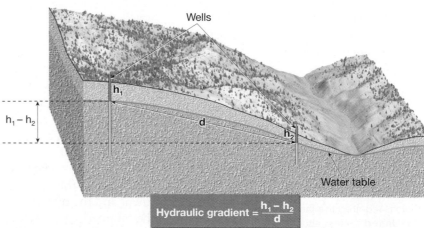

Hydraulic gradient = $\frac{h_1 - h_2}{d}$

FIGURE 10.9 The hydraulic gradient is determined by measuring the difference in elevation between two points on the water table ($h_1 - h_2$) divided by the distance between them, d. Wells are used to determine the height of the water table.

Darcy also discovered that the flow velocity varied with the permeability of the sediment—groundwater flows more rapidly through sediments having greater permeability than through materials having lower permeability. This factor is known as **hydraulic conductivity** and is a coefficient that takes into account the permeability of the aquifer and the viscosity of the fluid.

To determine discharge (Q)—that is, the actual volume of water that flows through an aquifer in a specified time—the following equation is used:

$$Q = \frac{KA(h_1 - h_2)}{d}$$

where $\frac{h_1 - h_2}{d}$ is the hydraulic gradient, K is the coefficient that represents hydraulic conductivity, and A is the cross-sectional area of the aquifer. This expression has come to be called *Darcy's law*.

CONCEPT CHECK 10.5

❶ Distinguish between porosity and permeability.

❷ What is the difference between an aquifer and an aquitard?

❸ What factors cause water to follow the paths shown in Figure 10.8?

❹ Relate groundwater movement to hydraulic gradient and hydraulic conductivity.

Springs

Groundwater

ESSENTIALS OF GEOLOGY Springs and Wells

Springs have aroused the curiosity and wonder of people for thousands of years. The fact that springs were, and to some people still are, rather mysterious phenomena is not difficult to understand, for here is water flowing freely from the ground in all kinds of weather in seemingly inexhaustible supply but with no obvious source.

Not until the middle of the 1600s did the French physicist Pierre Perrault invalidate the age-old assumption that precipitation could not adequately account for the amount of water emanating from springs and flowing in rivers. Over several years

FIGURE 10.10 Thousand Springs along the Snake River in Hagerman Valley, Idaho. *(Photo by David R. Frazier/Alamy)*

Perrault computed the quantity of water that fell on France's Seine River basin. He then calculated the mean annual runoff by measuring the river's discharge. After allowing for the loss of water by evaporation, he showed that there *was* sufficient water remaining to feed the springs. Thanks to Perrault's pioneering efforts and the measurements by many afterward, we now know that the source of springs is water from the zone of saturation and that the ultimate source of this water is precipitation.

Whenever the water table intersects Earth's surface, a natural outflow of groundwater results, which we call a **spring.** Springs such as the one pictured in **FIGURE 10.10** form when an aquitard blocks the downward movement of groundwater and forces it to move laterally. Where the permeable bed outcrops, a spring results. Another situation leading to the formation of a spring is illustrated in Figure 10.5. Here an aquitard is situated above the main water table. As water percolates downward, a portion of it is intercepted by the aquitard, thereby creating a localized zone of saturation called a **perched water table.**

Springs, however, are not confined to places where a perched water table creates a flow at the surface. Many geological situations lead to the formation of springs because subsurface conditions vary greatly from place to place. Even in areas underlain by impermeable crystalline rocks, permeable zones may exist in the form of fractures or solution channels. If these openings fill with water and intersect the ground surface along a slope, a spring will result.

CONCEPT CHECK 10.6

❶ Describe the circumstances that create springs such as those in Figure 10.5 and Figure 10.10.

Wells

Groundwater

ESSENTIALS OF GEOLOGY Springs and Wells

The most common device used by people for removing groundwater is the **well,** a hole bored into the zone of saturation (**FIGURE 10.11**). Wells serve as small reservoirs into which groundwater migrates and from which it can be pumped to the surface. The use of wells dates back many centuries and continues to be an important method of obtaining water today.

The water-table level may fluctuate considerably during the course of a year, dropping during dry seasons and rising following periods of rain. Therefore, to ensure a continuous supply of water, a well must penetrate below the water table. Often when water is withdrawn from a well, the water table around the well is lowered. This effect, termed **drawdown,** decreases with increasing distance from the well. The result is a depression in the water table, roughly conical in shape, known as a **cone of depression** (Figure 10.11). Because the cone of depression increases the slope of the water table near the well, groundwater will flow more rapidly toward the opening. For most small domestic wells, the cone of depression is negligible.

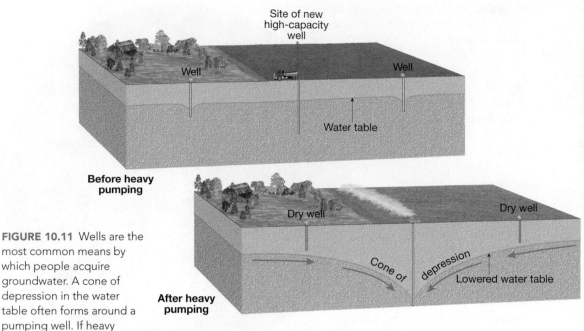

Site of new
high-capacity
well

Well

Well

Water table

**Before heavy
pumping**

Dry well

Dry well

Cone of depression

Lowered water table

**After heavy
pumping**

FIGURE 10.11 Wells are the most common means by which people acquire groundwater. A cone of depression in the water table often forms around a pumping well. If heavy pumping lowers the water table, some wells may be left dry. *(Photo by ASP/YPP/agefotostock)*

Well drilling rig

However, when wells are heavily pumped for irrigation or industrial purposes, the withdrawal of water can be great enough to create a very wide and steep cone of depression. This may substantially lower the water table in an area and cause nearby shallow wells to become "high and dry." Figure 10.11 also illustrates this situation.

Digging a successful well is a familiar problem for people in areas where groundwater is the primary source of supply. One well may be successful at a depth of 10 meters (33 feet), whereas a neighbor may have to go twice as deep to find an adequate supply. Still others may be forced to go deeper or try a different site altogether. When subsurface materials are heterogeneous, the amount of water a well is capable of providing may vary a great deal over short distances. For example, when two nearby wells are drilled to the

same level and only one is successful, it may be caused by the presence of a perched water table beneath one of them. Such a case is shown in Figure 10.5. Massive igneous and metamorphic rocks provide a second example. These crystalline rocks are usually not very permeable except where they are cut by many intersecting joints and fractures. Therefore, when a well drilled into such rock does not intersect an adequate network of fractures, it is likely to be unproductive.

CONCEPT CHECK 10.7

❶ How does a heavily pumping well affect the water table?

❷ In Figure 10.5, two wells are at the same level. Why was one successful and the other not?

Artesian Wells

 Groundwater

ESSENTIALS OF GEOLOGY **Springs and Wells**

In most wells, water cannot rise on its own. If water is first encountered at 30 meters depth, it remains at that level, fluctuating perhaps a meter or two with seasonal wet and dry periods. However, in some wells, water rises, sometimes overflowing at the

surface. Such wells are abundant in the *Artois* region of northern France, and so we call these self-rising wells *artesian*.

The term **artesian** is applied to any situation in which groundwater rises in a well above the level where it was initially encountered. For such a situation to occur, two conditions usually exist (**FIGURE 10.12**): (1) Water is confined to an aquifer that is inclined so that one end is exposed at the surface, where it can receive water; and (2) aquitards, both above and below the aquifer, must be present to prevent the water from escaping. Such an aquifer is called a **confined aquifer.** When such a layer is tapped, the pressure created by the weight of the water above will force the water to rise. If there were no friction, the water in the well would rise to the level of the water at the top of the aquifer. However, friction reduces the height of this pressure surface. The greater the distance from the recharge area (area where water enters the inclined aquifer), the greater the friction and the less the rise of water.

In Figure 10.12, Well 1 is a *nonflowing artesian well*, because at this location the pressure surface is below ground level. When the pressure surface is above the ground and a well is drilled into the aquifer, a *flowing artesian well* is created (Well 2). Not all artesian systems are wells.

FIGURE 10.12 Artesian systems occur when an inclined aquifer is surrounded by impermeable beds (aquitards). Such aquifers are called a *confined aquifers*. The photo shows a flowing artesian well. *(Photo by James E. Patterson)*

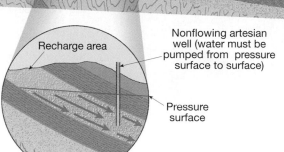

On a different scale, city water systems can be considered to be examples of artificial artesian systems (**FIGURE 10.13**). The water tower, into which water is pumped, can be considered the area of recharge; the pipes the confined aquifer; and the faucets in homes the flowing artesian wells.

CONCEPT CHECK 10.8

① Sketch a simple cross section of an artesian system with a flowing well. Label the aquitards, aquifer, and pressure surface.

This hot spring in Iceland is surrounded by a lava field. *(Photo by Markus Diouhy/Das Fotoarchiv/Photolibrary)*

Artesian springs also exist. Here groundwater may reach the surface by rising along a natural fracture such as a fault rather than through an artificially produced hole. In deserts, artesian springs are sometimes responsible for creating an oasis.

Artesian systems act as conduits, transmitting water from remote areas of recharge great distances to the points of discharge. In this manner, water that fell in central Wisconsin years ago is now taken from the ground and used by communities many kilometers away in Illinois. In South Dakota, such a system brings water from the Black Hills in the west eastward across the state.

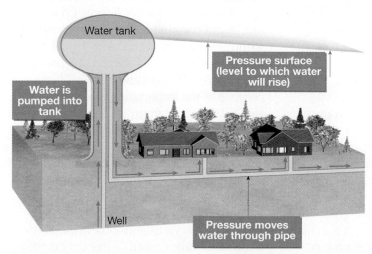

FIGURE 10.13 City water systems can be considered to be artificial artesian systems.

hot spring

Environmental Problems Associated with Groundwater

As with many of our valuable natural resources, groundwater is being exploited at an increasing rate. In some areas, overuse threatens the groundwater supply. In other places, groundwater withdrawal has caused the ground and everything resting upon it to sink. Still other localities are concerned with the possible contamination of their groundwater supply.

Treating Groundwater as a Nonrenewable Resource

Many natural systems tend to establish a condition of equilibrium. The groundwater system is no exception. The water table's height reflects a balance between the rate of water added by precipitation and the rate of water removed by discharge and withdrawal. Any imbalance will either raise or lower the water table. A long-term drop in the water table can occur if there is either a decrease in recharge due to a prolonged drought or an increase in groundwater discharge or withdrawal.

For many, groundwater appears to be an endlessly renewable resource because it is continually replenished by rainfall and melting snow. But in some regions groundwater has been and continues to be treated as a *nonrenewable resource*. Where this occurs, the water available to recharge the aquifer falls significantly short of the amount being withdrawn.

The High Plains aquifer provides one example (**FIGURE 10.14**). Underlying about 111 million acres (450,000 square kilometers or 174,000 square miles) in parts of eight western states, it is one of the largest and most agriculturally significant aquifers in the United States. It accounts for about 30 percent of all groundwater withdrawn for irrigation in the country. Mean annual precipitation is modest—ranging from about 40 centimeters (16 inches) in western portions to about 71 centimeters (28 inches) in eastern parts. Evaporation rates, on the other hand, are high—ranging from about 150 centimeters (60 inches) per year in the cooler northern parts of the region to 265 centimeters (105 inches) per year in the warmer southern parts. Because evaporation rates are high relative to precipitation, there is little rainwater to recharge the aquifer. Thus, in some parts of the region, where intense irrigation has been practiced for an extended period, depletion of groundwater has been severe.

Groundwater depletion has been a concern in the High Plains and other areas of the West for many years, but it is worth pointing out that the problem is not confined to this part of the country. Increased demands on groundwater resources have overstressed aquifers in many areas, not just in arid and semiarid regions.

Land Subsidence Caused by Groundwater Withdrawal

As you will see later in this chapter, surface subsidence can result from natural processes related to groundwater. However, the ground may also sink when water is pumped from wells faster than natural recharge processes can replace it. This effect is particularly pronounced in areas underlain by thick layers of loose sediment. As water is withdrawn, the weight of the overburden packs the sediment grains more tightly together and the ground subsides.

Many areas can be used to illustrate such land subsidence caused by excessive pumping of groundwater from relatively loose sediment. A classic example in the United States occurred in the San Joaquin Valley of California. This important agricultural region relies heavily on irrigation. Land subsidence due to groundwater withdrawal began in the valley in the mid-1920s and locally exceeded 8 meters (28 feet) by 1970 (**FIGURE 10.15**). Then, because of the importation of surface water and a decrease in groundwater pumping, water levels in the aquifer recovered and subsidence ceased.

However, during a drought from 1976 to 1977, heavy groundwater pumping led to renewed subsidence. This time, water levels dropped at a much faster rate than during the previous period because of the reduced storage capacity caused by earlier compaction of material in the aquifer. In all, more than 13,400 square kilometers

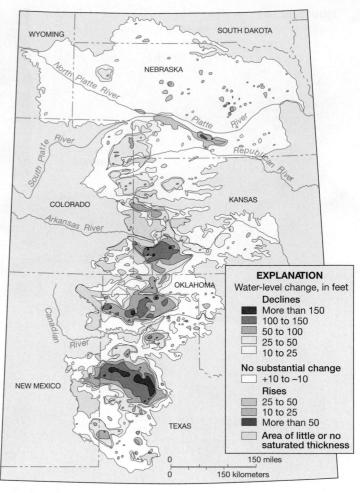

FIGURE 10.14 Changes in groundwater levels in the High Plains aquifer from predevelopment to 2005. Extensive pumping for irrigation has led to water-level declines in excess of 100 feet in parts of Kansas, Oklahoma, Texas, and New Mexico. Water level rises have occurred where surface water is used for irrigation, such as along the Plate River in Nebraska. *(After U.S. Geological Survey)*

DID YOU KNOW?
The U.S. Geological Survey estimates that during the past 50 to 60 years, water in storage in the High Plains aquifer declined about 200 million acre-feet (about 65 trillion gallons) with 62 percent of the total decline occurring in Texas.

In the low-lying coastal area between Houston and Galveston, land subsidence ranges from 1.5 to 3 meters (5 to 9 feet). The result is that about 78 square kilometers (30 square miles) are permanently flooded.

Outside the United States, one of the most spectacular examples of subsidence occurred in Mexico City, which is built on a former lake bed. In the first half of the 20th century, thousands of wells were sunk into the water-saturated sediments beneath the city. As water was withdrawn, portions of the city subsided by as much as 6 to 7 meters. In some places buildings have sunk to such a point that access to them from the street is located at what used to be the second-floor level!

Groundwater Contamination

The pollution of groundwater is a serious matter, particularly in areas where aquifers provide a large part of the water supply. One common source of groundwater pollution is sewage. Its sources include an ever-increasing number of septic tanks, as well as inadequate or broken sewer systems and farm waste.

If sewage water that is contaminated with bacteria enters the groundwater system, it may become purified through natural processes. The harmful bacteria may be mechanically filtered by the sediment through which the water percolates, destroyed by chemical oxidation, and/or assimilated by other organisms. For purification to occur, however, the aquifer must be of the correct composition. For example, extremely permeable aquifers (such as highly fractured crystalline rock, coarse gravel, or cavernous limestone) have such large openings that contaminated groundwater might travel long distances without being cleansed. In this case, the water flows too rapidly and is not in contact with the surrounding material long enough for purification to occur. This is the problem at Well 1 in **FIGURE 10.16**.

FIGURE 10.15 The shaded area on the map shows California's San Joaquin Valley. This important agricultural region relies heavily on irrigation. The marks on the utility pole in the photo indicate the level of the surrounding land in preceding years. Between 1925 and 1975, this part of the San Joaquin Valley subsided almost 9 meters (30 feet) because of the withdrawal of groundwater and the resulting compaction of sediments. *(Photo courtesy of U.S. Geological Survey, U.S. Department of the Interior)*

(5200 square miles) of irrigable land, half the entire valley, were affected by subsidence. Damage to structures, including highways, bridges, water lines, and wells, was extensive.

Many other cases of land subsidence due to groundwater pumping exist in the United States, including cases in Las Vegas, Nevada; New Orleans and Baton Rouge, Louisiana; portions of Southern Arizona; and the Houston–Galveston area of Texas.

DID YOU KNOW?
In the contiguous United States, the size of the area affected by land subsidence due to groundwater withdrawal amounts to an estimated 26,000 square kilometers (more than 10,000 square miles)—an area about the size of Massachusetts!

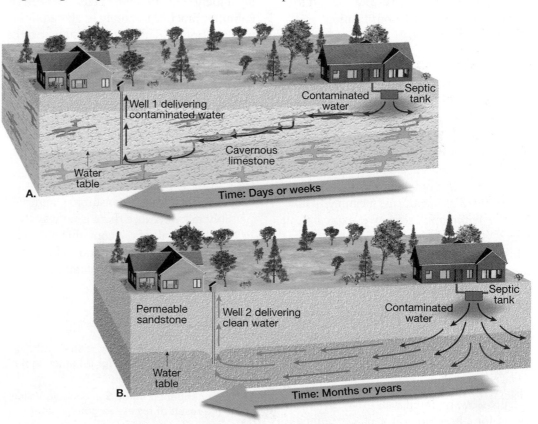

FIGURE 10.16 A. Although the contaminated water has traveled more than 100 meters before reaching Well 1, the water moves too rapidly through the cavernous limestone to be purified. **B.** As the discharge from the septic tank percolates through the permeable sandstone, it is purified in a relatively short distance.

A.

B.

FIGURE 10.17 Sometimes agricultural chemicals **A.** and materials leached from landfills **B.** find their way into the groundwater. These are two of the potential sources of groundwater contamination. *(Photo **A.** by Michael Collier; Photo **B.** by F. Rossotto/Corbis/The Stock Market)*

In contrast, when the aquifer is composed of sand or permeable sandstone, the water can sometimes be purified after traveling only a few dozen meters through it. The openings between sand grains are large enough to permit water movement, yet the movement of the water is slow enough to allow ample time for its purification (Well 2, Figure 10.16).

Other sources and types of contamination also threaten groundwater supplies. These include widely used substances such as highway salt, fertilizers that are spread across the land surface, and pesticides (**FIGURE 10.17**). In addition, a wide array of chemicals and industrial materials may leak from pipelines, storage tanks, landfills, and holding ponds. Some of these pollutants are classified as *hazardous*, meaning that they are flammable, corrosive, explosive, or toxic. As rainwater oozes through the refuse, it may dissolve a variety of potential contaminants. If the leached material reaches the water table, it will mix with the groundwater and contaminate the supply. Similar problems may result from leakage of shallow excavations called holding ponds into which various liquid wastes are disposed.

Because groundwater movement is usually slow, polluted water may go undetected for a long time. In fact, contamination is sometimes discovered only after drinking water has been affected and people become ill. By this time, the volume of polluted water may be very large, and

even if the source of contamination is removed immediately, the problem is not solved. Although the sources of groundwater contamination are numerous, there are relatively few solutions.

Once the source of the problem has been identified and eliminated, the most common practice is simply to abandon the water supply and allow the pollutants to be flushed away gradually. This is the least costly and easiest solution, but the aquifer must remain unused for many years. To accelerate this process, polluted water is sometimes pumped out and treated. Following removal of the tainted water, the aquifer is allowed to recharge naturally, or in some cases the treated water or other fresh water is pumped back in. This process is costly, time-consuming, and possibly risky because there is no way to be certain that all of the contamination has been removed. Clearly, the most effective solution to groundwater contamination is prevention.

CONCEPT CHECK 10.9

❶ Describe the problem associated with pumping groundwater for irrigation in the southern High Plains.

❷ What happened in the San Joaquin Valley as the result of excessive groundwater pumping?

❸ Which aquifer would be most effective in purifying polluted groundwater: coarse gravel, sand, or cavernous limestone?

Hot Springs and Geysers

By definition, the water in **hot springs** is 6 to 9 °C (11 to 16 °F) warmer than the mean annual air temperature for the localities where they occur. In the United States alone, there are well over 1000 such springs.

Temperatures in deep mines and oil wells usually rise with increasing depth, an average of about 2 °C per 100 meters (1 °F per 100 feet). Therefore, when groundwater circulates at great depths, it becomes heated. If it rises to the surface, the water may emerge as a hot spring. The water of some hot springs in the eastern United States is heated in this manner. The great majority (over 95 percent) of the hot springs (and geysers) in the United States are found in the West. The reason for such a distribution is that the source of heat for most hot springs is cooling igneous rock, and it is in the West that igneous activity has occurred most recently.

Geysers are intermittent hot springs or fountains where columns of water are ejected with great force at various intervals, often rising 30 to 60 meters (100 to 200 feet) into the air (**FIGURE 10.18**). After the jet of water ceases, a column of steam rushes out, usually with a thunderous roar. Perhaps the most famous geyser in the world is Old Faithful in Yellowstone National Park. The great abundance, diversity, and spectacular nature of Yellowstone's geysers and other thermal features undoubtedly were the primary reason for its becoming the first national park in the United States. Geysers are also found in other parts of the world, notably New Zealand and Iceland. In fact, the Icelandic word *geysa*, "to gush," gives us the name *geyser*.

Geysers occur where extensive underground chambers exist within hot igneous rocks. How they operate is shown in **FIGURE 10.19**. As relatively cool groundwater enters the chambers, it is heated by the surrounding rock. At the bottom of the chambers, the water is under great pressure because of the weight of the overlying water. This great pressure prevents the water from boiling at the normal surface temperature of 100 °C (212 °F).

For example, water at the bottom of a 300-meter (1000-foot) water-filled chamber must reach nearly 230 °C (450 °F) to boil. The heating causes the water to expand, with the result that some is forced out at the surface. This loss of water reduces the pressure on the remaining water in the chamber, which lowers the boiling point. A portion of the water deep within the chamber quickly turns to steam, and the geyser erupts. Following the eruption, cool groundwater again seeps into the chamber, and the cycle begins anew.

When groundwater from hot springs and geysers flows out at the surface, material in solution is often precipitated, producing an accumulation of chemical sedimentary rock. The material deposited at any given place commonly reflects the chemical make-up of the rock through which the water circulated. When the water contains dissolved silica, a material called *siliceous sinter* or *geyserite* is deposited around the spring. When the water contains dissolved calcium carbonate, a form of limestone called *travertine* or *calcareous tufa* is deposited. The latter term is used if the material is spongy and porous.

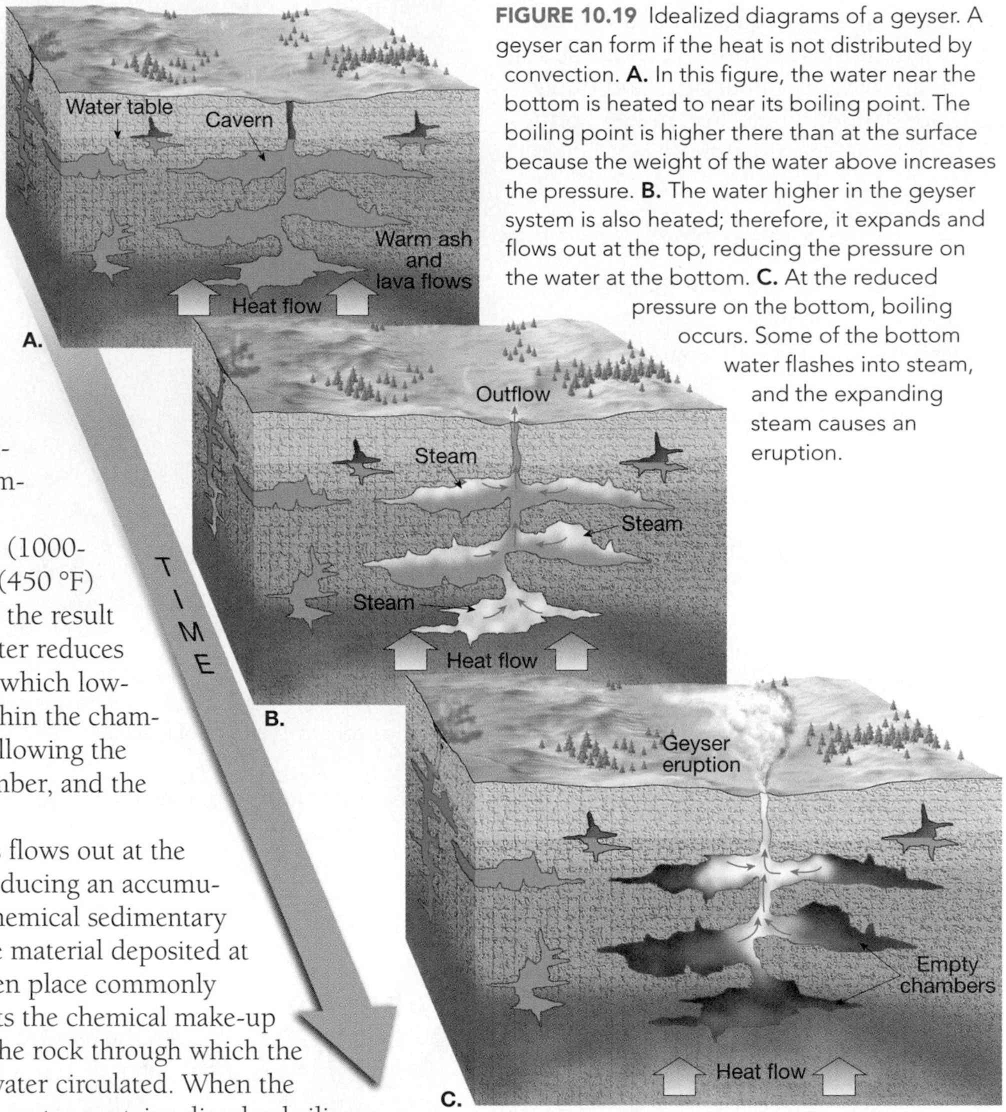

FIGURE 10.19 Idealized diagrams of a geyser. A geyser can form if the heat is not distributed by convection. **A.** In this figure, the water near the bottom is heated to near its boiling point. The boiling point is higher there than at the surface because the weight of the water above increases the pressure. **B.** The water higher in the geyser system is also heated; therefore, it expands and flows out at the top, reducing the pressure on the water at the bottom. **C.** At the reduced pressure on the bottom, boiling occurs. Some of the bottom water flashes into steam, and the expanding steam causes an eruption.

FIGURE 10.18 Old Faithful geyser in Yellowstone National Park, Wyoming. *(Photo by Jeff Vanuga/Corbis)*

The deposits at Mammoth Hot Springs in Yellowstone National Park are more spectacular than most (**FIGURE 10.20**). As the hot water flows upward through a series of channels and then out at the surface, the reduced pressure allows carbon dioxide to separate and escape from the water. The loss of carbon dioxide causes the water to become supersaturated with calcium carbonate, which then precipitates. In addition

DID YOU KNOW?
Many people think that Old Faithful erupts so reliably—every hour on the hour—that you can set your watch by it. So goes the legend, but it's not true. Time spans between eruptions vary from about 65 minutes to more than 90 minutes and have generally increased over the years thanks to changes in the geyser's plumbing.

to containing dissolved silica and calcium carbonate, some hot springs contain sulfur, which gives water a poor taste and unpleasant odor. Undoubtedly, Rotten Egg Spring, Nevada, is such a situation.

CONCEPT CHECK 10.10

❶ What is the source of heat for most hot springs and geysers?

❷ Describe what occurs to cause a geyser to erupt.

Geothermal Energy

Geothermal energy is harnessed by tapping natural underground reservoirs of steam and hot water. These occur where subsurface temperatures are high, owing to relatively recent volcanic activity. Geothermal energy is put to use in two ways: The steam and hot water are used for heating and to generate electricity.

Iceland is a large volcanic island with many active volcanoes (**FIGURE 10.21**). In Iceland's capital, Reykjavik, hot water is pumped into buildings throughout the city for space heating. (Reykjavik literally means "bay of steam.") It also warms greenhouses, where fruits and vegetables are grown year-round. In the United States, localities in several western states use hot water from geothermal sources for space heating.

As for generating electricity geothermally, the Italians were the first to do so in 1904, so the idea is not new. In 2009 more than 250 geothermal power plants in 24 countries were producing more than 10,000 megawatts (million watts). These plants provide power to more than 60 million people. The leading producers of geothermal power are listed in **TABLE 10.2**.

The first commercial geothermal power plant in the United States was built in 1960 at The Geysers, north of San Francisco (**FIGURE 10.22**). The Geysers remains the world's largest geothermal power plant, generating nearly 1000 megawatts. In addition to The Geysers, geothermal development is occurring elsewhere in the western United States, including Nevada,

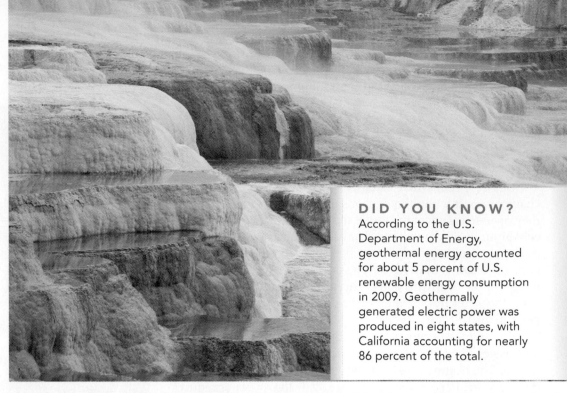

DID YOU KNOW?
According to the U.S. Department of Energy, geothermal energy accounted for about 5 percent of U.S. renewable energy consumption in 2009. Geothermally generated electric power was produced in eight states, with California accounting for nearly 86 percent of the total.

FIGURE 10.20 Mammoth Hot Springs in Yellowstone National Park. Although most of the deposits associated with geysers and hot springs in Yellowstone Park are silica-rich geyserite, the deposits at Mammoth Hot Springs consist of a form of limestone called travertine. *(Photo by Jamie & Judy Wild/Danita Delimont.com)*

Utah, and the Imperial Valley in southern California.

What geologic factors favor a geothermal reservoir of commercial value?

TABLE 10.2

Worldwide Geothermal Power Production, 2007	
Producing Country	**Megawatts**
United States	2687
Philippines	1970
Indonesia	992
Mexico	953
Italy	810
Japan	535
New Zealand	472
Iceland	421
El Salvador	204
Costa Rica	163
All others	525
Total	9732

Source: Geothermal Resources Council.

1. *A potent source of heat* such as a large magma chamber deep enough to ensure adequate pressure and slow cooling yet not so deep that the natural water circulation is inhibited. Such magma chambers are most likely in regions of recent volcanic activity.

2. *Large and porous reservoirs with channels connected to the heat source,* near which water can circulate and then be stored in the reservoir.

3. *A cap of low permeability rocks* that inhibits the flow of water and heat to the surface. A deep, well-insulated reservoir contains much more stored energy than does a similar but uninsulated reservoir.

As with other alternative methods of power production, geothermal sources are not expected to provide a high percentage of the world's growing energy needs. Nevertheless, in regions where its potential can be developed, its use will continue to grow.

CONCEPT CHECK 10.11

❶ List two ways that geothermal energy is put to use.

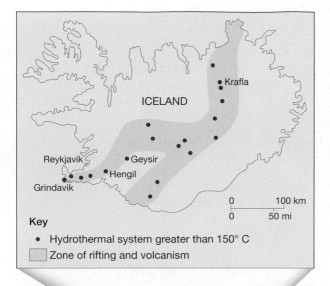

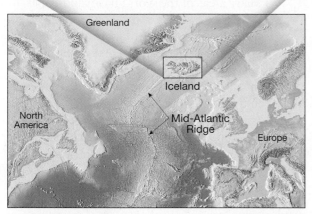

FIGURE 10.21 Iceland straddles the Mid-Atlantic Ridge. This divergent plate boundary is the site of numerous active volcanoes and geothermal systems. Because the entire country consists of geologically young volcanic rocks, warm water can be encountered in holes drilled almost anywhere. More than 45 percent of Iceland's energy comes from geothermal sources. The photo shows a power station in southwestern Iceland. The steam is used to generate electricity. Hot (83 °C) water from the plant is sent via an insulated pipeline to Reykjavik for space heating. *(Photo by Simon Fraser/Science Photo Library/ Photo Researchers, Inc.)*

The Geologic Work of Groundwater

Groundwater dissolves rock. This fact is the key to understanding how caverns and sinkholes form. Because soluble rocks, especially limestone, underlie millions of square kilometers of Earth's surface, it is here that groundwater carries on its important role as an erosional agent. Limestone is nearly insoluble in pure water, but it is quite easily dissolved by water containing small quantities of carbonic acid, and most groundwater contains this acid. It forms because rainwater readily dissolves carbon dioxide from the air and from decaying plants. Therefore, when groundwater comes in contact with limestone, the carbonic acid reacts with calcite (calcium carbonate) in the rocks to form calcium bicarbonate, a soluble material that is then carried away in solution.

Caverns

The most spectacular results of groundwater's erosional handiwork are limestone **caverns.** In the United States alone, about 17,000 caves have been discovered, and new ones are being found every year. Although most are modest, some have spectacular dimensions. Mammoth Cave in Kentucky and Carlsbad Caverns in southeastern New Mexico are two famous examples that are both national parks. The Mammoth Cave system is the most extensive in the world, with more than 557 kilometers (345 miles) of interconnected passages. The dimensions at Carlsbad Caverns are impressive in a different way. Here we find the largest and perhaps most spectacular single chamber. The Big Room at Carlsbad Caverns has an area equivalent to 14 football fields and enough height to accommodate the U.S. Capitol Building.

Most caverns are created at or just below the water table in the zone of saturation. Here acidic groundwater follows lines of weakness in the rock, such as joints and bedding planes. As time passes, the dissolving process slowly creates cavities and gradually enlarges them into caverns. The material dissolved by groundwater is carried away and discharged into streams and transported to the ocean.

In many caves, development has occurred at several levels, with the current cavern-forming activity occurring at the lowest elevation. This situation reflects the close relationship between the formation of major subterranean passages and the river valleys into which they drain. As streams cut their valleys deeper, the water table drops as the elevation of the river drops. Consequently, during periods when surface streams are rapidly eroding downward, surrounding groundwater levels drop rapidly and cave passages are abandoned by the water table while the passages are still relatively small in cross-sectional area. Conversely, when the entrenchment of streams is slow or negligible, there is time for large passages to form.

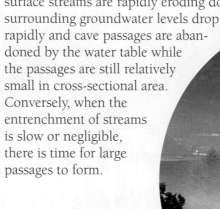

FIGURE 10.22 The Geysers, near the city of Santa Rosa in northern California, is the world's largest electricity-generating geothermal development. Most of the steam wells are about 3000 meters deep. *(AP Photo/Calpine)*

Certainly the features that arouse the greatest curiosity for most cavern visitors are the stone formations that give some caverns a wonderland appearance. These are not *erosional* features like the cavern itself, but are depositional features created by the seemingly endless dripping of water over great spans of time. The calcium carbonate that is left behind produces the limestone we call *travertine*. These cave deposits, however, are also commonly called *dripstone*, an obvious reference to their mode of origin. Although the formation of caverns takes place in the zone of saturation, the deposition of dripstone is not possible until the caverns are above the water table in the unsaturated zone. As soon as the chamber is filled with air, the stage is set for the decoration phase of cavern building to begin.

The various dripstone features found in caverns are collectively called **speleothems,** no two of which are exactly alike (**FIGURE 10.23**). Perhaps the most familiar speleothems are **stalactites**. These icicle-like pendants hang from the ceiling of the cavern and form where water seeps through cracks above. When the water reaches the air in the cave, some of the carbon dioxide in solution escapes from the drop and calcium carbonate precipitates. Deposition occurs as a ring around the edge of the water drop. As drop after drop follows, each leaves an infinitesimal trace of calcite behind, and a hollow limestone tube is created. Water then moves through the tube, remains suspended momentarily at the end, contributes a tiny ring of calcite, and falls to the cavern floor.

The stalactite just described is appropriately called a *soda straw* (Figure 10.23A). Often the hollow tube of the soda straw becomes plugged or its supply of water increases. In either case, the water is forced to flow, and hence deposit, along the outside of the tube. As deposition continues, the stalactite takes on the more common conical shape.

Speleothems that form on the floor of a cavern and reach upward toward the ceiling are called **stalagmites.** The water supplying the calcite for stalagmite growth falls from the ceiling and splatters over the surface. As a result, stalagmites do not have a central tube, and they are usually more massive in appearance and more rounded on their upper ends than stalactites. Given enough time, a downward-growing stalactite and an upward-growing stalagmite may join to form a *column.*

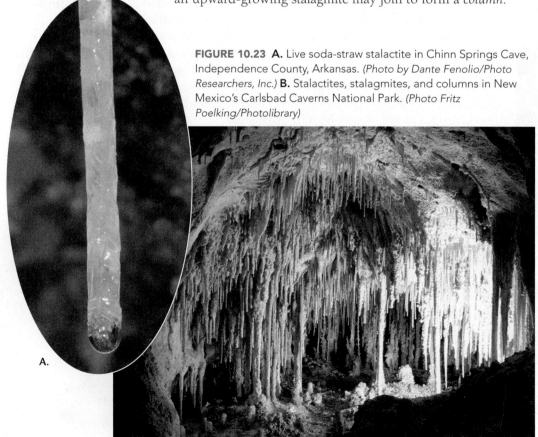

FIGURE 10.23 A. Live soda-straw stalactite in Chinn Springs Cave, Independence County, Arkansas. *(Photo by Dante Fenolio/Photo Researchers, Inc.)* **B.** Stalactites, stalagmites, and columns in New Mexico's Carlsbad Caverns National Park. *(Photo Fritz Poelking/Photolibrary)*

A.

B.

Karst Topography

Many areas of the world have landscapes that to a large extent have been shaped by the dissolving power of groundwater. Such areas are said to exhibit **karst topography,** named for the Krs Plateau, located along the northeastern shore of the Adriatic Sea in the border area between Slovenia and Italy where such topography is strikingly developed. In the United States, karst landscapes occur in many areas that are underlain by limestone, including portions of Kentucky, Tennessee, Alabama, southern Indiana, and central and northern Florida (**FIGURE 10.24**). Generally, arid and semiarid areas are too dry to develop karst topography. When solution features exist in such regions, they are likely to be remnants of a time when rainier conditions prevailed.

Karst areas typically have irregular terrain punctuated with many depressions, called **sinkholes** or **sinks** (**FIGURE 10.25**). In the limestone areas of Florida, Kentucky, and southern Indiana, there are literally tens of thousands of these depressions varying in depth from just a meter or two to a maximum of more than 50 meters.

Sinkholes commonly form in two ways. Some develop gradually over many years without any physical disturbance to the rock. In these situations, the limestone immediately below the soil is dissolved by downward-seeping rainwater that is freshly charged with carbon dioxide. With time, the bedrock surface is lowered and the fractures into which the water seeps are enlarged. As the fractures grow in size, soil subsides into the widening voids, from which it is removed by groundwater flowing in the passages below. These depressions are usually shallow and have gentle slopes.

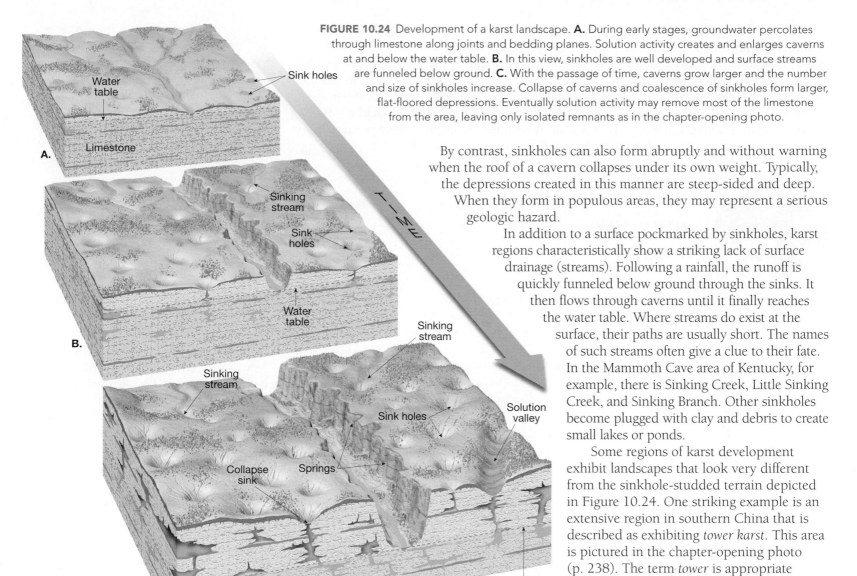

FIGURE 10.24 Development of a karst landscape. **A.** During early stages, groundwater percolates through limestone along joints and bedding planes. Solution activity creates and enlarges caverns at and below the water table. **B.** In this view, sinkholes are well developed and surface streams are funneled below ground. **C.** With the passage of time, caverns grow larger and the number and size of sinkholes increase. Collapse of caverns and coalescence of sinkholes form larger, flat-floored depressions. Eventually solution activity may remove most of the limestone from the area, leaving only isolated remnants as in the chapter-opening photo.

By contrast, sinkholes can also form abruptly and without warning when the roof of a cavern collapses under its own weight. Typically, the depressions created in this manner are steep-sided and deep. When they form in populous areas, they may represent a serious geologic hazard.

In addition to a surface pockmarked by sinkholes, karst regions characteristically show a striking lack of surface drainage (streams). Following a rainfall, the runoff is quickly funneled below ground through the sinks. It then flows through caverns until it finally reaches the water table. Where streams do exist at the surface, their paths are usually short. The names of such streams often give a clue to their fate. In the Mammoth Cave area of Kentucky, for example, there is Sinking Creek, Little Sinking Creek, and Sinking Branch. Other sinkholes become plugged with clay and debris to create small lakes or ponds.

Some regions of karst development exhibit landscapes that look very different from the sinkhole-studded terrain depicted in Figure 10.24. One striking example is an extensive region in southern China that is described as exhibiting *tower karst*. This area is pictured in the chapter-opening photo (p. 238). The term *tower* is appropriate because the landscape consists of a maze of isolated steepsided hills that rise abruptly

FIGURE 10.25 A. Groundwater was responsible for creating these sinkholes west of Timaru on New Zealand's South Island. *(Photo by David Wall/Alamy)* **B.** This small sinkhole formed suddenly in 1991 when the roof of a cavern collapsed, destroying this house in Frostproof, Florida. *(Photo by St. Petersburg Times/Getty Images/Liaison)*

A.

B.

from the ground. Each is riddled with interconnected caves and passageways. This type of karst topography forms in wet tropical and subtropical regions having thick beds of highly jointed limestone. Here groundwater has dissolved large volumes of limestone, leaving only these residual towers. Karst development is more rapid in tropical climates due to the abundant rainfall and the greater availability of carbon dioxide from the decay of lush tropical vegetation. The extra carbon dioxide in the soil means there is more carbonic acid for dissolving limestone. Other tropical areas of advanced karst development include portions of Puerto Rico, western Cuba, and northern Vietnam.

CHAPTER TEN
Groundwater in Review

◉ As a resource, *groundwater* represents the largest reservoir of fresh water that is readily available to humans. Geologically, the dissolving action of groundwater produces *caves* and *sinkholes*. Groundwater is also an equalizer of streamflow.

◉ Each day in the United States we use about 349 billion gallons of fresh water. Groundwater provides about 79 billion gallons or 23 percent of the total. More groundwater is used for irrigation than for all other uses combined.

◉ Groundwater is water that completely fills the pore spaces in sediment and rock in the subsurface *zone of saturation*. The upper limit of this zone is the *water table*. The *unsaturated zone* is above the water table where the soil, sediment, and rock are not saturated.

◉ The interaction between streams and groundwater takes place in one of three ways: Streams gain water from the inflow of groundwater (*gaining stream*); they lose water through the streambed to the groundwater system (*losing stream*); or they do both, gaining in some sections and losing in others.

◉ The quantity of water that can be stored in a material depends upon the material's *porosity* (the volume of open spaces). The *permeability* (the ability to transmit a fluid through interconnected pore spaces) of a material is a very important factor controlling the movement of groundwater.

◉ Materials with very small pore spaces (such as clay) hinder or prevent groundwater movement and are called *aquitards*. *Aquifers* consist of materials with larger pore spaces (such as sand) that are permeable and transmit groundwater freely.

◉ Groundwater moves in looping curves that are a compromise between the downward pull of gravity and the tendency of water to move toward areas of reduced pressure.

◉ The primary factors influencing the velocity of groundwater flow are the slope of the water table (*hydraulic gradient*) and the permeability of the aquifer (*hydraulic conductivity*).

◉ *Springs* occur whenever the water table intersects the land surface and a natural flow of groundwater results. *Wells*, openings bored into the zone of saturation, draw groundwater out and create roughly conical depressions in the water table known as *cones of depression*. *Artesian wells* occur when water rises above the level at which it was initially encountered.

◉ When groundwater circulates at great depths, it becomes heated. If it rises, the water may emerge as a *hot spring*. Geysers occur when groundwater is heated in underground chambers and expands and some water quickly changes to steam, causing the geyser to erupt. The source of heat for most hot springs and geysers is hot igneous rock. *Geothermal energy* is harnessed by tapping natural underground reservoirs of steam and hot water.

◉ Some of the current environmental problems involving groundwater include (1) *overuse* by intense irrigation; (2) *land subsidence* caused by groundwater withdrawal; and (3) contamination by pollutants.

◉ Most *caverns* form in limestone at or below the water table when acidic groundwater dissolves rock along lines of weakness, such as joints and bedding planes. The various *dripstone* features found in caverns are collectively called *speleothems*. Landscapes that to a large extent have been shaped by the dissolving power of groundwater exhibit *karst topography*, an irregular terrain punctuated with many depressions, called *sinkholes* or *sinks*.

Key Terms

GIVE IT SOME THOUGHT

❶ Imagine a water molecule that is part of a groundwater system in an area of gently rolling hills in the eastern United States. Describe some possible paths the molecule might take through the hydrologic cycle if:

  a. it were pumped from the ground to irrigate a farm field.

  b. there was a long period of heavy rainfall.

  c. the water table in the vicinity of the molecule developed a steep cone of depression due to heavy pumping from a nearby well.

Combine your understanding of the hydrologic cycle with your imagination and include possible short-term and long-term destinations and information as to how the molecule gets to these places via evaporation, transpiration, condensation, precipitation, infiltration, and runoff. Remember to consider possible interactions with streams, lakes, groundwater, the ocean, and the atmosphere.

❷ Identify a location in the United States where you might find (a) a gaining stream, and (b) a losing stream. Why did you choose each location? Describe a situation that would cause a gaining stream to become a losing stream. Also describe a situation in which a losing stream becomes a gaining stream.

❸ Which one of the three basic rock types (igneous, sedimentary, or metamorphic) has the greatest likelihood of being a good aquifer? Why?

❹ During a trip to the grocery store, your friend wants to buy some bottled water. Some brands promote the fact that their product is artesian. Other brands boast that their water comes from a spring. Your friend asks, "Is artesian water or spring water necessarily better than water from other sources?" How would you answer?

❺ What is the likely difference between an intermittent stream (one that flows off and on) and a stream that flows all the time, even during extended dry periods?

❻ Imagine you are an environmental scientist who has been hired to solve a groundwater contamination problem. Several homeowners have noticed that their well water has a funny smell and taste. Some think the contamination is coming from a landfill, but others believe it might be the cattle feedlot or a nearby chemical plant. Your first step was to gather data from wells in the area and prepare the map of the water table shown here.

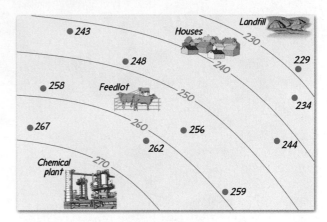

  a. Based on your map, can any of the three potential sources of contamination be eliminated? If so explain.

  b. What other steps would you take to determine the source of the contamination?

❼ An acquaintance is considering purchasing a large tract of productive irrigated farmland in West Texas. His intention is to continue growing crops on the land for years to come. If he asked your opinion about the area he selected, what figure in this chapter would you consult before you responded? How would this figure help your friend evaluate his potential purchase?

Companion Website

www.mygeoscienceplace.com

The *Essentials of Geology, 11e* companion Website contains numerous multimedia resources accompanied by assessments to aid in your study of the topics in this chapter. The use of this site's learning tools will help improve your understanding of geology. Utilizing the access code that accompanies this text, visit **www.mygeoscienceplace.com** in order to:

  • **Review** key chapter concepts.
  • **Read** with links to the Pearson eText and to chapter-specific web resources.
  • **Visualize** and comprehend challenging topics using the learning activities in *GEODe: Essentials of Geology* and the *Geoscience Animations Library*.
  • **Test** yourself with online quizzes.

Glaciers and Glaciation

C LIMATE HAS A STRONG INFLUENCE ON THE NATURE AND INTENSITY of Earth's external processes. This fact is dramatically illustrated in this chapter because the existence and extent of glaciers are largely controlled by Earth's changing climate.

Glaciers continue to sculpt the Swiss Alps.
(Arno Balzarini/epa/Corbis)

To assist you in learning the important concepts in this chapter, focus on the following questions:

⊙ What is a glacier? What are the different types of glaciers? Where are glaciers located?

⊙ How does glacial ice move, and at what rates does it move?

⊙ What determines whether the front of a glacier advances, retreats, or remains stationary?

⊙ What are the various processes of glacial erosion?

⊙ What are the features created by glacial erosion and deposition? What materials make up the depositional features?

⊙ What is the evidence for the Ice Age? What are some indirect effects of Ice Age glaciers?

⊙ What are some proposals that attempt to explain the causes of glacial ages?

FOCUS ON CONCEPTS

Glaciers: A Part of Two Basic Cycles

Glaciers and Glaciation

ESSENTIALS OF GEOLOGY **Introduction**

Like the running water and groundwater that were the focus of the preceding two chapters, glaciers represent a significant erosional process. These moving masses of ice are responsible for creating many unique landforms and are part of an important link in the rock cycle in which the products of weathering are transported and deposited as sediment.

Today, glaciers cover nearly 10 percent of Earth's land surface. However, in the recent geologic past, ice sheets were three times more extensive, covering vast areas with ice thousands of meters thick. Many regions still bear the mark of these glaciers. The basic character of such diverse places as the Alps, Cape Cod, and Yosemite valley was fashioned by now vanished masses of glacial ice. Moreover, Long Island, the Great Lakes, and the fiords of Norway and Alaska all owe their existence to glaciers. Glaciers, of course, are not just a phenomenon of the geologic past. As you will see, they are still modifying the physical landscapes of many regions today (**FIGURE 11.1**).

Glaciers are a part of two fundamental cycles in the Earth system—the hydrologic cycle and the rock cycle. Earlier you learned that the water of the hydrosphere is constantly cycled through the atmosphere, biosphere, and geosphere. Time and time again water is evaporated from the oceans into the atmosphere, precipitated upon the land, and carried by rivers and underground flow back to the sea. However, when precipitation falls at high elevations or high latitudes, the water may not immediately make its way toward the sea. Instead, it may become part of a glacier. Although the ice will eventually melt, allowing the water to continue its path to the sea, water can be stored as glacial ice for many tens, hundreds, or even thousands of years.

FIGURE 11.1 Hubbard Glacier is still eroding the Alaskan landscape. The dark stripe of sediment in the center of the glacier is a medial moraine. (*Photo by Michael Collier*)

DID YOU KNOW?
Approximately 160,000 glaciers presently occupy Earth's polar regions and high mountain environments. Today's glaciers cover only about one third of the area that was covered at the height of the most recent Ice Age.

A **glacier** is a thick ice mass that forms over hundreds or thousands of years. It originates on land from the accumulation, compaction, and recrystallization of snow. A glacier appears to be motionless, but it is not—glaciers move very slowly. Like running water, groundwater, wind, and waves, glaciers are dynamic erosional agents that accumulate, transport, and deposit sediment. As such, glaciers are among the processes that perform a very basic function in the rock cycle. Although glaciers are found in many parts of the world today, most are located in remote areas, either near Earth's poles or in high mountains.

Valley (Alpine) Glaciers

Literally thousands of relatively small glaciers exist in lofty mountain areas, where they usually follow valleys that were originally occupied by streams. Unlike the rivers that previously flowed in these valleys, the glaciers advance slowly, perhaps only a few centimeters per day. Because of their setting, these moving ice masses are termed **valley glaciers** or **alpine glaciers** (see Figure 11.1). Each glacier is a stream of ice, bounded by precipitous rock walls, that flows downvalley from a snow accumulation center near its head. Like rivers, valley glaciers can be long or short, wide or narrow, single or with branching tributaries. Generally, the widths of alpine glaciers are small compared to their lengths. Some extend for just a fraction of a kilometer, whereas others go on for many tens of kilometers. The west branch of the Hubbard Glacier, for example, runs through 112 kilometers (nearly 70 miles) of mountainous terrain in Alaska and the Yukon Territory.

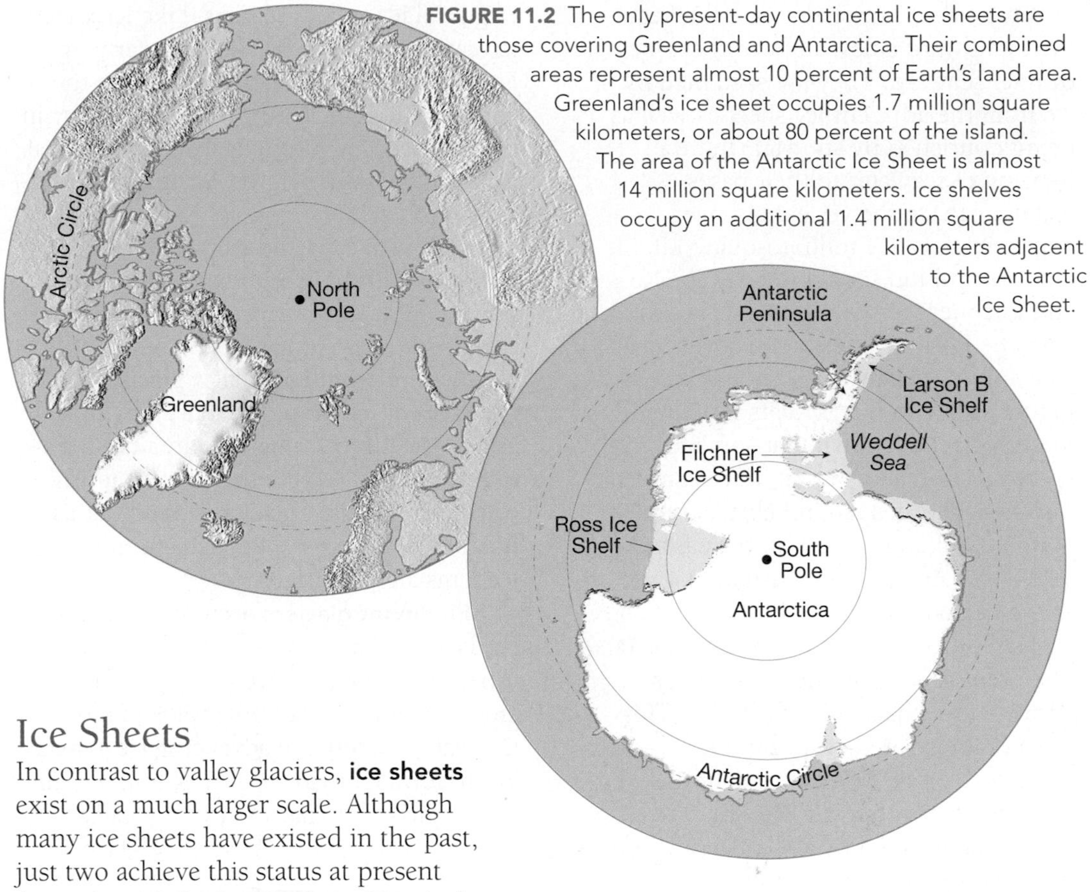

FIGURE 11.2 The only present-day continental ice sheets are those covering Greenland and Antarctica. Their combined areas represent almost 10 percent of Earth's land area. Greenland's ice sheet occupies 1.7 million square kilometers, or about 80 percent of the island. The area of the Antarctic Ice Sheet is almost 14 million square kilometers. Ice shelves occupy an additional 1.4 million square kilometers adjacent to the Antarctic Ice Sheet.

Ice Sheets

In contrast to valley glaciers, **ice sheets** exist on a much larger scale. Although many ice sheets have existed in the past, just two achieve this status at present (**FIGURE 11.2**). In the Northern Hemisphere, Greenland is covered by an imposing ice sheet that occupies 1.7 million square kilometers (0.7 million square miles) or about 80 percent of this large island. Averaging nearly 1500 meters (5000 feet) thick, in places the ice extends 3000 meters (10,000 feet) above the island's bedrock floor.

In the Southern Hemisphere, the huge Antarctic ice sheet attains a maximum thickness of almost 4300 meters (14,000 feet) and covers nearly the entire continent, an area of more than 13.6 million square kilometers (5.3 million square miles). Because of the proportions of these huge features, they are often called *continental ice sheets.* Indeed, the combined areas of present-day continental ice sheets represent almost 10 percent of Earth's land area.

These enormous masses flow out in all directions from one or more snow-accumulation centers and completely obscure all but the highest areas of underlying terrain. Even sharp variations in the topography beneath the glaciers usually appear as relatively subdued undulations on the surface of the ice. Such topographic differences, however, do affect the behavior of the ice sheets, especially near their margins, by guiding flow in certain directions and creating zones of faster and slower movement.

Along portions of the Antarctic coast, glacial ice flows into the adjacent ocean, creating features called **ice shelves**. They are large, relatively flat masses of floating ice that extend seaward from the coast but remain attached to the land along one

DID YOU KNOW?
If it's hard to imagine how large the Greenland ice sheet really is, try this. The ice sheet is long enough to extend from Key West, Florida, to 100 miles north of Portland, Maine. Its width could reach from Washington, D.C. to Indianapolis, Indiana. Put another way, the ice sheet is 80 percent as big as the entire United States east of the Mississippi River. The area of Antarctica's ice sheet is more than eight times as great!

or more sides. The shelves are thickest on their landward sides, and they become thinner seaward. They are sustained by ice from the adjacent ice sheet as well as being nourished by snowfall and the freezing of seawater to their bases. Antarctica's ice shelves extend over approximately 1.4 million square kilometers (0.6 million square miles). The Ross and Filchner ice shelves are the largest, with the Ross ice shelf alone covering an area approximately the size of Texas (see Figure 11.2). In recent years, satellite monitoring has shown that some ice shelves are breaking apart. For example, during a 35-day span in February and March 2002 an ice shelf on the eastern side of the Antarctic Peninsula, known as the Larsen B ice shelf, broke apart and separated from the continent (**FIGURE 11.3**). The event sent thousands of icebergs adrift in the adjacent Weddell Sea. This was not an isolated event but rather part of a trend. Since 1974 the extent of seven ice shelves surrounding the Antarctic Peninsula declined by about 13,500 square kilometers (nearly 5300 square miles). Scientists attribute the breakup of the ice shelves to a strong regional climate warming.

Other Types of Glaciers

In addition to valley glaciers and ice sheets, other types of glaciers are also identified. Covering some uplands and plateaus are masses

of glacial ice called **ice caps**. Like ice sheets, ice caps completely bury the underlying landscape but are much smaller than the continental-scale features. Ice caps occur in many places, including Iceland and several of the large islands in the Arctic Ocean (**FIGURE 11.4**).

Often ice caps and ice sheets feed **outlet glaciers**. These tongues of ice flow down valleys extending outward from the margins of these larger ice masses. The tongues are essentially valley glaciers that are avenues for ice movement from an ice cap or ice sheet through mountainous terrain to the sea. Where they encounter the ocean, some outlet glaciers spread out as floating ice shelves. Often large numbers of icebergs are produced.

Piedmont glaciers occupy broad lowlands at the bases of steep mountains and form when one or more valley glaciers emerge from the confining walls of mountain valleys. Here the advancing ice spreads out to form a broad sheet. The size of individual piedmont glaciers varies greatly. Among the largest is the broad Malaspina

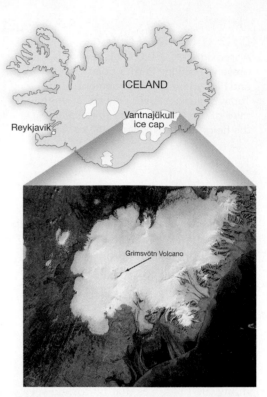

FIGURE 11.4 The ice cap in this satellite image is the Vantnajükull in southeastern Iceland (jükull means "ice cap" in Danish). In 1996 the Grimsvötn Volcano erupted beneath the ice cap, producing large quantities of melted glacial water that created floods. (*Landsat image from NASA*)

Glacier along the coast of southern Alaska. It covers more than 5000 square kilometers (2000 square miles) of the flat coastal plain at the foot of the lofty St. Elias Range (**FIGURE 11.5**).

What if the Ice Melted?

How much water is stored as glacial ice? Estimates by the U.S. Geological Survey indicate that only slightly more than 2 percent of the world's water is tied up in

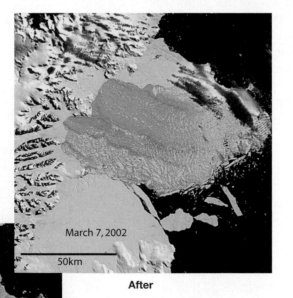

March 7, 2002

50km

After

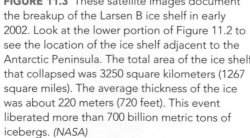

January 31, 2002

50km

Before

FIGURE 11.3 These satellite images document the breakup of the Larsen B ice shelf in early 2002. Look at the lower portion of Figure 11.2 to see the location of the ice shelf adjacent to the Antarctic Peninsula. The total area of the ice shelf that collapsed was 3250 square kilometers (1267 square miles). The average thickness of the ice was about 220 meters (720 feet). This event liberated more than 700 billion metric tons of icebergs. (*NASA*)

DID YOU KNOW?
Because Iceland has active volcanoes but is also a place with abundant glaciers, it is often called the "land of fire and ice." When volcanoes buried beneath the island's ice caps erupt, they trigger rapid melting that can cause major floods. Because the problem is so common here, there is an Icelandic word, *jükulhlaup* ("glacier flood"), to refer to the phenomenon. See Figure 11.4.

FIGURE 11.5 Malaspina Glacier in southeastern Alaska is a classic piedmont glacier. Piedmont glaciers occur where valley glaciers exit a mountain range onto broad lowlands, are no longer laterally confined, and spread to become wide lobes. Malaspina Glacier is actually a compound glacier, formed by the merger of several valley glaciers. Seen here are the most prominent: Agassiz Glacier (left) and Seward Glacier (right). In total, Malaspina Glacier is up to 65 kilometers (40 miles) wide and extends up to 45 kilometers (28 miles) from the mountain front nearly to the sea. *(Image from NASA/JPL)*

way. If the ice sheet were melted at a uniform rate, it could feed (1) the Mississippi River for more than 50,000 years, (2) all the rivers in the United States for about 17,000 years, (3) the Amazon River for approximately 5000 years, or (4) all the rivers of the world for about 750 years.

CONCEPT CHECK 11.1

❶ Where are glaciers found today, and what percentage of Earth's land surface do they cover?

❷ Describe how glaciers fit into the hydrologic cycle. What role do they play in the rock cycle?

❸ List and briefly distinguish among four types of glaciers.

Formation and Movement of Glacial Ice

Glaciers and Glaciation

Budget of a Glacier

glaciers. But even 2 percent of a vast quantity is still large. The total volume of just valley glaciers is about 210,000 cubic kilometers, comparable to the combined volume of the world's largest saline and freshwater lakes.

As for ice sheets, Antarctica's represents 80 percent of the world's ice and an estimated 65 percent of Earth's fresh water. It covers an area almost one and a half times the area of the entire United States. If this ice melted, sea level would rise an estimated 60 to 70 meters and the ocean would inundate many densely populated coastal areas (**FIGURE 11.6**).

The hydrologic importance of the Antarctic ice can be illustrated in another

Snow is the raw material from which glacial ice originates; therefore, glaciers form in areas where more snow falls in winter than melts during the summer. Glaciers develop in the high latitude polar realm because, even though annual snowfall totals are modest, temperatures are so low that little of the snow melts. Glaciers can form in mountains because temperatures drop with an increase in altitude. So even near the equator, glaciers may form at elevations above about 5000 meters (16,400 feet). For example, Tanzania's Mount Kilimanjaro, which is located practically astride the equator at an altitude of 5895 meters (19,336 feet), has glaciers at its summit. The elevation above which snow remains throughout the year is called the **snowline**. As you would expect, the elevation of the snowline varies with latitude. Near the equator, it occurs high in the mountains, whereas in the vicinity of the 60th parallel it is at or near sea level. Before a glacier is created, snow must be converted into glacial ice.

Glacial Ice Formation

When temperatures remain below freezing following a snowfall, the fluffy accumulation of delicate hexagonal crystals soon changes. As air infiltrates the spaces between the crystals, the extremities of the crystals evaporate and the water vapor condenses near the centers of the crystals. In this manner snowflakes become smaller, thicker, and more spherical, and the large pore spaces disappear. By this process air is forced out and what was once light, fluffy snow is recrystallized into a much denser mass of small grains having the consistency of coarse sand. This granular recrystallized snow is called **firn** and is commonly found making up old snowbanks near the end of winter. As more snow is added, the pressure on the lower layers gradually increases, compacting the ice grains at depth. Once the thickness of ice and snow exceeds 50 meters (160 feet), the weight is sufficient to fuse firn into a solid mass of interlocking ice crystals. Glacial ice has now been formed.

The rate at which this transformation occurs varies. In regions where the annual snow accumulation is great, burial is relatively rapid and snow may turn to glacial ice in a matter of a decade or less. Where the yearly addition of snow is less abundant, burial is slow and the transformation of snow to glacial ice may take hundreds of years.

FIGURE 11.6 This map of a portion of North America shows the present-day coastline compared to the coastline that would exist if the ice sheets on Greenland and Antarctica melted.

How Glaciers Move

The movement of glacial ice is generally referred to as *flow*. The fact that glacial movement is described in this way seems paradoxical—how can a solid flow? The way in which ice flows is complex and is of two basic types. The first of these, *plastic flow*, involves movement *within* the ice. Ice behaves as a brittle solid until the pressure upon it is equivalent to the weight of about 50 meters (165 feet) of ice. Once that load is surpassed, ice behaves as a plastic material, and flow begins. A second and often equally important mechanism of glacial movement consists of the entire ice mass slipping along the ground. The lowest portions of most glaciers are thought to move by this sliding process called *basal slip* (**FIGURE 11.7**).

The upper 50 meters or so of a glacier is not under sufficient pressure to exhibit plastic flow. Rather, the ice in this uppermost zone is brittle and is appropriately referred to as the *zone of fracture*. The ice in this zone is carried along "piggyback" style by the ice below. When the glacier moves over irregular terrain, the zone of fracture is subjected to tension, resulting in cracks called **crevasses** (**FIGURE 11.8**). These gaping cracks, which often make travel across glaciers dangerous, may extend to depths of 50 meters (165 feet). Beyond this depth, plastic flow seals them off.

Rates of Glacial Movement

Unlike streamflow, glacial movement is not obvious. If we could watch an alpine glacier move, we would see that like the water in a river, not all of the ice moves downvalley at an equal rate. Just as friction with the bedrock floor slows the movement of the ice at the bottom of the glacier, the drag created by the valley walls leads to the flow being greatest in the center of the glacier. This was first demonstrated by experiments during the 19th century, in which markers were carefully placed in a straight line across the top of a valley glacier. Periodically, the positions of the stakes were recorded, revealing the type of movement just described. More about these experiments may be found in Chapter 1, pages 10–11.

How rapidly does glacial ice move? Average velocities vary considerably from one glacier to another.* Some move so slowly that trees and other vegetation may become well established in the debris that has accumulated on the glacier's surface, whereas others may move at rates of up to several meters per day. For example, Byrd Glacier, an outlet glacier in Antarctica that was the subject of a 10-year study using satellite images, moved at an average rate of 750 to 800 meters per year (about 2 meters per day). Other glaciers in the study advanced at one fourth that rate.

The advance of some glaciers is characterized by periods of extremely rapid movements called *surges*. Glaciers that exhibit such movement may flow along in an apparently normal manner and then speed up for a relatively short time before returning to the

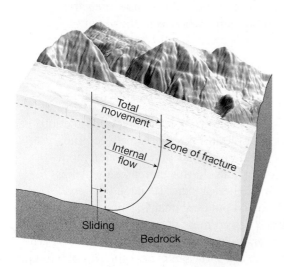

FIGURE 11.7 A vertical cross section through a glacier illustrates ice movement. Glacial movement is divided into two components. Below about 50 meters (160 feet), ice behaves plastically and flows. In addition, the entire mass of ice may slide along the ground. The ice in the zone of fracture is carried along "piggyback" style. Notice that the rate of movement is slowest at the base of the glacier, where frictional drag is greatest.

FIGURE 11.8 Crevasses form in the brittle ice of the zone of fracture. They can extend to depths of 50 meters and can obviously make travel across glaciers dangerous. *(Photo by Bo Tornvig/age footstock)*

*Specialized instruments aboard satellites allow us to monitor some glaciers from space. Figure 1.9, p. 8, provides one example of monitoring the movement of Antarctica's Lambert Glacier.

DID YOU KNOW?
The Kutiah Glacier in Pakistan holds the record for the fastest glacial surge ever recorded. In 1953, it raced more than 12 kilometers (7.4 miles) in three months, averaging about 130 meters (430 feet) per day.

August 1964

August 1965

Position before surge

Position after surge

normal rate again. The flow rates during surges are as much as 100 times the normal rate (**FIGURE 11.9**).

Budget of a Glacier

Snow is the raw material from which glacial ice originates; therefore, glaciers form in areas where more snow falls in winter than melts during the summer. Glaciers are constantly gaining and losing ice.

GLACIAL ZONES. Snow accumulation and ice formation occur in the **zone of accumulation**. Its outer limits are defined by the *snowline*. As noted earlier, the elevation of the snowline varies greatly, from sea level in polar regions to altitudes approaching 5000 meters near the equator. Above the snowline, in the zone of accumulation, the addition of snow thickens the glacier and promotes movement. Below the snowline is the **zone of wastage**. Here there is a net loss to the glacier as all of the snow from the previous winter melts, as does some of the glacial ice (**FIGURE 11.10**).

In addition to melting, glaciers also waste as large pieces of ice break off the front of the glacier in a process called **calving**. Calving creates *icebergs* in places where the glacier has reached the sea or a lake (**FIGURE 11.11**). Because icebergs are just slightly less dense than seawater, they float very low in the water, with about 80 percent of their mass submerged. Along the margins of Antarctica's ice shelves, calving is the primary means by which these masses lose ice. The relatively flat

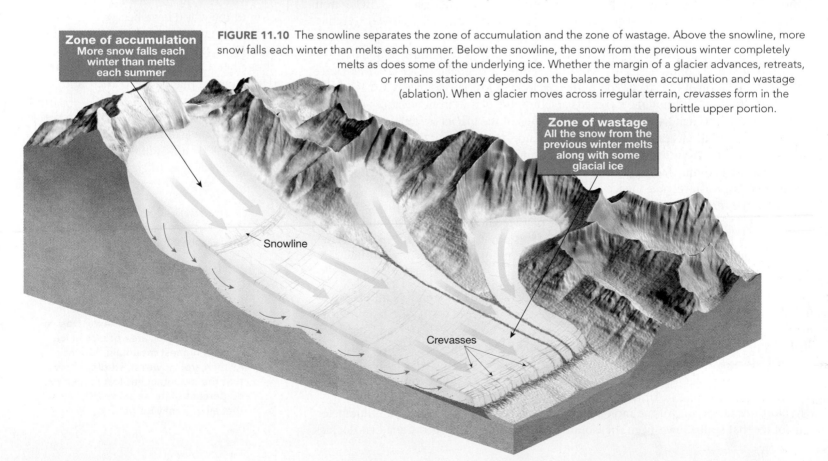

Zone of accumulation
More snow falls each winter than melts each summer

FIGURE 11.10 The snowline separates the zone of accumulation and the zone of wastage. Above the snowline, more snow falls each winter than melts each summer. Below the snowline, the snow from the previous winter completely melts as does some of the underlying ice. Whether the margin of a glacier advances, retreats, or remains stationary depends on the balance between accumulation and wastage (ablation). When a glacier moves across irregular terrain, *crevasses* form in the brittle upper portion.

Zone of wastage
All the snow from the previous winter melts along with some glacial ice

Snowline

Crevasses

FIGURE 11.12 Athabasca Glacier in Jasper National Park in the Canadian Rockies. This photo was taken in August 2005. The date marker shows where the snout of the glacier was in 1992. The glacier is currently receding at a rate of 2 to 3 meters per year. *(Photo by Hughrocks)*

Geologist's Sketch

FIGURE 11.11 Icebergs are created when large pieces calve from the front of a glacier after it reaches a water body. In the large image, ice is calving from the terminus of Hubbard Glacier in Wrangell-St. Elias National Park, Alaska. *(Photo by Bernhard Edmaier/Photo Researchers, Inc.)* As the drawing illustrates, only about 20 percent (or less) of an iceberg protrudes above the waterline. The smaller photo shows an iceberg in Canada's Labrador Sea. *(Photo by Radius Images/Photolibrary)*

icebergs produced here can be several kilometers across and 600 meters thick. By comparison, thousands of irregularly shaped icebergs are produced by outlet glaciers flowing from the margins of the Greenland ice sheet. Many drift southward and find their way into the North Atlantic, where they can pose a hazard to navigation.

Whether the margin of a glacier is advancing, retreating, or remaining stationary depends on the budget of the glacier. The **glacial budget** is the balance, or lack of balance, between accumulation at the upper end of the glacier and loss at the lower end. This loss is termed **ablation**. If ice accumulation exceeds ablation, the glacial front advances until the two factors balance. When this happens, the terminus of the glacier is stationary.

If a warming trend increases ablation and/or if a drop in snowfall decreases accumulation, the ice front will retreat. As the terminus of the glacier retreats, the extent of the zone of wastage diminishes. Therefore, in time a new balance will be reached between accumulation and wastage, and the ice front will again become stationary.

Whether the margin of a glacier is advancing, retreating, or stationary, the ice within the glacier continues to flow forward. In the case of a receding glacier, the ice still flows forward but not rapidly enough to offset ablation. This point is illustrated well in Figure 1.11B. As the line of stakes within the Rhone Glacier continued to move downvalley, the terminus of the glacier slowly retreated upvalley.

GLACIERS IN RETREAT. Glaciers are sensitive to changes in temperature and precipitation and therefore can provide clues about changes in climate. In 1991, the mummified remains of a Neolithic man were discovered by hikers in the Alps. The remains had been locked and preserved in glacial ice for more than 5000 years. The discovery and subsequent analysis provided a fascinating glimpse of a very different time and fascinated the world. The discovery also meant that the glacier had reached a 5000 year minimum. The shrinking glacier that held the "ice man" is not unique.

In North America, one of the most visited glaciers is Athabasca Glacier, a 7-kilometer tongue of ice that spills down from the Columbia Icefield in the rugged Canadian Rockies.

People who return to the area a few years after their first visit will likely notice a change in the position of the end of the glacier (**FIGURE 11.12**). Over the past 125 years, Athabasca Glacier has lost about half of its volume and receded more than 1.5 kilometers (nearly 1 mile), leaving a thick and uneven layer of rocky debris in its place.

With few exceptions, glaciers around the world have been retreating at unprecedented rates over the last century (**FIGURE 11.13**). Some have disappeared altogether. One study that used satellite data reported that between 1961 and 2004, glaciers around the world (outside of the Greenland and Antarctic ice sheets) lost an estimated 8000 cubic kilometers (1900 cubic miles) of ice.* That is approximately enough ice to cover a

---

*Mark B. Dyurgerov and Mark F. Meier. *Glaciers and the Changing Earth System,* Boulder, Colorado: INSTAAR Occasional Paper No. 58, 2005.

DID YOU KNOW?
Glaciers are found on all continents except Australia. Surprisingly, tropical Africa has a small area of glacial ice atop its highest mountain, Mount Kilimanjaro. However, studies show that the mountain has lost more than 80 percent of its ice since 1912 and that all of it may be gone by 2020.

FIGURE 11.13 Two images taken 63 years apart from the same spot in Alaska's Glacier Bay National Park. Muir Glacier, which is prominent in the 1941 photo, has retreated out of the field of view in the 2004 image. Also Riggs Glacier (upper right) has thinned and retreated significantly. *(Photos courtesy of National Snow and Ice Data Center)*

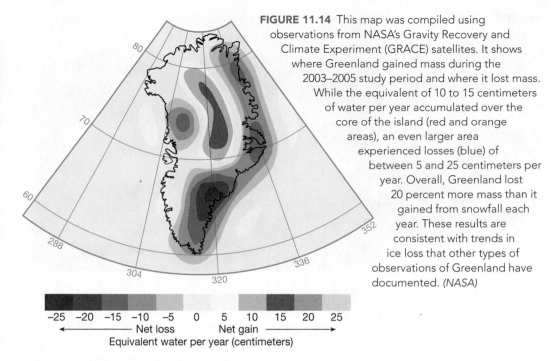

FIGURE 11.14 This map was compiled using observations from NASA's Gravity Recovery and Climate Experiment (GRACE) satellites. It shows where Greenland gained mass during the 2003–2005 study period and where it lost mass. While the equivalent of 10 to 15 centimeters of water per year accumulated over the core of the island (red and orange areas), an even larger area experienced losses (blue) of between 5 and 25 centimeters per year. Overall, Greenland lost 20 percent more mass than it gained from snowfall each year. These results are consistent with trends in ice loss that other types of observations of Greenland have documented. *(NASA)*

Net loss ← Net gain →
Equivalent water per year (centimeters)

2-kilometer wide swath of land between New York and Los Angeles with a glacier 1 kilometer thick.

GREENLAND'S GLACIAL BUDGET. Is the Greenland ice sheet growing or shrinking? Answering this relatively simple question is not a simple task and is much more difficult for ice sheets than for valley glaciers. Accurate answers require large quantities of data. Conducting field studies on an ice sheet poses significant challenges—sampling a huge area with an extreme climate in a remote part of the world over an extended time span. Data can only be collected for short periods each year at widely scattered sites. The information gathered by such studies, though useful, provided only a glimpse into what was happening on the Greenland ice sheet. However, in recent years satellite observations have started to provide scientists with the data they need

to begin to determine whether the margins of the ice sheet are melting faster than the interior is gaining ice (net loss) or whether the interior is adding new ice faster than the edges are wasting away (net gain).

The map in **FIGURE 11.14** was prepared using satellite data for the years 2003 to 2005. During that span, the island's coastal area lost 155 gigatons (41 cubic miles) of ice per year, while snow accumulation in the interior of the ice sheet was only 54 gigatons per year. Clearly, during this span, the processes resulting in a net loss of mass were dominant.

CONCEPT CHECK 11.2

❶ Describe the two components of glacial flow. How fast do glaciers move?

❷ Under what circumstances will the front of a glacier advance? Retreat? Remain stationary?

Glacial Erosion

Glaciers are capable of great erosion. For anyone who has observed the terminus of an alpine glacier, the evidence of its erosive force is clear (**FIGURE 11.15**). You can witness firsthand the release of rock fragments of various sizes from the ice as it melts. All

signs lead to the conclusion that the ice has scraped, scoured, and torn rock from the floor and walls of the valley and carried it downvalley. It should be pointed out, however, that in mountainous regions mass-wasting processes also make substantial contributions to the sediment load of a glacier. A glance back at the Chapter 8 opening photo on p. 196 provides a striking example.

Once rock debris is acquired by a glacier, the enormous competency of ice will not allow the debris to settle out like the load carried by a stream or by the wind. Indeed, as a medium of sediment transport, ice has no equal. Glaciers can transport huge blocks that no other erosional agent could possibly budge. Although today's glaciers are of limited importance as erosional agents, many landscapes that were modified by the widespread glaciers of the most recent Ice Age still reflect to a high degree the work of ice.

Glaciers erode the land primarily in two ways—*plucking* and *abrasion*. First, as a glacier flows over a fractured bedrock surface, it loosens and lifts blocks of rock and incorporates them into the ice. This process, known as **plucking**, occurs when meltwater penetrates the cracks and joints of bedrock beneath a glacier and freezes. When water freezes it expands, exerting tremendous leverage that pries the rock

FIGURE 11.15 Glaciers are capable of great erosion. As the terminus of this Alaskan glacier wastes away, it deposits large quantities of unsorted sediment called *till*. (Photo by Michael Collier)

loose. In this manner sediment of all sizes, ranging from particles as fine as flour to blocks as big as houses, becomes part of the glacier's load.

The second major erosional process is **abrasion** (FIGURE 11.16). As the ice and its load of rock fragments slide over bedrock, they function like sandpaper to smooth and polish the surface below. The pulverized rock produced by the glacial gristmill is appropriately called **rock flour**. So much rock flour may be produced that meltwater streams flowing out of a glacier often have the grayish appearance of skim milk and offer visible evidence of the grinding power of ice.

When the ice at the bottom of a glacier contains large fragments of rock, long scratches and grooves called **glacial striations** may be gouged into the bedrock. These linear grooves provide clues to the direction of ice flow. By mapping the striations over large areas, patterns of glacial flow can often be reconstructed.

In contrast, not all abrasive action produces striations. The rock surfaces over which the glacier moves may also become highly polished by the ice and its load of finer particles. The broad expanses of smoothly polished granite in Yosemite National Park provide an excellent example (Figure 11.16B).

As is the case with other agents of erosion, the rate of glacial erosion is highly variable. This differential erosion by ice is largely controlled by four factors: (1) rate of glacial movement; (2) thickness of the ice; (3) shape, abundance, and hardness of the rock fragments contained in the ice at the base of the glacier; and (4) the erodibility of the surface beneath the glacier. Variations in any or all of these factors from time to time and/or from place to place mean that the features, effects, and degree of landscape modification in glaciated regions can vary greatly.

CONCEPT CHECK 11.3

❶ How do glaciers accumulate their load of sediment?

❷ What factors influence the capability of a glacier to erode?

Landforms Created by Glacial Erosion

 Glaciers and Glaciation

ESSENTIALS OF GEOLOGY **Reviewing Glacial Features**

The erosional effects of valley glaciers and ice sheets are quite different. A visitor to a glaciated mountain region is likely to see a sharp and angular topography. The reason is that as alpine glaciers move downvalley, they tend to accentuate the irregularities of the mountain landscape by creating steeper canyon walls and making bold peaks even more jagged. By contrast, continental ice sheets generally override the terrain and hence subdue rather than accentuate the irregularities they encounter. Although the erosional potential of ice sheets is enormous, landforms carved by these huge ice masses usually do not inspire the same

A.

FIGURE 11.16
A. Glacial abrasion created the scratches and grooves in this bedrock. Cove Glacier, Prince William, Sound, northeast of Whittier, Alaska. **B.** Glacially polished granite in California's Yosemite National Park. (Photos by Michael Collier)

B.

wonderment and awe as do the erosional features created by valley glaciers. Much of the rugged mountain scenery so celebrated for its majestic beauty is the product of erosion by alpine glaciers. Take a moment to study **FIGURE 11.17**, which shows a hypothetical mountain area before, during, and after glaciation. You will refer to this often in the following discussion.

Glaciated Valleys

A hike up a glaciated valley reveals a number of striking ice-created features. The valley itself is often a dramatic sight. Unlike streams, which create their own valleys, glaciers take the path of least resistance by following the course of existing stream

valleys. Prior to glaciation, mountain valleys are characteristically narrow and V-shaped because streams are well above base level and are therefore downcutting. However, during glaciation these narrow valleys undergo a transformation as the glacier widens and deepens them, creating a U-shaped **glacial trough** (Figure 11.17C and **FIGURE 11.18**). In addition to producing a broader and deeper valley, the glacier also straightens the valley. As ice flows around sharp curves, its great erosional force removes the spurs of land that extend into the valley.

The amount of glacial erosion that takes place in different valleys in a mountainous area varies. Prior to glaciation, the mouths of tributary streams join the main (*trunk*) valley at the elevation of the stream in that valley. During glaciation, the amount of ice flowing through the main valley can be much greater than the amount advancing down each tributary. Consequently, the valley containing the trunk glacier is eroded deeper than the smaller valleys that feed it. Thus, when the glaciers eventually recede, the valleys of tributary glaciers are left standing above the main glacial trough

> **DID YOU KNOW?**
> There are many areas where scientists report that glaciers are wasting away. For example, 150 years ago there were 147 glaciers in Montana's Glacier National Park. Today only 37 remain, and these may vanish by 2030. Similarly, glaciers all across the Alps are retreating and disappearing every year.

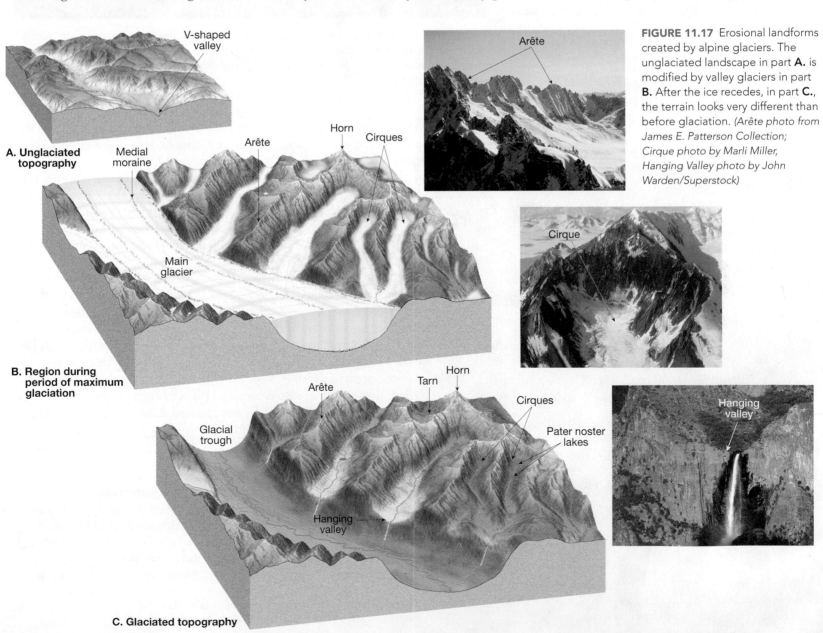

FIGURE 11.17 Erosional landforms created by alpine glaciers. The unglaciated landscape in part **A.** is modified by valley glaciers in part **B.** After the ice recedes, in part **C.**, the terrain looks very different than before glaciation. (*Arête photo from James E. Patterson Collection; Cirque photo by Marli Miller, Hanging Valley photo by John Warden/Superstock*)

and are termed **hanging valleys** (Figure 11.17C). Rivers flowing through hanging valleys can produce spectacular waterfalls, such as those in Yosemite National Park.

As hikers walk up a glacial trough, they may pass a series of bedrock depressions on the valley floor that were probably formed by plucking and scoured by the abrasive force of the ice. If these depressions are filled with water, they are called **pater noster lakes**. The Latin name means "our Father" and is a reference to a string of rosary beads.

At the head of a glacial valley is a very characteristic and often imposing feature associated with an alpine glacier called a **cirque**. As Figure 11.17 illustrates, this bowl-shaped depression has precipitous walls on three sides but is open on the downvalley side. The cirque is the focal point of the glacier's growth, because it is the area of snow accumulation and ice formation. It begins as an irregularity in the

FIGURE 11.18 Prior to glaciation, a mountain valley is typically narrow and V-shaped. During glaciation, an alpine glacier widens, deepens, and straightens the valley, creating the U-shaped glacial trough seen here. This glacial trough is near Carcross in Canada's Yukon Territory. *(Photo by Michael Collier)*

mountainside that is subsequently enlarged by frost wedging and plucking along the sides and bottom of the glacier. After the glacier has melted away, the cirque basin is often occupied by a small lake called a **tarn** (Figure 11.17C).

Before leaving the topic of glacial troughs and their associated features, one rather well-known feature should be discussed. **Fiords** are deep, often spectacular, steep-sided inlets of the sea that are present at high latitudes where mountains are adjacent to the ocean (**FIGURE 11.19**). They are drowned glacial troughs that became submerged as the ice left the valley and sea level rose following the Ice Age.

The depths of fiords can exceed 1000 meters (3300 feet). However, the great depths of these flooded troughs is only partly explained by the post–Ice Age rise in sea level. Unlike the situation governing the downward erosional work of rivers, sea level does not act as base level for glaciers. As a consequence, glaciers are capable of eroding their beds far below the surface of the sea. For example, a 300-meter-thick alpine glacier can carve its valley floor more than 250 meters below sea level before downward erosion ceases and the ice begins to float. Norway, British Columbia, Greenland, New Zealand, Chile, and Alaska all have coastlines characterized by fiords.

Arêtes and Horns

A visit to the Alps, the Northern Rockies, or many other scenic mountain landscapes carved by valley glaciers reveals not only glacial troughs, cirques, pater noster lakes, and the other related features just discussed. You are also likely to see sinuous, sharp-edged ridges called **arêtes** (French for *knife-edge*) and sharp, pyramidlike peaks called **horns** projecting above the surroundings. Both features can

This ice cave is located in Argentina's Los Glaciares National Park. *(Photo by Martin Harvey/Getty Images)*

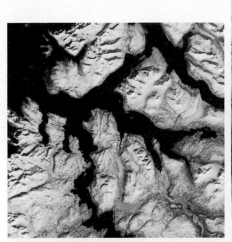

FIGURE 11.19 The coast of Norway is known for its many fiords. Frequently these ice-sculpted inlets of the sea are hundreds of meters deep. *(Satellite images courtesy of NASA; photo by Wolfgang Meier/ Photolibrary)*

FIGURE 11.20 Horns are sharp, pyramidlike peaks that are fashioned by alpine glaciers. This example is the famous Matterhorn in the Swiss Alps. *(Photo by Imagebroker/Alamy)*

originate from the same basic process, the enlargement of cirques produced by plucking and frost action. In the case of the spires of rock called horns, cirques around a single high mountain are responsible. As the cirques enlarge and converge, an isolated horn is produced. The most famous example is the Matterhorn in the Swiss Alps (**FIGURE 11.20**).

Arêtes can be formed in a similar manner, except that the cirques are not clustered around a point but rather exist on opposite sides of a divide. As the cirques grow, the divide separating them is reduced to a narrow knifelike partition. An arête, however, may also be created in another way. When two glaciers occupy parallel valleys, an arête can form when the divide separating the moving tongues of ice is progressively narrowed as the glaciers scour and widen their adjacent valleys.

Roches Moutonnées

In many glaciated landscapes, but most frequently where continental ice sheets have modified the terrain, the ice carves small streamlined hills from protruding bedrock knobs. Such an asymmetrical knob of bedrock is called a **roche moutonnée** (French for *sheep rock*). It is formed when glacial abrasion smoothes the gentle slope facing the oncoming ice sheet and plucking steepens the opposite side as the ice rides over the knob (**FIGURE 11.21**). Roches moutonnées indicate the direction of glacial flow, because the gentler slope is generally on the side from which the ice advanced.

CONCEPT CHECK 11.4

❶ How does a glaciated mountain valley differ in appearance from a mountain valley that was not glaciated?

❷ Describe the features created by glacial erosion that you might expect to see in an area where valley glaciers recently existed.

FIGURE 11.21 Roche moutonnée in Yosemite National Park, California. The gentle slope was abraded and the steep side was plucked. The ice moved from right to left. *(Photo by E. J. Tarbuck)*

Geologist's Sketch

Glacial Deposits

 Glaciers and Glaciation

ESSENTIALS OF GEOLOGY Reviewing Glacial Features

Glaciers pick up and transport huge loads of debris as they slowly advance across the land. Ultimately these materials are deposited when the ice melts. In regions where glacial sediment is deposited, it can play a truly significant role in forming the physical landscape. For example, in many areas once covered by the ice sheets of the recent Ice Age, the bedrock is rarely exposed because glacial deposits that are tens or even hundreds of meters thick completely mantle the terrain. The general effect of these deposits is to reduce the local relief and thus level the topography. Indeed, rural country scenes that are familiar to many of us—rocky pastures in New England, wheat fields in the Dakotas, rolling farmland in the Midwest—result directly from glacial deposition.

Types of Glacial Drift

Long before the theory of an extensive Ice Age was proposed, much of the soil and rock debris covering portions of Europe was recognized as coming from elsewhere. At the time, these foreign materials were believed to have been "drifted" into their present positions by floating ice during an ancient flood. As a consequence, the term *drift* was applied to this sediment. Although rooted in a concept that was not correct, this term was so well established by the time the true glacial origin of the debris became widely recognized that it remained in the basic glacial vocabulary. Today, **glacial drift** is an all-embracing term for sediments of glacial origin, no matter how, where, or in what form they were deposited.

Glacial drift is divided into two distinct types: (1) materials deposited directly by the glacier, which are known as *till*, and (2) sediments laid down by glacial meltwater, called *stratified drift*. Here is the difference: **Till** is deposited as glacial ice melts and drops its load of rock fragments. Unlike moving water and wind, ice cannot sort the sediment it carries. Therefore, deposits of till are characteristically

DID YOU KNOW?
The Arctic Ocean surrounds the North Pole, so the ice in this region is sea ice (frozen seawater), *not* glacial ice. Remember, glaciers form on land from the accumulation of snow.

unsorted mixtures of many particle sizes (**FIGURE 11.22**). A close examination of this sediment often shows that many of the pieces are scratched and polished as a result of being dragged along by the glacier. **Stratified drift** is sorted according to the size and weight of the fragments. Because ice is not capable of such sorting activity, these sediments are not deposited directly by the glacier. Rather, they reflect the sorting action of glacial meltwater.

Some deposits of stratified drift are made by streams issuing directly from the glacier. Other stratified deposits involve sediment that was originally laid down as till and later picked up, transported, and redeposited by meltwater beyond the margin of the ice. Accumulations of stratified drift often consist largely of sand and gravel, because the meltwater is not capable of moving larger material and because the finer rock flour remains suspended and is commonly carried far from the glacier. An indication that stratified drift consists primarily of sand and gravel can be seen in many areas where these deposits are actively mined as aggregate for road work and other construction projects.

Close up of cobble

FIGURE 11.22 Glacial till is an unsorted mixture of many different sediment sizes. *(Photos by E. J. Tarbuck)*

FIGURE 11.23 A large, glacially transported boulder in Denali National Park, Alaska. Such boulders are called *glacial erratics. (Photo by Michael Collier)*

When boulders are found in the till or lying free on the surface, they are called **glacial erratics** if they are different from the bedrock below (**FIGURE 11.23**). Of course, this means that they must have been derived from a source outside the area where they are found. Although the locality of origin for most erratics is unknown, the origin of some can be determined. In many cases, boulders were transported as far as 500 kilometers from their source area and, in a few instances, more than 1000 kilometers. Therefore, by studying glacial erratics as well as the mineral composition of the till, geologists can sometimes trace the path of a lobe of ice. In portions of New England, as well as other areas, erratics can be seen dotting pastures and farm fields. In some places, these rocks were cleared from fields and piled to make fences and walls.

Moraines, Outwash Plains, and Kettles

Perhaps the most widespread features created by glacial deposition are *moraines*, which are simply layers or ridges of till. Several types of moraines are identified; some are common only to mountain valleys, and others are associated with areas affected by either ice sheets or valley glaciers. Lateral and medial moraines fall in the first category, whereas end moraines and ground moraines are in the second.

LATERAL AND MEDIAL MORAINES. The sides of a valley glacier accumulate large quantities of debris from the valley walls. When the glacier wastes away, these materials are left as ridges, called **lateral moraines,** along the sides of the valley (**FIGURE 11.24**). **Medial moraines** are formed when two valley glaciers coalesce to form a single ice stream. The till that was once carried along the edges of each glacier joins to form a single dark stripe of debris within the newly enlarged glacier. The creation of these dark stripes within the ice stream is one obvious proof that glacial ice moves, because the medial moraine could not form if the ice did not flow downvalley (Figure 11.24). It is common to see several medial moraines within a large alpine glacier because a streak will form whenever a tributary glacier joins the main valley.

END AND GROUND MORAINES. An **end moraine** is a ridge of till that forms at the terminus of a glacier and is characteristic of ice sheets and valley glaciers alike. These relatively common landforms are deposited when a state of equilibrium is attained between ablation and ice accumulation. That is, the end moraine forms when the ice is melting and evaporating near the end of the glacier at a rate equal to the forward advance of the glacier from its region of nourishment. Although the terminus of the glacier is now stationary, the ice continues to flow forward, delivering a continuous supply of sediment in the same manner a conveyor belt delivers goods to the end of a production line. As the ice melts,

FIGURE 11.24 Medial moraines form when the lateral moraines of merging glaciers join. St. Elias National Park, Alaska. Another example of a medial moraine may be seen in Figure 11.1, p. 262 *(Photo by Tom Bean/Alamy)*

Medial moraines

Lateral moraine

Lateral moraines join to form medial moraine

Lateral moraine

Medial moraine

Geologist's Sketch

the till is dropped and the end moraine grows. The longer the ice front remains stable, the larger the ridge of till will become.

Eventually the time comes when ablation exceeds nourishment. At this point, the front of the glacier begins to recede in the direction from which it originally advanced. However, as the ice front retreats, the conveyor-belt action of the glacier continues to provide fresh supplies of sediment to the terminus. In this manner a large quantity of till is deposited as the ice melts away, creating a rock-strewn, undulating plain. This gently rolling layer of till deposited as the ice front recedes is termed **ground moraine.** It has a leveling effect, filling in low spots and clogging old stream channels, often leading to a derangement of the existing drainage system. In areas where this layer of till is still relatively fresh, such as the northern Great Lakes region, poorly drained swampy lands are quite common.

Periodically, a glacier will retreat to a point where ablation and nourishment once again balance. When this happens, the ice front stabilizes and a new end moraine forms.

The pattern of end moraine formation and ground moraine deposition may be repeated many times before the glacier has completely vanished. Such a pattern is illustrated in **FIGURE 11.25**. The very first end moraine to form signifies the farthest advance of the glacier and is called the *terminal end moraine.* Those end moraines that form as the ice front occasionally stabilizes during retreat are termed *recessional end moraines.* Terminal and recessional moraines are essentially alike; the only difference between them is their relative positions.

End moraines deposited by the most recent stage of Ice Age glaciation are prominent features in many parts of the Midwest and Northeast. In Wisconsin, the wooded, hilly terrain of the Kettle Moraine near Milwaukee is a particularly picturesque example. A well-known example in the Northeast is Long Island. This linear strip of glacial sediment that extends northeastward from New York City is part of an end moraine complex that stretches from eastern Pennsylvania to Cape Cod, Massachusetts (**FIGURE 11.26**).

FIGURE 11.27 represents a hypothetical area during glaciation and after the retreat

of ice sheets. It shows the end moraines that were just described as well as the depositional features that are discussed in the sections that follow. This figure depicts landscape features similar to what might be encountered if you were traveling in the upper Midwest or New England. As you read upcoming sections dealing with other glacial deposits, you will be referred to this figure several times.

OUTWASH PLAINS AND VALLEY TRAINS. At the same time that an end moraine is forming, water from the melting glacier cascades over the till, sweeping some of it out in front of the growing ridge of unsorted debris. Meltwater generally emerges from the ice in rapidly moving streams that are often choked with suspended material and carry a substantial bed load as well. As the water leaves the glacier, it moves onto the relatively flat surface beyond and rapidly loses velocity. As a consequence, much of its bed load is dropped and the meltwater begins weaving a complex pattern of braided channels (Figure 11.27). In this way, a broad, ramp-like surface composed of stratified drift is built adjacent to the downstream edge of most end moraines. When the feature is formed in association with an ice sheet it is termed an **outwash plain**; when it is largely confined to a mountain valley it is usually referred to as a **valley train**.

KETTLES. Often end moraines, outwash plains, and valley trains are pockmarked with basins or depressions known as **kettles** (Figure 11.27). Kettles form when blocks of stagnant ice become wholly or partly buried in drift and eventually melt, leaving pits in the glacial sediment. Although most kettles do not exceed 2 kilometers in diameter, some with diameters exceeding 10 kilometers occur in Minnesota. Likewise, the typical depth of most kettles is less than 10 meters, although the vertical dimensions of some approach 50 meters. In many cases water eventually fills the depression and forms a pond or lake. One well-known example is Walden Pond near Concord, Massachusetts. It is here that Henry David Thoreau lived alone for two years in the 1840s and about which he wrote his famous book *Walden,* or *Life in the Woods.*

FIGURE 11.25 End moraines of the Great Lakes region. Those deposited during the most recent (Wisconsinan) stage are most prominent.

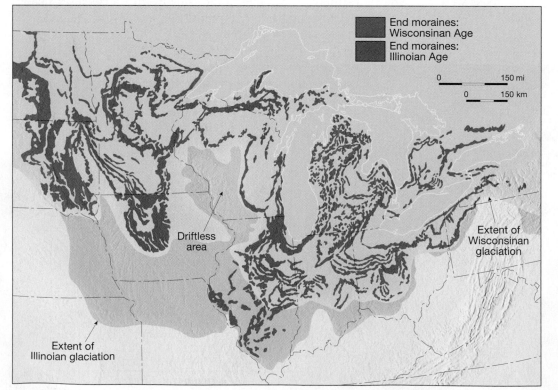

End moraines: Wisconsinan Age

End moraines: Illinoian Age

0   150 mi

0   150 km

Driftless area

Extent of Wisconsinan glaciation

Extent of Illinoian glaciation

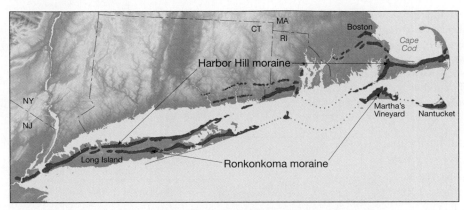

FIGURE 11.26 End moraines make up substantial parts of Long Island, Cape Cod, Martha's Vineyard, and Nantucket. Although portions are submerged, the Ronkonkoma moraine extends through central Long Island, Martha's Vineyard, and Nantucket. It was deposited about 20,000 years ago. The Harbor Hill moraine, which formed about 14,000 years ago, extends along the north shore of Long Island, through southern Rhode Island and Cape Cod.

Drumlins, Eskers, and Kames

Moraines are not the only landforms deposited by glaciers. Some landscapes are characterized by numerous elongate parallel hills made of till. Other areas exhibit conical hills and relatively narrow winding ridges composed largely of stratified drift.

DRUMLINS. Drumlins are streamlined, asymmetrical hills composed of till (Figure 11.27). They range in height from 15 to 60 meters (50 to 200 feet) and average 0.4 to 0.8 kilometer (0.25 to 0.50 mile) in length. The steep side of the hill faces the direction from which the ice advanced, whereas the gentler slope points in the direction the ice moved. Drumlins are not found singly but rather occur in clusters, called *drumlin fields*. One such cluster, east of Rochester, New York, is estimated to contain about 10,000 drumlins. Their streamlined shape indicates that they were molded in the zone of flow within an active glacier. It is thought that drumlins originate when glaciers advance over previously deposited drift and reshape the material.

ESKERS AND KAMES. In some areas that were once occupied by glaciers, sinuous ridges composed largely of sand and gravel can be found. Known as **eskers**, these ridges are deposited by meltwater rivers flowing within, on top of, and beneath a mass of motionless, stagnant glacial ice (Figure 11.27). Many sediment sizes are carried by the torrents of meltwater in the ice-banked channels, but only the coarser material can settle out of the turbulent stream. In some areas

DID YOU KNOW?
Studies have shown that the retreat of glaciers may reduce the stability of faults and hasten earthquake activity. When the weight of the glacier is removed, the ground rebounds. In tectonically active areas experiencing post-glacial rebound, earthquakes may occur sooner and/or be stronger than if the ice were present.

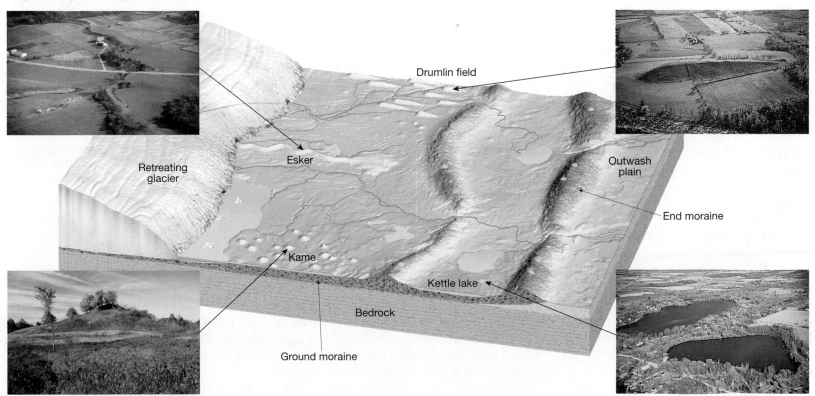

FIGURE 11.27 This hypothetical area illustrates many common depositional landforms. The outermost end moraine marks the limit of glacial advance and is called the *terminal end moraine*. End moraines that form as the ice front occasionally becomes stationary during retreat are called *recessional end moraines*. (Drumlin photo courtesy of Ward's Natural Science Establishment; Kame, Esker, and Kettle photos by Richard P. Jacobs/JLM Visuals)

they are mined for sand and gravel, and for this reason, eskers are disappearing in some localities.

Kames are steep-sided hills that, like eskers, are composed of sand and gravel (Figure 11.27). Kames originate when glacial meltwater washes sediment into openings and depressions in the stagnant wasting terminus of a glacier. When the ice eventually melts away, the stratified drift is left behind as mounds or hills.

CONCEPT CHECK 11.5

❶ What is the difference between till and stratified drift?

❷ Distinguish between a terminal end moraine and a recessional end moraine.

❸ Describe the formation of a medial moraine.

❹ List four depositional features other than moraines.

Other Effects of Ice Age Glaciers

In addition to the massive erosional and depositional work carried on by Ice-Age glaciers, the ice sheets had other effects, sometimes profound, on the landscape. For example, as the ice advanced and retreated, animals and plants were forced to migrate. This led to stresses that some organisms could not tolerate. Hence, a number of plants and animals became extinct. Other effects of Ice Age glaciers that are described in this section involve adjustments in Earth's crust due to the addition and removal of ice and sea-level changes associated with the formation and melting of ice sheets. The advance and retreat of ice sheets also led to significant changes in the routes taken by rivers. In some regions, glaciers acted as dams that created large lakes. When these ice dams failed, the effects on the landscape were profound. In areas that today are deserts, lakes of another type, called pluvial lakes, formed.

Crustal Subsidence and Rebound

In areas that were centers of ice accumulation, such as Scandinavia and the Canadian

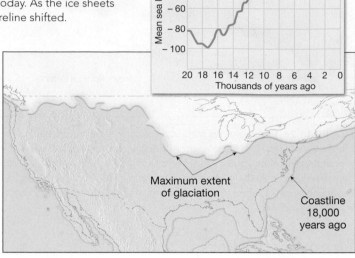

FIGURE 11.28 Changing sea level during the past 20,000 years. About 18,000 years ago, when the most recent ice advance was at a maximum, sea level was nearly 100 meters lower than at present. Thus, land that is presently covered by the ocean was exposed and the shoreline looked very different than it does today. As the ice sheets melted, sea level rose and the shoreline shifted.

Maximum extent of glaciation

Coastline 18,000 years ago

Shield, the land has been slowly rising over the past several thousand years. Uplifting of almost 300 meters (1000 feet) has occurred in the Hudson Bay region. This, too, is the result of the continental ice sheets, but how can glacial ice cause such vertical crustal movement? We now understand that the land is rising because the added weight of the 3-kilometer-thick (2-mile-thick) mass of ice caused downwarping of Earth's crust. Following the removal of this immense load, the crust has been adjusting by gradually rebounding upward ever since.*

Sea-Level Changes

One of the more interesting and perhaps dramatic effects of the Ice Age was the fall and rise of sea level that accompanied the advance and retreat of the glaciers. Today if all of Earth's glaciers melted, sea level would rise approximately 70 meters (230 feet) worldwide.

Although the total volume of glacial ice today is great, exceeding 25 million cubic kilometers, during the Ice Age the volume of glacial ice amounted to about 70 million cubic kilometers (45 million cubic kilometers more than at present). Because we know that the snow from which glaciers are made ultimately comes from the evaporation of ocean water, the growth of ice sheets must have caused a worldwide drop in sea level (**FIGURE 11.28**). Indeed, estimates suggest that sea level was as much as 100 meters lower than it is today. Thus, land that is presently flooded by the oceans was dry. The Atlantic Coast of the United States lay more than 100 kilometers to the east of New York City; France and Britain were joined where the famous English Channel is today; Alaska and Siberia were connected across the Bering Strait; and Southeast Asia was tied by dry land to the islands of Indonesia.

> **DID YOU KNOW?**
> Antarctica's ice sheet weighs so much that it depresses Earth's crust by an estimated 900 meters (3000 feet) or more.

Rivers Before and After the Ice Age

FIGURE 11.29A shows the familiar present-day pattern of rivers in the central United States, with the Missouri, Ohio, and Illinois rivers as major tributaries to the Mississippi. Figure 11.29B depicts drainage systems in this region prior to the Ice Age. The pattern is *very* different from the present. This remarkable transformation of river systems resulted from the advance and retreat of the ice sheets.

Notice that prior to the Ice Age, a significant part of the Missouri River drained north toward Hudson Bay. Moreover, the Mississippi River did not follow the present Iowa–Illinois boundary but rather flowed across west-central Illinois, where the lower Illinois River flows today. The preglacial Ohio River barely reached to the present-day state of Ohio,

*For a more complete discussion of this concept, see the section on *Isostasy* in Chapter 17, p. 432.

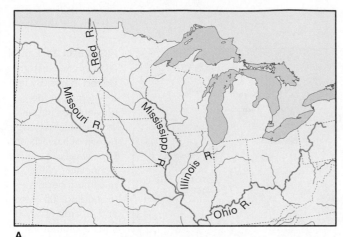

A.

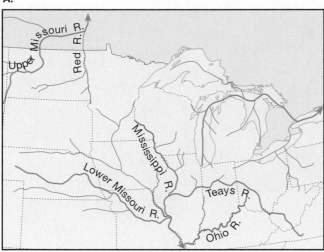

B.

FIGURE 11.29 A. This map shows the Great Lakes and the familiar present-day pattern of rivers in the central United States. Pleistocene ice sheets played a major role in creating this pattern. **B.** Reconstruction of drainage systems in the central United States prior to the Ice Age. The pattern was very different from today, and there were no Great Lakes.

and the rivers that today feed the Ohio in western Pennsylvania flowed north and drained into the North Atlantic. The Great Lakes were created by glacial erosion during the Ice Age. Prior to the Pleistocene era, the basins occupied by these huge lakes were lowlands with rivers that ran eastward to the Gulf of St. Lawrence.

The large Teays River was a significant feature prior to the Ice Age (**FIGURE 11.29B**). It flowed from West Virginia across Ohio, Indiana, and Illinois, where it discharged into the Mississippi River not far from present-day Peoria. This river valley, which would have rivaled the Mississippi in size,

Moraine Lake, Banff National Park, Canada. *(Photo by Steven Vidler/Eurasia Press/Corbis)*

was completely obliterated during the Pleistocene, buried by glacial deposits hundreds of feet thick. Today the sands and gravels in the buried Teays Valley make it an important aquifer.

Clearly, if we are to understand the present pattern of rivers in the central United States (and many other places as well), we must be aware of glacial history.

DID YOU KNOW?
The Great Lakes constitute the largest body of fresh water on Earth. Formed by glacial erosion between about 10,000 and 12,000 years ago, the lakes currently contain about 19 percent of Earth's surface fresh water.

Ice Dams Create Proglacial Lakes

Ice sheets and alpine glaciers can act as dams to create lakes by trapping glacial meltwater and blocking the flow of rivers. Some of these lakes are relatively small, short-lived impoundments. Others can be large and exist for hundreds or thousands of years.

FIGURE 11.30 is a map of Lake Agassiz—the largest lake to form during the Ice Age in North America. With the retreat of the ice sheet came enormous volumes of meltwater. The Great Plains generally slope upward to the west. As the terminus of the ice sheet receded northeastward, meltwater was trapped between the ice on one side and the sloping land on the other, causing Lake Agassiz to deepen and spread across the landscape. It came into existence about 12,000 years ago and lasted for about 4500 years. Such water bodies are termed *proglacial lakes*, referring to their position just beyond the outer limits of a glacier or ice sheet. The lake's history is complicated by the dynamics of the ice sheet, which at various times readvanced and affected lake levels and drainage systems. Where drainage occurred depended upon the water level of the lake and the position of the ice sheet.

Lake Agassiz left marks over a broad region. Former beaches, many kilometers from any water, mark former shorelines. Several modern river valleys, including the Red River and the Minnesota River, were originally cut by water entering or leaving the lake. Present-day remnants of Lake Agassiz include Lakes Winnipeg, Manitoba, Winnipegosis, and Lake of the Woods. The sediments of the former lake basin are now fertile agricultural land.

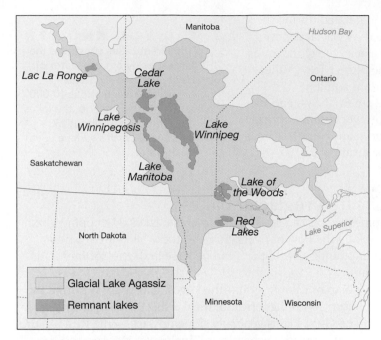

FIGURE 11.30 Map showing the extent of glacial Lake Agassiz. It was an immense feature—bigger than all of the present-day Great Lakes combined. The modern-day remnants of this proglacial water body are still major landscape features.

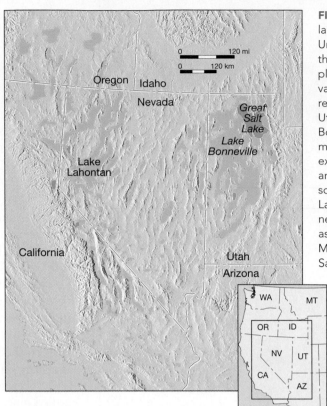

FIGURE 11.31 Pluvial lakes of the western United States. By far the largest of the pluvial lakes in the vast Basin and Range region of Nevada and Utah was Lake Bonneville. With maximum depths exceeding 300 meters and an area of 50,000 square kilometers, Lake Bonneville was nearly the same size as present-day Lake Michigan. The Great Salt Lake is a remnant of this huge pluvial lake. *(After R. F. Flint)*

Research shows that the shifting of glaciers and the failure of ice dams can cause the rapid release of huge volumes of water. Such events occurred during the history of Lake Agassiz and many other proglacial lakes. The erosional and depositional results of such megafloods were dramatic.

Pluvial Lakes

While the formation and growth of ice sheets was an obvious response to significant changes in climate, the existence of the glaciers themselves triggered important climatic changes in the regions beyond their margins. In arid and semiarid areas on all of the continents, temperatures were lower and thus evaporation rates were lower, but at the same time moderate precipitation totals were experienced. This cooler, wetter climate formed many **pluvial lakes** (from the Latin term *pluvis*, meaning *rain*). In North America the greatest concentration of pluvial lakes occurred in the vast Basin and Range region of Nevada and Utah (**FIGURE 11.31**). By far the largest of the lakes in this region was Lake Bonneville. With maximum depths exceeding 300 meters and an area of 50,000 square kilometers, Lake Bonneville was nearly the same size as present-day Lake Michigan. As the ice sheets waned, the climate again grew more arid, and the lake levels lowered in response. Although most of the lakes completely disappeared, a few small remnants of Lake Bonneville remain, the Great Salt Lake being the largest and best known.

CONCEPT CHECK 11.6

1. Describe at least four effects of Ice Age glaciers aside from the formation of major erosional and depositional features.

Glaciers of the Ice Age

At various points in the preceding pages we mentioned the Ice Age, a time when ice sheets and alpine glaciers were far more extensive than they are today. There was a time when the most popular explanation for what we now know to be glacial deposits was that the material had been drifted in by means of icebergs or perhaps simply swept across the landscape by a catastrophic flood. However, during the nineteenth century, field investigations by many scientists provided convincing proof that an extensive Ice Age was responsible for these deposits and for many other features.

By the beginning of the 20th century, geologists had largely determined the extent of Ice Age glaciation. Further, they discovered that many glaciated regions had not one layer of drift but rather several layers. Close examination of these older deposits showed well-developed zones of chemical weathering and soil formation as well as the remains of plants that require warm temperatures. The evidence was clear: There had been not just one glacial advance but many, each separated by extended periods when climates were as warm as or warmer than they are at present. The Ice Age had not simply been a time when the ice advanced over the land, lingered for a while, and then receded. Rather, the period was a very complex event characterized by a number of advances and withdrawals of glacial ice.

The glacial record on land is punctuated by many erosional gaps. This makes it difficult to reconstruct the episodes of the Ice Age clearly, but sediment on the ocean floor provides an uninterrupted record of climate cycles for this period. Studies of cores drilled from these seafloor sediments show that glacial–interglacial cycles have occurred about every 100,000 years. About 20 such cycles of cooling and warming were identified for the span we call the Ice Age.

RUSSIA

North Pole

*Arctic
Ocean*

Alps

Iceland

*Pacific
Ocean*

*Atlantic
Ocean*

UNITED STATES

Glacial ice

Sea ice

FIGURE 11.32 Maximum extent of glaciation in the Northern Hemisphere during the Ice Age.

During the glacial age, ice left its imprint on almost 30 percent of Earth's land area, including about 10 million square kilometers of North America, 5 million square kilometers of Europe, and 4 million square kilometers of Siberia (**FIGURE 11.32**). The amount of glacial ice in the Northern Hemisphere was roughly twice that in the Southern Hemisphere. The primary reason is that the Southern Hemisphere has little land in the middle latitudes, and therefore the southern polar ice could not spread far beyond the margins of Antarctica. By contrast, North America and Eurasia provided great expanses of land for the spread of ice sheets.

Today we know that the Ice Age began between 2 million and 3 million years ago. This means that most of the major glacial episodes occurred during a division of the geologic time scale called the **Pleistocene epoch**. Although the Pleistocene is commonly used as a synonym for the Ice Age, this epoch does not encompass it all. The Antarctic ice sheet, for example, formed at least 30 million years ago.

CONCEPT CHECK 11.7

❶ About what percentage of Earth's land surface was affected by Pleistocene glaciers?

Causes of Glaciation

A great deal is known about glaciers and glaciation. Much has been learned about glacier formation and movement, the extent of glaciers past and present, and the features created by glaciers, both erosional and depositional. However, the causes of glacial ages are not completely understood.

Although widespread glaciation has been rare in Earth's history, the Pleistocene Ice Age is not the only glacial period for which a record exists. Earlier glaciations are indicated by deposits called *tillite*, a sedimentary rock formed when glacial till becomes lithified. Such strata usually contain striated rock fragments, and some overlie grooved and polished rock surfaces or are associated with sandstones and conglomerates that show features indicating they were deposited as stratified drift. Two Precambrian glacial episodes have been identified in the geologic record, the first approximately 2 billion years ago and the second about 600 million years ago. Further, a well-documented record of an earlier glacial age is found in late Paleozoic rocks that are about 250 million years old and exist on several landmasses.

Any theory that attempts to explain the causes of glacial ages must successfully answer two basic questions: (1) *What causes the onset of glacial conditions?* For continental ice sheets to have formed, average temperatures must have been somewhat lower than at present and perhaps substantially lower than throughout much of geologic time. Thus, a successful theory would have to account for the cooling that finally leads to glacial conditions. (2) *What caused the alternation of glacial and interglacial stages that have been documented for the Pleistocene epoch?* The first question deals with long-term trends in temperature on a scale of millions of years, but this second question relates to much shorter-term changes.

Although the literature of science contains many hypotheses relating to the possible causes of glacial periods, we will discuss only a few major ideas to summarize current thought.

Plate Tectonics

Probably the most attractive proposal for explaining the fact that extensive glaciations have occurred only a few times in the geologic past comes from the theory of plate tectonics.* Because glaciers can form only on land, we know that landmasses must exist somewhere in the higher latitudes before an ice age can commence. Many scientists suggest that ice ages have occurred only when Earth's shifting crustal plates have carried the continents from tropical latitudes to more poleward positions.

*A brief overview of the theory appears in Chapter 1, and a more extensive discussion is presented in Chapter 15.

This iceberg is in Canada's Labrador Sea *(Photo by Radius Images/ Photolibrary)*.

Glacial features in present-day Africa, Australia, South America, and India indicate that these regions, which are now tropical or subtropical, experienced an ice age near the end of the Paleozoic era, about 250 million years ago. However, there is no evidence that ice sheets existed during that period in what are today the higher latitudes of North America and Eurasia. For many years this puzzled scientists. Was the climate in these relatively tropical latitudes once like it is today in Greenland and Antarctica? Why did glaciers not form in North America and Eurasia? Until the plate tectonics theory was formulated, there had been no reasonable explanation.

Today scientists understand that the areas containing these ancient glacial features were joined together as a single supercontinent (Pangaea) located at latitudes far to the south of their present positions. Later this landmass broke apart, and its pieces, each moving on a different plate, migrated toward their present locations (**FIGURE 11.33**). Now we know that during the geologic past, plate movements accounted for many dramatic climatic changes as landmasses shifted in relation to one another and moved to different latitudinal positions.

Changes in oceanic circulation also must have occurred, altering the transport of heat and moisture and consequently the climate as well. Because the rate of plate movement is very slow—a few centimeters per year—appreciable changes in the positions of the continents occur only over great spans of geologic time. Thus, climate changes brought about by shifting plates are extremely gradual and happen on a scale of millions of years.

Variations in Earth's Orbit

Because climatic changes brought about by moving plates are extremely gradual, the plate tectonics theory cannot be used to explain the alternation between glacial and interglacial climates that occurred during the Pleistocene epoch. Therefore, we must look to some other triggering mechanism that may cause climate change on a scale of thousands rather than millions of years. Today many scientists strongly suspect that the climatic oscillations that characterized

the Pleistocene may be linked to variations in Earth's orbit. This hypothesis was first developed and strongly advocated by the Serbian scientist Milutin Milankovitch and is based on the premise that variations in incoming solar radiation are a principal factor controlling Earth's climate.

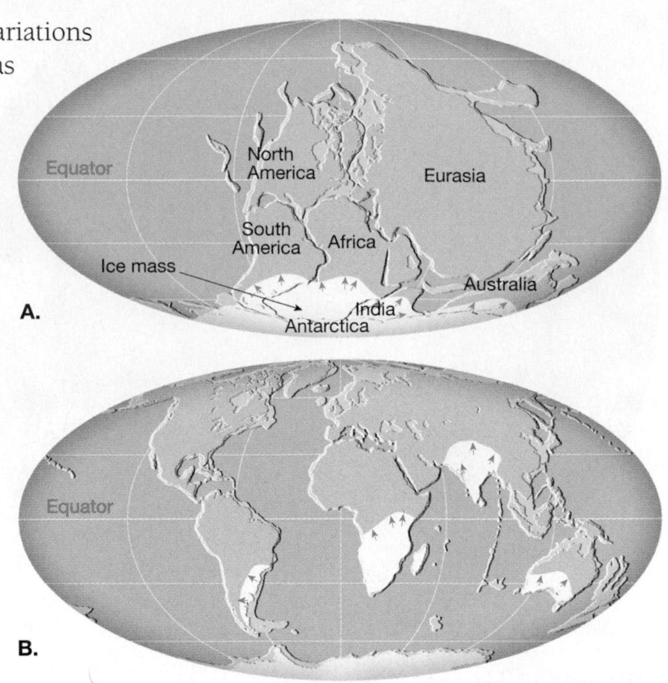

Milankovitch formulated a comprehensive mathematical model based on the following elements (**FIGURE 11.34**):

1. Variations in the shape (*eccentricity*) of Earth's orbit about the Sun;

2. Changes in *obliquity*; that is, changes in the angle that the axis makes with the plane of Earth's orbit; and

3. The wobbling of Earth's axis, called *precession*.

Using these factors, Milankovitch calculated variations in the receipt of solar energy and the corresponding surface temperature of Earth back into time in an attempt to correlate these changes with the climate fluctuations of the Pleistocene. It should be noted that these factors cause little or no variation in the *total* solar energy reaching the ground. Instead, their impact is felt because they change the degree of contrast between the seasons. Somewhat milder winters in the mid to high latitudes means greater snowfall totals, whereas cooler summers would bring a reduction in snowmelt.

Among the studies that have added credibility to the astronomical hypothesis of Milankovitch is one in which deep-sea sediments containing certain climatically sensitive microorganisms were analyzed to establish a chronology of temperature changes going back nearly 500,000 years.[*] This time scale of climatic change was then compared to astronomical calculations of eccentricity, obliquity, and precession to determine whether a correlation did indeed exist.

Although the study was very involved and mathematically complex, the conclusions were straightforward. The researchers found that major variations in climate over the past several hundred thousand years were closely associated with changes in the geometry of Earth's orbit; that is, cycles of climatic change were shown to correspond closely with the periods of obliquity, precession, and orbital eccentricity. More specifically, the authors stated: "It is concluded that changes in the earth's orbital geometry are the fundamental cause of the succession of Quaternary ice ages."[†]

Let us briefly summarize the ideas that were just described. The theory of plate tectonics provides us with an explanation for the widely spaced and nonperiodic onset of glacial conditions at various times in the geologic past; the astronomical model proposed by Milankovitch and supported by the work of J. D. Hays and his colleagues furnishes an explanation for the alternating glacial and interglacial episodes of the Pleistocene.

FIGURE 11.33 A. The supercontinent Pangaea showing the area covered by glacial ice about 300 million years ago. **B.** The continents as they are today. The white areas indicate where evidence of the old ice sheets exists.

---

[*]J. D. Hays, John Imbrie, and N. J. Shackelton, "Variations in the Earth's Orbit: Pacemaker of the Ice Ages," *Science* 194 (1976): 1121–32.

[†]J. D. Hays et al., ibid., p. 1131. The term *quaternary* refers to the period on the geologic time scale that encompasses the last 2.6 million years.

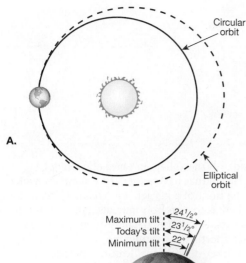

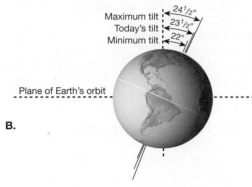

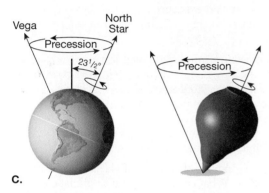

FIGURE 11.34 Orbital variations. **A.** The shape of Earth's orbit changes during a cycle that spans about 100,000 years. It gradually changes from nearly circular to more elliptical and then back again. This diagram greatly exaggerates the amount of change. **B.** Today the axis of rotation is tilted about 23.5° to the plane of Earth's orbit. During a cycle of 41,000 years, this angle varies from 21.5° to 24.5°. **C.** Precession: Earth's axis wobbles like a spinning top. Consequently, the axis points to different spots in the sky during a cycle of about 26,000 years.

Other Factors

Variations in Earth's orbit correlate closely with the timing of glacial–interglacial cycles. However, the variations in solar energy reaching Earth's surface caused by these orbital changes do not adequately explain the magnitude of the temperature changes that occurred during the most recent Ice Age. Other factors must also have contributed. One factor involves vari-

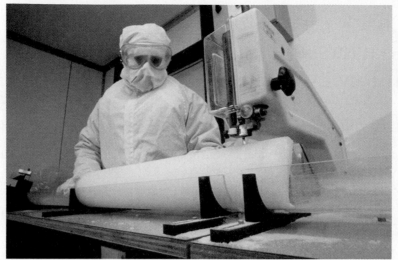

FIGURE 11.35 Scientist slicing an ice core sample from Antarctica for analysis. He is wearing protective clothing and a mask to minimize contamination of the sample. Chemical analysis of ice cores can provide important data about past climates. *(Photo by British Antarctic Survey/Photo Researchers, Inc.)*

ations in the chemical composition of the atmosphere. Other influences involve changes in both the reflectivity of Earth's surface and ocean circulation. Let's take a brief look at these factors.

Chemical analyses of air bubbles that become trapped in glacial ice at the time of ice formation indicate that the Ice-Age atmosphere contained less of the gases carbon dioxide and methane than the post Ice-Age atmosphere (**FIGURE 11.35**). Carbon dioxide and methane are important "greenhouse" gases, which means that they trap radiation emitted by Earth and contribute to the heating of the atmosphere. When the amount of carbon dioxide and methane in the atmosphere increases, global temperatures rise, and when there is a reduction in these gases, as occurred during the Ice Age, temperatures fall. Therefore, reductions in the concentrations of greenhouse gases help explain the magnitude of the temperature drop that occurred during glacial times. Although scientists know that concentrations of carbon dioxide and methane dropped, they do not know what caused the drop. As often occurs in science, observations gathered during one investigation yield information and raise questions that require further analysis and explanation.

Obviously, whenever Earth enters an ice age, extensive areas of land that were once ice free are covered with ice and snow. In addition, a colder climate causes the area covered by sea ice (frozen surface sea water) to expand as well. Ice and snow reflect a large portion of incoming solar energy back to space. Thus, energy that would have warmed Earth's surface and the

air above is lost, and global cooling is reinforced.*

Yet another factor that influences climate during glacial times relates to ocean currents. Research has shown that ocean circulation changes during ice ages. For example, studies suggest that the warm current that transports large amounts of heat from the tropics toward higher latitudes in the North Atlantic was significantly weaker during the Ice Age. This would lead to a colder climate in Europe, amplifying the cooling attributable to orbital variations.

In conclusion, we emphasize that the ideas just discussed do not represent the only possible explanations for glacial ages. Although interesting and attractive, these proposals are certainly not without critics, nor are they the only possibilities currently under study. Other factors may be, and probably are, involved.

CONCEPT CHECK 11.8

❶ How does the theory of Plate Tectonics help us understand the cause of ice ages?

❷ Does the theory of plate tectonics explain alternating glacial-interglacial climates during the Pleistocene? Why or why not?

*Recall from Chapter 1 that something that reinforces (adds to) the initial change is called a *positive feedback mechanism*. To review this idea, see the discussion on feedback mechanisms in the section on "Earth as a System" in Chapter 1.

CHAPTER ELEVEN
Glaciers and Glaciation in Review

⊙ A *glacier* is a thick mass of ice originating on the land from the compaction and recrystallization of snow, and it shows evidence of past or present flow. Today, *valley* or *alpine glaciers* are found in mountain areas where they usually follow valleys that were originally occupied by streams. *Ice sheets* exist on a much larger scale, covering most of Greenland and Antarctica.

⊙ Near the surface of a glacier, in the *zone of fracture*, ice is brittle. However, below about 50 meters, pressure is great, causing ice to *flow* like a *plastic material*. A second important mechanism of glacial movement consists of the entire ice mass *slipping* along the ground.

⊙ The average velocity of glacial movement is generally quite slow, but it varies considerably from one glacier to another. The advance of some glaciers is characterized by periods of extremely rapid movements called *surges*.

⊙ Glaciers form in areas where more snow falls in winter than melts during summer. Snow accumulation and ice formation occur in the *zone of accumulation*. Its outer limits are defined by the *snowline*. Beyond the snowline is the *zone of wastage*, where there is a net loss to the glacier. The *glacial budget* is the balance, or lack of balance, between accumulation at the upper end of the glacier, and loss, called *ablation*, at the lower end.

⊙ Glaciers erode land and acquire debris by *plucking* (lifting pieces of bedrock out of place) and *abrasion* (grinding and scraping of a rock surface). Mass-wasting processes also make significant contributions to the load of many alpine glaciers. Erosional features produced by valley glaciers include *glacial troughs, hanging valleys, pater noster lakes, fiords, cirques, arêtes, horns,* and *roches moutonnées.*

⊙ Any sediment of glacial origin is called *drift*. The two distinct types of glacial drift are (1) *till*, which is unsorted sediment deposited directly by the ice; and (2) *stratified drift*, which is relatively well-sorted sediment laid down by glacial meltwater.

⊙ The most widespread features created by glacial deposition are layers or ridges of till, called *moraines*. Associated with valley glaciers are *lateral moraines*, formed along the sides of the valley, and *medial moraines*, formed between two valley glaciers that have joined. *End moraines*, which mark the former position of the front of a glacier, and *ground moraine*, an undulating layer of till deposited as the ice front retreats, are common to both valley glaciers and ice sheets. An *outwash plain* is often associated with the end moraine of an ice sheet. A *valley train* may form when the glacier is confined to a valley. Other depositional features include *drumlins* (streamlined asymmetrical hills composed of till), *eskers* (sinuous ridges composed largely of sand and gravel deposited by streams flowing in tunnels beneath the ice near the terminus of a glacier), and *kames* (steep-sided hills composed of sand and gravel).

⊙ The *Ice Age*, which began 2 million to 3 million years ago, was a very complex period characterized by a number of advances and withdrawals of glacial ice. Most of the major glacial episodes occurred during a division of the geologic time scale called the *Pleistocene epoch*. Perhaps the most convincing evidence for the occurrence of several glacial advances during the Ice Age is the widespread existence of *multiple layers of drift* and an uninterrupted record of climate cycles preserved in *seafloor sediments*.

⊙ In addition to massive erosional and depositional work, other effects of Ice-Age glaciers include the *forced migration of organisms, changes in stream courses, formation of large proglacial lakes, adjustments of the crust* by rebounding after the removal of the immense load of ice, and *climate changes* caused by the existence of the glaciers themselves. In the sea, the most far-reaching effect of the Ice Age was the *worldwide change in sea level* that accompanied each advance and retreat of the ice sheets.

⊙ Any theory that attempts to explain the causes of glacial ages must answer two basic questions: (1) What causes the onset of glacial conditions? and (2) What caused the alternating glacial and interglacial stages that have been documented for the Pleistocene epoch? Two of the many hypotheses for the cause of glacial ages involve (1) plate tectonics and (2) variations in Earth's orbit. Other factors that are related to climate change during glacial ages include changes in atmospheric composition, variations in the amount of sunlight reflected by Earth's surface, and changes in ocean circulation.

Key Terms

ablation (p. 268)
abrasion (p. 270)
alpine glaciers (p. 263)
arêtes (p. 272)
calving (p. 267)
cirque (p. 272)
crevasses (p. 266)
drumlins (p. 277)
end moraine (p. 275)
eskers (p. 277)
fiords (p. 272)
firn (p. 265)

glacial budget (p. 268)
glacial drift (p. 274)
glacial erratics (p. 275)
glacial striations (p. 270)
glacial trough (p. 271)
glacier (p. 263)
ground moraine (p. 276)
hanging valleys (p. 272)
horns (p. 272)
ice caps (p. 264)
ice sheets (p. 263)
ice shelves (p. 263)

kames (p. 278)
kettles (p. 276)
lateral moraines (p. 275)
medial moraines (p. 275)
outlet glaciers (p. 264)
outwash plain (p. 276)
pater noster lakes (p. 272)
piedmont glaciers (p. 264)
Pleistocene epoch (p. 281)
plucking (p. 269)
pluvial lakes (p. 280)
roche moutonnée (p. 273)

rock flour (p. 270)
snowline (p. 265)
stratified drift (p. 274)
tarn (p. 272)
till (p. 274)
valley glaciers (p. 263)
valley train (p. 276)
zone of accumulation (p. 267)
zone of wastage (p. 267)

GIVE IT SOME THOUGHT

1 The accompanying diagram shows the results of a classic experiment used to determine how glacial ice moves in a mountain valley. The experiment occurred over an eight-year span. Refer to the diagram and answer the following:

   a. What was the average yearly rate at which ice in the center of the glacier advanced?

   b. About how fast was the center of the glacier advancing *per day*?

   c. Calculate the average rate at which ice along the sides of the glacier moved forward.

   d. Why was the rate at the center different than along the sides?

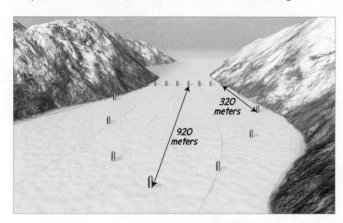

2 Studies have shown that during the Ice Age the margins of some ice sheets advanced southward from the Hudson Bay region at rates ranging from about 50 to 320 meters per year.

   a. Determine the maximum amount of time required for an ice sheet to move from the southern end of Hudson Bay to the south shore of present day Lake Erie, a distance of 1600 kilometers.

   b. Calculate the minimum number of years required for an ice sheet to move this distance.

3 If Earth were to experience another Ice Age, one hemisphere would have substantially more expansive ice sheets than the other. Would it be the Northern Hemisphere or the Southern Hemisphere? What is the reason for the large disparity?

4 While taking a break from a hike in the Northern Rockies with a fellow geology enthusiast, you notice that the boulder you are sitting on is part of a large deposit of sediment that consists of a jumbled mixture of many different sediment sizes. Since you are in an area that once had extensive valley glaciers, your colleague suggests that the deposit must be glacial till. Although you know this is certainly a likely possibility, you remind your companion that material deposited by landslides can also consist of an unsorted mix of many sizes. How might you and your friend determine whether this deposit is actually glacial till?

5 For each of the statements below, identify the type of glacier that is being described.

   a. A glacier that is often described as *continental*.

   b. A glacier that forms when one or more valley glaciers spreads out at the base of a steep mountain.

   c. Greenland is the only example of this type of glacier in the Northern Hemisphere.

   d. A glacier that may also be called an *alpine glacier*.

   e. This glacier is a stream of ice leading from the margin of an ice sheet through the mountains to the sea.

6 Glacial ice is classified as a metamorphic rock, yet glaciers are a basic part of the hydrologic cycle. Should glaciers be considered a part of the geosphere or do they belong to the hydrosphere? Explain.

7 If the budget of a valley glacier were balanced for an extended span, what feature would you expect to find at the terminus of the glacier? Now assume the glacier's budget changes so that ablation exceeds accumulation. How would the terminus of the glacier change? Describe the deposit you would expect to form under these conditions.

8 Assume you and a non-geologist friend are visiting Alaska's Hubbard Glacier shown in Figure 11.1. After studying the glacier for quite a long time, your friend asks, "Do these things really move?" How would you convince your companion that this glacier does indeed move using evidence that is clearly visible in this image?

Companion Website

12

Deserts and Wind

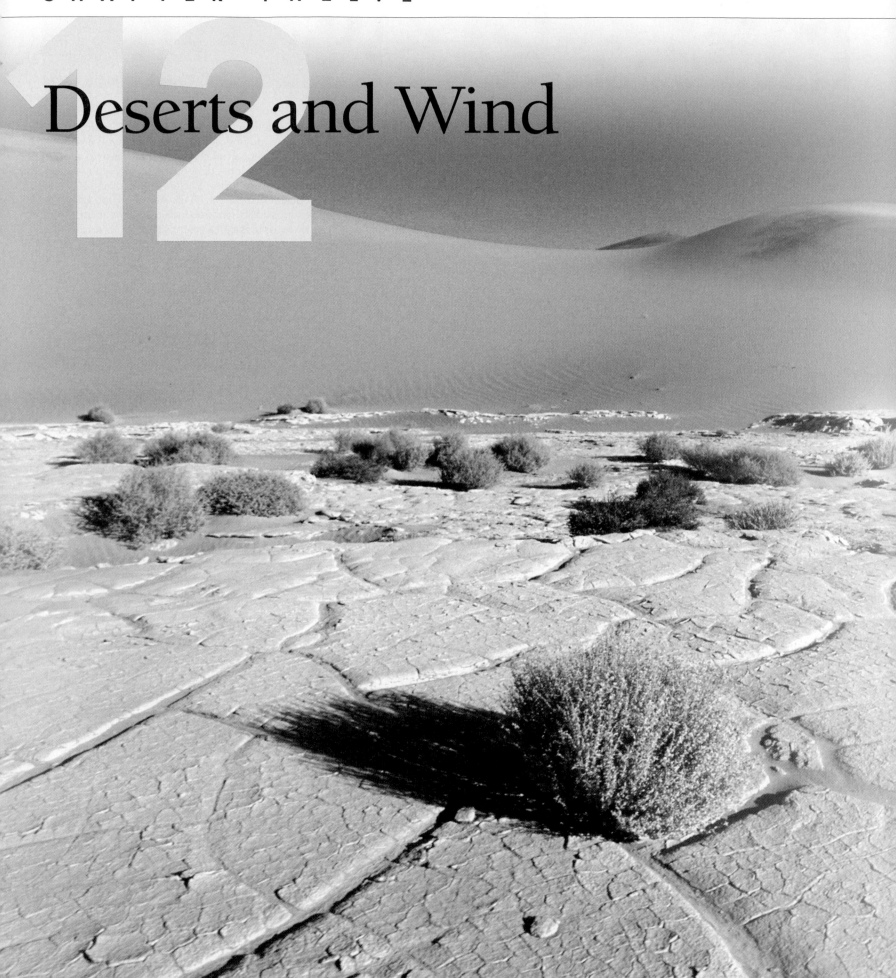

CLIMATE HAS A STRONG INFLUENCE ON THE NATURE AND INTENSITY

of Earth's external processes. This was clearly demonstrated in the preceding chapter on glaciers. Another excellent example of the strong link between climate and geology is seen when we examine the development of arid landscapes.

To assist you in learning the important concepts in this chapter, focus on the following questions:

- ⊙ What are the causes of deserts in the low and middle latitudes?
- ⊙ What are the roles of weathering, water, and wind in arid and semiarid climates?
- ⊙ How have many of the landscapes in the dry Basin and Range region of the United States evolved?
- ⊙ How does wind erode?
- ⊙ What are some depositional features produced by wind?

California's Death Valley is a desert in the rainshadow of the Sierra Nevada and exhibits many classic features associated with a mountainous desert landscape.
(Photo by Russ Bishop/Photolibrary)

FOCUS ON CONCEPTS

Distribution and Causes of Dry Lands

 Deserts and Wind

ESSENTIALS OF GEOLOGY **Distribution and Causes of Dry Lands**

The word *desert* literally means *deserted* or *unoccupied*. For many dry regions this is a very appropriate description, although where water is available in deserts, plants and animals thrive. Nevertheless, the world's dry regions are among the least familiar land areas on Earth outside the polar realm.

Desert landscapes frequently appear stark. Their profiles are not softened by a carpet of soil and abundant plant life. Instead, barren rocky outcrops with steep, angular slopes are common. At some places the rocks are tinted orange and red. At others they are gray and brown and streaked with black. For many visitors, desert scenery exhibits a striking beauty; to others the terrain seems bleak. No matter

which feeling is elicited, it is clear that deserts are very different from the more humid places where most people live.

As you will see, arid regions are not dominated by a single geologic process. Rather, the effects of tectonic forces, running water, and wind are all apparent. Because these processes combine in different ways from place to place, the appearance of desert landscapes varies a great deal as well (**FIGURE 12.1**).

We all recognize that deserts are dry places, but just what is meant by the term *dry*? That is, how much rain defines the boundary between humid and dry regions? Sometimes it is arbitrarily defined by a single rainfall figure—for example, 25 centimeters per year of precipitation. However, the concept of *dryness* is very relative; it refers to *any situation in which water deficiency exists*. Hence, climatologists define **dry climate** as one in which yearly precipitation is less than the potential loss of water by evaporation.

Dryness, then, is related not only to annual rainfall totals but is also a function of evaporation, which in turn closely depends upon temperature. As temperatures climb, potential evaporation also increases. Twenty-five centimeters of rain may support only a sparse vegetative cover in Nevada, whereas the same amount of precipitation falling in northern Scandinavia is sufficient to support forests.

Within these water-deficient regions, two climatic types are commonly recognized: **desert**, which is *arid*, and **steppe**, which is *semiarid*. The two share many features; their differences are primarily matters of degree. The steppe is a marginal and more humid variant of the desert and is a transition zone that surrounds the desert and separates it from bordering humid climates. The world map showing the distribution of desert and steppe regions reveals that dry lands are concentrated in the subtropics and in the mid-latitudes (**FIGURE 12.2**).

FIGURE 12.1 The appearance of desert landscapes varies a great deal from place to place. This scene is near the Al-'Ula oasis in northwestern Saudi Arabia. *(Photo by Aldo Pavan/DanitaDelimont.com)*

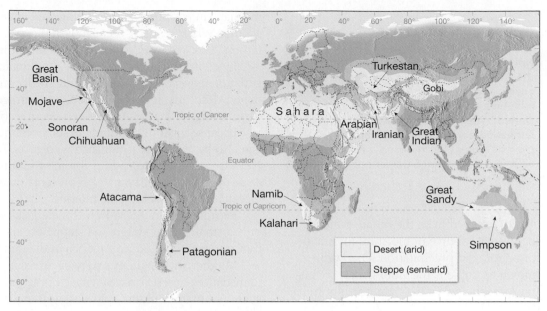

FIGURE 12.2 Arid and semiarid climates cover about 30 percent of Earth's land surface.

Low-Latitude Deserts

The heart of the low-latitude dry climates lies in the vicinities of the Tropics of Cancer and Capricorn. Figure 12.2 shows a virtually unbroken desert environment stretching for more than 9300 kilometers (nearly 5800 miles) from the Atlantic coast of North Africa to the dry lands of northwestern India. In addition to this single great expanse, the Northern Hemisphere contains another, much smaller area of subtropical desert and steppe in northern Mexico and the southwestern United States.

In the Southern Hemisphere, dry climates dominate Australia. Almost 40 percent of the continent is desert, and much of the remainder is steppe. In addition, arid and semiarid areas occur in southern Africa and make a limited appearance in coastal Chile and Peru.

What causes these bands of low-latitude desert? The answer is the global distribution of air pressure and winds. **FIGURE 12.3A**, an idealized diagram of Earth's general circulation, helps visualize the relationship. Heated air in the pressure belt known as the *equatorial low* rises to great heights (usually between 15 and 20 kilometers) and then spreads out. As the upper-level flow reaches 20° to 30° latitude, north or south, it sinks toward the surface. Air that rises through the atmosphere expands and cools, a process that leads to the development of clouds and precipitation. For this reason, the areas under the influence of the equatorial low are among the rainiest on Earth. Just the opposite is true for the regions in the vicinity of 30° north and south latitude, where high pressure predominates. Here, in the zones known as the *subtropical highs*, air is subsiding. When air sinks, it is compressed and warmed. Such conditions are just the opposite of what is needed to produce

clouds and precipitation. Consequently, these regions are known for their clear skies, sunshine, and ongoing drought (**FIGURE 12.3B**).

Middle-Latitude Deserts

Unlike their low-latitude counterparts, middle-latitude deserts and steppes are not controlled by the subsiding air masses associated with high pressure. Instead, these dry lands exist principally because they are sheltered in the deep interiors of large landmasses. They

FIGURE 12.3 A. Idealized diagram of Earth's general circulation. The deserts and steppes that are centered in the latitude belt between 20° and 30° north and south coincide with the subtropical high-pressure belts. Here, dry, subsiding air inhibits cloud formation and precipitation. By contrast, the pressure belt known as the equatorial low is associated with areas that are among the rainiest on Earth. **B.** In this view of Earth from space, North Africa's Sahara Desert, the adjacent Arabian Desert, and the Kalahari and Namib deserts in southern Africa are clearly visible as tan-colored, cloud-free zones. The band of clouds that extends across central Africa and the adjacent oceans coincides with the equatorial low-pressure belt. *(Photo courtesy of NASA)*

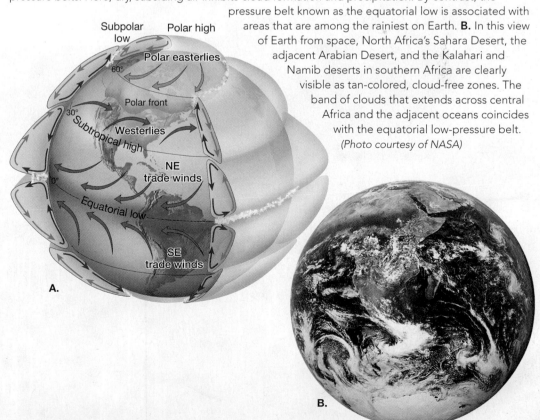

are far removed from the ocean, which is the ultimate source of moisture for cloud formation and precipitation. One well-known example is the Gobi Desert of central Asia, shown on the map north of India (Figure 12.2).

The presence of high mountains across the paths of prevailing winds further separates these areas from water-bearing maritime air masses. Also, the mountains force the air to lose much of its water. The mechanism is simple: As prevailing winds meet mountain barriers, the air is forced to ascend. When air rises, it expands and cools, a process that can produce clouds and precipitation. The windward sides of mountains, therefore, often have high precipitation.

By contrast, the leeward sides of mountains are usually much drier (**FIGURE 12.4**). This situation exists because air reaching the leeward side has lost much of its moisture, and if the air descends, it is compressed and warmed, making cloud formation even less likely. The dry region that results is often referred to as a **rainshadow desert**.

Because most mid-latitude deserts occupy sites on the leeward sides of mountains, they can also be classified as rainshadow deserts. In North America, the Coast Ranges, Sierra Nevada, and Cascades are the foremost mountain barriers to moisture from the Pacific. In Asia, the great Himalayan chain prevents the summertime monsoon flow of moist Indian Ocean air from reaching the interior.

Because the Southern Hemisphere lacks extensive land areas in the middle latitudes, only a small area of desert and steppe occurs in this latitude range, existing primarily near the southern tip of South America in the rainshadow of the towering Andes.

Middle-latitude deserts provide an example of how tectonic processes affect climate. Rainshadow deserts exist by virtue of the mountains produced when plates collide. Without such mountain-building episodes, wetter climates would prevail where many dry regions exist today.

CONCEPT CHECK 12.1

❶ How extensive are the desert and steppe regions of Earth?

❷ What is the primary cause of low-latitude deserts? Of middle-latitude deserts?

❸ In which hemisphere (Northern or Southern) are middle-latitude deserts more common? Explain.

Organ Pipe Cactus National Monument in Arizona's Sonoran Desert. *(Photo by Jeff Lapore/ Photo Researchers, Inc.)*

FIGURE 12.4 Many deserts in the middle latitudes are rainshadow deserts. As moving air meets a mountain barrier, it is forced to rise. Clouds and precipitation on the windward side often result. Air descending the leeward side is much drier. The mountains effectively cut the leeward side off from the sources of moisture, producing a rainshadow desert. The Great Basin desert is a rainshadow desert that covers nearly all of Nevada and portions of adjacent states. *(Photo on left by Dean Pennala/Shutterstock. Photo on right by Dennis Tasa)*

Geologic Processes in Arid Climates

Deserts and Wind

ESSENTIALS OF GEOLOGY **Common Misconceptions about Deserts**

The angular hills, the sheer canyon walls, and the desert surface of pebbles or sand contrast sharply with the rounded hills and curving slopes of more humid places. Indeed, to a visitor from a humid region, a desert landscape may seem to have been shaped by forces altogether different from those operating in well-watered areas. However, although the contrasts might be striking, they do not reflect different processes. They merely disclose the differ- ing effects of the same processes that oper- ate under contrasting climatic conditions.

Weathering

In humid regions, relatively well-developed soils support an almost continuous cover of vegetation. Here the slopes and rock edges are rounded, reflecting the strong influence of chemical weathering in a humid climate. By contrast, much of the weathered debris in deserts consists of unaltered rock and mineral fragments—the results of mechani- cal weathering processes. In dry lands, rock weathering of any type is greatly reduced because of the lack of moisture and the scarcity of organic acids from decaying plants. However, chemical weathering is not completely lacking in deserts. Over long spans of time, clays and thin soils do form, and many iron-bearing silicate miner- als oxidize, producing the rust-colored stain found tinting some desert landscapes.

The Role of Water

Permanent streams are normal in humid regions, but almost all desert stream beds are dry most of the time (**FIGURE 12.5A**). Deserts have **ephemeral streams**, which means that they carry water only in response to specific episodes of rainfall. A typical ephemeral stream might flow only a few days or perhaps just a few hours during the year. In some years, the channel might carry no water at all.

This fact is obvious even to the casual traveler who notices numerous bridges with no streams beneath them or numerous dips in the road where dry channels cross. However, when the rare heavy showers do come, so much

A.

B.

FIGURE 12.5 **A.** Most of the time, desert stream channels are dry. **B.** An ephemeral stream shortly after a heavy shower. Although such floods are short-lived, large amounts of erosion occur. *(Photos by E. J. Tarbuck)*

rain falls in such a short time that all of it cannot soak in (**FIGURE 12.6**). Because desert vegetative cover is sparse, runoff is largely unhindered and consequently rapid, often creating flash floods along valley floors (**FIGURE 12.5B**). These floods are quite unlike floods in humid regions. A flood on a river such as the Mississippi may take several days to reach its crest and then subside, but desert floods arrive suddenly and subside quickly. Because much of the surface material in a desert is not anchored by vegetation, the amount of erosional work that occurs during a single, short-lived rain event is impressive.

In the dry western United States, different names are used for ephemeral streams, including *wash* and *arroyo*. In other parts of the world, a dry stream may be a *wadi* (Arabia and North Africa), a *donga* (South Africa), or a *nullah* (India). The satellite images in **FIGURE 12.7** show a wadi in the Sahara Desert.

Humid regions are notable for their integrated drainage systems, but in arid regions streams usually lack an extensive system of tributaries. In fact, a basic characteristic of desert streams is that they are small and die out before reaching the sea. Because the water table is usually far below the surface, few desert streams can draw upon it as streams do in humid regions (see Figure 10.7C, p. 244). Without a steady supply of water, the combination of evaporation and infiltration soon depletes the stream.

The few permanent streams that do cross arid regions, such as the Colorado and Nile rivers, originate *outside* the desert, often in well-watered mountains. Here the water supply must be great to compensate for the

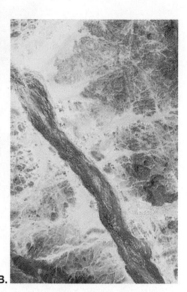

A.

B.

FIGURE 12.7 These two satellite images show how rain transformed a wadi in the North African country of Niger. **A.** The wadi in its usual dry state. **B.** Following a rainy period, freshly sprouted vegetation turns the wadi green. *(NASA)*

FIGURE 12.6 There are often many weeks, months, or occasionally even years separating periods of rain in the desert. When rains do occur, they are often heavy and of relatively short duration. Because the rainfall intensity is high, not all of the water can soak in, and rapid runoff and erosion result. *(Photo by Ed Darack/Getty Images)*

DID YOU KNOW?
Not all deserts are hot. Cold temperatures are experienced in middle-latitude deserts. For example, at Ulan Bator in Mongolia's Gobi Desert, the average *high* temperature on January days is only −19 °C (−2 °F)!

losses occurring as the stream crosses the desert. For example, after the Nile leaves its headwaters in the lakes and mountains of central Africa, it traverses almost 3000 kilometers (1900 miles) of the Sahara *without a single tributary*. By contrast, in humid regions the discharge of a river increases as it flows downstream because tributaries and groundwater contribute additional water along the way.

It should be emphasized that *running water, although infrequent, nevertheless does most of the erosional work in deserts* (**FIGURE 12.8**). This is contrary to a common belief that wind is the most important erosional agent sculpting desert landscapes. Although wind erosion is indeed more significant in dry areas than elsewhere, most desert landforms are carved by running water. As you will see shortly, the main role of wind is in the transportation and deposition of sediment, which creates and shapes the ridges and mounds we call *dunes*.

CONCEPT CHECK 12.2

❶ How does the rate of rock weathering in dry climates compare to the rate in humid regions?

❷ When a permanent stream such as the Nile River crosses a desert, does discharge increase or decrease? How does this compare to a river in a humid area?

❸ What is the most important erosional agent in deserts?

Basin and Range: The Evolution of a Mountainous Desert Landscape

Deserts and Winds

ESSENTIALS OF GEOLOGY **Reviewing Landforms and Landscapes**

Because arid regions typically lack permanent streams, they are characterized as having **interior drainage**. This means that they have a discontinuous pattern of intermittent streams that do not flow out of the desert to the ocean. In the United States, the dry Basin and Range region provides an excellent example. The region includes southern Oregon, all of Nevada, western Utah, southeastern California, southern Arizona, and southern New Mexico. The name Basin and Range is an apt description for this almost 800,000-square-kilometer (more than 300,000-square-mile) area, as it is characterized by more than 200 relatively small mountain ranges that rise 900 to 1500 meters (3000 to 5000 feet) above the basins that separate them.

In this region, as in others like it around the world, most erosion occurs without reference to the ocean (ultimate base level), because the interior drainage never reaches the sea. Even where permanent streams flow to the ocean, few tributaries exist, and thus only a narrow strip of land adjacent to the stream has sea level as its ultimate level of land reduction.

FIGURE 12.8 Desert rains are infrequent, but when they occur, erosion can be significant. The dry channels of ephemeral streams are conspicuous in this aerial view of a parched desert surface. *(Photo by Michael Collier)*

Arizona's Monument Valley.
(Photo by Fuse/Jupiter Images)

aridity

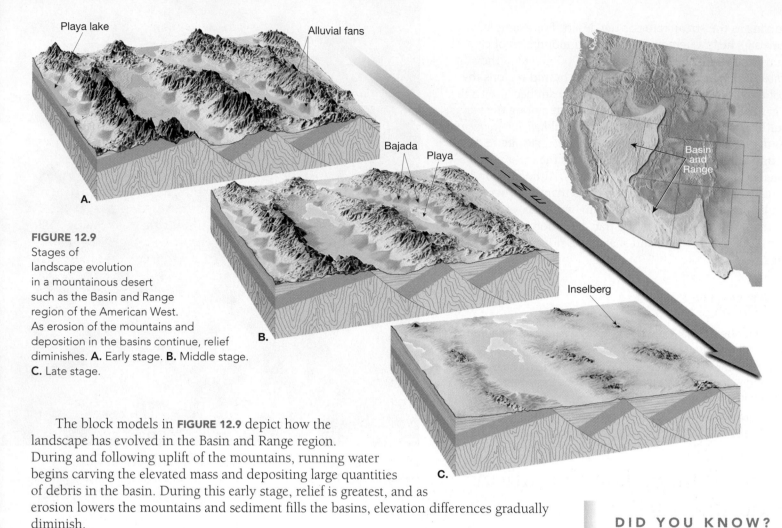

FIGURE 12.9
Stages of landscape evolution in a mountainous desert such as the Basin and Range region of the American West. As erosion of the mountains and deposition in the basins continue, relief diminishes. **A.** Early stage. **B.** Middle stage. **C.** Late stage.

The block models in **FIGURE 12.9** depict how the landscape has evolved in the Basin and Range region. During and following uplift of the mountains, running water begins carving the elevated mass and depositing large quantities of debris in the basin. During this early stage, relief is greatest, and as erosion lowers the mountains and sediment fills the basins, elevation differences gradually diminish.

When the occasional torrents of water produced by sporadic rains move down the mountain canyons, they are heavily loaded with sediment. Emerging from the confines of the canyon, the runoff spreads over the gentler slopes at the base of the mountains and quickly loses velocity. Consequently, most of its load is dumped within a short distance. The result is a cone of debris at the mouth of a canyon known as an **alluvial fan**. Because the coarsest material is dropped first, the head of the fan is steepest, having a slope of perhaps 10 to 15 degrees. Moving down the fan, the size of the sediment and the steepness of the slope decrease and merge imperceptibly with the basin floor. An examination of the fan's surface usually reveals a braided channel pattern because of the water shifting its course as successive channels became choked with sediment. Over the years a fan enlarges, eventually coalescing with fans from adjacent canyons to produce an apron of sediment called a **bajada** along the mountain front.

On the rare occasions of abundant rainfall, streams may flow across the bajada to the center of the basin, converting the basin floor into a shallow **playa lake**. Playa lakes are temporary features that last only a few days or at best a few weeks before evaporation and infiltration remove the water. The dry, flat lake bed that remains is called a **playa**. Playas are typically composed of fine silts and clays and are occasionally encrusted with salts precipitated during evaporation (see Figure 6.14, p. 160). These precipitated salts may be unusual. A case in point is the sodium borate (better known as borax) mined from ancient playa lake deposits in Death Valley, California.

With the ongoing erosion of the mountain mass and the accompanying sedimentation, the local relief continues to diminish. Eventually, nearly the entire mountain mass is gone. Thus, by the late stages of erosion, the mountain areas are reduced to a few large bedrock knobs projecting above the surrounding sediment-filled basin. These isolated erosional

DID YOU KNOW?
Not all low-latitude deserts are hot, sunny places with low humidity and cloudless skies. In West Coast subtropical deserts, such as the Atacama and Namib, cold ocean currents are responsible for cool temperatures, frequent fogs, and gloomy periods of dense (but rainless) low clouds.

remnants on a late-stage desert landscape are called **inselbergs**, a German word meaning "island mountains."

Each of the stages of landscape evolution in an arid climate depicted in Figure 12.9 can be observed in the Basin and Range region. Recently uplifted mountains in an early stage of erosion are found in southern Oregon and northern Nevada. Death Valley, California, and southern Nevada fit into the more advanced middle stage, whereas the late stage, with its inselbergs, can be seen in southern Arizona.

FIGURE 12.10 includes a satellite image (left) and an aerial view (right) of a portion of Death Valley. Many of the features that were just described are visible. The satellite image is from February 2005 and shows the area shortly after a rare heavy rain. As has occurred many times over thousands of years, a wide, shallow playa lake formed in the lowest spot. By May 2005, only three months after the storm, the valley floor had returned to being a dry, salt-encrusted playa.

CONCEPT CHECK 12.3

❶ Describe the features and characteristics associated with each stage in the evolution of a mountainous desert. Where in the United States can each stage be observed?

Transportation of Sediment by Wind

Moving air, like moving water, is turbulent and able to pick up loose debris and transport it to other locations. Just as in a stream, the velocity of wind increases with height above the surface. Also like a stream, wind transports fine particles in suspension while heavier ones are carried as bed load. However, the transport of sediment by wind differs from that by running water in two significant ways. First, wind's lower density compared to water renders it less capable of picking up and transporting coarse materials. Second, because wind is not confined to channels, it can spread sediment over large areas, as well as high into the atmosphere.

Bed Load

The **bed load** carried by wind consists of sand grains. Observations in the field and experiments using wind tunnels indicate

that windblown sand moves by skipping and bouncing along the surface—a process termed **saltation**. The term is not a reference to salt, but instead derives from the Latin word meaning "to jump."

The movement of sand grains begins when wind reaches a velocity sufficient to overcome the inertia of the resting particles. At first the sand rolls along the surface. When a moving sand grain strikes another grain, one or both of them may jump into the air. Once in the air, the grains are carried forward by the wind until gravity pulls them back toward the surface. When the sand hits the surface, it either bounces back into the air or dislodges other grains, which then jump upward. In this manner, a chain reaction is established, filling the air near the ground with saltating sand grains in a short period of time (**FIGURE 12.11**).

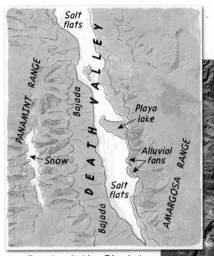

Geologist's Sketch

Geologist's Sketch

Death Valley

FIGURE 12.10 The image on the left is a satellite view of a portion of Death Valley, California, a classic Basin and Range landscape. Shortly before this image was taken in February 2005, heavy rains led to the formation of a playa lake— the pool of greenish water on the basin floor. By May 2005, the lake reverted to a salt-covered playa. *(NASA)* The aerial view on the right provides a close-up of one of Death Valley's many alluvial fans. *(Photo by Michael Collier)*

7808

Bouncing sand grains never travel far from the surface. Even when winds are very strong, the height of the saltating sand seldom exceeds a meter and usually is no greater than a half meter.

Suspended Load

Unlike sand, finer dust particles can be swept high into the atmosphere by the wind. Because dust is often composed of rather flat particles that have large surface areas compared to their weight, it is relatively easy for turbulent air to counterbalance the pull of gravity and keep these fine particles airborne for hours or even days. Although both silt and clay can be carried in suspension, silt commonly makes up the bulk of the **suspended load** because the reduced level of chemical weathering in deserts produces only small amounts of clay.

Fine particles are easily carried by the wind, but they are not so easily picked up to begin with. The reason is that the wind velocity is practically zero within a very thin layer close to the ground. Thus, the wind cannot lift the sediment by itself. Instead, the dust must be ejected or spattered into the moving air currents by bouncing sand grains or other distur-

bances. This idea is illustrated nicely by a dry unpaved country road on a windy day. Left undisturbed, little dust is raised by the wind. However, as a car or truck moves over the road, the layer of silt is kicked up, creating a thick cloud of dust.

Although the suspended load is usually deposited relatively near its source, high winds are capable of carrying large quantities of dust great distances. The images in FIGURE 12.12 show examples on two differ-

ent scales. In the 1930s, silt that was picked up in Kansas was transported to New England and beyond into the North Atlantic. Similarly, dust blown from the Sahara has been traced as far as the West Indies.

FIGURE 12.11 The bed load carried by wind consists of sand grains that move by bouncing along the surface. Sand never travels far from the surface, even when winds are very strong. *(Photo by Bernd Zoller/Photolibrary)*

CONCEPT CHECK 12.4

❶ Contrast how wind transports sand with how wind transports dust.

FIGURE 12.12 Two examples of suspended load. **A.** Dust blackens the sky on May 21, 1937, near Elkhart, Kansas. It was because of storms like this that portions of the Great Plains were called the Dust Bowl in the 1930s. *(Photo reproduced from the collection of the Library of Congress)* **B.** This satellite image shows thick plumes of dust from the Sahara Desert blowing across the Red Sea on June 30, 2009. Such dust storms are common in arid North Africa. In fact, this region is the largest dust source in the world. Satellites are excellent tools for studying the transport of dust on a global scale. They show us that dust storms can cover huge areas and that dust can be transported great distances. *(NASA)*

A.

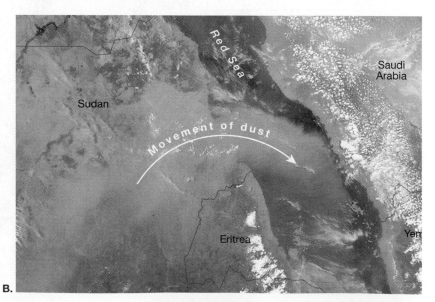

B.

15,241,216   3,375

DID YOU KNOW?
In some regions deserts are expanding. The problem is called *desertification*, a term that refers to the alteration of land to desertlike conditions as the result of human activities. It commonly takes place on the margins of deserts and results primarily from inappropriate land use. It is triggered when the modest natural vegetation in marginal areas is removed by plowing or grazing. When drought occurs, as it inevitably does, and the vegetative cover has been destroyed beyond the minimum to hold the soil against erosion, the destruction becomes irreversible. Desertification is occurring in many places but is particularly serious in the region south of the Sahara Desert known as the Sahel.

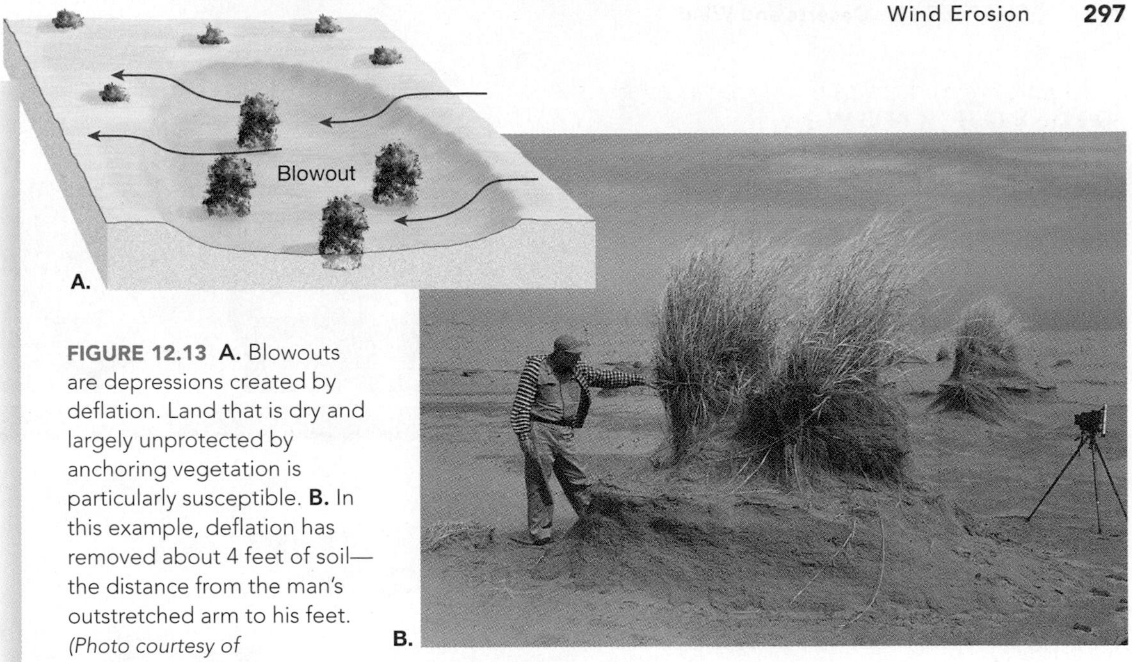

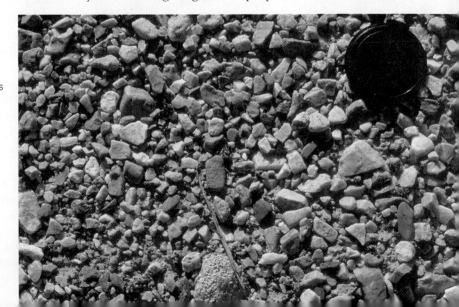

FIGURE 12.13 A. Blowouts are depressions created by deflation. Land that is dry and largely unprotected by anchoring vegetation is particularly susceptible. **B.** In this example, deflation has removed about 4 feet of soil—the distance from the man's outstretched arm to his feet. *(Photo courtesy of U.S.D.A./Natural Resources Conservation Service)*

Wind Erosion

Compared to running water and glaciers, wind is a relatively insignificant erosional agent. Recall that even in deserts, most erosion is performed by intermittent running water, not by the wind. Wind erosion is more effective in arid lands than in humid areas because in humid regions moisture binds particles together and vegetation anchors the soil. For wind to be effective, dryness and scanty vegetation are important prerequisites.

When such circumstances exist, wind may pick up, transport, and deposit great quantities of fine sediment. During the 1930s, parts of the Great Plains experienced great dust storms. The plowing under of the natural vegetative cover for farming, followed by severe drought, made the land ripe for wind erosion and led to the areas being labeled the Dust Bowl. There is more about this environmental disaster in Chapter 5.

Deflation and Blowouts

One way that wind erodes is by **deflation**, the lifting and removal of loose material. Although the effects of deflation are sometimes difficult to notice because the entire surface is being lowered at the same time, they can be significant. In portions of the 1930s Dust Bowl, vast areas of land were lowered by as much as a meter in only a few years.

The most noticeable results of deflation in some places are shallow depressions called **blowouts** (**FIGURE 12.13**). In the Great Plains region, from Texas in the south to Montana in the north, thousands of blowouts are visible on the landscape. They range from small dimples less than a meter deep and 3 meters wide to depressions that approach 50 meters in depth and several kilometers across. The factor that controls the depths of these basins (that is, acts as base level) is the local water table. When blowouts are lowered to the water table, damp ground and vegetation prevent further deflation.

DID YOU KNOW?
Deserts don't necessarily consist of mile after mile of drifting sand dunes. Surprisingly, sand accumulations represent only a small percentage of the total desert area. In the Sahara, dunes cover only one tenth of its area. The sandiest of all deserts, the Arabian, is one third sand covered.

Desert Pavement

In portions of many deserts, the surface consists of a closely packed layer of coarse particles. This veneer of pebbles and cobbles, called **desert pavement**, is only one or two stones thick (**FIGURE 12.14**). Beneath is a layer containing a significant proportion of silt and sand.

FIGURE 12.14 Desert pavement consists of a closely packed veneer of pebbles and cobbles that is only one or two stones thick. Beneath the pavement is material containing a significant proportion of finer particles. If left undisturbed, desert pavement will protect the surface from deflation. *(Photo by Bobbé Christopherson)*

When desert pavement is present, it is an important control on wind erosion because pavement stones are too large for deflation to remove. When this armor is disturbed, wind can easily erode the exposed fine silt.

For many years, the most common hypothesis for the formation of desert pavement was that it develops when wind removes sand and silt within poorly sorted surface deposits. As **FIGURE 12.15A** illustrates, the concentration of larger particles

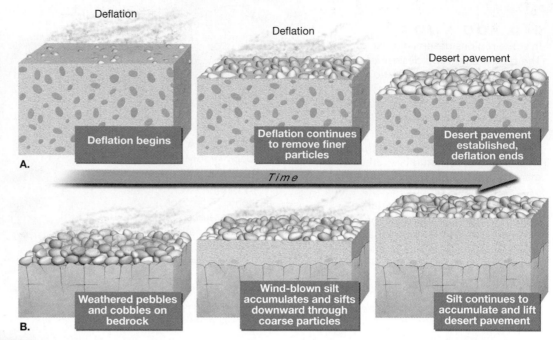

FIGURE 12.15 Formation of desert pavement. **A.** This model portrays an area with poorly sorted surface deposits. Coarse particles gradually become concentrated into a tightly packed layer as deflation lowers the surface by removing sand and silt. Here desert pavement is the result of wind erosion. **B.** This model shows the formation of desert pavement on a surface initially covered with coarse pebbles and cobbles. Windblown dust accumulates at the surface and gradually sifts downward through spaces between coarse particles. Infiltrating rainwater aids the process. This depositional process raises the surface and produces a layer of coarse pebbles and cobbles underlain by a substantial layer of fine sediment.

at the surface gradually increases as the finer particles are blown away. Eventually the surface is completely covered with pebbles and cobbles too large to be moved by the wind.

Studies have shown that the process depicted in Figure 12.15A is not an adequate explanation for all environments in which desert pavement exists. For example, in many places, desert pavement is underlain by a relatively thick layer of silt that contains few if any pebbles and cobbles. In such a setting, deflation of fine sediment could not leave behind a layer of coarse particles. Studies also showed that in some areas the pebbles and cobbles composing desert pavement have all been exposed at the surface for about the same length of time. This would not be the case for the process shown in Figure 12.15A. Here, the coarse particles that make up the pavement reach the surface over an extended time span as deflation gradually removes the fine material.

As a result, an alternate explanation for desert pavement was formulated (**FIGURE 12.15B**). This hypothesis suggests that pavement develops on a surface that initially consists of coarse particles. Over time, protruding cobbles trap fine, windblown grains that settle and sift downward through the spaces between the larger surface stones. The process is aided by infiltrating rainwater. In this model, the cobbles composing the pavement were never buried. Moreover, it successfully explains the lack of coarse particles beneath the desert pavement.

Ventifacts and Yardangs

Like glaciers and streams, wind also erodes by **abrasion.** In dry regions as well as along some beaches, windblown sand cuts and polishes exposed rock surfaces.

Great Sand Dunes National Park and Preserve *(Photo by K. D. McGraw-Rainbow/Science Faction/CORBIS)*

sand dunes

A.

B.

FIGURE 12.16 A. Ventifacts are rocks that are polished and shaped by sandblasting. *(Photo by Richard M. Busch)* **B.** Yardangs are usually small, wind-sculpted landforms that are aligned parallel with the wind. *(Photo by Peter M. Wilson/CORBIS)*

Abrasion sometimes creates interestingly shaped stones called **ventifacts** (**FIGURE 12.16A**). The side of the stone exposed to the prevailing wind is abraded, leaving it polished, pitted, and with sharp edges. If the wind is not consistently from one direction, or if the pebble becomes reoriented, it may have several faceted surfaces.

Unfortunately, abrasion is often given credit for accomplishments beyond its capabilities. Such features as balanced rocks that stand high atop narrow pedestals, and intricate detailing on tall pinnacles, are not the results of abrasion. Sand seldom travels more than a meter above the surface, so the wind's sandblasting effect is obviously limited in vertical extent.

In addition to ventifacts, wind erosion is responsible for creating much larger features, called yardangs (from the Turkistani word *yar* meaning "steep bank"). A **yardang** is a streamlined, wind-sculpted landform that is oriented parallel to the prevailing wind (**FIGURE 12.16B**). Individual yardangs are generally small features that stand less than 5 meters (16 feet) high and no more than about 10 meters (32 feet) long. Because the sandblasting effect of wind is greatest near the ground, these abraded bedrock remnants are usually narrower at their base. Sometimes yardangs are large features. Peru's Ica Valley contains yardangs that approach 100 meters (330 feet) in height and several kilometers in length. Some in the desert of Iran reach 150 meters (nearly 500 feet) in height.

CONCEPT CHECK 12.5

❶ Why is wind erosion relatively more important in arid regions than in humid areas?

❷ What factor limits the depths of blowouts?

❸ Briefly describe two hypotheses used to explain the formation of desert pavement.

Wind Deposits

Deserts and Wind
ESSENTIALS OF GEOLOGY **Reviewing Landforms and Landscapes**

Although wind is relatively unimportant in producing *erosional* landforms, significant *depositional* landforms are created by the wind in some regions. Accumulations of windblown sediment are particularly conspicuous in the world's dry lands and along many sandy coasts. Wind deposits are of two distinctive types: (1) mounds and ridges of sand from the wind's bed load, which we call *dunes*, and (2) extensive blankets of silt, called *loess*, that once were carried in suspension.

Sand Deposits

As is the case with running water, wind drops its load of sediment when its velocity falls and the energy available for transport diminishes. Thus, sand begins to accumulate wherever an obstruction across the path of the wind slows its movement. Unlike many deposits of silt, which form

blanketlike layers over large areas, winds commonly deposit sand in mounds or ridges called **dunes** (**FIGURE 12.17**).

As moving air encounters an object, such as a clump of vegetation or a rock, the wind sweeps around and over it, leaving a shadow of slower moving air behind the obstacle as well as a smaller zone of quieter air just in front of the obstacle. Some of the saltating sand grains moving with the wind come to rest in these wind shadows. As the accumulation of sand continues, it becomes a more imposing barrier to the wind and thus a more efficient trap for even more sand. If there is a sufficient supply of sand and the wind blows steadily for a long enough time, the mound of sand grows into a dune.

Many dunes have an asymmetrical profile with the leeward (sheltered) slope being steep and the windward slope being more gently inclined. The dunes in Figure 12.17 are a good example. Sand moves up the gentle slope on the windward side by saltation. Just beyond the crest of the dune, where the wind velocity is reduced, the sand accumulates. As more sand collects, the slope steepens, and eventually some of it slides or slumps under the pull of gravity. In this way, the leeward slope of the dune, called the **slip face**, maintains an angle of about 34 degrees, the angle of repose for loose dry sand (Figure 12.17B). (Recall from Chapter 8 that the angle of repose is the steepest angle at which loose material remains stable.) Continued sand accumulation, coupled with periodic slides down the slip face, results in the slow migration of the dune in the direction of air movement.

A.

B.

FIGURE 12.17 A. Dunes composed of gypsum sand at White Sands National Monument in southeastern New Mexico. These dunes are gradually migrating from the background (top) toward the foreground (bottom) of the photo. Strong winds move sand up the more gentle windward slopes. As sand accumulates near the dune crest, the slope steepens and some of the sand slides down the steeper *slip face*. **B.** Sand sliding down the steep slip face of a dune in White Sands National Monument, New Mexico. *(Photos by Michael Collier)*

DID YOU KNOW?
Migrating dunes can lead to costly problems. For example, to keep a portion of Highway 95 near Winnemucca, Nevada, open to traffic, sand must be removed about three times a year. Each time between 1500 and 4000 m^3 of sand are removed. Attempts at stabilizing the dunes with vegetation have failed.

As sand is deposited on the slip face, layers form that are inclined in the direction the wind is blowing. These sloping layers are called **cross beds**. When the dunes are eventually buried under other layers of sediment and become part of the sedimentary rock record, their asymmetrical shape is destroyed, but the cross beds remain as testimony to their origin. Nowhere is cross-bedding more prominent than in the sandstone walls of Zion Canyon in southern Utah (**FIGURE 12.18**).

Types of Sand Dunes

Dunes are not just random heaps of windblown sediment. Rather, they are accumulations that usually assume surprisingly consistent patterns. A broad assortment of dune forms exists, generally simplified to a few major types for discussion. Of course, gradations exist among different forms as well as irregularly shaped dunes that do not fit easily into any category. Several factors influence the

form and size that dunes ultimately assume. These include wind direction and velocity, availability of sand, and the amount of vegetation present. Six basic dune types are shown in **FIGURE 12.19**, with arrows indicating wind directions.

BARCHAN DUNES. Solitary sand dunes shaped like crescents with tips pointing downwind are called **barchan dunes** (Figure 12.19A). These dunes form where supplies of sand are limited and the surface is relatively flat, hard, and lacking vegetation. They migrate slowly with the wind at a rate of up to 15 meters per year. Their size is usually modest, with the largest barchans reaching heights of about 30 meters while the maximum spread between their horns approaches 300 meters. When the wind direction is nearly constant, the crescent form of these dunes is nearly symmetrical. However, when the wind direction is not perfectly fixed, one tip becomes larger than the other.

TRANSVERSE DUNES. In regions where the prevailing winds are steady, sand is plentiful, and vegetation is sparse or absent, the dunes form a series of long ridges that are separated by troughs and oriented at right angles to the prevailing wind. Because of this orientation, they are termed **transverse dunes** (Figure 12.19B). Typically, many coastal dunes are of this type. In addition, transverse dunes are common in many arid regions where the extensive surface of wavy sand is sometimes called a *sand sea*. In some parts of the Sahara and Arabian deserts, transverse dunes reach heights of 200 meters, are 1 to 3 kilometers across, and can extend for distances of 100 kilometers or more.

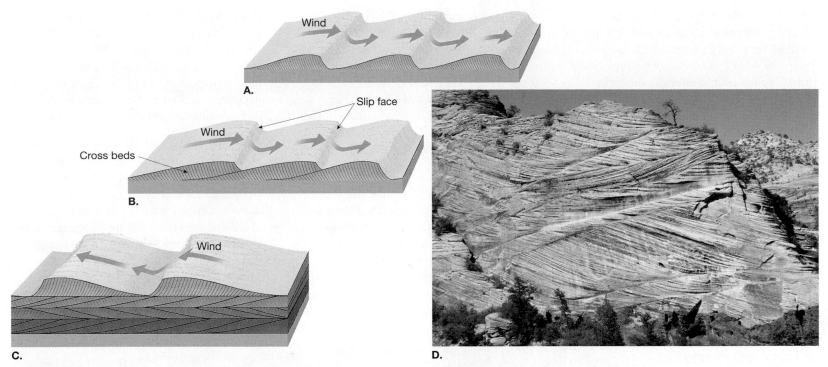

FIGURE 12.18 As parts **A.** and **B.** illustrate, dunes commonly have an asymmetrical shape. The steeper leeward side is called the slip face. Sand grains deposited on the slip face at the angle of repose create the cross-bedding of the dunes. **C.** A complex pattern develops in response to changes in wind direction. Also notice that when dunes are buried and become part of the sedimentary record, the cross-bedded structure is preserved. **D.** Cross beds are an obvious characteristic of the Navajo Sandstone in Zion National Park, Utah. *(Photo by Dennis Tasa)*

FIGURE 12.19 Sand dune types. **A.** Barchan dunes. **B.** Transverse dunes. **C.** Barchanoid dunes. **D.** Longitudinal dunes. **E.** Parabolic dunes. **F.** Star dunes.

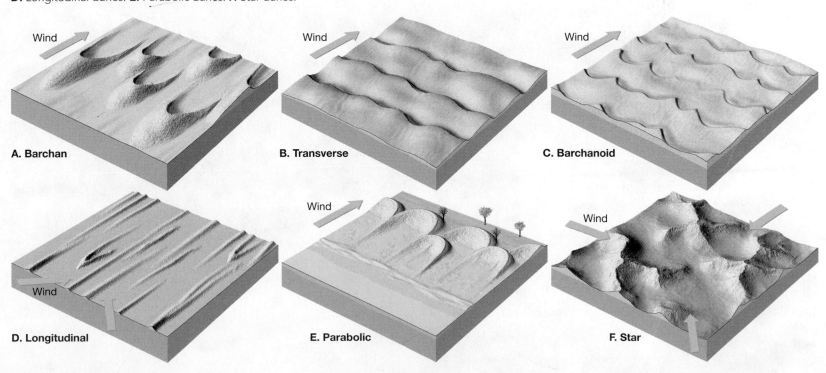

BARCHANOID DUNES. There is a relatively common dune form that is intermediate between isolated barchans and extensive waves of transverse dunes. Such dunes, called **barchanoid dunes**, form scalloped rows of sand oriented at right angles to the wind (Figure 12.19C). The rows resemble a series of barchans that have been positioned side by side. Visitors exploring the gypsum dunes at White Sands National Monument, New Mexico, will recognize this form (Figure 12.17A).

LONGITUDINAL DUNES. Longitudinal dunes are long ridges of sand that form more or less parallel to the prevailing wind where sand supplies are moderate (Figure 12.19D). Apparently the prevailing wind direction varies somewhat but remains in the same quadrant of the compass. Although the smaller types are only 3 or 4 meters high and several dozens of meters long, in some large deserts longitudinal dunes can reach great size. For example, in portions of North Africa, Arabia, and central Australia, these dunes can approach a height of 100 meters and extend for distances of more than 100 kilometers (62 miles).

PARABOLIC DUNES. Unlike the other dunes that have been described thus far, **parabolic dunes** form where vegetation partially covers the sand. The shape of these dunes resembles the shape of barchans except that their tips point into the wind rather than downwind (Figure 12.19E). Parabolic dunes often form along coasts where there are strong onshore winds and abundant sand. If the sand's sparse vegetative cover is disturbed at some spot, deflation creates a blowout. Sand is then transported out of the depression and deposited as a curved rim that grows higher as deflation enlarges the blowout.

STAR DUNES. Confined largely to parts of the Sahara and Arabian deserts, **star dunes** are isolated hills of sand that exhibit a complex form (Figure 12.19F). Their name is derived from the fact that the bases of these dunes resemble multipointed stars. Usually three or four sharp-crested ridges diverge from a central high point that in some cases may approach a height of 90 meters. As their form suggests, star dunes develop where wind directions are variable.

DID YOU KNOW?
The highest dunes in the world are located along the southwest coast of Africa in the Namib Desert. In places, these huge dunes reach heights of 300 to 350 meters (1000 to 1167 feet).

Dunes in Egypt encroaching on an irrigated field. *(Photo by Georg Gerster/Photo Researchers, Inc.)*

Loess (Silt) Deposits

In some parts of the world the surface topography is mantled with deposits of windblown silt, called **loess**. Over thousands of years dust storms deposited this material. When loess is breached by streams or road cuts, it tends to maintain vertical cliffs and lacks any visible layers, as you can see in **FIGURE 12.20**.

The distribution of loess worldwide indicates that there are two principal sources for this sediment: deserts and glacial deposits. The thickest and most extensive deposits of loess on Earth occur in western and northern China. They were blown there from the extensive desert basins of Central Asia. Accumulations of 30 meters are common, and thicknesses of more than 100 meters have been measured. It is this fine, buff colored sediment that gives the Yellow River (Hwang Ho) its name.

In the United States, deposits of loess are significant in many areas, including South Dakota, Nebraska, Iowa, Missouri, and Illinois, as well as portions of the Columbia Plateau in the Pacific Northwest. The correlation between the distribution of loess and important farming regions in the Midwest and eastern Washington State is not a coincidence, because soils derived from this wind-deposited sediment are among the most fertile in the world.

Unlike the deposits in China, which originated in deserts, the loess in the United States and Europe is an indirect product of glaciation. Its source is deposits of stratified drift. During the retreat of the ice sheets, many river valleys were choked with sediment deposited by meltwater. Strong westerly winds sweeping across the barren floodplains picked up the finer sediment and dropped it as a blanket on the eastern sides of the valleys. Such an origin is confirmed by the fact that loess deposits are thickest and coarsest on the lee side of such major glacial drainage outlets as the Mississippi and Illinois rivers and rapidly thin with increasing distance from the valleys.

A.

B.

C.

FIGURE 12.20 **A.** This vertical loess bluff near the Mississippi River in southern Illinois is about 3 meters high. *(Photo by James E. Patterson)* **B.** In parts of China, loess has sufficient structural strength to permit the excavation of dwellings. *(Photo by Christopher Liu/ChinaStock Photo Library)* **C.** This satellite image from March 13, 2003, shows streamers of windblown dust moving southward into the Gulf of Alaska. It illustrates a process similar to the one that created many loess deposits in the American Midwest during the Ice Age. Fine silt is produced by the grinding action of glaciers, then transported beyond the margin of the ice by running water and deposited. Later, the fine silt is picked up by strong winds and deposited as loess. *(NASA)*

CONCEPT CHECK 12.6

❶ How do sand dunes migrate?

❷ List and briefly distinguish among basic dune types.

❸ How is loess different from sand? How are some loess deposits related to glaciers?

CHAPTER TWELVE
Deserts and Wind in Review

⊙ The *concept of dryness is relative;* it refers to any situation in which a water deficiency exists. Dry regions encompass about 30 percent of Earth's land surface. Two climatic types are commonly recognized: *desert,* which is arid, and *steppe* (a marginal and more humid variant of desert), which is semiarid. *Low-latitude deserts* coincide with the zones of subtropical highs in lower latitudes. In contrast, *middle-latitude deserts* exist principally because of their positions in the deep interiors of large landmasses far removed from the ocean.

⊙ The same geologic processes that operate in humid regions also operate in deserts, but under contrasting climatic conditions. In dry lands *rock weathering of any type is greatly reduced* because of the lack of moisture and the scarcity of organic acids from decaying plants. Much of the weathered debris in deserts is the result of *mechanical weathering.* Practically all desert streams are dry most of the time and are said to be *ephemeral.* Stream courses in deserts are seldom well integrated and lack an extensive system of tributaries. Nevertheless, *running water is responsible for most of the erosional work in a desert.* Although wind erosion is more significant in dry areas than elsewhere, the main role of wind in a desert is in the transportation and deposition of sediment.

⊙ Because arid regions typically lack permanent streams, they are characterized as having *interior drainage.* Many of the landscapes of the Basin and Range region of the western and southwestern United States are the result of streams eroding uplifted mountain blocks and depositing the sediment in interior basins. *Alluvial fans, playas,* and *playa lakes* are features often associated with these landscapes. In the late stages of erosion, the mountain areas are reduced to a few large bedrock knobs, called *inselbergs,* projecting above the sediment-filled basin.

⊙ The transport of sediment by wind differs from that by running water in two ways. First, wind has a low density compared to water; thus, it is not capable of picking up and transporting coarse materials. Second, because wind is not confined to channels, it can spread sediment over large areas. The *bed load* of wind consists of sand grains skipping and bouncing along the surface in a process termed *saltation.* Fine dust particles are capable of being carried by the wind great distances as *suspended load.*

⊙ Compared to running water and glaciers, wind is a less significant erosional agent. *Deflation,* the lifting and removal of loose material, often produces shallow depressions called *blowouts.* Wind also erodes by *abrasion,* sometimes creating interestingly shaped stones termed *ventifacts. Yardangs* are narrow, streamlined, wind-sculpted landforms.

⊙ *Desert pavement* is a thin layer of coarse pebbles and cobbles that covers some desert surfaces. Once established, it protects the surface from deflation. Depending on circumstances, it may develop as a result of deflation or deposition of fine particles.

⊙ Wind deposits are of two distinct types: (1) *mounds and ridges of sand,* called *dunes,* that are formed from sediment carried as part of the wind's bed load; and (2) extensive *blankets of silt,* called *loess,* that once were carried by wind in *suspension.* The profiles of many dunes show an asymmetrical shape, with the leeward (sheltered) slope being steep and the windward slope being more gently inclined. The *types of sand dunes* include (1) *barchan dunes,* (2) *transverse dunes,* (3) *barchanoid dunes,* (4) *longitudinal dunes,* (5) *parabolic dunes,* and (6) *star dunes.* The thickest and most extensive deposits of loess occur in western and northern China. Unlike the deposits in China, which originated in deserts, the loess in the United States and Europe is an indirect product of glaciation.

Key Terms

abrasion (p. 298)
alluvial fan (p. 294)
bajada (p. 294)
barchan dunes (p. 300)
barchanoid dunes (p. 302)
bed load (p. 295)
blowouts (p. 297)
cross beds (p. 300)

deflation (p. 297)
desert (p. 288)
desert pavement (p. 297)
dry climate (p. 288)
dunes (p. 299)
ephemeral streams (p. 291)
interior drainage (p. 293)
inselbergs (p. 294)

loess (p. 303)
longitudinal dunes (p. 302)
parabolic dunes (p. 302)
playa (p. 294)
playa lake (p. 294)
rainshadow desert (p. 290)
saltation (p. 295)

slip face (p. 299)
star dunes (p. 302)
steppe (p. 288)
suspended load (p. 296)
transverse dunes (p. 300)
ventifacts (p. 299)
yardang (p. 299)

GIVE IT SOME THOUGHT

❶ Albuquerque, New Mexico, receives an average of 20.7 centimeters (8.07 inches) of rainfall annually. Albuquerque is considered a desert when the commonly used Köppen climate classification is applied. The Russian city of Verkhoyansk is located near the Arctic Circle in Siberia. The yearly precipitation at Verkhoyansk averages 15.5 centimeters (6.05 inches), about 5 centimeters less than Albuquerque, yet it is classified as a humid climate. Explain how this can occur.

❷ Compare and contrast the sediment deposited by a stream, the wind, and a glacier. Which deposit should have the most uniform grain size? Which one would exhibit the poorest sorting? Explain your choices.

❸ Is either of the following statements true? Are they both true? Explain your answer.

a. Wind is more effective as an agent of erosion in dry places than in humid places.

b. Wind is the most important agent of erosion in deserts.

❹ Examine the photo on p. 302 of sand dunes encroaching on irrigated fields. What type of dunes are these? From what direction (left or right) is the prevailing wind? How were you able to determine the wind direction?

❺ Name three deserts that are visible on this classic view of Earth from space. Briefly explain why these regions are so dry. Can you also explain the cause of the band of clouds in the region of the equator?

❻ "Deserts are hot, lifeless, and sand-covered landscapes shaped largely by the force of wind." The preceding statement summarizes most people's image of arid regions, especially those living in humid places. Is it an accurate perception? Explain why or why not.

Companion Website

Shorelines

THE RESTLESS WATERS
OF THE OCEAN ARE
CONSTANTLY IN MOTION.
Winds generate surface currents, gravity of the Moon and Sun produces tides, and density differences create deep-ocean circulation. Further, waves carry the energy from storms to distant shores, where their impact erodes the land.

Hatteras Island, North Carolina is a barrier island. When storm waves strike this beach, these homes are obviously vulnerable. *(Photo by Michael Collier)*

To assist you in learning the important concepts in this chapter, focus on the following questions:

- ◉ Why is the shoreline considered a dynamic interface?
- ◉ What are the various parts of the coastal zone?
- ◉ What factors influence the height, length, and period of a wave?
- ◉ What is the motion of water particles within a wave?
- ◉ How do waves erode?
- ◉ What are some typical features produced by wave erosion and from sediment deposited by beach drift and longshore currents?
- ◉ What are the local factors that influence shoreline erosion, and what are some basic responses to shoreline erosion problems?
- ◉ How do emergent and submergent coasts differ in their formation and characteristic features?
- ◉ How are tides produced?

The Shoreline: A Dynamic Interface

Shorelines are dynamic environments. Their topography, geologic make-up, and climate vary greatly from place to place. Continental and oceanic processes converge along coasts to create landscapes that frequently undergo rapid change. When it comes to the deposition of sediment, they are transition zones between marine and continental environments.

Nowhere is the restless nature of the ocean's water more noticeable than along the shore—the dynamic interface among air, land, and sea. An *interface* is a common boundary where different parts of a system interact. This is certainly an appropriate designation for the coastal zone. Here we can see the rhythmic rise and fall of tides and observe waves constantly rolling in and breaking. Sometimes the waves are low and gentle. At other times, they pound the shore with awesome fury (**FIGURE 13.1**).

Although it may not be obvious, the shoreline is constantly being shaped and modified by waves. For example, along Cape Cod, Massachusetts, wave activity is eroding cliffs of poorly consolidated glacial sediments so aggressively that the cliffs are retreating inland at up to 1 meter per year (**FIGURE 13.2A**). By contrast, at Point Reyes, California, the far more durable bedrock cliffs are less susceptible to wave attack and are therefore retreating much more slowly (**FIGURE 13.2B**). Along both coasts, wave activity moves sediment toward and away from the shore as well as along it. Such activity sometimes produces narrow sandbars that frequently change size and shape as storm waves come and go.

The nature of present-day shorelines is not just the result of the relentless attack on the land by the sea. The shore has a complex character that results from multiple geologic processes. For example, practically all coastal areas were affected by the worldwide rise in sea level that accompanied the melting of ice sheets at the close of the Pleistocene epoch (see Figure 11.28, p. 278). As the sea encroached landward, the shoreline retreated, becoming superimposed upon existing landscapes that had resulted from such diverse processes as stream erosion, glaciation, volcanic activity, and the forces of mountain building.

Today, the coastal zone is experiencing intensive human activity. Unfortunately, people often treat the shoreline as if it were a stable platform on which structures can safely be built. This attitude inevitably leads to conflicts between people and nature. As you will see, many coastal landforms, especially beaches and barrier islands, are relatively fragile, short-lived geological features that are inappropriate sites for development.

CONCEPT CHECK 13.1

❶ What is an interface?

❷ Which of Earth's spheres are represented in Figure 13.1?

FIGURE 13.1 Wind is responsible for creating the ocean waves that modify shorelines. These waves are crashing along the coast of the Hawaiian island of Oahu. (*Photo by Douglas Peebles/Corbis*)

FIGURE 13.2 **A.** This satellite image includes the familiar outline of Cape Cod. Boston is to the upper left. The two large islands off the south shore of Cape Cod are Martha's Vineyard (left) and Nantucket (right). Although the work of waves constantly modifies this coastal landscape, shoreline processes are not responsible for creating it. Rather, the present size and shape of Cape Cod result from the positioning of moraines and other glacial materials deposited during the Pleistocene epoch. *(Satellite Image courtesy of Earth Satellite Corporation/Science Photo Library/Photo Researchers, Inc.)* **B.** High-altitude image of the Point Reyes area north of San Francisco, California. The 5.5-kilometer-long south-facing cliffs at Point Reyes (bottom of photo) are exposed to the full force of the waves from the Pacific Ocean. Nevertheless, this promontory retreats slowly because the bedrock from which it formed is very resistant. *(High-altitude image courtesy of USDA-ASCS)*

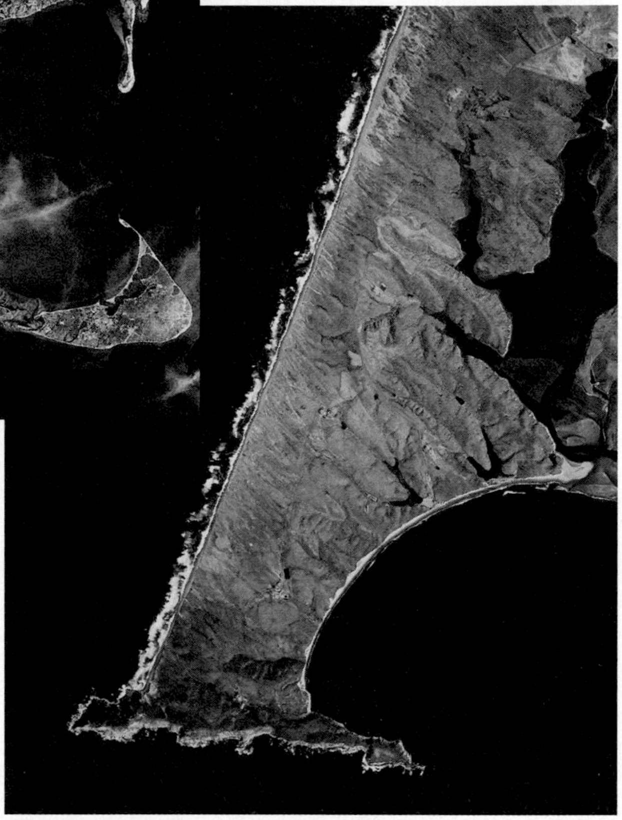

A.

B.

The Coastal Zone

In general conversation several terms are used when referring to the boundary between land and sea. In the preceding section, the terms *shore, shoreline, coastal zone,* and *coast* were all used. Moreover, when many think of the land–sea interface, the word *beach* comes to mind. Let's take a moment to clarify these terms and introduce some other terminology used by those who study the land–sea boundary zone. You will find it helpful to refer to **FIGURE 13.3**, which is an idealized profile of the coastal zone.

Basic Features

The **shoreline** is the line that marks the contact between land and sea. Each day, as tides rise and fall, the position of the shoreline migrates. Over longer time spans, the average position of the shoreline gradually shifts as sea level rises or falls.

FIGURE 13.3 The coastal zone consists of several parts. The beach is an accumulation of sediment on the landward margin of the ocean or a lake. It can be thought of as material in transit along the shore.

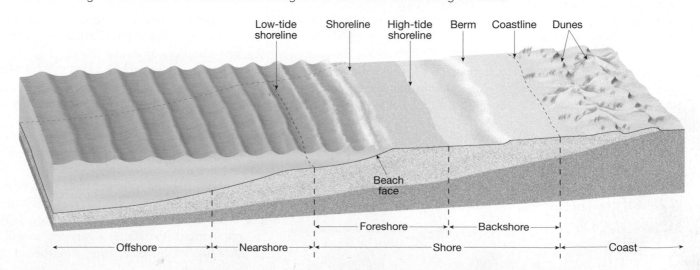

Low-tide shoreline | Shoreline | High-tide shoreline | Berm | Coastline | Dunes

Beach face

Offshore | Nearshore | Foreshore | Backshore | Shore | Coast

The **shore** is the area that extends between the lowest tide level and the highest elevation on land that is affected by storm waves. By contrast, the **coast** extends inland from the shore as far as ocean-related features can be found. The **coastline** marks the coast's seaward edge, whereas the inland boundary is not always obvious or easy to determine.

As Figure 13.3 illustrates, the shore is divided into the *foreshore* and the *backshore*. The **foreshore** is the area exposed when the tide is out (low tide) and submerged when the tide is in (high tide). The **backshore** is landward of the high-tide shoreline. It is usually dry, being affected by waves only during storms. Two other zones are commonly identified. The **nearshore zone** lies between the low-tide shoreline and the line where waves break at low tide. Seaward of the nearshore zone is the **offshore zone**.

Beaches

For many, a beach is the sandy area where people lie in the sun and walk along the water's edge. Technically, a **beach** is an accumulation of sediment found along the landward margin of the ocean or a lake. Along straight coasts, beaches may extend for tens or hundreds of kilometers. Where coasts are irregular, beach formation may be confined to the relatively quiet waters of bays.

Beaches consist of one or more **berms,** which are relatively flat platforms often composed of sand that are adjacent to coastal dunes or cliffs and marked by a change in slope at the seaward edge. Another part of the beach is the **beach face,** which is the wet sloping surface that extends from the berm to the shoreline. Where beaches are sandy, sunbathers usually prefer the berm, whereas joggers prefer the wet, hard-packed sand of the beach face.

Beaches are composed of whatever material is locally abundant. The sediment for some beaches is derived from the erosion of adjacent cliffs or nearby coastal mountains. Other beaches are built from sediment delivered to the coast by rivers.

Although the mineral makeup of many beaches is dominated by durable quartz grains, other minerals may be dominant. For example, in areas such as southern Florida where there are no mountains or other sources of rock-forming minerals nearby, most beaches are composed of shell fragments and the remains of organisms that live in coastal waters (see Figure 6.9A p. 157). Some beaches on volcanic islands in the open ocean are composed of weathered grains of the basaltic lava that comprise the islands or of coarse debris eroded from coral reefs that develop around islands in low latitudes (**FIGURE 13.4**).

Regardless of the composition, the material that comprises the beach does not stay in one place. Instead, crashing waves are constantly moving it. Thus, beaches can be thought of as material in transit along the shore.

CONCEPT CHECK 13.2

 Distinguish between shore, shoreline, coast, and coastline.

❷ What is a beach? Distinguish between beach face and berm.

FIGURE 13.4 The black sand at this beach is derived from dark volcanic rock. *(Photo by E. J. Tarbuck)*

Waves

Shorelines
ESSENTIALS OF GEOLOGY **Waves and Beaches**

Ocean waves are energy traveling along the interface between ocean and atmosphere, often transferring energy from a storm far out at sea over distances of several thousand kilometers. That is why even on calm days the ocean has waves that travel across its surface. When observing waves, always remember that you are watching *energy* travel though a medium (water). If you make waves by tossing a pebble into a pond, splashing in a pool, or blowing across the surface of a cup of coffee, you are imparting *energy* to the liquid, and the waves you see are just the visible evidence of the energy passing through.

Wind-generated waves provide most of the energy that shapes and modifies shorelines. Where the land and sea meet, waves that may have traveled unimpeded for hundreds or thousands of kilometers suddenly encounter a barrier that will not allow them to advance further. Stated another way, the shore is where a practi-

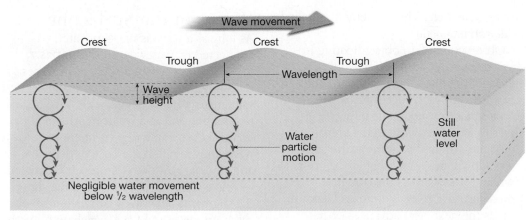

FIGURE 13.5 This diagram illustrates the basic parts of a wave as well as the movement of water particles with the passage of the wave. Negligible water movement occurs below a depth equal to one half the wavelength (the level of the dashed line).

cally irresistible force confronts an almost immovable object. The conflict that results is neverending and sometimes dramatic.

Wave Characteristics

Most ocean waves derive their energy and motion from the wind. When a breeze is less than 3 kilometers (2 miles) per hour, only small wavelets appear. At greater wind speeds, more stable waves gradually form and advance with the wind.

Characteristics of ocean waves are illustrated in **FIGURE 13.5**, which shows a simple, nonbreaking waveform. The tops of the waves are the *crests*, which are separated by *troughs*. Halfway between the crests and troughs is the *still water level*, which is the level the water would occupy if there were no waves. The vertical distance between trough and crest is called the **wave height**, and the horizontal distance between successive crests (or troughs) is the **wavelength.** The time it takes one full wave—one wavelength—to pass a fixed position is the **wave period.**

The height, length, and period that are eventually achieved by a wave depend upon three factors: (1) wind speed; (2) length of time the wind has blown; and (3) *fetch*, the distance that the wind has traveled across the open water. As the quantity of energy transferred from the wind to the water increases, the height and steepness of the waves increase as well. Eventually a critical point is reached where waves grow so tall that they topple over, forming ocean breakers called *whitecaps*.

For a particular wind speed, there is a maximum fetch and duration of wind beyond which waves will no longer increase in size. When the maximum fetch and duration are reached for a given wind velocity, the waves are said to be fully developed. The reason that waves can grow no further is that they are losing as much energy through the breaking of whitecaps as they are receiving from the wind.

When the wind stops or changes direction or the waves leave the stormy area where they were created, they continue on without relation to local winds. The waves also undergo a gradual change to *swells* that are lower in height and longer in length and may carry the storm's energy to distant shores. Because

Waves crashing at Big Sur, California. (Photo by Radius Images/ Photolibrary)

many independent wave systems exist at the same time, the sea surface acquires a complex and irregular pattern. Hence, the sea waves we watch from the shore are usually a mixture of swells from faraway storms and waves created by local winds.

Circular Orbital Motion

Waves can travel great distances across ocean basins. In one study, waves generated near Antarctica were tracked as they traveled through the Pacific Ocean basin. After more than 10,000 kilometers (over 6000 miles), the waves finally expended their energy a week later along the shoreline of the Aleutian Islands of Alaska. The water itself does not travel the entire distance, but the wave form does. As the wave travels, the water passes the energy along by

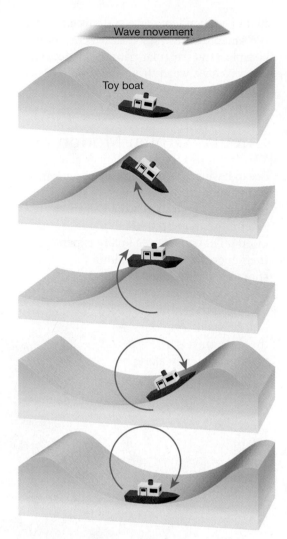

FIGURE 13.6 The movements of the toy boat show that the wave form advances but the water does not advance appreciably from its original position. In this sequence, the wave moves from left to right as the toy boat (and the water in which it is floating) rotates in an imaginary circle.

moving in a circle. This movement is called *circular orbital motion*.

Observation of an object floating in waves reveals that it moves not only up and down but also slightly forward and backward with each successive wave. **FIGURE 13.6** shows that a floating object moves up and backward as the crest approaches, up and forward as the crest passes, down and forward after the crest, down and backward as the trough approaches, and rises and moves backward again as the next crest advances. When the movement of the toy boat shown in Figure 13.6 is traced as a wave passes, it can be seen that the boat moves in a circle and it returns to essentially the same place. Circular orbital motion allows a waveform (the wave's shape) to move forward *through the water* while the individual water particles that transmit the wave move around in a circle. Wind moving across a field of wheat causes a similar phenomenon: The wheat itself doesn't travel across the field, but the waves do.

The energy contributed by the wind to the water is transmitted not only along the surface of the sea but also downward. However, beneath the surface the circular motion rapidly diminishes until, at a depth equal to one half the wavelength measured from still water level, the movement of water particles becomes negligible. This depth is known as the *wave base*. The dramatic decrease of wave energy with depth is shown by the rapidly diminishing diameters of water-particle orbits in Figure 13.5.

Waves in the Surf Zone

As long as a wave is in deep water, it is unaffected by water depth (**FIGURE 13.7**, left). However, when a wave approaches the shore, the water becomes shallower and influences wave behavior. The wave begins to "feel bottom" at a water depth equal to its wave base. Such depths interfere with water movement at the base of the wave and slow its advance (Figure 13.7, center).

As a wave advances toward the shore, the slightly faster waves farther out to sea catch up, decreasing the wavelength. As the speed and length of the wave diminish, the wave steadily grows higher. Finally, a critical point is reached when the wave is too steep to support itself and the wave front collapses, or *breaks* (Figure 13.7, right), causing water to advance up the shore.

The turbulent water created by breaking waves is called **surf.** On the landward margin of the surf zone the turbulent sheet of water from collapsing breakers, called *swash*, moves up the slope of the beach. When the energy of the swash has been expended, the water flows back down the beach toward the surf zone as *backwash*.

CONCEPT CHECK 13.3

❶ List three factors that determine the height, length, and period of a wave.

❷ Describe the motion of a floating object as a wave passes.

❸ How does a wave's speed, wavelength, and height change as it moves into shallow water and breaks?

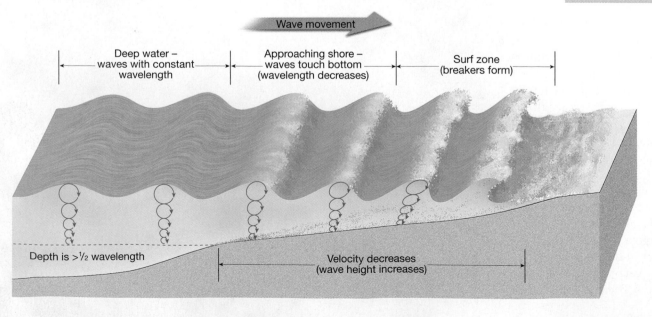

FIGURE 13.7 Changes that occur when a wave moves onto shore. The waves touch bottom as they encounter water depths less than half a wavelength. The wave speed decreases, and the waves stack up against the shore, causing the wavelength to decrease. This results in an increase in wave height to the point where the waves pitch forward and break in the surf zone.

Wave Erosion

 Shorelines

ESSENTIALS OF GEOLOGY **Wave Erosion**

During calm weather wave action is minimal. However, just as streams do most of their work during floods, so too do waves perform most of their work during storms. The impact of high, storm-induced waves against the shore can be awesome (**FIGURE 13.8**). Each breaking wave may hurl thousands of tons of water against the land, sometimes making the ground tremble. The pressures exerted by Atlantic waves in wintertime, for example, average nearly 10,000 kilograms per square meter (more than 2000 pounds per square foot). The force during storms is even greater.

It is no wonder that cracks and crevices are quickly opened in cliffs, coastal structures, and anything else that is subjected to these enormous shocks. Water is forced into every opening, greatly compressing air trapped in the cracks. When the wave subsides, the air expands rapidly, dislodging rock fragments and enlarging and extending fractures.

In addition to the erosion caused by wave impact and pressure, **abrasion,** the sawing and grinding action of water armed with rock fragments, is also important. In fact, abrasion is probably more intense in the surf zone than in any other environment. Smooth, rounded stones and pebbles along the shore are obvious reminders of the grinding action of rock

against rock in the surf zone (**FIGURE 13.9A**). Further, such fragments are used as tools by the waves as they cut horizontally into the land (**FIGURE 13.9B**).

CONCEPT CHECK 13.4

❶ Describe two ways in which waves cause erosion.

Sand Movement on the Beach

Beaches are sometimes called "rivers of sand." The reason is that the energy from breaking waves often causes large quantities of sand to move along the beach face and in the surf zone roughly parallel to the

FIGURE 13.8 When waves break against the shore, the force of the water can be powerful and the erosional work that is accomplished can be great. These storm waves are breaking along the coast of Wales. *(The Photo Library Wales/Alamy)*

A.

B.

FIGURE 13.9 A. Abrasion can be intense in the surf zone. Smooth, rounded stones along the shore are an obvious reminder of this fact. *(Photo by Michael Collier)* **B.** Sandstone cliff undercut by wave erosion at Gabriola Island, British Columbia, Canada. *(Photo by Fletcher and Baylis/Photo Researchers, Inc.)*

shoreline. Wave energy also causes sand to move perpendicular to (toward and away from) the shoreline.

Movement Perpendicular to the Shoreline

If you stand ankle deep in water at the beach, you will see that swash and backwash move sand toward and away from the shoreline. Whether there is a net loss or addition of sand depends on the level of wave activity. When wave activity is relatively light (less energetic waves), much of the swash soaks into the beach, which reduces the backwash. Consequently, the

swash dominates and causes a net movement of sand up the beach face toward the berm.

When high-energy waves prevail, the beach is saturated from previous waves, so much less of the swash soaks in. As a result, the berm erodes because backwash is strong and causes a net movement of sand down the beach face.

Along many beaches, light wave activity is the rule during the summer. Therefore, a wide sand berm gradually develops. During winter, when storms are frequent and more powerful, strong wave activity erodes and narrows the berm. A wide berm that may have taken months to build can be dramatically narrowed in just a few hours by the high-energy waves created by a strong winter storm.

Wave Refraction

The bending of waves, called **wave refraction**, plays an important part in shoreline processes (**FIGURE 13.10**). It affects the distribution of energy along the shore

and thus strongly influences where and to what degree erosion, sediment transport, and deposition will take place.

Waves seldom approach the shore truly straight on. Rather, most waves move toward the shore at an angle. However, when they reach the shallow water of a smoothly sloping bottom, they are bent and tend to become parallel to the shore. Such bending occurs because the part of the wave nearest the shore reaches shallow water and slows first, whereas the end that is still in deep water continues forward at its full speed. The net result is a wave front that may approach nearly parallel to the shore regardless of the original direction of the wave.

Because of refraction, wave impact is concentrated against the sides and ends of headlands that project into the water, whereas wave attack is weakened in bays. This differential wave attack along irregular coastlines is illustrated in Figure 13.10. Because the waves reach the shallow water in front of the headland before they reach adjacent bays, they are bent more nearly

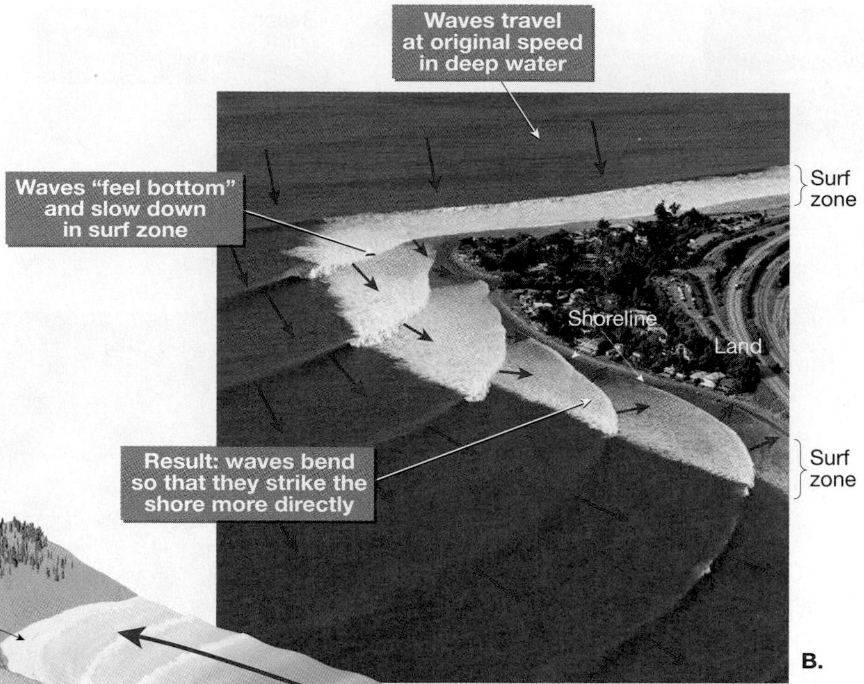

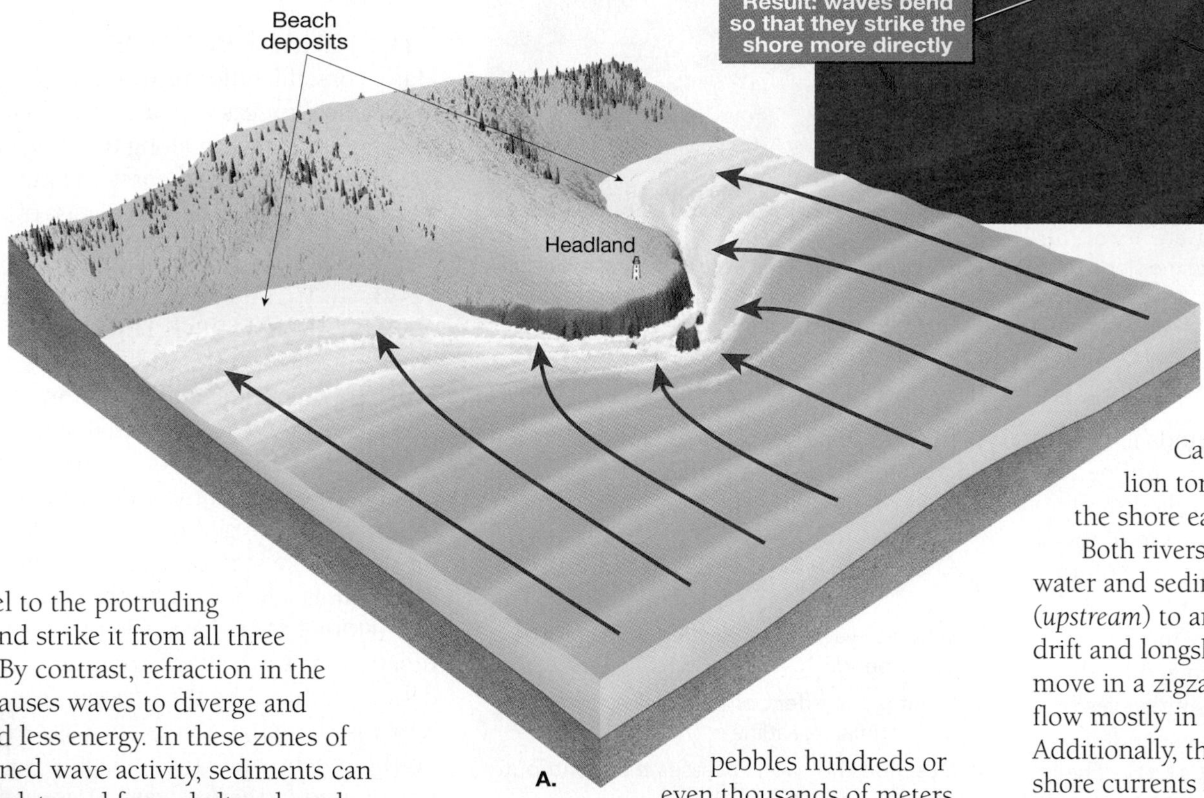

FIGURE 13.10 As waves first touch bottom in the shallows along an irregular coast, they are slowed, causing them to bend (refract) and align nearly parallel to the shoreline. **A.** In this diagram, waves approach nearly straight on. Refraction causes wave energy to be concentrated at headlands (resulting in erosion) and dispersed in bays (resulting in deposition). **B.** Wave refraction at Rincon Point, California. *(Photo by Woody Woodworth/Creation Captured)*

parallel to the protruding land and strike it from all three sides. By contrast, refraction in the bays causes waves to diverge and expend less energy. In these zones of weakened wave activity, sediments can accumulate and form sheltered sandy beaches. Over a long period, erosion of the headlands and deposition in the bays will straighten an irregular shoreline.

Longshore Transport

Although waves are refracted, most still reach the shore at an angle, however slight. Consequently, the uprush of water from each breaking wave (the swash) is not head on, but oblique. However, the backwash is straight down the slope of the beach. The effect of this pattern of water movement is to transport particles of sediment in a zigzag pattern along the beach face (**FIGURE 13.11**). This movement is called **beach drift**, and it can transport sand and

pebbles hundreds or even thousands of meters each day. However, a more typical rate is 5 to 10 meters per day.

Waves that approach the shore at an angle also produce currents within the surf zone that flow parallel to the shore and move substantially more sediment than beach drift does. Because the water here is turbulent, these **longshore currents** easily move the fine suspended sand and roll larger sand and gravel along the bottom. When the sediment transported by longshore currents is added to the quantity moved by beach drift, the total can be very large. At Sandy Hook, New Jersey, for example, the quantity of sand transported along the shore over a 48-year period averaged almost 750,000 tons annually. For a

10-year period at Oxnard, California, more than 1.5 million tons of sediment moved along the shore each year.

Both rivers and coastal zones move water and sediment from one area (*upstream*) to another (*downstream*). Beach drift and longshore currents, however, move in a zigzag pattern, whereas rivers flow mostly in a turbulent, swirling fashion. Additionally, the direction of flow of longshore currents along a shoreline can change, whereas rivers flow in the same direction (downhill). Longshore currents change direction because the direction that waves approach the beach changes seasonally. Nevertheless, longshore currents generally flow southward along both the Atlantic and Pacific shores of the United States.

Rip Currents

Rip currents are concentrated movements of water that flow in the *opposite* direction from breaking waves.* Most of the

*Sometimes rip currents are incorrectly called *rip tides*, although they are unrelated to tidal phenomena.

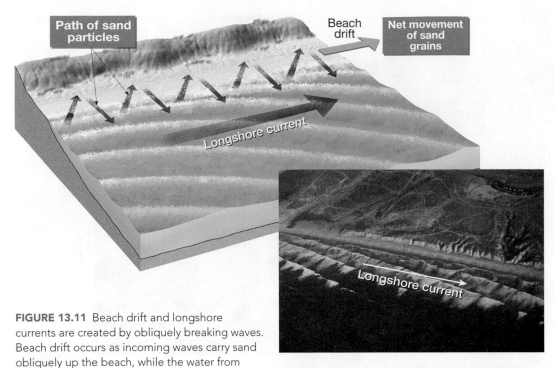

FIGURE 13.11 Beach drift and longshore currents are created by obliquely breaking waves. Beach drift occurs as incoming waves carry sand obliquely up the beach, while the water from spent waves carries it directly down the slope of the beach. Similar movements occur offshore in the surf zone to create the longshore current. These processes transport large quantities of material along the beach and in the surf zone. In the photo, waves approaching the beach at a slight angle near Oceanside, California, produce a longshore current moving from left to right. *(Photo by John S. Shelton)*

back-wash from spent waves finds its way back to the open ocean as an unconfined flow across the ocean bottom called *sheet flow*. However, sometimes a portion of the returning water moves seaward in the form of surface rip currents. These currents do not travel far beyond the surf zone before breaking up and can be recognized by the way they interfere with incoming waves or by the sediment that is often suspended within the rip current (**FIGURE 13.12**). They can be a hazard to swimmers, who, if caught in them, can be carried out away

from shore. The best strategy for exiting a rip current is to swim *parallel* to the shore for a few tens of meters.

CONCEPT CHECK 13.5

❶ Why do waves approaching the shoreline often bend?

❷ What is the effect of wave refraction along an irregular coastline?

❸ Describe the two processes that contribute to longshore transport.

FIGURE 13.12 A rip current (red arrow) extends outward from shore and interferes with incoming waves. As the warning sign indicates, rip currents can be dangerous. *(Photo by A. P. Trujillo/APT Photos)*

Shoreline Features

A fascinating assortment of shoreline features can be observed in the world's coastal regions. These shoreline features vary depending on the type of rocks exposed along the shore, the intensity of waves, the nature of coastal currents, and whether the coast is stable, sinking, or rising. Features that owe their origin primarily to the work of erosion are called *erosional features*; deposits of sediment produce *depositional features*.

Erosional Features

Many coastal landforms owe their origin to erosional processes. Such erosional features are common along the rugged and irregular New England coast and along the steep shorelines of the West Coast of the United States.

WAVE-CUT CLIFFS, WAVE-CUT PLATFORMS, AND MARINE TERRACES.
Wave-cut cliffs, as the name implies, originate due to the cutting action of the surf against the base of coastal land. As erosion progresses, rocks overhanging the notch at the base of the cliff crumble into the surf, and the cliff retreats. A relatively flat, benchlike surface, called a **wave-cut platform,** is left behind by the receding cliff (**FIGURE 13.13**, left). The platform broadens as wave attack continues. Some debris produced by the breaking waves remains along the water's edge as sediment on the beach, while the remainder is transported farther seaward. If a wave-cut platform is uplifted above sea level by tectonic forces, it becomes a **marine terrace** (Figure 13.13, right). Marine terraces are easily recognized by their gentle

DID YOU KNOW?
During a storm that struck the coast of Scotland, a 1350-ton portion of a steel-and-concrete breakwater was torn from the rest of the structure and moved toward shore. Five years later the 2600-ton unit that replaced the first met a similar fate.

FIGURE 13.13 Wave-cut platform and marine terrace. A wave-cut platform is exposed at low tide along the California coast at Bolinas Point near San Francisco. A wave-cut platform has been uplifted to create the marine terrace. *(Photo by John S. Shelton)*

seaward-sloping shape and are often desirable sites for coastal roads, buildings, or agriculture.

SEA ARCHES AND SEA STACKS.

Headlands that extend into the sea are vigorously attacked by waves because of refraction. The surf erodes the rock selectively, wearing away the softer or more highly fractured rock at the fastest rate. At first, sea caves may form. When two caves on opposite sides of a headland unite, a **sea arch** results (**FIGURE 13.14**). Eventually, the arch falls in, leaving an isolated remnant, or **sea stack,** on the wave-cut platform (Figure 13.14). In time, it too will be consumed by the action of the waves.

Depositional Features

Sediment eroded from the beach is transported along the shore and deposited in

areas where wave energy is low. Such processes produce a variety of depositional features.

SPITS, BARS, AND TOMBOLOS. Where beach drift and longshore currents are active, several features related to the movement of sediment along the shore may develop. A **spit** is an elongated ridge of sand that projects from the land into the mouth of an adjacent bay. Often the end in the water hooks landward in response to

the dominant direction of the longshore current (**FIGURE 13.15**). The term **baymouth bar** is applied to a sandbar that completely crosses a bay, sealing it off from the open ocean (Figure 13.15). Such a feature tends to form across bays where currents are weak, allowing a spit to extend to the other side. A **tombolo,** a ridge of sand that connects an island to the mainland or to another island, forms in much the same manner as a spit.

FIGURE 13.14 Sea arch and sea stack at the tip of Mexico's Baja Peninsula. *(Photo by Lew Robertson/Getty Images)*

BARRIER ISLANDS. The Atlantic and Gulf Coastal Plains are relatively flat and slope gently seaward. The shore zone is characterized by **barrier islands.** These low ridges of sand parallel the coast at distances from 3 to 30 kilometers offshore. From Cape Cod, Massachusetts, to Padre Island, Texas, nearly 300 barrier islands rim the coast (**FIGURE 13.16**).

Most barrier islands are 1 to 5 kilometers wide and between 15 and 30 kilometers long. The tallest features are sand dunes, which usually reach heights of 5 to 10 meters. The lagoons that separate these narrow islands from the shore are zones of relatively quiet water that allow small craft traveling between New York and northern Florida to avoid the rough waters of the North Atlantic.

Barrier islands probably form in several ways. Some originate as spits that were subsequently severed from the mainland by wave erosion or by the general rise in sea level following the last episode of glaciation. Others are created when turbulent waters in the line of breakers heap up sand that has been scoured from the bottom. Finally, some barrier islands may be former sand-dune ridges that originated along the shore during the last glacial period, when sea level was lower. As the ice sheets

Wind is not only responsible for creating waves, it also provides the force that drives the ocean's surface circulation. *(Photo by Novastock/F1ONLINE/agefotostock)*

FIGURE 13.15 A. High-altitude image of a well-developed spit and baymouth bar along the coast of Martha's Vineyard, Massachusetts. *(Image courtesy of USDA-ASCS)* **B.** This photograph, taken from the International Space Station, shows Provincetown Spit located at the tip of Cape Cod. Can you pick this feature out in the satellite image in Figure 13.2A? *(NASA image)*

Baymouth bar

Spit

Tidal delta

A.

Provincetown Spit

B.

FIGURE 13.16 Nearly 300 barrier islands rim the Gulf and Atlantic coasts. The islands along the coast of North Carolina are excellent examples. *(Photo by Michael Collier)*

deposition will eventually produce a straighter, more regular coast.

FIGURE 13.17 illustrates the evolution of an initially irregular coast. As waves erode the headlands, creating cliffs and a wave-cut platform, sediment is carried along the shore. Some material is deposited in the bays, whereas other debris is formed into spits and baymouth bars. At the same time, rivers fill the bays with sediment. Ultimately, a generally straight, smooth coast results.

CONCEPT CHECK 13.6

❶ Describe the formation of each labeled feature in parts B and C of Figure 13.17.

❷ List three ways that barrier islands may form.

Stabilizing the Shore

Shorelines

Waves and Beaches

Today the coastal zone teems with human activity. Unfortunately, people often treat the shoreline as if it were a stable platform on which structures can be built safely. This approach jeopardizes both people and the shoreline because many coastal landforms are relatively fragile, short-lived features that are easily damaged by development. As anyone who has endured a tsunami or a tropical storm knows, the shoreline is not always a safe place to live. We will examine this latter idea more closely in the section on "Hurricanes—The Ultimate Coastal Hazard."

Compared with natural hazards such as earthquakes, volcanic eruptions, and landslides, shoreline erosion appears to be a more continuous and predictable process that causes relatively modest damage to limited areas. In reality, the shoreline is a dynamic place that can change rapidly in response to natural forces. Exceptional storms are capable of eroding beaches and cliffs at rates that far exceed the long-term

melted, sea level rose and flooded the area behind the beach-dune complex.

The Evolving Shore

A shoreline continually undergoes modification regardless of its initial configuration. At first most coastlines are irregular, although the degree of, and reason for, the irregularity may differ considerably from place to place. Along a coastline of varied geology, the pounding surf may initially increase its irregularity because the waves will erode the weaker rocks more easily than the stronger ones. However, if a shoreline remains stable, marine erosion and

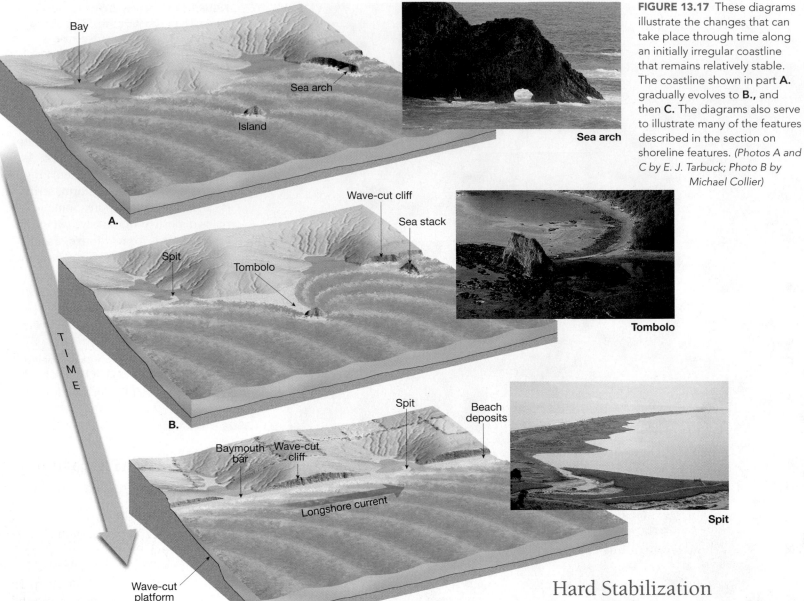

FIGURE 13.17 These diagrams illustrate the changes that can take place through time along an initially irregular coastline that remains relatively stable. The coastline shown in part **A.** gradually evolves to **B.**, and then **C.** The diagrams also serve to illustrate many of the features described in the section on shoreline features. *(Photos A and C by E. J. Tarbuck; Photo B by Michael Collier)*

average. Such bursts of accelerated erosion not only have a significant impact on the natural evolution of a coast but can also have a profound impact on people who reside in the coastal zone. Erosion along our coasts causes significant property damage. Huge sums are spent annually not only to repair damage but also to prevent or control erosion. Already a problem at many sites, shoreline erosion is certain to become increasingly serious as extensive coastal development continues.

Although the same processes cause change along every coast, not all coasts respond in the same way. Interactions among different processes and the relative importance of each process depend on local factors, which include (1) the proximity of a coast to sediment-laden rivers, (2) the degree of tectonic activity, (3) the topography and composition of the land, (4) prevailing winds and weather patterns, and (5) the configuration of the coastline and nearshore areas.

During the past 100 years, growing affluence and increasing demands for recreation have brought unprecedented development to many coastal areas. As both the number and the value of structures have increased, so too have efforts to protect property from storm waves by stabilizing the shore. Also, controlling the natural migration of sand is an ongoing struggle in many coastal areas. Such interference can result in unwanted changes that are difficult and expensive to correct.

Hard Stabilization

Structures built to protect a coast from erosion or to prevent the movement of sand along a beach are known as **hard stabilization.** Hard stabilization can take many forms and often results in predictable yet unwanted outcomes. Hard stabilization includes jetties, groins, breakwaters, and seawalls.

JETTIES. From relatively early in America's history a principal goal in coastal areas was the development and maintenance of harbors. In many cases, this involved the construction of jetty systems. **Jetties** are usually built in pairs and extend into the ocean at the entrances to rivers and harbors. With the flow of water confined to a narrow zone, the ebb and flow caused by the rise and fall of the tides keep the sand in motion and prevent deposition in

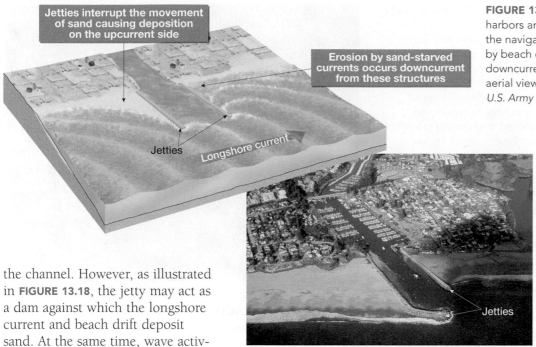

Jetties interrupt the movement of sand causing deposition on the upcurrent side

Erosion by sand-starved currents occurs downcurrent from these structures

Jetties

Longshore current

FIGURE 13.18 Jetties are built at the entrances to rivers and harbors and are intended to prevent deposition in the navigation channel. Jetties interrupt the movement of sand by beach drift and longshore currents. Beach erosion occurs downcurrent from the site of the structure. The photo is an aerial view of jetties at Santa Cruz Harbor, California *(Photo by U.S. Army Corps of Engineers)*

Jetties

the channel. However, as illustrated in **FIGURE 13.18**, the jetty may act as a dam against which the longshore current and beach drift deposit sand. At the same time, wave activity removes sand on the other side. Because the other side is not receiving any new sand, there is soon no beach at all.

GROINS. To maintain or widen beaches that are losing sand, groins are sometimes constructed. A **groin** is a barrier built at a right angle to the beach to trap sand that is moving parallel to the shore. Groins are usually constructed of large rocks but may also be composed of wood. These structures often do their job so effectively that the longshore current beyond the groin becomes sand starved. As a result, the current erodes sand from the beach on the downstream side of the groin.

To offset this effect, property owners downstream from the structure may erect a groin on their property. In this manner, the number of groins multiplies, resulting in a *groin field* (**FIGURE 13.19**). An example of such proliferation is the shoreline of New Jersey, where hundreds of these structures have been built. Because it has been shown that groins often do not provide a satisfactory solution, they are no longer the preferred method of keeping beach erosion in check.

BREAKWATERS AND SEAWALLS. Hard stabilization can also be built parallel to the shoreline. One such structure is a **breakwater,** the purpose of which is

to protect boats from the force of large breaking waves by creating a quiet water zone near the shoreline. However, when this is done, the reduced wave activity along the shore behind the structure may allow sand to accumulate. If this happens, the marina will eventually fill with sand while the downstream beach erodes and retreats. At Santa Monica, California, where the building of a breakwater created such a problem, the city had to install a dredge to remove sand from the protected quiet water zone and deposit it down the beach where longshore currents and beach drift could recirculate the sand (**FIGURE 13.20**).

Another type of hard stabilization built parallel to the shoreline is a **seawall;** which

FIGURE 13.19 A series of groins along the shoreline near Chichester, Sussex, England. *(Photo by Sandy Stockwell/London Aerial Photo Library/Corbis)*

is designed to armor the coast and defend property from the force of breaking waves. Waves expend much of their energy as they move across an open beach. Seawalls cut this process short by reflecting the force of unspent waves seaward. As a consequence, the beach to the seaward side of the seawall experiences significant erosion and may, in some instances, be eliminated entirely (**FIGURE 13.21**). Once the width of the beach is reduced, the seawall is subjected to even greater pounding by the waves. Eventually this battering will cause the wall to fail, and a larger, more expensive wall must be built to take its place.

The wisdom of building temporary protective structures along shorelines is increasingly questioned. The feelings of many coastal scientists and engineers are expressed in the following excerpt from a position paper that grew out of a conference on America's eroding shoreline:

It is now clear that halting the receding shoreline with protective structures benefits only a few and seriously degrades or destroys the natural beach and the value it holds for the majority. Protective structures divert the ocean's energy temporarily from private properties, but usually refocus the energy on the adjacent natural beaches. Many interrupt the natural sand flow in coastal currents, robbing many beaches of vital sand replacement.

Alternatives to Hard Stabilization

Armoring the coast with hard stabilization has several potential drawbacks, including the cost of the structure and the loss of sand on the beach. Alternatives to hard stabilization include beach nourishment and relocation.

*Strategy for Beach Preservation Proposed, *Geotimes 30* (No. 12, December 1985), 15.

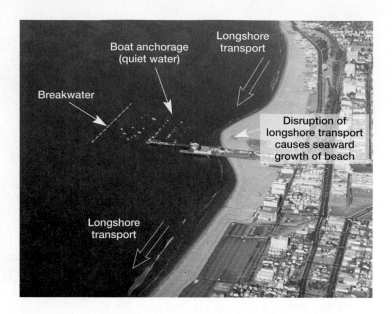

FIGURE 13.20 Aerial view of a breakwater at Santa Monica, California. The breakwater appears as a line in the water behind which many boats are anchored. The construction of the breakwater disrupted longshore transport and caused the seaward growth of the beach. *(Photo by John S. Shelton)*

BEACH NOURISHMENT. Beach nourishment represents an approach to stabilizing shoreline sands without hard stabilization. As the term implies, this practice simply involves the addition of large quantities of sand to the beach system (**FIGURE 13.22**). By building the beaches seaward storm protection is improved. Beach nourishment, however, is not a permanent solution to the problem of shrinking beaches. The same processes that removed the sand in the first place will eventually remove the replacement sand as well. In addition, beach nourishment is very expensive because huge volumes of sand must be transported to the beach from offshore areas, nearby rivers, or other source areas. Orrin Pilkey, a respected coasted scientist, described the situation this way:

Nourishment has been carried out on beaches on both sides of the continent, but by far the greatest effort in dollars and sand volume has been expended on the East Coast barrier islands between the southern shore of Long Island, New York, and southern Florida. Over this shoreline reach, communities have added a total of 500 million cubic yards of sand on 195 beaches in 680 separate instances. Some beaches…have been renourished more than 20 times each since 1965. Virginia Beach has been renourished more than 50 times. Nourished beaches typically cost between $2 million and $10 million per mile.

In some instances, beach nourishment can lead to unwanted environmental effects. For example, beach replenishment at Waikiki Beach, Hawaii, involved replac-

*"Beaches Awash with Politics," *Geotimes*, July 2005, 36–39.

FIGURE 13.21 Seabright in northern New Jersey once had a broad, sandy beach. A seawall 5 to 6 meters (16 to 20 feet) high and 8 kilometers (5 miles) long was built to protect the town and the railroad that brought tourists to the beach. As you can see, after the wall was built, the beach narrowed dramatically. *(Photo by Rafael Macia/Photo Researchers, Inc.)*

A.   **B.**

FIGURE 13.22 Miami Beach. **A.** Before beach nourishment and **B.** after beach nourishment. *(Courtesy of the U.S. Army Corps of Engineers, Vicksburg District)*

ing coarse calcareous sand with softer, muddier calcareous sand. Destruction of the soft beach sand by breaking waves increased the water's turbidity and killed offshore coral reefs. At Miami Beach, increased turbidity also damaged local coral communities.

Beach nourishment appears to be an economically viable long-range solution to the beach preservation problem only in areas where there exists dense development, large supplies of sand, relatively low wave energy, and reconcilable environmental issues. Unfortunately, few areas possess all these attributes.

RELOCATION. Instead of building structures such as groins and seawalls to hold the beach in place or adding sand to replenish eroding beaches, another option is available. Many coastal scientists and planners are calling for a policy shift from defending and rebuilding beaches and coastal property in high hazard areas to removing or relocating storm-damaged buildings in those places and letting nature reclaim the beach. This approach is similar to that adopted by the federal government for river floodplains following the devastating 1993 Mississippi River floods in which vulnerable structures are abandoned and relocated on higher, safer ground.

Such proposals, of course, are controversial. People with significant nearshore investments shudder at the thought of not rebuilding and not defending coastal developments from the erosional wrath of the sea. Others, however, argue that with sea level rising, the impact of coastal storms will only get worse in the decades to come. This group advocates that oft-damaged structures be abandoned or relocated to improve personal safety and to reduce costs.

DID YOU KNOW?

In 1870, a lighthouse was constructed on Cape Hatteras, North Carolina's famous barrier island (see the map in Figure 13.16). When it was built, the structure was about 460 meters (1500 feet) from the shoreline. Despite various attempts to stop shoreline retreat, the beach continued to get narrower and crashing waves eventually threatened the historic building. As a result, in 1999 the 21-story tall, 4800-ton lighthouse was moved to a safer location about 0.5 kilometer (1600 feet) from the shore at a cost of $12 million.

CONCEPT CHECK 13.7

❶ List at least three examples of *hard stabilization* and describe what each is intended to do. How does each affect distribution of sand on a beach?

❷ What are two alternatives to hard stabilization and potential problems associated with each?

Erosion Problems along U.S. Coasts

The shoreline along the Pacific Coast of the United States is strikingly different from that characterizing the Atlantic and Gulf coast regions. Some of the differences are related to plate tectonics. The West Coast represents the leading edge of the North American plate, and because of this it experiences active uplift and deformation. By contrast, the East Coast is a tectonically quiet region that is far from any active plate boundary. Because of this basic geological difference the nature of shoreline erosion problems along America's opposite coasts is different.

Atlantic and Gulf Coasts

Much of the development along the Atlantic and Gulf coasts has occurred on barrier islands. Typically, barrier islands consist of a wide beach that is backed by dunes and separated from the mainland by marshy lagoons. The broad expanses of sand and exposure to the ocean have made barrier islands exceedingly attractive sites for development. Unfortunately, development has taken place more rapidly than our understanding of barrier island dynamics has grown.

Because barrier islands face the open ocean, they receive the full force of major storms that strike the coast. When a storm occurs, the barriers absorb the energy of the waves primarily through the movement of sand. This process and the dilemma that results have been described as follows:

Waves may move sand from the beach to offshore areas or, conversely, into the dunes; they may erode the dunes, depositing sand onto the beach or carrying it out to sea; or they may carry sand from the beach and the dunes into the marshes behind the barrier, a process known as over-wash. The common factor is movement. Just as a flexible reed may survive a wind that destroys an oak tree, so the barriers survive hurricanes and nor'easters not through unyielding strength but by giving before the storm.

This picture changes when a barrier is developed for homes or a resort. Storm waves that previously rushed harmlessly through gaps between the dunes now encounter buildings and roadways. Moreover, because the dynamic nature of the barrier is readily perceived only during storms, homeowners tend to attribute damage to a particular storm, rather than to the basic mobility of coastal barriers. With their homes or investments at stake, local residents are more likely to seek to hold the sand in place and the waves at bay than to admit that development was improperly placed to begin with. *

Pacific Coast

In contrast to the broad, gently sloping coastal plains of the Atlantic and Gulf coasts, much of the Pacific Coast is characterized by relatively narrow beaches that are backed by steep cliffs and mountain ranges. Recall that America's western margin is a more rugged and tectonically active region than the eastern margin. Because uplift continues, a rise in sea level in the West is not so readily apparent. Nevertheless, like the shoreline erosion

*Frank Lowenstein, Beaches or Bedrooms—The Choice as Sea Level Rises, *Oceanus* 28 (No. 3, Fall 1985), 22.

problems facing the East's barrier islands, West Coast difficulties also stem largely from the alteration of a natural system by people.

A major problem facing the Pacific shoreline, and especially portions of Southern California, is a significant narrowing of many beaches. The bulk of the sand on many of these beaches is supplied by rivers that transport it from the mountainous regions to the coast. Over the years this natural flow of material to the coast has been interrupted by dams built for irrigation and flood control. The reservoirs effectively trap the sand that would otherwise nourish the beach environment. When the beaches were wider, they served to protect the cliffs behind them from the force of storm waves. Now, however, the waves move across the narrowed beaches without losing much energy and cause more rapid erosion of the sea cliffs.

Although the retreat of the cliffs provides material to replace some of the sand impounded behind dams, it also endangers homes and roads built on the bluffs. In addition, development atop the cliffs aggravates the problem. Urbanization increases runoff, which, if not carefully controlled, can result in serious bluff erosion. Watering lawns and gardens adds significant quantities of water to the slope. This water percolates downward toward the base of the cliff, where it may emerge in small seeps. This action reduces the slope's stability and facilitates mass wasting.

Shoreline erosion along the Pacific Coast varies considerably from one year to the next, largely because of the sporadic occurrence of storms. As a consequence, when the infrequent but serious episodes of erosion occur, the damage is often blamed on the unusual storms and not on coastal development or the sediment-trapping dams that may be great distances away. If, as predicted, sea level rises at an increasing rate in the years to come, increased shoreline erosion and sea-cliff retreat should be expected along many parts of the Pacific Coast. This issue is addressed in the section on "Sea-Level Rise" in Chapter 20, p. 516.

CONCEPT CHECK 13.8

❶ Briefly describe what happens when storm waves strike an undeveloped barrier island.
❷ How might building a dam on a river that flows to the sea affect a coastal beach?

Hurricanes—The Ultimate Coastal Hazard

The whirling tropical cyclones that occasionally have wind speeds exceeding 300 kilometers (185 miles) per hour are known in the United States as *hurricanes*—the greatest storms on Earth (**FIGURE 13.23**). In the western Pacific they are called *typhoons*, and in the Indian Ocean they are simply called *cyclones*. No matter which name is used, these storms are among the most destructive of natural disasters. When a hurricane reaches land, it is capable of annihilating coastal areas and killing tens of thousands of people.

Fortunately, when Hurricane Ike made landfall along the Texas coast near Galveston on September 13, 2008, loss of life was relatively small. Because forecasts were accurate and warnings were timely, tens of thousands were safely evacuated from the most vulnerable areas before the storm arrived. Nevertheless, damages to the coastal zone and inland areas were substantial and costly (**FIGURE 13.24**).

On the positive side, hurricanes provide essential rainfall over many areas they cross. Consequently, a resort owner along the Florida coast may dread the coming of hurricane season, whereas a farmer in Japan may welcome its arrival.

A.

B.

FIGURE 13.23 Satellite images of Hurricane Katrina in late August 2005 shortly before it devastated the Gulf Coast. **A.** Color-enhanced infrared image from the *GOES-East* satellite. The most intense activity is associated with red and orange. **B.** A more traditional satellite image from NASA's *Terra* satellite. *(NASA)*

Our coasts are vulnerable. People are flocking to live near the ocean. The proportion of the U.S. population residing within 75 kilometers (45 miles) of a coast in 2010 is well in excess of 50 percent. The concentration of such large numbers of people near the shoreline means that hurricanes place millions at risk. Moreover, the potential costs of property damage are incredible.

Profile of a Hurricane

Most hurricanes form between the latitudes of 5 degrees and 20 degrees over all the tropical oceans except those of the South Atlantic and eastern South Pacific (**FIGURE 13.26**). The North Pacific has the greatest number of storms, averaging 20 per year. Fortunately for those living in the coastal regions of the southern and eastern United States, fewer than five hurricanes, on the average, develop annually in the warm sector of the North Atlantic.

The vast majority of hurricane-related deaths and damage are caused by relatively infrequent yet powerful storms. The storm that pounded an unsuspecting Galveston, Texas, in 1900 was not just the deadliest U.S. hurricane ever, but also the deadliest natural disaster *of any kind* to affect the United States (**FIGURE 13.25**).* Of course, the deadliest and most costly storm in recent memory occurred in August 2005, when Hurricane Katrina devastated the Gulf Coast of Louisiana, Mississippi, and Alabama. Although hundreds of thousands fled before the storm made landfall, thousands of others were caught by the storm. In addition to the human suffering and tragic loss of life that were left in the wake of Hurricane Katrina, the financial losses caused by the storm are practically incalculable. Up until August 2005, the $25 billion in damages associated with Hurricane Andrew in 1992 represented the costliest natural disaster in U.S. history. This figure was exceeded many times over when Katrina's economic impact was calculated. Although imprecise, some suggest that the final accounting could exceed $100 billion.

Isaac's Storm—A Man, A Time, and the Deadliest Hurricane in History, by Eric Larson (Random House © 1999) is an engaging and readable account of the 1900 Galveston storm.

FIGURE 13.24 Crystal Beach, Texas, on September 16, 2008, three days after Hurricane Ike came ashore. At landfall the storm had sustained winds of 165 kilometers (105 miles) per hour. The extraordinary storm surge caused much of the damage pictured here. *(Photo by Earl Nottingham/Associated Press)*

FIGURE 13.25 Aftermath of the Galveston hurricane that struck an unsuspecting and unprepared city on September 8, 1900. It was the worst natural disaster in U.S. history. Entire blocks were swept clean, while mountains of debris accumulated around the few remaining buildings. (AP Photo)

As the inward rush of warm, moist surface air approaches the core of the storm, it turns upward and ascends in a ring of cumulonimbus towers (Figure 13.27). This doughnut-shaped wall of intense convective activity surrounding the center of the storm is called the **eye wall.** It is here that the greatest wind speeds and heaviest rainfall occur. Surrounding the eye wall are curved bands of clouds that trail away in a spiral fashion. Near the top of the hurricane, the airflow is outward, carrying the rising air away from the storm center, thereby providing room for more inward flow at the surface.

At the very center of the storm is the **eye** of the hurricane (Figure 13.27). This well-known feature is a zone about 20 kilometers (12.5 miles) in diameter where precipitation ceases and winds subside. It offers a brief but deceptive break from the extreme weather in the enormous curving wall clouds that surround it. The air within the eye gradually descends and heats by compression, making it the warmest part of the storm. Although many people believe that the eye is characterized by clear blue skies, such is usually not the case because the subsidence in the eye is seldom strong enough to produce cloudless conditions. Although the sky appears much brighter in the eye, scattered clouds at various levels are common.

Hurricane Destruction

The amount of damage caused by a hurricane depends on several factors, including the size and population density of the area affected and the shape of the ocean bottom near the shore. The most significant factor, of course, is the strength of the storm itself. By studying past storms, a scale has been established to rank the relative intensities of hurricanes. As **TABLE 13.1** indicates, a *category 5* storm is the worst possible, whereas a *category 1* hurricane is least severe.

During hurricane season it is common to hear scientists and reporters alike use the numbers from the *Saffir–Simpson Hurricane Scale.* When Hurricane Katrina made landfall, sustained winds were 225 kilometers (140 miles) per hour, making it a strong category 4 storm. Storms that fall into category 5 are rare. Only three

Although many tropical disturbances develop each year, only a few reach hurricane status. By international agreement, a hurricane has wind speeds in excess of 119 kilometers (74 miles) per hour and a rotary circulation. Mature hurricanes average 600 kilometers (375 miles) in diameter and often extend 12,000 meters (40,000 feet) above the ocean surface. The lowest pressures ever recorded in the Western Hemisphere are associated with these storms.

An extremely rapid change in air pressure generates the strong, inward-spiraling winds of a hurricane (**FIGURE 13.27**). As the air rushes toward the center of the storm, its velocity increases. This occurs for the same reason that skaters with their arms extended spin faster as they pull their arms in close to their bodies.

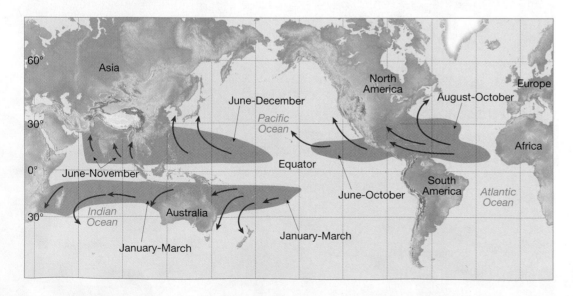

FIGURE 13.26 This world map shows the regions where most hurricanes form as well as their principal months of occurrence and the most common tracks they follow. Hurricanes do not develop within about 5° of the equator because the Coriolis effect (a force related to Earth's rotation that gives storms their "spin") is too weak. Because warm surface ocean temperatures are necessary for hurricane formation, they seldom form poleward of 20° latitude or over the cool waters of the South Atlantic and the eastern South Pacific.

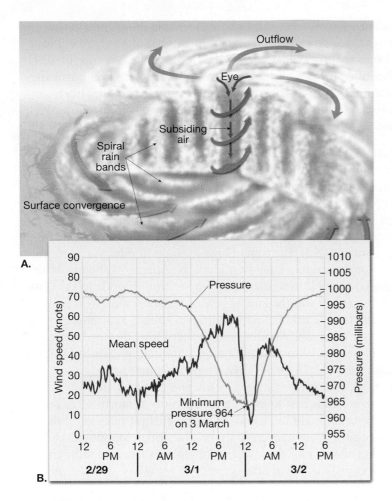

A.

B.

FIGURE 13.27 A. Cross section of a hurricane. Note that the vertical dimension is greatly exaggerated. The eye, the zone of relative calm at the center of the storm, is a distinctive hurricane feature. Sinking air in the eye warms by compression. Surrounding the eye is the eye wall, the zone where winds and rain are most intense. Tropical moisture spiraling inward creates rain bands that pinwheel around the storm center. Outflow of air at the top of the hurricane is important because it prevents the convergent flow at lower levels from "filling in" the storm. *(After NOAA)* **B.** Measurements of surface pressure and wind speed during the passage of Cyclone Monty at Mardie Station in Western Australia between February 29 and March 2, 2004. (Hurricanes are called "cyclones" in this part of the world.) The strongest winds are associated with the eye wall, and the weakest winds and lowest pressure are found in the eye. *(Data from World Meteorological Organization)*

storms this powerful are known to have hit the continental United States: Andrew struck Florida in 1992, Camille pounded Mississippi in 1969, and a Labor Day hurricane struck the Florida Keys in 1935. Damage caused by hurricanes can be divided into three categories: (1) storm surge, (2) wind damage, and (3) inland flooding.

STORM SURGE. Without question, the most devastating damage in the coastal zone is caused by the storm surge. It not only accounts for a large share of coastal property losses but is also responsible for a high percentage of all hurricane-caused deaths. A **storm surge** is a dome of water 65 to 80 kilometers (40 to 50 miles) wide that sweeps across the coast near the point where the eye makes landfall. If all wave activity were smoothed out, the storm surge would be the height of the water above normal tide level. In addition, tremendous wave activity is superimposed on the surge. We can easily imagine the damage that this surge of water could inflict on low-lying coastal areas (see Figure 13.24). The worst surges occur in places like the Gulf of Mexico, where the continental shelf is very shallow and gently sloping. In addition, local features such as bays and rivers can cause the surge height to double and increase in speed.

As a hurricane advances toward the coast in the Northern Hemisphere, storm surge is always most intense on the right side of the eye where winds are blowing *toward* the shore. In addition, on this side of the storm, the forward movement of the hurricane also contributes to the storm surge. In **FIGURE 13.28**, assume a hurricane with peak winds of 175 kilometers (109 miles) per hour is moving toward the shore at 50 kilometers (31 miles) per hour. In this case, the net wind speed on the right side of the advancing storm is 225 kilometers (140 miles) per hour. On the left side, the hurricane's winds are blowing opposite the direction of storm movement, so the net winds are *away* from the coast at 125 kilometers (78 miles) per hour. Along the shore facing the left side of the oncoming hurricane, the water level may actually decrease as the storm makes landfall.

> **DID YOU KNOW?**
> The difference between a *hurricane* and a *tropical storm* relates to intensity. Both are tropical cyclones—circular zones of low pressure with strong, inward-spiraling winds. When sustained winds are between 61 and 119 kilometers (37 to 74 miles) per hour, the cyclone is called a tropical storm. It is during this phase that a name is given (Andrew, Fran, Rita, and so on). When the cyclone's sustained winds exceed 119 kilometers (74 miles) per hour, it has reached hurricane status.

TABLE 13.1

Saffir–Simpson Hurricane Scale

Scale Number (Category)	Central Pressure (Millibars)	Wind Speed (KPH)	Wind Speed (MPH)	Storm Surge (Meters)	Storm Surge (Feet)	Damage
1	≥980	119–153	74–95	1.2–1.5	4–5	*Minimal*
2	965–979	154–177	96–110	1.6–2.4	6–8	*Moderate*
3	945–964	178–209	111–130	2.5–3.6	9–12	*Extensive*
4	920–944	210–250	131–155	3.7–5.4	13–18	*Extreme*
5	<920	>250	>155	>5.4	>18	*Catastrophic*

FIGURE 13.28 Winds associated with a Northern Hemisphere hurricane advancing toward the coast. This hypothetical storm, with peak winds of 175 kilometers (109 miles) per hour, is moving toward the coast at 50 kilometers (31 miles) per hour. On the right side of the advancing storm, the 175-kilometer-per-hour winds are in the same direction as the movement of the storm (50 kilometers per hour). Therefore, the *net* wind speed on the right side of the storm is 225 kilometers (140 miles) per hour. On the left side, the hurricane's winds are blowing opposite the direction of the storm movement, so the *net* winds of 125 kilometers (78 miles) per hour are away from the coast. Storm surge will be greatest along that part of the coast hit by the right side of the advancing hurricane.

WIND DAMAGE. Destruction caused by wind is perhaps the most obvious of the classes of hurricane damage. Debris such as signs, roofing materials, and small items left outside become dangerous flying missiles in hurricanes. For some structures the force of the wind is sufficient to cause total ruin. Mobile homes are particularly vulnerable. High-rise buildings are also susceptible to hurricane-force winds. Upper floors are most vulnerable because wind speeds usually increase with height. Recent research suggests that people should stay below the 10th floor but remain above floors that are at risk for flooding. In regions with good building codes, wind damage is usually not as catastrophic as storm-surge damage. However, hurricane-force winds affect a much larger area than storm surge and can cause huge economic losses. For example, in 1992 it was largely the winds associated with Hurricane Andrew that produced more than $25 billion of damage in southern Florida and Louisiana.

Hurricanes sometimes produce tornadoes that contribute to the storm's destructive power. Studies have shown that more than half of the hurricanes that make landfall produce at least one tornado. In 2004 the number of tornadoes associated with tropical storms and hurricanes was extraordinary. Tropical Storm Bonnie and five landfalling hurricanes—Charley, Frances, Gaston, Ivan, and Jeanne—produced nearly 300 tornadoes that affected the southeast and mid-Atlantic states.

HEAVY RAINS AND INLAND FLOODING. The torrential rains that accompany most hurricanes represent a third significant threat: flooding. Whereas the effects of storm surge and strong winds are concentrated in coastal areas, heavy rains may affect places hundreds of kilometers from the coast for up to several days after the storm has lost its hurricane-force winds.

In September 1999, Hurricane Floyd brought flooding rains, high winds, and rough seas to a large portion of the Atlantic seaboard. More than 2.5 million people evacuated their homes from Florida to the Carolinas and beyond. It was the largest peacetime evacuation in U.S. history up to that time. Torrential rains falling on already saturated ground created devastating inland flooding. Altogether Floyd dumped more than 48 centimeters (19 inches) of rain on Wilmington, North Carolina; 33.98 centimeters (13.38 inches) in a single 24-hour span.

Another well-known example is Hurricane Camille (1969). Although this storm is best known for its exceptional storm surge and the devastation it brought to coastal areas, the greatest number of deaths associated with this storm occurred in the Blue Ridge Mountains of Virginia two days after Camille's landfall. Many places received more than 25 centimeters (10 inches) of rain.

SUMMARY. To summarize, extensive damage and loss of life in the coastal zone can result from storm surge, strong winds, and torrential rains. When loss of life occurs, it is commonly caused by the storm surge, which can devastate entire barrier islands or zones within a few blocks of the coast. Although wind damage is usually not as catastrophic as the storm surge, it affects a much larger area. Where building codes are inadequate, economic losses can be especially severe. Because hurricanes weaken as they move inland, most wind damage occurs within 200 kilometers (125 miles) of the coast. Far from the coast, a weakening storm can produce extensive flooding long after the winds have diminished below hurricane levels. Sometimes the damage from inland flooding exceeds storm-surge destruction.

CONCEPT CHECK 13.9

❶ List the three categories of hurricane damage. Which one is responsible for the greatest number of hurricane-related deaths?

Coastal Classification

The great variety of shorelines demonstrates their complexity. Indeed, to understand any particular coastal area, many factors must be considered, including rock types, size and direction of waves, frequency of storms, range between high and low tides, and offshore topography. In addition, practically all coastal areas were affected by the worldwide rise in sea level that accompanied the melting of Ice Age glaciers at the close of the Pleistocene epoch. Finally, tectonic events that elevate or drop the land or change the volume of ocean basins must be taken into account. The large number of factors that influence coastal areas make shoreline classification difficult.

Many geologists classify coasts based on changes that have occurred with respect to sea level. This commonly used classification divides coasts into two very general categories: emergent and submergent. **Emergent coasts** develop either because an area experiences uplift or as a result of a drop in sea level. Conversely, **submergent coasts** are created when sea level rises or the land adjacent to the sea subsides.

Emergent Coasts

In some areas the coast is clearly emergent because rising land or a falling water level exposes wave-cut cliffs and platforms above sea level. Excellent examples include portions of coastal California where uplift has occurred in the recent geological past. The marine terrace shown in Figure 13.13, p. 317, illustrates this. In the case of the Palos Verdes Hills, south of Los Angeles, seven different terrace levels exist, indicating seven episodes of uplift. The ever-persistent sea is now cutting a new platform at the base of the cliff. If uplift follows, it too will become an elevated marine terrace.

Other examples of emergent coasts include regions that were once buried beneath great ice sheets. When glaciers were present, their weight depressed the crust, and when the ice melted, the crust began gradually to spring back. Consequently, prehistoric shoreline features today are found high above sea level. The Hudson Bay region of Canada is one such area, portions of which are still rising at a rate of more than 1 centimeter per year.

Submergent Coasts

In contrast to the preceding examples, other coastal areas show definite signs of submergence. Shorelines that have been submerged in the relatively recent past are often highly irregular because the sea typically floods the lower reaches of river valleys flowing into the ocean. The ridges separating the valleys, however, remain above sea level and project into the sea as headlands. These drowned river mouths, which are called **estuaries**, characterize many coasts today. Along the Atlantic coastline, the Chesapeake and Delaware bays are examples of estuaries created by submergence (**FIGURE 13.29**).

FIGURE 13.29 Estuaries along the East Coast of the United States. The lower portions of many river valleys were submerged by the rise in sea level that followed the end of the Ice Age. Chesapeake Bay and Delaware Bay are especially prominent examples.

Wave erosion caused by strong storms forced abandonment of this highly developed shoreline area in Long Island, New York. Coastal areas are dynamic places that can change rapidly in response to natural forces. *(Photo by Mark Wexler/Woodfin Camp & Associates)*

DID YOU KNOW?
Tides can be used to generate electrical power when a dam is constructed across the mouth of a bay or estuary in a coastal area having a large tidal range. The narrow opening between the bay and the open ocean magnifies the variations in water level as tides rise and fall and the strong in-and-out flow is used to drive turbines. The largest plant yet constructed is along the coast of France.

The picturesque coast of Maine, particularly in the vicinity of Acadia National Park, is another excellent example of an area that was flooded by the postglacial rise in sea level and transformed into a highly irregular submerged coastline.

Keep in mind that most coasts have a complicated geologic history. With respect to sea level, many have at various times emerged and then submerged again. Each time, they retain some of the features created during the previous situation.

CONCEPT CHECK 13.10

❶ What observable features would lead you to classify a coastal area as emergent?

❷ Are estuaries associated with submergent or emergent coasts? Explain.

Tides

Tides are daily changes in the elevation of the ocean surface. Their rhythmic rise and fall along coastlines have been known since antiquity. Other than waves, they are the easiest ocean movements to observe. An exceptional example of extreme daily tides is shown in **FIGURE 13.30**.

Although known for centuries, tides were not explained satisfactorily until Sir Isaac Newton applied the law of gravitation to them. Newton showed that there is a

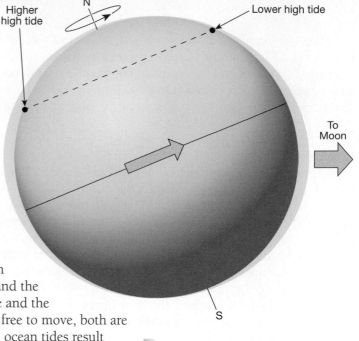

FIGURE 13.31 Idealized tidal bulges on Earth caused by the Moon. If Earth were covered to a uniform depth with water, there would be two tidal bulges: one on the side of Earth facing the Moon (right) and the other on the opposite side of Earth (left). Depending on the Moon's position, tidal bulges may be inclined relative to Earth's equator. In this situation, Earth's rotation causes an observer to experience two unequal high tides during a day.

mutual attractive force between two bodies, as between Earth and the Moon. Because the atmosphere and the ocean both are fluids and thus free to move, both are deformed by this force. Hence, ocean tides result from the gravitational attraction exerted upon Earth by the Moon and, to a lesser extent, by the Sun.

Causes of Tides

To illustrate how tides are produced, we will assume Earth is a rotating sphere covered to a uniform depth with water (**FIGURE 13.31**). It is easy to see how the Moon's gravitational force can cause the water to bulge on the side of Earth nearer the Moon. In addition, however, an equally large tidal bulge is produced on the side of Earth directly opposite the Moon.

Both tidal bulges are caused, as Newton discovered, by the pull of gravity. Gravity is inversely proportional to the square of the distance between two objects, meaning simply that it quickly weakens with distance. In this case, the two objects are the Moon and Earth. Because the force of gravity decreases with distance, the Moon's gravitational pull on Earth is slightly greater on the near side of Earth than on the far side. The result of this differential pulling is to stretch (elongate) the "solid" Earth very slightly. In contrast, the world ocean, which is mobile, is deformed quite dramatically by this effect, producing the two opposing tidal bulges.

Because the position of the Moon changes only moderately in a single day, the tidal bulges remain in place while Earth rotates through them. For this reason, if you stand on the seashore for 24 hours, Earth will rotate you through alternating areas of deeper and shallower water. As you are carried into each tidal bulge, the tide rises, and as you are carried into the intervening troughs between the tidal bulges, the tide falls. Therefore, most places on Earth experience two high tides and two low tides each tidal day.

Further, the tidal bulges migrate as the Moon revolves around Earth every 29 days. As a result, the tides, like the time of moonrise, shift about 50 minutes later each day. After 29 days the cycle is complete and a new one begins.

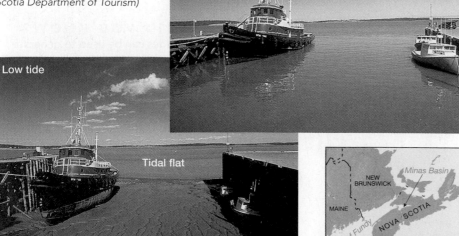

FIGURE 13.30 High tide and low tide on Nova Scotia's Minas Basin in the Bay of Fundy. Tidal flats are exposed during low tide. *(Photos courtesy of Nova Scotia Department of Tourism)*

DID YOU KNOW?

The world's largest tidal range (difference between successive high and low tides) is found in the northern end of Nova Scotia's Bay of Fundy. Here the maximum spring tidal range is about 17 meters (56 feet). This leaves boats "high and dry" during low tide (see Figure 13.30).

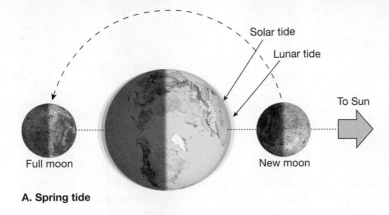

A. Spring tide

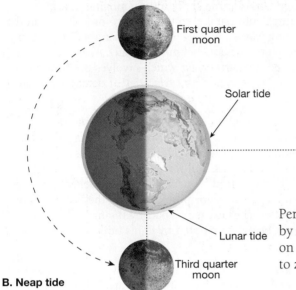

B. Neap tide

FIGURE 13.32 Earth–Moon–Sun positions and the tides. **A.** When the Moon is in the full or new position, the tidal bulges created by the Sun and Moon are aligned, there is a large tidal range on Earth, and spring tides are experienced. **B.** When the Moon is in the first- or third-quarter position, the tidal bulges produced by the Moon are at right angles to the bulges created by the Sun. Tidal ranges are smaller, and neap tides are experienced.

In many locations, there may be an inequality between the high tides during a given day. Depending on the Moon's position, the tidal bulges may be inclined to the equator as in Figure 13.31. This figure illustrates that one high tide experienced by an observer in the Northern Hemisphere is considerably higher than the high tide half a day later. In contrast, a Southern Hemisphere observer would experience the opposite effect.

Monthly Tidal Cycle

The primary body that influences the tides is the Moon, which makes one complete revolution around Earth every 29 and a half days. The Sun, however, also influences the tides. It is far larger than the Moon, but because it is much farther away, its effect is considerably less. In fact, the Sun's tide-generating effect is only about 46 percent that of the Moon's.

Near the times of new and full moons, the Sun and Moon are aligned and their forces are added together (**FIGURE 13.32A**). Accordingly, the combined gravity of these two tide-producing bodies causes larger tidal bulges (higher high tides) and deeper tidal troughs (lower low tides), producing a large tidal range. These are called the **spring tides,** which have no connection with the spring season but occur twice a month during the time when the Earth–Moon–Sun system is aligned. Conversely, at about the time of the first and third quarters of the Moon, the gravitational forces of the Moon and Sun act on Earth at right angles, and each partially offsets the influence of the other (**FIGURE 13.32B**). As a result, the daily tidal range is less. These are called **neap tides,** which also occur twice each month. Each month, then, there are two spring tides and two neap tides, each about one week apart.

Tidal Currents

Tidal current is the term used to describe the *horizontal* flow of water accompanying the rise and fall of the tide. These water movements induced by tidal forces can be important in some coastal areas. Tidal currents that advance into the coastal zone as the tide rises are called *flood currents.* As the tide falls, seaward-moving water generates *ebb currents.* Periods of little or no current, called *slack water,* separate flood and ebb. The areas affected by these alternating tidal currents are called **tidal flats** (see Figure 13.30, left). Depending on the nature of the coastal zone, tidal flats vary from narrow strips seaward of the beach to zones that may extend for several kilometers.

Although tidal currents are not important in the open sea, they can be rapid in bays, river estuaries, straits, and other narrow places. Off the coast of Brittany in France, for example, tidal currents that accompany a high tide of 12 meters (40 feet) may attain a speed of 20 kilometers (12 miles) per hour. Although tidal currents are not generally considered to be major agents of erosion and sediment transport, notable exceptions occur where tides move through narrow inlets. Here they scour the small entrances to many good harbors that would otherwise be blocked.

Sometimes deposits called **tidal deltas** are created by tidal currents (**FIGURE 13.33**). They may develop either as *flood deltas* landward of an inlet or as *ebb deltas* on the seaward side of an inlet. Because wave activity and longshore currents are reduced on the sheltered landward side, flood deltas are more common and actually more prominent (see Figure 13.15). They form after the tidal current moves rapidly through an inlet. As the current emerges into more open waters from the narrow passage, it slows and deposits its load of sediment.

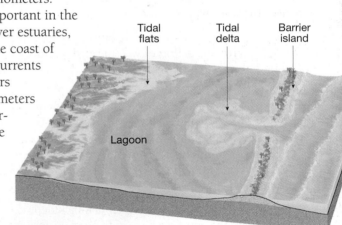

FIGURE 13.33 As a rapidly moving tidal current moves through a barrier island's inlet into the quiet waters of the lagoon, the current slows and deposits sediment, creating a tidal delta. Because this tidal delta has developed on the landward side of the inlet, it is called a *flood delta.*

CONCEPT CHECK 13.11

❶ Explain why an observer can experience two unequal high tides during one day.

❷ Distinguish between *neap tides* and *spring tides.*

❸ Contrast *flood current* and *ebb current.*

CHAPTER THIRTEEN

Shorelines in Review

⊙ The *shore* is the area extending between the lowest tide level and the highest elevation on land that is affected by storm waves. The *coast* extends inland from the shore as far as ocean-related features can be found. The shore is divided into the *foreshore* and *backshore*. Seaward of the foreshore are the *nearshore* and *offshore* zones.

⊙ A *beach* is an accumulation of sediment found along the landward margin of the ocean or a lake. Among its parts are one or more *berms* and the *beach face*. Beaches are composed of whatever material is locally abundant and should be thought of as material in transit along the shore.

⊙ *Waves* are *moving energy* and most *ocean waves are initiated by the wind*. The three factors that influence the *height*, *wavelength*, and *period* of a wave are (1) *wind speed*, (2) *length of time the wind has blown*, and (3) *fetch*, the distance the wind has traveled across open water. Once waves leave a storm area, they are termed *swells*, which are symmetrical, longer-wavelength waves.

⊙ As waves travel, *water particles transmit energy by circular orbital motion*, which extends to a depth equal to one half the wavelength. When a wave travels into shallow water, it experiences physical changes that can cause the wave to collapse, or *break*, and form *surf*.

⊙ Wave erosion is caused by *wave impact pressure* and *abrasion* (the sawing and grinding action of water armed with rock fragments). The bending of waves is called *wave refraction*. Owing to refraction, wave impact is concentrated against the sides and ends of headlands.

⊙ Most waves reach the shore at an angle. The uprush (swash) and backwash of water from each breaking wave moves the sediment in a zigzag pattern along the beach. This movement, called *beach drift*, can transport sand hundreds or even thousands of meters each day. Oblique waves also produce *longshore currents* within the surf zone that flow parallel to the shore.

⊙ Features produced by *shoreline erosion* include *wave-cut cliffs* (which originate from the cutting action of the surf against the base of coastal land), *wave-cut platforms* (relatively flat, benchlike surfaces left behind by receding cliffs), *sea arches* (formed when a headland is eroded and two caves from opposite sides unite), and *sea stacks* (formed when the roof of a sea arch collapses).

⊙ Some of the depositional features formed when sediment is moved by beach drift and longshore currents are *spits* (elongated ridges of sand that project from the land into the mouth of an adjacent bay), *baymouth bars* (sandbars that completely cross a bay), and *tombolos* (ridges of sand that connect an island to the mainland or to another island). Along the Atlantic and Gulf coastal plains, the shore zone is characterized by *barrier islands*, low ridges of sand that parallel the coast at distances of 3 to 30 kilometers offshore.

⊙ Local factors that influence shoreline erosion are (1) the *proximity of a coast to sediment-laden rivers*, (2) the *degree of tectonic activity*, (3) the *topography* and *composition of the land*, (4) *prevailing winds* and *weather patterns*, and (5) the *configuration of the coastline* and *nearshore areas*.

⊙ *Hard stabilization* involves building solid, massive structures in an attempt to protect a coast from erosion or prevent the movement of sand along the beach. Hard stabilization includes *groins* (short walls constructed at a right angle to the shore to trap moving sand), *breakwaters* (structures built parallel to the shore to protect it from the force of large breaking waves), and *seawalls* (armoring the coast to prevent waves from reaching the area behind the wall). *Alternatives to hard stabilization* include *beach nourishment*, which involves the addition of sand to replenish eroding beaches, and *relocation* of damaged or threatened buildings.

⊙ Because of basic geological differences, the *nature of shoreline erosion problems along America's Pacific and Atlantic coasts is very different*. Much of the development along the Atlantic and Gulf coasts has occurred on barrier islands, which receive the full force of major storms. Much of the Pacific Coast is characterized by narrow beaches backed by steep cliffs and mountain ranges. A major problem facing the Pacific shoreline is a narrowing of beaches caused in part by the natural flow of materials to the coast being interrupted by dams built for irrigation and flood control.

⊙ Although damages caused by a hurricane depend on several factors, including the size and population density of the area affected and the nearshore bottom configuration, the most significant factor is the strength of the storm itself. The *Saffir–Simpson* scale ranks the relative intensities of hurricanes. A category 5 storm is most severe and a category 1 storm is least severe. Damage caused by hurricanes can be divided into three categories: (1) *storm surge*, which is most intense on the right side of the eye where winds are blowing toward the shore, occurs when a dome of water sweeps across the coast near the point where the eye makes landfall; (2) *wind damage*; and (3) *inland flooding*, which is caused by torrential rains that accompany most hurricanes.

⊙ One common classification of coasts is based upon changes that have occurred with respect to sea level. Emergent coasts, often with wave-cut cliffs and wave-cut platforms above sea level, develop either because an area experiences uplift or as a result of a drop in sea level. Conversely, submergent coasts, with their drowned river mouths, called estuaries, are created when sea level rises or the land adjacent to the sea subsides.

⊙ Tides, the daily rise and fall in the elevation of the ocean surface, are caused by the *gravitational attraction* of the Moon and, to a lesser extent, by the Sun. Near the times of new and full moons, the Sun and Moon are aligned, and their gravitational forces are added together to produce especially high and low tides. These are called the *spring tides*. Conversely, at about the times of the first and third quarters of the Moon, when the gravitational forces of the Moon and Sun are at right angles, the daily tidal range is less. These are called *neap tides*.

⊙ *Tidal currents* are horizontal movements of water that accompany the rise and fall of tides. *Tidal flats* are the areas that are affected by the advancing and retreating tidal currents. When tidal currents slow after emerging from narrow inlets, they deposit sediment that may eventually create *tidal deltas*.

Key Terms

abrasion (p. 313)
backshore (p. 310)
barrier islands (p. 318)
baymouth bar (p. 317)
beach (p. 310)
beach drift (p. 315)
beach face (p. 310)
beach nourishment (p. 322)
berms (p. 310)
breakwater (p. 321)
coast (p. 310)
coastline (p. 310)

emergent coasts (p. 329)
estuaries (p. 329)
eye (p. 326)
eye wall (p. 326)
foreshore (p. 310)
groin (p. 321)
hard stabilization (p. 320)
jetties (p. 320)
longshore currents (p. 315)
marine terrace (p. 316)
neap tides (p. 331)
nearshore zone (p. 310)

offshore zone (p. 310)
rip currents (p. 315)
sea arch (p. 317)
sea stack (p. 317)
seawall (p. 321)
shore (p. 310)
shoreline (p. 309)
spit (p. 317)
spring tides (p. 331)
storm surge (p. 327)
submergent coasts (p. 329)
surf (p. 312)

tidal current (p. 331)
tidal deltas (p. 331)
tidal flats (p. 331)
tides (p. 330)
tombolo (p. 317)
wave-cut cliffs (p. 316)
wave-cut platform (p. 316)
wave height (p. 311)
wavelength (p. 311)
wave period (p. 311)
wave refraction (p. 314)

GIVE IT SOME THOUGHT

1 Examine Figure 13.19. In what direction are beach drift and longshore currents moving sand—toward the top or toward the bottom of the photo? How can you tell?

2 During a visit to the beach, you get in a small rubber raft and paddle out *beyond* the surf zone. Tiring, you stop and take a rest. Describe the movement of your raft during your rest. How does this movement differ, if at all, from what you would have experienced if you had stopped paddling while *in* the surf zone?

3 Refer to Figure 13.23A, a color-enhanced infrared image of Hurricane Katrina. Assume that the eye makes landfall exactly along the 90th meridian (the vertical green line extending through the storm). Which of the following statements is most accurate? Explain your answer.

 a. The storm surge will be greater west of the 90th meridian.

 b. The storm surge will be greater east of the 90th meridian.

 c. There is no way to make an educated guess about which side will have the bigger storm surge.

4 You and a friend set up an umbrella and chairs at a beach. Your friend then goes into the surf zone to play Frisbee with another person. Several minutes later, your friend looks back toward the beach and is surprised to see that she is no longer near where the umbrella and chairs were set up. Although she is still in the surf zone, she is 30 or 40 yards away from where she started. How would you explain to your friend why she moved along the shore?

5 A friend wants to purchase a vacation home on a barrier island. If consulted, what advise would you give your friend?

6 Great damage and significant loss of life can take place a day or more *after* a hurricane has moved ashore and weakened. When this occurs, what is the likely cause?

7 The force of gravity plays a critical role in creating ocean tides. The more massive an object, the stronger its pull of gravity. Explain why the Sun's influence is only about half that of the Moon's, even though the Sun is much more massive than the Moon.

Companion Website

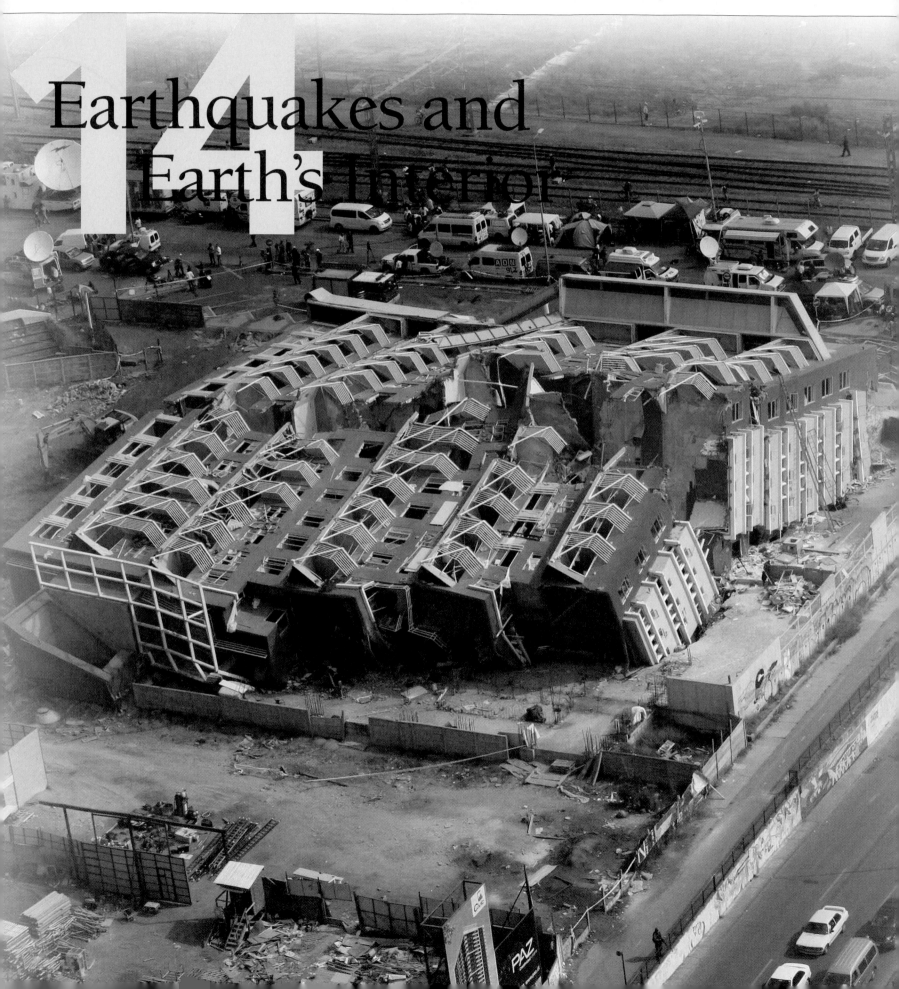

Earthquakes and Earth's Interior

O n February 27, 2010, at 3:34 am local time, an earthquake measuring 8.8 occurred approximately 11 kilometers (7 miles) off the coast of Chile, resulting in 486 fatalities and widespread property damage. This powerful quake prompted seismologists to estimate that the length of the day was shortened by 1.26 microseconds and Earth's axis moved by 8 centimeters (3 inches). In addition, the event laterally displaced cities, including Concepción by more than 3 meters (10 feet) and Santiago approximately 24 centimeters (10 inches).

The Borde Rio apartment building in Concepción, Chile, was toppled by an 8.8-magnitude earthquake on February 27, 2010 *(Photo by Pilar Olivares/Reuters)*

To assist you in learning the important concepts in this chapter, focus on the following questions:

- ◉ What is an earthquake?
- ◉ What are the types of earthquake waves?
- ◉ How is the epicenter of an earthquake determined?
- ◉ Where are the principal earthquake zones on Earth?
- ◉ How is earthquake strength expressed?
- ◉ What are the main factors that affect the amount of destruction caused by seismic shaking?
- ◉ What is a tsunami?
- ◉ How did Earth acquire its layered structure?
- ◉ What are the major zones of Earth's interior?
- ◉ How do continental crust and oceanic crust differ?

An Earthquake Disaster In Haiti

On Tuesday, January 12, 2010, an estimated 230,000 people lost their lives when a magnitude 7.0 earthquake struck the small Caribbean nation of Haiti, the poorest country in the Western Hemisphere (**FIGURE 14.1**).

The quake originated only 25 kilometers (15 miles) from the country's densely populated capital city of Port-au-Prince. It occurred along a San Andreas-like fault at a shallow depth of just 10 kilometers (6 miles). Because of the quake's shallow depth, ground shaking was exceptional for an event of this magnitude. Other factors also contributed to the Port-au-Prince disaster including the city's geologic setting and the nature of its buildings. The city is not built on solid bedrock, but rather on sediment, which is more susceptible to ground shaking by earthquake waves. More importantly, inadequate or nonexistent building codes meant that buildings collapsed far more readily than they should have. At least 52 aftershocks, measuring 4.5 or greater, jolted the area and added to the trauma survivors experienced for days after the original quake.

The death toll of 230,000 rivals the loss of life in the tragic 2004 Indonesian tsunami, an earthquake-generated event that is described later in the chapter. Both of these natural disasters produced death tolls equivalent to the populations of cities the size of Madison, Wisconsin, or Orlando, Florida. The loss of life associated with the 2010 Haitian event is even more extraordinary when it is compared with the 1989 Loma Prieta earthquake in southern California, which had a similar magnitude (7.1) but claimed just 67 lives. In addition to the staggering death toll, there were more than 300,000 injuries, and 250,000 residences were destroyed. Nearly a million people were left homeless.

Relief agencies from around the globe stepped in to distribute food and water and to provide for the enormous medical, security, and social needs of the injured and displaced. For a lengthy period following the quake, inadequate infrastructure coupled with devastating damages combined to inhibit the timely delivery of essential services. Such a human crisis can easily continue for an extended period due to the spread of disease and complications from minimally or untreated injuries. The rebuilding of Haiti will likely take a decade or more and billions of dollars in assistance.

CONCEPT CHECK 14.1

① List three factors that contributed to making the Haiti earthquake a serious natural disaster.

What Is an Earthquake?

 Earthquakes

ESSENTIALS OF GEOLOGY What Is an Earthquake?

Earthquakes are natural geologic phenomena caused by the sudden and rapid movement of a large volume of rock. The violent shaking and destruction caused by earthquakes are the result of rupture and slippage along fractures in Earth's crust called **faults.** Larger quakes result from the rupture of larger fault segments. The origin of an earthquake occurs at depths between 5 and 700 kilometers, at the **focus**. The point at the surface directly above the focus is called the **epicenter** (**FIGURE 14.2**).

During large earthquakes, a massive amount of energy is released as **seismic waves**—a form of elastic energy that causes vibrations in the material that transmits them. Seismic waves are analogous to waves produced when a stone is dropped

FIGURE 14.1 A Haitian girl stands amid the rubble of a destroyed building following the 7.0 magnitude earthquake that devastated Port-au-Prince, Haiti, on January 12, 2010. (*Photo by Orlando Barria/Corbis*)

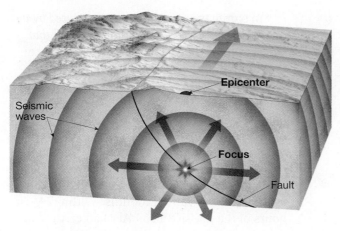

FIGURE 14.2 Earthquake focus and epicenter. The *focus* is the zone within Earth where the initial displacement occurs. The *epicenter* is the surface location directly above the focus.

Thousands of earthquakes occur around the world every day. Fortunately, most are so small that they can only be detected by sensitive instruments. Of these, only about 75 strong quakes are recorded each year and many of these occur in remote regions. Occasionally, a large earthquake is triggered near a major population center. Such events are among the most destructive natural forces on Earth. The shaking of the ground, coupled with the liquefaction of soils, wreaks havoc on buildings, roadways, and other structures. In addition, when a quake occurs in a populated area, power and gas lines are often ruptured, causing numerous fires. In the famous 1906 San Francisco earthquake, much of the damage was caused by fires that became uncontrollable when broken water mains left firefighters with only trickles of water (**FIGURE 14.3**).

into a calm pond. Just as the impact of the stone creates a pattern of waves in motion, an earthquake generates waves that radiate outward in all directions from the focus. Even though seismic energy dissipates rapidly with increasing distance, sensitive instruments located around the world detect and record these events.

Discovering the Causes of Earthquakes

The energy released by atomic explosions or by the movement of magma in Earth's crust can generate earthquakelike waves, but these events are generally quite weak. What mechanism produces a destructive earthquake? As you have learned, Earth is not a static planet. We know that large sections of Earth's crust have been thrust upward, because fossils of marine organisms have been discovered thousands of meters above sea level. Other regions exhibit evidence of extensive subsidence. In addition to these vertical displacements, offsets

FIGURE 14.3 San Francisco in flames after the 1906 earthquake. (Reproduced from the collection of the Library of Congress) Inset photo shows fire triggered when a gas line ruptured during the Northridge earthquake in Southern California in 1994. *(AFP/Getty Images)*

FIGURE 14.4 Slippage along a fault produced an offset in this orange grove east of Calexico, California. *(Photo by John S. Shelton)* Inset photo shows a fence offset 2.5 meters (8.5 feet) during the 1906 San Francisco earthquake. *(Photo by G. K. Gilbert, U.S. Geological Survey)*

In summary, *earthquakes are produced by the rapid release of elastic energy stored in rock that has been deformed by differential stresses.* Once the strength of the rock is exceeded, it suddenly ruptures, causing the vibrations of an earthquake.

Aftershocks and Foreshocks

Strong earthquakes are followed by numerous smaller tremors, called **aftershocks**, that gradually diminish in frequency and intensity over a period of several months. Within 24 hours of the massive 1964 Alaskan earthquake,

in fence lines, roads, and other structures indicate that horizontal movements are also common (**FIGURE 14.4**).

The actual mechanism of earthquake generation eluded geologists until H. F. Reid of Johns Hopkins University conducted a study following the great 1906 San Francisco earthquake. The earthquake was accompanied by horizontal surface displacements of several meters along the northern portion of the San Andreas Fault. Field studies determined that during this single earthquake, the Pacific plate lurched as much as 4.7 meters (15 feet) northward past the adjacent North American plate.

What Reid concluded from his investigations is illustrated in **FIGURE 14.5**. Tectonic stresses acting over tens to hundreds of years slowly deform the crustal rocks on both sides of a fault. When deformed by differential stress, rocks bend and store elastic energy, much like a wooden stick does if bent (Figure 14.5B). Eventually, the frictional resistance holding the rocks in place is overcome. Slippage allows the deformed (strained) rock to "snap back" to its original, stress-free shape (Figure 14.5C, D). The "springing back" was termed **elastic rebound** by Reid because the rock behaves elastically, much like a stretched rubber band does when it is released. The vibrations we know as an earthquake are generated by the rock elastically returning to its original shape.

FIGURE 14.5 Elastic rebound. As rock is deformed, it bends, storing elastic energy. Once strained beyond its breaking point, the rock cracks, releasing the stored-up energy in the form of earthquake waves.

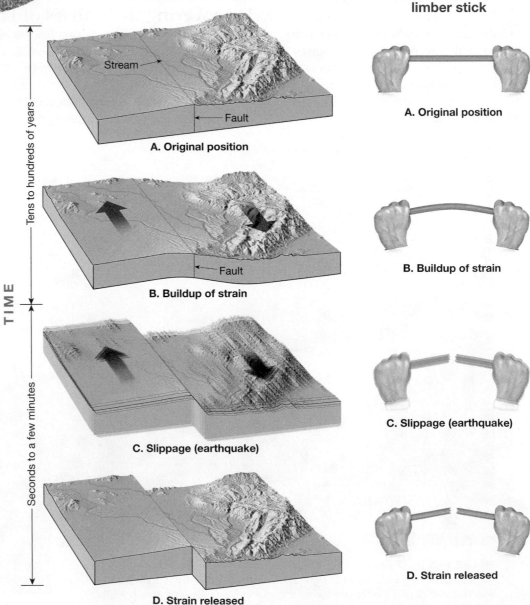

Deformation of rocks

A. Original position

B. Buildup of strain

C. Slippage (earthquake)

D. Strain released

Deformation of a limber stick

A. Original position

B. Buildup of strain

C. Slippage (earthquake)

D. Strain released

28 aftershocks were recorded, 10 of which had magnitudes that exceeded 6. More than 10,000 aftershocks with magnitudes of 3.5 or above occurred in the following 69 days and thousands of minor tremors were recorded over a span of 18 months. Because aftershocks happen mainly on the section of the fault that has slipped, they provide geologists with data that is useful in establishing the dimensions of the rupture surface.

Although aftershocks are weaker than the main earthquake, they can trigger the destruction of already weakened structures. This occurred in northwestern Armenia (1988) where many people lived in large apartment buildings constructed of brick and concrete slabs. After a moderate earthquake of magnitude 6.9 weakened the buildings, a strong aftershock of magnitude 5.8 completed the demolition.

In contrast to aftershocks, small earthquakes called **foreshocks** often precede a major earthquake by days or in some cases years. Monitoring of foreshocks to predict forthcoming earthquakes has been attempted with limited success.

CONCEPT CHECK 14.2

❶ What is an *earthquake*? Under what circumstances do earthquakes occur?

❷ How are *faults*, *foci*, and *epicenters* related?

❸ Who was first to explain the actual mechanism by which earthquakes are generated?

❹ Sketch or explain what is meant by *elastic rebound*.

Earthquakes and Faults

Earthquakes take place along faults both new and old that occur in places where differential stresses have ruptured Earth's crust. Some faults are large and capable of generating major earthquakes. One example is the San Andreas Fault, which is the transform fault boundary that separates two great sections of Earth's lithosphere: the North American plate and the Pacific plate. Other faults are small and capable of producing only minor earthquakes.

Most of the displacement that occurs along faults can be satisfactorily explained by the plate tectonics theory, which states that large slabs of Earth's lithosphere are in continual slow motion. These mobile plates interact with neighboring plates, straining and deforming the rocks at their margins. Faults associated with plate boundaries are the source of most large earthquakes.

Large faults are not perfectly straight or continuous; instead, they consist of numerous branches and smaller fractures that display kinks and offsets. Such a pattern is displayed in **FIGURE 14.6**, which shows the San Andreas Fault as a system of several faults of various sizes.

The San Andreas is undoubtedly the most studied fault system in the world. Over the years, research has shown that displacement occurs along discrete segments that behave somewhat differently from one another. A few sections of the San Andreas exhibit a slow, gradual displacement known as *fault creep*, which occurs without the buildup of significant strain. These sections produce only minor seismic shaking. Other segments slip at regular intervals, producing small to moderate earthquakes. Still other segments remain locked and store energy for a few hundred years before rupturing in great earthquakes. Earthquakes that occur along locked segments of the San Andreas Fault tend to

be repetitive. As soon as one is over, the continuous motion of the plates begins building strain anew. Decades or centuries later, the fault fails again.

CONCEPT CHECK 14.3

❶ Faults that are experiencing no active creep may be considered safe. Rebut or defend this statement.

❷ Why is the San Andreas sometimes referred to as a *fault system*?

Seismology: The Study of Earthquake Waves

 Earthquakes

ESSENTIALS OF GEOLOGY **Seismology**

The study of earthquake waves, **seismology,** dates back to attempts made by the Chinese almost 2000 years ago to determine the direction from which these waves originated. Modern **seismographs,** instruments that record earthquake waves, are similar to the instruments used by the Chinese. Seismographs have a weight freely suspended from a support that is

Aftermath of a devastating tsunami at the coastal town of Banda Aceh on the Indonesian island of Sumatra, December 26, 2004. *(Mark Pearson/Alamy)*

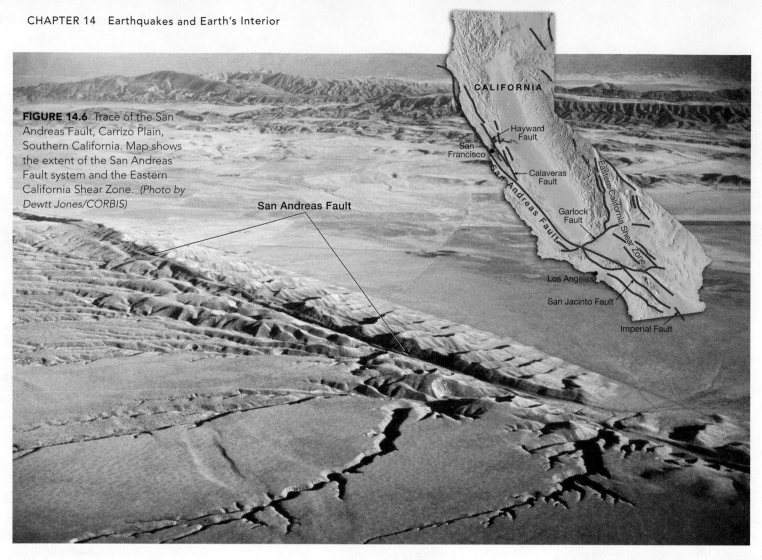

FIGURE 14.6 Trace of the San Andreas Fault, Carrizo Plain, Southern California. Map shows the extent of the San Andreas Fault system and the Eastern California Shear Zone. *(Photo by Dewtt Jones/CORBIS)*

securely attached to bedrock (**FIGURE 14.7**). When vibrations from an earthquake reach the instrument, the inertia of the weight keeps it relatively stationary while Earth and the support move. (*Inertia* is the tendency of objects at rest to stay at rest and objects in motion to remain in motion.)

To detect very weak earthquakes, or a great earthquake that has occurred in another part of the world, most seismographs are designed to amplify ground motion. Other instruments are designed to withstand the violent shaking that occurs very near the focus.

The records obtained from seismographs, called **seismograms**, provide useful information about the nature of seismic waves. Seismograms reveal that two main groups of seismic waves are generated by

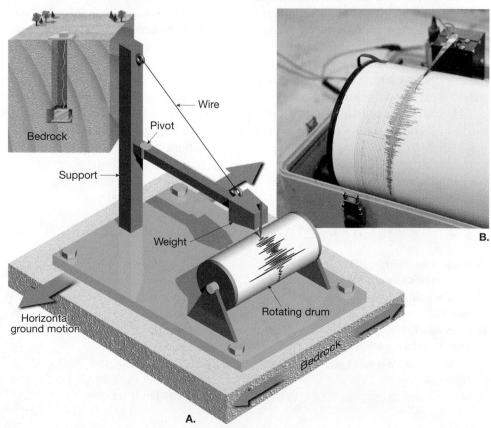

FIGURE 14.7 Principle of the seismograph. **A.** The inertia of the suspended weight tends to keep it motionless while the recording drum, which is anchored to bedrock, vibrates in response to seismic waves. The stationary weight provides a reference point from which to measure the amount of displacement occurring as a seismic wave passes through the ground. **B.** Seismograph recording earthquake tremors. *(Photo courtesy of Zephyr/Photo Researchers, Inc.)*

the slippage of a rock mass. Some are called **surface waves**, because their motion is restricted to near Earth's surface. Others travel through Earth's interior and are called **body waves**. Body waves are divided into two types—**primary (P) waves** and **secondary (S) waves.**

Body waves are identified by their mode of travel through intervening materials. P waves are "push–pull" waves—they momentarily push (squeeze) and pull (stretch) rocks in the direction the wave is traveling (**FIGURE 14.8A**). This wave motion is similar to that generated by human vocal cords as they move air to create sound. Solids, liquids, and gases resist a change in volume when compressed and will elastically spring back once the force is removed. Therefore, P waves can travel through all three of these materials.

On the other hand, S waves "shake" the particles at right angles to their direction of travel. This can be illustrated by fastening one end of a rope and shaking the other end, as shown in Figure 14.8B. Unlike P waves, which temporarily change the *volume* of intervening material by alternately squeezing and stretching it, S waves change the *shape* of the material that transmits them. Because fluids (gases and liquids) do not resist stresses that cause changes in shape—meaning fluids will not return to their original shape once the stress is removed—they will not transmit S waves.

The motion of surface waves is somewhat more complex. As surface waves travel along the ground, they cause the ground and anything resting upon it to move, much like ocean swells toss a ship (Figure 14.8C). In addition to their up-and-down motion, surface waves have a side-to-side motion similar to an S wave oriented in a horizontal plane (Figure

14.8D). This latter motion is particularly damaging to the foundations of structures.

By examining the "typical" seismic record shown in **FIGURE 14.9**, you can see a major difference among seismic waves—their speed of travel. P waves are the first to arrive at a recording station, then S waves, and finally surface waves. The velocity of P waves through the crustal rock granite is about 6 kilometers per second and increases to nearly 13 kilometers per

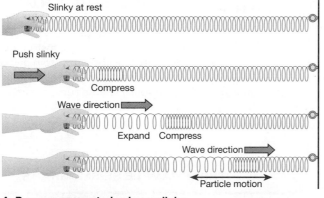

A. P waves generated using a slinky

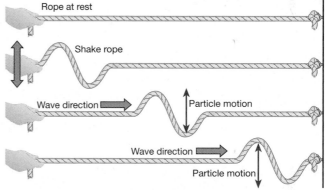

B. S waves generated using a rope

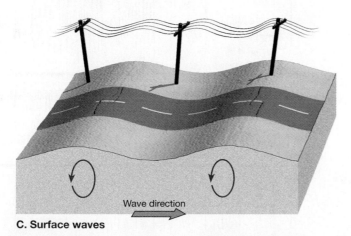

C. Surface waves

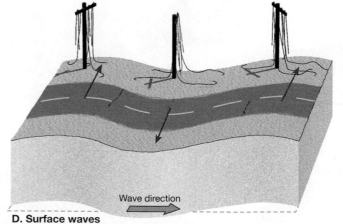

D. Surface waves

FIGURE 14.8 Types of seismic waves and their characteristic motion. (Note that during a strong earthquake, ground shaking consists of a combination of various kinds of seismic waves.) **A.** As illustrated by a slinky, P waves are compressional waves that alternately compress and expand the material through which they pass. **B.** S waves cause material to oscillate at right angles to the direction of wave motion. **C.** One type of surface wave travels along Earth's surface similar to rolling ocean waves. The red arrows show the elliptical movement of rock as the wave passes. **D.** Another type of surface wave moves the ground from side to side and can be particularly damaging to the foundations of buildings.

second at the base of the mantle. P waves can pass through Earth's mantle in about 20 minutes. Generally, in any solid material, P waves travel about 1.7 times faster than S waves, and surface waves are roughly 10 percent slower than S waves.

In addition to velocity differences, notice in Figure 14.9 that the height, or *amplitude,* of these wave types also varies. S waves have slightly greater amplitudes than P waves, while surface waves exhibit even greater amplitudes. Surface waves also retain their maximum amplitude longer than P and S waves. As a result, surface waves tend to cause greater ground shaking, and hence greater destruction, than either P or S waves.

Seismic waves are useful in determining the location and magnitude of earthquakes. In addition, seismic waves provide an important tool for probing Earth's interior.

CONCEPT CHECK 14.4

❶ Describe the principle of a *seismograph.*

❷ List the major differences between *P, S,* and *surface waves.*

❸ Which type of seismic wave causes the greatest destruction to buildings?

Locating the Source of an Earthquake

Earthquakes

ESSENTIALS OF GEOLOGY **Locating The Source of an Earthquake**

When analyzing an earthquake, the first task seismologists undertake is determining its *epicenter,* the point on Earth's surface directly above the focus (see Figure 14.2). The method used for locating an earthquake's epicenter relies on the fact that P waves travel faster than S waves.

The method is analogous to the results of a race between two autos, one faster than the other. The first P wave, like the faster auto, always wins the race, arriving ahead of the first S wave. The greater the length of the race, the greater the difference in their arrival times at the finish line (the seismic station). Therefore, the greater the interval between the arrival of the first P wave and the arrival of the first S wave, the greater the distance to the epicenter. **FIGURE 14.10** shows three simplified seismograms for the same earthquake. Based on the P–S interval, which city—Nagpur, Darwin, or Paris—is farthest from the epicenter?

The system for locating earthquake epicenters was developed by using seismograms from earthquakes whose epicenters

could be easily pinpointed from physical evidence. From these seismograms, travel-time graphs were constructed (**FIGURE 14.11**). Using the sample seismogram for Nagpur, India in Figure 14.10A and the travel-time curve in Figure 14.11, we can determine the distance separating the recording station from the earthquake in two steps: (1) Using the seismogram, determine the time interval between the arrival of the first P wave and the arrival of the first S wave, and (2) using the travel-time graph, find the P–S interval on the vertical axis and use that information to determine the distance to the epicenter on the horizontal axis. Following this procedure, we can determine that the earthquake occurred 3400 kilometers (2100 miles) from the recording instrument in Nagpur, India.

DID YOU KNOW?

Although seismographs were developed to record earthquake waves, they are sensitive instruments that record vibrations from any source, including underground nuclear tests, volcanic eruptions, or simply waves beating on a nearby shore.

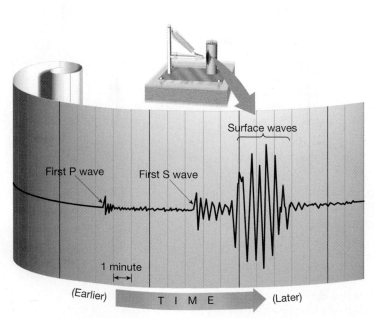

FIGURE 14.9 Typical seismogram. Note the time interval (about 5 minutes) between the arrival of the first P wave and the arrival of the first S wave.

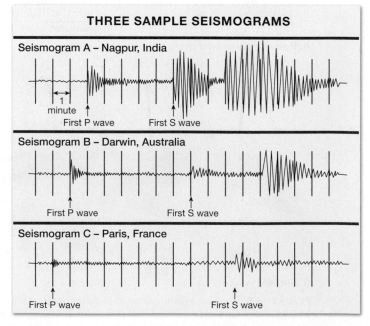

FIGURE 14.10 Simplified seismograms of the same earthquake recorded in three different cities. **A.** Nagpur, India. **B.** Darwin, Australia. **C.** Paris, France.

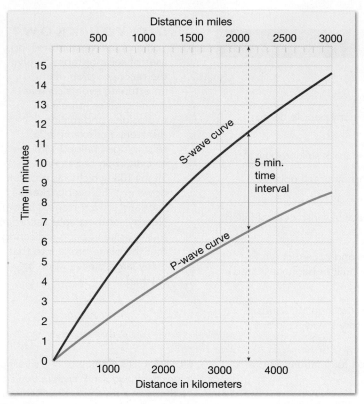

FIGURE 14.11 A travel-time graph is used to determine the distance to an earthquake epicenter. The difference in arrival time between the first P wave and the first S wave in the example is 5 minutes. Thus, the epicenter is roughly 3400 kilometers (2100 miles) away.

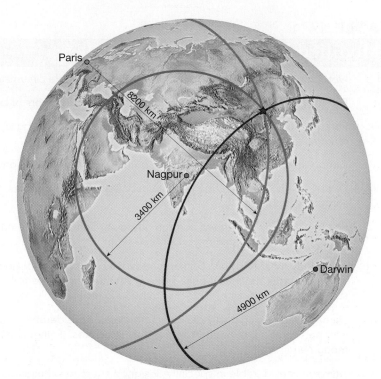

FIGURE 14.12 Determining an earthquake epicenter using the distances obtained from three or more seismic stations—a method called *triangulation*.

Now we know the *distance,* but what about *direction*? The epicenter could be in any direction from the seismic station. Using a method called *triangulation,* the precise location can be determined using the distance from three or more seismic stations (**FIGURE 14.12**). On a globe, a circle is drawn around each seismic station. The radius of these circles is equal to the distance from the seismic station to the epicenter. The point where the three circles intersect is the epicenter of the quake.

CONCEPT CHECK 14.5

❶ What information does a travel-time graph provide?

❷ Briefly describe the *triangulation* method used to determine the epicenter of an earthquake.

Measuring the Size of Earthquakes

Historically, seismologists have employed a variety of methods to determine two fundamentally different measures that describe the size of an earthquake—intensity and magnitude. The first of these to be used was **intensity**—a measure of the degree of earthquake shaking at a given locale based on observed effects. Later, with the development of seismographs, it became possible to measure ground motion using instruments. This quantitative measurement, called **magnitude,** relies on data gleaned from seismic records to estimate the amount of energy released at an earthquake's source.

Intensity and magnitude provide useful, though different, information about earthquake strength. Consequently, both measures are used to describe earthquake severity.

Modified Mercalli Intensity Scale

Numerous intensity scales have been developed over the last 150 years. The one widely used is the **Modified Mercalli Intensity Scale**—named after Giuseppe Mercalli, who initially developed it in 1902 (**TABLE 14.1**). This intensity scale is divided into twelve levels of severity based on observed effects such as people awakening from sleep, furniture moving, plaster cracking and falling, and finally—total destruction. As Table 14.1 illustrates, the lower numbers on the Mercalli scale (I-V) refer to what people in various locations felt during the quake, whereas the higher numbers (VI-XII) are based on observable damage to buildings and other structures. **FIGURE 14.13** shows shaking intensity maps for two San Francisco Bay area quakes—1989 Loma Prieta and 1906 San Francisco earthquakes. Although the 1989 Loma Prieta quake caused billions of dollars in damage and claimed more than 60 lives, these shaking maps show that a repeat of the 1906 San Francisco earthquake would certainly be more catastrophic.

TABLE 14.1

Modified Mercalli Intensity Scale

I	Not felt except by a very few under especially favorable circumstances.
II	Felt only by a few persons at rest, especially on upper floors of buildings.
III	Felt quite noticeably indoors, especially on upper floors of buildings, but many people do not recognize it as an earthquake.
IV	During the day felt indoors by many, outdoors by few. Sensation like heavy truck striking building.
V	Felt by nearly everyone, many awakened. Disturbances of trees, poles, and other tall objects sometimes noticed.
VI	Felt by all; many frightened and run outdoors. Some heavy furniture moved; few instances of fallen plaster or damaged chimneys. Damage slight.
VII	Everybody runs outdoors. Damage negligible in buildings of good design and construction; slight to moderate in well-built ordinary structures; considerable in poorly built or badly designed structures.
VIII	Damage slight in specially designed structures; considerable in ordinary substantial buildings with partial collapse; great in poorly built structures. (Fall of chimneys, factory stacks, columns, monuments, walls.)
IX	Damage considerable in specially designed structures. Buildings shifted off foundations. Ground cracked conspicuously.
X	Some well-built wooden structures destroyed. Most masonry and frame structures destroyed. Ground badly cracked.
XI	Few, if any, (masonry) structures remain standing. Bridges destroyed. Broad fissures in ground.
XII	Damage total. Waves seen on ground surfaces. Objects thrown upward into air.

DID YOU KNOW?
Some of the most interesting uses of seismographs involve the reconstruction of unfortunate events such as airline crashes, pipeline explosions, and mining disasters. For example, a seismologist helped with the investigation of Pan Am Flight 103, which was brought down in 1988 over Lockerbie, Scotland, by a terrorist's bomb. Nearby seismographs recorded six separate impacts indicating that the plane broke into that many large pieces when it crashed.

FIGURE 14.13 Comparison of shaking intensity for two San Francisco Bay area earthquakes—the 1989 Loma Prieta and the 1906 San Francisco earthquakes. The colors on the map represent levels of shaking based on the Modified Mercalli Intensity Scale. *(Data from U.S. Geological Survey)*

Despite their usefulness in providing a tool to compare earthquake severity, intensity scales have significant drawbacks. These scales are based on effects (largely destruction) that depend not only on the severity of ground shaking but also on factors such as building design and the nature of surface materials. For example, the modest 7.0 magnitude 2010 Haiti earthquake mentioned earlier was extremely destructive, mainly because of inferior building practices. Thus, the destruction wrought by an earthquake is frequently not a good measure of the amount of energy that was unleashed.

Magnitude Scales

In order to more accurately compare earthquakes across the globe, a measure was needed that does not rely on parameters that vary considerably from one part of the world to another. As a consequence, several magnitude scales were developed.

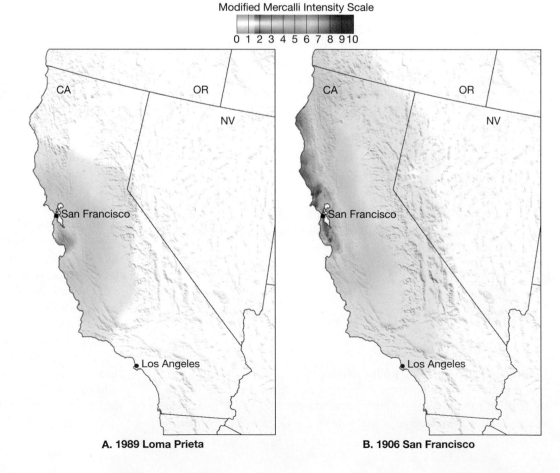

Modified Mercalli Intensity Scale

0 1 2 3 4 5 6 7 8 9 10

A. 1989 Loma Prieta

B. 1906 San Francisco

RICHTER MAGNITUDE. In 1935 Charles Richter of the California Institute of Technology developed the first magnitude scale using seismic records. As shown in **FIGURE 14.14** (top), the **Richter scale** is based on the amplitude of the largest seismic wave (P, S, or surface wave) recorded on a seismogram. Because seismic waves weaken as the distance between the focus and the seismograph increases, Richter developed a method that accounts for the decrease in wave amplitude with increasing distance. Theoretically, as long as equivalent instruments are used, monitoring stations at various locations will obtain the same Richter magnitude for each recorded earthquake. In practice, however, different recording stations often obtain slightly different Richter magnitudes for the same earthquake—a consequence of the variations in rock types through which the waves travel.

Earthquakes vary enormously in strength, and great earthquakes produce wave amplitudes that are thousands of times larger than those generated by weak

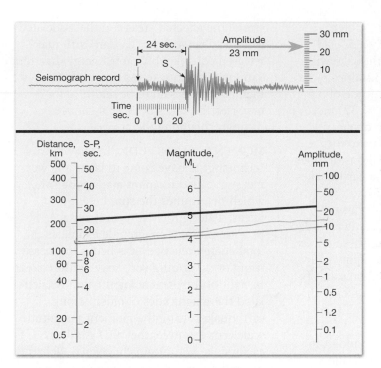

FIGURE 14.14 Illustration showing how the Richter magnitude of an earthquake can be determined graphically using a seismograph record from a Wood-Anderson instrument. First, measure the height (amplitude) of the largest wave on the seismogram (23 mm) and then the distance to the epicenter using the time interval between S and P waves (24 seconds). Next, draw a line between the distance scale (left) and the wave amplitude scale (right). By doing this, you should obtain the Richter magnitude (M_L) of 5. (*Data from California Institute of Technology*)

tremors (**FIGURE 14.15**). To accommodate this wide variation, Richter used a *logarithmic scale* to express magnitude, in which a *tenfold* increase in wave amplitude corresponds to an increase of 1 on the magnitude scale. Thus, the degree of ground shaking for a 5-magnitude earthquake is 10 times greater than that produced by an earthquake having a Richter magnitude of 4.

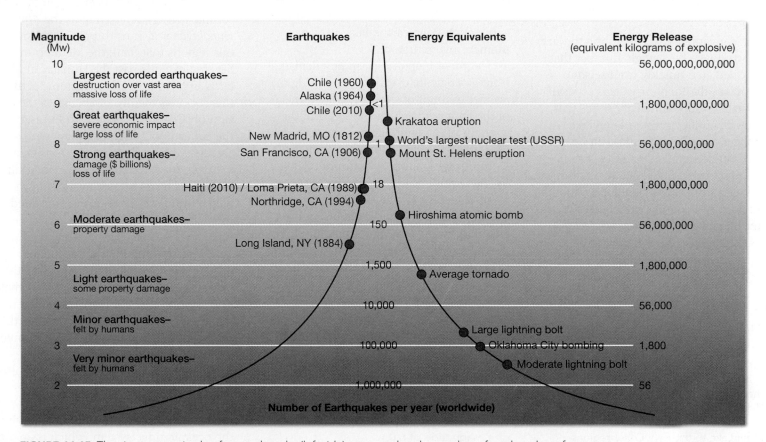

FIGURE 14.15 The size or magnitude of an earthquake (left side) compared to the number of earthquakes of various magnitudes that occur worldwide each year. The largest earthquakes occur less than once a year, whereas strong earthquakes happen more than once a month; weak quakes, those less than magnitude 2, occur hundreds of times per day. (*Data from IRIS Consortium, www.iris.edu*)

In addition, each unit of Richter magnitude equates to roughly a *32-fold energy increase.* Thus, an earthquake with a magnitude of 6.5 releases 32 times more energy than one with a magnitude of 5.5, and roughly 1000 times (32 × 32) more energy than a 4.5-magnitude quake. Furthermore, a major earthquake with a magnitude of 8.5 releases millions of times more energy than the smallest earthquakes felt by humans (Figure 14.15).

Although the Richter scale has no upper limit, the largest magnitude recorded was 8.9. Great shocks such as these release an amount of energy that is roughly equivalent to the detonation of 1 billion tons of explosives. Conversely, earthquakes with a Richter magnitude of less than 2.0 are generally not felt by humans.

Richter's original goal was modest in that he only attempted to rank shallow earthquakes in southern California into groups of large, medium, and small magnitude. Hence, Richter magnitude was designed to classify relatively local earthquakes and is designated by the symbol (M_L)—where M is for *magnitude* and L is for *local.*

The convenience of describing the size of an earthquake by a single number that can be calculated quickly from seismograms makes the Richter scale a powerful tool. Further, unlike intensity scales that can only be applied to populated areas of the globe, Richter magnitudes can be assigned to earthquakes in more remote regions and even to events that occur in the ocean basins. In time, seismologists modified Richter's work and developed new Richter-like magnitude scales.

Despite its usefulness, the Richter scale is not adequate for describing very large earthquakes. For example, the 1906 San Francisco earthquake and the 1964 Alaskan earthquake had roughly the same Richter magnitudes. However, based on the relative

size of the affected areas and the associated tectonic changes, the Alaskan earthquake released considerably more energy than the San Francisco quake. As a result, the Richter scale is said to be *saturated* for major earthquakes because it cannot distinguish among them.

MOMENT MAGNITUDE. In recent years, seismologists have come to favor a newer measure called **moment magnitude** (M_W), which determines the strain energy released along the entire fault surface. Because moment magnitude estimates the total energy released, it is better for measuring or describing very large earthquakes. In light of this, seismologists have recalculated the magnitudes of older, strong earthquakes using the moment magnitude scale. For example, the 1964 Alaskan earthquake was originally given a Richter magnitude of 8.3, but a recent recalculation using the moment magnitude scale resulted in an upgrade to 9.2. Similarly, the 1906 San Francisco earthquake, which had a Richter magnitude of 8.3, was downgraded to a M_W 7.9. The strongest earthquake on record is the 1960 Chilean subduction zone earthquake, with a moment magnitude of 9.5.

Moment magnitude can be calculated from geologic fieldwork by measuring the average amount of slip on the fault, the area of the fault surface that slipped, and strength of the faulted rock. The area of the fault plane can be roughly calculated by multiplying the surface-rupture length by the depth of the aftershocks. This method is most effective for determining the magnitude of large earthquakes generated along large faults in which the ruptures reach the surface. Moment magnitude can also be calculated using data from seismograms.

CONCEPT CHECK 14.6

1 What information does the Modified Mercalli Intensity Scale provide about an earthquake?

2 An earthquake measuring 7.0 on the Richter scale releases about _____ times more energy than an earthquake with a magnitude of 6.0.

3 Why is the *moment magnitude scale* favored over the Richter scale?

Earthquake Belts and Plate Boundaries

About 95 percent of the energy released by earthquakes originates in the few relatively narrow zones shown in **FIGURE 14.16**. The zone of greatest seismic activity, called the *circum-Pacific belt,* encompasses the coastal regions of Chile, Central America, Indonesia, Japan, and Alaska, including the Aleutian Islands (Figure 14.16). Most earthquakes in the circum-Pacific belt occur along convergent plate boundaries where one plate slides at a low angle beneath another. The zone of contact between the subducting and overlying plates forms a huge fault called a *megathrust,* along which Earth's largest earthquakes are generated. Because subduction zone earthquakes usually happen beneath the ocean they may also generate destructive waves called *tsunami.* For example, the 2004 quake off the coast of Sumatra produced a tsunami that claimed an estimated 230,000 lives.

Another major concentration of strong seismic activity, referred to as the *Alpine-Himalayan belt,* runs through the mountainous regions that flank the Mediterranean Sea and extends past the Himalayan Mountains (see Figure 14.16). Tectonic activity in this region is mainly attributed to the collision of the African plate with Eurasia and the collision of the Indian plate with southeast Asia. These plate interactions created many faults that remain active. In addition, numerous faults located away from these plate boundaries have been reactivated as India continues its northward advance into Asia. For example, slippage on a complex fault system in 2008 in the Sichuan Province of China killed at least 70,000 people and left 1.5 million others homeless. The "culprit" is the Indian subcontinent, which shoves the Tibetan Plateau northeastward against the rocks of the Sichuan Basin.

Figure 14.16 shows another continuous earthquake belt that extends for thousands of kilometers through the world's oceans. This zone coincides with the oceanic ridge system, which is an area of frequent but low-intensity seismic activity. As tensional forces pull the plates apart during seafloor spreading, displacement

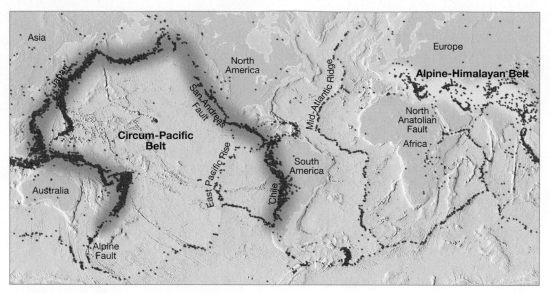

FIGURE 14.16 Distribution of nearly 15,000 earthquakes with magnitudes equal to or greater than 5 for a 10-year period. *(Data from Dewitt Jones/CORBIS)*

along normal faults generates most of the earthquakes. The remaining seismic activity in this zone is associated with slippage along transform faults located between ridge segments.

Transform faults also run through continental crust where they may generate large earthquakes that tend to occur on a cyclical basis. Examples include California's San Andreas Fault, New Zealand's Alpine Fault, and Turkey's North Anatolian Fault that produced a deadly earthquake in 1999.

CONCEPT CHECK 14.7

❶ Name the zone in which most strong earthquakes occur.

❷ Name another major concentration of strong earthquake activity.

Earthquake Destruction

The most violent earthquake ever recorded in North America—the Good Friday Alaskan earthquake—occurred at 5:36 PM on March 27, 1964. Felt throughout that state, the earthquake had a moment magnitude (M_W) of 9.2 and lasted 3 to 4 minutes. This event left 131 people dead, thousands homeless, and the economy of the state badly disrupted. Had schools and business districts been open, the toll surely would have been higher. Within 24 hours of the initial shock, 28 aftershocks were recorded, 10 of which exceeded a magnitude of 6.0. The location of the epicenter and the towns that were hardest hit by the quake are shown in **FIGURE 14.17**.

Destruction from Seismic Vibrations

The 1964 Alaskan earthquake provided geologists with insights into the role of ground shaking as a destructive force. As the energy released by an earthquake travels along Earth's surface, it causes the ground to vibrate in a complex manner by moving up and down as

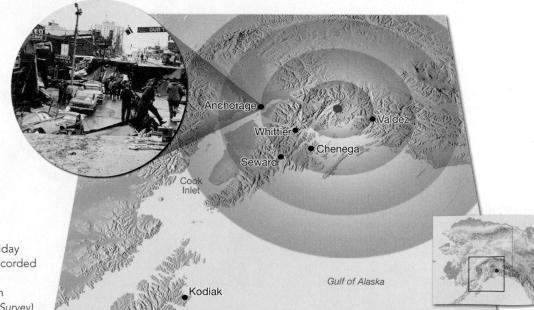

FIGURE 14.17 Region most affected by the Good Friday earthquake of 1964, the strongest earthquake ever recorded in North America. Note the location of the epicenter (red dot). Inset photo shows the collapse of a street in Anchorage, Alaska. *(Photo courtesy of U.S. Geological Survey)*

well as from side to side. The amount of damage to man-made structures attributable to the vibrations depends on several factors, including (1) the intensity and (2) the duration of the shaking, (3) the nature of the material upon which the structure rests, and (4) the nature of building materials and the construction practices of the region.

All of the multistory structures in Anchorage were damaged by the vibrations in 1964. The more flexible wood-frame residential buildings fared best. However, many homes were destroyed when the ground failed. A striking example of how construction variations affect earthquake damage is shown in **FIGURE 14.18**. You can see that the steel-frame building on the left withstood the vibrations, whereas the poorly designed J.C. Penney building was badly damaged. Engineers have learned that buildings built of blocks and bricks and not reinforced with steel rods are the most serious safety threats in earthquakes.

Most large structures in Anchorage were damaged, even though they were built according to the earthquake provisions of the Uniform Building Code. Perhaps some of that destruction can be attributed to the unusually long duration of this earthquake. Most quakes involve tremors that last less than a minute. For example, the 1994 Northridge earthquake was felt for about 40 seconds, and the strong vibrations of the 1989 Loma Prieta earthquake lasted less than 15 seconds, but the Alaska quake reverberated for 3 to 4 minutes.

AMPLIFICATION OF SEISMIC WAVES. Although the region near the epicenter

will experience about the same intensity of ground shaking, destruction may vary considerably within this area. Such differences are usually attributable to the nature of the ground on which the structures are built. Soft sediments, for example, generally amplify the vibrations more than solid bedrock. Thus, the buildings in Anchorage that were situated on unconsolidated sediments experienced heavy structural damage. By contrast, most of the town of Whittier, although much nearer the epicenter, rested on a firm foundation of solid bedrock and suffered much less damage from seismic shaking. Following the quake, however, Whittier was damaged by a tsunami—a phenomenon that will be described later in the chapter.

LIQUEFACTION. In areas where unconsolidated materials are saturated with water, earthquake vibrations can turn stable soil into a mobile fluid, a phenomenon known as **liquefaction.** As a result, the ground is not capable of supporting buildings, and underground storage tanks and sewer lines may literally float toward the surface. During the 1989 Loma Prieta earthquake, in San Francisco's Marina District, foundations failed and geysers of sand and water shot from the ground, indicating that liquefaction had occurred (**FIGURE 14.19**).

Landslides and Ground Subsidence

The greatest damage to structures is often caused by landslides and ground subsidence triggered by earthquake vibrations.

FIGURE 14.18 Damage caused to the five-story J.C. Penney Co. building, Anchorage, Alaska. Very little structural damage was incurred by the adjacent building. (Courtesy of NOAA/Seattle)

A.

B.

FIGURE 14.19 Liquefaction. **A.** These "mud volcanoes" were produced the Loma Prieta earthquake of 1989. They formed when geysers of sand and water shot from the ground, an indication that liquefaction occurred *(Photo by Richard Hilton, courtesy of Dennis Fox)* **B.** Students experiencing nature of liquefaction. *(Photo by Marli Miller)*

This was the case during the 1964 Alaskan earthquake in Valdez and Seward, where the violent shaking caused deltaic sediments to slump, carrying both waterfronts away. In Valdez, 31 people died on a dock when it slid into the sea. Because of the threat of recurrence, the entire town of Valdez was relocated to more stable ground about 7 kilometers away.

Much of the damage in the city of Anchorage was attributed to landslides. Homes were destroyed in Turnagain Heights when a layer of clay lost its strength and more than 200 acres of land slid toward the ocean (**FIGURE 14.20**). A portion of this spectacular landslide was left in its natural condition as a reminder of this destructive event. The site was appropriately named "Earthquake Park." Downtown Anchorage was also disrupted as sections of the main business district dropped by as much as 3 meters (10 feet).

Fire

More than 100 years ago, San Francisco was the economic center of the western United States, largely because of gold and silver mining. Then, at dawn on April 18, 1906, a violent earthquake struck unexpectedly, triggering an enormous firestorm. Much of the city was reduced to ashes and ruins. It is estimated that 3000 people died and 225,000 of the city's 400,000 residents were left homeless.

That historic earthquake reminds us of the formidable threat of fire. The central city contained mostly large, older wooden structures and brick buildings. Although many of the unreinforced brick buildings were extensively damaged by vibrations, the greatest destruction was caused by fires, which started when gas and electrical lines were severed. The fires raged out of control for three days and devastated more than 500 blocks of the city. The initial ground shaking, which broke the city's water lines into hundreds of disconnected pieces, made controlling the fires virtually impossible.

The fires were finally contained when buildings were dynamited along a wide boulevard to provide a fire break, similar to the strategy used in fighting forest fires. Only a few deaths were attributed to the San Francisco fires, but other earthquake-initiated fires have been more destructive

Effects of liquefaction. This building rested on unconsolidated sediment that behaved like quicksand during the 1985 Mexican earthquake. *(Photo by James L. Beck)*

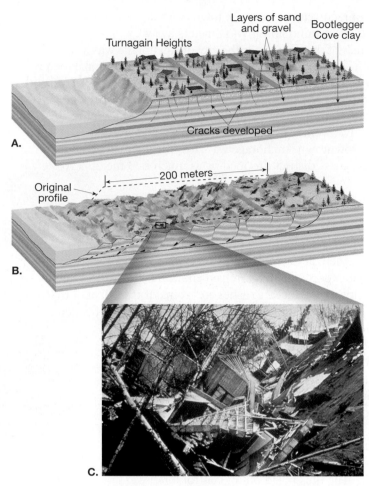

FIGURE 14.20 Turnagain Heights slide caused by the 1964 Alaskan earthquake. **A.** Vibrations from the earthquake caused cracks to appear near the edge of the bluff. **B.** Within seconds blocks of land began to slide toward the sea on a weak layer of clay. In less than 5 minutes, as much as 200 meters of the Turnagain Heights bluff area had been destroyed. **C.** Photo of a small portion of the Turnagain Heights slide. *(Photo courtesy of U.S. Geological Survey)*

and claimed many more lives. For example, a 1923 earthquake in Japan triggered an estimated 250 fires, which devastated the city of Yokohama and destroyed more than half the homes in Tokyo. More than 100,000 deaths were attributed to the fires, which were driven by unusually high winds.

What Is a Tsunami?

Large undersea earthquakes occasionally set in motion massive waves that scientists call **seismic sea waves.** You may be more familiar with the Japanese term **tsunami,** which is frequently used to describe these destructive phenomena. Because of Japan's location

along the circum-Pacific belt and its expansive coastline, it is especially vulnerable to tsunami destruction.

Most tsunami are caused by the vertical displacement of a slab of seafloor along a fault on the ocean floor or less often by a large submarine landslide triggered by an earthquake (**FIGURE 14.21**). Once generated, a tsunami resembles the ripples formed when a pebble is dropped into a pond. In contrast to ripples, tsunami advance across the ocean at amazing speeds, between 500 and 950 kilometers per hour. Despite this striking characteristic, a tsunami in the open ocean can pass undetected because its height (amplitude) is usually less than 1 meter and the distance between wave crests is great, ranging from 100 to 700 kilometers. However, upon entering shallow coastal waters, these destructive waves "feel bottom" and slow, causing the water to pile up (see Figure 14.21). A few exceptional tsunami have reached 30 meters (100 feet) in height. As the crest of a tsunami approaches the shore, it appears as a rapid rise in sea level with a turbulent and chaotic surface (**FIGURE 14.22A**).

The first warning of an approaching tsunami is a rapid withdrawal of water from beaches. Some inhabitants of the Pacific basin have learned to heed this warning and move to higher ground. Approximately 5 to 30 minutes after the retreat of water, a surge capable of extending hundreds of meters inland occurs. In a successive fashion, each surge is followed by a rapid oceanward retreat of the sea.

TSUNAMI DAMAGE FROM THE 2004 INDONESIAN EARTHQUAKE. A massive undersea earthquake of moment magnitude 9.1 occurred near the island of Sumatra on December 26, 2004, and sent waves of water racing across the Indian Ocean and Bay of Bengal. It was one of the deadliest natural disasters of any kind in modern times, claiming more than 230,000 lives. As water surged several kilometers inland, cars and trucks were flung around like toys in a bathtub, and fishing boats were rammed into homes. In some locations, the backwash of water dragged bodies and huge amounts of debris out to sea.

The destruction was indiscriminate, destroying luxury resorts and poor fishing hamlets on the Indian Ocean coast (see **FIGURE 14.22B**). Damages were reported as far away as the Somalia coast of Africa, 4100 kilometers (2500 miles) west of the earthquake epicenter.

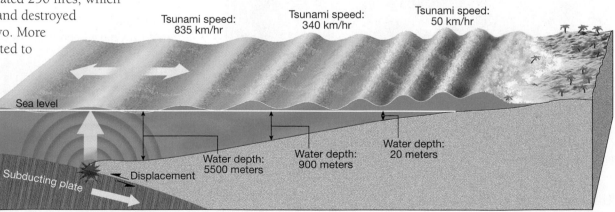

FIGURE 14.21 Schematic drawing of a tsunami generated by displacement of the ocean floor. The speed of a wave correlates with ocean depth. Waves moving in deep water advance at speeds in excess of 800 kilometers per hour. Speed gradually slows to 50 kilometers per hour at depths of 20 meters. Decreasing depth slows the movement of the wave. As waves slow in shallow water, they grow in height until they topple and rush onto shore with tremendous force. The size and spacing of these swells are not to scale.

The killer waves generated by this massive quake achieved heights as great as 10 meters (33 feet) and struck many unprepared areas during a three-hour span following the earthquake. Although the Pacific basin had a tsunami warning system in place, the Indian Ocean unfortunately did not. The rarity of tsunami in the Indian Ocean also contributed to a lack of preparedness. It should come as no surprise that a tsunami warning system for the Indian Ocean was subsequently established.

TSUNAMI WARNING SYSTEM. In 1946, a large tsunami struck the Hawaiian Islands without warning. A wave more than 15 meters (50 feet) high left several coastal villages in shambles. This destruction motivated the U.S. Coast and Geodetic Survey to establish a tsunami warning system for coastal areas of the Pacific. Seismic observatories throughout the region report large earthquakes to the Tsunami Warning Center in Honolulu. Scientists at the Center use deep-sea buoys equipped with pressure sensors to detect energy released by an earthquake. In addition, tidal gauges measure the rise and fall in sea level that accompany tsunami, resulting in warnings issued within the hour. Although tsunami travel very rapidly, there is sufficient time to evacuate all but the areas nearest the epicenter. For example, a tsunami generated off the coast of Chile in 2010 took about 15 hours to reach the Hawaiian Islands (**FIGURE 14.23**).

FIGURE 14.22 A massive earthquake (M$_W$ 9.1) off the Indonesian island of Sumatra sent a tsunami racing across the Indian Ocean and Bay of Bengal on December 26, 2004. **A.** Unsuspecting foreign tourists, who at first walked on the sand after the water receded, now rush toward shore as the first of six tsunami roll toward Hat Rai Lay Beach near Krabi in southern Thailand. *(AFP/Getty Images Inc.)* **B.** Tsunami survivors walk among the debris from this earthquake-triggered event. *(Photo by Kimmasa Mayama/Reuters/CORBIS)*

CONCEPT CHECK 14.8

1. List four factors that affect the amount of destruction caused by seismic vibrations.
2. In addition to the destruction created directly by seismic vibrations, list three other types of destruction associated with earthquakes.
3. What is a *tsunami*? How is one generated?
4. Cite at least three reasons an earthquake with a moderate magnitude might cause more extensive damage than a quake with a high magnitude.

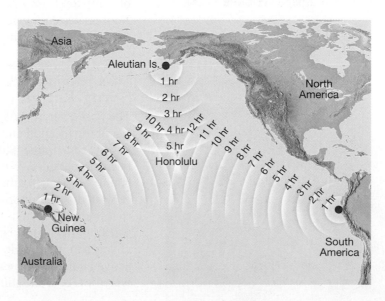

FIGURE 14.23 Tsunami travel times to Honolulu, Hawaii, from selected locations throughout the Pacific. *(Data from NOAA)*

Damaging Earthquakes East of the Rockies

When you think "earthquake," you probably think of California and Japan. However, six major earthquakes have occurred in the central and eastern United States since colonial times. Three of these had estimated Richter magnitudes of 7.5, 7.3, and 7.8, and they were centered near the Mississippi River valley in southeastern Missouri. Occurring on December 16, 1811; January 23, 1812; and February 7, 1812, these earthquakes, plus numerous smaller tremors, destroyed the town of New Madrid, Missouri, triggered massive landslides, and caused damage over a six-state area. The course of the Mississippi River was altered, and Tennessee's Reelfoot Lake was enlarged. The distances over which these earthquakes were felt are truly remarkable. Chimneys were reported downed in Cincinnati, Ohio, and Richmond, Virginia, while Boston residents, located 1770 kilometers (1100 miles) to the northeast, felt the tremor.

Despite the history of the New Madrid earthquake, Memphis, Tennessee, the largest population center in the area, does not have adequate earthquake provisions in its building code. Further, because Memphis is located on unconsolidated floodplain deposits, buildings are more susceptible to damage than are similar structures built on bedrock. It has been estimated that if an earthquake the size of the 1811–1812 New Madrid event were to strike in the next decade, it would result in casualties in the thousands and damages in tens of billions of dollars. Damaging earthquakes that occurred in Aurora, Illinois (1909), and Valentine, Texas (1931), remind us that other areas in the central United States are vulnerable.

The greatest historical earthquake in the eastern states occurred on August 31, 1886, in Charleston, South Carolina. The event, which spanned 1 minute, caused 60 deaths, numerous injuries, and great economic loss within a radius of 200 kilometers (120 miles) of Charleston. Within 8 minutes, effects were felt as far away as Chicago and St. Louis, where strong vibrations shook the upper floors of buildings, causing people to rush outdoors. In Charleston alone, more than 100 buildings were destroyed and 90 percent of the remaining structures damaged (**FIGURE 14.24**).

FIGURE 14.24 Damage to Charleston, South Carolina, caused by the August 31, 1886 earthquake. Damage ranged from toppled chimneys and broken plaster to total collapse. *(Photo courtesy of U.S. Geological Survey)*

Numerous other strong earthquakes have been recorded in the eastern United States. New England and adjacent areas have experienced sizable shocks ever since colonial times. The first reported earthquake in the Northeast took place in Plymouth, Massachusetts, in 1683 and was followed in 1755 by the destructive Cambridge, Massachusetts, quake. Moreover, from the time that records have been kept, New York state alone has experienced more than 300 earthquakes large enough to be felt.

Earthquakes in the central and eastern United States occur far less frequently than in California, yet history indicates that the East is vulnerable. Further, these shocks east of the Rockies have generally produced structural damage over a larger area than counterparts of similar magnitude in California. The reason is that the underlying bedrock in the central and eastern United States is older and more rigid. As a result, seismic waves are able to travel greater distances with less attenuation than in the western United States. It is estimated that for earthquakes of similar magnitude, the region of maximum ground motion in the East may be up to 10 times larger than in the West. Consequently, the higher rate of earthquake occurrence in the western United States is balanced somewhat by the fact that central and eastern U.S. quakes can damage larger areas.

CONCEPT CHECK 14.9

❶ List two reasons a repeat of the 1811–1812 New Madrid earthquakes could be destructive in the Memphis, Tennessee, metropolitan area.

❷ Explain why an earthquake east of the Rockies may produce damage over a larger area than one of similar magnitude in California.

Can Earthquakes Be Predicted?

The vibrations that shook Northridge, California, in 1994 caused 60 deaths and more than $40 billion in damage (**FIGURE 14.25**). This level of destruction was the result of an earthquake of moderate intensity (M_W 6.7). Seismologists warn that other earthquakes of comparable or greater strength can be expected along the San Andreas system, which cuts a 1300-kilometer (800-mile) path through the state. The obvious question is: Can these earthquakes be predicted?

Short-Range Predictions

The goal of short-range earthquake prediction is to provide a warning of the location and magnitude of a large earthquake within a narrow time frame. Substantial efforts to achieve this objective have been attempted in Japan, the United States, China, and Russia—countries where earthquake risks are high. This research has concentrated on monitoring possible *precursors*—events or changes that precede a forthcoming earthquake and thus may provide a warning. In California, for example, seismologists are monitoring changes in ground elevation and variations in strain levels near active faults. Other researchers are measuring changes in groundwater levels, while still others are trying to predict earthquakes

FIGURE 14.25 Damage to Interstate 5 caused by the January 17, 1994, Northridge earthquake. *(Photo by Tom McHugh/Photo Researchers, Inc.)*

based on an increase in the frequency of foreshocks that precede some, but not all, earthquakes.

One claim of a successful short-range prediction was made by the Chinese government after the February 4, 1975, earthquake in Liaoning Province. According to reports, very few people were killed— even though more than 1 million lived near the epicenter—because the earthquake was predicted and the residents were evacuated. Some Western seismologists have questioned this claim and suggest instead that an intense swarm of foreshocks, which began 24 hours before the main earthquake, may have caused many people to evacuate on their own accord.

One year after the Liaoning earthquake, an estimated 240,000 people died in the Tangshan, China, earthquake, which was *not* predicted (**TABLE 14.2**). There were no foreshocks. Predictions can also lead to false alarms. In a province near Hong Kong, people reportedly evacuated their dwellings for more than a month, but no earthquake followed.

In order for a short-range prediction scheme to warrant general acceptance, it must be both accurate and reliable. Thus, *it must have a small range of uncertainty in regard to location and timing, and it must produce few failures or false alarms.* Can you imagine the debate that would precede an order to evacuate a large city in the United States, such as Los Angeles or San Francisco? The cost of evacuating millions of people, arranging for living accommodations, and providing for their lost work time and wages would be staggering.

Currently, no reliable method exists for making short-range earthquake predictions. In fact, except for a brief period of optimism during the 1970s, the leading seismologists of the past 100 years have generally concluded that short-range earthquake prediction is not feasible.

Long-Range Forecasts

In contrast to short-range predictions, which aim to predict earthquakes within a time frame of hours, or at most days, long-range forecasts give the probability of a certain magnitude earthquake occurring on a time scale of 30 to 100 years or more. These forecasts give statistical estimates of the expected intensity of ground motion fora given area, over a specified time frame. Although long-range forecasts are not as informative as we might like, these data are useful for providing important guides for building codes so that buildings, dams, and roadways are constructed to withstand expected levels of ground shaking.

For example, in the 1970s, before the 800-mile-long Trans-Alaskan oil pipeline was built, geologists did a hazards study of the Denali Fault system—a major tectonic structure across Alaska. It was determined that during a magnitude 8 earthquake on the Denali Fault, it would experience a 6-meter (20-foot) horizontal displacement. As a result of this investigation, the pipeline was designed to allow it to slide horizontally without breaking (**FIGURE 14.26**). In 2002, the Denali Fault ruptured producing a 7.9 magnitude earthquake. Although the total displacement along the fault was about 5 meters (18 feet), there was no oil

FIGURE 14.26 The Trans-Alaskan oil pipeline was designed and built to withstand several meters of horizontal displacement where it crosses the Denali Fault. During a magnitude 7.9 earthquake in 2002, the pipeline moved as predicted and no oil spill occurred. This illustrates the importance of estimating potential ground motion and designing structures to mitigate the risks. *(Photo courtesy of USGS)*

TABLE 14.2

Some Notable Earthquakes

Year	Location	Deaths (est.)	Magnitude[†]	Comments
1556	Shensi, China	830,000		Possibly the greatest natural disaster.
1755	Lisbon, Portugal	70,000		Tsunami damage extensive.
*1811–1812	New Madrid, Missouri	Few	7.9	Three major earthquakes.
*1886	Charleston, South Carolina	60		Greatest historical earthquake in the eastern United States.
*1906	San Francisco, California	3,000	7.8	Fires caused extensive damage.
1908	Messina, Italy	120,000		
1923	Tokyo, Japan	143,000	7.9	Fire caused extensive destruction.
1960	Southern Chile	5,700	9.5	The largest-magnitude earthquake ever recorded.
*1964	Alaska	131	9.2	Greatest North American earthquake.
1970	Peru	70,000	7.9	Great rockslide.
*1971	San Fernando, California	65	6.5	Damage exceeded $1 billion.
1975	Liaoning Province, China	1,328	7.5	First major earthquake to be predicted.
1976	Tangshan, China	255,000	7.5	Not predicted.
1985	Mexico City	9,500	8.1	Major damage occurred 400 km from epicenter.
1988	Armenia	25,000	6.9	Poor construction practices.
*1989	San Francisco Bay area	62	7.1	Damages exceeded $6 billion.
1990	Iran	50,000	7.4	Landslides and poor construction practices caused great damage.
1993	Latur, India	10,000	6.4	Located in stable continental interior.
*1994	Northridge, California	60	6.7	Damages in excess of $15 billion.
1995	Kobe, Japan	5,472	6.9	Damages estimated to exceed $100 billion.
1999	Izmit, Turkey	17,127	7.4	Nearly 44,000 injured and more than 250,000 displaced.
1999	Chi-Chi, Taiwan	2,300	7.6	Severe destruction; 8,700 injuries.
2001	Bhuj, India	25,000+	7.9	Millions homeless.
2003	Bam, Iran	41,000+	6.6	Ancient city with poor construction.
2004	Indian Ocean (Sumatra)	230,000	9.1	Devastating tsunami damage.
2005	Pakistan/Kashmir	86,000	7.6	Many landslides; 4 million homeless.
2008	Sichuan, China	70,000	7.9	Millions homeless, some towns will not be rebuilt.
2010	Port-au-Prince, Haiti	230,000	7.0	More than 300,000 injured and a million homeless.
2010	Maule, Chile	486	8.8	One of the ten largest earthquakes by magnitude.

Source: U.S. Geological Survey
*U.S. earthquakes.
[†]Widely differing magnitudes have been estimated for some of these earthquakes. When available, moment magnitudes are used.

spill. The Trans-Alaskan pipeline carries nearly 20 percent of the domestic oil supply of the United States—roughly 600,000 barrels per day—with a degree of scientific reassurance that it will withstand future displacement.

Long-range forecasts are based on evidence that many large faults break repeatedly, producing similar quakes at roughly similar intervals. In other words, as soon as a section of a fault ruptures, the continuing motions of Earth's plates begin to build strain in the rocks again until they fail once more. As a result seismologists study historical records of earthquakes to see if there are any discernible patterns so that the probability of recurrence may be established.

With this concept in mind, seismologists plot the distribution of rupture zones associated with great earthquakes around the globe. The maps reveal that individual rupture zones tend to occur adjacent to one another without appreciable overlap, thereby tracing out a plate boundary. Because plates are moving at known velocities, the rate at which strain builds can also be estimated.

Researchers' study of historical records led to the discovery that some seismic zones had not produced a large earthquake in more than a century, and in some locations, for several centuries. These quiet zones, called **seismic gaps**, are believed to be inactive zones that are storing strain for future major quakes.

An area of recent interest to seismologists is the northern edge of the Indian plate, which is colliding with Asia (**FIGURE 14.27**). Although this area had historically been seismically quiet, four major earthquakes have struck the plate

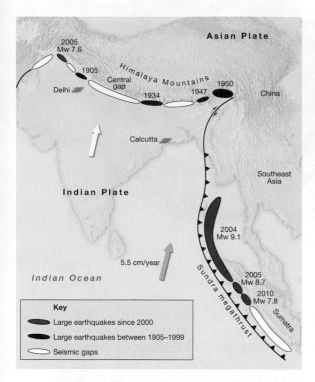

FIGURE 14.27 Seismic gaps. Map of the northern boundary of the Indian plate, where it is moving toward Asia. Shown in red are the rupture zones for the three large earthquakes that have occurred along this plate boundary since 2000. Large earthquakes that occurred between 1905 and 1999 are shown in black, and seismic gaps are shown in white. Seismic gaps are "quiet zones" thought to be inactive zones that are storing elastic strain that will eventually produce major earthquakes.

boundary since 2004. The most destructive was the previously described December 2004 Sumatra earthquake (M_W 9.1). Then, in March 2005, and again in April 2010, two strong earthquakes (M_W 8.6) struck Indonesia on the same fault system directly south of the deadly 2004 event. Fortunately, the later events were much less destructive, because no substantial tsunami was generated.

In October 2005, the Pakistan/Kashmir earthquake struck, claiming 86,000 lives. The severity of the destruction caused by this quake was attributed to severe thrusting, coupled with poor construction practices (**FIGURE 14.28**).

Regrettably, as the map in Figure 14.27 illustrates, several mature seismic gaps (shown in white) are located along this plate margin. One of these lies on the

FIGURE 14.28 Destruction caused by the Pakistan/Kashmir earthquake (M_W 7.6) that struck the region in October 2005. The severity of the destruction was attributed to severe thrusting, coupled with poor construction practices. More than 86,000 fatalities occurred. *(Photo by AP Wide World Photos)*

Sunda megathrust, just south of the March 2005/April 2010 ruptures. Did the displacement in 2005 transfer sufficient stress to nudge the neighboring region toward failure?

Other seismic gaps are located within the continent along the margins of the Himalayan Mountains. One of these sites is located adjacent to the area of slippage that produced the October 2005 Pakistan/Kashmir quake. Another is a 600-kilometer-long region on the central Himalaya that has apparently not ruptured since 1505.

In summary, *the best prospects for making useful earthquake predictions involve forecasting magnitudes and locations on time scales of years or perhaps even decades. These forecasts are important because they provide information that can be used in the design of structures and to assist in land-use planning in order to reduce injuries and loss of life and property.*

CONCEPT CHECK 14.10

❶ Are accurate, short-range earthquake predictions possible using modern seismic instruments?

❷ What is the value of long-range earthquake forecasts?

Earth's Interior

 Earth's Interior

ESSENTIALS OF GEOLOGY **Earth's Layered Structure**

If you could slice Earth in half, the first thing you would notice is that it has three distinct layers. The heaviest materials (metals) would be in the center. Lighter solids (rocks) would be in the middle, and liquids and gases would be on top. Within Earth we know these layers as the iron core, the rocky mantle and crust, the liquid ocean, and the gaseous atmosphere. More than 95 percent of the variations in composition and temperature within Earth are due to layering. However, this is not the end of the story. If it were, Earth would be a dead, lifeless cinder floating in space.

There are also variations in composition and temperature with depth that indicate the interior of our planet is very dynamic. The rocks of the mantle and crust are in constant motion, not only moving about through plate tectonics, but also continuously recycling between the surface and the deep interior. Furthermore, it is from Earth's deep interior that the water and air of our oceans and atmosphere are replenished, allowing life to exist at the surface.

Probing Earth's Interior: "Seeing" Seismic Waves

Discovering the structure and properties of Earth's deep interior has not been easy. Light does not travel through rock, so we must find other ways to "see" into our planet. The best way to learn about Earth's interior is to dig or drill a hole and examine it directly. Unfortunately, this is only possible at shallow depths. The deepest a drilling rig has ever penetrated is 12.3 kilometers (8 miles), which is about 1/500 of the way to Earth's center! Even this was an extraordinary accomplishment because temperature and pressure increase rapidly with depth.

Fortunately, many earthquakes are large enough that their seismic waves travel all the way through Earth and can be recorded on the other side (**FIGURE 14.29**). This means that the seismic waves act like medical x-rays used to take images of a person's insides. There are about 100 to 200 earthquakes each year that are large enough (about $M_w > 6$) to be well recorded by seismographs all around the globe. These large earthquakes provide the means to "see" into our planet and have been the source of most of the data that have allowed us to figure out the nature of Earth's interior.

Interpreting the waves recorded on seismograms in order to identify Earth structure is challenging because seismic waves do not travel along straight paths. Instead, seismic waves are *reflected*, *refracted*, and *diffracted* as they pass through our planet. They reflect off boundaries between different layers, they refract (or bend) when passing from one layer to another layer, and they diffract around any obstacles they encounter (Figure 14.29). These different wave behav-

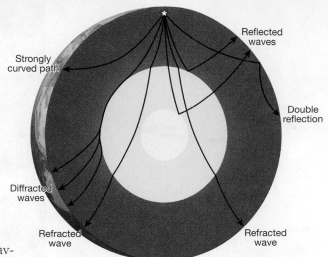

FIGURE 14.29 Slice through Earth's interior showing some of the ray paths that seismic waves from an earthquake would take. Notice that in the mantle, the rays follow curved (refracting) paths rather than straight paths because the seismic velocity of rocks increases with depth, a result of increasing pressure with depth.

iors have been used to identify the boundaries that exist within Earth.

One of the most noticeable behaviors of seismic waves is that they follow strongly curved paths (Figure 14.29). This occurs because the velocity of seismic waves generally increases with depth. In addition, seismic waves travel faster when rock is stiffer or less compressible. These properties of stiffness and compressibility are then used to interpret the composition and temperature of the rock. For instance, when·rock is hotter, it becomes less stiff (imagine taking a frozen chocolate bar and then heating it up!), and waves travel more slowly. Waves also travel at different speeds through rocks of different compositions. Thus, the speed that seismic waves travel can help determine both the kind of rock that is inside Earth and how hot it is.

Formation of Earth's Layered Structure

As material accumulated to form Earth (and for a short period afterward), the high-velocity impact of nebular debris and the decay of radioactive elements caused the temperature of our planet to steadily increase. During this time of intense heating, Earth became hot enough that iron and nickel began to melt. Melting produced liquid blobs of heavy metal that sank toward the center of the planet. This process occurred rapidly on the scale of geologic time and produced Earth's dense iron-rich core.

The early period of heating resulted in another process of chemical differentiation, whereby melting formed buoyant masses of molten rock that rose toward the surface, where they solidified to produce a primitive crust. These rocky materials were rich in oxygen and "oxygen-seeking" elements, particularly silicon and aluminum, along with lesser amounts of calcium, sodium, potassium, iron, and magnesium. In addition, some heavy metals such as gold,

lead, and uranium, which have low melting points or were highly soluble in the ascending molten masses, were scavenged from Earth's interior and concentrated in the developing crust. This early period of chemical segregation established the three basic divisions of Earth's interior—the iron-rich *core;* the thin *primitive crust;* and Earth's largest layer, called the *mantle,* which is located between the core and crust (**FIGURE 14.30**).

Earth's Internal Structure

In addition to these three compositionally distinct layers, Earth can be divided into layers based on physical properties. The physical properties used to define such zones include whether the layer is solid or liquid and how weak or strong it is. Knowledge of both types of layers is essential to our understanding of basic geologic processes, such as volcanism, earthquakes, and mountain building (Figure 14.30).

EARTH'S CRUST. The **crust,** Earth's relatively thin, rocky outer skin, is of two types—continental crust and oceanic crust. Both share the word *crust,* but the similarity ends there. The oceanic crust is roughly 7 kilometers (4 miles) thick and composed of the dark igneous rock *basalt.* By contrast, the continental crust averages 35 to 40 kilometers (22 to 25 miles) thick but may exceed 70 kilometers (40 miles) in some mountainous regions such as the Rockies and Himalayas. Unlike the oceanic crust, which has a relatively homogeneous chemical composition, the continental crust consists of many rock types. Although the upper crust has an average composition of a *granitic rock* called *granodiorite,* it varies considerably from place to place.

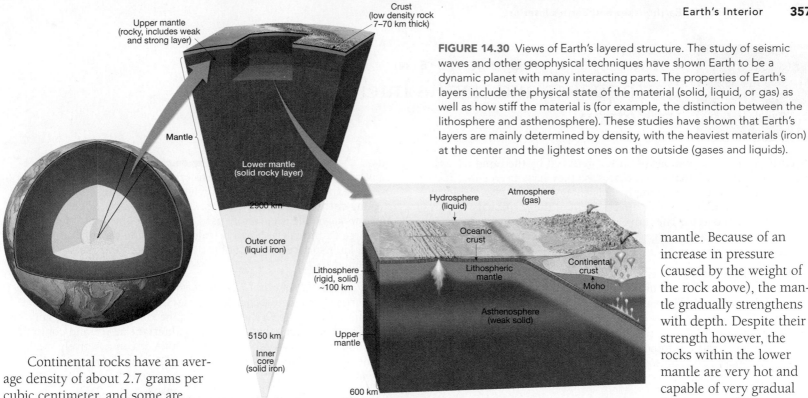

Upper mantle (rocky, includes weak and strong layer)

Mantle

Lower mantle (solid rocky layer)

2900 km

Outer core (liquid iron)

5150 km

Inner core (solid iron)

6, 371 km

Crust (low density rock 7–70 km thick)

Hydrosphere (liquid)   Atmosphere (gas)

Oceanic crust

Continental crust

Lithospheric mantle

Moho

Lithosphere (rigid, solid) ~100 km

Asthenosphere (weak solid)

Upper mantle

600 km

FIGURE 14.30 Views of Earth's layered structure. The study of seismic waves and other geophysical techniques have shown Earth to be a dynamic planet with many interacting parts. The properties of Earth's layers include the physical state of the material (solid, liquid, or gas) as well as how stiff the material is (for example, the distinction between the lithosphere and asthenosphere). These studies have shown that Earth's layers are mainly determined by density, with the heaviest materials (iron) at the center and the lightest ones on the outside (gases and liquids).

Continental rocks have an average density of about 2.7 grams per cubic centimeter, and some are 4 billion years old. The rocks of the oceanic crust are younger (180 million years or less) and denser (about 3.0 grams per cubic centimeter) than continental rocks.*

EARTH'S MANTLE. More than 82 percent of Earth's volume is contained in the **mantle,** a solid, rocky shell that extends to a depth of about 2900 kilometers (1800 miles). The boundary between the crust and mantle represents a marked change in chemical composition. The dominant rock type in the uppermost mantle is *peridotite,* which is richer in the metals magnesium and iron than the minerals found in either the continental or oceanic crust.

The upper mantle extends from the crust–mantle boundary down to a depth of about 660 kilometers (410 miles). The upper mantle can be divided into two different parts. The top portion of the upper mantle is part of the stiff *lithosphere,* and beneath that is the weaker *asthenosphere.* The **lithosphere** (sphere of rock) consists of the entire crust and uppermost mantle and forms Earth's relatively cool, rigid outer shell. Averaging about 100 kilometers (62 miles) in thickness, the lithosphere is more than 250 kilometers (155 miles) thick below the oldest portions of the continents

*Liquid water has a density of 1 gram per cubic centimeter; therefore, the density of basalt is three times that of water.

(see Figure 14.30). Beneath this stiff layer to a depth of about 350 kilometers (217 miles) lies a soft, comparatively weak layer known as the **asthenosphere** (weak sphere). The top portion of the asthenosphere has a temperature/pressure regime that results in a small amount of melting. Within this very weak zone, the lithosphere is mechanically detached from the layer below. The result is that the lithosphere is able to move independently of the asthenosphere, a fact we will consider in the next chapter.

It is important to emphasize that the strength of various Earth materials is a function of both their composition and the temperature and pressure of their environment. The entire lithosphere does *not* behave like a brittle solid similar to rocks found on the surface. Rather, the rocks of the lithosphere get progressively hotter and weaker (more easily deformed) with increasing depth. At the depth of the uppermost asthenosphere, the rocks are close enough to their melting temperature that they are very easily deformed, and some melting may actually occur. Thus, the uppermost asthenosphere is weak because it is near its melting point, just as hot wax is weaker than cold wax.

From 660 kilometers (410 miles) deep to the top of the core, at a depth of 2900 kilometers (1800 miles), is the lower

mantle. Because of an increase in pressure (caused by the weight of the rock above), the mantle gradually strengthens with depth. Despite their strength however, the rocks within the lower mantle are very hot and capable of very gradual flow.

EARTH'S CORE. The composition of the **core** is thought to be an iron–nickel alloy with minor amounts of oxygen, silicon, and sulfur—elements that readily form compounds with iron. At the extreme pressure found in the core, this iron-rich material has an average density of nearly 11 grams per cubic centimeter and approaches 14 times the density of water at Earth's center.

The core is divided into two regions that exhibit very different mechanical strengths. The **outer core** is a *liquid layer* 2270 kilometers (1410 miles) thick. It is the movement of metallic iron within this zone that generates Earth's magnetic field. The **inner core** is a sphere with a radius of 1216 kilometers (754 miles). Despite its higher temperature, the iron in the inner core is *solid* due to the immense pressures that exist in the center of the planet.

CONCEPT CHECK 14.11

❶ Briefly describe how seismic waves are used to probe Earth's interior.

❷ How did Earth acquire its layered structure?

❸ Contrast the physical make-up of the *asthenosphere* and the *lithosphere.*

❹ How do continental crust and oceanic crust differ?

CHAPTER FOURTEEN
Earthquakes and Earth's Interior in Review

⦿ *Earthquakes* are vibrations of Earth produced by the rapid release of energy from rocks that rupture because they have been subjected to stresses that exceed their strength. This energy, which takes the form of *seismic waves,* radiates in all directions from the earthquake's source, called the *focus.* The movements that produce most large earthquakes occur along large fractures, called *faults,* that are usually associated with plate boundaries.

⦿ Along a fault, rocks store energy as they are bent. As slippage occurs at the weakest point (the focus), displacement will exert stress farther along a fault, where additional slippage will occur until most of the built-up strain is released. An earthquake occurs as the rock elastically returns to its original shape. The "springing back" of the rock is termed *elastic rebound.* Small earthquakes, called *foreshocks,* often precede a major earthquake. The adjustments that follow a major earthquake often generate smaller earthquakes called *aftershocks.*

⦿ Two main types of *seismic waves* are generated during an earthquake: (1) *surface waves,* which travel along the outer layer of Earth, and (2) *body waves,* which travel through Earth's interior. Body waves are further divided into *primary (P) waves,* which push (squeeze) and pull (stretch) rocks in the direction the wave is traveling, and *secondary (S) waves,* which "shake" the particles in rock at right angles to their direction of travel. P waves can travel through solids, liquids, and gases. Fluids (gases and liquids) will not transmit S waves. In any solid material, P waves travel about 1.7 times faster than S waves.

⦿ The location on Earth's surface directly above the focus of an earthquake is the *epicenter.* Using the difference in arrival times between P and S waves, the distance separating a recording station from the earthquake epicenter can be determined. When the distances are known from three or more seismic stations, the epicenter can be located using a method called *triangulation.*

⦿ Seismologists use two fundamentally different measures to describe the size of an earthquake—intensity and magnitude. *Intensity* is a measure of the degree of ground shaking at a given locale based on the observed effects. The *Modified Mercalli Intensity Scale* is divided into 12 levels of severity based on observed effects such as people awakened from sleep, furniture moving, plaster cracking and falling, and finally—total destruction. *Magnitude* is calculated from seismic records and estimates the amount of energy released at the source of an earthquake. Using the *Richter scale,* the magnitude of an earth-

quake is estimated by measuring the *amplitude* (maximum displacement) of the largest seismic wave recorded. A logarithmic scale is used to express magnitude, in which a tenfold increase in ground shaking corresponds to an increase of 1 on the magnitude scale. *Moment magnitude* is currently used to estimate the size of moderate and large earthquakes. It can be calculated using the amount of slip on the fault surface, the area of the fault surface, and the strength of the faulted rock.

⦿ *A close correlation exists between earthquake epicenters and plate boundaries.* The greatest energy is released by earthquakes along the margin of the Pacific Ocean, known as the *circum-Pacific belt,* and the mountainous regions that flank the Mediterranean Sea and continue past the Himalayan complex. Another zone of comparatively weak seismicity runs through the world's oceans along the *oceanic ridge system.*

⦿ The primary factors that determine the amount of destruction accompanying an earthquake are the magnitude of the earthquake and the proximity of the quake to a populated area. Structural damage attributable to ground shaking depends on several factors, including (1) the intensity and (2) the duration of ground shaking, (3) the nature of the material upon which the structure rests, and (4) the design of the structure. Secondary effects of earthquakes include *landslides, ground subsidence, fire,* and *tsunami damage.*

⦿ Substantial research to predict earthquakes is underway in Japan, the United States, China, and Russia—countries where earthquake risk is high. No reliable method of short-range prediction has yet been devised. Long-range forecasts are based on the premise that earthquakes are repetitive or cyclical. Seismologists study the history of earthquakes for patterns so their occurrences may be predicted. Long-range forecasts are important because they provide information used to develop the Uniform Building Code and to assist in land-use planning.

⦿ As indicated by the behavior of P and S waves as they travel through Earth, the four major zones of Earth's interior are: (1) *crust* (the very thin outer layer, 5 to 40 kilometers); (2) *mantle* (a rocky layer located below the crust with a thickness of 2900 kilometers); (3) *outer core* (a layer about 2270 kilometers thick, which exhibits the characteristics of a mobile liquid); and (4) *inner core* (a solid metallic sphere with a radius of about 1216 kilometers).

Key Terms

GIVE IT SOME THOUGHT

❶ Briefly describe the concept of *elastic rebound*. Develop an analogy other than a rubber band to illustrate this concept.

❷ The accompanying map shows the locations of the 15 largest earthquakes in the world since 1900. Refer to the map of Earth's plates (Figure 15.9, p. 370) and determine which type of plate boundary is most often associated with these events.

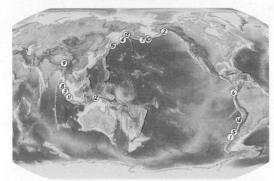

❸ Use the accompanying seismogram to answer the following questions:

   a. Which of the three types of seismic waves reached the seismograph first?

   b. What is the time interval between the arrival of the first P wave and the arrival of the first S wave?

   c. Using your answer from question b., and Figure 14.11, determine the distance from the seismic station to the epicenter.

   d. Which of the three types of seismic waves had the highest amplitude when they reached the seismic station?

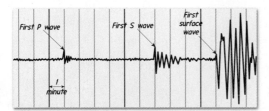

❹ You go for a jog on a beach and choose to run near the water where the sand is well packed and solid under your feet. With each step, you notice that your footprint quickly fills with water, but not water coming in from the ocean. What is this water's source? For what earthquake-related hazard is this phenomenon a good analogy?

❺ Explain, in your own words, why a tsunami often causes a rapid withdrawal of water from beaches before the first surge.

❻ Why is it possible to issue a tsunami warning, but not provide a warning for an earthquake?

❼ Using the accompanying map of the San Andreas Fault, answer the following questions:

   a. Which of the four segments of the San Andreas Fault do you think has the best chance of experiencing a major earthquake in the foreseeable future?

   b. Which segment do you think is experiencing *fault creep*?

   c. If major earthquakes occur along active segments of the San Andreas Fault about every 200 years, when should the section that ruptured during the Fort Tejon quake be expected to generate another major event?

   d. Do you think San Francisco or Los Angeles has the greatest risk for experiencing major earthquake damage in the near future? Defend your selection.

❽ Describe the two different ways that Earth's layers are defined.

❾ S waves temporarily change the shape of the material that transmits them. Can you identify a place in Earth's interior that would not transmit S waves? Why?

❿ Based on the properties of Earth's layers, and the mode of travel of body waves, predict where in Earth's interior waves should (a) travel fastest, and (b) travel slowest. Is there an exception for these generalities? Explain your answers.

Companion Website   my**geoscience** place

www.mygeoscienceplace.com

The *Essentials of Geology, 11e* companion Website contains numerous multimedia resources accompanied by assessments to aid in your study of the topics in this chapter. The use of this site's learning tools will help improve your understanding of geology. Utilizing the access code that accompanies this text, visit **www.mygeoscienceplace.com** in order to:

• **Review** key chapter concepts.
• **Read** with links to the Pearson eText and to chapter-specific web resources.
• **Visualize** and comprehend challenging topics using the learning activities in *GEODe: Essentials of Geology* and the *Geoscience Animations Library*.
• **Test** yourself with online quizzes.

Plate Tectonics: A Scientific Revolution Unfolds

P LATE TECTONICS IS THE FIRST THEORY TO PROVIDE A COMPREHENSIVE VIEW of the processes that produced Earth's major surface features, including the continents and ocean basins. Within the framework of this theory, geologists have found explanations for the basic causes and distribution of earthquakes, volcanoes, and mountain belts. Further, we are now better able to explain the distribution of plants and animals in the geologic past, as well as the distribution of economically significant mineral deposits.

To assist you in learning the important concepts in this chapter, focus on the following questions:

- ◉ What evidence was used to support the continental drift hypothesis?
- ◉ What was one of the main objections to the continental drift hypothesis?
- ◉ What is the theory of plate tectonics?
- ◉ In what major way does the plate tectonics theory depart from the continental drift hypothesis?
- ◉ What are the three types of plate boundaries?
- ◉ Where does new lithosphere form?
- ◉ How do mountain systems such as the Himalayas form?
- ◉ What type of plate motion occurs along a transform fault boundary?
- ◉ What evidence is used to support the plate tectonics theory?
- ◉ What are the major driving forces for plate tectonics?
- ◉ What models have been proposed to explain the driving mechanism for plate motion?

This hiker is walking across glacial debris in Pakistan's rugged and remote Karakoram Range, a part of the **Himalayas.** *(Photo by Bill Stevenson)*

FOCUS ON CONCEPTS

From Continental Drift To Plate Tectonics

Prior to the 1960s most geologists held the view that the ocean basins and continents had fixed geographic positions and were of great antiquity. Less than a decade later researchers came to realize that Earth's continents are not static, instead they gradually migrate across the globe. Because of these movements, blocks of continental material collide, deforming the intervening crust, thereby creating Earth's great mountain chains (**FIGURE 15.1**). Furthermore, landmasses occasionally split apart. As the continental blocks separate, a new ocean basin emerges between them. Meanwhile, other portions of the seafloor plunge into the mantle. In short, a dramatically different model of Earth's tectonic processes emerged.*

This profound reversal in scientific thought has been appropriately described as a *scientific revolution*. The revolution began early in the 20th century as a relatively straightforward proposal called *continental drift*. For more than 50 years the idea that continents were capable of movement was categorically rejected by the scientific establishment. Continental drift was particularly distasteful to North American geologists, perhaps because much of the supporting evidence had been gathered from the continents of Africa, South America, and Australia, with which most North American geologists were unfamiliar.

Following World War II, modern instruments replaced rock hammers as the tools of choice for many researchers.

*Tectonic processes are those that deform Earth's crust to create major structural features such as mountains, continents, and ocean basins.

FIGURE 15.1 Climbers camping on a sheer rock face of a mountain known as K7 in Pakistan's Karakoram, a part of the Himalayas. These mountains formed as India collided with Eurasia. *(Photo by Jimmy Chin/National Geographic Stock)*

Armed with these more advanced tools, geologists and a new breed of researchers, including *geophysicists* and *geochemists,* made several surprising discoveries that began to rekindle interest in the drift hypothesis. By 1968 these developments led to the unfolding of a far more encompassing explanation known as the *theory of plate tectonics.*

In this chapter, we will examine the events that led to this dramatic reversal of scientific opinion in an attempt to provide insight into how science works. We will also briefly trace the development of the *continental drift hypothesis*, examine why it was first rejected, and consider the evidence that finally led to the acceptance of its direct descendant—the theory of plate tectonics.

CONCEPT CHECK 15.1

❶ Briefly contrast the view held by most geologists regarding ocean basins and continents prior to the 1960s with their perspective a decade later.

Continental Drift: An Idea before Its Time

The idea that continents, particularly South America and Africa, fit together like pieces of a jigsaw puzzle came about during the 1600s as better world maps became available. However, little significance was given to this notion until 1915, when Alfred Wegener (1880–1930), a German meteorologist and geophysicist, wrote *The Origin of Continents and Oceans*. This book, published in several editions, set forth the basic outline of Wegener's hypothesis called **continental drift**—which dared to challenge the long-held assumption that the continents and ocean basins had fixed geographic positions.

Wegener suggested that a single **supercontinent** consisting of all Earth's landmasses once existed.* He named this giant landmass **Pangaea** (pronounced Pan-jee-ah; meaning "all lands") (**FIGURE 15.2**). Wegener further hypothesized that about

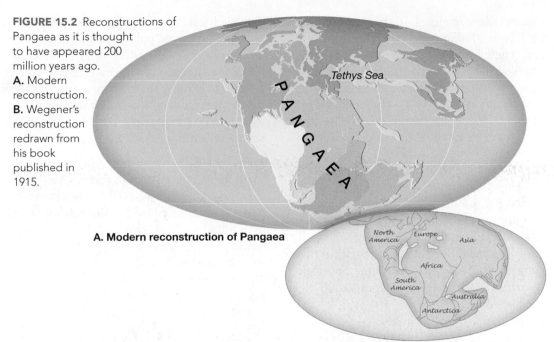

FIGURE 15.2 Reconstructions of Pangaea as it is thought to have appeared 200 million years ago. **A.** Modern reconstruction. **B.** Wegener's reconstruction redrawn from his book published in 1915.

A. Modern reconstruction of Pangaea

B. Wegener's Pangaea

200 million years ago, during the early part of the Mesozoic era, this supercontinent began to fragment into smaller landmasses. These continental blocks then "drifted" to their present positions over a span of millions of years. The inspiration for continental drift is believed to have come to Wegener when he observed the break-up of sea ice during a Danish-led expedition to Greenland.

Wegener and others who advocated the continental drift hypothesis collected substantial evidence to support their point of view. The fit of South America and Africa and the geographic distribution of fossils and ancient climates all seemed to buttress the idea that these now separate landmasses were once joined. Let us examine some of this evidence.

DID YOU KNOW?
Although Alfred Wegener is rightfully credited with formulating the continental drift hypothesis, he was not the first to suggest continental mobility. An American geologist, F. B. Taylor, published the first paper to outline this important idea. However, Taylor's paper provided little supporting evidence, whereas Wegener spent much of his professional life trying to substantiate his views.

Evidence: The Continental Jigsaw Puzzle

Like a few others before him, Wegener suspected that the continents might once have been joined when he noticed the remarkable similarity between the coastlines on opposite sides of the Atlantic Ocean. However, Wegener's use of present-day shorelines to fit these continents together was challenged immediately by other Earth scientists. These opponents correctly argued that shorelines are continually modified by wave erosion and depositional processes. Even if continental displacement had taken place, a good fit today would be unlikely. Because Wegener's original jigsaw fit of the continents was crude, it is assumed that he was aware of this problem (see **FIGURE 15.2B**).

Scientists later determined that a much better approximation of the outer boundary of a continent is the seaward edge of its continental shelf, which lies submerged a few hundred

*Wegener was not the first person to conceive of a long-vanished supercontinent. Edward Suess (1831–1914), a distinguished 19th-century geologist, pieced together evidence for a giant landmass consisting of the continents of South America, Africa, India, and Australia.

meters below sea level. In the early 1960s, Sir Edward Bullard and two associates constructed a map that pieced together the edges of the continental shelves of South America and Africa at a depth of about 900 meters (**FIGURE 15.3**). The remarkable fit that was obtained was more precise than even these researchers had expected. As shown in Figure 15.3 there are a few places where the continents overlap. Some of these overlaps are related to the process of stretching and thinning of the continental margins as they drifted apart. Others can be explained by the work of major river systems. For example, since the break-up of Pangaea the Niger River has built an extensive delta that enlarged the continental shelf of Africa.

Evidence: Fossils Match across the Seas

Although the seed for Wegener's hypothesis came from the remarkable similarities of the continental margins on opposite sides of the Atlantic, it was when he learned that identical fossil organisms had been discovered in rocks from both South America and Africa that his pursuit of continental drift became more focused. Through a review of the literature, Wegener learned that most paleontologists (scientists who study the fossilized remains of ancient organisms) were in agreement that some type of land connection was needed to explain the existence of similar Mesozoic age life forms on widely separated landmasses. Just as modern life forms native to North America are quite different from those of Africa and Australia, one would expect that during the Mesozoic era, organisms on widely separated continents would be distinct.

MESOSAURUS. To add credibility to his argument, Wegener documented cases of several fossil organisms that were found on different landmasses despite the unlikely possibility that their living forms could have crossed the vast ocean presently separating them (**FIGURE 15.4**). A classic example is *Mesosaurus,* an aquatic fish-catching reptile whose fossil remains are limited to black shales of the Permian period (about 260 million years ago) in eastern South America and southwestern Africa. If *Mesosaurus* had been able to make the long journey across the South Atlantic, its remains would likely be more widely distributed. As this is not the case, Wegener asserted that South America and Africa must have been joined during that period of Earth history.

How did opponents of continental drift explain the existence of identical fossil organisms in places separated by thousands of kilometers of open ocean? Rafting, transoceanic land bridges (isthmian links), and island stepping stones were the most widely invoked explanations for these migrations (**FIGURE 15.5**). We know, for example, that during the Ice Age that ended about 8,000 years ago the lowering of sea level allowed mammals (including humans) to cross the narrow Bering Strait that separates Russia and Alaska. Was it possible that land bridges once connected Africa and South America but later subsided below sea level? Modern maps of the seafloor substantiate Wegener's contention that if land bridges of this magnitude once existed, their remnants would still lie below sea level.

FIGURE 15.3 Drawing that shows the best fit of South America and Africa along the continental slope at a depth of 500 fathoms (about 900 meters). The areas where continental blocks overlap appear in orange. (*After A. G. Smith, "Continental Drift," in* Understanding the Earth, *edited by I. G. Gass.*)

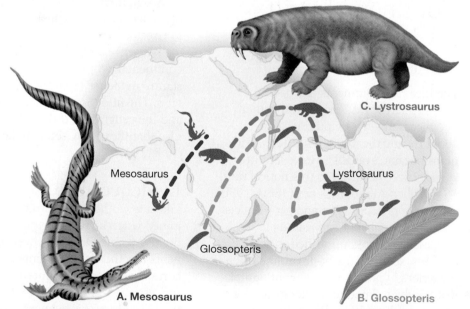

FIGURE 15.4 Fossil evidence supporting continental drift. **A.** Fossils of *Mesosaurus* are found only in nonmarine deposits in eastern South America and western Africa. *Mesosaurus* was a freshwater reptile incapable of swimming the 5000 kilometers of open ocean that now separate these continents. **B.** Remains of *Glossopteris* and related flora are found in Australia, Africa, South America, Antarctica, and India, landmasses which currently have quite varied climates. However, when *Glossopteris* inhabited these regions during the late Paleozoic era, their climates were all subpolar. **C.** Fossils of *Lystrosaurus,* a land-dwelling reptile, are also found on three of these landmasses.

FIGURE 15.5 These sketches by John Holden illustrate various explanations for the occurrence of similar species on landmasses that are presently separated by vast oceans. *(Reprinted with permission of John Holden)*

GLOSSOPTERIS. Wegener also cited the distribution of the fossil "seed fern" *Glossopteris* as evidence for the existence of Pangaea (see Figure 15.4). This plant, identified by its tongue-shaped leaves and seeds that were too large to be carried by the wind, was known to be widely dispersed among Africa, Australia, India, and South America. Later, fossil remains of *Glossopteris* were also discovered in Antarctica.* Wegener also learned that these seed ferns and associated flora grew only in a subpolar climate. Therefore, he concluded that when these landmasses were joined, they were located much closer to the South Pole.

Evidence: Rock Types and Geologic Features

Anyone who has worked a jigsaw puzzle knows that its successful completion requires that you fit the pieces together while maintaining the continuity of the picture. The "picture" that must match in the "continental drift puzzle" is one of rock types and geologic features such as mountain belts. If the continents were once together, the rocks found in a particular region on one continent should closely match in age and type those found in adjacent positions on the once adjoining

*In 1912 Captain Robert Scott and two companions froze to death lying beside 35 pounds of rock on their return from a failed attempt to be the first to reach the South Pole. These samples, collected on the moraines of Beardmore Glacier, contained fossil remains of *Glossopteris*.

continent. Wegener found evidence of 2.2-billion-year-old igneous rocks in Brazil that closely resembled similarly aged rocks in Africa.

Similar evidence can be found in mountain belts that terminate at one coastline, only to reappear on landmasses across the ocean. For instance, the moun-

tain belt that includes the Appalachians trends northeastward through the eastern United States and disappears off the coast of Newfoundland (**FIGURE 15.6A**). Mountains of comparable age and structure are found in the British Isles, and Scandinavia. When these landmasses are reassembled, as in **FIGURE 15.6B**, the mountain chains form a nearly continuous belt.

Wegener described how the similarities in geologic features on both sides of the Atlantic

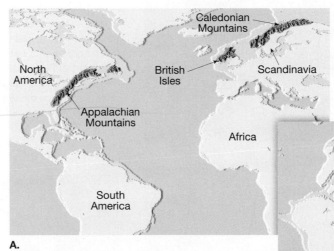

A.

FIGURE 15.6 Matching mountain ranges across the North Atlantic. **A.** The Appalachian Mountains trend along the eastern flank of North America and disappear off the coast of Newfoundland. Mountains of comparable age and structure are found in the British Isles and Scandinavia. **B.** When these landmasses are placed in their pre-drift locations, these ancient mountain chains form a nearly continuous belt.

DID YOU KNOW?

A group of scientists proposed an interesting although incorrect explanation for the cause of continental drift. Their proposal suggested that early in Earth's history, our planet was only about half its current diameter and completely covered by continental crust. Through time Earth expanded, causing the continents to split into their current configurations, while new seafloor "filled in" the spaces as they drifted apart.

linked these landmasses when he said, "It is just as if we were to refit the torn pieces of a newspaper by matching their edges and then check whether the lines of print run smoothly across. If they do, there is nothing left but to conclude that the pieces were in fact joined in this way."*

*Alfred Wegener, *The Origin of Continents and Oceans,* translated from the 4th revised German ed. of 1929 by J. Birman (London: Methuen, 1966).

Evidence: Ancient Climates

Because Alfred Wegener was a student of world climates, he suspected that paleoclimatic (*paleo* = ancient, *climatic* = climate) data might also support the idea of mobile continents. His assertion was bolstered when he learned that evidence for a glacial period that dated to the late Paleozoic had been discovered in southern Africa, South America, Australia, and India (**FIGURE 15.7A**). This meant that about 300 million years ago, vast ice sheets covered extensive portions of the Southern Hemisphere as well as India (**FIGURE 15.7B**). Much of the land area that contains evidence of this period of Paleozoic glaciation presently lies within 30 degrees of the equator in subtropical or tropical climates.

How could extensive ice sheets form near the Equator? One proposal suggested that our planet experienced a period of extreme global cooling. Wegener rejected this explanation because during the same span of geologic time, large tropical

swamps existed in several locations in the Northern Hemisphere. The lush vegetation in these swamps was eventually buried and converted to coal. Today these deposits comprise major coal fields in the eastern United States, Northern Europe, and Asia. Many of the fossils found in these coal-bearing rocks were produced by tree ferns that possessed large fronds; a fact consistent with a warm, moist climate. Furthermore, these fern trees lacked growth rings, a characteristic of tropical plants that grow in regions having minimal yearly fluctuations in temperature. By contrast, trees that inhabit the middle latitudes, like those found in most of the United States, develop multiple tree rings—one for each growing season.

Wegener suggested that a more plausible explanation for the late Paleozoic glaciation was provided by the supercontinent of Pangaea. In this configuration the southern continents are joined together and located near the South Pole (**FIGURE 15.7C**). This would account for the conditions necessary

A.

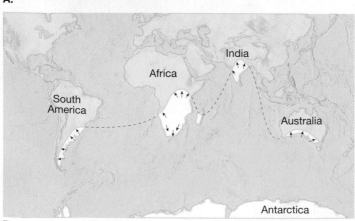

B.

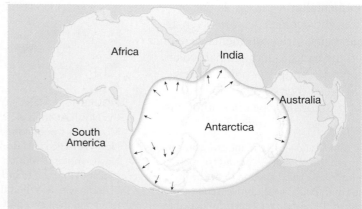

C.

FIGURE 15.7 Paleoclimatic evidence for continental drift. **A.** Glacial striations (scratches) and grooves like these are produced as glaciers drag rock debris across the underlying bedrock. The direction of glacial movement can be deduced from the distinctive patterns of aligned scratches and grooves. *(Photo by Gregory S. Springer)* **B.** Near the end of the Paleozoic era (about 300 million years ago) ice sheets covered extensive areas of the Southern Hemisphere and India. Arrows show the direction of ice movement that can be inferred from the pattern of glacial striations and grooves found in the bedrock. **C.** The continents restored to their pre-drift positions when they were part of Pangaea. This configuration accounts for the conditions necessary to generate a vast ice sheet and also explains the directions of ice movement that radiated away from an area near the present position of the South Pole.

to generate extensive expanses of glacial ice over much of these landmasses. At the same time, this geography would place today's northern continents nearer the equator and account for the tropical swamps that generated the vast coal deposits. Wegener was so convinced that his explanation was correct that he wrote, "This evidence is so compelling that by comparison all other criteria must take a back seat."

How does a glacier develop in hot, arid central Australia? How do land animals migrate across wide expanses of the ocean? As compelling as this evidence may have been, 50 years passed before most of the scientific community accepted the concept of continental drift and the logical conclusions to which it led.

CONCEPT CHECK 15.2

① What was the first line of evidence that led early investigators to suspect the continents were once connected?

② Describe the four kinds of evidence that Wegener and his supporters gathered to substantiate the continental drift hypothesis.

③ Explain why the discovery of the fossil remains of *Mesosaurus* in both South America and Africa, but nowhere else, supports the continental drift hypothesis.

④ Early in the 20th century, what was the prevailing view of how land animals migrated across vast expanses of open ocean?

⑤ How did Wegener account for the existence of glaciers in the southern landmasses at a time when areas in North America, Europe, and Asia supported lush tropical swamps?

The Great Debate

Wegener's proposal did not attract much open criticism until 1924, when his book was translated into English, French, Spanish, and Russian. From that point until his death in 1930, the drift hypothesis encountered a great deal of hostile criticism. The respected American geologist R.T. Chamberlain stated, "Wegener's hypothesis in general is of the foot-loose type, in that it takes considerable liberty with our globe, and is less bound by restrictions or tied down by awkward, ugly facts than most of its rival theories. Its appeal seems to lie in the fact that it plays a game in which there are few restrictive rules and no sharply drawn code of conduct."

One of the main objections to Wegener's hypothesis stemmed from his inability to identify a credible mechanism for continental drift. Wegener proposed

The movement of Earth's tectonic plates is the cause of our most destructive earthquakes. Pisco, Peru, following a powerful earthquake on August 16, 2007. *(Sergio Erday/epa/Corbis)*

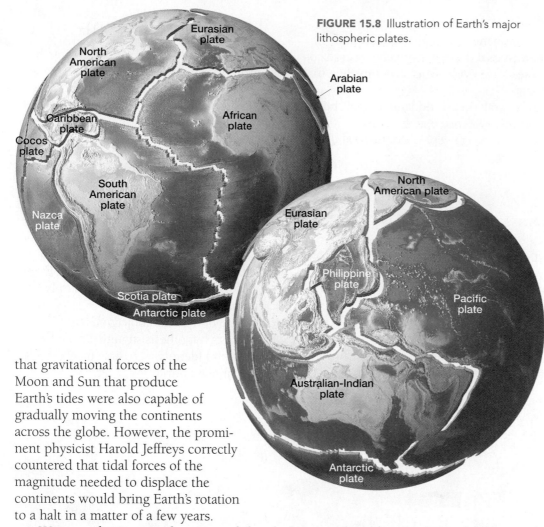

FIGURE 15.8 Illustration of Earth's major lithospheric plates.

that gravitational forces of the Moon and Sun that produce Earth's tides were also capable of gradually moving the continents across the globe. However, the prominent physicist Harold Jeffreys correctly countered that tidal forces of the magnitude needed to displace the continents would bring Earth's rotation to a halt in a matter of a few years.

Wegener also incorrectly suggested that the larger and sturdier continents broke through thinner oceanic crust, much like ice breakers cut through ice. However, no evidence existed to suggest that the ocean floor was weak enough to permit passage of the continents without the continents being appreciably deformed in the process.

In 1930 Wegener made his fourth and final trip to the Greenland ice sheet. Although the primary focus of this expedition was to study the harsh winter polar climate on the ice-covered island, Wegener continued to test his continental drift hypothesis. As in earlier expeditions, he used astronomical methods in an attempt to verify that Greenland had drifted westward with respect to Europe. While returning from Eismitte (an experimental station located in the center of Greenland), Wegener perished along with a companion. His intriguing idea, however, did not die.

What went wrong? Why was Wegener unable to overturn the established scientific views of his day? Foremost was the fact that, although the central theme of Wegener's drift hypothesis was correct, it contained some incorrect details. For example, continents do not break through the ocean floor, and tidal energy is much too weak to cause continents to be displaced. Moreover, in order for any comprehensive scientific theory to gain wide acceptance, it must stand up to critical testing from all areas of science. Wegener's great contribution to our understanding of Earth notwithstanding, not *all* of the evidence supported the continental drift hypothesis as he had formulated it.

Although many of Wegener's contemporaries opposed his views, even to the point of open ridicule, some considered his ideas plausible. For those geologists who continued the search, the exciting concept of continents adrift held their interest. Others viewed continental drift as a solution to previously unexplainable observations. Nevertheless, most of the scientific community, particularly in North America, either categorically rejected continental drift or at least treated it with considerable skepticism.

CONCEPT CHECK 15.3

❶ To which two aspects of Wegener's continental drift hypothesis did most Earth scientists object?

Plate Tectonics

 **PLATE TECTONICS**

ESSENTIALS OF GEOLOGY **Introduction**

Following World War II, oceanographers equipped with new marine tools and ample funding from the U.S. Office of Naval Research embarked on an unprecedented period of oceanographic exploration. Over the next two decades a much better picture of large expanses of the seafloor slowly and painstakingly began to emerge. From this work came the discovery of a global **oceanic ridge system** that winds through all of the major oceans in a manner similar to the seams on a baseball.

In other parts of the ocean, new discoveries were also being made. Earthquake studies conducted in the western Pacific demonstrated that tectonic activity was occurring at great depths beneath deep-ocean trenches. Of equal importance was the fact that dredging of the seafloor did not bring up any oceanic crust that was older than 180 million years. Further, sediment accumulations in the deep-ocean basins were found to be thin, not the thousands of meters that were predicted.

By 1968, these developments, among others, led to the unfolding of a far more encompassing theory than continental drift, known as **plate tectonics**. According to the plate tectonics model, the uppermost mantle and the overlying crust behave as a strong, rigid layer, known as the **lithosphere**, which is broken into segments commonly referred to as *plates* (**FIGURE 15.8**). The lithosphere is thinnest in the oceans where it varies from as little as a few kilometers along the axis of the oceanic ridge system to about 100 kilometers (60 miles) in the deep-ocean basins. By contrast, continental lithosphere is generally thicker than 100 kilometers and may extend to a depth of 200 to 300 kilometers beneath stable continental cratons.

The lithosphere, in turn, overlies a weak region in the mantle known as the **asthenosphere**. The temperatures and pressures in the upper asthenosphere (100 to 200 kilometers in depth) are such that the rocks there are very near their melting temperatures and, hence, respond to stress by flowing. As a result, Earth's rigid outer shell is effectively detached from the layers below, which permits it to move independently.

Earth's Major Plates

The lithosphere is composed of about two dozen segments having irregular sizes and shapes called **lithospheric plates** or **tectonic plates** that are in constant motion with respect to one another. As shown in **FIGURE 15.9**, seven major lithospheric plates are recognized. These plates, which account for 94 percent of Earth's surface area, include the *North American, South American, Pacific, African, Eurasian, Australian-Indian,* and *Antarctic plates.* The largest is the Pacific plate, which encompasses a significant portion of the Pacific Ocean basin. The six other large plates include an entire continent plus a significant amount of ocean floor. Notice in Figure 15.9 that the South American plate encompasses almost all of South America and about one half of the floor of the South Atlantic. This is a major departure from Wegener's continental drift hypothesis, which proposed that the continents moved through the ocean floor, not with it. Note also that none of the plates are defined entirely by the margins of a single continent.

Intermediate-sized plates include the *Caribbean, Nazca, Philippine, Arabian, Cocos, Scotia,* and *Juan de Fuca plates.* These plates, with the exception of the Arabian plate, are composed mostly of oceanic lithosphere. In addition, there are several smaller plates (*microplates*) that have been identified but are not shown in Figure 15.9.

Plate Boundaries

One of the main tenets of the plate tectonics theory is that plates move as semicoherent units relative to all other plates. As plates move, the distance between two locations on different plates, such as New York and London, gradually changes whereas the distance between sites on the same plate—New York and Denver, for example—remains relatively constant.

Because plates are in constant motion relative to each other, most major interactions among them (and, therefore, most deformation) occur along their *boundaries.* In fact, plate boundaries were first established by plotting the locations of earthquakes and volcanoes. Plates are bounded by three distinct types of boundaries, which are differentiated by the type of movement they exhibit. These boundaries are depicted at the bottom of Figure 15.9 and are briefly described here:

1. **Divergent boundaries** (*constructive margins*)—where two plates move apart, resulting in upwelling of hot material from the mantle to create new seafloor (**FIGURE 15.9A**).

2. **Convergent boundaries** (*destructive margins*)—where two plates move together, resulting in oceanic lithosphere descending beneath an overriding plate, eventually to be reabsorbed into the mantle or possibly in the collision of two continental blocks to create a mountain system (**FIGURE 15.9B**).

3. **Transform fault boundaries** (*conservative margins*)—where two plates grind past each other without the production or destruction of lithosphere (**FIGURE 15.9C**).

Divergent and convergent plate boundaries each account for about 40 percent of all plate boundaries. Transform faults account for the remaining 20 percent. In the following sections we will summarize the nature of the three types of plate boundaries.

CONCEPT CHECK 15.4

❶ Compare and contrast the lithosphere and the asthenosphere.

❷ List the three types of plate boundaries, and describe the relative motion at each of them.

Divergent Boundaries

 PLATE TECTONICS

ESSENTIALS OF GEOLOGY Divergent Boundaries

Most **divergent boundaries** are located along the crests of oceanic ridges and can be thought of as *constructive plate margins* because this is where new ocean floor is generated (**FIGURE 15.10**). Divergent boundaries are also called **spreading centers**, because seafloor spreading occurs at these boundaries. Here, two adjacent plates are moving away from each other, producing long, narrow fractures in the ocean crust. As a result, hot rock from the mantle below migrates upward to fill the voids left as the crust is being ripped apart. This molten material gradually cools to produce new slivers of seafloor. In a slow, yet unending manner, adjacent plates spread apart and new oceanic lithosphere forms between them.

Oceanic Ridges and Seafloor Spreading

Most divergent plate boundaries are associated with *oceanic ridges:* elevated areas of the seafloor that are characterized by high heat flow and volcanism. The global ridge system is the longest topographic feature on Earth's surface, exceeding 70,000 kilometers (43,000 miles) in length. As shown in Figure 15.9 various segments of the global ridge system have been named, including the Mid-Atlantic Ridge, East Pacific Rise, and Mid-Indian Ridge.

Representing 20 percent of Earth's surface, the oceanic ridge system winds through all major ocean basins like the seam on a baseball. Although the crest of the oceanic ridge is commonly 2 to 3 kilometers higher than the adjacent ocean

DID YOU KNOW?

An observer on another planet would notice, after only a few million years, that all the continents and ocean basins on Earth are indeed moving. The Moon, on the other hand, is tectonically dead, so it would look virtually unchanged millions of years into the future.

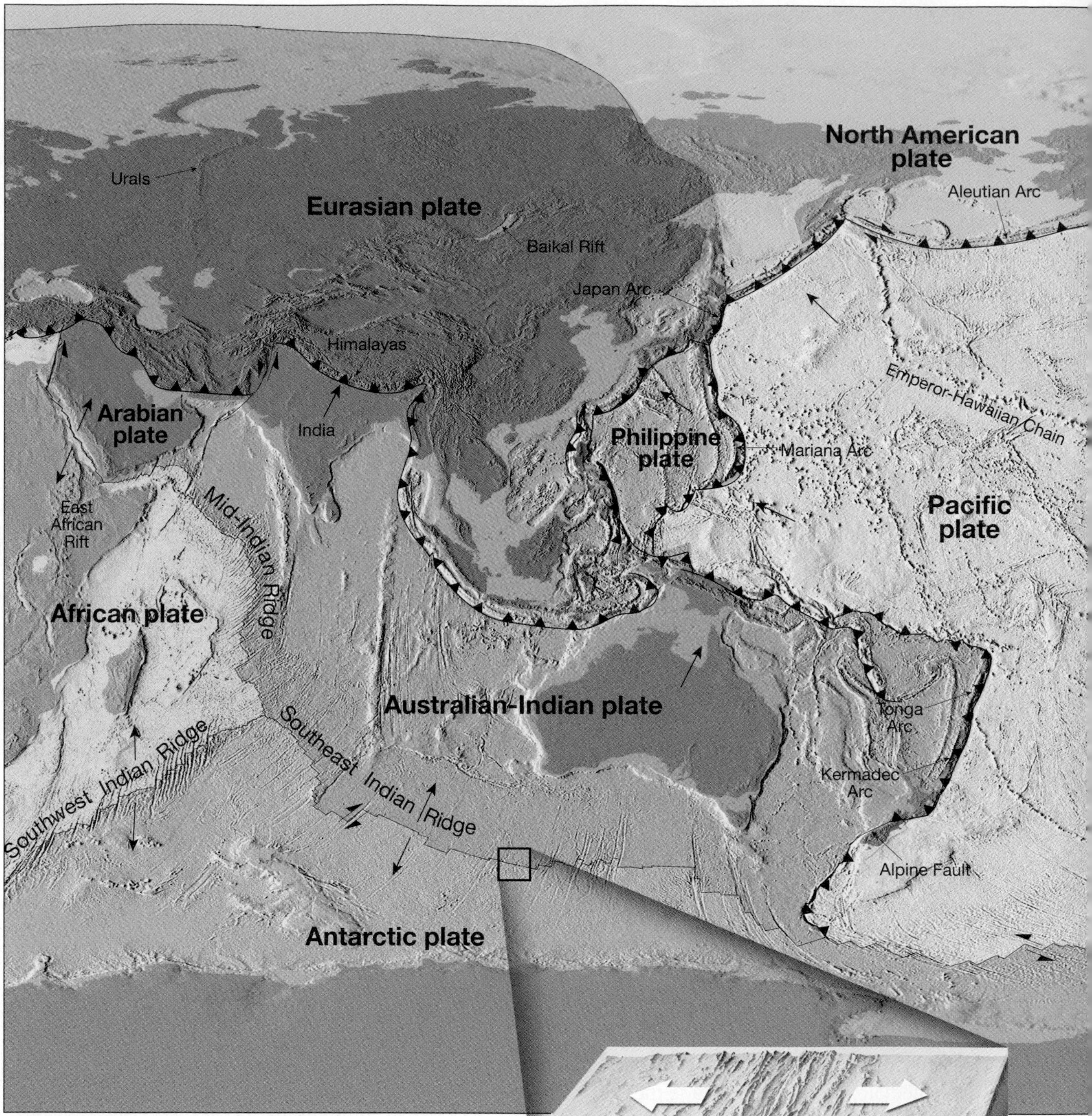

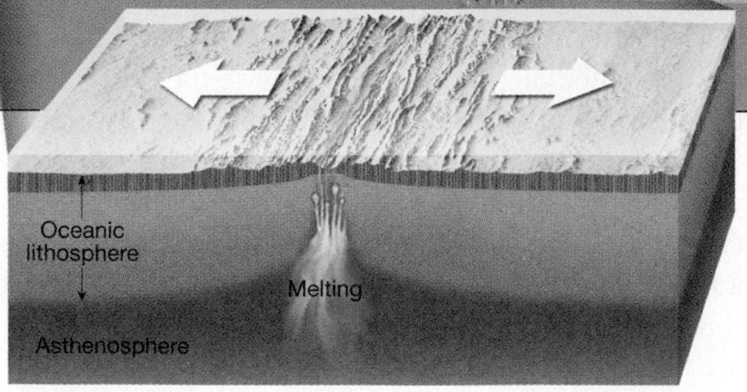

FIGURE 15.9 A mosaic of rigid plates constitutes Earth's outer shell. (*After W. B. Hamilton, U.S. Geological Survey*)

A. Divergent boundary

basins, the term "ridge" may be misleading because this feature is not narrow but has widths that vary from 1000 to more than 4000 kilometers. Further, along the axis of some ridge segments is a deep downfaulted structure called a **rift valley**. This structure is evidence that tensional forces are actively pulling the ocean crust apart at the ridge crest.

The mechanism that operates along the oceanic ridge system to create new seafloor is appropriately called **seafloor spreading**. Typical rates of spreading average around 5 centimeters (2 inches) per year. This is roughly the same rate at which human fingernails grow. Comparatively slow spreading rates of 2 centimeters per year are found along the Mid-Atlantic Ridge, whereas spreading rates exceeding 15 centimeters (6 inches) per year have been measured along sections of the East Pacific Rise. Although these rates of seafloor production are slow on a human time scale, they are nevertheless rapid enough to have generated all of Earth's ocean basins within the last 200 million years. In fact, none of the ocean floor that has been dated thus far exceeds 180 million years in age.

The primary reason for the elevated position of the oceanic ridge is that newly created oceanic crust is hot, making it less dense than cooler rocks found away from the ridge axis. As soon as new lithosphere forms, it is slowly yet continually displaced away from the zone of upwelling. Thus, it begins to cool and contract, thereby increasing in density. This thermal contraction accounts for the increase in ocean depths away from the ridge crest. It takes about 80 million years for the temperature of the crust to stabilize and contraction to cease. By this time, rock that was once part of the elevated oceanic ridge system is located in the deep-ocean basin, where it may be buried by substantial accumulations of sediment.

In addition, cooling strengthens the hot material directly below the oceanic crust, thereby adding to the plate's thickness.

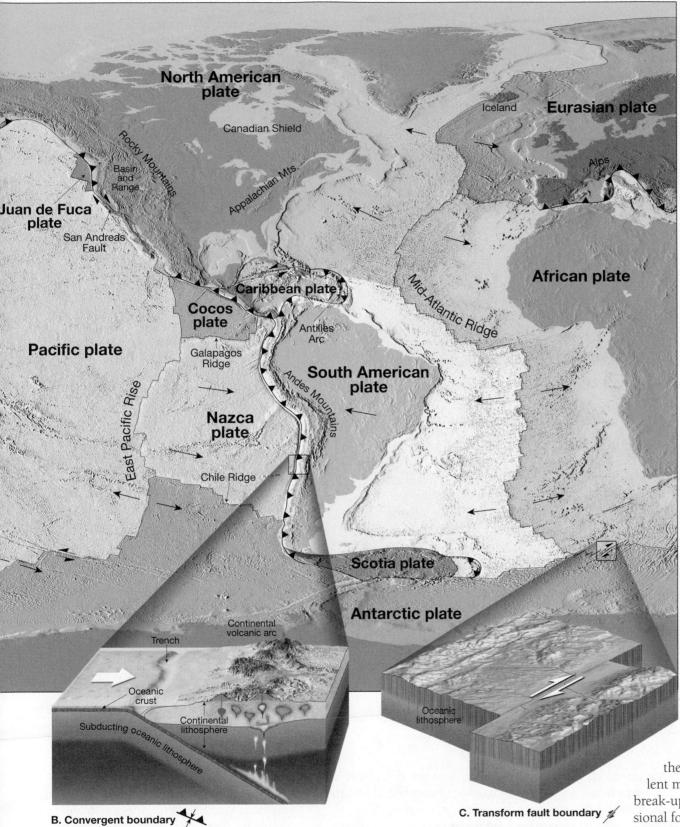

North American plate

Canadian Shield

Rocky Mountains

Basin and Range

Juan de Fuca plate

San Andreas Fault

Appalachian Mts.

Iceland

Eurasian plate

Alps

African plate

Mid-Atlantic Ridge

Caribbean plate

Cocos plate

Antilles Arc

Pacific plate

Galapagos Ridge

East Pacific Rise

Andes Mountains

South American plate

Nazca plate

Chile Ridge

Scotia plate

Antarctic plate

Continental volcanic arc

Trench

Oceanic crust

Subducting oceanic lithosphere

Continental lithosphere

B. Convergent boundary

Oceanic lithosphere

C. Transform fault boundary

Continental Rifting

Divergent boundaries can also develop within a continent, in which case the landmass may split into two or more smaller segments separated by an ocean basin. Continental rifting occurs where opposing tectonic forces act to pull the lithosphere apart. The initial stage of rifting tends to include mantle upwelling that is associated with broad upwarping of the overlying lithosphere (**FIGURE 15.11A**). As a result, the lithosphere is stretched, causing the brittle crustal rocks to break into large slabs. As the tectonic forces continue to pull the crust apart, these crustal fragments sink, generating an elongated depression called a **continental rift** (**FIGURE 15.11B**).

A modern example of an active continental rift is the East African Rift (see Figure 15.9, left). Whether this rift will eventually result in the break-up of Africa is a topic of continued research. Nevertheless, the East African Rift is an excellent model of the initial stage in the break-up of a continent. Here, tensional forces have stretched and thinned the crust, allowing molten rock to ascend from the mantle. Evidence for recent volcanic activity includes several large volcanic mountains including Mount Kilimanjaro and Mount Kenya, the tallest

Stated another way, the thickness of oceanic lithosphere is age-dependent. The older (cooler) it is, the greater its thickness.

Oceanic lithosphere that exceeds 80 million years in age is about 100 kilometers thick: about its maximum thickness.

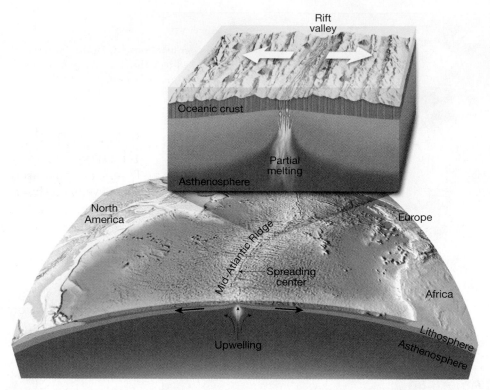

FIGURE 15.10 Most divergent plate boundaries are situated along the crests of oceanic ridges.

peaks in Africa. Research suggests that if rifting continues, the rift valley will lengthen and deepen, eventually extending out to the margin of the landmass (**FIGURE 15.11C**). At this point, the rift will become a narrow sea with an outlet to the ocean. The Red Sea, which formed when the Arabian Peninsula split from Africa, is a modern example of such a feature. Consequently, the Red Sea provides us with a view of how the Atlantic Ocean may have looked in its infancy (**FIGURE 15.11D**).

CONCEPT CHECK 15.5

❶ Sketch or describe how two plates move in relation to each other along divergent plate boundaries.

❷ List four facts that characterize the oceanic ridge system.

❸ Briefly describe the process of continental rifting. Where is it occurring today?

Convergent Boundaries

 PLATE TECTONICS

ESSENTIALS OF GEOLOGY **Convergent Boundaries**

New lithosphere is constantly being produced at the oceanic ridges; however, our planet is not growing larger—its total surface area remains constant. A balance is maintained because older, denser portions of oceanic lithosphere descend into the mantle at a rate equal to seafloor production. This activity occurs along **convergent boundaries**, where two plates move toward each other and the leading edge of one is bent downward, as it slides beneath the other.

Convergent boundaries are also called **subduction zones**, because they are sites where lithosphere is descending (being subducted) into the mantle. Subduction occurs because the density of the descending tectonic plate is greater than the density of the underlying asthenosphere. In general, oceanic lithosphere is more dense than the asthenosphere, whereas continental lithosphere is less dense and resists subduction. As a consequence, only oceanic lithosphere will subduct to great depths.

Deep-ocean trenches are the surface manifestations produced as oceanic lithosphere descends into the mantle. These large linear depressions are remarkably long and deep. The Peru–Chile trench along the west coast of South America is more than 4500 kilometers (3000 miles) in length, and its base is as much as 8 kilometers (5 miles) below sea level. The trenches in the western Pacific, including the Mariana and Tonga trenches, tend to be even deeper than those of the eastern Pacific.

Slabs of oceanic lithosphere descend into the mantle at angles that vary from a few degrees to nearly vertical (90 degrees). The angle at which oceanic lithosphere descends depends largely on its density. For example, when a spreading center is located near a subduction zone, as is the case along the Peru–Chile trench, the subducting lithosphere is young and, therefore, warm and buoyant. Because of this, the angle of descent is small, which results in considerable interaction between the descending slab and the overriding plate. Consequently, the region around the Peru–Chile trench experiences great earthquakes, including the 2010 Chilean earthquake—one of the 10 largest on record.

As oceanic lithosphere ages (gets farther from the spreading center), it gradually cools, which causes it to thicken and increase in density. In parts of the western Pacific, some oceanic lithosphere is 180 million years old. This is the thickest and densest in today's oceans. The very dense slabs in this region typically plunge into the mantle at angles approaching 90 degrees. This largely explains the fact that most trenches in the western Pacific are deeper than trenches in the eastern Pacific.

Although all convergent zones have the same basic characteristics, they are highly variable features. Each is controlled by the type of crustal material involved and the tectonic setting. Convergent boundaries can form between *two oceanic plates, one oceanic and one continental plate,* or *two continental plates.*

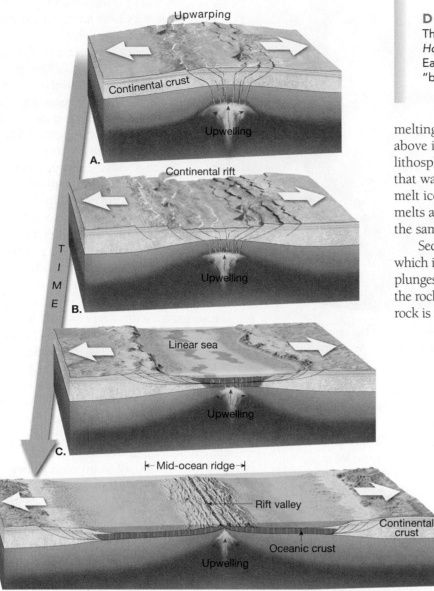

melting is triggered within the wedge of hot asthenosphere that lies above it. But how does the subduction of a cool slab of oceanic lithosphere cause mantle rock to melt? The answer lies in the fact that water contained in the descending plates acts like salt does to melt ice. That is, "wet" rock in a high-pressure environment melts at substantially lower temperatures than does "dry" rock of the same composition.

Sediments and oceanic crust contain a large amount of water, which is carried to great depths by a subducting plate. As the plate plunges downward, heat and pressure drive water from the voids in the rock. At a depth of roughly 100 kilometers, the wedge of mantle rock is sufficiently hot that the introduction of water from the

FIGURE 15.11 Continental rifting and the formation of a new ocean basin. **A.** The initial stage of continental rifting tends to include upwelling in the mantle that is associated with broad upwarping of the lithosphere. Tensional forces and buoyant uplifting of the heated lithosphere cause the crust to be broken into large slabs. **B.** As the crust is pulled apart, these large blocks sink, generating a continental rift valley. **C.** Further spreading generates a narrow sea similar to the present-day Red Sea. **D.** Eventually, an expansive deep-ocean basin and oceanic ridge are created.

Oceanic–Continental Convergence

Whenever the leading edge of a plate capped with continental crust converges with a slab of oceanic lithosphere, the buoyant continental block remains "floating," while the denser oceanic slab sinks into the mantle (**FIGURE 15.12A**). When a descending oceanic slab reaches a depth of about 100 kilometers (60 miles),

Alfred Wegener shown waiting out the 1912–13 Arctic winter during an expedition to Greenland, where he made a 1200-kilometer traverse across the widest part of the island's ice sheet. *(Photo courtesy of Archive of Alfred Wegener Institute)*

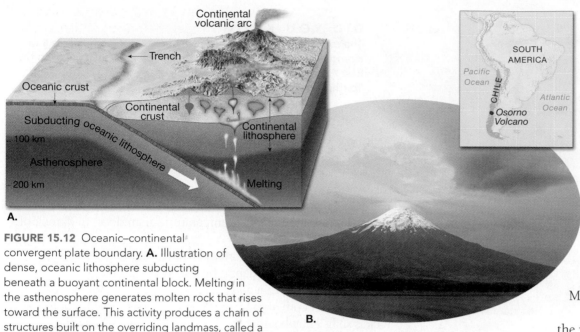

FIGURE 15.12 Oceanic–continental convergent plate boundary. **A.** Illustration of dense, oceanic lithosphere subducting beneath a buoyant continental block. Melting in the asthenosphere generates molten rock that rises toward the surface. This activity produces a chain of structures built on the overriding landmass, called a continental volcanic arc. **B.** Osorno Volcano is one of the most active volcanoes of the southern Chilean Andes, having erupted 11 times between 1575 and 1869. Located on the shore of Lake Llanquihue, Osorno is similar in appearance to Mount Fuji, Japan. *(Photo by Michael Collier)*

slab below leads to some melting. This process, called **partial melting**, is thought to generate about 10 percent molten material, which is intermixed with unmelted mantle rock. Being less dense than the surrounding mantle, this hot mobile material gradually rises toward the surface. Depending on the environment, these mantle-derived masses of molten rock may ascend through the crust and give rise to a volcanic eruption. However, much of this material never reaches the surface; rather, it solidifies at depth—a process that thickens the crust.

The volcanoes of the towering Andes are the product of molten rock generated by the subduction of the Nazca plate beneath the South American continent (**FIGURE 15.12B**). Mountain systems, such as the Andes, which

are produced in part by volcanic activity associated with the subduction of oceanic lithosphere, are called **continental volcanic arcs**. The Cascade Range in Washington, Oregon, and California is another that consists of several well-known volcanic mountains, including Mount Rainier, Mount Shasta, and Mount St. Helens. This active volcanic arc also extends into Canada, where it includes Mount Garibaldi, Mount Silverthrone, and others.

Oceanic–Oceanic Convergence

An *oceanic–oceanic convergent boundary* has many features in common with oceanic–continental plate margins. Where two oceanic slabs converge, one descends beneath the other, initiating volcanic activity by the same mechanism that operates at all subduction zones (**FIGURE 15.13A**). Water squeezed from the subducting slab of oceanic lithosphere triggers melting in the hot wedge of mantle rock above. In this setting, volcanoes grow up from the ocean floor, rather than upon a continental platform. When subduction is sustained, it will

*More on these volcanic events is found in Chapter 4.

eventually build a chain of volcanic structures large enough to emerge as islands. The volcanic islands tend to be spaced at intervals of about 80 kilometers (50 miles). This newly formed land consisting of an arc-shaped chain of volcanic islands is called a **volcanic island arc**, or simply an **island arc** (**FIGURE 15.13B**).

The Aleutian, Mariana, and Tonga islands are examples of relatively young volcanic island arcs. Island arcs are generally located 100 to 300 kilometers (60 to 200 miles) from a deep-ocean trench. Located adjacent to the island arcs just mentioned are the Aleutian trench, the Mariana trench, and the Tonga trench.

Most volcanic island arcs are located in the western Pacific. Only two are located in the Atlantic—the Lesser Antilles arc, on the eastern margin of the Caribbean Sea and the Sandwich Islands located off the tip of South America. The Lesser Antilles are a product of the subduction of the Atlantic seafloor beneath the Caribbean plate. Located within this volcanic arc are the United States and British Virgin Islands as well as the island of Martinique, where Mount Pelée erupted in 1902, destroying the town of St. Pierre and killing an estimated 28,000 people. This chain of islands also includes Montserrat, where there has been recent volcanic activity.*

Relatively young island arcs are fairly simple structures made of numerous volcanic cones that are underlain by oceanic crust that is generally less than 20 kilometers (12 miles) thick. By contrast, older island arcs are more complex and are underlain by highly deformed crust that may reach 35 kilometers in thickness. Examples include the islands that make up the countries of Japan, Indonesia, and the Philippines. These island arcs are built upon material generated by earlier episodes of subduction or on small slivers of continental crust.

Continental–Continental Convergence

The third type of convergent boundary results when one landmass moves toward the margin of another because of subduction of the intervening seafloor

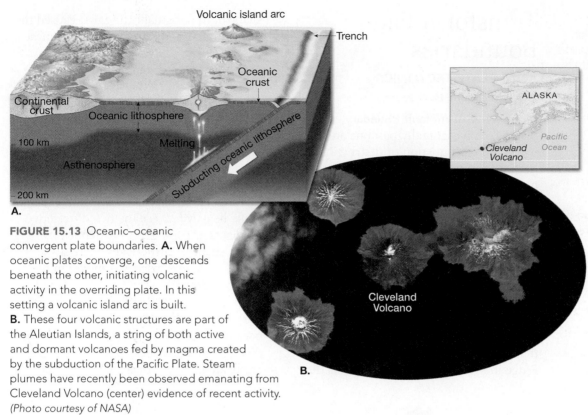

Volcanic island arc

Trench

Oceanic crust

Continental crust

Oceanic lithosphere

100 km

Melting

Asthenosphere

Subducting oceanic lithosphere

200 km

A.

FIGURE 15.13 Oceanic–oceanic convergent plate boundaries. **A.** When oceanic plates converge, one descends beneath the other, initiating volcanic activity in the overriding plate. In this setting a volcanic island arc is built. **B.** These four volcanic structures are part of the Aleutian Islands, a string of both active and dormant volcanoes fed by magma created by the subduction of the Pacific Plate. Steam plumes have recently been observed emanating from Cleveland Volcano (center) evidence of recent activity. *(Photo courtesy of NASA)*

ALASKA

Cleveland Volcano

Pacific Ocean

Cleveland Volcano

B.

(**FIGURE 15.14A**). Whereas oceanic lithosphere tends to be dense and sink into the mantle, the buoyancy of continental material inhibits it from being subducted. Consequently, a collision between two converging continental fragments ensues (**FIGURE 15.14C**). This event folds and deforms the accumulation of sediments and sedimentary rocks along the continental margins as if they had been placed in a gigantic vise. The result is the formation of a new mountain range composed of deformed sedimentary and metamorphic rocks that often contain slivers of oceanic crust.

Such a collision began about 50 million years ago when the subcontinent of India "rammed" into Asia, producing the Himalayas—the most

FIGURE 15.14 The ongoing collision of India and Asia began about 50 million years ago—producing the majestic Himalayas. **A.** As India migrated northward the intervening ocean closed as the seafloor was subducted beneath Eurasia. **B.** Position of India in relation to Eurasia at various times. *(Modified after Peter Molnar)* **C.** Eventually the two landmasses collided, deforming and elevating the sediments that had been deposited along their continental margins. In addition, slices of crustal rocks were thrust onto the colliding plates.

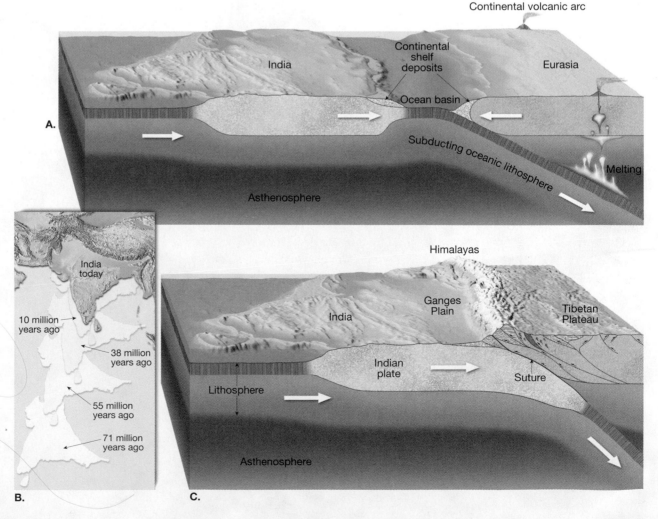

Continental volcanic arc

India

Continental shelf deposits

Eurasia

Ocean basin

Subducting oceanic lithosphere

Melting

A.

Asthenosphere

India today

10 million years ago

38 million years ago

55 million years ago

71 million years ago

B.

Himalayas

Ganges Plain

India

Tibetan Plateau

Indian plate

Suture

Lithosphere

Asthenosphere

C.

spectacular mountain range on Earth (Figure 15.14C). During this collision, the continental crust buckled and fractured and was generally shortened and thickened. In addition to the Himalayas, several other major mountain systems, including the Alps, Appalachians, and Urals, formed as continental fragments collided. (This topic will be considered further in Chapter 17.)

CONCEPT CHECK 15.6

1. Why does the rate of lithosphere production roughly balance with the rate at which it is destroyed?

2. Compare a continental volcanic arc and a volcanic island arc.

3. Why does oceanic lithosphere subduct while continental lithosphere does not?

4. Briefly describe how mountain systems such as the Himalayas form.

Transform Fault Boundaries

 PLATE TECTONICS

ESSENTIALS OF GEOLOGY **Transform Fault Boundaries**

Along a **transform fault boundary**, plates slide horizontally past one another without the production or destruction of lithosphere (*conservative plate margins*). The nature of transform faults was discovered in 1965 by Canadian geologist J. Tuzo Wilson, who proposed that these large faults connected two spreading centers (divergent boundaries) or less commonly two trenches (convergent boundaries). Most transform faults are found on the ocean floor (**FIGURE 15.15A**). Here they offset segments of the oceanic ridge system, producing a steplike plate margin. Notice that the zigzag shape of the Mid-Atlantic Ridge in Figure 15.9 roughly reflects the

shape of the original rifting that caused the break-up of the supercontinent of Pangaea. (Compare the shapes of the continental margins of the landmasses on both sides of the Atlantic with the shape of the Mid-Atlantic Ridge.)

Typically, transform faults are part of prominent linear breaks in the seafloor known as **fracture zones**, which include both the active transform faults as well as their inactive extensions into the plate interior (**FIGURE 15.15B**). Active transform faults lie *only between* the two offset ridge segments and are generally defined by weak, shallow earthquakes. Here seafloor produced at one ridge axis moves in the opposite direction of seafloor produced at an opposing ridge segment. Thus, between the ridge segments these adjacent slabs of oceanic crust are grinding past each other along a transform fault. Beyond the ridge crests are inactive zones, where the frac-

The Alps were created by the collision of the African and Eurasian plates. (*Photo by Gareth McCormack/Alamy*)

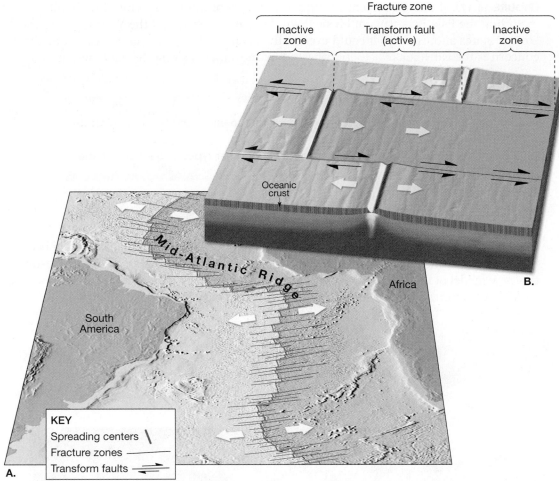

KEY
Spreading centers \
Fracture zones ——
Transform faults ⇄

A.

FIGURE 15.15 Transform fault boundaries. **A.** Most transform faults offset segments of a spreading center, producing a steplike plate margin. The zigzag shape of the Mid-Atlantic Ridge roughly reflects the shape of the zone of rifting that produced the break-up of Pangaea. **B.** Fracture zones are long, narrow fractures in the seafloor that are nearly perpendicular to the offset ridge segments. They include both the active transform fault and its "fossilized" trace, where oceanic crust of different ages is juxtaposed. The offsets between ridge segments (transform faults) do not change in length over time.

Like the Mendocino Fault, most transform fault boundaries are located within the ocean basins; however, a few cut through continental crust. Two examples are the earthquake-prone San Andreas Fault of California and New Zealand's Alpine Fault. Notice in Figure 15.16 that the San Andreas Fault connects a spreading center located in the Gulf of California to the Cascadia subduction zone and the Mendocino Fault located along the northwest coast of the United States. Along the San Andreas Fault, the Pacific plate is moving toward the northwest, past the North American plate

tures are preserved as linear topographic depressions. The trend of these fracture zones roughly parallels the direction of plate motion at the time of their formation. Thus, these structures are useful in mapping the direction of plate motion in the geologic past.

In another role, transform faults provide the means by which the oceanic crust created at ridge crests can be transported to a site of destruction—the deep-ocean trenches. **FIGURE 15.16** illustrates this situation. Notice that the Juan de Fuca plate moves in a southeasterly direction, eventually being subducted under the west coast of the United States. The southern end of this plate is bounded by a transform fault called the Mendocino Fault. This transform boundary connects the Juan de Fuca ridge to the Cascadia subduction zone. Therefore, it facilitates the movement of the crustal material created at the Juan de Fuca ridge to its destination beneath the North American continent.

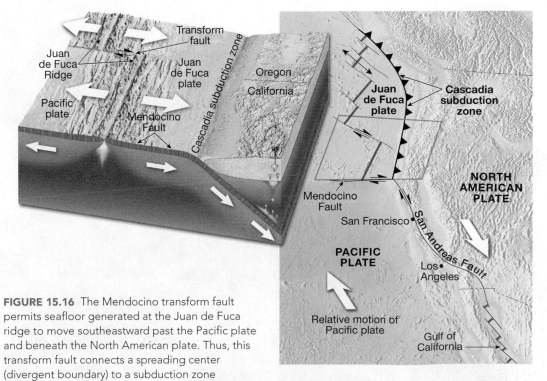

FIGURE 15.16 The Mendocino transform fault permits seafloor generated at the Juan de Fuca ridge to move southeastward past the Pacific plate and beneath the North American plate. Thus, this transform fault connects a spreading center (divergent boundary) to a subduction zone (convergent boundary). Furthermore, the San Andreas Fault, also a transform fault, connects two spreading centers: the Juan de Fuca ridge and a divergent zone located in the Gulf of California.

DID YOU KNOW?
Olympus Mons is a huge volcano on Mars that strongly resembles the Hawaiian shield volcanoes. Rising 25 kilometers above the surrounding plains, Olympus Mons owes its massiveness to the fact that plate tectonics does not operate on Mars. Consequently, instead of being carried away from the hot spot by plate motion, as occurred with the Hawaiian volcanoes, Olympus Mons remained fixed and grew to a gigantic size.

(**FIGURE 15.17**). If this movement continues, that part of California west of the fault zone, including the Baja Peninsula of Mexico, will become an island off the West Coast of the United States and Canada. It could eventually reach Alaska. However, a more immediate concern is the earthquake activity triggered by movements along this fault system.

CONCEPT CHECK 15.7

❶ Sketch or describe how two plates move in relation to each other along a transform plate boundary.

❷ Differentiate between transform faults and the two other types of plate boundaries.

Testing the Plate Tectonics Model

Some of the evidence supporting continental drift and seafloor spreading has already been presented. With the development of the theory of plate tectonics, researchers began testing this new model of how Earth works. Although new supporting data were obtained, it was often new interpretations of already existing data that swayed the tide of opinion.

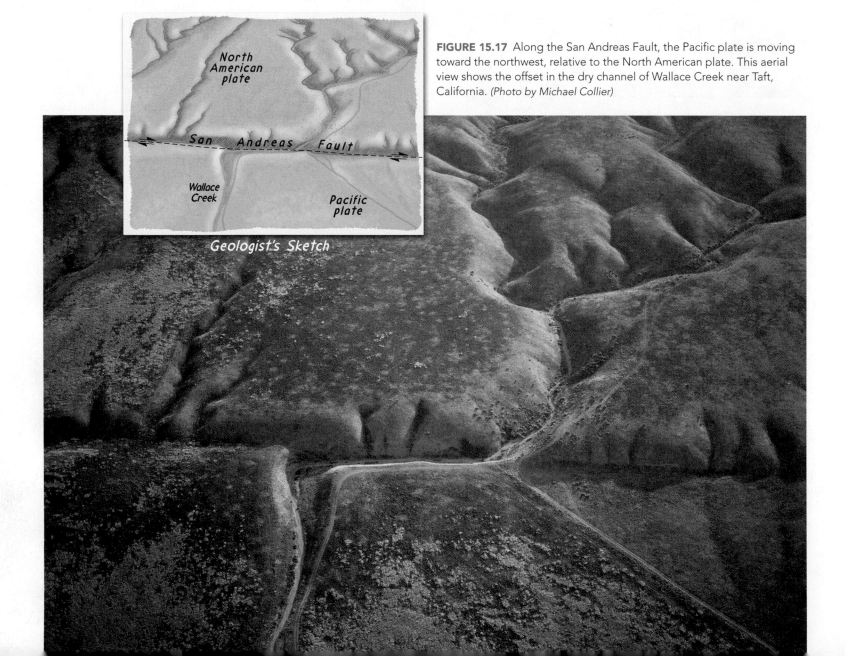

FIGURE 15.17 Along the San Andreas Fault, the Pacific plate is moving toward the northwest, relative to the North American plate. This aerial view shows the offset in the dry channel of Wallace Creek near Taft, California. *(Photo by Michael Collier)*

Evidence: Ocean Drilling

Some of the most convincing evidence for seafloor spreading came from the Deep Sea Drilling Project, which operated from 1968 until 1983. One of the early goals was to gather samples of the ocean floor in order to establish its age. To accomplish this, the *Glomar Challenger,* a drilling ship capable of working in water thousands of meters deep, was built. Hundreds of holes were drilled through the layers of sediments that blanket the ocean crust, as well as into the basaltic rocks below. Rather than radiometrically dating the crustal rocks, researchers used the fossil remains of microorganisms found in the sediments resting directly on the crust to date the seafloor at each site.*

When the oldest sediment from each drill site was plotted against its distance from the ridge crest, the plot showed that the sediments increased in age with increasing distance from the ridge (**FIGURE 15.18**). This finding supported the seafloor-spreading hypothesis, which predicted that the youngest oceanic crust would be found at the ridge crest, the site of seafloor production, and the oldest oceanic crust would be located adjacent to the continents.

The data collected by the Deep Sea Drilling Project also reinforced the idea that the ocean basins are geologically young because no seafloor with an age in excess of 180 million years was ever found. By comparison, most continental crust exceeds several hundred million years in age and some has been located that exceeds 4 billion years in age.

The thickness of ocean-floor sediments provided additional verification of seafloor spreading. Drill cores from the *Glomar Challenger* revealed that sediments are almost entirely absent on the ridge crest and that sediment thickness increases with increasing distance from the ridge (Figure 15.18). This pattern of sediment distribution should be expected if the seafloor-spreading hypothesis is correct.

The Ocean Drilling Program, the successor to the Deep Sea Drilling Project, employed a more technologically advanced

*Radiometric dates of the ocean crust itself are unreliable because of the alteration of basalt by seawater.

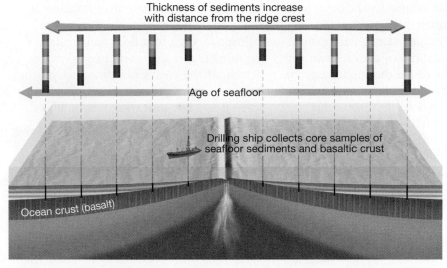

FIGURE 15.18 Since 1968 drilling ships have gathered core samples of seafloor sediment and crustal rocks at hundreds of sites. Results from these efforts showed that the ocean floor is indeed youngest at the ridge axis. This was the first direct evidence supporting the seafloor spreading hypothesis and the broader theory of plate tectonics.

drilling ship, the *JOIDES Resolution,* to continue the work of the *Glomar Challenger*. While the Deep Sea Drilling Project validated many of the major tenets of the theory of plate tectonics, the *JOIDES Resolution* was able to probe deeper into the oceanic crust. This allowed for the study of earthquake-generating zones at convergent plate margins and for the direct examination of oceanic plateaus and seamounts. Sediment cores from the Ocean Drilling Program have also extended our knowledge of long- and short-term climatic changes.

In October 2003, the *JOIDES Resolution* became part of a new program, the Integrated Ocean Drilling Program (IODP). This new international effort uses multiple vessels for exploration, including the massive 210-meter-long (nearly 770-foot-long) *Chikyu,* (meaning "planet Earth" in Japanese) which began operations in 2007. One of the goals of the IODP is to recover a complete section of the ocean crust, from top to bottom.

Hot-spot volcanism, Kilauea, Hawaii. *(U.S. Geological Survey)*

Evidence: Hot Spots

Mapping volcanic islands and seamounts (submarine volcanoes) in the Pacific Ocean revealed several linear chains of volcanic structures. One of the most studied chains consists of at least 129 volcanoes that extend

hot-spot volcanism

from the Hawaiian Islands to Midway Island and continue northward toward the Aleutian trench (**FIGURE 15.19**). Radiometric dating of this structure, called the *Hawaiian Island–Emperor Seamount chain,* showed that the volcanoes increase in age with increasing distance from the "big island" of Hawaii. The youngest volcanic island in the chain (Hawaii) rose from the ocean floor less than a million years ago, whereas Midway Island is 27 million years old, and Suiko Seamount, near the Aleutian trench, is about 65 million years old (Figure 15.19).

Most researchers are in agreement that a cylindrically shaped upwelling of hot rock, called a **mantle plume**, is located beneath the island of Hawaii. As the hot, rocky plume ascends through the mantle, the confining pressure drops, which triggers partial melting. (This process, called *decompression melting,* is discussed in Chapter 3.) The surface manifestation of this activity is a **hot-spot**, an area of volcanism, high heat flow, and crustal uplifting that is a few hundred kilometers across. As the Pacific plate moved over the hot spot, a chain of volcanic structures known as a **hot-spot track** was built. As shown in Figure 15.19, the age of each volcano indicates how much time has elapsed since it was situated over the mantle plume.

Taking a closer look at the five largest Hawaiian Islands, we see a similar pattern of ages from the volcanically active island of Hawaii, to the inactive volcanoes that make up the oldest island, Kauai (see Figure 15.19). Five million years ago, when Kauai was positioned over the hot spot, it was the *only* Hawaiian Island in existence. Visible evidence of the age of Kauai can be seen by examining its extinct volcanoes, which have been eroded into jagged peaks and vast canyons. By contrast, the relatively young island of Hawaii exhibits many fresh lava flows, and one of its five major volcanoes, Kilauea, remains active today.

Research suggests that at least some mantle plumes originate at great depth, perhaps at the core–mantle boundary. Others, however, may have a much shallower origin. Of the 40 or so hot spots that have been identified worldwide, more than a dozen are located near spreading centers. For example, the mantle plume located beneath Iceland is responsible for the vast accumulation of volcanic rocks found along this exposed section of the Mid-Atlantic Ridge.

Evidence: Paleomagnetism

Anyone who has used a compass to find direction knows that Earth's magnetic field has a north and south magnetic pole. Today these magnetic poles align closely, but not exactly, with the geographic poles. (The geographic poles are located where Earth's rotational axis intersects the surface.) Earth's magnetic field is similar to that produced by a simple bar magnet. Invisible lines of force pass through the planet and extend from one magnetic pole to the other (**FIGURE 15.20**). A compass needle, itself a small magnet free to rotate on an axis, becomes aligned with the magnetic lines of force and points to the magnetic poles.

Unlike the pull of gravity, we cannot feel Earth's magnetic field, yet its presence is revealed because it deflects a compass needle. In addition, some naturally occurring minerals are magnetic and hence are influenced by Earth's magnetic field. One of the most common is the iron-rich mineral *magnetite,* which is abun-

FIGURE 15.19 The chain of islands and seamounts that extends from Hawaii to the Aleutian trench was generated as the Pacific plate moved over a mantle plume (hot spot). Radiometric dating of the Hawaiian Islands shows that the volcanic activity increases in age moving away from the "big island" of Hawaii.

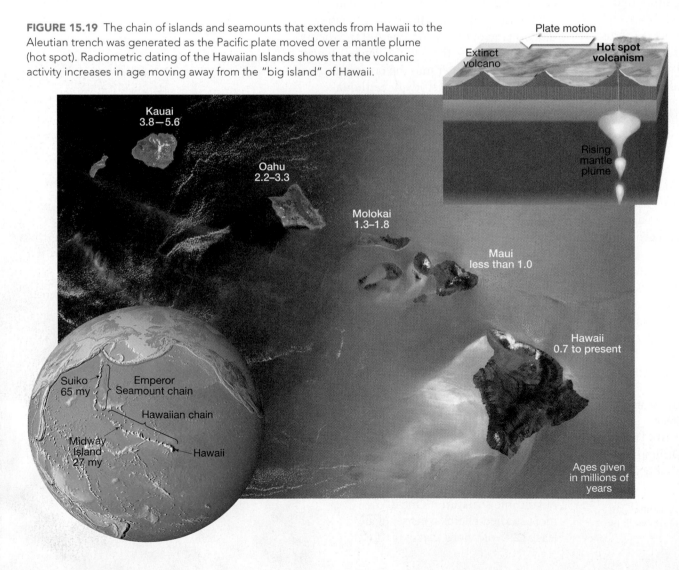

Plate motion

Extinct volcano

Hot spot volcanism

Rising mantle plume

Kauai
3.8—5.6

Oahu
2.2–3.3

Molokai
1.3–1.8

Maui
less than 1.0

Hawaii
0.7 to present

Ages given in millions of years

Suiko
65 my

Emperor Seamount chain

Hawaiian chain

Midway Island
27 my

Hawaii

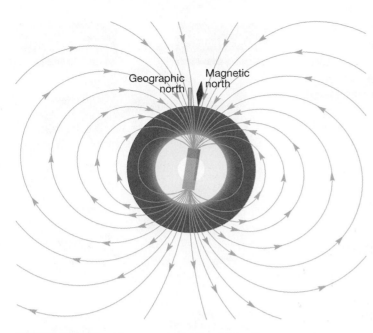

FIGURE 15.20 Earth's magnetic field consists of lines of force much like those a giant bar magnet would produce if placed at the center of Earth.

dant in lava flows of basaltic composition.* Basaltic lavas erupt at the surface at temperatures greater than 1000 °C, exceeding a threshold temperature for magnetism known as the **Curie point** (about 585 °C). Consequently, the magnetite grains in molten lava are nonmagnetic. However, as the lava cools these iron-rich grains become magnetized and align themselves in the direction of the existing magnetic lines of force. Once the minerals solidify, the magnetism they possess will usually remain "frozen" in this position. Thus, they act like a compass needle because they "point" toward the position of the magnetic poles at the time of their formation. Rocks that formed thousands or millions of years ago and contain a

DID YOU KNOW?
When all of the continents were joined to form Pangaea, the rest of Earth's surface was covered with a huge ocean called *Panthalassa* (*pan* = all, *thalassa* = sea). Today all that remains of Panthalassa is the Pacific Ocean, which has been decreasing in size since the break-up of Pangaea.

*Some sediments and sedimentary rocks contain enough iron-bearing mineral grains to acquire a measurable amount of magnetization.

The *JOIDES Resolution*, one of the drilling ships of the Integrated Ocean Drilling Program. (Ocean Drilling Program) Inset photo shows the newly launched Japanese vessel *Chikyu*, the world's most advanced drilling ship. (*AP/Wide World*)

"record" of the direction of the magnetic poles at the time of their formation are said to possess **fossil magnetism**, or **paleomagnetism**. During the 1950s paleomagnetic data was collected from lava flows around the globe.

Apparent Polar Wandering

A study of rock magnetism conducted in Europe led to an interesting discovery. The magnetic alignment of iron-rich minerals in lava flows of different ages indicated that the position of the paleomagnetic poles had changed through time. A plot of the location of the magnetic north pole with respect to Europe revealed that during the past 500 million years, the pole had gradually wandered from a location near Hawaii northward to its present location near the North Pole (**FIGURE 15.21A**). This was strong evidence that either the magnetic poles had migrated, an idea known as *polar wandering,* or that the lava flows moved—in other words, Europe had drifted in relation to the poles.

Although the magnetic poles are known to move in an erratic path around the geographic poles, studies of paleomagnetism from numerous locations show that the positions of the magnetic poles, averaged over thousands of years, correspond closely to the positions of the geographic poles. Therefore, a more acceptable explanation for the apparent polar wandering paths was provided by Wegener's hypothesis. If the magnetic poles remain stationary, their *apparent movement* is produced by continental drift.

Further evidence for continental drift came a few years later when a polar-wandering path was constructed for North America (Figure 15.21A). For the first 300 million years or so, the paths for North America and Europe were found to be similar in direction, but separated by about 5000 kilometers (3000 miles). Then, during the middle of the Mesozoic era (180 million years ago) they began to converge on the present North Pole. Because the North Atlantic Ocean is about 5000 kilometers wide, the explanation for these curves was that North America and Europe were joined until the Mesozoic,

when the Atlantic began to open. From this time forward, these continents continuously moved apart. When North America and Europe are moved back to their pre-drift positions, as shown in **FIGURE 15.21B**, these apparent wandering paths coincide. This is evidence that North America and Europe were once joined and moved relative to the poles as part of the same continent.

Magnetic Reversals and Seafloor Spreading

Another discovery came when geophysicists learned that over periods of hundreds of thousands of years, Earth's magnetic field periodically reverses polarity. During a **magnetic reversal** the north magnetic pole becomes the south magnetic pole, and vice versa. Lava solidifying during a period of reverse polarity will be magnetized with the polarity opposite that of volcanic rocks being formed today. When rocks exhibit the same magnetism as the present magnetic field, they are said to possess **normal polarity**, whereas rocks exhibiting the opposite magnetism are said to have **reverse polarity**.

Once the concept of magnetic reversals was confirmed, researchers set out to establish a time scale for these occurrences. The task was to measure the magnetic polarity of hundreds of lava flows and use radiometric dating techniques to establish the age of each flow. **FIGURE 15.22** shows the **magnetic time scale** established for the past few million years. The major divisions of the magnetic time scale are called *chrons* and last for roughly 1 million years. As more measurements became available, researchers realized that several, short-lived reversals (less than 200,000 years long) often occurred during a single chron.

Meanwhile, oceanographers had begun to do magnetic surveys of the ocean floor in conjunction with their efforts to construct detailed maps of seafloor topography. These magnetic surveys were accomplished by towing very sensitive instruments, called **magnetometers**, behind research vessels. The goal of these geophysical surveys was to map variations in the strength of Earth's magnetic field that arise from differences in

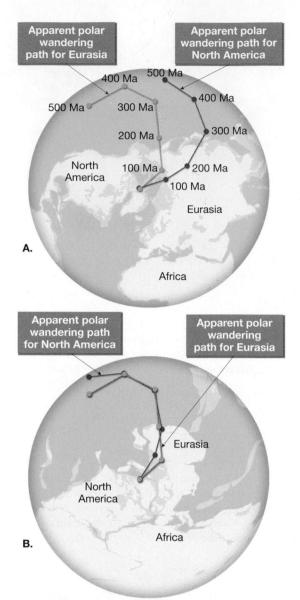

FIGURE 15.21 Simplified apparent polar-wandering paths as established from North American and Eurasian paleomagnetic data. **A.** The more westerly path determined from North American data was caused by the westward movement of North America from Eurasia during the break-up of Pangaea. **B.** The positions of the wandering paths when the landmasses are reassembled.

the magnetic properties of the underlying crustal rocks.

The first comprehensive study of this type was carried out off the Pacific Coast of North America and had an unexpected outcome. Researchers discovered alternating stripes of high- and low-intensity magnetism as shown in **FIGURE 15.23**. This relatively simple pattern of magnetic variation defied explanation until 1963, when Fred Vine and D. H. Matthews demonstrated that the

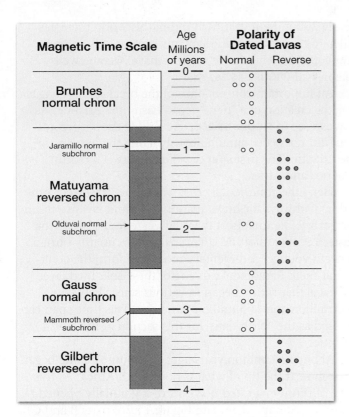

Magnetic Time Scale		Age Millions of years	Polarity of Dated Lavas	
			Normal	Reverse
Brunhes normal chron		—0—		
Jaramillo normal subchron		—1—		
Matuyama reversed chron				
Olduvai normal subchron		—2—		
Gauss normal chron		—3—		
Mammoth reversed subchron				
Gilbert reversed chron		—4—		

FIGURE 15.22 Time scale of Earth's magnetic field in the recent past. This time scale was developed by establishing the magnetic polarity for lava flows of known age. *(Data from Allen Cox and G. B. Dalrymple)*

high- and low-intensity stripes supported the concept of seafloor spreading. Vine and Matthews suggested that the stripes of high-intensity magnetism are regions where the paleomagnetism of the ocean crust exhibits normal polarity (**FIGURE 15.24**). Consequently, these rocks *enhance* (reinforce) Earth's magnetic field. Conversely, the low-intensity stripes are regions where the ocean crust is polarized in the reverse direction and therefore *weaken* the existing magnetic field. But how do parallel stripes of normally and reversely magnetized rock become distributed across the ocean floor?

Vine and Matthews reasoned that as magma solidifies along narrow rifts at the crest of an oceanic ridge, it is magnetized with the polarity of the existing magnetic field (**FIGURE 15.25**). Because of seafloor spreading, this strip of magnetized crust would gradually increase in width. When Earth's magnetic field reverses polarity, any newly formed seafloor (having the opposite polarity) would form in the middle of the old strip. Gradually, the two halves of the old strip are carried in opposite directions away from the ridge crest. Subsequent reversals would build a pattern of normal and reverse magnetic stripes as shown in Figure 15.25. Because new rock is added in equal amounts to both trailing edges of the spreading ocean floor, we should expect the pattern of stripes (size and polarity) found on one side of an oceanic ridge to be a mirror image of the other side. A few years later a survey across the Mid-Atlantic Ridge just south of Iceland revealed a pattern of magnetic stripes exhibiting a remarkable degree of symmetry to the ridge axis.

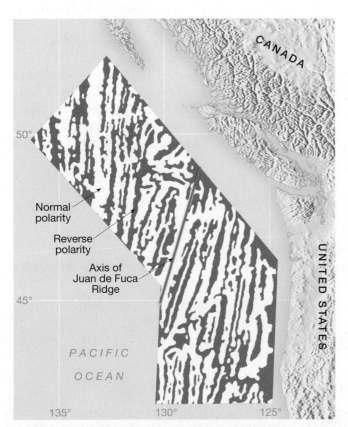

FIGURE 15.23 Pattern of alternating stripes of high- and low-intensity magnetism discovered off the Pacific Coast of North America.

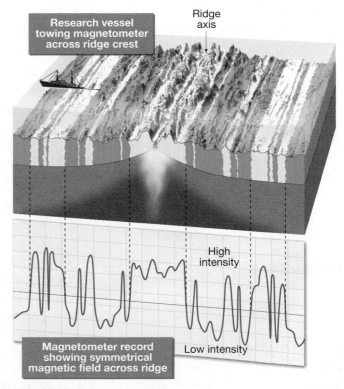

FIGURE 15.24 The ocean floor as a magnetic tape recorder. Magnetic intensities are recorded as a magnetometer is towed across a segment of the oceanic ridge. Notice the symmetrical stripes of low- and high-intensity magnetism that parallel the ridge crest. Vine and Matthews suggested that the stripes of high-intensity magnetism occur where normally magnetized oceanic rocks enhanced the existing magnetic field. Conversely, the low-intensity stripes are regions where the crust is polarized in the reverse direction, which weakens the existing magnetic field.

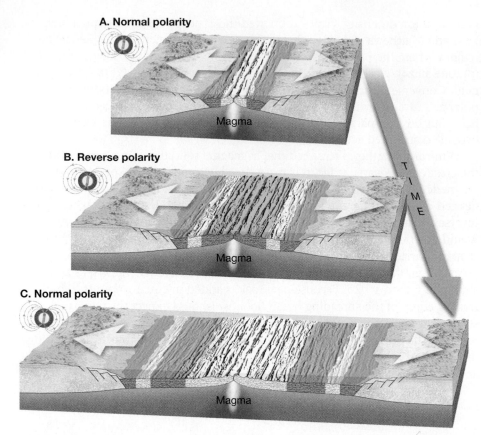

A. Normal polarity

Magma

B. Reverse polarity

Magma

C. Normal polarity

Magma

TIME

FIGURE 15.25 As new basalt is added to the ocean floor at mid-ocean ridges, it is magnetized according to Earth's existing magnetic field. Hence, oceanic crust behaves much like a tape recorder as it records each reversal of our planet's magnetic field.

CONCEPT CHECK 15.8

❶ What is the age of the oldest sediments recovered by deep-ocean drilling? How do the ages of these sediments compare to the ages of the oldest continental rocks?

❷ Assuming hot spots remain fixed, in what direction was the Pacific plate moving while the Hawaiian Islands were forming? When Suiko Seamount was forming?

❸ Describe how Fred Vine and D. H. Matthews related the seafloor-spreading hypothesis to magnetic reversals.

The Breakup of Pangaea

Wegener used evidence from fossils, rock types, and ancient climates to create a jigsaw-puzzle fit of the continents—thereby creating his supercontinent of Pangaea. In a similar manner, but employing modern tools not available to Wegener, geologists have recreated the steps in the break-up of this supercontinent, an event that began

nearly 200 million years ago. From this work, the dates when individual crustal fragments separated from one another and their relative motions have become well established (**FIGURE 15.26**).

An important consequence of the break-up of Pangaea was the creation of a "new" ocean basin: the Atlantic. As you can see in **FIGURE 15.26B**, splitting of the supercontinent did not occur simultaneously along the margins of the Atlantic. The first split developed between North America and Africa.

Here, the continental crust was highly fractured, providing pathways for huge quantities of fluid lavas to reach the surface. Remnants of these lavas are found along the Eastern Seaboard of the United States—primarily buried beneath younger sedimentary rocks that form the continental shelf. Radiometric dating of these solidified lavas indicates that rifting began in various stages between 180 million and 165 million years ago. This time span can be used as the "birth date" for this section of the North Atlantic.

About 130 million years ago, the South Atlantic began to open near the tip of what is now South Africa. As this zone of rifting migrated northward, it gradually opened the South Atlantic (compare Figure 15.26, parts B and C). Continued break-up in the Southern Hemisphere led to the separation of Africa and Antarctica and sent India on a northward journey. By the early Cenozoic, about 50 million years ago, Australia had separated from Antarctica, and the South Atlantic had emerged as a full-fledged ocean (**FIGURE 15.26D**).

A modern map (**FIGURE 15.26F**) shows that India eventually collided with Asia, an event that began about 50 million years ago and created the Himalayas as well as the Tibetan Highlands. About the same time, the separation of Greenland from Eurasia completed the break-up of the northern landmass. During the last 20 million years or so of Earth history, Arabia rifted from Africa to form the Red Sea, and Baja California separated from Mexico to form the Gulf of California (**FIGURE 15.26E**). Meanwhile, a sliver of land (now known as Central America) was trapped between North America and South America to produce our globe's familiar, modern appearance.

CONCEPT CHECK 15.9

❶ When did the supercontinent of Pangaea begin to break apart?

❷ What two continents were the first to separate?

❸ During the break-up of Pangaea, which continent was actually growing in size through the accretion of other landmasses?

How Is Plate Motion Measured?

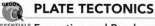

PLATE TECTONICS

ESSENTIALS OF GEOLOGY **Formation and Break-up of Pangaea**

A number of methods have been employed to establish the direction and rate of plate motion. Paleomagnetism stored in the rocks of the ocean floor is one method used to determine the speeds at which plates move away from the ridge axes where they were generated. In addition, hot spot tracks, such as the Hawaiian Island–Emperor Seamount chain, trace the speed and direction of plate movement relative to the hot plume embedded in the mantle below.

Mantle Plumes and Plate Motions

By measuring the length of a hot spot track and the time interval between the formation of its oldest and youngest volcanic structures, an average rate of plate motion can be calculated. For example, that portion of the Hawaiian Island–Emperor Seamount chain that extends from Hawaii to Suiko Seamount is roughly 6000 kilometers in length and formed over the past 65 million years. Thus, the average rate of movement of the Pacific plate, relative to the mantle plume, was about 9 centimeters (4 inches) per year.

Hot spot tracks can also be useful when establishing the direction a plate is moving. Notice in Figure 15.19 that there is a bend in the Hawaiian Island–Emperor Seamount chain. This bend occurred about 50 million years ago when the motion of the Pacific plate changed from one that was nearly due north to a more northwesterly path. Similarly, hot spots found on the floor of the Atlantic have increased our understanding of the migration of landmasses following the break-up of Pangaea.

The existence of mantle plumes and their association with hot spots is well documented. Most mantle plumes are long-lived features that appear to maintain relatively fixed positions within the mantle. However, recent evidence has shown that some hot spots may slowly migrate. Preliminary results suggest that the Hawaiian hotspot may have migrated southward by as much as 20 degrees latitude. If this is the case, models of past plate motion that were based on a "fixed hot spot" frame of reference will need to be reevaluated.

Measuring Plate Motion from Space

Plates are not flat surfaces; instead they are curved sections of a sphere, which greatly complicates how plate motion is described. In addition, plates usually exhibit some degree of rotational motion, which can cause two locations on the same plate to move at different speeds and in different directions. The latter fact can be illustrated by rotating your dinner plate in a clockwise matter. When doing so you will notice that the items on the left side of the plate move away from you (divergence) as the items on the right side move toward you (convergence). Items in the center will rotate, but their position relative to yours will not change. The complex nature of plate motion makes the task of describing plate motions more difficult than simply establishing the relative motion between two plates along the boundary that separates them. Fortunately, using space-age technology, researchers have recently been able to accurately calculate the absolute motion of hundreds of locations across the globe.

You may be familiar with the Global Positioning System (GPS), which is part of the navigation system used in automobiles to locate one's position and to provide directions to some other location. The Global Positioning System employs two dozen satellites that send radio signals that are intercepted by GPS receivers located at Earth's surface. The exact position of the receiver is determined by simultaneously establishing the distance from the receiver to four or more satellites. Researchers use specifically designed equipment that is able to locate the position of a point on Earth to within a few millimeters (about the diameter of a small pea). To establish plate motion, a particular site is surveyed repeatedly over a number of years.

Data obtained from these and other similar techniques are shown in **FIGURE 15.27**. Calculations show that Hawaii is moving in a northwesterly direction and approaching Japan at 8.3 centimeters per year. A site located in Maryland is retreating from one in England at a speed of 1.7 centimeters per year—a value that is close to the 2.0-centimeters-per-year spreading rate that was established from paleomagnetic

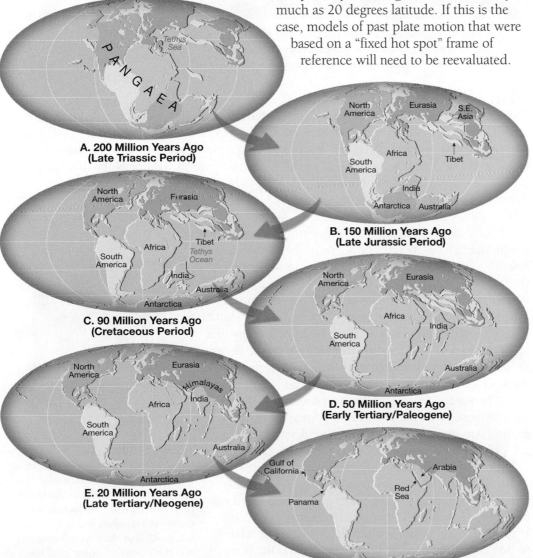

A. 200 Million Years Ago (Late Triassic Period)

B. 150 Million Years Ago (Late Jurassic Period)

C. 90 Million Years Ago (Cretaceous Period)

D. 50 Million Years Ago (Early Tertiary/Paleogene)

E. 20 Million Years Ago (Late Tertiary/Neogene)

F. Present

FIGURE 15.26 Several views of the break-up of Pangaea over a period of 200 million years.

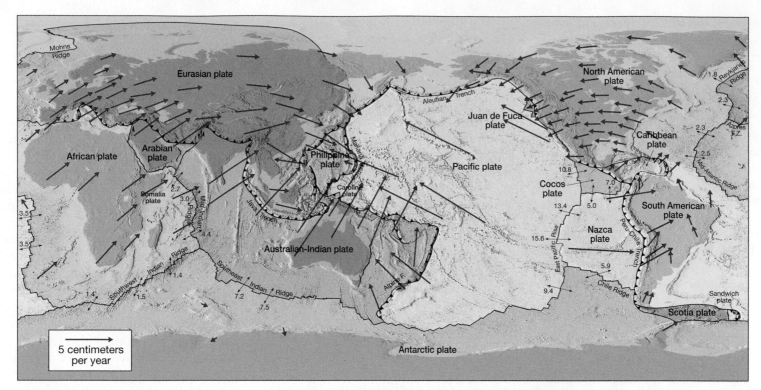

FIGURE 15.27 This map illustrates directions and rates of plate motion in centimeters per year. The red arrows show plate motion at selected locations based on GPS data. The small black arrows and labels show seafloor spreading velocities. *(Seafloor data from DeMets and others; GPS data from Jet Propulsion Laboratory)*

evidence. Techniques using GPS devices have also been useful in establishing small-scale crustal movements such as those that occur along faults in regions known to be tectonically active.

CONCEPT CHECK 15.10

❶ Briefly describe how hot-spot tracks can be used to determine the rate of plate motion.

❷ Refer to Figure 15.27 and determine which three plates appear to exhibit the highest rates of motion.

What Drives Plate Motions?

The plate tectonics theory *describes* plate motion and the role that this motion plays in generating and modifying the major features of Earth's crust. Therefore, acceptance of plate tectonics does not rely on knowing precisely what drives plate motion. This is fortunate, because none of the models yet proposed can account for all major facets of plate tectonics.

Plate–Mantle Convection

From geophysical evidence, we have learned that although the mantle consists almost entirely of solid rock, it is hot and weak enough to exhibit fluidlike convective flow. The simplest type of convection is analogous to heating a pot of water on a stove (**FIGURE 15.28**). Heating the base causes the material to rise in relatively thin sheets or blobs that spread out at the surface and cool. Eventually, the surface layer thickens (increases in density) and sinks back to the bottom where it is reheated until it achieves enough buoyancy to rise again.

Mantle convection is considerably more complex than the model just described. The shape of the mantle does not resemble that of a cooking pot. Rather it is a spherically shaped zone with a much larger upper boundary (Earth's surface) than lower boundary (core–mantle boundary). Furthermore, mantle convection is driven by a combination of three thermal processes: heating at the bottom by heat loss from Earth's core; heating from within by the decay of radioactive isotopes; and cooling from the top that creates thick, cold lithospheric slabs that sink into the mantle.

When seafloor spreading was first introduced, geologists proposed that the main driving force for plate motion was upwelling that came deep in the mantle. Upon reaching the base of the lithosphere, this flow was thought to spread laterally and drag the plates along. Thus, plates were viewed as being carried passively by convective flow in the mantle. Based on physical evidence, however, it became clear that upwelling beneath oceanic ridges is quite shallow and not related to deep circulation in the lower mantle. It is the horizontal movement of lithospheric plates away from the

FIGURE 15.28 Convection is a type of heat transfer that involves the actual movement of a substance. Here the stove warms the water in the bottom of a cooking pot. The heated water expands, becomes less dense (more buoyant), and rises. Simultaneously, the cooler, denser water near the top sinks.

ridge that causes mantle upwelling, not the other way around. Thus, modern models have plates being an integral part of mantle convection and perhaps even its most active component.

Although convection in the mantle is still poorly understood, researchers generally agree on the following:

1. Convective flow in the rocky 2900-kilometer-thick mantle—in which warm, buoyant rock rises and cooler, denser material sinks—is the underlying driving force for plate movement.

2. Mantle convection and plate tectonics are part of the same system. Subducting oceanic plates drive the cold downward-moving portion of convective flow while shallow upwelling of hot rock along the oceanic ridge and buoyant mantle plumes are the upward-flowing arms of the convective mechanism.

3. Convective flow in the mantle is the primary mechanism for transporting heat away from Earth's interior to the surface where it is eventually radiated into space.

What is not known with any high degree of certainty is the exact structure of this convective flow. First, we will look at some of the forces that contribute to plate motion, and then we will examine two models that have been proposed to describe plate–mantle convection.

Forces That Drive Plate Motion

There is general agreement that the subduction of cold, dense slabs of oceanic lithosphere is a major driving force of plate motion (**FIGURE 15.29**). As these slabs sink into the asthenosphere, they "pull" the trailing plate along. This phenomenon, called **slab pull**, occurs because cold slabs of oceanic lithosphere are more dense than the underlying asthenosphere and hence "sink like a rock."

Another important driving force is **ridge push** (Figure 15.29). This gravity-driven mechanism results from the elevated position of the oceanic ridge, which causes slabs of lithosphere to "slide" down the flanks of the ridge. Ridge push appears to contribute far less to plate motions than slab

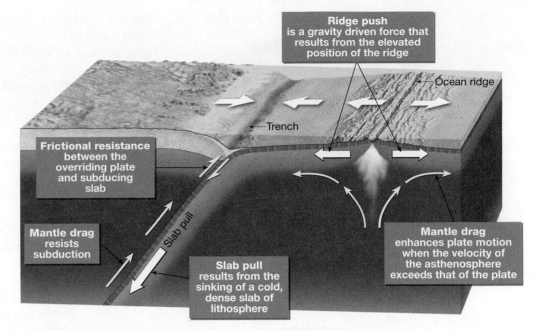

FIGURE 15.29 Illustration of some of the forces that act on tectonic plates.

pull. The primary evidence for this comes from comparing rates of seafloor spreading along ridge segments having different elevations. For example, despite its greater average height above the seafloor, spreading rates along the Mid-Atlantic Ridge are considerably less than spreading rates along the less steep East Pacific Rise (see Figure 15.27). In addition, fast-moving plates are being subducted along a larger percentage of their margins than slow-moving plates. This fact supports the notion that slab pull is a more significant driving force than ridge push. Examples of fast-moving plates that have extensive subduction zones along their margins include the Pacific, Nazca, and Cocos plates.

Although slab pull and ridge push appear to be the dominant forces acting on plates, they are not the only forces that influence plate motion. Beneath plates, convective flow in the mantle exerts a force, perhaps best described as "mantle drag" (Figure 15.29). When flow in the asthenosphere is moving at a velocity that exceeds that of the plate, mantle drag enhances plate motion. However, if the asthenosphere is moving more slowly than the plate, or in the opposite direction, this force tends to resist plate motion. Another type of resistance to plate motion occurs along subduction zones. Here friction between the overriding plate and the descending slab generates significant earthquake activity.

Models of Plate–Mantle Convection

Any acceptable model for plate–mantle convection must explain compositional variations known to exist in the mantle. For example, the basaltic lavas that erupt along oceanic ridges, as well as those that are generated by hot-spot volcanism, such as those found in Hawaii, have mantle sources. Yet ocean ridge basalts are very uniform in composition and depleted in certain elements. Hot-spot eruptions, on the other hand, have high concentrations of these elements and tend to have varied compositions. Because basaltic lavas that arise from different tectonic settings have different compositions, they are assumed to be derived from chemically distinct mantle reservoirs.

LAYERING AT 660 KILOMETERS. Some researchers argue that the mantle resembles a "giant layer cake" divided at a depth of 660 kilometers. As shown in **FIGURE 15.30A**, this layered model has two zones of convection—a thin, dynamic layer in the upper mantle and a thick, sluggish one located below. This model successfully explains why basaltic lavas that erupt along the oceanic ridges have a different chemical make-up than those that erupt in Hawaii as a result of hot-spot activity. The mid-ocean ridge basalts come from the upper convective layer, which is well mixed, whereas the mantle plume that feeds the Hawaiian

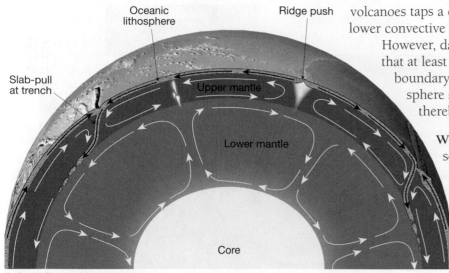

A. Layering at 660 kilometers

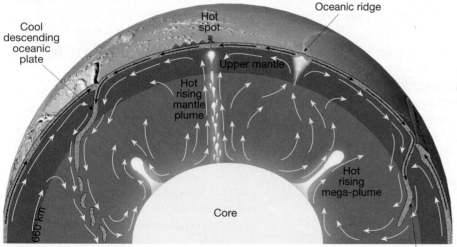

B. Whole mantle convection

FIGURE 15.30 Proposed models for mantle convection. **A.** The "layer cake" model consists of two convection layers—a thin, convective layer above 660 kilometers and a thick one below. **B.** In this whole-mantle convection model, cold oceanic lithosphere descends into the lowermost mantle, while hot mantle plumes transport heat toward the surface.

volcanoes taps a deeper, more primitive magma source that resides in the lower convective layer.

However, data gathered from the study of earthquake waves have shown that at least some subducting oceanic slabs penetrate the 660-kilometer boundary and descend deep into the mantle. The subducting lithosphere should serve to mix the upper and lower layers together, thereby destroying the layered structure proposed in this model.

WHOLE-MANTLE CONVECTION. Other researchers favor some type of *whole-mantle convection* in which cold oceanic lithosphere sinks to great depths and stirs the entire mantle (**FIGURE 15.30B**). One whole-mantle model suggests that the burial ground for subducting slabs is the core-mantle boundary. Over time, this material melts and buoyantly rises toward the surface as a mantle plume, thereby transporting hot material toward the surface (Figure 15.30B).

Recent work has predicted that whole-mantle convection would cause the entire mantle to completely mix in a matter of a few hundred million years. This, in turn, would eliminate chemically distinct magma sources—those that are observed in hot-spot volcanism and those associated with volcanic activity along oceanic ridges. Thus, the whole-mantle model also has shortcomings.

Although there is still much to be learned about the mechanisms that cause Earth's tectonic plates to migrate across the globe, one thing is clear. The unequal distribution of heat in Earth's interior generates some type of thermal convection that ultimately drives plate–mantle motion.

CONCEPT CHECK 15.11

❶ Describe slab pull and ridge push. Which of these forces appears to contribute more to plate motion?

❷ What role are mantle plumes thought to play in the convective flow of the mantle?

❸ Briefly describe the two models proposed for mantle–plate convection. What is lacking in each of these models?

Plate Tectonics in the Future

Geologists have extrapolated present-day plate movements into the future. **FIGURE 15.31** illustrates where Earth's landmasses may be 50 million years from now if present plate movements persist during this time span.

In North America we see that the Baja Peninsula and the portion of southern California that lies west of the San Andreas Fault will have slid past the North American plate. If this northward migration takes place, Los Angeles and San Francisco will pass each other in about 10 million years, and in about 60 million years Los Angeles will begin to descend into the Aleutian Trench.

If Africa continues on a northward path, it will collide with Eurasia, closing the Mediterranean and initiating a major mountain-building episode (Figure 15.31). In other parts of the world, Australia will be astride the equator and, along with New Guinea, will be on a collision course with Asia. Meanwhile, North and South America will begin to separate, while the Atlantic and Indian Oceans continue to grow at the expense of the Pacific Ocean.

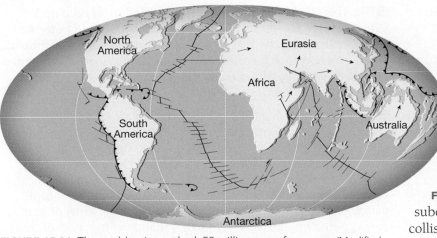

FIGURE 15.31 The world as it may look 50 million years from now. *(Modified after Robert S. Dietz, John C. Holden, C. Scotese, and others)*

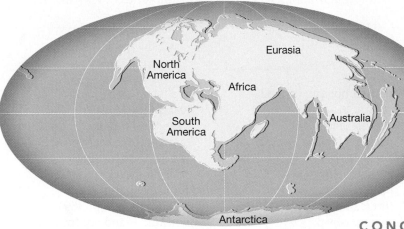

FIGURE 15.32 Reconstruction of Earth as it may appear 250 million years into the future.

A few geologists have even speculated on the nature of the globe 250 million years into the future. As shown in **FIGURE 15.32**, the next supercontinent may form as a result of subduction of the floor of the Atlantic Ocean, resulting in the collision of the Americas with the Eurasian–African landmass. Support for the possible closing of the Atlantic comes from a similar event when the proto-Atlantic closed to form the Appalachian and Caledonian mountains. During the next 250 million years, Australia is also projected to collide with Southeast Asia. If this scenario is accurate, the dispersal of Pangaea will end when the continents reorganize into the next supercontinent.

Such projections, although interesting, must be viewed with considerable skepticism because many assumptions must be correct for these events to unfold as just described. Nevertheless, changes in the shapes and positions of continents that are equally profound will undoubtedly occur for many hundreds of millions of years to come. Only after much more of Earth's internal heat has been lost will the engine that drives plate motions cease.

CONCEPT CHECK 15.12

❶ Briefly describe some major changes to the globe when we extrapolate present-day plate movements 50 million years into the future.

CHAPTER FIFTEEN
Plate Tectonics in Review

⊙ In the early 1900s *Alfred Wegener* set forth his *continental drift* hypothesis. One of its major tenets was that a supercontinent called *Pangaea* began breaking apart about 200 million years ago. The rifted continental fragments then "drifted" to their present positions. To support his hypothesis, Wegener used the *fit of South America and Africa, fossil evidence, rock types and structures,* and *ancient climates.* One of the main objections to the continental drift hypothesis was its inability to provide an acceptable mechanism for the movement of continents.

⊙ By 1968, continental drift was replaced by a far more encompassing theory known as *plate tectonics.* According to plate tectonics, Earth's rigid outer layer (*lithosphere*) overlies a weaker region called the *asthenosphere.* Further, the lithosphere is broken into several large and numerous smaller segments, called *plates,* that are in motion and continually changing in shape and size. Plates move as relatively coherent units and are deformed mainly along their boundaries.

⊙ *Divergent plate boundaries* occur where plates move apart, resulting in upwelling of material from the mantle to create new seafloor. Most divergent boundaries occur along the axis of the oceanic ridge system and are associated with seafloor spreading. New divergent boundaries may form within a continent (for example, the East African Rift Valleys), where they may fragment a landmass and develop a new ocean basin.

⊙ *Convergent plate boundaries* occur where plates move together, resulting in the subduction of oceanic lithosphere into the mantle along a deep-ocean trench. Convergence of an oceanic and continental block results in subduction of the oceanic slab and the formation of a *continental volcanic arc* such as the Andes of South America. Oceanic–oceanic convergence results in an arc-shaped chain of volcanic islands called a *volcanic island arc*. When two plates carrying continental crust converge, the buoyant continental blocks collide, resulting in the formation of a mountain belt as exemplified by the Himalayas.

⊙ *Transform fault boundaries* occur where plates grind past each other without the production or destruction of lithosphere. Most transform faults join two segments of a mid-ocean ridge where they provide the means by which oceanic crust created at a ridge crest can be transported to its site of destruction—a deep-ocean trench. Still others, like the San Andreas Fault, cut through continental crust.

⊙ The theory of plate tectonics is supported by (1) the ages of *sediments* from the floors of the deep-ocean basins; (2) the existence of island groups that formed over *hot spots* and that provide a frame of reference for tracing the direction of plate motion; and (3) *Paleomagnetism*, the direction and intensity of Earth's magnetism in the geologic past.

⊙ Two basic models for mantle convection are currently being evaluated. Mechanisms that contribute to this convective flow are slab pull and ridge push. *Slab pull* occurs where cold, dense oceanic lithosphere is subducted and pulls the trailing lithosphere along. *Ridge push* results when gravity sets the elevated slabs astride oceanic ridges in motion. Hot, buoyant *mantle plumes* are considered the upward flowing arms of mantle convection. One model suggests that mantle convection occurs in two layers separated at a depth of 660 kilometers (410 miles). Another model proposes whole-mantle convection that stirs the entire 2900-kilometer-thick (1800-mile-thick) rocky mantle.

Key Terms

asthenosphere (p. 369)
continental drift (p. 363)
continental rift (p. 371)
continental volcanic
  arcs (p. 374)
convergent boundaries (p. 369)
Curie point (p. 381)
deep-ocean trenches (p. 372)
divergent boundaries (p. 369)
fossil magnetism (p. 382)

fracture zones (p. 376)
hot-spot (p. 380)
hot-spot track (p. 380)
island arc (p. 374)
lithosphere (p. 368)
lithospheric plates (p. 369)
magnetometers (p. 382)
magnetic reversal (p. 382)
magnetic time scale (p. 382)
mantle plume (p. 380)

normal polarity (p. 382)
oceanic ridge
  system (p. 368)
paleomagnetism (p. 382)
Pangaea (p. 363)
partial melting (p. 374)
plate tectonics (p. 368)
reverse polarity (p. 382)
ridge push (p. 387)
rift valley (p. 370)

seafloor spreading (p. 370)
slab pull (p. 387)
spreading centers (p. 369)
subduction zones (p. 372)
supercontinent (p. 363)
tectonic plates (p. 369)
transform fault
  boundaries (p. 369)
volcanic island arc (p. 374)

GIVE IT SOME THOUGHT

❶ After referring to the section in Chapter 1 entitled "The Nature of Scientific Inquiry" answer the following:

  a. What observation led Alfred Wegener to develop his continental drift hypothesis?

  b. What evidence did he gather to support his proposal?

  c. Why was the continental drift hypothesis rejected by the majority of the scientific community?

  d. Do you think Wegener followed the basic principles of scientific inquiry? Support your answer.

❷ Referring to the accompanying diagrams that illustrate the three types of convergent plate boundaries, complete the following:

  a. Identify each type of convergent boundary.

  b. Volcanic island arcs develop on what type of crust?

  c. Why are volcanoes largely absent where two continental blocks collide?

  d. Describe two ways that oceanic–oceanic convergent boundaries are different from oceanic–continental boundaries? How are they similar?

A.

B.

C.

❸ Some predict that California will sink into the ocean. Is this idea consistent with the theory of plate tectonics? Explain.

❹ Refer to the accompanying hypothetical plate map to answer the following questions:

a. Portions of how many plates are shown?

b. Are continents A, B, and C moving toward or away from each other? How did you determine your answer?

c. Explain why active volcanoes are found on both continent A and continent B.

d. Why does continent C lack active volcanoes? Provide at least one scenario in which volcanic activity might be triggered on this continent.

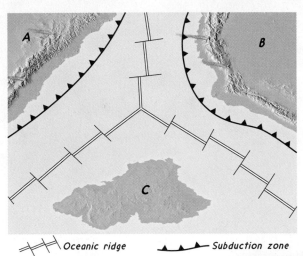

❺ Volcanoes, such as the Hawaiian chain, that form over mantle plumes are some of the largest on Earth. However, several volcanoes on Mars are gigantic compared to those on Earth. What does this difference tell us about how, or if, the process of plate motion operates on Mars? Explain.

❻ Imagine you are studying seafloor spreading along two different oceanic ridges. Along the first ridge the magnetic stripes are uniformly narrow. Along the second ridge they are wide near the ridge crest, but they become narrower as you move away from the crest. What can you say about the history of motion in each example?

❼ Australian marsupials (kangaroos, koala bears, etc.) have direct fossil links to marsupial opossums found in the Americas. Yet the modern marsupials in Australia are markedly different from their American relatives. How does the break-up of Pangaea help to explain these differences (see Figure 15.26)?

❽ Density is a key component in the behavior of Earth materials and is especially important in understanding key aspects of plate tectonics. Describe three different ways that density and /or density differences play a role in plate tectonics.

❾ Refer to the accompanying map to complete the following:

a. List the cities (in pairs) that are moving farther apart as a result of plate motion.

b. List the cities (in pairs) that are moving closer together as a result of plate motion.

c. List the cities (in pairs) that are presently not moving relative to each other.

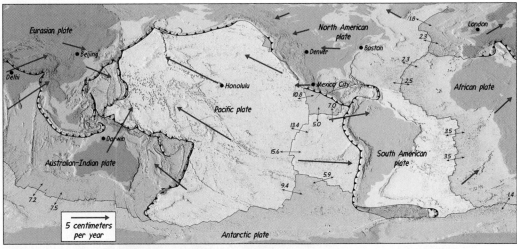

Companion Website   mygeoscience place

www.mygeoscienceplace.com

The *Essentials of Geology, 11e* companion Website contains numerous multimedia resources accompanied by assessments to aid in your study of the topics in this chapter. The use of this site's learning tools will help improve your understanding of geology. Utilizing the access code that accompanies this text, visit **www.mygeoscienceplace.com** in order to:

- **Review** key chapter concepts.
- **Read** with links to the Pearson eText and to chapter-specific web resources.
- **Visualize** and comprehend challenging topics using the learning activities in *GEODe: Essentials of Geology* and the *Geoscience Animations Library.*
- **Test** yourself with online quizzes.

Origin and Evolution of the Ocean Floor

THE OCEAN IS THE LARGEST FEATURE ON EARTH, COVERING MORE than 70 percent of our planet's surface. In this chapter, we will examine the topography of the ocean floor and look at the processes that produced its varied features. You will also learn about the composition, structure, and origin of oceanic crust.

Pacific Ocean, Cabo San Lucas, Mexico.
(Stuart Westmoreland/Corbis)

To assist you in learning the important concepts in this chapter, focus on the following questions:

⊙ How is the ocean floor mapped?

⊙ What are the three major topographic provinces of the ocean floor and the important features associated with each?

⊙ What is the nature and origin of the oceanic ridge system?

⊙ How does oceanic crust form, and how is it different from continental crust?

⊙ What is the relationship between continental rifting and new ocean basins?

⊙ What is the supercontinent cycle?

FOCUS ON CONCEPTS

An Emerging Picture of the Ocean Floor

Divergent Boundaries

Mapping the Ocean Floor

Prior to World War II, information about the ocean floor was extremely limited. Recall that Wegener's continental drift hypothesis was initially rejected by the scientific community, in part because little was known about the ocean floor. Until the 20th century, weighted lines were used to measure water depth, a task that took hours to perform and could be wildly inaccurate, especially in deep water.

With the development of modern instruments our understanding of the diverse topography of the ocean floor improved dramatically. Particularly significant was the discovery of the global oceanic ridge system. This broad elevated landform, which stands 2 to 3 kilometers higher than the adjacent deep-ocean basins, is the longest topographic feature on Earth.

Today we know that oceanic ridges mark divergent plate margins where new oceanic lithosphere is born. Oceanographic studies also discovered deep-ocean trenches, where oceanic lithosphere descends into the mantle. Because the processes of plate tectonics continuously create oceanic crust at mid-ocean ridges and consume it at subduction zones, oceanic crust is perpetually renewed and recycled.

Mapping the Seafloor

If all water could be drained from the ocean basins, a great variety of features would be observed, including volcanic peaks, deep trenches, extensive plains, linear ridges, and large plateaus. In fact, the topography would be nearly as diverse as that on the continents.

The complex nature of ocean-floor topography did not unfold until the historic three-and-a-half-year voyage of the HMS *Challenger* (**FIGURE 16.1**). From December 1872 to May 1876, the *Challenger* expedition made the first comprehensive study of the global ocean ever attempted. During the 127,500-kilometer (79,200-mile) voyage, the ship and its crew of scientists traveled to every ocean except the Arctic. Throughout the voyage, they sampled a multitude of ocean properties, including water depth, which was accomplished by laboriously lowering long weighted lines overboard. The knowledge gained by the *Challenger* of

the ocean's great depth and varied topography expanded with the laying of transatlantic telegraph cables. A far better understanding of the seafloor emerged with the development of modern instruments that measure ocean depths.

MODERN BATHYMETRIC TECHNIQUES. **Bathymetry** (*bathos* = depth, *metry* = measurement) is the measurement of ocean depths and the charting of the shape or topography of the ocean floor. Today, sound energy is used to measure water depths. The basic approach employs **sonar**, an acronym for *so*und *na*vigation and *r*anging. The first devices that used sound to measure water depth, called **echo sounders**, were developed early in the 20th century. Echo sounders work by transmitting a sound wave (called a *ping*) into the water in order to produce an echo when it bounces off any object, such as a large marine organism or the ocean floor (**FIGURE 16.2A**). A sensitive receiver intercepts the reflected echo, and a clock precisely measures the travel time to fractions of a second. By knowing the speed of sound waves in water—about 1500 meters (4900 feet) per second—and the time required for the energy pulse to reach the ocean floor and return, depth can be calculated. Depths determined from continuous monitoring of these echoes are plotted to obtain a profile of the ocean floor. By laboriously combining profiles, a chart of the seafloor was produced.

Following World War II, the U.S. Navy developed *sidescan sonar* to look for explosive devices that had been deployed in shipping lanes (**FIGURE 16.2B**). These torpedo-shaped instruments are towed behind ships and send out a fan of sound extending to either side of the ship's path. By combining swaths of sidescan sonar data, oceanographers produced the first photographlike images of the seafloor. Although sidescan sonar provides valuable

FIGURE 16.1 The first systematic bathymetric measurements of the ocean were made aboard the HMS *Challenger* during its historic three-and-a-half-year voyage which departed England in December of 1872 and returned in May 1876. *(From C. W. Thompson and Sir John Murray, Report on the Scientific Results of the Voyage of the HMS Challenger, Vol. 1, Great Britain: Challenger Office, 1895, Plate 1. Library of Congress)*

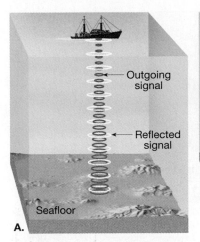
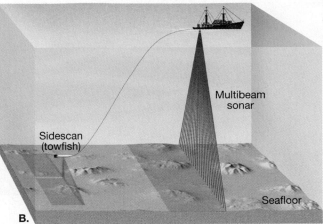

FIGURE 16.2 Various types of sonar. **A.** An echo sounder determines the water depth by measuring the time interval required for an acoustic wave to travel from a ship to the seafloor and back. The speed of sound in water is 1500 m/sec. Therefore, depth $= \frac{1}{2}$ (1500 m/sec $\times$ echo travel time) **B.** Modern multibeam sonar and sidescan sonar obtain an "image" of a narrow swath of seafloor every few seconds.

views of the seafloor, it does not provide bathymetric (water depth) data.

This drawback was resolved in the 1990s with the development of *high-resolution multibeam* instruments. These systems use hull-mounted sound sources that send out sound in several directions, then record reflections from the seafloor through a set of narrowly focused receivers aimed at different angles. Rather than obtaining the depth of a single point every few seconds, this technique allows a survey ship to map a swath of ocean floor tens of kilometers wide. In addition, these systems collect bathymetric data of such high resolution that they can distinguish depths that differ by less than a meter. When multibeam sonar is used to map sections of seafloor, the ship travels in a regularly spaced back-and-forth pattern known as "mowing the lawn."

Despite their greater efficiency and enhanced detail, research vessels equipped with multibeam sonar travel at a mere 10 to 20 kilometers (6 to 12 miles) per hour. It would take at least 100 vessels outfitted with this equipment hundreds of years to map the entire seafloor. This explains why only about 5 percent of the seafloor has been mapped in detail—and why large areas of the seafloor have not yet been mapped with sonar at all.

SEISMIC REFLECTION PROFILES. Marine geologists are also interested in viewing the rock structure beneath the

sediments that blanket much of the seafloor. This is accomplished by making a **seismic reflection profile.** To construct such a profile strong, low-frequency sounds are produced by explosions (depth charges) or air guns. The sound waves penetrate the seafloor and reflect off the boundaries between rock layers and fault surfaces. **FIGURE 16.3** shows a seismic profile of a portion of the Madeira abyssal plain in the eastern Atlantic. Although the seafloor is flat, the image allows us to see the irregular ocean crust buried by a thick accumulation of sediments.

Viewing the Ocean Floor from Space

Another technological breakthrough that led to an enhanced understanding of the seafloor involves measuring the shape of the ocean surface from space. After compensating for waves, tides, currents, and atmospheric effects, it was discovered that the water's surface is not perfectly "flat." Because massive structures such as seamounts and ridges exert stronger-than-average gravitational attraction, they produce elevated areas on the ocean surface. Conversely, canyons and trenches create slight depressions.

Satellites equipped with *radar altimeters* are able to measure subtle differences in sea level by bouncing microwaves off the sea surface. These devices can measure variations as small as a few centimeters. Such data have added greatly to our knowledge of ocean-floor topography. Combined with traditional sonar depth measurements, the data are used to produce detailed ocean-floor maps, such as the one in Figure 1.27 (pp. 26–27).

DID YOU KNOW?
The U.S. Navy utilizes the biological sonar of bottlenose dolphins to help defend its ships and facilities. Specifically, dolphins have been trained to detect mines intended to blow up ships. When Navy ships enter mine-infested waters, the trained dolphins are released and use their sonar to locate the deadly mines.

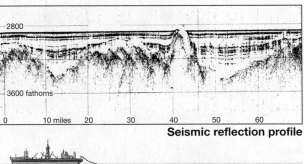

Seismic reflection profile

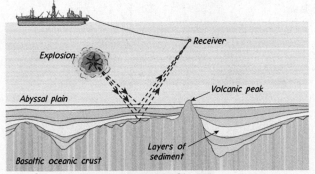

Geologist's Sketch

FIGURE 16.3 Seismic cross-section and matching sketch across a portion of the Madeira abyssal plain in the eastern Atlantic Ocean, showing the irregular oceanic crust buried by sediments. *(Image courtesy of Charles Hollister, Woods Hole Oceanographic Institution)*

Provinces of the Ocean Floor

Oceanographers studying the topography of the ocean floor identify three major areas: *continental margins, deep-ocean basins,* and *oceanic (mid-ocean) ridges.* The map in **FIGURE 16.4** outlines these provinces for the North Atlantic Ocean, and the profile at the bottom shows the varied topography. Profiles of this type usually have their vertical dimension exaggerated many times (40 times in this case) to make topographic features more conspicuous. Vertical exaggeration, however, makes slopes appear *much* steeper than they actually are.

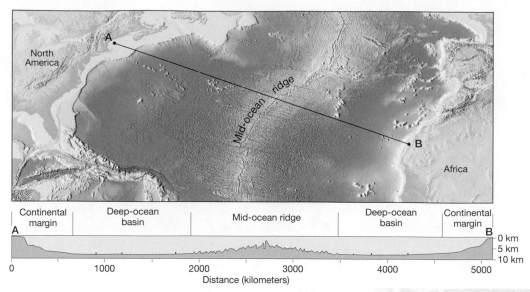

FIGURE 16.4 Major topographic divisions of the North Atlantic and a profile from New England to the coast of North Africa.

DID YOU KNOW?

The salinity of water in the Dead Sea is nearly 10 times greater than the salinity of normal seawater. As a result, the water is so dense and buoyant that people and objects easily float.

The University of Hawaii's research vessel, *Kilo Moana*, arriving at Kodiak, Alaska. *(Marshalena Delaney/AP Photo/U.S. Coast Guard)*

CONCEPT CHECK 16.1

❶ Define *bathymetry*.

❷ Assuming that the average speed of sound waves in water is 1500 meters per second, determine the water depth if the signal sent out by an echo sounder requires 6 seconds to strike bottom and return to the recorder.

❸ Describe how satellites orbiting Earth can determine features on the seafloor without being able to directly observe them beneath several kilometers of seawater.

❹ What are the three major topographic provinces of the ocean floor?

Continental Margins

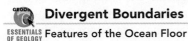

Divergent Boundaries

ESSENTIALS OF GEOLOGY Features of the Ocean Floor

Two types of **continental margins** have been identified—*passive* and *active*. Passive margins are found along most of the coastal areas that surround the Atlantic and Indian oceans, including the east coasts of North and South America, as well as the coastal areas of Europe and Africa. Passive margins consist of continental crust capped with weathered materials eroded from adjacent landmasses.

By contrast, active continental margins occur where oceanic lithosphere subducts into the mantle beneath the edge of a continent. As a result, active margins are narrow and consist of highly deformed sediments that were scraped from the descending lithospheric slab and plastered against the margin of the overriding continent. Active continental margins are common around the Pacific Rim, where they parallel deep-ocean trenches.

Passive Continental Margins

The features comprising **passive continental margins** include the continental shelf, the continental slope, and the continental rise (**FIGURE 16.5**).

CONTINENTAL SHELF. The **continental shelf** is a gently sloping, submerged surface extending from the shoreline toward the deep-ocean basin. Because it is underlain by continental crust, it is clearly a flooded extension of the continents.

The continental shelf varies greatly in width. Although almost nonexistent along some continents, the shelf extends seaward more than 1500 kilometers (930 miles) along others. On average, the continental shelf is about 80 kilometers (50 miles) wide, and its seaward edge is about 130 meters (425 feet) deep. As a result, the

average inclination of the continental shelf is only about one tenth of 1 degree, a slope so slight that it would appear to an observer to be a horizontal surface.

The continental shelf tends to be relatively featureless; however, some areas are mantled by extensive glacial deposits and thus are quite rugged. In addition, some continental shelves are dissected by large valleys running from the coastline into deeper waters. Many of these *shelf valleys* are the seaward extensions of river valleys on the adjacent landmass. They were eroded during the last Ice Age (Pleistocene epoch) when enormous quantities of water were stored in vast ice sheets on the continents causing sea levels to drop at least 100 meters (330 feet). Because of this drop, rivers extended their courses, and land-dwelling plants and animals migrated to the newly exposed portions of the continents. Dredging off the coast of North America has retrieved the ancient remains of numerous land dwellers, including mammoths, mastodons, and horses; further evidence that portions of the continental shelves were once above sea level.

Although continental shelves represent only 7.5 percent of the total ocean area, they have economic and political significance because they contain important mineral deposits and support important fishing grounds (**FIGURE 16.6**). According to the United Nations Convention on the Law of the Sea, all countries that ratified the treaty had until 2009 to claim any extension of their continental shelf beyond the normal 200 nautical miles. The extension could be no more than 350 nautical miles from land and required scientific evidence to show

FIGURE 16.5 Schematic view showing the major features of a passive continental margin. Note that the slopes shown for the continental shelf and continental slope are greatly exaggerated. The continental shelf has an average slope of one tenth of 1 degree, while the continental slope has an average slope of about 5 degrees.

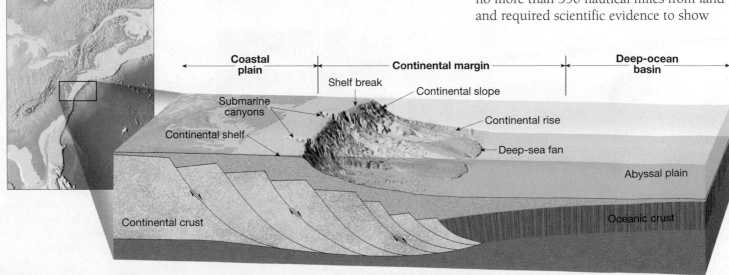

FIGURE 16.6 Offshore drilling rigs are used to tap the oil and natural gas reserves of the continental shelf. This platform is in the North Sea. *(Photo by Peter Bowater/Photo Researchers, Inc.)*

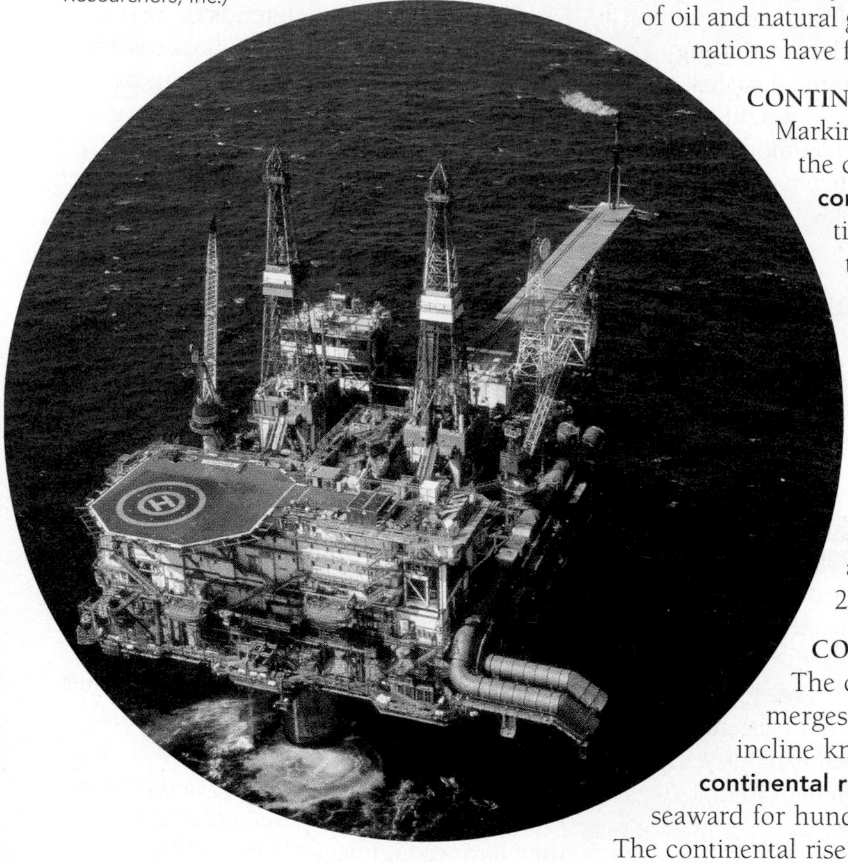

that the seafloor in question was, indeed, continental shelf. In 2001, Russia was the first country to submit a continental-shelf claim, based mainly on known reservoirs of oil and natural gas. Many other nations have followed suit.

CONTINENTAL SLOPE. Marking the seaward edge of the continental shelf is the **continental slope**, a relatively steep structure that marks the boundary between continental crust and oceanic crust. Although the inclination of the continental slope varies greatly from place to place, it averages about 5 degrees and in places exceeds 25 degrees.

CONTINENTAL RISE. The continental slope merges into a more gradual incline known as the **continental rise** that may extend seaward for hundreds of kilometers. The continental rise consists of a thick accumulation of sediment that has moved down the continental slope and onto deep-ocean floor. Most of the sediments are delivered to the seafloor by *turbidity currents* that periodically flow down *submarine canyons* (see Figure 6.21, p. 167). When these muddy slurries emerge from the mouth of a canyon onto the relatively flat ocean floor, they deposit sediment that forms a **deep-sea fan** (see Figure 16.5). As fans from adjacent submarine canyons grow, they merge laterally to produce a continuous wedge of sediment at the base of the continental slope forming the continental rise.

Active Continental Margins

Along active continental margins the continental shelf is very narrow, if it exists at all, and the continental slope descends abruptly into a deep-ocean trench. In these settings, the landward wall of a trench and the continental slope are essentially the same feature.

Active continental margins are located primarily around the Pacific Ocean in areas where oceanic lithosphere is being subducted beneath the leading edge of a continent (**FIGURE 16.7**). In such settings, sediments from the ocean floor and pieces of oceanic crust are scraped from the descending oceanic plate and plastered against the edge of the overriding continent. This chaotic accumulation of deformed sediment and scraps of oceanic crust is called an **accretionary wedge** (*ad* = toward, *crescere* = to grow). Prolonged plate subduction can produce massive accumulations of sediment along active continental margins.

Some active margins have little or no sediment accumulation, indicating that material is being carried into the mantle with the subducting plate. This tends to occur where old oceanic lithosphere is subducting nearly vertically into the mantle. In these locations the continental margin is very narrow, as the trench may lie a mere 50 kilometers (31 miles) offshore.

DID YOU KNOW?

Faced with a shortage of land along its heavily populated coast, Japan has constructed artificial islands on the adjacent continental shelf. One of these islands is located about 3 miles off the coast of Osaka, where the Kansai Airport was built.

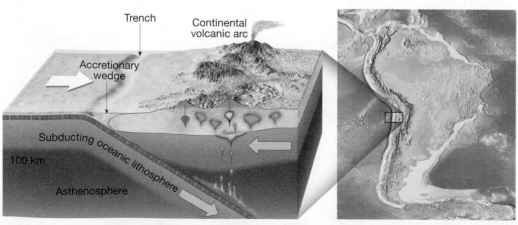

FIGURE 16.7 Active continental margin. Sediments from the ocean floor are scraped from the descending plate and added to the continental crust as an accretionary wedge.

CONCEPT CHECK 16.2

1. List the three major features that comprise a passive continental margin. Which of these features is considered a flooded extension of the continent? Which one has the steepest slope?
2. Describe the differences between active and passive continental margins. Include how various features relate to plate tectonics, and give a geographic example of each type of margin.

Features of Deep-Ocean Basins

Divergent Boundaries

ESSENTIALS OF GEOLOGY **Features of the Ocean Floor**

Between the continental margin and the oceanic ridge lies the **deep-ocean basin** (see Figure 16.4). The size of this region—almost 30 percent of Earth's surface—is roughly comparable to the percentage of land above sea level. This region includes *deep-ocean trenches,* which are extremely deep linear depressions in the ocean floor; remarkably flat areas known as *abyssal plains;* tall volcanic peaks called *seamounts* and *guyots;* and large flood basalt provinces called *oceanic plateaus.*

Deep-Ocean Trenches

Deep-ocean trenches are long, relatively narrow creases in the seafloor that represent the deepest parts of the ocean floor (**TABLE 16.1**). Most trenches are located along the margins of the Pacific Ocean (**FIGURE 16.8**), where many exceed 10 kilometers (6 miles) in depth. A portion of one trench—the Challenger Deep in the Mariana Trench—has been measured at 11,022 meters (36,163 feet) below sea level, making it the deepest

known part of the world ocean. Two trenches are located in the Atlantic—the Puerto Rico Trench adjacent to the Lesser Antilles arc and the South Sandwich Trench.

Trenches are sites of plate convergence where slabs of oceanic lithosphere subduct and plunge back into the mantle. In addition

TABLE 16.1

Dimensions of Some Deep-Ocean Trenches

Trench	Depth (kilometers)	Average Width (kilometers)	Length (kilometers)
Aleutian	7.7	50	3700
Central America	6.7	40	2800
Japan	8.4	100	800
Java	7.5	80	4500
Kurile–Kamchatka	10.5	120	2200
Mariana	11.0	70	2550
Peru–Chile	8.1	100	5900
Philippine	10.5	60	1400
Puerto Rico	8.4	120	1550
South Sandwich	8.4	90	1450
Tonga	10.8	55	1400

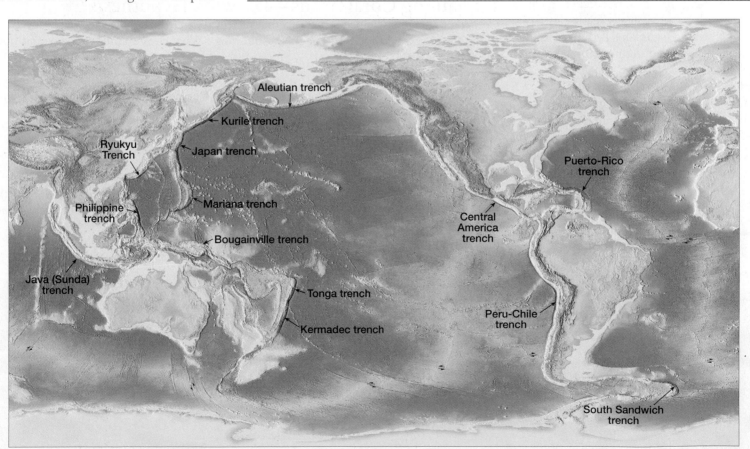

FIGURE 16.8 Distribution of the world's deep-ocean trenches.

to earthquakes being created as one plate "scrapes" against another, volcanic activity is also associated with these regions. Thus, trenches are often paralleled by an arc-shaped row of active volcanoes called a *volcanic island arc*. Furthermore, *continental volcanic arcs*, such as those making up portions of the Andes and Cascades, are located parallel to trenches that lie adjacent to continental margins. The volcanic activity associated with the trenches that surround the Pacific Ocean explains why this region is called the *Ring of Fire*.

Abyssal Plains

Abyssal plains (*a* = without, *byssus* = bottom) are deep, flat features—in fact, they are likely the most level places on Earth. The abyssal plain found off the coast of Argentina, for example, has less than 3 meters (10 feet) of relief over a distance exceeding 1300 kilometers (800 miles). The monotonous topography of abyssal plains is occasionally interrupted by the protruding summit of a partially buried volcanic peak.

Using *seismic profilers,* researchers have determined that the relatively featureless topography of abyssal plains is due to thick accumulations of sediment that have buried an otherwise rugged ocean floor (see Figure 16.4). The nature of the sediment indicates that these plains consist primarily of fine sediments transported far out to sea by turbidity currents, deposits that have precipitated out of seawater, and shells and skeletons of microscopic marine organisms.

Abyssal plains are found in all oceans. However, the Atlantic Ocean has the most extensive abyssal plains because it has few trenches to act as traps for sediment carried down the continental slope.

Seamounts, Guyots, and Oceanic Plateaus

Dotting the ocean floor are submarine volcanoes called **seamounts,** which may rise hundreds of meters above the surrounding topography. It is estimated that more than a million exist. Some grow large enough to become oceanic islands, but most do not have a sufficiently long eruptive history to build a structure above sea level. Although seamounts are found on the floors of all the oceans, they are most common in the Pacific.

Some, like the Hawaiian Island–Emperor Seamount chain, which stretches from the Hawaiian Islands to the Aleutian trench, form over volcanic hot spots in association with mantle plumes (see Figure 15.19 p. 380). Others are born near oceanic ridges. If the volcano is large enough before it is carried from the magma source by plate movement, the structure may emerge as an island. Examples in the Atlantic include Azores, Ascension, Tristan da Cunha, and St. Helena.

During the time they exist as islands, some of these volcanic structures are lowered to near sea level by the forces of weathering and erosion. In addition, islands gradually sink and disappear below the water surface as the moving plate slowly carries them away from the elevated oceanic ridge or hot spot where they originated. Submerged, flat-topped seamounts that formed in this manner are called **guyots** or **tablemounts.***

The ocean floor also contains several massive **oceanic plateaus,** which resemble flood basalt provinces on the continents. Oceanic plateaus, which in some cases are more than 30 kilometers thick, were generated from vast outpourings of fluid basaltic lavas. Some oceanic plateaus appear to have formed quickly in geologic terms. Examples include the Ontong Java, which formed in less than 3 million years, and the Kerguelen Plateau, which formed in 4.5 million years.

Explaining Coral Atolls—Darwin's Hypothesis

Coral *atolls* are ring-shaped structures that often extend from slightly above sea level to depths of several thousand meters (**FIGURE 16.9**). What causes atolls to form, and how do they attain such thicknesses?

*The term *guyot* is named after Princeton University's first geology professor. It is pronounced "GEE-oh" with a hard *g* as in "give."

FIGURE 16.9 An aerial view of Tetiaroa Atoll in the Pacific. The light blue waters of the relatively shallow lagoon contrast with the dark blue color of the deep ocean surrounding the atoll. *(Photo by Douglas Peebles Photography/Alamy)*

DID YOU KNOW?
Today, coral reefs cover about 230,000 square miles of the world's marine areas. Many different stresses can destroy coral reefs, including human activity and rising sea levels due to global warming.

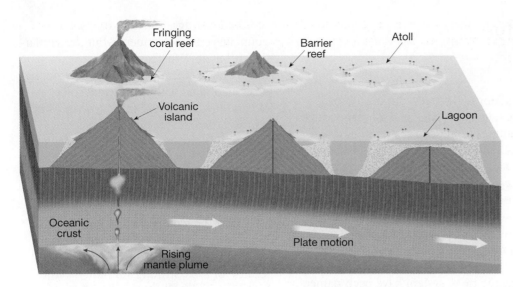

FIGURE 16.10 Formation of a coral atoll due to the gradual sinking of oceanic crust and upward growth of the coral reef. A fringing coral reef forms around an active volcanic island generated by a mantle plume. As the volcanic island moves away from the region of hot-spot activity it sinks, and the fringing reef gradually becomes a barrier reef. Eventually, the volcano is completely submerged, and a coral atoll remains.

Corals are tiny animals that generally appear in large numbers and when linked form colonies. Most corals create a hard external skeleton made of calcium carbonate. Some build large calcium carbonate structures, called *reefs*, where new colonies grow atop the strong skeletons of previous colonies. Sponges and algae may attach to the reef, enlarging it further.

Reef-building corals grow best in waters with an average annual temperature of about 24 °C (75 °F). They cannot survive prolonged exposure to temperatures below 18 °C (64 °F) or above 30 °C (86 °F). In addition, reef-builders require clear, sunlit water. Consequently, the depth of most active reef growth is limited to no more than about 45 meters (150 feet).

The strict environmental conditions required for coral growth create an interesting paradox: How can corals—which require warm, shallow, sunlit water no deeper than a few dozen meters—create thick structures such as coral atolls that extend to great depths?

The naturalist Charles Darwin was one of the first to formulate a hypothesis on the origin of ringed-shaped atolls. From 1831 to 1836 he sailed aboard the British ship HMS *Beagle* during its famous global circumnavigation. In various places that Darwin visited, he noticed a progression of stages in coral reef development from (1) a *fringing reef* along the margins of a volcano to (2) a *barrier reef* with a volcano in the middle to (3) an *atoll*, consisting of a continuous or broken ring of coral reef surrounding a central lagoon. The essence of Darwin's hypothesis, illustrated in **FIGURE 16.10**, was that as a volcanic island slowly sinks, corals continue to build the reef complex upward. During Darwin's time, however, there was no plausible mechanism to account for how an island might sink.

Currently, plate tectonics helps explain how volcanic islands become extinct and sink to great depths over long periods of time. Some volcanic islands form over a relatively stationary mantle plume, which causes the lithosphere to be buoyantly uplifted. Over a span of millions of years, these volcanic islands become inactive and gradually sink as the moving plate carries them away from the region of hot-spot volcanism (Figure 16.10).

CONCEPT CHECK 16.3

❶ Explain how deep-ocean trenches are related to plate boundaries.

❷ Why are abyssal plains more extensive on the floor of the Atlantic than on the floor of the Pacific?

❸ How does a flat-topped *seamount*, or *guyot*, form?

The deep-diving submersible *Alvin* is 7.6 meters long, weights 16 tons, has a cruising speed of 1 knot, and can reach depths as great as 4000 meters (13,000 feet). A pilot and two scientific observers are along during a normal 6- to 10-hour dive. *(Courtesy of Rod Catanach/ Woods Hole Oceanographic Institution)*

submersible *Alvin*

Anatomy of the Oceanic Ridge

Divergent Boundaries

Oceanic Ridges and Seafloor Spreading

Along well-developed divergent plate boundaries, the seafloor is elevated, forming a broad linear swell called the **oceanic ridge**, or **mid-ocean ridge.** Our knowledge of the oceanic ridge system comes from soundings of the ocean floor, core samples from deep-sea drilling, visual inspection using deep-diving submersibles, and even firsthand inspection of slices of ocean floor that have been thrust onto dry land during continental collisions. At oceanic ridges we find extensive normal and strike-slip faulting, earthquakes, high heat flow, and volcanism.

The oceanic ridge system winds through all major oceans in a manner similar to the seam on a baseball and is the longest topographic feature on Earth, exceeding 70,000 kilometers (43,000 miles) in length (**FIGURE 16.11**). The crest of the ridge typically stands 2 to 3 kilometers above the adjacent deep-ocean basins and marks the plate boundary where new oceanic crust is created.

Notice in Figure 16.11 that large sections of the oceanic ridge system have been named based on their locations within the various ocean basins. Some ridges run through the middle of ocean basins, where they are appropriately called *mid-ocean* ridges. The Mid-Atlantic Ridge and the Mid-Indian Ridge are examples. By contrast, the East Pacific Rise is *not* a "mid-ocean" feature. Rather, as its name implies, it is located in the eastern Pacific, far from the center of the ocean.

The term *ridge* is somewhat misleading, because these features are not narrow and steep as the term implies, but rather have widths of between 1000 and 4000 kilometers and the appearance of broad, elongated swells that exhibit varying degrees of ruggedness. Furthermore, the ridge system is broken into segments that range from a few tens to hundreds of kilometers in length. Each segment is offset from the adjacent segment by a transform fault.

Oceanic ridges are as high as some mountains on the continents, but the similarities end there. Whereas most mountain ranges on land form when the compressional forces associated with continental collisions fold and metamorphose thick sequences of sedimentary rocks, oceanic ridges form where upwelling from the mantle generates new oceanic crust. Oceanic ridges consist of layers and piles of newly formed basaltic rocks that are buoyantly uplifted by the hot mantle rocks from which they formed.

Along the axis of some segments of the oceanic ridge system are deep, down-faulted structures called **rift valleys** because of their striking similarity to the continental rift valleys found in East Africa (**FIGURE 16.12**). Some rift valleys, including those along the rugged Mid-Atlantic Ridge, are typically 30 to 50 kilometers wide and have walls that tower 500 to 2500 meters above the valley floor. This makes them comparable to the deepest and widest part of Arizona's Grand Canyon.

CONCEPT CHECK 16.4

❶ Briefly describe the oceanic ridge system.

❷ Although oceanic ridges can be as tall as some mountains found on the continents, how are these features different?

DID YOU KNOW?
In January 1960, U.S. Navy Lt. Don Walsh and explorer Jacques Piccard descended to the bottom of the Challenger Deep region of the Mariana Trench. More than five hours after leaving the surface, they reached the bottom at 10,912 meters (35,800 feet)—a record depth of human descent that has never been surpassed.

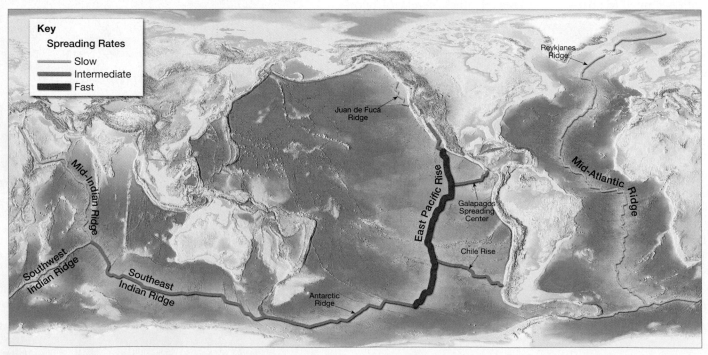

FIGURE 16.11 Distribution of the oceanic ridge system. The map shows ridge segments that exhibit slow, intermediate, and fast spreading rates.

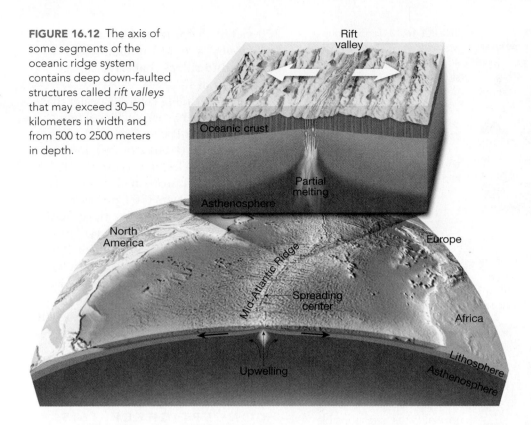

FIGURE 16.12 The axis of some segments of the oceanic ridge system contains deep down-faulted structures called *rift valleys* that may exceed 30–50 kilometers in width and from 500 to 2500 meters in depth.

Oceanic Ridges and Seafloor Spreading

 Divergent Boundaries

ESSENTIALS OF GEOLOGY **Oceanic Ridges and Seafloor Spreading**

The greatest volume of magma (more than 60 percent of Earth's total yearly output) is produced along the oceanic ridge system in association with seafloor spreading. As plates diverge, fractures created in the oceanic crust fill with molten rock that gradually wells up from the hot mantle below. This molten material slowly cools and crystallizes, producing new slivers of seafloor. This process repeats itself in episodic bursts, generating new lithosphere that moves away from the ridge crest in a conveyor belt fashion.

Seafloor Spreading

Harry Hess of Princeton University formulated the concept of seafloor spreading in the early 1960s. Later, geologists were able to verify Hess's view that seafloor spreading occurs along the crests of oceanic ridges where hot mantle rock rises to replace the material that has shifted horizontally. Recall from Chapter 3 that as rock rises it experiences a decrease in confining pressure that may lead to *decompression melting*.

Partial melting of mantle rock produces basaltic magma that has a surprisingly consistent chemical composition. The newly formed melt separates from the mantle rock and rises toward the surface. Along some ridge segments, the melt collects in small elongated reservoirs located just beneath the ridge crest. Eventually, about 10 percent migrates upward along fissures to erupt as lava flows on the ocean floor (Figure 16.12). This activity continuously adds new basaltic rock to diverging plate margins, temporarily welding them together, only to be broken as spreading continues. Along some ridges, outpourings of bulbous lavas build submerged shield volcanoes (seamounts) as well as elongated lava ridges. At other locations, more voluminous lava flows create a relatively subdued topography.

Why Are Oceanic Ridges Elevated?

The primary reason for the elevated position of the ridge system is that newly created oceanic lithosphere is hot and therefore less dense than cooler rocks of the deep-ocean basin. As the newly formed basaltic crust travels away from the ridge crest, it is cooled from above as seawater circulates through the pore spaces and fractures in the rock. In addition, it cools because it gets farther and farther from the zone of hot mantle upwelling. As a result, the lithosphere gradually cools, contracts, and becomes more dense. This thermal contraction accounts for the greater ocean depths that occur away from the ridge. It takes almost 80 million years of cooling and contraction for rock that was once part of an elevated ocean-ridge system to relocate to the deep-ocean basin.

Gas hydrates are natural gas reservoirs in icelike crystalline solids found in submarine sediments. *(Photo courtesy of GEOMAR Research Center)*

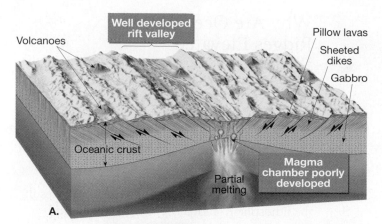

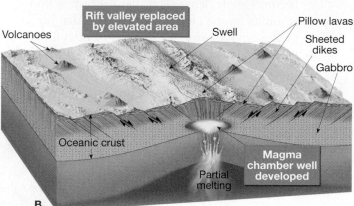

FIGURE 16.13 Topography of the crest of an oceanic ridge. **A.** Where spreading rates are slow, a prominent rift valley develops along the ridge crest, and the topography is typically rugged. **B.** Where spreading rates are rapid, no median rift valleys develop, and the topography is comparatively smooth.

Spreading Rates and Ridge Topography

When researchers studied various segments of the oceanic ridge system, it was clear that there were topographic differences. Many appear to be controlled by spreading rates—which largely determine the amount of melt generated at a rift zone. At fast spreading centers more magma wells up from the mantle than at slow spreading centers. This difference in output causes differences in the structure and topography of various ridge segments.

Oceanic ridges that exhibit slow spreading rates from 1 to 5 centimeters per year have prominent rift valleys and rugged topography (**FIGURE 16.13A**). The Mid-Atlantic and Mid-Indian ridges are examples. The vertical displacement of large slabs of oceanic crust along normal faults is responsible for the steep walls of the rift valleys. Furthermore, volcanism produces numerous cones in the rift valley that enhance the rugged topography of the ridge crest.

By contrast, along the Galápagos ridge, an intermediate spreading rate of 5 to 9 centimeters per year is the norm. As a result, the rift valleys that develop are relatively shallow—often less than 200 meters deep. In addition, their topography is more subdued when compared to ridges that have slower spreading rates.

At fast spreading centers (greater than 9 centimeters per year), such as along much of the East Pacific Rise, rift valleys are generally absent (**FIGURE 16.13B**). Instead, the ridge axis is elevated. These elevated structures, called *swells*, are built from lava flows up to 10 meters (30 feet) thick that have incrementally paved the ridge crest with volcanic rocks (**FIGURE 16.14**). In addition, because the depth of the ocean depends largely on the age of the seafloor, ridge segments that exhibit faster spreading rates tend to have more gradual profiles than ridges that have slower spreading rates (**FIGURE 16.15**). Because of these differences in topography, the gently sloping, less rugged portions of fast spreading ridges are called *rises*.

CONCEPT CHECK 16.5

❶ What is the source of magma for seafloor spreading?

❷ What is the primary reason for the elevated position of the oceanic ridge system?

❸ Compare and contrast a slow spreading center such as the Mid-Atlantic Ridge with one that exhibits a faster spreading rate, such as the East Pacific Rise.

DID YOU KNOW?
If you were thinking about buying an island hideaway, the Indian Ocean might be the place to look. Numerous small coral islands, including the Maldives Islands, are scattered about the Indian Ocean. The Maldives include more than 1000 islands, most of which are uninhabited.

FIGURE 16.14 False-color sonar image of a segment of the East Pacific Rise. The linear pink area is the swell that formed above the ridge axis. Note also the large volcanic cone in the lower left portion of the image. *(Courtesy of Dr. Ken C. Macdonald)*

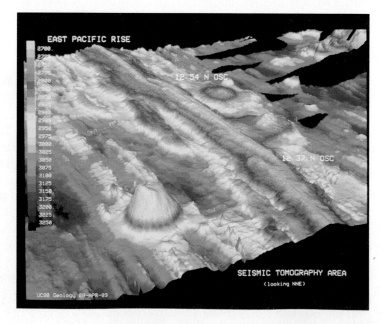

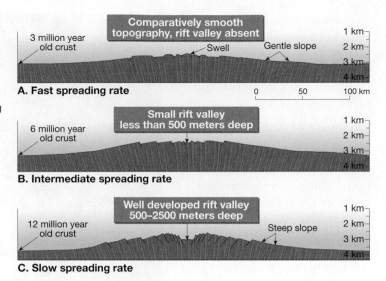

FIGURE 16.15 Schematic of ridge segments that exhibit fast, intermediate, and slow spreading rates. Fast spreading centers have gentle slopes and lack a rift valley. By contrast, ridges that have slow spreading rates have well-developed rift valleys and steep flanks. The slopes of all these profiles are greatly exaggerated.

feed into a few dozen larger, elongated channels, perhaps 100 meters (300 feet) or more wide. These structures, in turn, feed lens-shaped magma chambers located directly beneath the ridge crest. With the addition of melt from below, the pressure inside the chambers steadily increases. As a result, the rocks above these reservoirs periodically fracture, allowing the melt to ascend into the young oceanic crust above.

The molten rock surges upward along numerous vertical fractures that develop in the ocean crust. Some cools and solidifies to form dikes. New dikes intrude older dikes, which are still warm and weak, to form a **sheeted dike complex.** This portion of the oceanic crust is usually 1 to 2 kilometers thick.

Roughly 10 percent of the melt eventually erupts on the ocean floor. Because the surface of a submarine lava flow is chilled quickly by seawater, it generally travels no more than a few kilometers before completely solidifying. The forward motion occurs as lava accumulates behind the

The Nature of Oceanic Crust

An interesting aspect of oceanic crust is that its thickness and structure are remarkably consistent throughout the entire ocean basin. Seismic soundings indicate that its thickness averages only about 7 kilometers (5 miles). Furthermore, it is composed almost entirely of mafic (basaltic) rocks that are underlain by a layer of the ultramafic rock peridotite, which forms the lithospheric mantle.

Although most oceanic crust forms out of view, far below sea level, geologists have been able to examine the structure of the ocean floor firsthand. In such locations as Newfoundland, Cyprus, Oman, and California, slivers of oceanic crust have been thrust high above sea level. From these exposures, and core samples collected by deep-sea drilling ships, researchers conclude that the ocean crust consists of four distinct layers (**FIGURE 16.16**):

- Layer 1: The upper layer is a sequence of unconsolidated sediments. Sediments are very thin near the axes of oceanic ridges but may be several kilometers thick next to continents.

- Layer 2: Below the layer of sediments is a rock unit composed mainly of basaltic lavas that contain abundant pillowlike structures called *pillow basalts.*

- Layer 3: The middle, rocky layer is made up of numerous interconnected dikes having a nearly vertical orientation, called the *sheeted dike complex.* These dikes are former pathways where magma rose to feed lava flows on the ocean floor.

- Layer 4: The lowest unit is mainly *gabbro,* the coarse-grained equivalent of basalt, which crystallized deeper in the crust without erupting.

This sequence of layers composing the oceanic crust is called an **ophiolite complex** (see Figure 16.16). From studies of various ophiolite complexes around the globe and related data, geologists have pieced together a scenario for the formation of the ocean floor.

How Does Oceanic Crust Form?

The molten rock that goes into the making of new oceanic crust originates from partial melting of the mantle rock peridotite at depths greater than 40 kilometers. This process generates a melt having the composition of basalt, which is less dense than the surrounding solid rock. The newly formed melt rises through the upper mantle along thousands of tiny conduits that

FIGURE 16.16 The four layers that make up a typical section of oceanic crust—based on data obtained from ophiolite complexes, seismic profiling, and core samples obtained from deep-sea drilling expeditions.

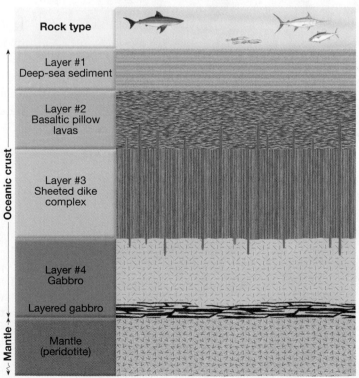

congealed margin and then breaks through. This process occurs repeatedly, as molten basalt is extruded—like toothpaste from a tightly squeezed tube (**FIGURE 16.17**). The result is protuberances resembling large bed pillows stacked one atop the other, hence the name **pillow basalts** (**FIGURE 16.18**).

In some settings, pillow lavas may build volcano-size mounds that resemble shield volcanoes, whereas in other situations they form elongated ridges tens of kilometers long. These structures are eventually separated from their supply of magma as they are carried away from the ridge crest by seafloor spreading.

The lowest unit of the ocean crust develops from crystallization within the central magma chamber itself. The first minerals to crystallize are olivine, pyroxene, and occasionally chromite (chromium oxide), which settle through the magma to form a layered zone near the floor of the reservoir. The remaining melt tends to cool along the walls of the chamber to form massive amounts of coarse-grained gabbro. This portion accounts for up to 5 of the 7 kilometers of ocean crust thickness.

Although molten rock rises continuously from the asthenosphere toward the surface, seafloor spreading occurs in pulselike bursts. As the melt begins to accumulate in the lens-shaped reservoirs, it is blocked from continuing upward by the stiff overlying rocks. As the amount of melt entering the magma reservoirs increases, pressure rises. Periodically, the pressure exceeds the strength of the overlying rocks, which fracture and initiate a short episode of seafloor spreading.

FIGURE 16.17 A photograph taken from the *Alvin* during Project FAMOUS shows lava extrusions in the rift valley of the Mid-Atlantic Ridge. Large toothpastelike extrusions such as this were common features. A mechanical arm is sampling an adjacent blisterlike extrusion. *(Photo courtesy of Woods Hole Oceanographic Institution)*

Interactions between Seawater and Oceanic Crust

In addition to serving as a mechanism for the dissipation of Earth's internal heat, the interaction between seawater and the newly formed basaltic crust alters both the seawater and the crust. The permeable and highly fractured lava of the upper oceanic crust allows seawater to penetrate to depths of 2 to 3 kilometers. As seawater circulates through the hot crust, it is heated and chemically reacts with the

black smoker

Black smoker chimney. *(Photo courtesy of P. Arona)*

FIGURE 16.18 Pillow lava exposed along a sea cliff, Cape Wanbrow, New Zealand. Notice that each "pillow" shows an outer, rapidly cooled, dark glassy layer enclosing a dark gray basalt interior. *(Photo by G.R. Roberts/Photo Researchers, Inc.)*

basaltic rock by a process called *hydrothermal* (hot water) *metamorphism* (see Chapter 7, p. 189). This alteration causes the dark silicates (olivine and pyroxene) in basalt to form new metamorphic minerals such as chlorite and serpentine. Simultaneously, the hot seawater dissolves ions of silica, iron, copper, and occasionally silver and gold from the hot basalts. When the water temperature reaches a few hundred degrees Celsius, these mineral-rich fluids buoyantly rise along fractures and eventually spew out on the ocean floor.

Studies conducted by submersibles along the Juan de Fuca Ridge have photographed these metallic-rich solutions as they gushed from the seafloor to form particle-filled clouds called **black smokers.** As the hot liquid (up to 400 °C) mixes with the cold, mineral-laden seawater, the dissolved minerals precipitate to form massive, metallic sulfide deposits, some of which are economically important. Occasionally these deposits grow upward to form underwater chimneylike structures equivalent in height to skyscrapers.

CONCEPT CHECK 16.6

❶ Briefly describe the four layers of the ocean crust.

❷ How does the *sheeted dike complex* form?

❸ How does hydrothermal metamorphism alter the basaltic rocks that make up the seafloor? How is seawater changed during this process?

❹ What is a black smoker?

DID YOU KNOW?
The circulation of seawater through the oceanic crust extracts heat from the volcanically produced material and is the main method by which the crust is cooled.

Continental Rifting: The Birth of a New Ocean Basin

The reason Pangaea began to split apart nearly 200 million years ago is a subject still debated among geoscientists. Nevertheless, the event illustrates that ocean basins originate when continents break apart. This was, undoubtedly, the case for the Atlantic Ocean, which formed as the Americas drifted from Europe and Africa.

Evolution of an Ocean Basin

The opening of a new ocean basin begins with the formation of a **continental rift**, an elongated depression along which the entire lithosphere is stretched and thinned. Where the lithosphere is thick, cool, and strong, rifts tend to be narrow—often less than a few hundred kilometers wide. Modern examples of narrow continental rifts include the East African Rift, the Baikal Rift (south central Siberia), and the Rhine Valley (northwestern Europe).

Iceland's largest fishing port was extensively damaged in 1973 by the eruption of the Heimaey Volcano. *(Photo by Bettman/Corbis)*

By contrast, where the crust is thin, hot, and weak, rifts can be more than 1000 kilometers wide, as exemplified by the Basin and Range Province in the western United States.

In those settings where rifting continues, the rift system evolves into a young, narrow ocean basin, such as the present-day Red Sea. Continued seafloor spreading eventually results in the formation of a mature ocean basin bordered by rifted continental margins. The Atlantic Ocean is such a feature. What follows is an overview of ocean basin evolution using modern examples to represent the various stages of rifting.

EAST AFRICAN RIFT. The East African Rift is a continental rift that extends through eastern Africa for approximately 3000 kilometers (2000 miles). It consists of several interconnected rift valleys that split into eastern and western sections around Lake Victoria (**FIGURE 16.19**). Whether this rift will eventually develop into a spreading center, with the Somali subplate separating from the continent of Africa, is uncertain.

The most recent period of rifting began about 20 million years ago as upwelling in the mantle intruded the base of the lithosphere (**FIGURE 16.20A**). Buoyant uplifting of the heated lithosphere led to doming and stretching of the crust. Consequently, the upper crust was broken along high-angle normal faults, producing downfaulted blocks, or *grabens,* while the lower crust deformed by ductile stretching (**FIGURE 16.20B**).

In the early stages of rifting, magma generated by decompression melting of the rising mantle rocks intruded the crust. Occasionally, some of the magma migrated upward along fractures and erupted at the surface. This activity produced extensive basaltic flows within the rift as well as volcanic cones—some forming more than 100 kilometers from the rift axis. Examples include Mount Kenya and Mount Kilimanjaro, the highest point in Africa, rising almost 6000 meters (20,000 feet) above the Serengeti Plain.

RED SEA. Research suggests that if spreading continues, a rift valley will lengthen and deepen, eventually extending to the margin of the continent (**FIGURE 16.20C**). At this point, the continental rift becomes a narrow linear sea with an outlet to the ocean, similar to the Red Sea.

The Red Sea formed when the Arabian Peninsula rifted from Africa beginning about 30 million years ago. Steep fault scarps that rise as much as 3 kilometers above sea level flank the margins of this water body. Thus, the escarpments surrounding the Red Sea are similar to the steep cliffs that border the East African Rift. Although the Red Sea reaches oceanic depths (up to 5 kilometers) in only a few locations, symmetrical magnetic stripes indicate that typical seafloor spreading has been occurring for at least the past 5 million years.

ATLANTIC OCEAN. If spreading continues, the Red Sea will grow wider and develop an elevated oceanic ridge similar to the Mid-Atlantic ridge (**FIGURE 16.20D**). As new oceanic crust is added to the diverging plates, the rifted continental margins gradually recede from the region of upwelling. As a result, they cool, contract, and sink.

Over time, continental margins subside below sea level and material eroded from the adjacent highlands blanket this once-rugged topography. The result is a *passive continental margin* consisting of a rifted continental crust that has been covered by a thick wedge of relatively undisturbed sediment and sedimentary rock.

Not all continental rift valleys develop into full-fledged spreading centers. In the central United States a failed rift extends from Lake Superior into central Kansas. This once-active rift valley is filled with sediments and volcanic rock that was extruded onto the crust more than a billion years ago. Why one rift valley develops into a full-fledged active spreading center while others are abandoned is not fully understood.

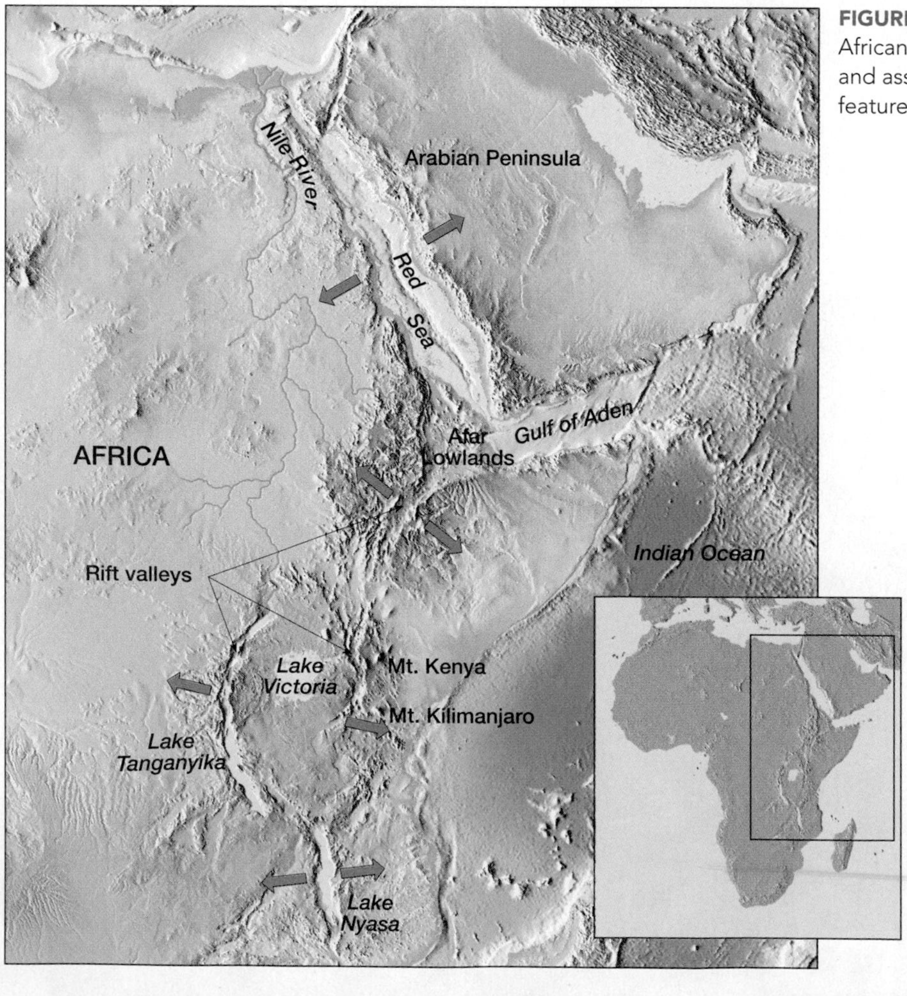

FIGURE 16.19 East African Rift valleys and associated features.

CONCEPT CHECK 16.7

❶ Name a modern example of a continental rift.

❷ Briefly describe each of the four stages in the evolution of an ocean basin.

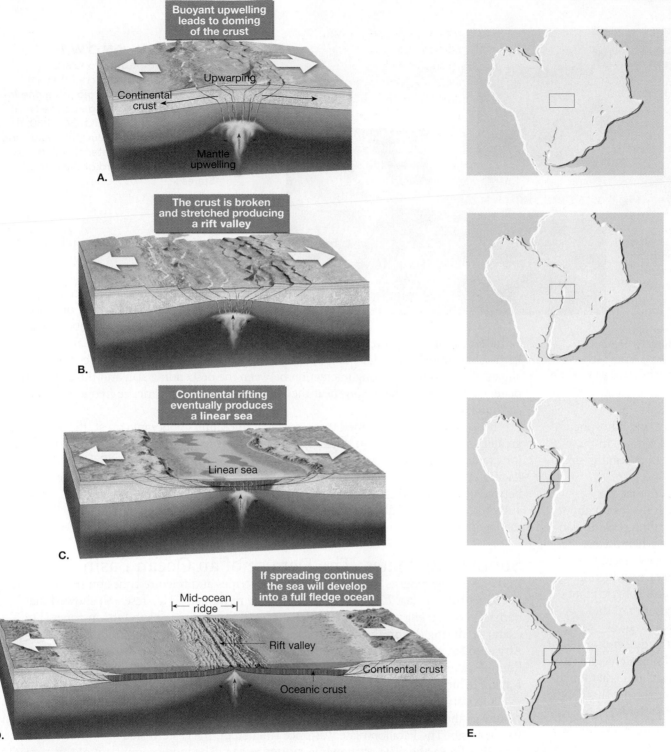

A. Buoyant upwelling leads to doming of the crust

Upwarping

Continental crust

Mantle upwelling

B. The crust is broken and stretched producing a rift valley

C. Continental rifting eventually produces a linear sea

Linear sea

D. If spreading continues the sea will develop into a full fledge ocean

Mid-ocean ridge

Rift valley

Continental crust

Oceanic crust

FIGURE 16.20
Formation of an ocean basin. **A.** Tensional forces and buoyant uplifting of the heated lithosphere cause the upper crust to be broken along normal faults, while the lower crust deforms by ductile stretching. **B.** As the crust is pulled apart, large slabs of rock sink, generating a rift zone. **C.** Further spreading generates a narrow sea. **D.** Eventually, an expansive ocean basin and ridge system are created. **E.** These four diagrams illustrate the separation of South America and Africa to form the South Atlantic.

E.

Destruction of Oceanic Lithosphere

Although new lithosphere is continually being produced at divergent plate boundaries, Earth's surface area is not growing larger. In order to balance the amount of newly created lithosphere, there must be a process whereby plates are destroyed.

Why Oceanic Lithosphere Subducts

The process of plate subduction is complex, and the ultimate fate of oceanic lithosphere is still being debated. What is known with some certainty is that oceanic lithosphere will resist subduction unless its overall density is greater than that of the underlying mantle.

It takes about 15 million years for a young slab of oceanic lithosphere to become cooler and denser than the supporting asthenosphere. In parts of the western Pacific, some oceanic lithosphere is nearly 180 million years old, the thickest and densest in today's oceans. The subducting slabs in this region typically descend into the mantle at angles approaching 90 degrees (**FIGURE 16.21A**). By contrast, when a spreading center is located near a

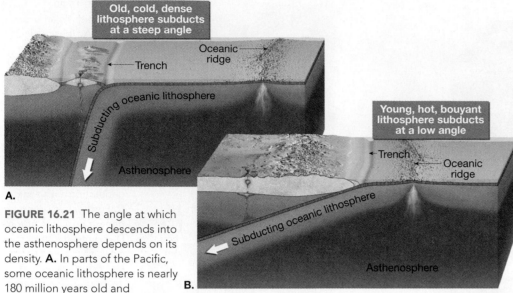

FIGURE 16.21 The angle at which oceanic lithosphere descends into the asthenosphere depends on its density. **A.** In parts of the Pacific, some oceanic lithosphere is nearly 180 million years old and typically descends into the mantle at angles approaching 90 degrees. **B.** Young oceanic lithosphere is warm and buoyant, hence it tends to subduct at a low angle.

subduction zone, the oceanic lithosphere is still young, warm, and buoyant. In these settings, the slab's angle of descent is small (**FIGURE 16.21B**).

It is important to note that the *lithospheric mantle,* which makes up about 80 percent of the descending oceanic slab, drives subduction. Even when the *oceanic crust* is quite old, its density is less than the underlying asthenosphere. Subduction, therefore, depends on lithospheric mantle, which is colder and denser than the asthenosphere that supports it.

When an oceanic slab descends to about 400 kilometers, mineral phase changes (the transition from low-density mineral to high-density), enhance subduction. At this depth, the transition from olivine (low-density) to spinel (its high-density form) increases the

density of the slab, which helps pull the plate into the subduction zone.

At some locations, the oceanic crust is unusually thick and buoyant because it is capped by large outpourings of basaltic lava, or other thick crustal fragments. In these settings subduction may be modified, or even prevented. This appears to be the situation in two areas along the Peru–Chile Trench, where the angle of descent is shallow—about 10 to 15 degrees. Low dip angles often result in a strong interaction between the descending slab and the overriding plate. Consequently, the regions near the Peru–Chile Trench experience frequent, great earthquakes.

It has also been determined that unusually thick units of oceanic crust, those that are greater than 30 kilometers in thickness, are not likely to subduct. The Ontong Java Plateau, for example, is a thick oceanic plateau, about the size of Alaska, located in the western Pacific. About 20 million years ago this plateau reached the trench that forms the boundary between the subducting Pacific plate and the overriding Australian–Indian plate. Apparently too buoyant to subduct, the Ontong Java Plateau clogged the trench. We will consider the fate of crustal fragments that are too buoyant to subduct in the next chapter.

Subducting Plates: The Demise of an Ocean Basin

In the 1970s, geologists began using magnetic stripes and fracture zones on the ocean floor to reconstruct the last 200 million years of plate movement. This research showed that parts of, or even entire, ocean basins have been destroyed along subduction zones. For example, during the break-up of Pangaea shown in Figure 15.26 (p. 385), notice that the African plate moves northward eventually colliding with Eurasia. During this event, the floor of the intervening Tethys Ocean was almost entirely consumed into the mantle, leaving behind a small remnant—the Mediterranean Sea.

Reconstructions of the break-up of Pangaea also helped investigators understand the demise of the Farallon plate—a large oceanic plate that once occupied much of the eastern Pacific basin. The Farallon plate was once situated on the eastern side of a spreading center opposite the Pacific plate as shown in **FIGURE 16.22A**. The modern remnant of this spreading center, which generated both the Farallon and Pacific plates, is the East Pacific Rise.

Beginning about 180 million years ago, the Americas were propelled westward by seafloor spreading in the Atlantic. Therefore, the Farallon plate, which was subducting beneath the Americas faster than it was being generated, decreased in size (**FIGURE 16.22B**). As its surface area shrank, it broke into smaller pieces, some of which subducted entirely. The remaining fragments of the once-extensive Farallon plate are the Juan de Fuca, Cocos, and Nazca plates.

FIGURE 16.22
Simplified illustration of the demise of the Farallon plate, which once ran along the western margin of the Americas. Because the Farallon plate was subducting faster than it was being generated, it got smaller and smaller. The remaining fragments of the once mighty Farallon plate are the Juan de Fuca, Cocos, and Nazca plates.

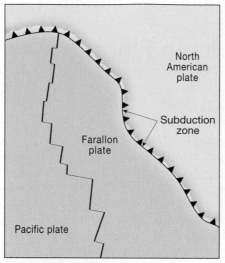

A. 56 million years ago

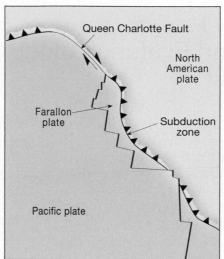

B. 37 million years ago

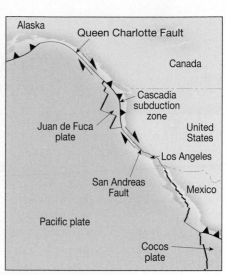

C. Today

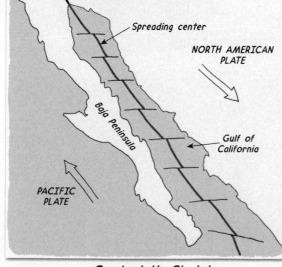

Geologist's Sketch

FIGURE 16.23 Satellite image showing the separation between the Baja Peninsula and North America. *(Image courtesy of NASA)*

The westward migration of North America also caused a section of the East Pacific Rise to enter the subduction zone that once lay off the coast of California (Figure 16.22B). As this spreading center subducted, it was destroyed and replaced by a newly generated transform fault system that currently accommodates the differential motion between the North American and Pacific plates. As more of the ridge subducted, the transform fault system, which we now call the San Andreas Fault, increased in length (**FIGURE 16.22C**). A similar event generated the Queen Charlotte transform fault located off the west coast of Canada and southern Alaska.

Today, the southern end of the San Andreas Fault connects to a young spreading center that is generating the Gulf of California (**FIGURE 16.23**). Because of this change in plate geometry, the Pacific plate has captured a sliver of North America (the Baja Peninsula and a portion of southern California) and carries it northwestward toward Alaska at a rate of about 6 centimeters per year.

CONCEPT CHECK 16.8

❶ Explain why oceanic lithosphere subducts even though the oceanic crust is less dense than the underlying asthenosphere.

❷ Why does the lithosphere thicken as it moves away from the ridge as a result of seafloor spreading?

❸ What happened to the Farallon plate?

DID YOU KNOW?
Sea turtles, salmon, and other animals travel great distances across the open ocean so that they can lay their eggs in the same location where they were hatched. Recent research suggests that these animals may use Earth's magnetic field for navigation. Interestingly, many animals that have "homing" abilities also contain minuscule amounts of the magnetic mineral magnetite.

CHAPTER SIXTEEN
Origin and Evolution of the Ocean Floor in Review

⊙ *Ocean bathymetry is determined using echo sounders and multibeam sonars,* which bounce sonic signals off the ocean floor. Ship-based receivers record the reflected echoes and accurately measure the time interval of the signals. With this information, ocean depths are calculated and plotted to produce maps of ocean floor topography. Recently, *satellite measurements* of the ocean surface have provided a new type of data that can be used to map the ocean floor.

⊙ Oceanographers studying the topography of ocean basins have delineated three major units: *continental margins, deep-ocean basins,* and *oceanic (mid-ocean) ridges.*

⊙ The zones that collectively make up a *passive continental margin* include the *continental shelf* (a gently sloping, submerged surface extending from the shoreline toward the deep-ocean basin; the *continental slope* (the true edge of the continent, which has a steep slope that leads from the continental shelf into deep water); and the *continental rise* (a gradual incline composed of sediments that have moved downslope from the continental shelf to the deep-ocean floor).

⊙ Most *active continental margins* are located around the Pacific Ocean in areas where the leading edge of a continent is overrunning oceanic lithosphere. At these sites, sediment scraped from the descending oceanic plate is plastered against the continent to form a collection of sediments called an *accretionary wedge.* An active continental margin generally has a narrow continental shelf, which grades into a deep-ocean trench.

⊙ The deep-ocean basin lies between the continental margin and the oceanic ridge system. Its features include *deep-ocean trenches* (long, narrow depressions that are the deepest parts of the ocean and are located where moving crustal plates descend back into the mantle); *abyssal plains* (among the most level places on Earth, consisting of thick accumulations of sediments that were deposited atop the low, rough portions of the ocean floor by turbidity currents); *seamounts* (volcanic peaks on the ocean floor that originate near oceanic ridges or in association with volcanic hot spots); and *oceanic plateaus* (large, thick, flood basalt provinces similar to those found on the continents).

⊙ *Oceanic (mid-ocean) ridges,* the sites of seafloor spreading, are found in all major oceans and represent more than 20 percent of Earth's surface. They are the most prominent features in the oceans and form an almost continuous swell that rises 2 to 3 kilometers above the adjacent ocean basin floor. Ridges are characterized by an *elevated position, extensive faulting,* and *volcanic structures* that have developed on newly formed oceanic crust. Most of the geologic activity associated with ridges occurs along a narrow region on the ridge crest, called the *rift zone,* where magma from the asthenosphere moves upward to create new slivers of oceanic crust. The topography of the oceanic ridge is controlled by the rate of seafloor spreading.

⊙ New oceanic crust is formed in a continuous manner by the process of seafloor spreading. The upper crust is composed of *pillow lavas* of basaltic composition. Below this layer are numerous interconnected dikes (*sheeted dike complex*) that are underlain by a thick layer of gabbro. This entire sequence is called an *ophiolite complex.*

⊙ The development of a new ocean basin begins with the formation of a *continental rift* similar to the East African Rift. In those settings where rifting continues, a narrow ocean basin develops, exemplified by the Red Sea. Eventually, seafloor spreading creates an ocean basin bordered by rifted continental margins similar to the present-day Atlantic Ocean. Mechanisms that drive continental rifting include: hot mantle plumes, upwelling from shallow levels in the mantle, and forces that arise from plate motions.

⊙ Oceanic lithosphere subducts because its overall density is greater than the underlying asthenosphere. The subduction of oceanic lithosphere may result in the destruction of sections of or entire ocean basins. A classic example is the Farallon plate, most of which subducted beneath the Americas as these continents were displaced westward by seafloor spreading in the Atlantic.

Key Terms

abyssal plains (p. 400)
accretionary wedge (p. 398)
active continental
   margins (p. 398)
bathymetry (p. 394)
black smokers (p. 407)
continental margins (p. 397)
continental rift (p. 407)

continental rise (p. 398)
continental shelf (p. 397)
continental slope (p. 398)
deep-ocean basin (p. 399)
deep-ocean trenches (p. 399)
deep-sea fan (p. 398)
echo sounders (p. 394)
guyots (p. 400)

mid-ocean ridge (p. 402)
oceanic plateaus (p. 400)
oceanic ridge (p. 402)
ophiolite complex (p. 405)
passive continental
   margins (p. 397)
pillow basalts (p. 406)
rift valleys (p. 402)

seamounts (p. 400)
seismic reflection
   profile (p. 395)
sheeted dike
   complex (p. 405)
sonar (p. 394)
tablemounts (p. 400)

GIVE IT SOME THOUGHT

① Refer to the accompanying map showing the Eastern Seaboard of the United States to complete the following:

  a. Which letter is associated with each of the following: continental shelf; continental rise; and shelf-break?

  b. How does the size of the continental shelf that surrounds the state of Florida compare with the size of the Florida peninsula?

  c. Why are there no deep-ocean trenches on this map?

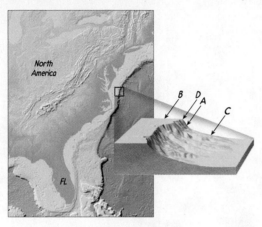

② In your own words, explain Darwin's hypothesis for the formation of coral atolls.

③ Do an internet search for "Mataiva," a small atoll in Tuomotus, French Polynesia. Using the map on the Wikipedia site, determine the approximate number of atolls in this small region of the western Pacific. What conclusion can you draw about the abundance of atolls in the western Pacific?

④ Referring to Figure 16.13, compare and contrast the topography of the crest of an oceanic ridge that exhibits a slow spreading rate with one that exhibits a fast spreading rate. Give examples of each.

⑤ Explain why the ocean floor generally gets deeper the farther one travels from the ridge crest.

⑥ Name a location where a new ocean basin may develop in the future. Explain why you chose that site.

⑦ Propose one or two ideas to explain why some continental rifts evolve into active plate boundaries while others do not.

⑧ Refer to Figure 16.8. At which of the following locations would you expect to have the highest angle of subduction, the Peru–Chile trench or the Mariana trench? Explain.

⑨ Refer to Figure 16.22. Predict the fate of the Juan de Fuca plate. What type of boundary might the Cascadia subduction zone become in the future? Explain.

⑩ Explain this Statement: The oceans are 4 billion years old, but the oldest ocean basin is only about 180 million years old.

Companion Website

www.mygeoscienceplace.com

The *Essentials of Geology, 11e* companion Website contains numerous multimedia resources accompanied by assessments to aid in your study of the topics in this chapter. The use of this site's learning tools will help improve your understanding of geology. Utilizing the access code that accompanies this text, visit **www.mygeoscienceplace.com** in order to:

  • **Review** key chapter concepts.

  • **Read** with links to the Pearson eText and to chapter-specific web resources.

  • **Visualize** and comprehend challenging topics using the learning activities in *GEODe: Essentials of Geology* and the *Geoscience Animations Library*.

  • **Test** yourself with online quizzes.

Crustal Deformation and Mountain Building

M OUNTAINS PROVIDE SOME OF THE MOST SPECTACULAR scenery on our planet. This splendor has been captured by poets, painters, and songwriters alike. Geologists understand that at some time all continental regions were mountainous masses. They have determined that continents grow by the addition of mountains to their flanks. Consequently, as geologists unravel the secrets of mountain formation, they also gain a deeper understanding of the evolution of Earth's continents.

Folded sedimentary layers exposed on the face of Mount Kidd, Alberta, Canada. (Photo by Tom Neversely/Photolibrary)

To assist you in learning the important concepts in this chapter, focus on the following questions:

⊙ What is rock deformation? What factors influence how rock deforms?

⊙ How are elastic, brittle, and ductile deformation different?

⊙ What structures form as a result of brittle deformation?

⊙ What structures form as a result of ductile deformation?

⊙ What are the most common types of faults? How does each type form?

⊙ How are Aleutian- and Andean-type mountain building similar? How are they different?

⊙ How does continental accretion relate to mountain building?

⊙ What are the stages in the formation of a major mountain belt?

⊙ How is the concept of isostasy related to mountain building?

⊙ What is meant by gravitational collapse?

Crustal Deformation

Crustal Deformation

ESSENTIALS OF GEOLOGY Deformation

Earth is a dynamic planet. Shifting lithospheric plates continually change the face of our planet by moving continents across the globe. The results of this tectonic activity are perhaps most strikingly apparent in Earth's major mountain belts (**FIGURE 17.1**). Rocks containing fossils of marine organisms are found thousands of meters above sea level, and massive rock units are bent, contorted, overturned, and sometimes rife with fractures.

We begin our look at mountain building by examining the process of rock deformation and the structures that result. Every mass of rock, no matter how strong, has a point at which it will fracture or flow. **Deformation** is a general term that refers to all changes in the original shape, size (volume), or orientation of a rock body. Geologists use the term **stress** to describe the forces that deform rocks. Stress that squeezes and shortens a rock mass is called *compressional stress*, whereas stress that pulls apart or elongates a rock body is known as *tensional stress*. Stress can also cause a rock to *shear*, which is similar to the slippage that occurs between individual playing cards when the top of the deck is moved relative to the bottom.

When rocks are subjected to stresses greater than their strength, they begin to deform, usually by flowing or fracturing. It is easy to visualize how rocks break, because we normally think of them as being brittle. But how can masses of rock be *bent* into intricate folds without fracturing in the process? To determine this, geologists performed laboratory experiments in which rocks were subjected to differential stresses under conditions that simulate those existing at various depths within the crust.

Elastic, Brittle, and Ductile Deformation

Although each rock type deforms somewhat differently, the general characteristics of rock deformation were determined from such experiments. Geologists learned that when stress is gradually applied, rocks first respond by deforming elastically. Changes that result from **elastic deformation** are recoverable; that is, like a rubber band, the rock will return to nearly its original size and shape when the stress is removed. During elastic deformation the chemical bonds of the minerals within a rock are stretched but do not break. As you saw in Chapter 15, the energy for most earth-

quakes comes from stored elastic energy that is released as rock snaps back to its original shape.

Once the elastic limit (strength) of a rock is surpassed, it either flows or fractures. Rocks that break into smaller pieces exhibit **brittle deformation**. From our everyday experience, we know that glass objects, wooden pencils, china plates, and even our bones exhibit brittle failure once their strength is surpassed. Brittle deformation occurs when stress causes the chemical bonds that hold a material together to break.

Ductile deformation, on the other hand, is a type of solid-state flow that produces a change in the shape of an object without fracturing. Ordinary objects that display ductile behavior include modeling clay, beeswax, taffy, and some metals. For example, a copper penny placed on a railroad track will be flattened and deformed (without breaking) by the force applied by a passing train. In rocks, ductile deformation is the result of some chemical bonds breaking, while others are forming, allowing minerals to change shape.

Factors That Affect Rock Strength

The major factors that influence the strength of a rock and how it will deform include temperature, confining pressure, rock type, and time.

FIGURE 17.1 A portion of the Sneffels Range near Ouray, Colorado *(Photo by James Hagar/Robert Harding)*

TEMPERATURE. The effect of temperature on the strength of a material can be easily demonstrated with a piece of glass tubing commonly found in a chemistry lab. If the tubing is dropped on a hard surface, it will shatter. However, if the tubing is heated over a Bunsen burner, it can be easily bent into a variety of shapes. Rocks respond similarly to heat. Where temperatures are high (deep in Earth's crust), rocks tend to deform ductilely and flow. Likewise, where temperatures are low (at or near the surface), rocks tend to behave like brittle solids and fracture.

CONFINING PRESSURE. Recall from Chapter 7 that pressure, like temperature, increases with depth as the thickness of the overlying rock increases. Buried rocks are subjected to confining pressure, which is much like water pressure, where the forces are applied equally in all directions. The deeper you go in the ocean, the greater the confining pressure. The same is true for rock that is buried. Confining pressure "squeezes" the materials in Earth's crust. Therefore, rocks that are deeply buried are "held together" by the immense pressure and tend to flow, rather than fracture.

ROCK TYPE. In addition to the physical environment, the mineral composition and texture of rock greatly influence how it will deform. For example, crystalline rocks composed of minerals that have strong internal molecular bonds tend to fail by brittle fracture. By contrast, sedimentary rocks that are weakly cemented, or metamorphic rocks that contain zones of weakness such as foliation, are more susceptible to ductile deformation.

Weak rocks that are most likely to behave in a ductile manner (flow or fold) when subjected to differential stress, include rock salt, shale, limestone, and schist. Igneous and some metamorphic rocks tend to be strong and brittle. In a near-surface environment, strong, brittle rocks will fail by fracturing when subjected to stresses that exceed their strength. At increasing depths, however, the strength of all rock types decreases significantly.

TIME. One key factor that researchers are unable to duplicate in the laboratory is how rocks respond to small stresses applied gradually over long spans of *geologic time*. However, insights into the effects of time on deformation are provided in everyday settings. For example, marble benches have been known to sag under their own weight over a span of 100 years or so, and wooden bookshelves may bend after being loaded with books for a relatively short period.

In general, when tectonic forces are applied slowly over long time spans, rocks tend to display ductile behavior and deform by flowing and folding. An analogous situation occurs when you take a taffy bar and slowly move the two ends together. The taffy will deform by folding. However, if you swiftly hit the taffy against the edge of a table, it will break into two or more pieces, exhibiting brittle failure.

Likewise, rocks tend to deform in a ductile manner by folding when stress builds gradually. These same rocks may fracture if the stress increases suddenly. As a consequence, folding and faulting may occur simultaneously in the same rock body (**FIGURE 17.2**).

To review, the processes by which rocks deform occur along a continuum that ranges from brittle fracture at one end to ductile flow at the other. The processes of deformation generate geologic changes on many scales. At one extreme are Earth's major mountain systems. At the other extreme are minor fractures in bedrock created by highly localized stresses. All of these phenomena, from the largest folds in the Alps to the smallest fractures in a slab of rock, are considered to be *rock* or *tectonic structures*.

FIGURE 17.2 Deformed sedimentary strata exposed in a road cut near Palmdale, California. In addition to the obvious folding, light-colored beds are offset along a fault located on the right side of the photograph. *(Photo by E. J. Tarbuck)*

CONCEPT CHECK 17.1

1. What is rock deformation? How might a rock body change during deformation?
2. Describe elastic deformation.
3. How is brittle deformation different from ductile deformation?
4. List and describe the four factors that affect rock strength.

DID YOU KNOW?
The tallest mountain in North America is the 20,321-foot-high Mount McKinley. The area surrounding the mountain was established as Mount McKinley National Park in 1917. In 1980 the park was enlarged and renamed Denali National Park and Preserve. Denali, the "High One" is the name used by native Athabascan people for this massive peak that crowns the 600-mile-long Alaska Range.

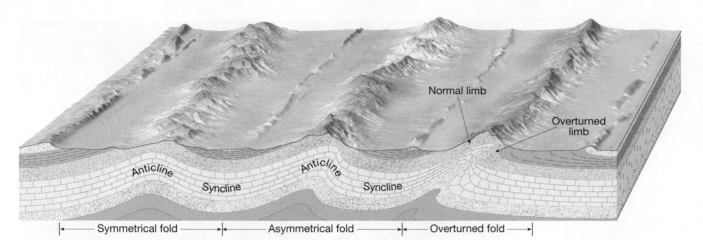

FIGURE 17.3 Block diagram of principal types of folded strata. The upfolded, or arched, structures are *anticlines*. The downfolds, or troughs, are *synclines*. Notice that the limb of an anticline is also the limb of the adjacent syncline.

Structures Formed by Ductile Deformation

Crustal Deformation

ESSENTIALS OF GEOLOGY Folds

Because folds are common features of deformed sedimentary rocks, we know rocks can bend without breaking. Ductile deformation is often accomplished by gradual slippage along planes of weakness within the atomic structure of mineral grains (see Figure 7.6, p. 183). This microscopic form of gradual solid-state flow involves slippage facilitated by chemical bonds breaking in one location as new ones form at another site. Rocks that display evidence of ductile flow usually were deformed at great depth and may exhibit contorted folds that give the impression that the strength of the rock was akin to soft putty.

Folds

During mountain building, flat-lying sedimentary and volcanic rocks are often bent into a series of wavelike undulations called **folds**. Folds in sedimentary strata are much like those that would form if you were to hold the ends of a sheet of paper and then push them together. In nature, folds come in a wide variety of sizes and configurations. Some folds are broad flexures in which strata hundreds of meters thick have been slightly warped. Others are very tight microscopic structures found in metamorphic rocks. Size differences notwithstanding, most folds are the result of *compressional stresses that result in a shortening and thickening of the crust.*

ANTICLINES AND SYNCLINES. The two most common types of folds are anticlines and synclines (**FIGURE 17.3**). **Anticlines** usually arise by upfolding, or arching, of sedimentary layers and are sometimes spectacularly displayed along highways that have been cut through deformed strata (**FIGURE 17.4A**).* Almost always found in association with anticlines are downfolds, or troughs, called **synclines** (**FIGURE 17.4B**). Notice in Figure 17.3 that the limb of an anticline is also a limb of the adjacent syncline.

Depending on their orientation, these basic folds are described as *symmetrical* when the limbs are mirror images of each other and *asymmetrical* when they are not. An asymmetrical fold is said to be

*By strict definition, an anticline is a structure in which the oldest strata are found in the center. A syncline is a structure in which the youngest strata are found in the center.

FIGURE 17.4 Anticline and syncline. **A.** An asymmetrical anticline in which one limb dips more steeply than the other. **B.** A nearly symmetrical syncline formed in limestone and siltstone strata. *(Photos by E. J. Tarbuck)*

A.

B.

FIGURE 17.5 Overturned fold, East Fork of Toklat River, Alaska. Overturned folds have one or both limbs tilted beyond vertical. *(Photo by Michael Collier)*

Overturned folds have one or both limbs tilted beyond vertical

Limb

Limb

East Fork of Toklat River, Alaska

Geologist's Sketch

overturned if one or both limbs are tilted beyond the vertical (**FIGURE 17.5**). An overturned fold can also "lie on its side" so a plane extending through the axis of the fold is horizontal. These *recumbent* folds are common in highly deformed mountainous regions such as the Alps.

Folds do not continue forever; rather their ends die out much like the wrinkles in cloth. Some folds *plunge* because the axis of the fold penetrates the ground. **FIGURE 17.6** shows an example of a plunging anticline and the pattern produced when erosion removes the upper layers of the structure and exposes its interior. Note that the outcrop pattern of an anticline points in the direction it is plunging. The opposite is true for a syncline. A good example of the kind of topography that results when erosional forces attack folded sedimentary strata is found in the Valley and Ridge Province of the Appalachians (see Figure 17.24).

It is important to realize that ridges are not necessarily associated with anticlines, nor are valleys related to synclines. Rather, ridges and valleys result because of differential weathering and erosion. For example, in the Valley and Ridge Province resistant sandstone beds remain as imposing ridges separated by valleys cut into more easily eroded shale or limestone beds.

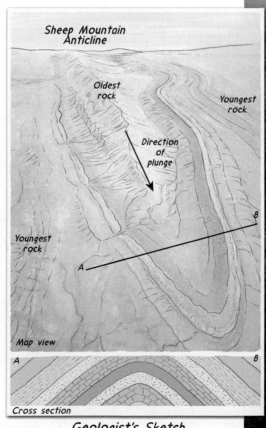

Sheep Mountain Anticline

Oldest rock

Youngest rock

Direction of plunge

B

Youngest rock

A

Map view

A   B

Cross section

Geologist's Sketch

FIGURE 17.6 Plunging anticline, Sheep Mountain, Wyoming. In a plunging anticline the outcrop pattern "points" in the direction of plunge, the opposite is true of plunging synclines. *(Photo by Michael Collier)*

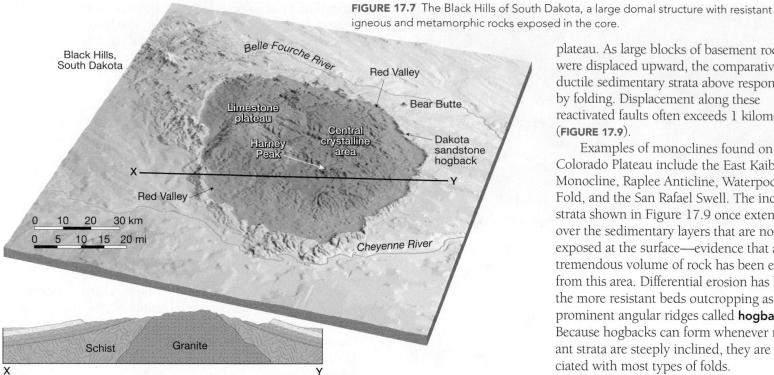

FIGURE 17.7 The Black Hills of South Dakota, a large domal structure with resistant igneous and metamorphic rocks exposed in the core.

plateau. As large blocks of basement rock were displaced upward, the comparatively ductile sedimentary strata above responded by folding. Displacement along these reactivated faults often exceeds 1 kilometer (**FIGURE 17.9**).

Examples of monoclines found on the Colorado Plateau include the East Kaibab Monocline, Raplee Anticline, Waterpocket Fold, and the San Rafael Swell. The inclined strata shown in Figure 17.9 once extended over the sedimentary layers that are now exposed at the surface—evidence that a tremendous volume of rock has been eroded from this area. Differential erosion has left the more resistant beds outcropping as prominent angular ridges called **hogbacks**. Because hogbacks can form whenever resistant strata are steeply inclined, they are associated with most types of folds.

CONCEPT CHECK 17.2

❶ Distinguish between anticlines and synclines, domes and basins, anticlines and domes.

❷ The Black Hills of South Dakota is a good example of what type of structural feature?

❸ Describe the formation of a monocline.

DOMES AND BASINS. Broad upwarps in basement rock may deform the overlying cover of sedimentary strata and generate large folds. When this upwarping produces a circular or slightly elongated structure, the feature is called a **dome.** Downwarped structures having a similar shape are termed **basins.**

The Black Hills of western South Dakota is a large domed structure generated by upwarping. Here erosion has stripped away the highest portions of the overlying sedimentary beds, exposing older igneous and metamorphic rocks in the center (**FIGURE 17.7**).

Several large basins exist in the United States (**FIGURE 17.8**). The basins of Michigan and Illinois have gently sloping beds similar to saucers. These basins are thought to be the result of large accumulations of sediment, whose weight caused the crust to subside (see section on isostasy later in this chapter). A few structural basins may have been the result of giant asteroid impacts.

Because large basins contain sedimentary beds sloping at low angles, they are usually identified by the age of the rocks composing them. The youngest rocks are found near the center, and the oldest rocks are at the flanks. This is just the opposite order of a domed structure, such as the Black Hills, where the oldest rocks form the core.

MONOCLINES. Although we have separated our discussion of folds and faults, in the real world folds can be intimately coupled with faults. Examples of this close association are broad, regional features called *monoclines*. Particularly prominent features of the Colorado Plateau, **monoclines** are large, step-like folds in otherwise horizontal sedimentary strata. These folds appear to be the result of the reactivation of ancient, steep, dipping faults located in basement rocks beneath the

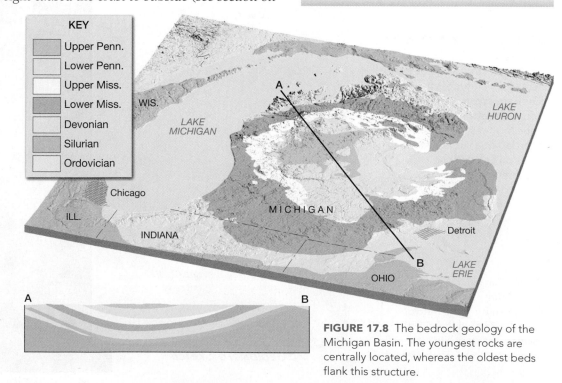

KEY

	Upper Penn.
	Lower Penn.
	Upper Miss.
	Lower Miss.
	Devonian
	Silurian
	Ordovician

FIGURE 17.8 The bedrock geology of the Michigan Basin. The youngest rocks are centrally located, whereas the oldest beds flank this structure.

FIGURE 17.9 The East Kaibab Monocline in northen Arizona. This monocline consists of bent sedimentary beds that were deformed by faulting in the bedrock below. The thrust fault in this sketch is called a *blind thrust* because it does not reach the surface. *(Photo by Michael Collier)*

Monocline

Hogback composed of Kaibab limestone

Sedimentary strata removed by erosion

East Kaibab Monocline, Arizona

Geologist's Sketch

Cross section

Fault in basement rock

Structures Formed by Brittle Deformation

 Crustal Deformation

ESSENTIALS OF GEOLOGY Faults and Fractures

You have probably seen a drinking glass drop on a hard surface, shattering into pieces. What you witnessed is analogous to brittle deformation that occurs in Earth's crust. In nature, brittle deformation occurs when stresses exceed the strength of a rock, causing it to break or fracture.

Faults

In the upper crust, to depths of about 10 to 15 kilometers (6 to 10 miles), rocks tend to exhibit brittle behavior by fracturing or by faulting. **Faults** are fractures in the crust along which appreciable displacement has taken place. Occasionally, small faults can be recognized in road cuts where sedimentary beds have been offset a few meters, as shown in **FIGURE 17.10**. Faults of this scale usually occur as single discrete breaks. By contrast, large faults, like the San Andreas Fault in California, have displacements of hundreds of kilometers and consist of many interconnecting fault surfaces. These structures, best described as *fault zones,* can be several kilometers wide and are often easier to identify from high-altitude photographs than at ground level.

Aerial view of the surface expression of the San Andreas Fault. *(Photo by D. Parker/Photo Researchers)*

San Andreas Fault

FIGURE 17.10 Faulting caused the vertical displacement of these strata. The rock immediately above a fault surface is the *hanging wall block,* and that below is called the *footwall block. (Photo by credit John S. Shelton)*

Dip-Slip Faults

Faults in which movement is primarily parallel to the *dip* (or inclination) of the fault surface are called **dip-slip faults**. It has become common practice to call the rock surface that is immediately above the fault the *hanging wall block* and to call the rock surface below, the *footwall block* (Figure 17.10). This nomenclature arose from prospectors and miners who excavated shafts and tunnels along fault zones that were sites of ore deposits. In these tunnels, the miners would walk on the rocks below the mineralized fault zone (the footwall block) and hang their lanterns on the rocks above (the hanging wall block).

NORMAL FAULTS. Dip-slip faults are classified as **normal faults** when the hanging wall block moves down relative to the footwall block (**FIGURE 17.11A**). Because of the downward motion of the hanging wall block, normal faults accommodate lengthening, or extension, of the crust.

Normal faults are found in a variety of sizes; some are small, having displacements of only a meter or so, like the one shown in the road cut in Figure 17.10. Others extend for tens of kilometers where they may sinuously trace the boundary of a mountain front. Most large, normal faults have relatively steep dips that tend to flatten out with depth. Vertical displacements along dip-slip faults may produce long, low cliffs called **fault scarps**.

In the western United States, large normal faults are associated with structures called *fault-block mountains.* Excellent examples of fault-block mountains are found in the Basin and Range Province, a region that encompasses Nevada and portions of the surrounding states (**FIGURE 17.12**). Here the crust has been elongated and broken to create more than 200 relatively small mountain ranges. Averaging about 80 kilometers in length, the ranges rise 900 to 1500 meters above the adjacent down-faulted basins.

The topography of the Basin and Range Province evolved in association with a system of roughly north–south trending normal faults. Movements along these faults produced alternating uplifted fault blocks called **horsts** and down-dropped blocks called **grabens**. Horsts generate elevated topography, whereas grabens form basins. As Figure 17.12 illustrates, structures called **half-grabens**, which are titled fault blocks, also contribute to the alternating topographic highs and lows in the Basin and Range Province. The horsts and higher ends of the tilted fault blocks are the source of sediments that have accumulated in the basins created by the grabens and lower ends of the tilted blocks.

Notice in Figure 17.12 that the slopes of the normal faults decrease with depth and eventually join to form a nearly horizontal fault called a *detachment fault.* These faults form a major boundary between the rocks below, which exhibit ductile deformation, and the rocks above, which exhibit brittle deformation.

Fault motion provides geologists with a method of determining the nature of the tectonic forces at work within Earth. Normal faults

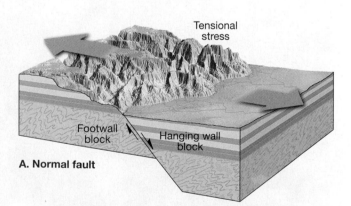

A. Normal fault

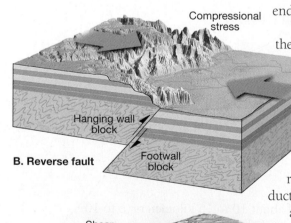

B. Reverse fault

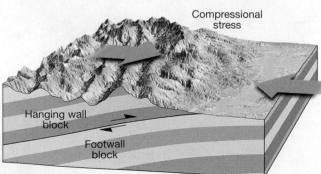

C. Thrust fault

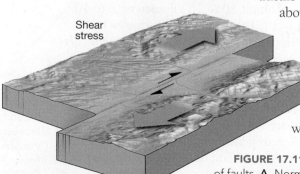

D. Strike-slip fault

FIGURE 17.11 Block diagrams of four types of faults. **A.** Normal fault **B.** Reverse fault **C.** Thrust fault **D.** Strike-slip fault

FIGURE 17.12 Normal faulting in the Basin and Range Province. Here, tensional stresses have elongated and fractured the crust into numerous blocks. Movement along these fractures has tilted the blocks, producing parallel mountain ranges called fault-block mountains. The down-faulted blocks (grabens) form basins, whereas the upfaulted blocks (horsts) erode to form rugged mountainous topography. In addition, numerous tilted blocks (half-grabens) form both basins and mountains. *(Photo by Michael Collier)*

DID YOU KNOW?
Many faults do not reach the surface and hence do not produce a fault scarp during an earthquake. Faults of this type are termed *blind* or *hidden faults*. The 1994 Northridge earthquake in the Los Angeles area occurred along a blind thrust fault.

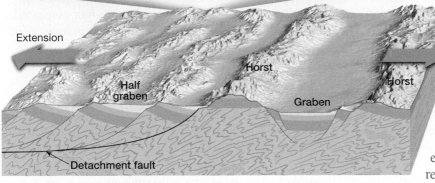

Extension

Half graben

Horst

Horst

Graben

Detachment fault

are associated with tensional forces that pull the crust apart. This "pulling apart" can be accomplished either by uplifting that causes the surface to stretch and break or by opposing horizontal forces.

REVERSE AND THRUST FAULTS. Reverse faults are dip-slip faults in which the hanging wall block moves up relative to the footwall block (**FIGURE 17.11B**). **Thrust faults** are reverse faults having dips less than 45°, so the overlying block moves nearly horizontally over the underlying block (**FIGURE 17.11C**).

Whereas normal faults occur in tensional environments, reverse faults result from strong compressional stresses. Because the hanging wall block moves up and over the footwall block, reverse and thrust faults accommodate horizontal shortening of the crust.

Most high-angle reverse faults are small and accommodate local displacements in regions dominated by other types of faulting. Thrust faults, on the other hand, exist at all scales with some large thrust faults having displacements on the order of tens to hundreds of kilometers.

Thrust faulting is most pronounced along convergent plate boundaries. Compressional forces associated with colliding plates generally create folds as well as

thrust faults that thicken and shorten the crust to produce mountainous topography (see Figure 17.21).

Strike-Slip Faults

A fault in which the dominant displacement is horizontal and parallel to the *trend* or *strike* of the fault surface is called a **strike-slip fault** (**FIGURE 17.11D**). The earliest scientific records of strike-slip faulting were made following surface ruptures that produced large earthquakes. One of the most noteworthy of these was the great San Francisco earthquake of 1906. During this strong earthquake, structures such as fences that were built across the San Andreas Fault were displaced as much as 4.7 meters (15 feet). Because movement along the San Andreas causes the crustal block on the opposite side of the fault to move to the right as you face the fault, it is called a *right-lateral* strike-slip fault (**FIGURE 17.13**).

The Great Glen Fault in Scotland is a well-known example of a *left-lateral* strike-slip fault, which exhibits the opposite sense of displacement. The total displacement along the Great Glen Fault is estimated to exceed 100 kilometers (60 miles). Also associated with this fault trace are numerous lakes, including Loch Ness, home of the legendary monster.

Some strike-slip faults cut through the lithosphere and accommodate motion between two large tectonic plates. Recall this special kind of strike-slip fault is called a **transform fault.** Numerous transform faults cut the oceanic lithosphere and link spread-

This fault scarp formed north of Landers, California, during an earthquake in 1992. *(Photo by Roger Ressmeyer/CORBIS)*

fault scarp

ing oceanic ridges. Others accommodate displacement between continental plates that move horizontally with respect to each other. One of the best-known transform faults is California's San Andreas Fault. This plate-bounding fault can be traced for about 950 kilometers (600 miles) from the Gulf of California to a point along the Pacific Coast north of San Francisco, where it heads out to sea. Since its formation about 30 million years ago, displacement along the San Andreas Fault has exceeded 560 kilometers. This movement has accommodated the northward displacement of southwestern California and the Baja Peninsula of Mexico in relation to the remainder of North America.

Joints

Among the most common rock structures are fractures called joints. Unlike faults, **joints** are fractures along which no appreciable displacement has occurred. Although some joints have a random orientation, most occur in roughly parallel groups (**FIGURE 17.14**).

We have already considered two types of joints. In Chapter 3 we learned that *columnar joints* form when igneous rocks cool and develop shrinkage fractures that produce elongated, pillar-like columns (see Figure 3.27). Also recall from Chapter 5 that sheeting produces a pattern of gently curved joints that develop more or less parallel to the surface of large exposed igneous bodies such as batholiths. Here the jointing results from the gradual expansion that occurs when erosion removes the overlying load (see Figure 5.5, p. 127).

Most joints are produced when rocks in the outermost crust are deformed as tensional stresses cause the rock to fail by brittle fracture. Extensive joint patterns often develop in response to relatively subtle and often barely perceptible regional upwarping and downwarping of the crust. In many cases, the cause for jointing at a particular locale is not readily apparent.

Many rocks are broken by two or even three sets of intersecting joints that slice the rock into numerous regularly shaped blocks. These joint sets often exert a strong influence on other geologic processes. For example, chemical weathering tends to be concentrated along joints, and in many areas groundwater movement and the resulting dissolution in soluble rocks is controlled by the joint pattern. Moreover, a system of joints can influence the direction stream courses follow. The rectangular drainage pattern described in Chapter 9 is such a case.

Highly jointed rocks present a risk to the construction of engineering projects, including highways and dams. On June 5, 1976, 14 lives were lost and nearly $1 billion in property damage occurred when the Teton Dam in Idaho failed. This earthen dam was constructed of very erodible clays and silts and was situated on highly jointed volcanic rocks. Although attempts were made to fill the voids in the jointed rock, water gradually penetrated the subsurface fractures and undermined the dam's foundation. Eventually the moving water cut a tunnel into the easily erodible clays and silts. Within minutes the dam failed, sending a 20-meter-high wall of water down the Teton and Snake rivers.

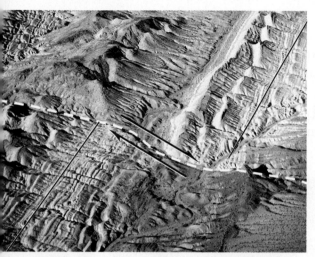

FIGURE 17.13 Aerial view of strike-slip (right-lateral) fault in southern Nevada. Notice the ridge of white rock in the top right portion of the photo was offset to the right relative to the portion of the same ridge that appears on the bottom left of the image. *(Photo by Marli Miller)*

CONCEPT CHECK 17.3

❶ Contrast the movements that occur along normal and reverse faults. What type of stress is indicated by each fault?

❷ What type of faults are associated with fault-block mountains?

❸ How are reverse faults different from thrust faults? In what way are they the same?

❹ Describe the relative movement along a strike-slip fault.

❺ How are joints different from faults?

FIGURE 17.14 Aerial view of highly jointed rock in Utah's Canyonlands National Park. Erosion has enlarged the network of joints *(Photo by Bernhard Edmaier/Photo Researches, Inc.)*

Mountain Building

Mountain building has occurred in the recent geologic past at several locations around the world (**FIGURE 17.15**). Young mountain belts include the American Cordillera, which runs along the western margin of the Americas from Cape Horn at the tip of South America to Alaska and includes the Andes and Rocky Mountains; the Alpine–Himalaya chain, that extends along the margin of the Mediterranean, through Iran to northern India, and into Indochina; and the mountainous terrains of the western Pacific, which include volcanic island arcs that comprise Japan, the Philippines, and Sumatra. Most of these young mountain belts have come into existence within the last 100 million years. Some, including the Himalayas, began their growth as recently as 50 million years ago.

In addition to these young mountain belts, there are several chains of Paleozoic-age mountains found on Earth. Although these older structures are deeply eroded and topographically less prominent, they exhibit the same structural features found in younger mountains. The Appalachians in the eastern United States and the Urals in Russia are classic examples of this group of older and well-worn mountain belts.

The term for the processes that collectively produce a mountain belt is **orogenesis.** Most major mountain belts display striking visual evidence of great horizontal forces that have shortened and thickened the crust. These **compressional mountains** contain large quantities of preexisting sedimentary and crystalline rocks that have been faulted and contorted into a series of folds. Although folding and thrust faulting are often the most conspicuous signs of orogenesis, varying degrees of metamorphism and igneous activity are always present.

How do mountain belts form? As early as the ancient Greeks, this question has intrigued some of the greatest philosophers and scientists. One early proposal suggested that mountains are simply wrinkles in Earth's crust, produced as the planet cooled from its original semimolten state. According to this idea, Earth contracted and shrank as it lost heat, which caused the crust to deform in a manner similar to how an orange peel wrinkles as the fruit dries out. However, neither this nor any other early hypothesis withstood scientific scrutiny.

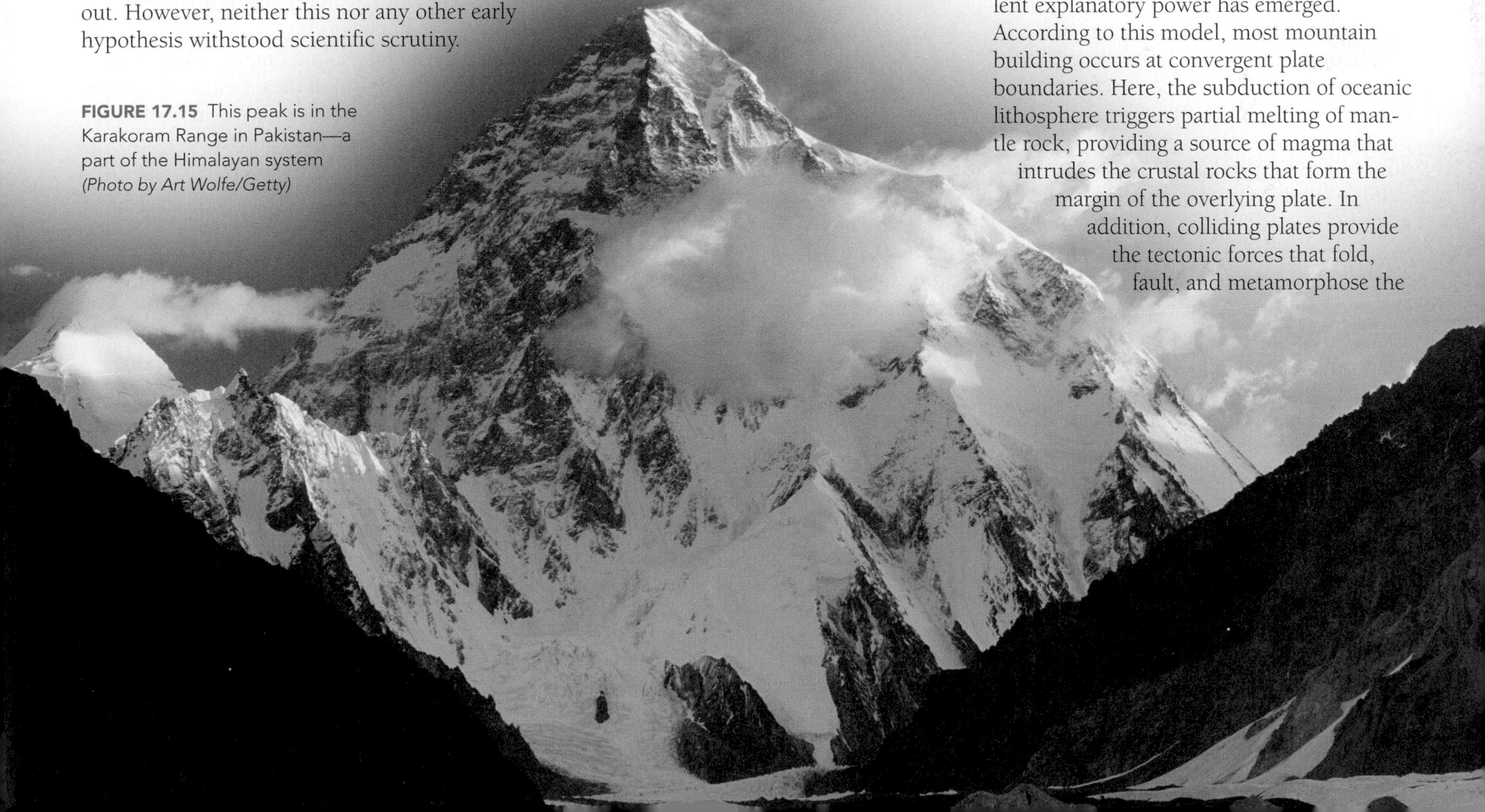

FIGURE 17.15 This peak is in the Karakoram Range in Pakistan—a part of the Himalayan system *(Photo by Art Wolfe/Getty)*

DID YOU KNOW?
The lake that is the home of the legendary Loch Ness Monster is located along the Great Glen Fault that divides northern Scotland. Strike-slip movement along this fault zone shattered a wide zone of rock, which glacial erosion subsequently removed, to produce the elongated valley in which Loch Ness is situated. The broken rock was easily eroded, a fact that accounts for the great depth of Loch Ness—as much as 600 feet below sea level.

CONCEPT CHECK 17.4

1. Define *orogenesis*.
2. In the plate tectonics model, which type of plate boundary is most directly associated with Earth's major mountain belts?

Mountain Building at Subduction Zones

With the development of the theory of plate tectonics, a model for orogenesis with excellent explanatory power has emerged. According to this model, most mountain building occurs at convergent plate boundaries. Here, the subduction of oceanic lithosphere triggers partial melting of mantle rock, providing a source of magma that intrudes the crustal rocks that form the margin of the overlying plate. In addition, colliding plates provide the tectonic forces that fold, fault, and metamorphose the

Volcanic Island Arcs

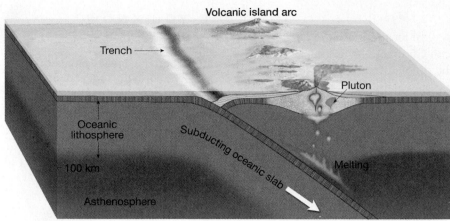

Island arcs result from the steady subduction of oceanic lithosphere, which may last for 200 million years or more (**FIGURE 17.16**). Periodic volcanic activity, the emplacement of igneous plutons at depth, and the accumulation of sediment that is scraped from the subducting plate gradually increase the volume of crustal material capping the upper plate. Some large volcanic island arcs, such as Japan, owe their size to having been built upon a preexisting fragment of continental crust.

The continued growth of a volcanic island arc can result in the formation of mountainous topography consisting of belts of igneous and metamorphic rocks. This activity, however, is viewed as just one phase in the development of a major mountain belt. As you will see later, some volcanic arcs are carried by a subducting plate to the margin of a large continental block, where they become involved in a large-scale mountain-building episode.

FIGURE 17.16 The development of a volcanic arc by the convergence of two oceanic plates. Continuous subduction results in the development of thick units of continental-type crust.

thick accumulations of sediments that have been deposited along the flanks of landmasses. Together, these processes thicken and shorten the continental crust, thereby elevating rocks that may have formed near the ocean floor to lofty heights.

To unravel the events that produce mountains, researchers examine ancient mountain structures as well as sites where orogenesis is currently active. Of particular interest are active subduction zones, where lithospheric plates are converging. Here the subduction of oceanic lithosphere generates Earth's strongest earthquakes and most explosive volcanic eruptions, as well as playing a pivotal role in generating many of Earth's mountain belts.

The subduction of oceanic lithosphere gives rise to two different types of tectonic structures. Where oceanic lithosphere subducts beneath an oceanic plate, a **volcanic island arc** and related tectonic features develop. Subduction beneath a continental block, on the other hand, results in the formation of a volcanic arc along the margin of a continent. Plate boundaries that generate continental volcanic arcs are referred to as **Andean-type plate margins**.

Mountain Building along Andean-type Margins

The first stage in the development of an Andean-type mountain belt occurs along a **passive continental margin** prior to the formation of the subduction zone. The East Coast of the United States provides a modern example of a passive continental margin where sedimentation has produced a thick platform of shallow-water sandstones, limestones, and shales (**FIGURE 17.17A**). At some point, the forces that drive plate motions change and a subduction zone develops along the margin of the continent. It is along these **active continental margins** that the structural elements of a developing mountain belt gradually take form.

A good place to examine an active continental margin is the west coast of South America. Here the Nazca plate is being subducted beneath the South American plate along the Peru–Chile trench. This subduction zone probably formed prior to the break-up of the supercontinent of Pangaea.

In an idealized Andean-type subduction, convergence of the continental block and the subducting oceanic plate leads to deformation and metamorphism of the continental margin. Once the oceanic plate descends to about 100 kilometers (60 miles), partial melting of mantle rock above the subducting slab generates magma that migrates upward (**FIGURE 17.17B**).

Thick continental crust greatly impedes the ascent of magma. Consequently, a high percentage of the magma that intrudes the crust never reaches the surface. Instead, it crystallizes at depth to form plutons. Eventually, uplifting and erosion exhume these igneous bodies and associated metamorphic rocks. Once they are exposed at the surface, these massive structures are called *batholiths* (**FIGURE 17.17C**). Composed of numerous plutons, batholiths form the core of the Sierra Nevada in California and are prevalent in the Peruvian Andes.

During the development of this continental volcanic arc, sediment derived from the land and scraped from the subducting plate is plastered against the landward side of the trench like piles of dirt in front of a bulldozer. This chaotic accumulation of sedimentary and metamorphic rocks with occasional scraps of ocean crust is called an **accretionary wedge** (Figure 17.17B). Prolonged subduction can build an accretionary wedge that is large enough to stand above sea level (Figure 17.17C).

Andean-type mountain belts are composed of two roughly parallel zones. The volcanic arc develops on the continental block. It consists of volcanoes and large intrusive bodies intermixed with high-temperature metamorphic rocks. The seaward segment is the accretionary wedge. It consists of folded and faulted sedimentary and metamorphic rocks (Figure 17.17C).

SIERRA NEVADA AND COAST RANGES. One of the best examples of an inactive Andean-type orogenic belt is found in the western United States. It includes the Sierra Nevada and the Coast Ranges in California. These parallel mountain belts were produced

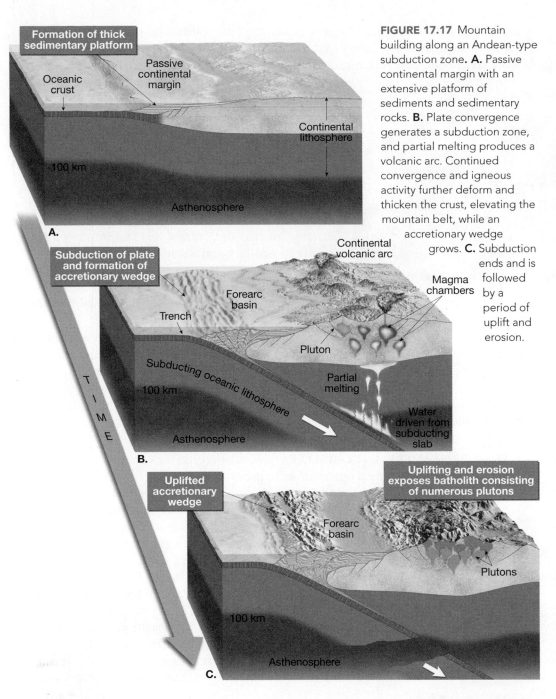

FIGURE 17.17 Mountain building along an Andean-type subduction zone. **A.** Passive continental margin with an extensive platform of sediments and sedimentary rocks. **B.** Plate convergence generates a subduction zone, and partial melting produces a volcanic arc. Continued convergence and igneous activity further deform and thicken the crust, elevating the mountain belt, while an accretionary wedge grows. **C.** Subduction ends and is followed by a period of uplift and erosion.

CONCEPT CHECK 17.5

❶ The formation of mountainous topography at a volcanic island arc is considered just one phase in the development of a major mountain belt. Explain.

❷ In what ways are the Sierra Nevada and the Andes similar?

❸ What is a passive margin? Give an example. Provide an example of an active continental margin.

❹ What is an accretionary wedge? Briefly describe its formation.

❺ In what way are the Coast Ranges of California related to subduction of oceanic lithosphere?

Collisional Mountain Belts

Convergent Boundaries

ESSENTIALS OF GEOLOGY **Continental Collisions**
Crustal Fragments and Mountain Building

Most major mountain belts are generated when one or more buoyant crustal fragments collide with a continental margin as a result of subduction. Oceanic lithosphere, which is relatively dense, readily subducts. Continental lithosphere, which contains significant amounts of low-density crustal rocks, is too buoyant to undergo subduction. Consequently, the arrival of a crustal fragment at a trench results in a collision with the margin of the adjacent continental block and an end to subduction.

Terranes and Mountain Building

The process of collision and accretion (joining together) of comparatively small crustal fragments to a continental margin has generated many of the mountainous regions that rim the Pacific. Geologists refer to these accreted crustal blocks as terranes. **Terrane** refers to any crustal fragment that has a geologic history distinct from that of the adjoining terranes.

THE NATURE OF TERRANES. What is the nature of these crustal fragments, and where did they originate? Research suggests that prior to their accretion to a continental block, some of these fragments

---

by the subduction of a portion of the Pacific Basin under the western edge of the North American plate. The Sierra Nevada batholith is a remnant of a portion of the continental volcanic arc that was produced by several surges of magma over tens of millions of years. Subsequent uplifting and erosion have removed most of the evidence of past volcanic activity and exposed a core of crystalline, igneous, and associated metamorphic rocks.

In the trench region, sediments scraped from the subducting plate, plus those provided by the eroding continental volcanic arc, were intensely folded and faulted into an accretionary wedge. This chaotic mixture of rocks presently constitutes the Franciscan Formation of California's Coast Ranges. Uplifting of the Coast Ranges took place only recently, as evidenced by the young, unconsolidated sediments that still mantle portions of these highlands.

In summary, *the growth of mountain belts at subduction zones is a response to crustal thickening caused by the addition of mantle-derived igneous rocks and sediments scraped from the descending oceanic slab.*

may have been **microcontinents** similar to the modern-day island of Madagascar, located east of Africa in the Indian Ocean.

Many others were island arcs similar to Japan, the Philippines, and the Aleutian Islands. Still others may have been submerged oceanic plateaus created by massive outpourings of basaltic lavas associated with mantle plumes (**FIGURE 17.18**). More than 100 of these relatively small crustal fragments are presently known to exist.

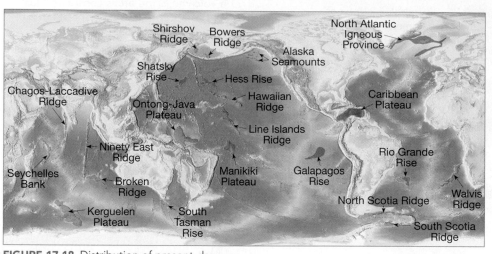

FIGURE 17.18 Distribution of present-day oceanic plateaus and other submerged crustal fragments. *(Data from Ben-Avraham and others)*

ACCRETION AND OROGENESIS. As oceanic plates move, they carry embedded oceanic plateaus, volcanic island arcs, and microcontinents to an Andean-type subduction zone. When an oceanic plate contains small seamounts, these structures are generally subducted along with the descending oceanic slab. However, large thick units of oceanic crust, such as the Ontong Java Plateau, which is the size of Alaska, or an island arc composed of abundant "light" igneous rocks, render the oceanic lithosphere too buoyant to subduct. In these situations, a collision between the crustal fragment and the continental margin occurs.

The sequence of events that happen when an island arc reaches an Andean-type margin is shown in **FIGURE 17.19**. Rather than subduct, the upper crustal layers of these thickened zones are peeled from the descending plate and thrust in relatively thin sheets upon the adjacent continental block. Because subduction often continues for 100 million years or longer, several crustal fragments can be transported to the continental margin. Each collision displaces earlier accreted terranes further inland, adding to the zone of deformation as well as to the thickness and lateral extent of the continental margin.

THE NORTH AMERICAN CORDILLERA. The relationship between mountain building and the accretion of crustal fragments arose primarily from studies conducted in the North American Cordillera (**FIGURE 17.20**). Researchers determined that some of the rocks in the orogenic belts of Alaska and British Columbia contained fossil and paleomagnetic evidence that indicated these strata previously lay much closer to the equator.

It is now known that many of the terranes that make up the North American Cordillera were scattered throughout the eastern Pacific, like the island arcs and oceanic plateaus currently distributed in the western Pacific. During the break-up of Pangaea, the eastern portion of the Pacific basin (Farallon plate) began to subduct under the western margin of North America. This activity resulted in the piecemeal addition of crustal fragments to the entire Pacific margin of the continent—from Mexico's Baja Peninsula to northern Alaska (Figure 17.20). Geologists expect that many modern microcontinents will likewise be accreted to active continental margins, producing new orogenic belts.

Continental Collisions

The Himalayas, Appalachians, Urals, and Alps represent mountain belts that were formed by the closure of major ocean basins. Continental collisions result in the development of mountains characterized by shortened and thickened crust achieved through folding and faulting. Some mountainous regions have crustal thicknesses that exceed 70 kilometers (40 miles).

Next, we will take a closer look at two examples of collision mountains—the Himalayas and the Appalachians. The Himalayas are the youngest collision mountains on Earth and are still rising. The Appalachians are a much older mountain belt, in which active mountain building ceased about 250 million years ago.

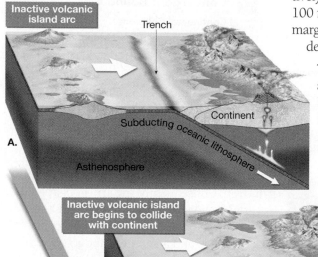

FIGURE 17.19 Sequence of events showing the collision and accretion of an island arc to a continental margin.

The Himalayas

The mountain-building episode that created the Himalayas began roughly 50 million years ago when India began to collide with Asia. Prior to the break-up of Pangaea, India was located between Africa and Antarctica in the Southern Hemisphere (see Figure 15.26, p. 385). As Pangaea fragmented, India moved rapidly, geologically speaking, a few thousand kilometers in a northward direction.

The subduction zone that facilitated India's northward migration was near the southern margin of Asia (**FIGURE 17.21A**). Continued subduction along Asia's margin created an Andean-type plate margin that contained a well-developed continental volcanic arc and an accretionary wedge. India's northern margin, on the other hand, was a passive continental margin consisting of a thick platform of shallow-water sediments and sedimentary rocks.

Geologists have determined that one or more small continental fragments were positioned on the subducting plate somewhere between India and Asia. During the closing of the intervening ocean basin, a small crustal fragment, which now forms southern Tibet, reached the trench. This event was followed by the docking of India itself. The tectonic forces involved in the collision of India with Asia were immense and caused the more deformable materials located on the seaward edges of these landmasses to become highly folded and faulted (see **FIGURE 17.21B**). The shortening and thickening of the crust elevated great quantities of crustal material, thereby generating the spectacular Himalayan mountains.

In addition to uplift, crustal shortening caused rocks at the "bottom of the pile" to become deeply buried—an environment where they experienced elevated temperatures and pressures (Figure 17.21B). Partial melting within the deepest and most deformed region of the developing mountain belt produced magmas that intruded the overlying rocks. It is in these environments that the metamorphic and igneous cores of collisional mountains are generated.

The formation of the Himalayas was followed by a period of uplift that raised the Tibetan Plateau. Seismic evidence suggests that a portion of the Indian subcontinent was thrust beneath Tibet a distance of perhaps 400 kilometers. If so, the added crustal thickness

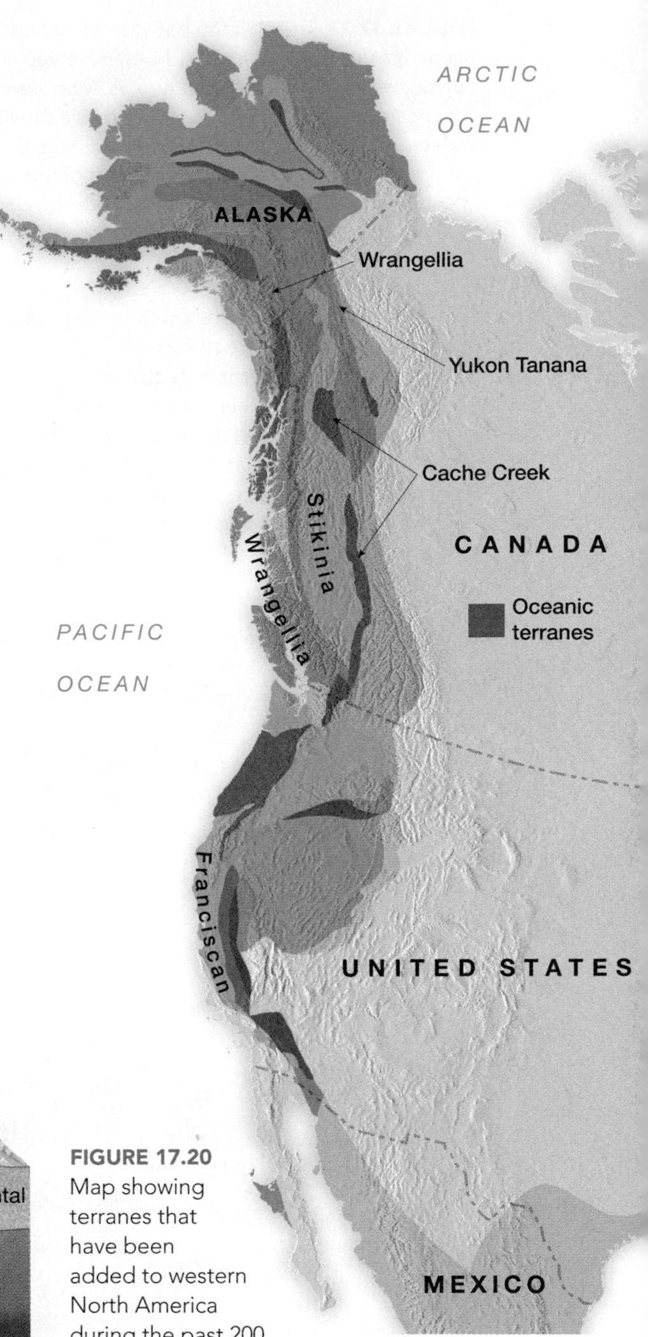

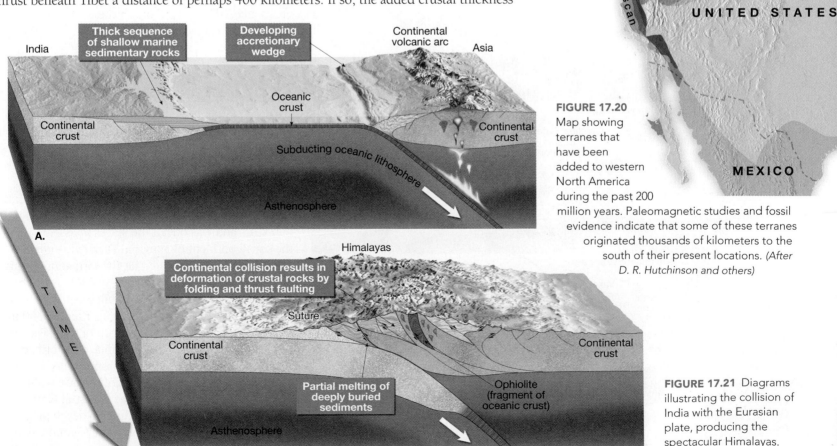

FIGURE 17.20 Map showing terranes that have been added to western North America during the past 200 million years. Paleomagnetic studies and fossil evidence indicate that some of these terranes originated thousands of kilometers to the south of their present locations. *(After D. R. Hutchinson and others)*

FIGURE 17.21 Diagrams illustrating the collision of India with the Eurasian plate, producing the spectacular Himalayas.

FIGURE 17.22 The collision between India and Asia that generated the Himalayas and Tibetan Plateau also severely deformed much of Southeast Asia. **A.** Map view of some of the major structural features of Southeast Asia thought to be related to this episode of mountain building. **B.** Re-creation of the deformation of Asia, with a rigid block representing India pushed into a mass of deformable modeling clay.

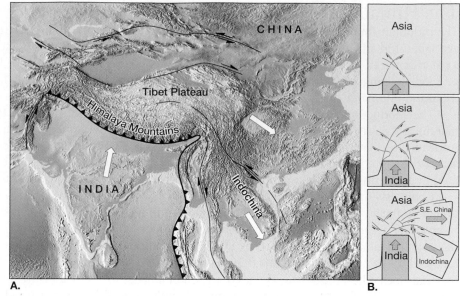

DID YOU KNOW?

The term *terrane* is used to designate a distinct and recognizable series of rock formations that have been transported by plate tectonic processes. Don't confuse this with the term *terrain*, which describes the shape of the surface topography or "lay of the land." In other words, you might observe a terrane as part of the terrain in the Appalachian Mountains.

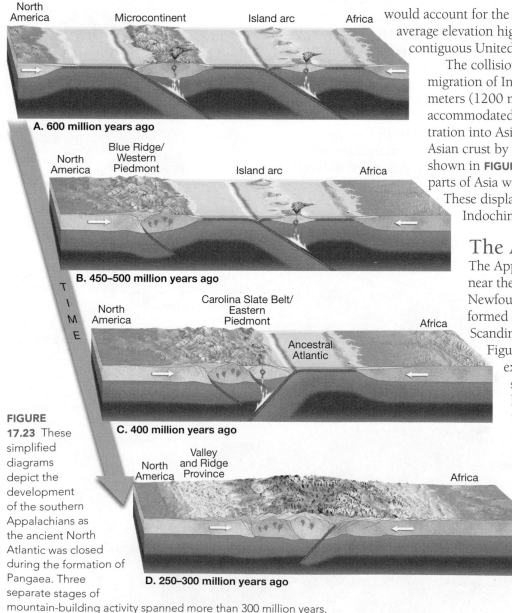

FIGURE 17.23 These simplified diagrams depict the development of the southern Appalachians as the ancient North Atlantic was closed during the formation of Pangaea. Three separate stages of mountain-building activity spanned more than 300 million years. (*After Zve Ben-Avraham, Jack Oliver, Larry Brown, and Frederick Cook*)

A. 600 million years ago

B. 450–500 million years ago

C. 400 million years ago

D. 250–300 million years ago

would account for the lofty landscape of southern Tibet, which has an average elevation higher than Mount Whitney, the highest point in the contiguous United States.

The collision with Asia slowed but did not stop the northward migration of India, which has since penetrated at least 2000 kilometers (1200 miles) into the mainland of Asia. Crustal shortening accommodated some of this motion. Much of the remaining penetration into Asia caused lateral displacement of large blocks of the Asian crust by a mechanism described as *continental escape*. As shown in **FIGURE 17.22**, when India continued its northward trek, parts of Asia were "squeezed" eastward out of the collision zone. These displaced crustal blocks include much of present-day Indochina and sections of mainland China.

The Appalachians

The Appalachian Mountains provide great scenic beauty near the eastern margin of North America from Alabama to Newfoundland. In addition, mountains of similar origin that formed during the same period are found in the British Isles, Scandinavia, north-western Africa, and Greenland (see Figure 15.6, p. 365). The orogeny that generated this extensive mountain system that presently lies on both sides of the North Atlantic lasted a few hundred million years and was one of the stages in assembling the supercontinent of Pangaea. Detailed studies of the Appalachians indicate that the formation of this mountain belt was complex and resulted from three distinct episodes of mountain building.

Our simplified overview begins roughly 750 million years ago with the break-up of a pre-Pangaea supercontinent called Rodinia, which rifted North America from Europe and Africa. This episode of continental rifting and seafloor spreading generated the ancestral North Atlantic. Located within this developing ocean basin was a fragment of continental crust that had been rifted from North America (**FIGURE 17.23A**).

FIGURE 17.24 The Valley and Ridge Province. This portion of the Appalachian Mountains consists of folded and faulted sedimentary strata that were displaced landward with the closing of the proto-Atlantic. *(LANDSAT image, courtesy of Phillips Petroleum Company, Exploration Projects Section)*

About 600 million years ago, plate motion dramatically changed, and the ancestral North Atlantic began to close. Two subduction zones likely formed—one located seaward of the coast of Africa that gave rise to a volcanic arc and another that developed along the margin of the continental fragment that lay off the coast of North America (Figure 17.23A).

Between 450 and 500 million years ago, the marginal sea between the crustal fragment and North America began to close. The collision that ensued, called the *Taconic Orogeny,* deformed the continental shelf and sutured the crustal fragment to the North American plate. The metamorphosed remnants of the continental fragment are recognized today as the crystalline rocks of the Blue Ridge and western Piedmont regions of the Appalachians (**FIGURE 17.23B**). In addition to the pervasive regional metamorphism, numerous magma bodies intruded the crustal rocks along the entire continental margin, particularly in what we know today as New England.

A second episode of mountain building, called the *Acadian Orogeny,* began about 400 million years ago. The continued closing of the ancestral North Atlantic resulted in the collision of the developing island arc with North America (**FIGURE 17.23C**). Evidence for this event is visible in the Carolina Slate Belt of the eastern Piedmont, which contains metamorphosed sedimentary and volcanic rocks characteristic of an island arc.

The final orogeny occurred between 250 and 300 million years ago, when Africa collided with North America. The result was landward displacement of the Blue Ridge and Piedmont provinces by as much as 250 kilometers (155 miles). This event also displaced and further deformed the shelf sediments and sedimentary rocks that had once flanked the eastern margin of North America (**FIGURE 17.23D**). Today these folded and thrust-faulted sandstones, limestones, and shales make up the largely unmetamorphosed rocks of the Valley and Ridge Province. Outcrops of the folded and thrust-faulted structures that characterize collision mountains are found as far inland as central Pennsylvania and western Virginia (**FIGURE 17.24**).

Following the collision of Africa and North America, the Appalachians lay in the interior of Pangaea. About 180 million years ago, this newly formed supercontinent began to break into smaller fragments, a process that ultimately created the modern Atlantic Ocean. Because this new zone of rifting occurred east of the suture that formed when Africa and North America collided, remnants of Africa remain "welded" to the North American plate (Figure 17.23D).

Other mountain ranges that exhibit evidence of continental collisions include the Alps and the Urals. The Alps formed as Africa and Europe collided during the closing of the Tethys Sea. Similarly, the Urals were uplifted during the assembly of Pangaea when northern Europe and northern Asia collided forming a major portion of Eurasia.

CONCEPT CHECK 17.6

❶ How is *terrane* different from *terrain?*

❷ In addition to microcontinents, what other structures are carried by the oceanic lithosphere and eventually accreted to a continent?

❸ How does the plate tectonics theory help explain the existence of fossil marine life in rocks atop compressional mountains?

❹ In your own words, briefly describe the stages in the formation of a major mountain belt according to the plate tectonics model.

Fault-Block Mountains

Most mountain belts form in compressional environments, as evidenced by the predominance of large thrust faults and folded strata. However, other tectonic processes, such as continental rifting, can also produce topographic mountains.

The mountains that form in these settings, termed **fault-block mountains**, are bounded by high-angle normal faults that gradually flatten with depth. Most fault-block mountains form in response to broad uplifting, which causes elongation and faulting.

The Teton Range in western Wyoming is an excellent example of fault-block mountains. This lofty structure was faulted and uplifted along its eastern flank as the block tilted downward to the west. Looking west from Jackson Hole, Wyoming, the eastern front of this mountain rises more than 2 kilometers above the valley making it one of the most imposing mountain fronts in the United States (**FIGURE 17.25**).

DID YOU KNOW?
During the assembly of Pangaea, the landmasses of Europe and Siberia collided to produce the Ural Mountains. Long before the discovery of plate tectonics, this extensively eroded mountain chain was regarded as the boundary between Europe and Asia.

Basin and Range Province

Located between the Sierra Nevada and Rocky Mountains is one of Earth's largest regions of fault-block mountains—the Basin and Range Province (see Figure 17.12). This region extends in a roughly north-south direction for nearly 3000 kilometers (2000 miles) and encompasses all of Nevada and portions of the surrounding states, as well as parts of southern Canada and western Mexico. In this region, the brittle upper crust has literally been broken into hundreds of fault blocks. Tilting of these faulted structures called *half-grabens*, gave rise to nearly parallel mountain ranges, averaging about 80 kilometers in length, which rise above adjacent sediment-filled basins (see Figure 17.12).

Extension in the Basin and Range Province began about 20 million years ago and appears to have "stretched" the crust as much as twice its original width. High heat flow in the region and several episodes of volcanism provide strong evidence that mantle upwelling caused doming of the crust, which in turn contributed to extension in the region.

CONCEPT CHECK 17.7

❶ Compare the processes that generate fault-block mountains to those associated with most other major mountain belts.

Vertical Movements of the Crust

Gradual up-and-down motions of the continental crust occur at many locations around the globe. Unfortunately, the reasons for these changes are not easily determined.

Isostasy

During the 1840s, researchers discovered that Earth's low-density crust floats on top of the high-density, deformable rocks of the mantle. The concept of a floating crust in gravitational balance is called **isostasy**. One way to explore this concept is to envision a series of wooden blocks of different heights floating in water, as shown in **FIGURE 17.26**. Note that the thicker wooden blocks float higher than the thinner blocks. Similarly, compressional mountains stand high above the surrounding terrain because crustal thickening creates buoyant crustal "roots" that extend deep into the supporting material below. Thus, lofty mountains behave much like the thicker wooden blocks shown in Figure 17.26.

ISOSTATIC ADJUSTMENT. Visualize what would happen if another small block of wood were placed atop one of the blocks in Figure 17.26. The combined block would sink until it reached a new isostatic (gravitational) balance. At this point, the top of the combined block would be higher than

before, and the bottom would be lower. This process of establishing a new level of gravitational equilibrium is called **isostatic adjustment**.

Applying the concept of isostatic adjustment, we should expect that when weight is added to the crust, it will respond by subsiding, and when weight is removed, it will rebound. (Visualize what happens when a ship's cargo is loaded or unloaded.) Evidence for crustal subsidence followed by crustal rebound is provided by Ice Age glaciers. When continental ice sheets occupied portions of North America during the Pleistocene epoch, the added weight of 3-kilometer-thick masses of ice caused downwarping of Earth's crust by hundreds of meters. In the 8000 years since this last ice sheet melted, uplifting of as much as 330 meters (1000 feet) has occurred in Canada's Hudson Bay region, where the thickest ice had accumulated.

One of the consequences of isostatic adjustment is that, as erosion lowers the summits of mountains, the crust rises in response to the reduced load (**FIGURE 17.27**). The processes of uplift and erosion continue until the mountain block reaches "normal" crustal thickness. When this occurs, these once-elevated structures will be near sea level, and the once-deeply buried interior of the mountain will be exposed at the surface.

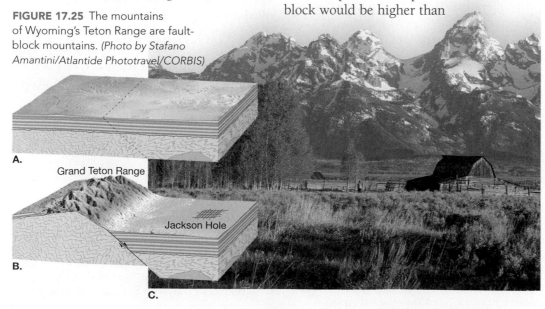

FIGURE 17.25 The mountains of Wyoming's Teton Range are fault-block mountains. *(Photo by Stafano Amantini/Atlantide Phototravel/CORBIS)*

A.

Grand Teton Range

B.

Jackson Hole

C.

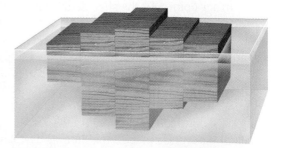

FIGURE 17.26 This drawing illustrates how wooden blocks of different thicknesses float in water. In a similar manner, thick sections of crustal material float higher than thinner crustal slabs.

Young mountain range

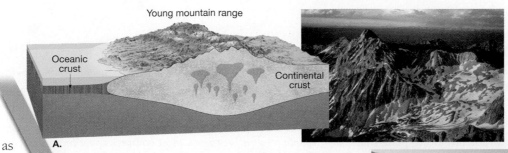

FIGURE 17.27 This sequence illustrates how the combined effects of erosion and isostatic adjustment result in a thinning of the crust in mountainous regions. **A.** When mountains are young, the continental crust is thickest. (*Photo by Michael Collier*) **B.** As erosion lowers the mountains, the crust rises in response to the reduced load. (*Photo by Mark Karrass/CORBIS*) **C.** Erosion and uplift continue until the mountains reach "normal" crustal thickness. (*Photo by Pat & Chuck Blackley/Alamy*)

In addition, as mountains are worn down, the eroded sediment is deposited on adjacent landscapes, causing these areas to subside (Figure 17.27).

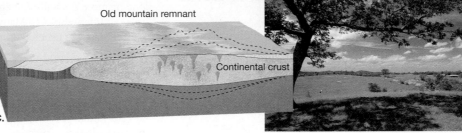

HOW HIGH IS TOO HIGH? Where compressional forces are great, such as those driving India into Asia, lofty mountains such as the Himalayas result. Is there a limit on how high a mountain can rise? As mountaintops are elevated, gravity-driven processes such as erosion and mass wasting are accelerated, carving the deformed strata into rugged landscapes. Equally important, however, is the fact that gravity also acts on the rocks within these massive structures. The higher the mountain, the greater the downward force on the rocks near the base. Eventually, the rocks deep within the developing mountain, which are comparatively warm and weak, begin to flow laterally, as shown in **FIGURE 17.28**. This is analogous to what happens when a ladle of very thick pancake batter is poured on a hot griddle. Similarly, mountains are altered by a process called **gravitational collapse**, which involves ductile spreading at depth and normal faulting and subsidence in the upper, brittle portion of the crust.

Considering these factors, a seemingly logical question follows—"What keeps the Himalayas standing?" Simply, the horizontal compressional forces that are driving India into Asia are greater than the vertical force of gravity. However, once India's northward trek ends, the downward pull of gravity, as well as weathering and erosion, will become the dominant forces acting on this mountainous region.

CONCEPT CHECK 17.8

① Define *isostasy*.

② Give one example of evidence that supports the concept of crustal uplift.

③ What happens to a floating object when weight is added? Subtracted?

④ Briefly describe how the principle of isostatic adjustment applies to changes in the elevations of mountains.

DID YOU KNOW?

Since the first successful ascent to the summit of Mount Everest in 1953, thousands have set out to match this feat. Unfortunately, more than 100 people have perished in their effort. In addition, debris has become a major problem because waste, such as empty oxygen bottles, does not disintegrate at high altitudes and is impossible to bury.

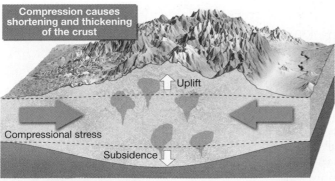

A. Horizontal compressional forces dominate

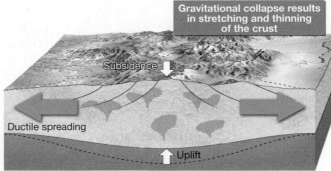

B. Gravitational forces dominate

FIGURE 17.28 Block diagram of a mountain belt that is collapsing under its own "weight." Gravitational collapse involves normal faulting in the upper, brittle portion of the crust and ductile spreading in the warm, weak rocks at depth.

Crustal Deformation and Mountain Building in Review

⊙ *Deformation* refers to changes in the shape, size (volume) or orientation of a rock body and is most pronounced along plate margins. Rocks deform differently depending on the environment (temperature and confining pressure); the composition and texture of the rock; and the length of time stress is maintained. Rocks first respond by deforming *elastically* and will return to their original shape when the stress is removed. Once their elastic limit (strength) is surpassed, rocks either deform by ductile flow or they fracture. *Ductile deformation* is a solid-state flow that changes the shape of an object without fracturing. Ductile flow may be accomplished by gradual slippage and recrystallization along planes of weakness within the crystal lattice of mineral grains. Ductile deformation tends to occur in a high-temperature/high-pressure environment. In a near-surface environment, most rocks deform by *brittle failure*.

⊙ Among the most basic geologic structures associated with rock deformation are *folds* (flat-lying sedimentary and volcanic rocks bent into a series of wavelike undulations). The two most common types of folds are *anticlines,* formed by the upfolding, or arching, of rock layers, and *synclines,* which are downfolds. Most folds are the result of horizontal *compressional stresses.* Folds can be *symmetrical, asymmetrical,* or, if one limb has been tilted beyond the vertical, *overturned. Domes* (upwarped structures) and *basins* (downwarped structures) are circular or somewhat elongated folds formed by vertical displacements of strata.

⊙ *Faults* are fractures in the crust along which appreciable displacement has occurred. Faults in which the movement is primarily vertical are called *dip-slip faults.* Dip-slip faults include both *normal* and *reverse faults.* Low-angle reverse faults are called *thrust faults.* Normal faults indicate *tensional stresses* that pull the crust apart. Reverse and thrust faulting indicate that *compressional forces* are at work. Large *thrust faults* are associated with subduction zones and other convergent boundaries where plates collide. *Strike-slip faults* exhibit mainly horizontal displacement parallel to the strike of the fault surface. Large strike-slip faults, called *transform faults,* accommodate displacement between plate boundaries.

⊙ *Joints* are fractures along which no appreciable displacement has occurred. Joints generally occur in groups with roughly parallel orientations and are the result of brittle failure of rock units located in the outermost crust.

⊙ The name for the processes that collectively produce a *compressional mountain belt* is *orogenesis.* Most compressional mountains consist of folded and faulted sedimentary and volcanic rocks, portions of which have been strongly metamorphosed and intruded by younger igneous bodies.

⊙ Subduction of oceanic lithosphere beneath a continental block gives rise to an *Andean-type plate margin* that is characterized by a continental volcanic arc and associated igneous plutons. In addition, sediment derived from the land, as well as material scraped from the subducting plate, becomes plastered against the landward side of the trench, forming an *accretionary wedge.*

⊙ Mountain belts can develop as a result of the collision and merger of one or more crustal fragments, including island arcs and oceanic plateaus to a continental block. Many of the mountain belts of the North American Cordillera were generated in this manner.

⊙ Continued subduction of oceanic lithosphere beneath an Andean-type continental margin will eventually close an ocean basin. The result will be a *continental collision* and the development of compressional mountains that are characterized by shortened and thickened crust. The development of a major mountain belt is often complex, involving two or more distinct episodes of mountain building. Continental collisions have generated most of Earth's major mountain belts, including the Himalayas, Alps, Urals, and Appalachians.

⊙ Although most mountains form along convergent plate boundaries, other tectonic processes, such as continental rifting, can produce topographic mountains. The *fault-block mountains* that form in these settings are bounded by high-angle normal faults. The Basin and Range Province in the western United States consists of hundreds of tilted, faulted blocks that give rise to nearly parallel mountain ranges that stand above sediment-laden basins.

⊙ Earth's crust floats on top of the deformable rocks of the mantle, much like wooden blocks float in water. The concept of a floating crust in gravitational balance is called *isostasy.* Most mountains consist of crustal rocks that have been shortened and thickened producing deep crustal roots that isostatically support them. As erosion lowers the peaks, *isostatic adjustment* gradually raises the mountains in response. The processes of uplift and erosion will continue until the mountain block reaches "normal" crustal thickness. Gravity can also cause these elevated structures to collapse under their own "weight."

Key Terms

accretionary wedge (p. 426)
active continental
   margins (p. 426)
Andean-type plate
   margins (p. 426)
anticlines (p. 418)
basins (p. 420)
brittle deformation (p. 416)
compressional
   mountains (p. 425)
deformation (p. 416)

dip-slip faults (p. 422)
dome (p. 420)
ductile deformation (p. 416)
elastic deformation (p. 416)
faults (p. 421)
fault-block mountains (p. 431)
fault scarps (p. 422)
folds (p. 418)
grabens (p. 422)
gravitational collapse (p. 433)
half-grabens (p. 422)

hogbacks (p. 420)
horsts (p. 422)
isostasy (p. 432)
isostatic adjustment (p. 432)
joints (p. 424)
microcontinents (p. 428)
monoclines (p. 420)
normal faults (p. 422)
orogenesis (p. 425)
passive continental
   margin (p. 426)

reverse fault (p. 423)
stress (p. 416)
strike-slip fault (p. 423)
synclines (p. 418)
terrane (p. 427)
thrust faults (p. 423)
transform fault (p. 423)
volcanic island arc (p. 426)

GIVE IT SOME THOUGHT

❶ Describe an environment within Earth in which you might expect rocks to experience ductile deformation. Suggest a scenario in which brittle rather than ductile deformation might occur.

❷ Is *granite* or *mica schist* more likely to fold or flow rather than fracture when subjected to differential stress? Explain.

❸ Refer to the accompanying diagrams to answer the following:

   a. What type of dip-slip fault is shown in Diagram 1? Were the dominant forces during faulting tensional, compressional, or shear?

   b. What type of dip-slip fault is shown in Diagram 2? Were the dominant forces during faulting tensional, compressional, or shear?

   c. Match the correct pair of arrows in Diagram 3 to the faults in Diagrams 1 and 2.

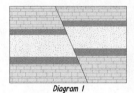 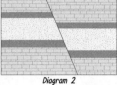

Diagram 1   Diagram 2   Diagram 3

❹ Refer to Figure 17.18 showing the distribution of present-day oceanic plateaus and other submerged crustal fragments. Is the *Galapagos Rise* or the *Rio Grande Rise* more likely to end up accreted to a continent? Explain your choice. In the future, how might a geologist determine that this accreted terrane is distinct from the continental crust to which it accreted?

❺ Suppose a sliver of oceanic crust was discovered in the interior of a continent. Would this support or refute the theory of plate tectonics? Explain.

❻ The Ural Mountains exhibit a north–south orientation through Eurasia (see Figure 1.27 pag. 27). How does the theory of plate tectonics explain the existence of this mountain belt in the interior of an expansive landmass?

❼ Explain why the Appalachian Mountains, which are similar in age to the Urals, are found along the margin of North America and not in the interior.

❽ Briefly describe the major differences between the evolution of the Appalachian Mountains and the North American Cordillera.

❾ What processes (besides formation and melting of large ice sheets) could cause isostatic adjustments?

Companion Website   mygeoscience place

www.mygeoscienceplace.com

The *Essentials of Geology, 11e* companion Website contains numerous multimedia resources accompanied by assessments to aid in your study of the topics in this chapter. The use of this site's learning tools will help improve your understanding of geology. Utilizing the access code that accompanies this text, visit **www.mygeoscienceplace.com** in order to:

- **Review** key chapter concepts.
- **Read** with links to the Pearson eText and to chapter-specific web resources.
- **Visualize** and comprehend challenging topics using the learning activities in *GEODe: Essentials of Geology* and the *Geoscience Animations Library*.
- **Test** yourself with online quizzes.

Geologic Time

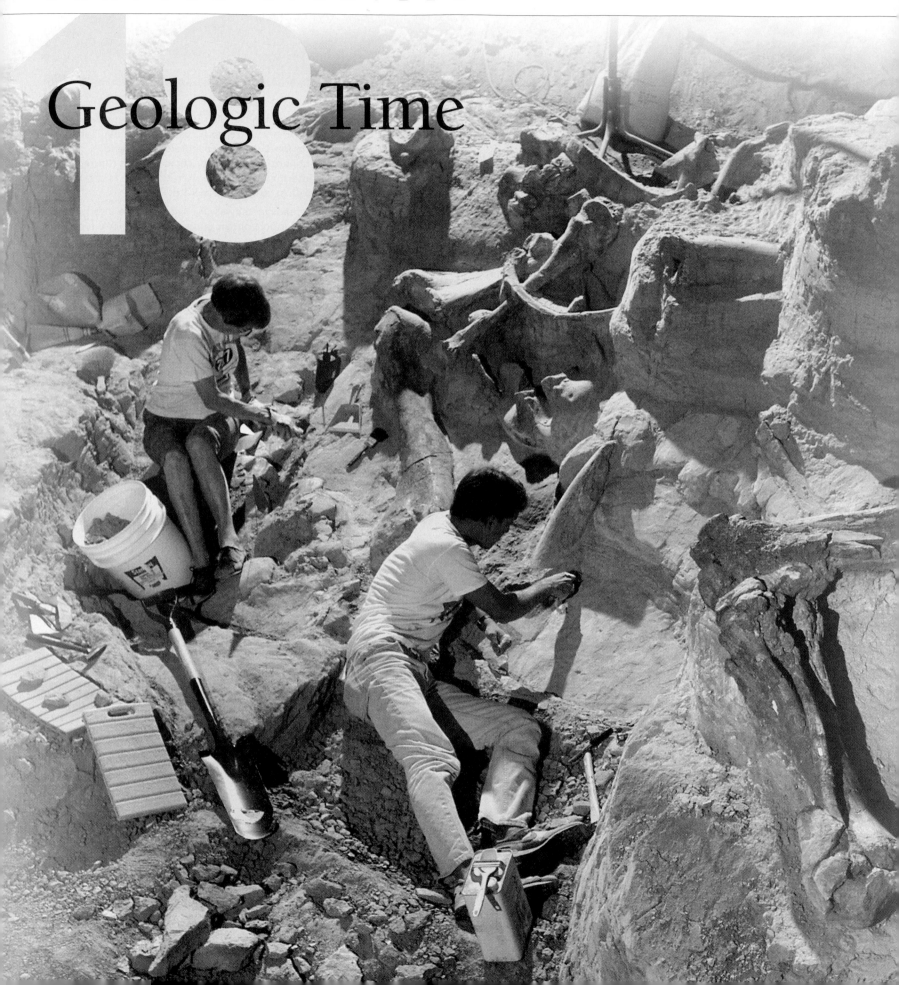

GEOLOGICAL EVENTS BY THEMSELVES HAVE LITTLE MEANING until they are put into a time perspective. Studying history, whether it be the Civil War or the Age of Dinosaurs, requires a calendar. Among geology's major contributions to human knowledge is the *geologic time scale* and the discovery that Earth history is exceedingly long.

To assist you in learning the important concepts in this chapter, focus on the following questions:

⊙ What are the two types of dates used by geologists to interpret Earth history?

⊙ What are the laws, principles, and techniques used to establish relative dates?

⊙ What are fossils? What are the conditions that favor the preservation of organisms as fossils?

⊙ How are fossils used to correlate rocks of similar ages that are in different places?

⊙ What is radioactivity, and how are radioactive isotopes used in radiometric dating?

⊙ What is the geologic time scale, and what are its principal subdivisions?

⊙ Why is it difficult to assign reliable numerical dates to samples of sedimentary rock?

Paleontologists excavating the remains of an Ice Age mammoth, a prehistoric relative of modern elephants, near Hot Springs in the Black Hills of South Dakota.
(Photo by Phil Degginger/Alamy)

Geology Needs a Time Scale

In the late 18th century, James Hutton recognized the immensity of Earth history and the importance of time as a component in all geological processes. In the 19th century, others effectively demonstrated that Earth had experienced many episodes of mountain building and erosion, which must have required great spans of geologic time. Although these pioneering scientists understood that Earth was very old, they had no way of knowing its true age. Was it tens of millions, hundreds of millions, or even billions of years old? Rather, a geologic time scale was developed that showed the sequence of events based on relative dating principles. What are these principles? What part do fossils play? With the discovery of radioactivity and radiometric dating techniques, geologists can now assign fairly accurate dates to many of the events in Earth history. What is radioactivity? Why is it a good "clock" for dating the geologic past?

In 1869, John Wesley Powell, who was later to head the U.S. Geological Survey, led a pioneering expedition down the Colorado River and through the Grand Canyon (**FIGURE 18.1**). Writing about the rock layers that were exposed by the downcutting of the river, Powell said, "the canyons of this region would be a Book of Revelations in the rock-leaved Bible of geology." He was undoubtedly impressed with the millions of years of Earth history exposed along the walls of the Grand Canyon and the other canyons of the Colorado Plateau (**FIGURE 18.2**).

Powell realized that the evidence for an ancient Earth is concealed in its rocks. Like the pages in a long and complicated history book, rocks record the geological events and changing life forms of the past. The book, however, is not complete. Many pages, especially in the early chapters, are missing. Others are tattered, torn, or smudged. Yet enough of the book remains to allow much of the story to be deciphered.

Interpreting Earth history is a prime goal of the science of geology. Like a modern-day sleuth, the geologist must interpret clues found preserved in the rocks. By studying rocks, especially sedimentary rocks, and the features they contain, geologists can unravel the complexities of the past.

CONCEPT CHECK 18.1

❶ Why is a geologic time scale a useful tool?

Relative Dating—Key Principles

 Geologic Time

ESSENTIALS OF GEOLOGY **Relative Dating—Key Principles**

The geologists who developed the geologic time scale revolutionized the way people think about time and perceive our planet. They learned that Earth is much older than anyone had previously imagined and that its surface and interior

DID YOU KNOW?

Early attempts at determining Earth's age proved to be unreliable. One method reasoned that if the rate at which sediment accumulates could be determined as well as the total thickness of sedimentary rock that had been deposited during Earth history, an estimate of Earth's age could be made. All that was necessary was to divide the rate of sediment accumulation into the total thickness of sedimentary rock. This method was riddled with difficulties. Can you think of some?

FIGURE 18.1 The start of the Powell expedition from Green River Station, Wyoming, is depicted in this drawing from Powell's 1875 book. The inset photo is of Major John Wesley Powell, pioneering geologist and the second director of the U.S. Geological Survey. (*Courtesy of the U.S. Geological Survey, Denver*)

have been changed over and over again by the same geological processes that operate today.

During the late 1800s and early 1900s, various attempts were made to determine Earth's age. Although some of the methods appeared promising at the time, none proved to be reliable. What these scientists were seeking was a **numerical date.** Such dates specify the actual number of years that have passed since an event occurred. Today our understanding of radioactivity allows us to accurately determine numerical dates for rocks that represent important events in Earth's distant past. We will study radioactivity later in this chapter. Prior to the discovery of radioactivity, geologists had no accurate and dependable method of numerical dating and had to rely solely on relative dating.

FIGURE 18.2 The Colorado River winding through Canyonlands National Park near Moab, Utah. The strata exposed in the canyon walls contain clues to millions of years of Earth history. *(Photo by John T. Parkinson/Mira.com)*

FIGURE 18.3 Applying the law of superposition to these layers exposed in the upper portion of the Grand Canyon, the Supai Group is oldest, and the Kaibab Limestone is youngest. *(Photo by E. J. Tarbuck)*

Geologist's Sketch

- *Kaibab Limestone—shallow marine limestone that rims much of the canyon*
- *Toroweap Formation—shallow marine, thin-to-medium bedded sandy limestone*
- *Coconino Sandstone—cliff-forming cross-bedded sandstone*
- *Hermit Shale—red, slope-forming thinly-bedded shales and siltstones*
- *Supai Group—alternating layers of sandstone, siltstone and shale*

Relative dating means placing rocks in their proper *sequence of formation*, first, second, third, and so on. Relative dating cannot tell us how long ago something took place, only that it followed one event and preceded another. The relative dating techniques that were developed are valuable and still widely used. Numerical dating methods did not replace these techniques; they simply supplemented them. To establish a relative time scale, a few basic principles or rules had to be discovered and applied. Although they may seem obvious to us today, they were major breakthroughs in thinking at the time, and their discovery and acceptance was an important scientific achievement.

Law of Superposition

Nicolaus Steno, a Danish anatomist, geologist, and priest (1636–1686), is credited with being the first to recognize a sequence of historical events in an outcrop of sedimentary rock layers. Working in the mountains of western Italy, Steno applied a very simple rule that has come to be the most basic principle of relative dating—the **law of superposition.** The law simply states that in an underformed sequence of sedimentary rocks, each bed is older than the one above it and younger than the one below. Although it may seem obvious that a rock layer could not be deposited unless it had something older beneath it for support, it was not until 1669 that Steno clearly stated the principle.

This rule also applies to other surface-deposited materials, such as lava flows and beds of ash from volcanic eruptions. Applying the law of superposition to the beds exposed in the upper portion of the Grand Canyon (**FIGURE 18.3**), you can easily place the layers in their proper order. Among those that are shown, the sedimentary rocks in the Supai Group must be the

oldest, followed in order by the Hermit Shale, Coconino Sandstone, Toroweap Formation, and Kaibab Limestone.

Principle of Original Horizontality

Steno is also credited with recognizing the importance of another basic principle, called the **principle of original horizontality.** Simply stated, it means that layers of sediment are generally deposited in a

horizontal position. Thus, if we observe rock layers that are flat, it means they have not been disturbed and they still have their *original* horizontality. The layers in Canyonlands National Park (Figure 18.2) and in the Grand Canyon (Figure 18.3) illustrate this. However, if they are folded or inclined at a steep angle, they must have been moved into that position by crustal disturbances sometime *after* their deposition (**FIGURE 18.4**).

FIGURE 18.4 Most layers of sediment are deposited in a nearly horizontal position. Thus, when we see rock layers that are folded or tilted we can assume that they were moved into that position by crustal disturbances after their deposition. These folds are at Agio Pavlos on the Mediterranean island of Crete. *(Photo by Marco Simoni/Robert Harding)*

Principle of Cross-Cutting Relationships

When a fault cuts through other rocks, or when magma intrudes and crystallizes, we can assume that the fault or intrusion is younger than the rocks affected. For example, in **FIGURE 18.5**, the faults and dikes clearly must have occurred *after* the sedimentary layers were deposited.

This is the **principle of cross-cutting relationships.** By applying the cross-cutting principle, you can see that fault *A* occurred *after* the sandstone layer was deposited, because it "broke" the layer. However, fault *A* occurred

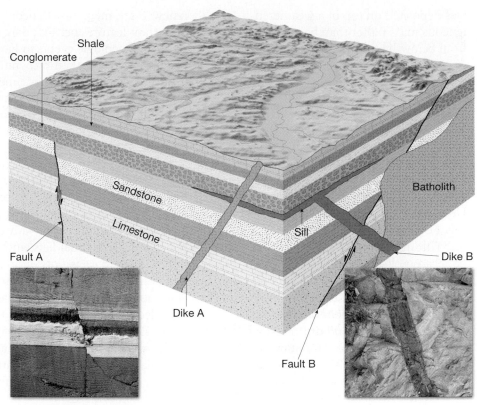

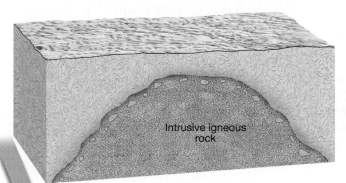

A. Intrusive igneous rock

FIGURE 18.5 Cross-cutting relationships are an important principle used in relative dating. An intrusive rock body is younger than the rocks it intrudes. A fault is younger than the rock layers it cuts.

FIGURE 18.6
These diagrams illustrate two ways that inclusions can form, as well as a type of unconformity termed a *nonconformity*. In part **A.**, the inclusions in the igneous mass represent unmelted remnants of the surrounding host rock that were broken off and incorporated at the time the magma was intruded. In part **C.**, the

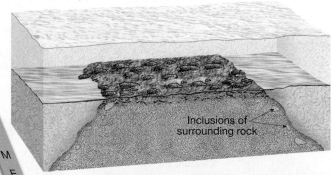

C. Deposition of sedimentary layers

igneous rock must be older than the overlying sedimentary beds because the sedimentary beds contain inclusions of the igneous rock. When older intrusive igneous rocks are overlain by younger sedimentary layers, a nonconformity is said to exist. The small inset in part **C.** shows an inclusion of dark igneous rock in a lighter-colored and younger host rock.

before the conglomerate was laid down, because that layer is unbroken.

We can also state that dike *B* and its associated sill are older than dike *A*, because dike *A* cuts the sill. In the same manner, we know that the batholith was emplaced after movement occurred along fault *B*, but before dike *B* was formed. This is true because the batholith cuts across fault *B* and dike *B* cuts across the batholith.

Inclusions

Sometimes inclusions can aid the relative dating process. **Inclusions** are pieces of one rock unit that are contained within another. The basic principle is logical and straightforward. The rock mass adjacent to the one containing the inclusions must have been there first in order to provide the rock fragments. Therefore, the rock mass containing inclusions is the younger of the two. **FIGURE 18.6** provides an example. Here the inclusions of intrusive igneous rock in the adjacent sedimentary layer indicate that the sedimentary layer was deposited on top of a weathered igneous mass rather than being intruded from below by magma that later crystallized.

Unconformities

When we observe layers of rock that have been deposited essentially without interruption, we call them **conformable.** Particular sites exhibit conformable beds representing certain spans of geologic time. However, no place on Earth has a complete set of conformable strata.

Throughout Earth history, the deposition of sediment has been interrupted again and again. All such breaks in the rock record are termed *unconformities*. An **unconformity** represents a long period during which deposition ceased, erosion removed previously formed rocks, and then deposition resumed. In each case, uplift and erosion are followed by subsidence and

renewed sedimentation. Unconformities are important features because they represent significant geologic events in Earth history. Moreover, their recognition helps us identify what intervals of time are not represented by strata and thus are missing from the geologic record.

The rocks exposed in the Grand Canyon of the Colorado River represent a tremendous span of geologic history. It is a wonderful place to take a trip through time. The canyon's colorful strata record a long history of sedimentation in a variety of environments—advancing seas, rivers and deltas, tidal flats, and sand dunes. The record is not continuous, however. Unconformities represent vast amounts of time that have not been recorded in the canyon's layers. **FIGURE 18.7** is a geologic cross-section of the Grand Canyon. Refer to it as you read about the three basic types of unconformities: angular unconformities, disconformities, and nonconformities.

ANGULAR UNCONFORMITY. Perhaps the most easily recognized unconformity is an **angular unconformity**. It consists of tilted or folded sedimentary rocks that are overlain by younger, more flat-lying strata. An angular unconformity indicates that during the pause in deposition, a period of deformation (folding or tilting) and erosion occurred (**FIGURE 18.8**).

When James Hutton studied an angular unconformity in Scotland more than 200 years ago, it was clear to him that it represented a major episode of geologic activity (**FIGURE 18.9**).* He also appreciated the immense time span implied by such relationships. When a companion later wrote of their visit to the site, he stated that "the mind seemed to grow giddy by looking so far into the abyss of time."

*This pioneering geologist is discussed in the section on "The Birth of Modern Geology" in Chapter 1.

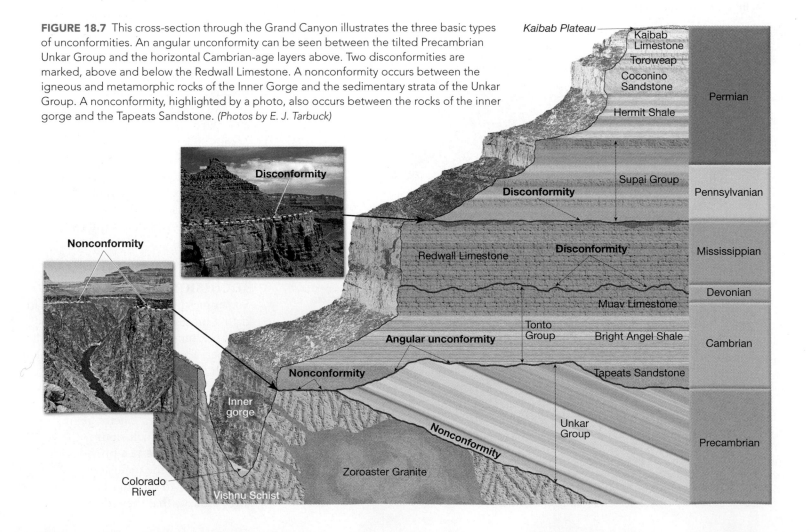

FIGURE 18.7 This cross-section through the Grand Canyon illustrates the three basic types of unconformities. An angular unconformity can be seen between the tilted Precambrian Unkar Group and the horizontal Cambrian-age layers above. Two disconformities are marked, above and below the Redwall Limestone. A nonconformity occurs between the igneous and metamorphic rocks of the Inner Gorge and the sedimentary strata of the Unkar Group. A nonconformity, highlighted by a photo, also occurs between the rocks of the inner gorge and the Tapeats Sandstone. *(Photos by E. J. Tarbuck)*

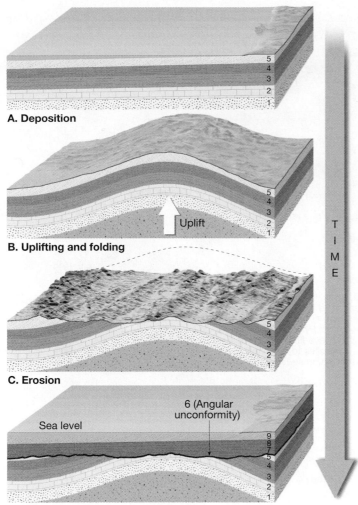

A. Deposition

Uplift

B. Uplifting and folding

TIME

C. Erosion

6 (Angular unconformity)

Sea level

D. Subsidence and renewed deposition

FIGURE 18.8 Formation of an angular unconformity. An angular unconformity represents an extended period during which deformation and erosion occurred.

Above the unconformity lie gently dipping beds of reddish sandstone and conglomerate

Angular unconformity

Rock hammer

Below the unconformity lie nearly vertical sandstones and shales

Geologist's Sketch

FIGURE 18.9 This angular unconformity at Siccar Point, Scotland, was first described by James Hutton in the late 18th century. *(Photo by Marli Miller)*

DISCONFORMITY. When contrasted with angular unconformities, **disconformities** are more common but usually far less conspicuous because the strata on either side are essentially parallel. For example, look at the disconformities in the cross-section of the Grand Canyon in Figure 18.7. Many disconformities are difficult to identify because the rocks above and below are similar and there is little evidence of erosion. Such a break often resembles an ordinary bedding plane. Other disconformities are easier to identify because the ancient erosion surface is cut deeply into the older rocks below.

NONCONFORMITY. The third basic type of unconformity is a **nonconformity.** Here the break separates older metamorphic or intrusive igneous rocks from younger sedimentary strata (Figures 18.6 and 18.7). Just as angular unconformities and disconformities imply crustal movements, so too do nonconformities. Intrusive igneous masses and metamorphic rocks originate far below the surface. Thus, for a nonconformity to develop, there must be a period of uplift and the erosion of overlying rocks. Once exposed at the surface, the igneous or metamorphic rocks are subjected to

weathering and erosion prior to subsidence and the renewal of sedimentation.

Using Relative Dating Principles

If you apply the principles of relative dating to the hypothetical geologic cross-section in **FIGURE 18.10**, you can place in proper sequence the rocks and the events they represent. The statements within the figure summarize the logic used to interpret the cross-section.

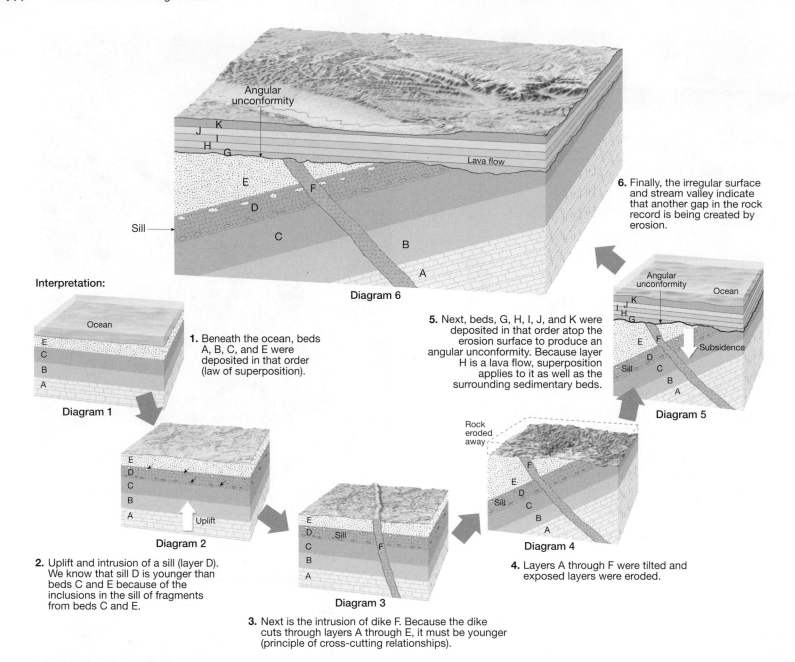

FIGURE 18.10 Interpreting a geologic cross-section of a hypothetical region.

In the diagram:

Diagram 6

6. Finally, the irregular surface and stream valley indicate that another gap in the rock record is being created by erosion.

Interpretation:

Diagram 1

1. Beneath the ocean, beds A, B, C, and E were deposited in that order (law of superposition).

Diagram 2

2. Uplift and intrusion of a sill (layer D). We know that sill D is younger than beds C and E because of the inclusions in the sill of fragments from beds C and E.

Diagram 3

3. Next is the intrusion of dike F. Because the dike cuts through layers A through E, it must be younger (principle of cross-cutting relationships).

Diagram 4

4. Layers A through F were tilted and exposed layers were eroded.

Diagram 5

5. Next, beds, G, H, I, J, and K were deposited in that order atop the erosion surface to produce an angular unconformity. Because layer H is a lava flow, superposition applies to it as well as the surrounding sedimentary beds.

In this example, we establish a relative time scale for the rocks and events in the area of the cross-section. Remember that this method gives us no idea of how many years of Earth history are represented, for we have no numerical dates. Nor do we know how this area compares to any other.

CONCEPT CHECK 18.2

❶ Distinguish between numerical dates and relative dates.

❷ Sketch and label three simple diagrams that illustrate each of the following: superposition, original horizontality, and cross-cutting relationships

❸ What is the significance of an unconformity?

❹ Distinguish among angular unconformity, disconformity, and nonconformity.

Correlation of Rock Layers

To develop a geologic time scale that is applicable to the entire Earth, rocks of similar age in different regions must be matched up. Such a task is referred to as **correlation.**

Within a limited area, correlating the rocks of one locality with those of another may be done simply by walking along the outcropping edges. However, this may not be possible when the rocks are mostly concealed by soil and vegetation. Correla-

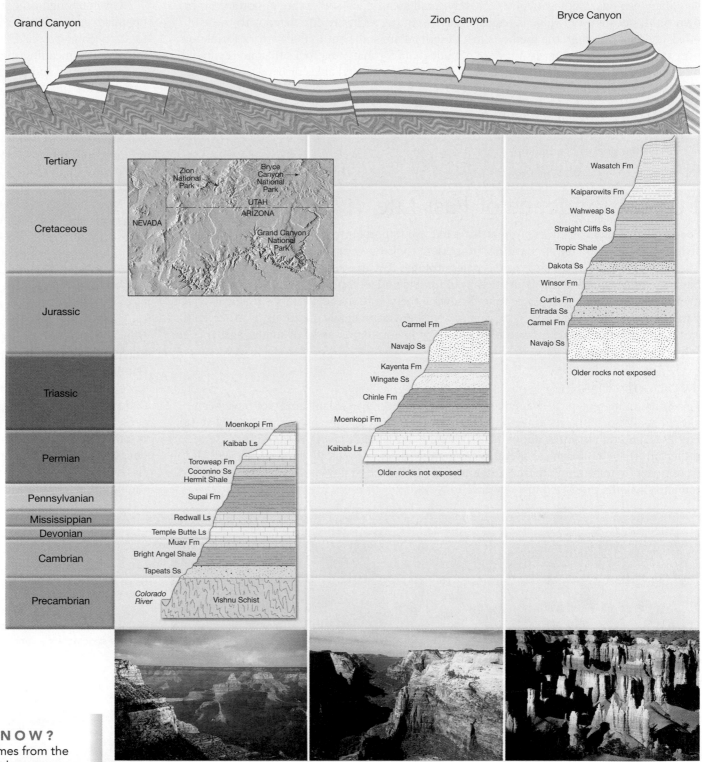

FIGURE 18.11 Correlation of strata at three locations on the Colorado Plateau provides a more complete picture of sedimentary rocks in the region. The diagram at the top is a geologic cross-section of the region. *(After U.S. Geological Survey; Photos by E. J. Tarbuck)*

Grand Canyon National Park   **Zion National Park**   **Bryce Canyon National Park**

DID YOU KNOW?
The word *fossil* comes from the Latin *fossilium*, which means "dug up from the ground." As originally used by medieval writers, a fossil was *any* stone, ore, or gem that came from an underground source. In fact, many early books on mineralogy are called books of fossils. The current meaning of fossil came about during the 1700s.

tion over short distances is often achieved by noting the position of a distinctive rock layer in a sequence of strata. Alternatively, a layer may be identified in another location if it is composed of very distinctive or uncommon minerals.

By correlating the rocks from one place to another, a more comprehensive view of the geologic history of a region is possible. **FIGURE 18.11**, for example, shows the correlation of strata at three sites on the Colorado Plateau in southern Utah and northern Arizona. No single locale exhibits the entire sequence, but correlation reveals a more complete picture of the sedimentary rock record.

Many geologic studies involve relatively small areas. Such studies are important in their own right, but their full value is realized only when the rocks are correlated with those of other regions. Although the methods just described are sufficient to trace a rock formation over relatively short distances, they are not adequate for matching rocks that are separated by great distances. When correlation between widely separated areas or between continents is the objective, geologists must rely on fossils.

CONCEPT CHECK 18.3

❶ What is the goal of correlation?

Fossils: Evidence of Past Life

Fossils, the remains or traces of prehistoric life, are important inclusions in sediment and sedimentary rocks. They are basic and important tools for interpreting the geologic past. The scientific study of fossils is called **paleontology.** It is an interdisciplinary science that blends geology and biology in an attempt to understand all aspects of the succession of life over the vast expanse of geologic time. Knowing the nature of the life-forms that existed at a particular time helps researchers understand past environmental conditions. Further, fossils are important time indicators and play a key role in correlating rocks of similar ages that are from different places.

Types of Fossils

Fossils are of many types. The remains of relatively recent organisms may not have been altered at all. Such objects as teeth, bones, and shells are common examples (**FIGURE 18.12**). Far less common are entire animals, flesh included, that have been preserved because of rather unusual circumstances. Remains of prehistoric elephants called *mammoths*, which were frozen in the Arctic tundra of Siberia and Alaska, are examples, as are the mummified remains of sloths preserved in a dry cave in Nevada.

Given enough time, the remains of an organism are likely to be modified. Often fossils become *petrified* (literally, "turned into stone"), meaning that the small internal cavities and pores of the original structure are filled with precipitated mineral matter (**FIGURE 18.13A**). In other instances *replacement* may occur. Here the cell walls and other solid material are removed and replaced with mineral matter. Sometimes the microscopic details of the replaced structure are faithfully retained.

Molds and casts constitute another common class of fossils. When a shell or other structure is buried in sediment and then dissolved by underground water, a *mold* is created. The mold faithfully reflects only the shape and surface marking of the organism; it does not reveal any information concerning its internal structure. If these hollow spaces are subsequently filled with mineral matter, *casts* are created (**FIGURE 18.13B**).

A type of fossilization called *carbonization* is particularly effective in preserving leaves and delicate animal forms. It occurs when fine sediment encases the remains of an organism. As time passes, pressure squeezes out the liquid and gaseous components

FIGURE 18.12 A. Excavating bones from Pit 91 at the La Brea tar pits in Los Angeles. It is a site rich in the unaltered remains of Ice Age organisms. Scientists have been digging here since 1915. *(AP/Wide World Photo)* **B**. Fossils of many relatively recent organisms are unaltered remains. The skeleton of this mammoth from the La Brea tar pits is a spectacular example. *(Alamy Images)*

A.

B.

FIGURE 18.13 There are many types of fossilization. Six examples are shown here. **A.** Petrified wood in Petrified Forest National Park, Arizona. **B.** This trilobite photo illustrates mold and cast. **C.** A fossil bee preserved as a thin carbon film. **D.** Impressions are common fossils and often show considerable detail. **E.** Insect in amber. **F.** Coprolite is fossil dung. *(Photo A by Bernhard Edmaier/Photo Researchers, Inc.; Photos B, D and F by E. J. Tarbuck; Photo C courtesy of Florissant Fossil Beds National Monument; Photo E by Collin Keates/Dorling Kindersly Media Library)*

and leaves behind a thin residue of carbon (**FIGURE 18.13C**). Black shales deposited as organic-rich mud in oxygen-poor environments often contain abundant carbonized remains. If the film of carbon is lost from a fossil preserved in fine-grained sediment, a replica of the surface, called an *impression*, may still show considerable detail (**FIGURE 18.13D**).

Delicate organisms, such as insects, are difficult to preserve and consequently are relatively rare in the fossil record. Not only must they be protected from decay, but they must also not be subjected to any pressure that would crush them. One way in which some insects have been preserved is in *amber*, the hardened resin of ancient trees. The fly in **FIGURE 18.13E** was pre-

served after being trapped in a drop of sticky resin. Resin sealed off the insect from the atmosphere and protected the remains from damage by water and air. As the resin hardened, a protective, pressure-resistant case was formed.

In addition to the fossils already mentioned, there are numerous other types, many of them only traces of prehistoric life. Examples of such *trace fossils* include:

1. Tracks—animal footprints made in soft sediment that was later lithified (see Figure 19.35B, p. 486.)

2. Burrows—tubes in sediment, wood, or rock made by an animal. These holes may later become filled with mineral matter and preserved. Some of the

oldest-known fossils are believed to be worm burrows.

3. Coprolites—fossil dung and stomach contents that can provide useful information pertaining to food habits of organisms (**FIGURE 18.13F**).

4. Gastroliths—highly polished stomach stones that were used in the grinding of food by some extinct reptiles.

Conditions Favoring Preservation

Only a tiny fraction of the organisms that have lived during the geologic past have been preserved as fossils. Normally, the remains of an animal or plant are destroyed. Under what circumstances are they preserved? Two special conditions appear to be necessary: rapid burial and the possession of hard parts.

When an organism perishes, its soft parts usually are quickly eaten by scavengers or decomposed by bacteria. Occasionally, however, the remains are buried by sediment. When this occurs, the remains are protected from the environment where destructive processes operate. Rapid burial therefore is an important condition favoring preservation.

In addition, animals and plants have a much better chance of being preserved as part of the fossil record if they have hard parts. Although traces and imprints of soft-bodied animals such as jellyfish, worms, and insects exist, they are not common. Flesh usually decays so rapidly that preservation is exceedingly unlikely. Hard parts such as shells, bones, and teeth predominate in the record of past life.

Because preservation is contingent on special conditions, the record of life in the geologic past is biased. The fossil record of those organisms with hard parts that lived in areas of sedimentation is quite abundant. However, we get only an occasional glimpse of the vast array of other life-forms that did not meet the special conditions favoring preservation.

Paleontologist working with dinosaur remains. *(Photo by Rich Frishman/Getty Images)*

Fossils and Correlation

The existence of fossils had been known for centuries, yet it was not until the late 1700s and early 1800s that their significance as geologic tools was made evident. During this period an English engineer and canal builder, William Smith, discovered that each rock formation in the canals he worked on contained fossils unlike those in the beds either above or below. Further, he noted that sedimentary strata in widely separated areas could be identified and correlated by their distinctive fossil content.

Based on Smith's classic observations and the findings of many geologists who followed, one of the most important and basic principles in historical geology was formulated: *Fossil organisms succeed one another in a definite and determinable order, and therefore any time period can be recognized by its fossil content.* This has come to be known as the **principle of fossil succession.** In other words, when fossils are arranged according to their age by applying the law of superposition to the rocks in which they are found they do not present a random or haphazard picture. To the contrary, fossils document the evolution of life through time.

For example, an Age of Trilobites is recognized quite early in the fossil record. Then, in succession, paleontologists recognize an Age of Fishes, an Age of Coal Swamps, an Age of Reptiles, and an Age of Mammals. These "ages" pertain to groups that were especially plentiful and characteristic during particular time periods. Within each of the "ages" there are many subdivisions based, for example, on certain species of trilobites and certain types of fish, reptiles, and so on. This same succession of dominant organisms, never out of order, is found on every continent.

Once fossils were recognized as time indicators, they became the most useful means of correlating rocks of similar age in different regions. Geologists pay particular attention to certain fossils called **index fossils.** These fossils are widespread geographically and are limited to a short span of geologic time, so their presence provides an important method of matching rocks of the same age Rock formations, however, do not always contain a specific index fossil. In such situations, groups of fossils are used to

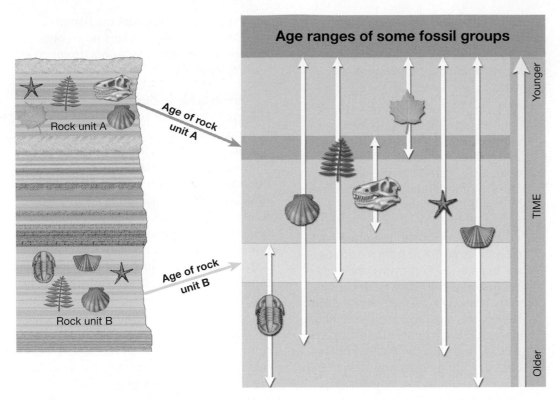

Age ranges of some fossil groups

Rock unit A

Age of rock unit A

Rock unit B

Age of rock unit B

Younger

TIME

Older

FIGURE 18.14 Overlapping ranges of fossils help date rocks more exactly than using a single fossil.

CONCEPT CHECK 18.4

❶ Describe several ways that an animal or plant can be preserved as a fossil.

❷ List two examples of trace fossils.

❸ What conditions favor the preservation of an organism as a fossil?

Dating with Radioactivity

 Geologic Time

ESSENTIALS OF GEOLOGY **Dating with Radioactivity**

In addition to establishing relative dates by using the principles described in the preceding sections, it is also possible to obtain reliable numerical dates for events in the geologic past. For example, we know that Earth is about 4.6 billion years old and that the dinosaurs became extinct about 65 million years ago. Dates that are expressed in millions and billions of years truly stretch our imagination because our personal calendars involve time measured in hours, weeks, and years. Nevertheless, the vast expanse of geologic time is a reality, and it is radiometric dating that allows us to measure it. In this section you will learn about radioactivity and its application in radiometric dating.

Reviewing Basic Atomic Structure

Recall from Chapter 2 that each atom has a *nucleus* containing protons and neutrons and that the nucleus is orbited by electrons. *Electrons* have a negative electrical charge, and *protons* have a positive charge. A *neutron* is actually a proton and an electron combined, so it has no charge (it is neutral).

The *atomic number* (each element's identifying number) is the number of protons in the nucleus. Every element has a different number of protons and thus a different atomic number (hydrogen = 1, carbon = 6, oxygen = 8, uranium = 92, etc.) Atoms of the same element always have the same number of protons, so the atomic number stays constant.

establish the age of the bed. **FIGURE 18.14** illustrates how a group of fossils can be used to date rocks more precisely than could be accomplished by the use of only one of the fossils.

In addition to being important and often essential tools for correlation, fossils are important environmental indicators. Although much can be deduced about past environments by studying the nature and characteristics of sedimentary rocks, a close examination of any fossils present can usually provide a great deal more information.

DID YOU KNOW?

Even when organisms die and their tissues decay, the organic compounds (hydrocarbons) of which they were made may survive in sediments. These are called *chemical fossils*. Commonly, these hydrocarbons form oil and gas, but some residues can persist in the rock record and be analyzed to determine the kind of organisms from which they are derived.

For example, when the remains of certain clam shells are found in limestone, the geologist can assume that the region was once covered by a shallow sea, because that is where clams live today. Also, by using what we know of living organisms, we can conclude that fossil animals with thick shells capable of withstanding pounding and surging waves must have inhabited shorelines. In contrast, the remains of animals with thin, delicate shells probably indicate deep, calm offshore waters. Hence, by noting carefully the types of fossils, the approximate position of an ancient shoreline may be identified.

Further, fossils can indicate the former temperature of the water. Certain present-day corals require warm and shallow tropical seas like those around Florida and the Bahamas. When similar corals are found in ancient limestones, they indicate that a Florida-like marine environment must have existed when they were alive. These examples illustrate how fossils can help unravel the complex story of Earth history.

Practically all of an atom's mass (99.9 percent) is in the nucleus, indicating that electrons have virtually no mass at all. By adding the protons and neutrons in an atom's nucleus, we derive the atom's *mass number*. The number of neutrons can vary, and these variants, or *isotopes*, have different mass numbers.

To summarize with an example, uranium's nucleus always has 92 protons, so its atomic number is always 92. Its neutron population varies, so uranium has three isotopes: uranium-234 (protons + neutrons = 234), uranium-235, and uranium-238. All three isotopes are mixed in nature. They look the same and behave the same in chemical reactions.

Radioactivity

The forces that bind protons and neutrons together in the nucleus usually are strong. However, in some isotopes, the nuclei are unstable because the forces binding protons and neutrons together are not strong enough. As a result, the nuclei spontaneously break apart (decay), a process called **radioactivity.**

What happens when unstable nuclei break apart? Three common types of radioactive decay are illustrated in **FIGURE 18.15** and are summarized as follows:

1. Alpha particles (α particles) may be emitted from the nucleus. An alpha particle is composed of 2 protons and 2 neutrons. Consequently, the emission of an alpha particle means that the mass number of the isotope is reduced by 4 and the atomic number is decreased by 2.
2. When a beta particle (β particle) or electron is given off from a nucleus, the mass number remains unchanged because electrons have practically no mass. However, because the electron has come from a neutron (remember, a neutron is a combination of a proton and an electron), the nucleus contains one more proton than before. Therefore, the atomic number increases by 1.
3. Sometimes an electron is captured by the nucleus. The electron combines with a proton and forms an additional neutron. As in the last example, the

mass number remains unchanged. However, as the nucleus now contains one less proton, the atomic number decreases by 1.

An unstable (radioactive) isotope of an element is called the *parent*. The isotopes resulting from the decay of the parent are the *daughter products*. **FIGURE 18.16** provides an example of radioactive decay. Here it can be seen that when the radioactive parent, uranium-238 (atomic number 92, mass number 238) decays, it follows a number of steps, emitting 8 alpha particles and 6 beta particles before finally becoming the stable daughter product lead-206 (atomic number 82, mass number 206). One of the unstable daughter products produced during this decay series is radon.

Certainly among the most important results of the discovery of radioactivity is that it provided a reliable means of calculating the ages of rocks and minerals that contain particular radioactive isotopes. The procedure is called **radiometric dating.** Why is radiometric dating reliable? The rates of decay for many isotopes have been precisely measured and do not vary under

The strata exposed in the Grand Canyon contain clues to hundreds of millions of years of Earth history. This view is from the North Rim. *(Photo by Lee Foster/Alamy)*

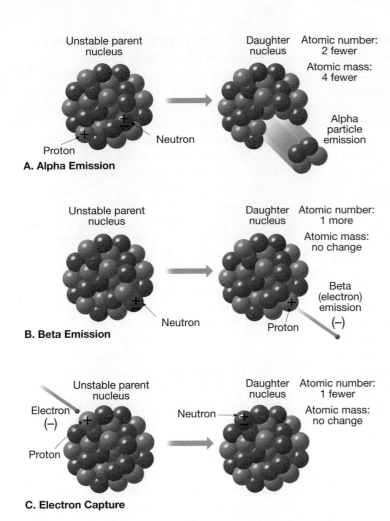

FIGURE 18.15 Common types of radioactive decay. Notice that in each case the number of protons (atomic number) in the nucleus changes, thus producing a different element.

the physical conditions that exist in Earth's outer layers. Therefore, each radioactive isotope used for dating has been decaying at a fixed rate ever since the formation of the rocks in which it occurs, and the products of decay have been accumulating at a corresponding rate. For example, when uranium is incorporated into a mineral that crystallizes from magma, there is no lead (the stable daughter product) from previous decay. The radiometric "clock" starts at this point. As the uranium in this newly formed mineral disintegrates, atoms of the daughter product are trapped, and measurable amounts of lead gradually accumulate.

Half-Life

The time required for half of the nuclei in a sample to decay is called the **half-life** of the isotope. Half-life is a common way of expressing the rate of radioactive disintegration. **FIGURE 18.17** illustrates what occurs when a radioactive parent decays directly into its stable daughter product. When the quantities of parent and daughters are equal (ratio 1:1), we know that one half-life has transpired. When one quarter of the original parent atoms remain and three quarters have decayed to the daughter product, the parent/daughter ratio is 1:3, and we know that two half-lives have passed. After

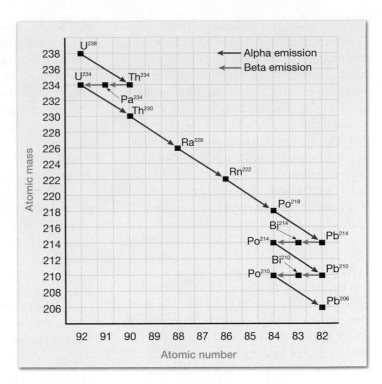

FIGURE 18.16 The most common isotope of uranium (U-238) is an example of a radioactive decay series. Before the stable end product (Pb-206) is reached, many different isotopes are produced as intermediate steps.

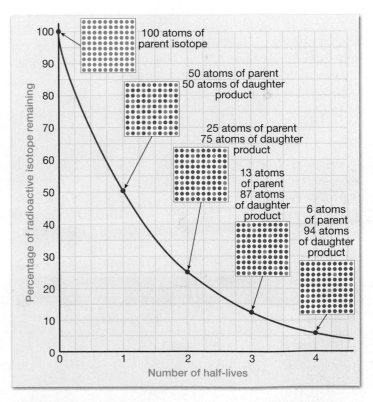

FIGURE 18.17 The radioactive-decay curve shows change that is exponential. Half of the radioactive parent remains after one half-life. After a second half-life, one quarter of the parent remains, and so forth.

three half-lives, the ratio of parent atoms to daughter atoms is 1:7 (one parent for every seven daughter atoms).

If the half-life of a radioactive isotope is known and the parent/daughter ratio can be measured, the age of the sample can be calculated. For example, assume that the half-life of a hypothetical unstable isotope is 1 million years and the parent/daughter ratio in a sample is 1:15. Such a ratio indicates that four half-lives have passed and that the sample must be 4 million years old.

Radiometric Dating

Notice that the *percentage* of radioactive atoms that decay during one half-life is always the same: 50 percent. However, the *actual number* of atoms that decay with the passing of each half-life continually decreases. Thus, as the percentage of radioactive parent atoms declines, the proportion of stable daughter atoms rises, with the increase in daughter atoms just matching the drop in parent atoms. This fact is the key to radiometric dating.

Of the many radioactive isotopes that exist in nature, five have proved particularly useful in providing radiometric ages for ancient rocks (**TABLE 18.1**). Rubidium-87, thorium-232, and the two isotopes of uranium are used only for dating rocks that are millions of years old, but potassium-40 is more versatile.

DID YOU KNOW?
Numerical dates of the fossil record show that life began in the ocean approximately 3.8 billion years ago.

POTASSIUM-ARGON. Although the half-life of potassium-40 is 1.3 billion years, analytical techniques make possible the detection of tiny amounts of its stable daughter product, argon-40, in some rocks that are younger than 100,000 years. Another important reason for its frequent use is that potassium is an abundant constituent of many common minerals, particularly micas and feldspars.

Although potassium (K) has three natural isotopes—^{39}K, ^{40}K, and ^{41}K—only ^{40}K is radioactive. When ^{40}K decays, it does so in two ways. About 11 percent changes to argon-40 (^{40}Ar) by means of electron capture (see Figure 18.15C). The remaining 89 percent of ^{40}K decays to calcium-40 (^{40}Ca) by beta emission (see Figure 18.15B). The decay of ^{40}K to ^{40}Ca however, is not useful for radiometric dating, because the ^{40}Ca produced by radioactive disintegration cannot be distinguished from calcium that may have been present when the rock formed.

The potassium-argon clock begins when potassium-bearing minerals crystallize from a magma or form within a metamorphic rock. At this point the new minerals will contain ^{40}K but will be free of ^{39}Ar because this element is an inert gas that does not chemically combine with other elements.

As time passes, the ^{40}K steadily decays by electron capture. The ^{40}Ar produced by this process remains trapped within the mineral's crystal lattice. Because no ^{40}Ar was present when the mineral formed, all of the daughter atoms trapped in the mineral must have come from the decay of ^{40}K. To determine a sample's age, the ^{40}K/^{40}Ar ratio is measured precisely and the known half-life for ^{40}K applied.

SOURCES OF ERROR. It is important to realize that an accurate radiometric date can be obtained only if the mineral remained a closed system during the entire period since its formation. A correct date is not possible unless there was neither addition nor loss of parent or daughter isotopes, and this is not always the case. In fact, an important limitation of the potassium-argon method arises from the fact that argon is a gas and it may leak from minerals, throwing off measurements. Indeed, losses can be significant if the rock is subjected to relatively high temperatures.

Of course, a reduction in the amount of ^{40}Ar leads to an underestimation of the rock's age. Sometimes temperatures are high enough for a sufficiently long period that all argon escapes. When this happens, the potassium-argon clock is reset, and dating the sample will give only the time of thermal resetting, not the true age of the rock. For other radiometric clocks, a loss of daughter atoms can occur if the rock has been subjected to weathering or leaching. To avoid such a problem, one simple safeguard is to use only fresh, unweathered material and not samples that may have been chemically altered.

Dating with Carbon-14

To date very recent events, carbon-14 is used. Carbon-14 is the radioactive isotope of carbon. The process is often called **radiocarbon dating.** Because the half-life of carbon-14 is only 5730 years, it can be used for dating events from the historic past as well as those from very recent geologic history. In some cases carbon-14 can be used to date events as far back as 70,000 years.

Carbon-14 is continuously produced in the upper atmosphere as a consequence of cosmic-ray bombardment. Cosmic rays, which are high-energy particles, shatter the nuclei of gas atoms, releasing neutrons. Some of the neutrons are absorbed by nitrogen atoms (atomic number 7), causing their nuclei to emit a proton. As a result, the atomic number decreases by 1 (to 6), and a different element, carbon-14, is created

TABLE 18.1

Radioactive Isotopes Frequently Used in Radiometric Dating

Radioactive Parent	Stable Daughter Product	Currently Accepted Half-Life Values
Uranium-238	Lead-206	4.5 billion years
Uranium-235	Lead-207	713 million years
Thorium-232	Lead-208	14.1 billion years
Rubidium-87	Strontium-87	47.0 billion years
Potassium-40	Argon-40	1.3 billion years

DID YOU KNOW?
Although movies and cartoons have depicted humans and dinosaurs living side by side, this was never the case. Dinosaurs flourished during the Mesozoic era and became extinct about 65 million years ago. Humans and their close ancestors did not appear on the scene until the late Cenozoic, more than 60 million years *after* the demise of dinosaurs.

FIGURE 18.18 **A**. Production and **B**. decay of carbon-14. These sketches represent the nuclei of the respective atoms.

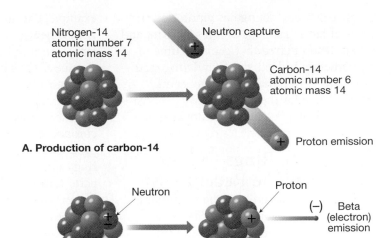

A. Production of carbon-14

B. Decay of carbon-14

(**FIGURE 18.18A**). This isotope of carbon quickly becomes incorporated into carbon dioxide, which circulates in the atmosphere and is absorbed by living matter. As a result, all organisms contain a small amount of carbon-14, including you.

While an organism is alive, the decaying radiocarbon is continually replaced, and the proportions of carbon-14 and carbon-12 remain constant. Carbon-12 is the stable and most common isotope of carbon. However, when any plant or animal dies, the amount of carbon-14 gradually decreases as it decays to nitrogen-14 by beta emission (**FIGURE 18.18B**). By comparing the proportions of carbon-14 and carbon-12 in a sample, radiocarbon dates can be determined. It is important to emphasize that carbon-14 is only useful in dating organic materials, such as wood, charcoal, bones, flesh, and even cloth made of cotton fibers.

Although carbon-14 is useful in dating only the last small fraction of geologic time, it has become a very valuable tool for anthropologists, archeologists, and historians, as well as for geologists who study very recent Earth history. In fact, the development of radiocarbon dating was considered so important that the chemist who discovered this application, Willard F. Libby, received a Nobel Prize in 1960.

Importance of Radiometric Dating
Although the basic principle of radiometric dating is simple, the actual procedure is quite complex. The chemical analysis that determines the quantities of parent and daughter must be painstakingly precise. In

addition, some radioactive materials do not decay directly into the stable daughter product. As you saw in Figure 18.16, uranium-238 produces 13 intermediate unstable daughter products before the 14th and final daughter product, the stable isotope lead-206, is produced.

Radiometric dating methods have produced literally thousands of dates for events in Earth history. Rocks exceeding 3.5 billion years in age are found on all of the continents. Earth's oldest rocks (so far) are gneisses from northern Canada near Great Slave Lake that have been dated at 4.03 billion years (b.y.). Rocks from western Greenland have been dated at 3.7 to 3.8 b.y., and rocks nearly as old are found in the Minnesota River valley and northern Michigan (3.5 to 3.7 b.y.), in southern Africa (3.4 to 3.5 b.y.), and in western Australia (3.4 to 3.6 b.y.). It is important to point out that these ancient rocks are not from any sort of "primordial crust" but originated as lava flows, igneous intrusions, and sediments deposited in shallow water—an indication that Earth history began *before* these rocks formed. Even older mineral grains have been dated. Tiny crystals of the mineral zircon having radiometric ages as old as 4.3 b.y. have been found in younger sedimentary rocks in western Australia. The source rocks for these tiny durable grains either no longer exist or have not yet been found.

DID YOU KNOW?
To guard against error in radiometric dating scientists often use cross-checks. Often this simply involves subjecting a sample to two different methods. If the two dates agree, the likelihood is high that the date is reliable. If an appreciable difference is found, other cross-checks must be employed to determine which, if either, is correct.

Natural casts of shelled invertebrates. (*Photo by E. J. Tarbuck*)

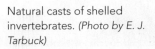

Radiometric dating has vindicated the ideas of James Hutton, Charles Darwin, and others who inferred that geologic time must be immense. Indeed, modern dating methods have proved that there has been enough time for the processes we observe to have accomplished tremendous tasks.

Using Tree Rings to Date and Study the Recent Past

If you look at the top of a tree stump or at the end of a log, you will see that it is composed of a series of concentric rings (**FIGURE 18.19**). Every year in temperate regions, trees add a layer of new wood under the bark. Characteristics of each tree ring, such as size and density, reflect the environmental conditions (especially climate) that prevailed during the year when the ring formed. Favorable growth conditions produce a wide ring; unfavorable ones produce a narrow ring. Trees growing at the same time in the same region show similar tree-ring patterns.

Because a single growth ring is usually added each year, the age of the tree when it was cut can be determined by counting the rings. If the year of cutting is known, the age of the tree and the year in which each ring formed can be determined by counting back from the outside ring. This procedure can be used to determine the dates of recent geologic events, for example, the minimum number of years since a new land surface was created by a landslide or a flood. The dating and study of annual rings in trees is called *dendrochronology*.

To make the most effective use of tree rings, extended patterns known as ring chronologies are established. They are produced by comparing the patterns of rings among trees in an area. If the same pattern can be identified in two samples, one of which has been dated, the second sample can be dated from the first by matching the ring pattern common to both. This technique, called *cross-dating*, is illustrated in **FIGURE 18.20**. Tree-ring chronologies extending back for thousands of years have been established for some regions. To date a timber sample of unknown age, its ring pattern is matched against the reference chronology.

Tree-ring chronologies are unique archives of environmental history and have important applications in such disciplines as climate, geology, ecology, and archaeology. For example, tree rings are used to reconstruct climate variations within a region for spans of thousands of years prior to human historical records. Knowledge of such long-term variations is of great value in making judgments regarding the recent record of climate change. Climate change is the main focus of Chapter 20.

In summary, dendrochronology provides useful numerical dates for events in the historic and recent prehistoric past. Moreover, because tree rings are a storehouse of data, they are a valuable tool in the reconstruction of past environments.

CONCEPT CHECK 18.5

1. List three types of radioactivity decay. For each type, describe how the atomic number and atomic mass change.
2. Sketch a simple diagram as a way of explaining the idea of half-life.
3. Why is radiometric dating a reliable method for determining numerical ages?
4. Briefly explain why tree rings are useful in studying the recent geologic past.

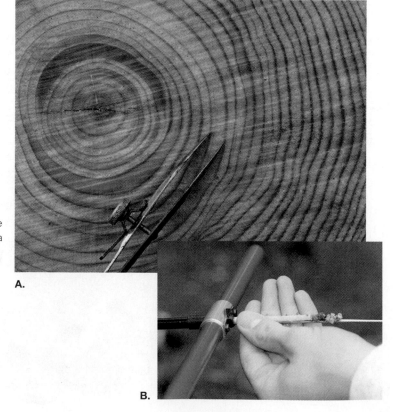

FIGURE 18.19 A. Each year a growing tree produces a layer of new cells beneath the bark. Because the amount of growth (thickness of a ring) depends on precipitation and temperature, tree rings are useful recorders of past climates. (*Photo by Stephen J. Krasemann/Photo Researchers, Inc.*) **B.** Scientists are not limited to working with trees that have been cut down. Small, nondestructive core samples can be taken from living trees. (*Photo by Gregory K. Scott/Photo Researchers, Inc.*).

A.

B.

DID YOU KNOW?
Dating with carbon-14 is useful to archaeologists and historians as well as geologists. For example, University of Arizona researchers used carbon-14 to determine the age of the Dead Sea Scrolls, considered among the great archaeological discoveries of the 20th century. Parchment from the scrolls dates between 150 BC and 5 BC. Portions of the scrolls contain dates that match those determined by the carbon-14 measurements.

The Geologic Time Scale

Geologic Time

Geologic Time Scale

Geologists have divided the whole of geologic history into units of varying magnitude. Together they comprise the **geologic time scale** of Earth history (**FIGURE 18.21**). The major units of the time scale were delineated during the 19th century, principally by workers in western Europe and Great Britain. Because radiometric dating was unavailable at that time, the entire time scale was created using methods of relative dating. It was only in the 20th century that radiometric dating permitted numerical dates to be added.

Structure of the Time Scale

The geologic time scale subdivides the 4.6-billion-year history of Earth into many different units and provides a meaningful time frame within which the events of the geologic past are arranged. As shown in Figure 18.21, **eons** represent the greatest expanses of time. The eon that began about 542 million years ago is the **Phanerozoic,** a term derived from Greek words meaning *visible life.* It is an appropriate description because the rocks and deposits of the Phanerozoic eon contain abundant fossils that document major evolutionary trends.

Another glance at the time scale reveals that eons are divided into **eras.** The three eras within the Phanerozoic are the **Paleozoic** ("ancient life"), the **Mesozoic** ("middle life"), and the **Cenozoic** ("recent life"). As the names imply, these eras are bound by profound worldwide changes in life forms. Major changes in life forms are discussed in Chapter 19 "Earth's Evolution through Geologic Time". Each era of the Phanerozoic eon is subdivided into **periods.** Each period is characterized by a somewhat less profound change in life-forms as compared with the eras.

Each period is divided into still smaller units called **epochs.** As you can see in Figure 18.21, seven epochs have been named for the periods of the Cenozoic. The epochs of other periods usually are simply termed *early, middle,* and *late.*

Precambrian Time

Notice that the detail of the geologic time scale does not begin until about 542 million years ago, the date for the beginning of the Cambrian period. The nearly 4 billion years prior to the Cambrian are divided into three eons, the **Hadean,** the **Archean,** and the **Proterozoic.** These, in turn, are divided into seven eras. It is also common for this vast expanse of time to simply be referred to as the **Precambrian.*** Although it represents about 88 percent of Earth history, the Precambrian is not divided into nearly as many smaller time units as the Phanerozoic eon.

Why is the huge expanse of Precambrian time not divided into numerous eras, periods, and epochs? The reason is that Precambrian history is not known in great enough detail. The quantity of information that geologists have deciphered about Earth's past is somewhat analogous to the detail of human history. The further back we go, the less is known. Certainly more data and information exist about the past 10 years than for the first decade of the 20th century; the events of the 19th century have been documented much better than the events of the 1st century AD; and so on. So it is with Earth history. The more recent past has the freshest, least disturbed, and more observable record. The further back in time the geologist goes, the more fragmented the record and clues become.

Terminology and the Geologic Time Scale

There are some terms associated with the geologic time scale that are not "officially" recognized as part of it. The best known, and most common, example is *Precambrian*—the informal name for the eons that came before the current Phanerozoic eon. Although the term *Precambrian* has no formal status on the geologic time scale, it has been traditionally used as though it did.

Hadean is another informal term that is found on some versions of the geologic time scale and is used by some geologists. It refers to the earliest interval (eon) of Earth history—before the oldest known rocks. When the term was coined in 1972, the age of Earth's oldest known rocks was about 3.8 billion years. Today that number stands at slightly greater than 4 billion and, of course, is subject to revision. The name *Hadean* derives from *Hades,* Greek for *underworld*—a reference to the "hellish" conditions that prevailed on Earth early in its history.

Effective communication in the geosciences requires that the geologic time scale consist of

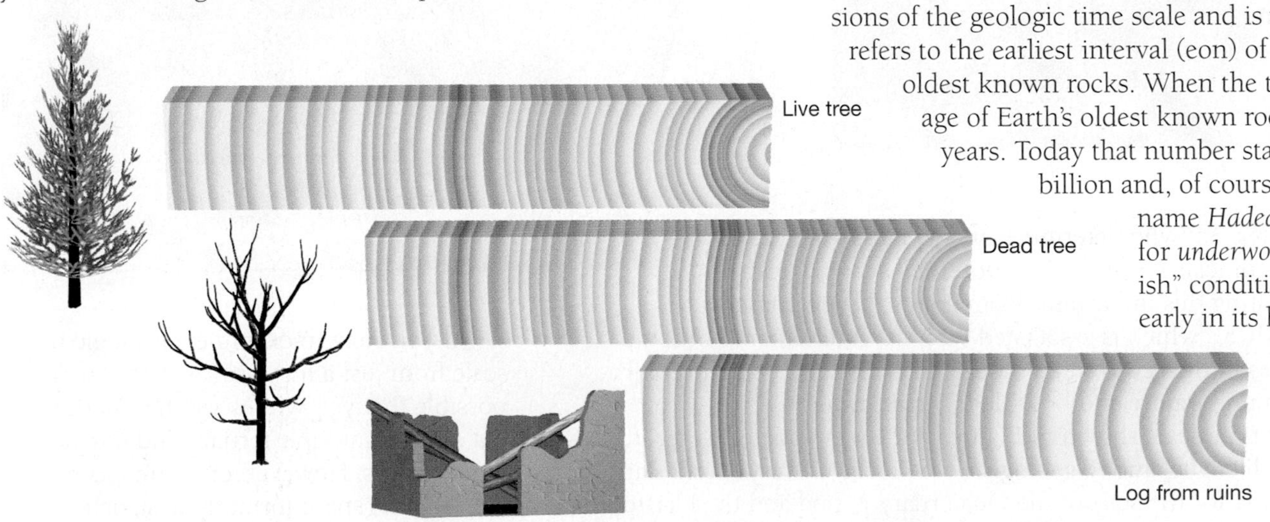

FIGURE 18.20 Cross-dating is a basic principle in dendrochronology. Here it was used to date an archaeological site by correlating tree-ring patterns for wood from trees of three different ages. First, a tree-ring chronology for the area is established using cores extracted from living trees. This chronology is extended further back in time by matching overlapping patterns from older, dead trees. Finally, cores taken from beams inside the ruin are dated using the chronology established from the other two sites.

Live tree

Dead tree

Log from ruins

*The terms Hadean and Precambrian are informal names that are in common use. There is more about this in the section on "Terminology and the Geologic Time Scale."

FIGURE 18.21 The geologic time scale. Numbers on the time scale represent time in millions of years before the present. The numerical dates were added long after the time scale had been established using relative dating techniques. The Precambrian accounts for more than 88 percent of geologic time. The time scale is a dynamic tool. Advances in the geosciences require that updates be made from time to time. This version was updated in July 2009.

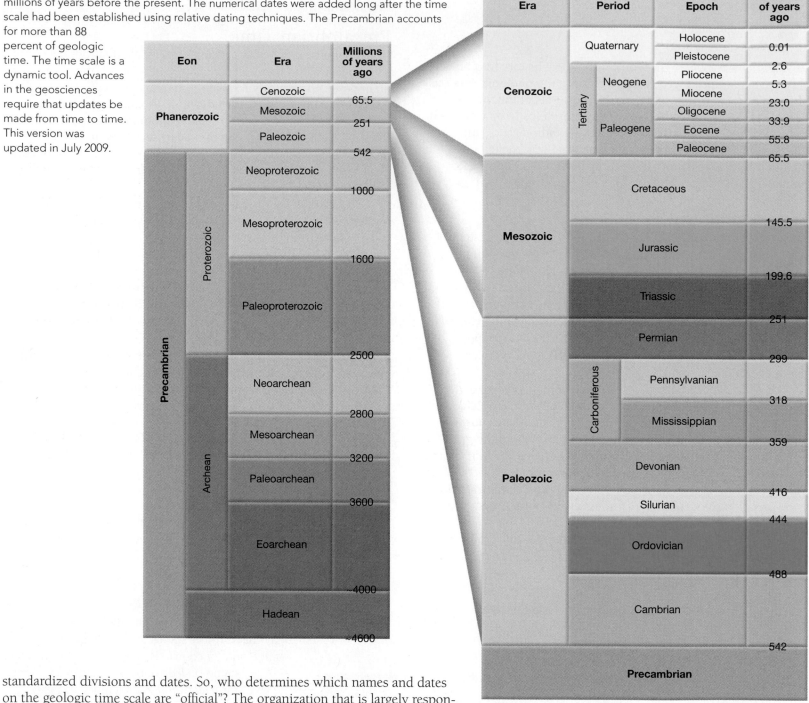

standardized divisions and dates. So, who determines which names and dates on the geologic time scale are "official"? The organization that is largely responsible for maintaining and updating this important document is the International Commission on Stratigraphy (ICS), which is associated with the International Union of Geological Sciences.* Advances in the geosciences require that the scale be periodically updated to include changes in unit names and boundary age estimates.

For example, the geologic time scale shown in Figure 18.21 was updated as recently as July 2009. After considerable dialogue among geologists who focus on very recent Earth history, the ICS changed the date for the start of the Quaternary period and the Pleistocene epoch from 1.8 million to 2.6 million years ago. Who knows, perhaps by the time you read this, other changes will have been made.

Stratigraphy is the branch of geology that studies rock layers (strata) and layering (stratification), thus its primary focus is sedimentary and layered volcanic rocks.

If you were to examine a geologic time scale from just a few years ago, it is quite possible that you would see the Cenozoic era divided into the Tertiary and Quaternary periods. However, on more recent versions the space formerly designated as Tertiary is divided into the Paleogene and Neogene periods. As our understanding of this time span changed, so too did its designation on the geologic time scale. Today, the Tertiary period is considered as a

FIGURE 18.22 A useful numerical date for this conglomerate is not possible, because the gravel particles that compose it were derived from rocks of diverse ages. *(Photo by E. J. Tarbuck)*

"historic" name and is given no official status on the ICS version of the time scale. Many time scales still contain references to the Tertiary period, including Figure 18.21. One reason for this is that a great deal of past (and some current) geological literature uses this name.

For those who study historical geology, it is important to realize that the geologic time scale is a dynamic tool that continues to be refined as our knowledge and understanding of Earth history evolves.

CONCEPT CHECK 18.6

❶ List the four basic units that make up the geologic time scale.

❷ Why is "zoic" part of so many names on the geologic time scale?

❸ What term applies to *all* of geologic time prior to the Phanerozoic eon? Why is this span *not* divided into as many smaller time units as the Phanerozoic eon?

Difficulties in Dating the Geologic Time Scale

Although reasonably accurate numerical dates have been worked out for the periods of the geologic time scale (see Figure 18.21), the task is not without difficulty. The primary problem in assigning numerical dates to units of time is the fact that not all rocks can be dated by radiometric methods. Recall that for a radiometric date to be useful, all minerals in the rock must have formed at about the same time. For this reason, radioactive isotopes can be used to determine when minerals in an igneous rock crystallized and when pres-

sure and heat created new minerals in a metamorphic rock.

However, samples of sedimentary rock can only rarely be dated directly by radiometric means. A sedimentary rock might include particles that contain radioactive isotopes, but the rock's age cannot be accurately determined, because the grains making up the rock are not the same age as the rock in which they occur. Rather, the sediments have been weathered from rocks of diverse ages. (**FIGURE 18.22**).

Radiometric dates obtained from metamorphic rocks might also be difficult to interpret because the age of a particular mineral in a metamorphic rock does not necessarily represent the time when the rock initially formed. Instead, the date could indicate any one of a number of subsequent metamorphic phases.

If samples of sedimentary rocks rarely yield reliable radiometric ages, how can numerical dates be assigned to sedimentary layers? Usually the geologist must relate them to datable igneous masses, as in **FIGURE 18.23**. In this example, radiometric dating has determined the ages of the

volcanic ash bed within the Morrison Formation and the dike cutting the Mancos Shale and Mesaverde Formation. The sedimentary beds below the ash are obviously older than the ash, and all the layers above the ash are younger (principle of superposition). The dike is younger than the Mancos Shale and the Mesaverde Formation but older than the Wasatch Formation because the dike does not intrude the Tertiary rocks (cross-cutting relationships).

*For more about this see "Are we now living in the Anthropocene?" in *GSA Today*, Vol. 18, No. 2, February 2008, pp. 4–8. *GSA Today* is a monthly publication of the Geological Society of America.

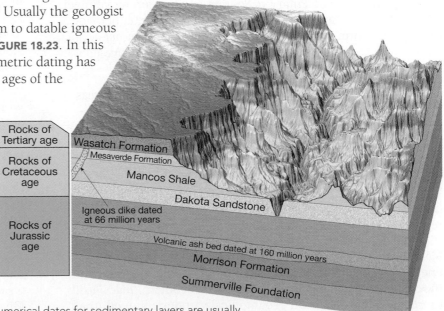

FIGURE 18.23 Numerical dates for sedimentary layers are usually determined by examining their relationship to igneous rocks. *(After U.S. Geological Survey)*

From this kind of evidence, geologists estimate that a part of the Morrison Formation was deposited about 160 million years ago, as indicated by the ash bed. Further, they conclude that the Tertiary period began after the intrusion of the dike, 66 million years ago. This is one example of literally thousands that illustrate how datable materials are used to *bracket* the various episodes in Earth history within specific time periods. It shows the necessity of combining laboratory dating methods with field observations of rocks.

CHAPTER EIGHTEEN
Geologic Time in Review

◉ The two types of dates used by geologists to interpret Earth history are (1) *relative dates*, which put events in their *proper sequence of formation*, and (2) *numerical dates*, which pinpoint the *time in years* when an event occurred.

◉ Relative dates can be established using the *law of superposition* (in an undeformed sequence of sedimentary rocks or surface-deposited igneous rocks, each bed is older than the one above and younger than the one below), the *principle of original horizontality* (most layers are deposited in a horizontal position), the *principle of cross-cutting relationships* (when a fault or intrusion cuts through another rock, the fault or intrusion is younger than the rocks cut through), and *inclusions* (the rock mass containing the inclusion is younger than the rock that provided the inclusion).

◉ *Unconformities* are gaps in the rock record. Each represents a long period during which deposition ceased, erosion removed previously formed rocks, and then deposition resumed. The three basic types of unconformities are *angular unconformities* (tilted or folded sedimentary rocks that are overlain by younger, more flat-lying strata), *disconformities* (the strata on either side of the unconformity are essentially parallel), and *nonconformities* (where a break separates older metamorphic or intrusive igneous rocks from younger sedimentary strata).

◉ *Correlation*, matching rocks of similar age in different areas, is used to develop a geologic time scale that applies to the entire Earth.

◉ *Fossils* are the remains or traces of prehistoric life. The special conditions that favor preservation are *rapid burial* and the possession of *hard parts* such as shells, bones, or teeth.

◉ Fossils are used to *correlate* sedimentary rocks from different regions by using the rocks' distinctive fossil content and applying the *principle of fossil succession*. It is based on the work of *William Smith* in the late 1700s and states that fossil organisms succeed one another in a definite and determinable order; therefore any time period can be recognized by its fossil content. *Index fossils*, those that are widespread geographically and are limited to a short span of geologic time, provide an important tool for matching rocks of the same age.

◉ Each atom has a nucleus containing *protons* (positively charged particles) and *neutrons* (neutral particles). Orbiting the nucleus are negatively charged *electrons*. The *atomic number* of an atom is the number of protons in the nucleus. The *mass number* is the number of protons plus the number of neutrons in an atom's nucleus. *Isotopes* are variants of the same atom but with a different number of neutrons and hence a different mass number.

◉ *Radioactivity* is the spontaneous breaking apart (decay) of certain unstable atomic nuclei. Three common types of radioactive decay are (1) emission of *alpha particles* from the nucleus, (2) emission of *beta particles* (or electrons) from the nucleus, and (3) *capture of electrons* by the nucleus.

◉ An unstable *radioactive isotope*, called the *parent*, will decay and form stable *daughter products*. The length of time for half of the nuclei of a radioactive isotope to decay is called the *half-life* of the isotope. If the half-life of the isotope is known, and the parent/daughter ratio can be measured, the age of a sample can be calculated. An accurate radiometric date can only be obtained if the mineral containing the radioactive isotope remained a closed system during the entire period since its formation.

◉ The *geologic time scale* divides Earth's history into units of varying magnitude. It is commonly presented in chart form, with the oldest time and event at the bottom and the youngest at the top. The principal subdivisions of the geologic time scale, called *eons*, include the *Archean*, *Proterozoic* (together, these two eons are commonly referred to as the *Precambrian*), and, beginning about 542 million years ago, the *Phanerozoic*. The Phanerozoic (meaning "visible life") eon is divided into the following *eras*: *Paleozoic* ("ancient life"), *Mesozoic* ("middle life"), and *Cenozoic* ("recent life").

◉ A significant problem in assigning numerical dates to units of time is that *not all rocks can be radiometrically dated*. A sedimentary rock may contain particles of many ages that have been weathered from different rocks that formed at various times. One way geologists assign numerical dates to sedimentary rocks is to relate them to datable igneous masses, such as dikes and volcanic ash beds.

Key Terms

GIVE IT SOME THOUGHT

❶ Refer to Figure 18.5 and answer the following questions. Briefly explain each answer.
   a. Is fault A older or younger than the sandstone layer?
   b. Is dike A older or younger than the sandstone layer?
   c. Was the conglomerate deposited before or after fault A?
   d. Was the conglomerate deposited before or after fault B?
   e. Which fault is older, A or B?
   f. Is dike A older or younger than the batholith?

❷ A mass of granite is in contact with a layer of sandstone. Using a principle described in this chapter, explain how you might determine whether the sandstone was deposited on top of the granite, or whether the magma that formed the granite was intruded after the sandstone was deposited.

❸ Refer to Figure 18.9 and consider the angular unconformity at Scotland's Siccar Point for the following exercises.
   a. Describe in general what occurred to produce this feature.
   b. Suggest ways in which all four spheres of the Earth system could have been involved.
   c. The Earth system is powered by energy from two sources. How are both sources represented in the Siccar Point unconformity?

❹ If a radioactive isotope of thorium (atomic number 90, mass number 232) emits 6 alpha particles and 4 beta particles during the course of radioactive decay, what is the atomic number and mass number of the stable daughter product?

❺ A hypothetical radioactive isotope has a half-life of 10,000 years. If the ratio of radioactive parent to stable daughter product is 1:3, how old is the rock containing the radioactive material?

❻ Solve the problems below that relate to the magnitude of Earth history. To make calculations easier, round Earth's age to 5 billion years.
   a. What percentage of geologic time is represented by recorded history? (Assume 5000 years for the length of recorded history)
   b. Humanlike ancestors (hominids) have been around for roughly 5 million years. What percentage of geologic time is represented by the existence of these ancestors?
   c. The first abundant fossil evidence does not appear until the beginning of the Cambrian period, about 540 million years ago. What percentage of geologic time is represented by abundant fossil evidence?

❼ The accompanying diagram is a cross-section of a hypothetical area. Place the lettered features in the proper sequence, from oldest to youngest. Where in the sequence of events can you identify an unconformity?

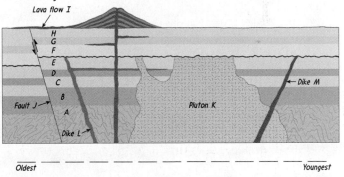

Oldest — — — — — — — — — — — — — Youngest

Companion Website

mygeoscience place PEARSON

www.mygeoscienceplace.com

Earth's Evolution through Geologic Time

EARTH HAS A LONG AND COMPLEX HISTORY. TIME AND AGAIN, THE splitting and colliding of continents has resulted in the formation of new ocean basins and the creation of great mountain ranges. Furthermore, the nature of life on our planet has experienced dramatic changes through time.

To assist you in learning the important concepts in this chapter, focus on the following questions:

- ◉ What makes Earth unique among the planets?
- ◉ How did Earth's atmosphere and oceans form and evolve through time?
- ◉ How did the first continents form?
- ◉ What is the supercontinent cycle?
- ◉ What are some of the important geological events of Earth's history?
- ◉ What was the Cambrian explosion?
- ◉ How has life changed through geologic time?

FOCUS ON CONCEPTS

Is Earth Unique?

There is only one place in the universe, as far as we know, that can support life—a modest-sized planet called Earth that orbits an average-sized star, the Sun. Life on Earth is ubiquitous; it is found in boiling mudpots and hot springs, in the deep abyss of the ocean, and even under the Antarctic ice sheet. Living space on our planet, however, is significantly limited when we consider the needs of individual organisms, particularly humans. The global ocean covers 71 percent of Earth's surface, but only a few hundred meters below the water's surface pressures are so intense that human lungs cannot function. In addition, many continental areas are too steep, too high, or too cold for us to inhabit (**FIGURE 19.1**). Nevertheless, based on what we know about other bodies in the solar system—and the hundreds of planets recently discovered orbiting around other stars—Earth is still, by far, the most accommodating.

What fortuitous events produced a planet so hospitable to life? Earth was not always as we find it today. During its formative years, our planet became hot enough to support a magma ocean. It also survived a several-hundred-million-year period of extreme bombardment, to which the heavily cratered surfaces of Mars, the Moon, and asteroids testify. The oxygen-rich atmosphere that makes higher life-forms possible developed relatively recently. Serendipitously, Earth seems to be the right planet, in the right location, at the right time.

The Right Planet

What are some of the characteristics that make Earth unique among the planets? Consider the following:

1. If Earth were considerably larger (more massive) its force of gravity would be proportionately greater. Like the giant planets, Earth would have retained a thick, hostile atmosphere consisting of ammonia and methane, and possibly hydrogen and helium.

2. If Earth were much smaller, oxygen, water vapor, and other volatiles would escape into space and be lost forever. Thus, like the Moon and Mercury, both of which lack an atmosphere, Earth would be void of life.

3. If Earth did not have a rigid lithosphere overlaying a weak asthenosphere, plate tectonics would not operate. The continental crust (Earth's "highlands") would not have formed without the recycling of plates. Consequently, the entire planet would likely be covered by an ocean a few kilometers deep. As the author Bill Bryson so aptly stated, "There might be life in that lonesome ocean, but there certainly wouldn't be baseball."*

4. Most surprisingly, perhaps, is the fact that if our planet did not have a molten metallic core, most of the life-forms on Earth would not exist. Fundamentally, without the flow of iron in the core, Earth could not support a magnetic field. It is the magnetic field which prevents lethal cosmic rays (the solar wind) from showering Earth's surface.

*A Short History of Nearly Everything, Broadway Books, 2003.

FIGURE 19.1 Climbers near the top of Mount Everest. At this altitude the level of oxygen is only one third the amount available at sea level. (*Photo courtesy of Woodfin Camp and Associates*)

The Right Location

One of the primary factors that determine whether a planet is suitable for higher life-forms is its location in the solar system. The following scenarios substantiate Earth's right position:

1. If Earth were about 10 percent closer to the Sun, like Venus, our atmosphere would consist mainly of the greenhouse gas carbon dioxide. As a result, Earth's surface temperature would be too hot to support higher life-forms.

2. If Earth were about 10 percent farther from the Sun, the problem would be reversed—it would be too cold. The oceans would freeze over and Earth's active water cycle would not exist. Without liquid water all life would perish.

3. Earth is near a star of modest size. Stars like the Sun have a life span of roughly 10 billion years. During most of this time, radiant energy is emitted at a fairly constant level. Giant stars, on the other hand, consume their nuclear fuel at very high rates and "burn out" in a few hundred million years. Therefore, Earth's proximity to a modest-sized star allowed enough time for the evolution of humans, who first appeared on this planet only a few million years ago.

The Right Time

The last, but certainly not the least, fortuitous factor is timing. The first organisms to inhabit Earth were extremely primitive and came into existence roughly 3.8 billion years ago. From this point in Earth's history innumerable changes occurred—life-forms came and went along with changes in the physical environment of our planet. Two of many timely, Earth-altering events include:

1. The development of our modern atmosphere. Earth's primitive atmosphere is thought to have been composed mostly of water vapor and carbon dioxide, with small amounts of other gases, but no free oxygen. Fortunately, microorganisms evolved that released oxygen into the atmosphere by the process of *photosynthesis*. About 2.2 billion years ago an atmosphere with free oxygen came into existence. The result was the evolution of the forebearers of the vast array of organisms that occupy Earth today.

2. About 65 million years ago our planet was struck by an asteroid 10 kilometers in diameter. This impact caused a mass extinction during which nearly three-quarters of all plant and animal species were obliterated—including dinosaurs (**FIGURE 19.2**). Although this may not seem fortuitous, the extinction of dinosaurs opened new habitats for small mammals that survived the impact. These habitats, along with evolutionary forces, led to the development of many large mammals that occupy our modern world. Without this event, mammals may not have evolved beyond the small rodentlike creatures that live in burrows.

As various observers have noted, Earth developed under "just right" conditions to support higher life-forms. Astronomers refer to this as the *Goldilocks scenario*. Like the classic Goldilocks and the Three Bears fable, Venus is too hot (Papa Bear's porridge), Mars is too cold (Mama Bear's porridge), but Earth is just right (Baby Bear's porridge). Do these "just right" conditions exist purely by chance as some researchers suggest, or might Earth's hospitable environment have developed for the evolution and survival of higher life-forms as others have argued?

The remainder of this chapter will focus on the origin and evolution of planet Earth—the one place in the Universe we know fosters life. As you learned in Chapter 18, researchers utilize many tools to interpret the clues about Earth's past. Using these tools, and clues contained in the rock record, scientists continue to unravel many complex events of the geologic past. The goal of this chapter is to provide a brief overview of the history of our planet and its life-forms—a journey that takes us back about 4.6 billion years to the formation of Earth and its atmosphere. Later, we will consider how our physical world assumed its present state and how Earth's inhabitants changed through time. We suggest that you reacquaint yourself with the *geologic time scale* presented in **FIGURE 19.3** and refer to it throughout the chapter.

CONCEPT CHECK 19.1

❶ Explain why Earth is just the right size.

❷ Why is Earth's molten, metallic core important to humans living today?

❸ Why is Earth's location in the solar system ideal for the development of higher life-forms?

FIGURE 19.2 This paleontologist is excavating the skeleton of a hadrosaurid or duck-billed dinosaur of late Cretaceous age in Coahuila State, Mexico. *(Photo by Francois Gohier/Photo Researchers, Inc.)*

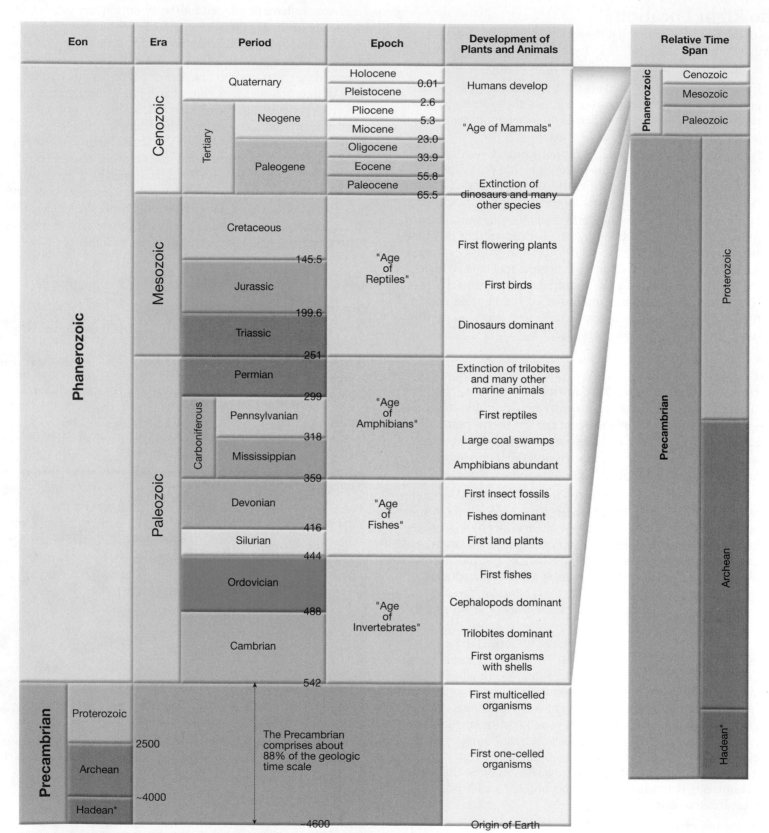

Eon	Era	Period		Epoch	Development of Plants and Animals
Phanerozoic	Cenozoic	Quaternary		Holocene 0.01	Humans develop
				Pleistocene 2.6	
		Tertiary	Neogene	Pliocene 5.3	"Age of Mammals"
				Miocene 23.0	
			Paleogene	Oligocene 33.9	
				Eocene 55.8	
				Paleocene 65.5	Extinction of dinosaurs and many other species
	Mesozoic	Cretaceous		"Age of Reptiles"	First flowering plants
		145.5			First birds
		Jurassic			
		199.6			Dinosaurs dominant
		Triassic			
		251			
	Paleozoic	Permian		"Age of Amphibians"	Extinction of trilobites and many other marine animals
		299			
		Carboniferous	Pennsylvanian		First reptiles
		318			Large coal swamps
			Mississippian		Amphibians abundant
		359			
		Devonian		"Age of Fishes"	First insect fossils
					Fishes dominant
		416			
		Silurian			First land plants
		444			
		Ordovician		"Age of Invertebrates"	First fishes
					Cephalopods dominant
		488			
		Cambrian			Trilobites dominant
					First organisms with shells
		542			
Precambrian	Proterozoic				First multicelled organisms
	2500				
	Archean		The Precambrian comprises about 88% of the geologic time scale		First one-celled organisms
	~4000				
	Hadean*				
	~4600				Origin of Earth

Relative Time Span

Phanerozoic: Cenozoic / Mesozoic / Paleozoic

Precambrian: Proterozoic / Archean / Hadean*

* Hadean is the informal name for the span that begins at Earth's formation and ends with Earth's earliest-known rocks.

FIGURE 19.3 The geologic time scale. Numbers represent time in millions of years before the present. These dates were added long after the time scale had been established using relative dating techniques. The Precambrian accounts for about 88 percent of geologic time.

Birth of a Planet

According to the Big Bang theory, the formation of our planet began about 13.7 billion years ago with a cataclysmic explosion that created all matter and space (**FIGURE 19.4**). Initially, atomic particles (protons, neutrons, and electrons) formed. Later, as this debris cooled, atoms of hydrogen and helium, the two lightest elements, began to form. Within a few hundred million years, clouds of these gases condensed and coalesced into stars that compose the galactic systems we now observe.

As these gases contracted to become the first stars, heating triggered the process of *nuclear fusion*. Within stars' interiors, hydrogen atoms convert to helium atoms, releasing enormous amounts of radiant energy (heat, light, cosmic rays). Astronomers have determined that in stars more massive than our Sun, other thermonuclear reactions occur that generate all the elements on the periodic table up to number 26, iron. The heaviest elements (beyond number 26) are only created at extreme temperatures during the explosive death of a star perhaps 10 to 20 times more massive than the Sun. During these cataclysmic **supernova** events, exploding stars produce all of the elements heavier than iron and spew them into interstellar space. It is from such debris that our Sun and solar system formed. According to the Big Bang sce-

nario, atoms in your body were produced billions of years ago in the hot interior of now defunct stars, and the formation of the gold in your jewelry was triggered by a supernova explosion that occurred trillions of miles away.

From Planetesimals to Protoplanets

Recall that the solar system, including Earth, formed about 4.6 billion years ago from the **solar nebula**, a large rotating cloud of interstellar dust and gas (see Figure 1.23, page 21). As the solar nebula contracted, most of the matter collected in the center to create the hot *protosun*, while the remainder became a flattened spinning disk. Within this spinning disk, matter gradually coalesced into clumps that collided and stuck together to become asteroid-size objects called **planetesimals**. The composition of each planetesimal was determined largely by its distance from the hot protosun.

Near the orbit of Mercury only metallic grains condensed from the solar nebula. Near Earth's orbit, metallic as well as rocky substances condensed, and beyond Mars ices of water, carbon dioxide, methane, and ammonia formed. It was from these clumps of matter that the planetesimals were made and, through repeated collisions and accretion (sticking together), grew into eight **protoplanets** and their moons (Figure 19.4).

At some point in Earth's evolution a giant impact occurred between a Mars-sized planetesimal and a young, semi-molten Earth. This collision ejected huge amounts of debris into space, some of which coalesced to form the Moon (Figure 19.4 J, K, L).

Earth's Early Evolution

As material continued to accumulate, the high-velocity impact of interplanetary debris (planetesimals) and the decay of

radioactive elements caused the temperature of our planet to steadily increase. During this period of intense heating, Earth became hot enough that iron and nickel began to melt. Melting produced liquid blobs of heavy metal that sank under their own weight. This process occurred rapidly on the scale of geologic time and produced Earth's dense iron-rich core. The formation of a molten iron core was the first of many stages of chemical differentiation in which Earth converted from a homogeneous body, with roughly the same matter at all depths, to a layered planet with material sorted by density.

This early period of heating also resulted in a magma ocean, perhaps several hundred kilometers deep. Within the magma ocean, buoyant masses of molten rock rose toward the surface and eventually solidified to produce a thin, primitive crust. Earth's first crust was likely basaltic in composition, similar to modern oceanic crust.

This period of chemical differentiation established the three major divisions of Earth's interior—the iron-rich *core,* the thin *primitive crust,* and Earth's thickest layer, the *mantle,* located between the core and the crust. In addition, the lightest materials, including water vapor, carbon dioxide, and other gases, escaped to form a primitive atmosphere and shortly thereafter the oceans (**FIGURE 19.5**).

CONCEPT CHECK 19.2

❶ What two elements made up most of the very early universe?

❷ What is the name of the cataclysmic event in which an exploding star produces all of the elements heavier than iron?

❸ Briefly describe the formation of the planets from the solar nebula.

A. Big Bang
13.7 Ga

B. Hydrogen and
helium atoms
created

C. Our galaxy forms
10 Ga

D. Heavy elements
synthesized by
supernova
explosions

G. Accretion of planetesimals
to form Earth and the
other planets

Earth

F. As material collects
to form the protosun
rotation flattens nebula

E. Solar nebula
begins to contract
4.7 Ga

H. Continual bombardment
and the decay of radioactive elements
produces magma ocean

I. Chemical differentiation
produces Earth's
layered structure

J. Mars-size
object impacts
young Earth
4.6 Ga

K. Debris orbits
Earth and
accretes

M. Outgassing produces Earth's
primitive atmosphere and ocean

L. Formation of Earth Moon system
4.5 Ga

FIGURE 19.4 Major events that led to the formation of early Earth.

FIGURE 19.5 Artistic depiction of Earth more than 4 billion years ago. This was a time of intense volcanic activity that produced Earth's primitive atmosphere and oceans, while early life-forms produced moundlike structures called stromatolites.

Origin of the Atmosphere and Oceans

We can be thankful for our atmosphere; without it there would be no greenhouse effect and Earth would be nearly 60 °F colder. Earth's water bodies would be frozen and the hydrologic cycle would be nonexistent.

The air we breathe is a stable mixture of 78 percent nitrogen, 21 percent oxygen, about 1 percent argon (an inert gas), and small amounts of gases such as carbon dioxide and water vapor. However, our planet's original atmosphere 4.6 billion years ago was substantially different.

Earth's Primitive Atmosphere

Early in Earth's formation, its atmosphere likely consisted of gases most common in the early solar system—hydrogen, helium, methane, ammonia, carbon dioxide, and water vapor. The lightest of these gases, hydrogen and helium, apparently escaped into space because Earth's gravity was too weak to hold them. Most of the remaining gases were probably scattered into space by strong *solar winds* (a vast stream of particles) from a young, active Sun. (All stars, including the Sun, apparently experience a highly active stage early in their evolution known as the *T-Tauri phase,* during which their solar winds are very intense.)

Earth's first enduring atmosphere was generated by a process called **outgassing,** through which gases trapped in the planet's interior are released. Outgassing from hundreds of active volcanoes still remains an important planetary function worldwide (**FIGURE 19.6**). However, early in Earth's history, when massive heating and fluid-like motion occurred in the mantle, the gas output must have been immense. Based on our understanding of modern volcanic eruptions, Earth's primitive atmosphere probably consisted of mostly water vapor, carbon dioxide, and

sulfur dioxide with minor amounts of other gases, and minimal nitrogen. Most important, free oxygen was not present.

Oxygen in the Atmosphere

As Earth cooled, water vapor condensed to form clouds, and torrential rains began to fill low-lying areas, which became the oceans. In those oceans, nearly 3.5 billion years ago, photosynthesizing bacteria began to release oxygen into the water. During *photosynthesis,* the Sun's energy is used by organisms to produce organic material (energetic molecules of sugar containing hydrogen and carbon) from carbon dioxide (CO_2) and water (H_2O). The first bacteria probably used hydrogen sulfide (H_2S) as the source of hydrogen rather than water. One of the earliest bacteria, *cyanobacteria* (once called blue-green algae), began to produce oxygen as a by-product of photosynthesis.

Initially, the newly released oxygen was readily consumed by chemical reactions with other atoms and molecules (particularly iron) in the ocean. It seems that large quantities of iron were released into the early ocean through submarine volcanism and associated hydrothermal vents. Iron has tremendous affinity for oxygen. When these two elements join, they become iron oxide (rust). As it accumulated on the seafloor, these early iron oxide deposits created alternating layers of iron-rich rocks and chert, called **banded iron formations** (**FIGURE 19.7**). Most banded iron deposits accumulated in the Precambrian between 3.5 and 2 billion years ago and represent the world's most important reservoir of iron ore.

As the number of oxygen-generating organisms increased, oxygen began to build in the atmosphere. Chemical analyses of rocks suggest that a significant amount of oxygen appeared in the atmosphere as early as 2.2 billion years ago and increased steadily until it reached

FIGURE 19.6 Earth's first enduring atmosphere was formed by a process called outgassing, which continues today from hundreds of active volcanoes worldwide. *(Photo by Game McGimsey/CORBIS)*

FIGURE 19.7 These layered, iron-rich rocks, called banded iron formations, were deposited during the Precambrian. Much of the oxygen generated as a by-product of photosynthesis was readily consumed by chemical reactions with iron to produce these rocks. *(Courtesy Spencer R. Titley)*

stable levels about 1.5 billion years ago. Obviously, the availability of free oxygen had a positive impact on the development of life.

Another significant benefit of the "oxygen explosion" is that oxygen (O_2) molecules readily absorb ultraviolet radiation and rearrange themselves to form *ozone* (O_3). Today ozone is concentrated above the surface in a layer called the *stratosphere* where it absorbs much of the ultraviolet radiation that strikes the upper atmosphere. For the first time, Earth's surface was protected from this type of solar

DID YOU KNOW?
The earliest evidence for life on Earth comes from hydrocarbon residues of bacteria (prokaryotes) in metamorphosed sedimentary rocks in Greenland and date from about 3.8 billion years ago.

radiation, which is particularly harmful to DNA. Marine organisms had always been shielded from ultraviolet radiation by the oceans, but the development of the atmosphere's protective ozone layer made the continents more hospitable.

Evolution of the Oceans

About 4 billion years ago, as much as 90 percent of the current volume of seawater was contained in the ocean basins. Because the primitive atmosphere was rich in carbon dioxide, sulfur dioxide, and hydrogen sulfide, the earliest rainwater was highly acidic—to an even greater degree than the acid rain that damaged lakes and streams in eastern North America during the latter part of the 20th century. Consequently, Earth's rocky surface weathered at an accelerated rate. The products released by chemical weathering included atoms and molecules of various substances, including sodium, calcium, potassium, and silica, that were carried into the newly formed oceans. Some of these dissolved substances precipitated to become chemical sediment that mantled the ocean floor. Other substances formed soluble salts, which increased the salinity of seawater. Research suggests that the salinity of the oceans increased rapidly at first but has remained constant over the last two billion years.

Earth's oceans also serve as a depository for tremendous volumes of carbon dioxide, a major constituent in the primitive atmosphere. This is significant because carbon dioxide is a greenhouse gas that strongly influences the heating of the atmosphere. Venus, once thought to be very similar to Earth, has an atmosphere composed of 97 percent carbon dioxide that produced a "runaway" greenhouse effect. As a result, its surface temperature is 475 °C (900 °F).

Carbon dioxide is readily soluble in seawater where it often joins other atoms or molecules to produce various chemical precipitates. The most common compound generated by this process is calcium carbonate ($CaCO_3$), which makes up limestone, the most abundant chemical sedimentary rock. About 542 million years ago, marine organisms began to extract calcium carbonate from seawater to make their shells and other hard parts.

Included were trillions of tiny marine organisms such as foraminifera, whose shells were deposited on the seafloor at the end of their life-cycle. Today, some of these deposits can be observed in the chalk beds exposed along the White Cliffs of Dover, England, shown in **FIGURE 19.8**. By "locking up" carbon dioxide, these limestone deposits store this greenhouse gas so it cannot easily reenter the atmosphere.

CONCEPT CHECK 19.3

❶ What is meant by outgassing, and what modern phenomenon serves that role today?

❷ Outgassing produced Earth's early atmosphere, which was rich in what two gases?

❸ Why is the evolution of a type of bacteria that employed photosynthesis to produce food important to most modern organisms?

❹ What was the source of water for the first oceans?

❺ How does the ocean remove carbon dioxide from the atmosphere? What role do tiny marine organisms, such as foraminifera, play?

Precambrian History: The Formation of Earth's Continents

Earth's first 4 billion years are encompassed in the time span called the *Precambrian*. Representing nearly 90 percent of Earth's history, the Precambrian is divided into the *Archean eon* ("ancient age") and the *Proterozoic eon* ("early life"). Our knowledge of this ancient time is limited because much of the early rock record has been obscured by the very Earth processes you have been studying, especially plate tectonics, erosion, and deposition. Most Precambrian rocks lack

DID YOU KNOW?
Fossil evidence suggests that the first organisms with relatively complex cells containing a nucleus (called eukaryotes) appeared about 2.7 billion years ago.

fossils, which hinders correlation of rock units. In addition, rocks this old are metamorphosed and deformed, extensively eroded, and frequently concealed by younger strata. Indeed, Precambrian history is written in scattered, speculative episodes, like a long book with many missing chapters.

Earth's First Continents

More than 95 percent of Earth's human population lives on the continents—the other 5 percent are people living on volcanic islands such as the Hawaiian Islands and Iceland. These islanders inhabit pieces of oceanic crust that were thick enough to rise above sea level.

What differentiates continental crust from oceanic crust? Recall that oceanic crust is a relatively dense (3.0 g/cm^3), homogeneous layer of basaltic rocks derived from partial melting of the rocky, upper mantle. In addition, oceanic crust is thin, averaging only 7 kilometers in thickness. Continental crust, on the other hand, is composed of a variety of rock types, has an average thickness of nearly 40 kilometers, and contains a large percentage of low-density (2.7 g/cm^3), silica-rich rocks such as granite.

The significance of these differences cannot be overstated in a review of Earth's geologic evolution. Oceanic crust, because it is relatively thin and dense, is found several kilometers below sea level—unless, of course, it has been pushed onto a landmass by tectonic forces. Continental crust, because of its great thickness and lower density, extends well above sea level. Also, recall that oceanic crust of normal thickness readily subducts, whereas thick, buoyant blocks of continental crust resist being recycled into the mantle.

MAKING CONTINENTAL CRUST. Earth's first crust was probably basalt, similar to that generated at modern oceanic ridges, but because physical evidence no longer exists we are not certain. The hot, turbulent mantle that most likely existed during the Archean eon recycled most of this material back into the mantle. In fact, it may have been continuously recycled, in much the same way that the "crust" that forms on a lava lake is repeatedly replaced with fresh lava from below (**FIGURE 19.9**).

The oldest preserved continental rocks occur as small, highly deformed terranes, which are incorporated within somewhat younger blocks of continental crust

(**FIGURE 19.10**). The oldest of these is the 4 billion-year-old Acasta gneiss located in the Slave Province of Canada's Northwest Territories.

The formation of continental crust is simply a continuation of the gravitational segregation of Earth materials that began during the final accretionary stage of our planet. After the metallic core and rocky mantle evolved, low density, silica-rich minerals were gradually extracted from the mantle to form continental crust. This is an ongoing, multistage process during which partial melting of ultramafic mantle rocks (peridotite) generates basaltic rocks. Next, the melting of basaltic rocks produces magmas that crystallize to form felsic, quartz-bearing rocks (see Chapter 3). However, little is known about the details of the mechanisms that generated these silica-rich rocks during the Archean.

Most geologists agree that some type of platelike motion operated early in Earth's history. In addition, hot spot volcanism likely played a role as well. However, because the mantle was hotter in the Archean than it is today, both of these phenomena would have progressed at higher rates than their modern counterparts. Hot spot volcanism is thought to have created immense shield volcanoes as

FIGURE 19.8 Prominent chalk deposit, the White Chalk Cliffs, Sussex, England. Similar deposits are also found in northern France. *(Photo by Jon Arnold/Getty Images)*

FIGURE 19.9 Rift pattern on lava lake. The crust covering this lava lake is continually being replaced with fresh lava from below, similar to how Earth's crust was recycled early in its history. *(Photo by Juerg Alean/www.stromboli.net)*

FIGURE 19.10 These rocks at Isua, Greenland, some of the world's oldest, have been dated at 3.8 billion years. *(Photo courtesy of James L. Amos/CORBIS)*

well as oceanic plateaus. Simultaneously, subduction of oceanic crust generated volcanic island arcs. Collectively, these relatively small crustal fragments represent the first phase in creating stable, continent-size landmasses.

FROM CONTINENTAL CRUST TO CONTINENTS. According to one model, the growth of large continental masses was accomplished through collision and accretion of various types of terranes as illustrated in **FIGURE 19.11**. This type of collision tectonics deformed and metamorphosed sediments caught between converging crustal fragments, thereby shortening and thickening the developing crust. Within the deepest regions of these collision zones, partial melting of the thickened crust generated silica-rich magmas that ascended and intruded the rocks above. The result was the formation of large crustal provinces that, in turn, accreted with others to form even larger crustal blocks called **cratons.** (The portion of a modern craton that is exposed at the surface is referred to as a *shield.*) The assembly of a large craton involves the accretion of several crustal blocks that cause major mountain-building episodes similar to India's collision with Asia. **FIGURE 19.12** shows the extent of crustal material that was produced during the Archean and Proterozoic eons. This was accomplished by the collision and accretion of many thin, highly mobile terranes into nearly recognizable continental masses.

Although the Precambrian was a time when much of Earth's continental crust was generated, a substantial amount of crustal material was destroyed as well. Crust can be lost by either weathering and erosion or by direct reincorporation into the mantle through subduction. Evidence suggests that during much of the Archean, thin slabs of continental crust were eliminated, mainly by subduction into the mantle. However, by about 3 billion years ago, cratons grew sufficiently large and thick to resist subduction. After that time, weathering and erosion became the primary processes of crustal destruction. By the close of the Precambrian, an estimated 85 percent of the modern continental crust had formed.

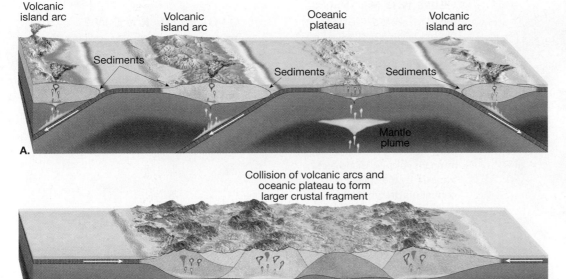

FIGURE 19.11 The growth of large continental masses was accomplished through the collision and accretion of various types of terranes.

The Making of North America

North America provides an excellent example of the development of continental crust and its piecemeal assembly into a continent. Notice in **FIGURE 19.13** that very little continental crust older than 3.5 billion years remains. In the late Archean, between 3 and 2.5 billion years ago, there was a period of major continental growth. During this span, the accretion of numerous island arcs and other fragments generated several large crustal provinces. North America contains some of these crustal units, including the Superior and Hearne/Rae cratons shown in Figure 19.13. It remains unknown where these ancient continental blocks formed.

About 1.9 billion years ago these crustal provinces collided to produce the Trans-Hudson mountain belt (Figure 19.13). (This mountain-building episode was not restricted

FIGURE 19.12 Illustration showing the extent of crustal material remaining from the Archean and Proterozoic eons.

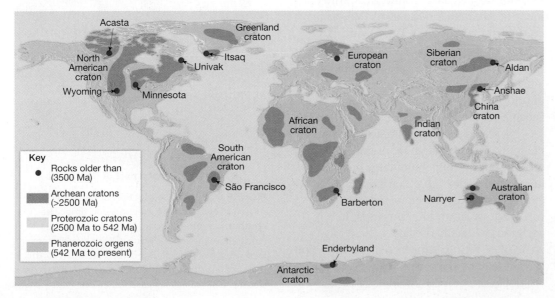

to North America, because ancient deformed strata of similar age are also found on other continents.) This event built the North American craton, around which several large and numerous small crustal fragments were later added. These late arrivals include Blue Ridge and Piedmont provinces of the Appalachians. Additionally, several terranes were added to the western margin of North America during the Mesozoic and Cenozoic eras to generate the mountainous North American Cordillera.

Supercontinents of the Precambrian

Supercontinents are large landmasses that contain all, or nearly all, the existing continents. Pangaea was the most recent, but certainly not the only, supercontinent to exist in the geologic past. The earliest well-documented supercontinent, *Rodinia*, formed during the Proterozoic eon about 1.1 billion years ago. Although its reconstruction is still being researched, it is clear that Rodinia's configuration was quite different from Pangaea's (**FIGURE 19.14**). One obvious distinction is that North America was located near the center of this ancient landmass.

Between 800 and 600 million years ago, Rodinia gradually split apart. By the end of the Precambrian, many of the fragments reassembled producing a large landmass in the Southern Hemisphere called *Gondwana*, comprised mainly of present-day South America, Africa, India, Australia, and Antarctica (**FIGURE 19.15**). Other continental fragments also developed—North America, Siberia, and northern Europe. We will consider the fate of these Precambrian landmasses later in the chapter.

SUPERCONTINENT CYCLE. The idea that rifting and dispersal of one supercontinent is followed by a long period during which the fragments are gradually reassembled into a new supercontinent with a different configuration is called the **supercontinent cycle.** The assembly and dispersal of supercontinents had a profound impact on the evolution of Earth's continents. In addition, this phenomenon greatly influenced global climates and contributed to periodic episodes of rising and falling sea level.

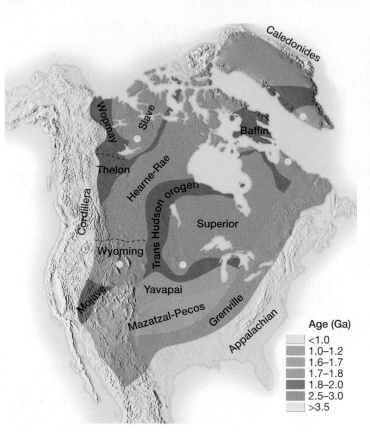

FIGURE 19.13 Map showing the major geological provinces of North America and their ages in billions of years (Ga). It appears that North America was assembled from crustal blocks that were joined by processes very similar to modern plate tectonics. These ancient collisions produced mountainous belts that include remnant volcanic island arcs trapped by the colliding continental fragments.

Age (Ga)	
	<1.0
	1.0–1.2
	1.6–1.7
	1.7–1.8
	1.8–2.0
	2.5–3.0
	>3.5

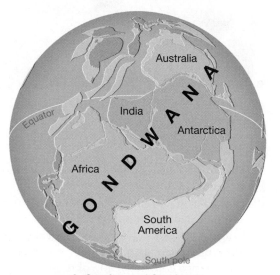

A. Continent of Gondwana

SUPERCONTINENTS AND CLIMATE. As continents move, the patterns of ocean currents and global winds change, which influences the global distribution of temperature and precipitation. One example of how a supercontinent's dispersal influenced climate is the formation of the Antarctic ice sheet. Although eastern Antarctica remained over the South Pole for more than 100 million years, it was not glaciated until about 25 million years ago. Prior to this period of glaciation, South America was connected to the Antarctic Peninsula. This arrangement of landmasses helped maintain a circulation pattern in which warm ocean currents reached the coast of Antarctica as shown in **FIGURE 19.16A**. This is similar to the way in which the modern Gulf Stream keeps Iceland mostly ice free, despite its name. However, as South America separated from Antarctica, it moved northward, permitting ocean circulation to flow from west to east around the entire continent of Antarctica (**FIGURE 19.16B**). This cold current, called the West Wind Drift, effectively isolated the entire Antarctic coast from the warm, poleward-directed currents in the southern oceans. As a result, most of the Antarctic landmass became covered with glacial ice.

Local and regional climates have also been impacted by large mountain systems created by the collision of large cratons. Because of their high elevations, mountains exhibit markedly lower average temperatures than surrounding lowlands. In addition, when air rises over these lofty structures, lifting "squeezes" moisture from the air, leaving the region downwind relatively dry. A modern analogy is the wet,

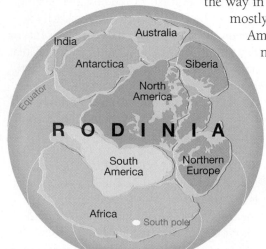

FIGURE 19.14 Simplified drawing showing one of several possible configurations of the supercontinent Rodinia. For clarity, the continents are drawn with somewhat modern shapes, not the shapes of 1 billion years ago. (*After P. Hoffman, J. Rogers, and others*)

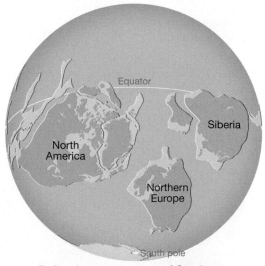

B. Continents not a part of Gondwana

FIGURE 19.15 Reconstruction of Earth as it may have appeared in late Precambrian time. The southern continents were joined into a single landmass called Gondwana. Other landmasses that were not part of Gondwana include North America, northwestern Europe, and northern Asia. (*After P. Hoffman, J. Rogers, and others*)

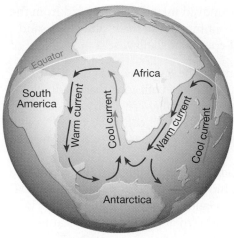 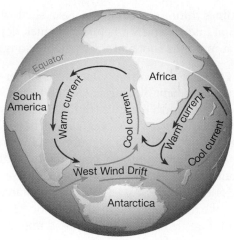

A. 50 million years ago   **B. Present**

FIGURE 19.16 Comparison of the oceanic circulation pattern 50 million years ago with that of the present. When South America separated from Antarctica, the West Wind Drift developed, which effectively isolated the entire Antarctic coast from the warm, poleward-directed currents in the southern oceans. This led to the eventual covering of much of Antarctica with glacial ice.

Geologic History of the Phanerozoic: The Formation of Earth's Modern Continents

The time span since the close of the Precambrian, called the *Phanerozoic eon*, encompasses 542 million years and is divided into three eras: *Paleozoic, Mesozoic,* and *Cenozoic.* The beginning of the Phanerozoic is marked by the appearance of the first life-forms with hard parts such as shells, scales, bones, or teeth—all of which greatly enhance the possibility of an organism being preserved in the fossil record (**FIGURE 19.17**).* Thus, the study of Phanerozoic

*For more about this see the discussion on "Conditions Favoring Preservation" in Chapter 18, p. 448.

heavily forested western slopes of the Sierra Nevada compared to the dry climate of the Great Basin desert that lies directly downwind (see Figure 12.4, p. 291)

SUPERCONTINENTS AND SEA LEVEL CHANGES. Significant and numerous sea level changes have been documented in geologic history, many of which appear related to the assembly and dispersal of supercontinents. If sea level rises, shallow seas advance onto the continents. Evidence for periods when the seas advanced onto the continents include thick sequences of ancient marine sedimentary rocks that blanket large areas of modern landmasses— including much of the eastern two thirds of the United States.

The supercontinent cycle and sea level changes are directly related to rates of *seafloor spreading*. When the rate of spreading is rapid, as it is along the East Pacific Rise today, the production of warm oceanic crust is also high. Because warm oceanic crust is less dense (takes up more space) than cold crust, fast spreading ridges occupy more volume in the ocean basins than do slow spreading centers. (Think of getting into a tub filled with water.) As a result, when the rates of seafloor spreading increase, sea level rises. This, in turn, causes shallow seas to advance onto the low-lying portions of the continents.

CONCEPT CHECK 19.4

❶ Explain why Precambrian history is more difficult to decipher than more recent geological history.

❷ Briefly describe how cratons come into being.

❸ What is the supercontinent cycle?

❹ How can the movement of continents trigger climate change?

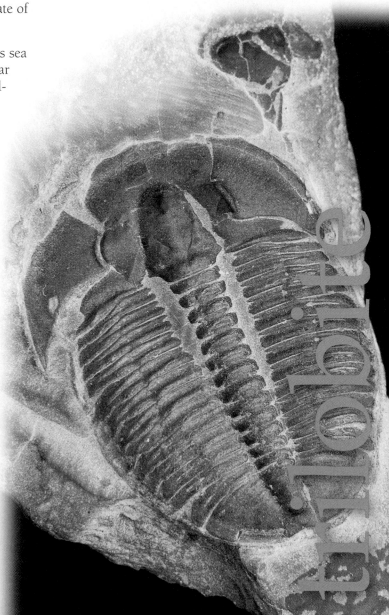

FIGURE 19.17 Trilobite fossil. Trilobites were very common Paleozoic life-forms. *(Photo by Ed Reschke/Photolibrary)*

crustal history was aided by the availability of fossils, which improved our ability to date and correlate geologic events. Moreover, because every organism is associated with its own particular niche, the greatly improved fossil record provided invaluable information for deciphering ancient environments.

Paleozoic History

As the Paleozoic era opened, North America hosted no living things, neither plant nor animal. There were no Appalachian or Rocky Mountains; the continent was largely a barren lowland.

Several times during the early Paleozoic, shallow seas moved inland then receded from the continental interior and left behind the thick deposits of limestone, shale, and clean sandstone that mark the shorelines of these previous mid-continent shallow seas.

FORMATION OF PANGAEA. One of the major events of the Paleozoic was the formation of the supercontinent of Pangaea, which began with a series of collisions that gradually joined together North America, Europe, Siberia, and other smaller crustal fragments (**FIGURE 19.18**). These events eventually generated a large northern continent called *Laurasia*. This tropical landmass supported warm wet conditions that led to the formation of vast swamps that eventually converted to coal.

Simultaneously, the vast southern continent of *Gondwana* encompassed five continents—South America, Africa, Australia, Antarctica, India, and perhaps portions of China. Evidence of extensive continental glaciation places this landmass near the South Pole. By the end of the Paleozoic, Gondwana had migrated northward and collided with Laurasia, culminating in the formation of the supercontinent *Pangaea.*

The accretion of Pangaea spans more than 300 million years and resulted in the formation of several mountain belts. The collision of northern Europe (mainly Norway) with Greenland produced the Caledonian Mountains, whereas the joining of northern Asia (Siberia) and Europe created the Ural Mountains. Northern China is also thought to have accreted to Asia by the end of the Paleozoic, whereas southern China may not have become part of Asia until after Pangaea had begun to rift. (Recall that India did not begin to accrete to Asia until about 50 million years ago.)

Pangaea reached its maximum size about 250 million years ago as Africa collided with North America (Figure 19.18D). This event marked the final episode of growth in the long history of the Appalachian Mountains (see Chapter 17).

Mesozoic History

Spanning about 186 million years, the Mesozoic era is divided into three periods: the *Triassic, Jurassic,* and *Cretaceous.* Major geologic events of the Mesozoic include the break-up of Pangaea and the evolution of our modern ocean basins.

The Mesozoic era began with much of the world's continents above sea level. The exposed Triassic strata are primarily red sandstones and mudstones that lack marine fossils, features that indicate a terrestrial environment. (The red color in sandstone comes from the oxidation of iron.)

As the Jurrasic period opened, the sea invaded western North America. Adjacent to this shallow sea, extensive continental sediments were deposited on what is now the Colorado Plateau. The most prominent is the Navajo Sandstone, a cross-bedded, quartz-rich layer, that in some places approaches 300 meters (1000 feet) thick. These remnants of massive dunes indicate that an enormous desert occupied much of the American Southwest during early Jurassic times (**FIGURE 19.19**). Another well-known Jurassic deposit is the Morrison Formation—the world's richest storehouse of dinosaur fossils. Included are the fossilized bones of massive dinosaurs such as Apatosaurus (formerly Brontosaurus), Brachiosaurus, and Stegosaurus.

As the Jurassic period gave way to the Cretaceous, shallow seas again encroached upon much of western North America, as

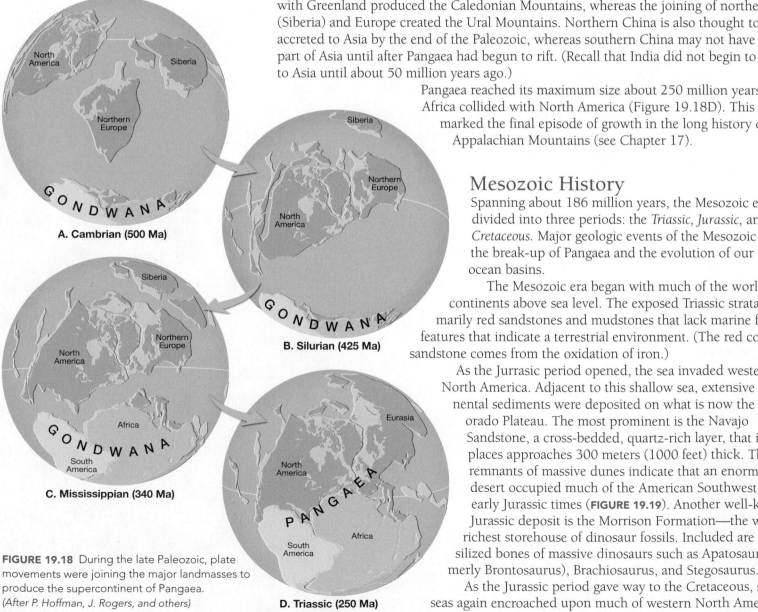

FIGURE 19.18 During the late Paleozoic, plate movements were joining the major landmasses to produce the supercontinent of Pangaea. *(After P. Hoffman, J. Rogers, and others)*

A. Cambrian (500 Ma)

B. Silurian (425 Ma)

C. Mississippian (340 Ma)

D. Triassic (250 Ma)

well as the Atlantic and Gulf coastal regions. This led to the formation of "coal swamps" similar to those of the Paleozoic era. Today, the Cretaceous coal deposits in the western United States and Canada are economically important. For example, on the Crow Native American reservation in Montana, there are nearly 20 billion tons of high-quality, Cretaceous-age coal.

Another major event of the Mesozoic era was the break-up of Pangaea. About 165 million years ago a rift developed between what is now North America and western Africa, which marked the birth of the Atlantic Ocean. As Pangaea gradually broke apart, the westward-moving North American plate began to override the Pacific basin. This tectonic event triggered a continuous wave of deformation that moved inland along the entire western margin of North America. By Jurassic times, subduction of the Farallon plate had begun to produce the chaotic mixture of rocks

that exist today in the Coast Ranges of California. Further inland, igneous activity was widespread, and for more than 100 million years volcanism was rampant as huge masses of magma rose within a few miles of Earth's surface. The remnants of this activity include the granitic plutons of the Sierra Nevada, as well as the Idaho batholith, and British Columbia's Coast Range batholith.

Tectonic activity initiated in the Jurassic continued throughout the Cretaceous. Compressional forces moved huge rock units in a shingle-like fashion toward the east. Across much of North America's western margin, older rocks were thrust eastward over younger strata, for distances exceeding 150 kilometers (90 miles). Ultimately, this activity was responsible for creating the vast Northern Rockies that extend from Wyoming to Alaska.

Toward the end of the Mesozoic, the southern portions of the Rocky Mountains

developed. This mountain-building event, called the *Laramide Orogeny,* occurred when large blocks of deeply buried Precambrian rocks were lifted nearly vertically along steeply dipping faults, upwarping the overlying younger sedimentary strata. The mountain ranges produced by the Laramide Orogeny include Colorado's Front Range, the Sangre de Cristo of New Mexico and Colorado, and the Bighorns of Wyoming.

Cenozoic History

The Cenozoic era, or "era of recent life," encompasses the last 65.5 million years of Earth history. It was during this span that the physical landscapes and life-forms of our modern world came into existence. The Cenozoic era represents a considerably smaller fraction of geologic time than either the Paleozoic or the Mesozoic. Nevertheless, much more is known about this time

FIGURE 19.19 These massive, cross-bedded sandstone cliffs in Zion National Park are the remnants of ancient sand dunes. *(Photo by Rick Whitacre/Shutterstock)*

span because the rock formations are more widespread and less disturbed than those of any preceding era.

Most of North America was above sea level during the Cenozoic era. However, the eastern and western margins of the continent experienced markedly contrasting events because of their different plate boundary relationships. The Atlantic and Gulf coastal regions, far removed from an active plate boundary, were tectonically stable. By contrast, western North America was the leading edge of the North American plate. As a result, plate interactions during the Cenozoic account for many events of mountain building, volcanism, and earthquakes.

EASTERN NORTH AMERICA. The stable continental margin of eastern North America was the site of abundant marine sedimentation. The most extensive deposition surrounded the Gulf of Mexico, from the Yucatan Peninsula to Florida, where a massive buildup of sediment caused the crust to downwarp. In many instances, faulting created structures in which oil and natural gas accumulated. Today, these and other petroleum traps are the Gulf Coast's most economically important resource, evidenced by numerous offshore drilling platforms.

By early Cenozoic time, most of the original Appalachians had eroded to create a low plain. Later, isostatic adjustments again raised the region and rejuvenated its

rivers. Streams eroded with renewed vigor, gradually sculpting the surface into its present-day topography. Sediments from this erosion were deposited along the eastern continental margin, where they accumulated to a thickness of many kilometers. Today, portions of the strata deposited during the Cenozoic are exposed as the gently sloping Atlantic and Gulf coastal plains, where a large percentage of the eastern and southeastern United States population resides.

WESTERN NORTH AMERICA. In the West, the Laramide Orogeny responsible for building the southern Rocky Mountains was coming to an end. As erosional forces lowered the mountains, the basins between uplifted ranges began to fill with sediment. East of the Rockies, a large wedge of sediment from the eroding mountains created the Great Plains.

Beginning in the Miocene epoch about 20 million years ago, a broad region from northern Nevada into Mexico experienced crustal extension that created more than 150 fault-block mountain ranges. Today, they rise abruptly above the adjacent basins, creating the Basin and Range Province (see Chapter 17).

As the Basin and Range Province was developing, the entire western interior of the continent gradually uplifted. This event reelevated the Rockies and rejuvenated many of the West's major rivers. As the rivers became incised, many spectacular gorges were created, including the

Grand Canyon of the Colorado River, the Grand Canyon of the Snake River, and the Black Canyon of the Gunnison River.

Volcanic activity was also common in the West during much of the Cenozoic. Beginning in the Miocene epoch, great volumes of fluid basaltic lava flowed from fissures in portions of present-day Washington, Oregon, and Idaho. These eruptions built the extensive (1.3 million square miles) Columbia Plateau. Immediately west of the Columbia Plateau, volcanic activity was different in character. Here, more viscous magmas with higher silica content erupted explosively, creating the Cascades, a chain of stratovolcanoes extending from northern California into Canada, some of which are still active (**FIGURE 19.20**).

A final episode of deformation occurred in the late Cenozoic, creating the Coast Ranges that stretch along the Pacific Coast. Meanwhile, the Sierra Nevada were faulted and uplifted along their eastern flank, forming the impressive mountains we see today.

As the Cenozoic was drawing to a close, the effects of mountain building, volcanic activity, isostatic adjustments, and extensive erosion and sedimentation created the physical landscape we know today. All that remained of Cenozoic time was the final 2.6-million-year episode called the Quaternary period. During this most recent, and

FIGURE 19.20 Mount Hood, Oregon. This volcano is one of several large composite cones that comprise the Cascade Range. (*Photo by John M. Roberts/CORBIS/Stock Market*)

ongoing, phase of Earth's history, in which humans evolved, the action of glacial ice, wind, and running water added the finishing touches to our planet's long, complex geologic history.

CONCEPT CHECK 19.5

❶ During which period of geologic history did the supercontinent of Pangaea come into existence?

❷ During which period of geologic history did Pangaea begin to break apart?

❸ Describe the climate during the early Jurassic period.

Earth's First Life

The oldest fossils provide evidence that life on Earth was established at least 3.5 billion years ago. Microscopic fossils similar to modern cyanobacteria (formerly known as blue-green algae) have been found in silica-rich chert deposits worldwide. Notable examples include southern Africa, where rocks date to more than 3.1 billion years, and the Lake Superior region of western Ontario and northern Minnesota, where the Gunflint Chert contains some fossils that are older than 2 billion years. Chemical traces of organic matter in rocks of greater age have led paleontologists to strongly suggest that life may have existed 3.8 billion years ago.

How did life begin? This question sparks considerable debate, and hypotheses abound. Requirements for life, assuming the presence of a hospitable environment, include the chemical raw materials that form the essential molecules of DNA, RNA, and proteins. These substances require organic compounds called *amino acids*. The first amino acids may have been synthesized from methane and ammonia, both of which were plentiful in Earth's primitive atmosphere. Some scientists suspect these gases could have been easily reorganized into useful organic molecules by ultraviolet light, while others consider lightning to have been the impetus, as the well-known experiments conducted by Stanley Miller and Harold Urey attempted to demonstrate.

Other researchers suggest that amino acids arrived "ready-made," delivered by asteroids or comets that collided with a young Earth. A group of meteorites (debris from asteroids and comets that strike Earth) called *carbonaceous chrondrites* known to contain amino acid-like organic compounds, led some to hypothesize that early life may have had an extraterrestrial beginning.

Yet another hypothesis proposes that the organic material needed for life came from the methane and hydrogen sulfide that spews from deep-sea hydrothermal vents (black smokers). Is it also possible that life originated within a hot spring similar to those in Yellowstone National Park? Some origin-

of-life researchers consider this scenario highly improbable as the scalding temperatures would have destroyed any early types of self-replicating molecules. Their supposition is that life would have first appeared along sheltered stretches of ancient beaches, where waves and tides brought together various organic materials formed in the Precambrian oceans.

Regardless of where or how life originated, it is clear that the journey from "then" to "now" involved change (**FIGURE 19.21**). The first known organisms were single-cell bacteria called **prokaryotes,** which means their genetic material (DNA) is *not separated* from the rest of the cell by a nucleus. Because oxygen was absent from Earth's early atmosphere and oceans, the first organisms employed anaerobic (without oxygen) metabolism to extract energy from "food." Their food source was likely organic molecules in their surroundings, but the supply was very limited. Later, bacteria evolved that used solar energy to synthesize organic compounds (sugars). This event was an important turning point in evolution—for the first time organisms had the capability of producing food for themselves as well as for other life-forms.

Recall that photosynthesis by ancient cyanobacteria, a type of prokaryote, contributed to the gradual rise in the level of oxygen, first in the ocean and later in the atmosphere. Thus, these early organisms dramatically transformed our planet. Fossil evidence for the existence of these microscopic bacteria includes distinctively layered mounds of calcium carbonate, called **stromatolites** (**FIGURE 19.22A**). Stromatolites are limestone mats built up by lime-accreting bacteria. What is known about these ancient fossils comes mainly from modern stromatolites like those found in Shark Bay, Australia (**FIGURE 19.22B**).

The oldest fossils of more advanced organisms, called **eukaryotes,** are about 2.1 billion years old. The first eukaryotes were microscopic, water-dwelling organisms. Unlike prokaryotes, the cellular structures of eukaryotes contain nuclei. This distinctive structure is what all multicellular organisms that now inhabit our planet—trees, birds, fishes, reptiles, and humans—have in common.

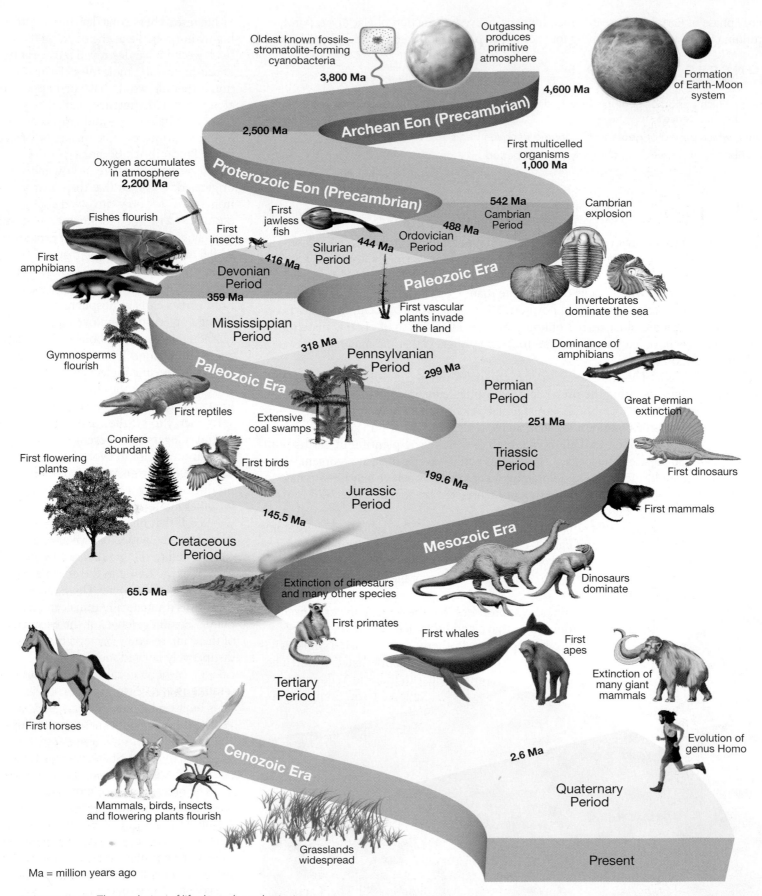

Oldest known fossils–
stromatolite-forming
cyanobacteria
3,800 Ma

Outgassing
produces
primitive
atmosphere

4,600 Ma

Formation
of Earth-Moon
system

2,500 Ma

Archean Eon (Precambrian)

First multicelled
organisms
1,000 Ma

Oxygen accumulates
in atmosphere
2,200 Ma

Proterozoic Eon (Precambrian)

542 Ma
Cambrian
Period

Cambrian
explosion

Fishes flourish

First
insects

First
jawless
fish

488 Ma
Ordovician
Period

444 Ma

416 Ma

Silurian
Period

Paleozoic Era

First
amphibians

Devonian
Period

359 Ma

First vascular
plants invade
the land

Invertebrates
dominate the sea

Mississippian
Period

318 Ma

Pennsylvanian
Period

299 Ma

Dominance of
amphibians

Gymnosperms
flourish

Paleozoic Era

Permian
Period

251 Ma

Great Permian
extinction

First reptiles

Extensive
coal swamps

Triassic
Period

First dinosaurs

Conifers
abundant

First birds

199.6 Ma

First flowering
plants

Jurassic
Period

First mammals

145.5 Ma

Mesozoic Era

Cretaceous
Period

Dinosaurs
dominate

65.5 Ma

Extinction of dinosaurs
and many other species

First primates

First whales

First
apes

Extinction of
many giant
mammals

Tertiary
Period

First horses

Cenozoic Era

2.6 Ma

Evolution of
genus Homo

Quaternary
Period

Mammals, birds, insects
and flowering plants flourish

Grasslands
widespread

Present

Ma = million years ago

FIGURE 19.21 The evolution of life through geologic time.

A.   **B.**

FIGURE 19.22 Stromatolites are among the most common Precambrian fossils. **A.** Precambrian fossil stromatolites composed of calcium carbonate deposited by algae. *(Photo by Sinclair Stammers/Photo Researchers, Inc.)* **B.** Modern stromatolites growing in shallow seas, western Australia. *(Photo by Bill Bachman/Photo Researchers, Inc.)*

During much of the Precambrian, life consisted exclusively of single-celled organisms. It wasn't until perhaps 1.5 billion years ago that multicelled eukaryotes evolved. Green algae, one of the first multicelled organisms, contained chloroplasts (used in photosynthesis) and were the ancestors of modern plants. The first primitive marine animals did not appear until somewhat later, perhaps 600 million years ago (**FIGURE 19.23**).

FIGURE 19.23 Ediacaran fossil. The Ediacarans are a group of sea-dwelling animals that may have come into existence about 600 million years ago. These soft-bodied organisms were up to one meter in length and are the oldest animal fossils so far discovered. *(Photo courtesy of the South Australian Museum)*

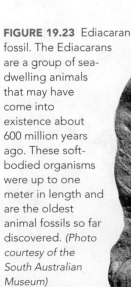

Fossil evidence suggests that organic evolution progressed at an excruciatingly slow pace until the end of the Precambrian. At this time, Earth's continents were barren, and the oceans were populated primarily by organisms too small to be seen with the naked eye. Nevertheless, the stage was set for the evolution of larger and more complex plants and animals.

CONCEPT CHECK 19.6

❶ Why do some researchers think that asteroids, or *carbonaceous chrondrites*, played an important role in the development of life on Earth?

❷ Compare *prokaryotes* with *eukaryotes*. Within which of these two groups do all multicelled organisms belong?

Paleozoic Era: Life Explodes

The Cambrian period marks the beginning of the Paleozoic era, a time span that saw the emergence of a spectacular variety of new life-forms. All major invertebrate (animals lacking backbones) groups made their appearance, including jellyfish, sponges, worms, mollusks (clams

and snails), and arthropods (insects and crabs). This huge expansion in biodiversity is often referred to as the *Cambrian explosion*.

But did the Cambrian explosion really happen? Recent research suggests that these life-forms may have gradually diversified late in the Precambrian but were not preserved in the fossil record. Considering that the Cambrian period marked the first time organisms developed hard parts, is it possible that the Cambrian event was simply an explosion of animal forms that grew in size and became "hard" enough to be fossilized?

Paleontologists may never answer that question definitively. They know, however, that hard parts clearly served many useful purposes and aided lifestyle adaptations. Sponges, for example, developed a network of fine, interwoven silica spicules that allowed them to grow larger and more erect, and thus capable of extending above the seafloor in search of food. Clams and snails secreted external shells of calcium carbonate that provided protection and allowed bodies to function in a more controlled environment. The successful trilobites developed a flexible exoskeleton of a protein called chitin (similar to a human fingernail), which permitted them to be mobile and search for food by burrowing through soft sediment (see Figure 19.17, page 473).

Early Paleozoic Life-Forms

The Cambrian period was the golden age of *trilobites*. More than 600 genera of these mud-burrowing scavengers flourished worldwide. The Ordovician marked the appearance of abundant cephalopods—

mobile, highly developed mollusks that became the major predators of their time (**FIGURE 19.24**). Descendants of these cephalopods include the squid, octopus, and chambered nautilus that inhabit our modern oceans. Cephalopods were the first truly large organisms on Earth, including one species that reached a length of nearly 10 meters (30 feet).

The early diversification of animals was driven, in part, by the emergence of predatory lifestyles. The larger mobile cephalopods preyed on trilobites that were typically smaller than a child's hand. The evolution of efficient movement was often associated with the development of greater sensory capabilities and more complex nervous systems. These early animals developed sensory devices for detecting light, odor, and touch.

Approximately 400 million years ago, green algae that had adapted to survive at the water's edge gave rise to the first multicellular land plants. The primary difficulty in sustaining plant life on land

was obtaining water and staying upright despite gravity and winds. These earliest land plants were leafless, vertical spikes about the size of a human index finger (**FIGURE 19.25**). The fossil record indicates that by the end of the Devonian, 40 million years later, there were forests with trees tens of meters tall—clear evidence that evolutionary processes were in full swing.

In the ocean, fishes perfected an internal skeleton as a new form of support and were the first creatures to have jaws. Armor-plated fishes that evolved during the Ordovician continued to adapt. Their armor plates thinned to lightweight scales that increased their speed and mobility.

FIGURE 19.24 During the Ordovician period (488–444 million years ago), the shallow waters of an inland sea over central North America contained an abundance of marine invertebrates. Shown in this reconstruction are straight-shelled cephalopods, trilobites, brachiopods, snails, and corals. (© The Field Museum, Neg. # GEO80820c, Chicago)

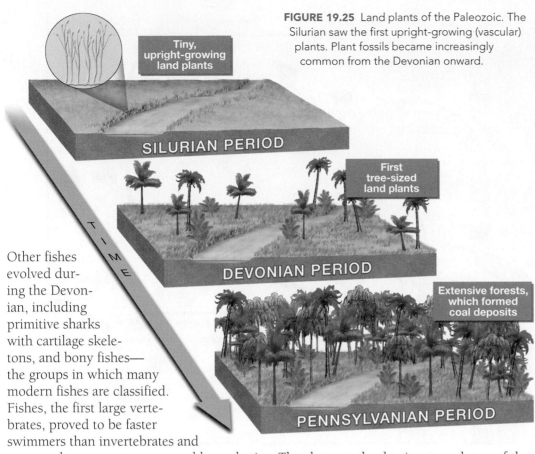

FIGURE 19.25 Land plants of the Paleozoic. The Silurian saw the first upright-growing (vascular) plants. Plant fossils became increasingly common from the Devonian onward.

Tiny, upright-growing land plants

SILURIAN PERIOD

First tree-sized land plants

DEVONIAN PERIOD

Extensive forests, which formed coal deposits

PENNSYLVANIAN PERIOD

Other fishes evolved during the Devonian, including primitive sharks with cartilage skeletons, and bony fishes—the groups in which many modern fishes are classified. Fishes, the first large vertebrates, proved to be faster swimmers than invertebrates and possessed more acute senses and larger brains. They became the dominant predators of the sea, which is why the Devonian period is often referred to as the "Age of the Fishes."

Vertebrates Move to Land

During the Devonian, a group of fishes called the *lobe-finned fish* began to adapt to terrestrial environments (**FIGURE 19.26**). Lobe-finned fishes had sacks that could be filled with air to supplement their "breathing" through gills. The first lobe-finned fish probably occupied freshwater tidal flats or small ponds near the ocean. Some began to use their fins to move from one pond to another in search of food, or to evacuate a deteriorating pond. This favored the evolution of a group of animals able to stay out of water longer and move on land more efficiently. By the late Devonian, lobe-finned fish had evolved into air-breathing amphibians with strong legs, yet retained a fish-like head and tail (**FIGURE 19.27**).

Modern amphibians, such as frogs, toads, and salamanders, are small and occupy limited biological niches. However, conditions during the late Paleozoic were ideal for these newcomers to land. Large tropical swamps that were teeming with large insects and millipedes extended across North America, Europe, and Siberia (**FIGURE 19.28**). With virtually no predatory risks, amphibians

A. Lobe-finned fish

B. Early amphibian

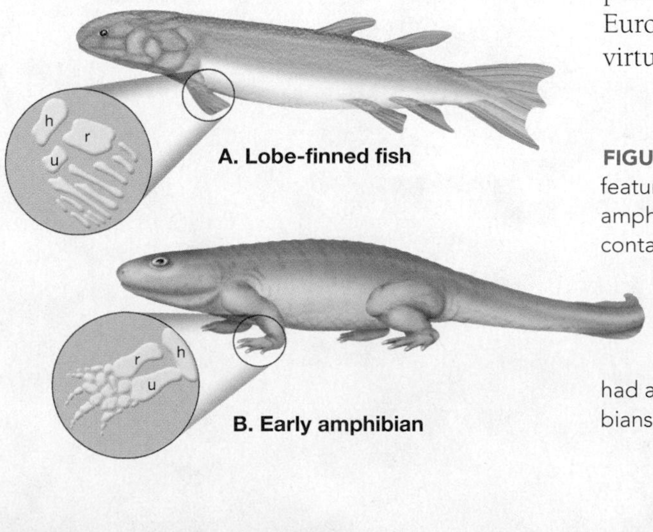

FIGURE 19.26 Comparison of the anatomical features of the lobe-finned fish and early amphibians. **A.** The fins on the lobe-finned fish contained the same basic elements (*h*, humerus, or upper arm; *r*, radius; and *u*, ulna, or lower arm) as those of the amphibians. **B.** This amphibian is shown with the standard five toes, but early amphibians had as many as eight toes. Eventually the amphibians evolved to have a standard toe count of five.

diversified rapidly. Some even took on lifestyles and forms similar to modern reptiles such as crocodiles.

Despite their success, amphibians were not fully adapted to life out of water. In fact, amphibian means "double life," because these animals need both the water from where they came and the land to which they moved. Amphibians are born in water, as exemplified by tadpoles, complete with gills and tails. These features disappear during the maturation process, resulting in air-breathing adults with legs.

The Great Permian Extinction

By the close of the Permian period, a mass extinction destroyed 70 percent of all land-dwelling vertebrate species, and perhaps 90 percent of all marine organisms. The late Permian extinction was the most significant of at least five mass extinctions that occurred over the past 500 million years. Each extinction wreaked havoc on the existing biosphere, wiping out large numbers of species. In each case, however, survivors entered new biological communities that were ultimately more diverse. Therefore, mass extinctions actually invigorated life on Earth, as the few hardy survivors eventually filled more environmental niches than those left behind by the victims.

Several mechanisms have been proposed to explain these ancient mass extinctions. Initially, paleontologists believed they were gradual events caused by a combination of climate change and biological forces, such as predation and competition. In the 1980s, a research team proposed the mass extinction of 65 million years ago was caused by the explosive impact of a large asteroid. This event, which caused the swift extinction of dinosaurs, is described later in the chapter.

Was the Permian extinction caused by a giant impact? For many years, researchers thought this was a possible explanation. However, scant evidence has been found to substantiate this hypothesis.

Another possible mechanism to explain the Permian extinction was voluminous eruptions of basaltic lavas, which began about 250 million years

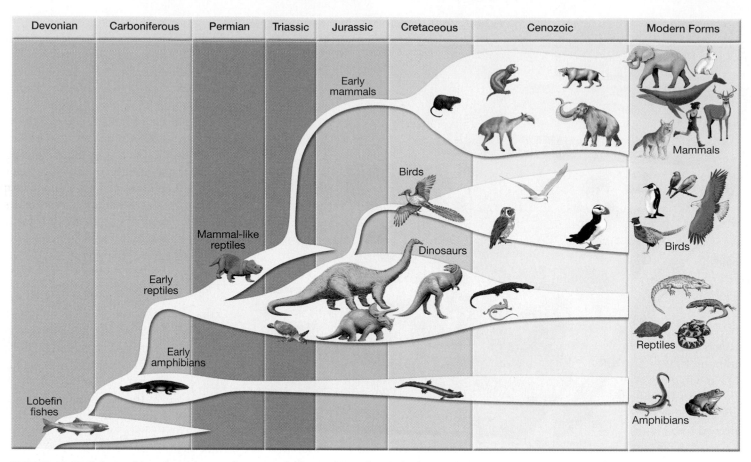

FIGURE 19.27 Relationships of various vertebrates and their evolution from a fish-like ancestor.

FIGURE 19.28 Restoration of a Pennsylvanian-age coal swamp (318 million to 299 million years ago). Shown are scale trees (left), seed ferns (lower left), and scouring rushes (right). Also note the large dragonfly. (© The Field Museum, Neg. # GEO85637c, Chicago. Photographer John Weinstein)

ago and covered thousands of square kilometers of the continents. (This period of volcanism produced the Siberian Traps located in northern Russia.) The release of carbon dioxide would certainly have enhanced greenhouse warming, and the emissions of sulfur dioxide probably resulted in copious amounts of acid rain.

A group of researchers have taken the global warming hypothesis still further. Although they agree that rapid greenhouse warming occurred, it alone would not have destroyed most plants because they are heat tolerant and consume CO_2 in photosynthesis. They contend, instead, that the trouble began in the ocean rather than on land.

Most organisms, including humans, use oxygen to metabolize food. However, some forms of bacteria employ *anaerobic* (without oxygen) metabolism. Under normal conditions, oxygen from the atmosphere is readily dissolved in seawater and is then evenly distributed to all depths by deep-water currents. This "oxygen-rich" water relegates "oxygen-hating" anaerobic bacteria to anoxic (oxygen-free) environments found in deep-water sediments.

An intense period of greenhouse warming caused by an extreme episode of volcanism would have heated the ocean surface, thereby significantly reducing the amount of oxygen absorbed by seawater (**FIGURE 19.29**). This condition favors deep-sea anaerobic bacteria, which generate toxic hydrogen sulfide as a waste gas. As these organisms proliferated, the amount of hydrogen sulfide dissolved in seawater would have steadily increased. Eventually, the concentration of hydrogen sulfide would reach a critical threshold and release large toxic bubbles into the atmosphere. On land, hydrogen sulfide is lethal to both plants and animals, but it is most destructive to oxygen-breathing marine life.

How plausible is this scenario? Remember that these ideas represent a *hypothesis*, a tentative explanation regarding a particular set of observations. Additional research about this and other hypotheses that relate to the Permian extinction continues.

CONCEPT CHECK 19.7

1. What is the *Cambrian explosion*?
2. What animal group was dominant in Cambrian seas?
3. What did plants have to overcome to move onto land?
4. What group of animals is thought to have left the ocean to become the first amphibians?
5. Why are amphibians not considered "true" land animals?

Mesozoic Era: Age of the Dinosaurs

As the Mesozoic era dawned, its life-forms were the survivors of the great Permian extinction. These organisms diversified in many ways to fill the biological voids created at the close of the Paleozoic. On land, conditions favored those that could adapt to drier climates. Among plants, gymnosperms were one such group. Unlike the first plants to invade the land, seed-bearing gymnosperms did not depend on free-standing water for fertilization. Consequently, these plants were not restricted to a life near the water's edge.

The gymnosperms quickly became the dominant trees of the Mesozoic. They included: cycads that resembled a large pineapple plant; ginkgoes that had fan-shaped leaves, much like their modern relatives; and the largest plants, the conifers, whose modern descendants include the pines, firs, and junipers. The best-known fossil occurrence of these ancient trees is in northern Arizona's Petrified Forest National Park. Here, huge petrified logs lie exposed at the surface, having been weathered from rocks of the Triassic Chinle Formation (**FIGURE 19.30**).

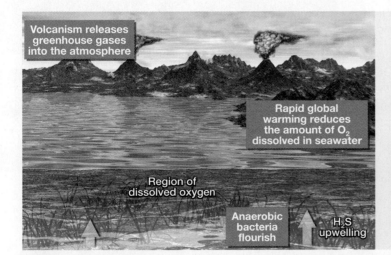
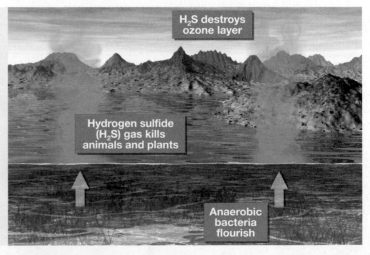

FIGURE 19.29 Model for the "Great Permian Extinction." Extensive volcanism released greenhouse gases, which resulted in extreme global warming. This condition reduced the amount of oxygen dissolved by seawater. This, in turn, favored "oxygen-hating" anaerobic bacteria, which generated toxic hydrogen sulfide as a waste gas. Eventually, the concentration of hydrogen sulfide reached a critical threshold and great bubbles of this toxin exploded into the atmosphere, wreaking havoc on organisms on land; but oxygen-breathing marine life was hit the hardest.

FIGURE 19.30 Petrified log of Triassic age in Arizona's Petrified Forest National Park. *(Photo by Michael Collier)*

Reptiles: The First True Terrestrial Vertebrates

Among the animals, reptiles readily adapted to the drier Mesozoic environment, thereby relegating amphibians to the wetlands where most remain today. Reptiles were the first true terrestrial animals with improved lungs for active lifestyles and "waterproof" skin that helped prevent the loss of body fluids. Most importantly, reptiles developed shell-covered eggs laid on land. The elimination of a water-dwelling stage (like the tadpole stage in frogs) was an important evolutionary advancement.

Of interest is the fact that the watery fluid within the reptilian egg closely resembles seawater in chemical composition. Because the reptile embryo develops in this watery environment, the shelled egg has been characterized as a "private aquarium" in which the embryos of these land vertebrates spend their water-dwelling stage of life. With this "sturdy egg," the remaining ties to the oceans were broken, and reptiles moved inland.

The first reptiles were small, but larger forms evolved rapidly, particularly the dinosaurs. One of the largest was *Apatosaurus,* which weighed more than 30 tons and measured over 25 meters (80 feet) from head to tail. Some of the largest dinosaurs were carnivorous (*Tyrannosaurus*), whereas others were herbivorous (like ponderous *Apatosaurus*).

Reptiles made perhaps the most spectacular adaptive radiations in all of Earth history. One group, the pterosaurs, became air-borne. These "dragons of the sky" possessed huge membranous wings that allowed them rudimentary flight. How the largest pterosaurs (some had wingspans of 26 feet and weighed 200 pounds) took flight is still unknown. Another group, exemplified by the fossil *Archaeopteryx,* led to more successful flyers—birds (**FIGURE 19.31**). Other reptiles returned to the sea, including fish-eating plesiosaurs and ichthyosaurs (**FIGURE 19.32**). These reptiles became proficient swimmers but retained their reptilian teeth and breathed by means of lungs.

For nearly 160 million years, dinosaurs reigned supreme. However, by the close of the Mesozoic, like many reptiles they became extinct (**FIGURE 19.33**). Select reptile groups survived to recent times, including turtles, snakes, crocodiles, and lizards. The huge, land-dwelling dinosaurs, the marine plesiosaurs, and the flying pterosaurs are known only through the fossil record. What caused this great extinction?

FIGURE 19.31 Paleontologists think that flying reptiles similar to *Archaeopteryx* were the ancestors of modern birds. Fossil evidence indicates that *Archaeopteryx* was capable of powerful flight but retained many characteristics of nonflight reptiles. Lower right is an artist's reconstruction of *Archaeopteryx,* perhaps the first birds. *(Photo by Michael Collier)*

Tail feathers (bird feature)

Long tail with vertebrae (reptilian feature)

Wing claws (reptilian feature)

Airfoil wings with feathers (bird feature)

Archaeopteryx

Geologist's Sketch

FIGURE 19.32 Marine reptiles such as this *Ichthyosaur* were the most spectacular of sea animals. *(Photo by Chip Clark)*

FIGURE 19.33 Fossil skull of a huge crocodile of Mesozoic age. *(Photo courtesy of Project Exploration)*

Demise of the Dinosaurs

The boundaries between divisions on the geologic time scale represent times of significant geological and/or biological change. Of special interest is the boundary between the Mesozoic era ("middle life") and the Cenozoic era ("recent life"), about 65 million years ago. During this transition about three quarters of all plant and animal species died out in a *mass extinction.* This boundary marks the end of the era in which dinosaurs and other reptiles dominated the landscape and the beginning of the era when mammals assumed that role (**FIGURE 19.34**). Because the last period of the Mesozoic is the Cretaceous (abbreviated K to avoid confusion with other "C" periods), and the first period of the Cenozoic is the Tertiary (abbreviated T), the time of this mass extinction is called the *Cretaceous–Tertiary,* or *KT, boundary.**

The extinction of the dinosaurs is generally attributed to their collective inability to adapt to radical change in environmental conditions. What event could have triggered the rapid extinction of one of the most successful groups of land animals? The most strongly supported hypothesis proposes that about 65 million years ago our planet was struck by a large carbonaceous meteorite, a relic from the formation of the solar system. The errant mass of rock was approximately 10 kilometers in diameter and was traveling at about 90,000 kilometers per hour at the time of impact. It collided with the southern portion of North America in a shallow tropical sea—now Mexico's Yucatan Peninsula (**FIGURE 19.35**). The energy released by the impact is estimated to have been equivalent to 100 million megatons (*mega* = million) of explosives.

Following the impact, suspended dust greatly reduced the amount of sunlight reaching Earth's surface, which resulted in global cooling ("impact winter"), and inhibited photosynthesis, disrupting food production. Long after the dust settled, carbon dioxide, water vapor, and sulfur oxides, added to the atmosphere by the blast, remained.

---

*The term *Paleogene* is now the preferred name for the first period of the Cenozoic era (see Figure 19.3). However, when this discovery was made, *Tertiary* was still widely used.

Mesozoic reptile

B.

Sulfate aerosols, because of their high reflectivity perpetuated the cooler surface temperatures for a few more years. Eventually, sulfate aerosols dissipated from the atmosphere as acid precipitation. With the aerosols gone, but carbon dioxide still present in large quantities, an enhanced greenhouse effect led to a long-term rise in average global temperatures. The likely result was that some of the plant and animal life that survived the initial impact eventually fell victim to stresses associated with global cooling, followed by acid precipitation and global warming.

FIGURE 19.35 Chicxulub crater is a giant impact crater that formed about 65 million years ago and has since been filled with sediments. About 180 kilometers (110 miles) in diameter, Chicxulub crater is the likely impact site that resulted in the demise of the dinosaurs.

The extinction of the dinosaurs opened habitats for the small mammals that survived. These new habitats, along with evolutionary forces, led to the development of the large mammals that occupy our modern world.

What evidence points to such a catastrophic collision 65 million years ago? First, a thin layer of sediment nearly 1 centimeter thick has been discovered at the KT boundary worldwide. This sediment contains a high level of the element *iridium*, rare in Earth's crust but found in high proportions in stony meteorites. Could this layer be the scattered remains of the meteorite responsible for the environmental changes that led to the demise of many reptile groups?

Despite substantial evidence and significant scientific support, some researchers disagree with the impact hypothesis. They suggest instead that huge volcanic eruptions may have led to a breakdown in the food chain. To support this hypothesis, they cite enormous outpourings of lavas in the Deccan Plateau of northern India about 65 million years ago. Regardless of what caused the KT extinction, its outcomes provide

DID YOU KNOW?

The turkey that is the focus of many Thanksgiving dinners is a descendant of reptiles, in particular, dinosaurs. The first evidence from the fossil record for the reptile–bird link came from a layer of 150-million-year-old Jurassic limestone that contained a nearly complete imprint of *Archaeopteryx*, a transitional fossil that exhibited wings and feathers but also had teeth and a skeleton similar to a small dinosaur (see Figure 19.31). Modern birds and reptiles share many traits, including the laying of eggs, similar skeletal structures, and the possession of scales, although in birds the scales are limited to their legs.

valuable lessons in understanding the role that catastrophic events play in shaping our planet's physical landscape and life-forms.

Cenozoic Era: Age of Mammals

During the Cenozoic, mammals replaced reptiles as the dominant land animals. At nearly the same time, angiosperms (flowering plants with covered seeds) replaced gymnosperms as the dominant plants. The Cenozoic is often called the "Age of Mammals" but can also be considered the "Age of Flowering Plants" because, in the plant world, angiosperms enjoy a status similar to that of mammals in the animal world.

The development of flowering plants strongly influenced the evolution of both birds and mammals that feed on seeds and fruits. During the middle of the Cenozoic, another type of angiosperm, grasses, developed rapidly and spread over the plains (**FIGURE 19.36**).

A.

B.

FIGURE 19.36 Angiosperms, commonly known as flowering plants, are seed-plants that have reproductive structures called flowers and fruits. **A.** The most diverse and widespread of modern plants, many angiosperms display easily recognizable flowers. *(Photo by WDG Photo/Shutterstock)* **B.** Some angiosperms, including grasses, have very tiny flowers. The expansion of the grasslands during the Cenozoic era greatly increased the diversity of grazing mammals and the predators that feed on them. *(Photo by Torleif/CORBIS)*

This fostered the emergence of herbivorous (plant-eating) mammals, which, in turn, provided the evolutionary foundation for large, predatory mammals.

During the Cenozoic, the ocean was teeming with modern fish such as tuna, swordfish, and barracuda. In addition, some mammals, including seals, whales, and walruses returned to the sea.

From Reptiles to Mammals

The earliest mammals coexisted with dinosaurs for nearly 100 million years but were small rodentlike creatures that gathered food at night when dinosaurs were less active. Then, about 65 million years ago, fate intervened when a large asteroid collided with Earth and dealt a crashing blow to the reign of the dinosaurs. This transition, during which one dominant group is replaced by another, is clearly visible in the fossil record.

Mammals are distinct from reptiles in that they give birth to live young that suckle on milk and are warm-blooded. This latter adaptation allowed mammals to lead more active lives and to occupy more diverse habitats than reptiles because they could survive in cold regions. (Most modern reptiles are dormant during cold weather.) Other mammalian adaptations included the development of insulating body hair and more efficient organs such as hearts and lungs.

With the demise of the large Mesozoic reptiles, Cenozoic mammals diversified rapidly. The many forms that exist today evolved from small primitive mammals that were characterized by short legs; flat, five-toed feet; and small brains. Their development and specialization took four

DID YOU KNOW?
The La Brea tar pits, located in downtown Los Angeles, are famous because they contain a rich and very well preserved assemblage of mammals that roamed southern California from 40,000 to 80,000 years ago. Tar pits form when crude oil seeps to the surface and the light portion evaporates, leaving behind sticky pools of heavy tar. See Figure 18.12, p. 447.

principal directions: (1) increase in size, (2) increase in brain capacity, (3) specialization of teeth to better accommodate their diet, and (4) specialization of limbs to be better equipped for a particular lifestyle or environment.

MARSUPIAL AND PLACENTAL MAMMALS. Two groups of mammals, the marsupials and the placentals, evolved and diversified during the Cenozoic. The groups differ principally in their modes of reproduction. Young marsupials are born live at a very early stage of development. At birth, the tiny and immature young enter the mother's pouch to suckle and complete their development. Today, marsupials are found primarily in Australia, where they underwent a separate evolutionary expansion largely isolated from placental mammals. Modern marsupials include kangaroos, opossums, and koalas (**FIGURE 19.37**). Placental mammals, conversely, develop within the mother's body for a much longer period, so birth occurs when the young are comparatively mature. Most modern mammals, including humans, are placental.

In South America, primitive marsupials and placentals coexisted in isolation for about 40 million years after the break-up of Pangaea. Evolution and specialization of both groups continued undisturbed until about 3 million years ago when the Panamanian land-bridge connected the two American continents. This event permitted the exchange of fauna between the two continents—monkeys, armadillos, sloths, and opossums arrived in North America, while various types of horses, bears, rhinos, camels, and wolves migrated southward. Many animals that had been unique to South America disappeared completely after this event, including hoofed mammals, rhino-sized rodents, and a number of carnivorous marsupials. Because this period of extinction coincided with the formation of the Panamanian land-bridge, it was thought that advanced carnivores from North America were responsible. However, recent research suggests that other factors, including climatic changes, may have played a significant role.

FIGURE 19.37 After the break-up of Pangaea, the Australian marsupials evolved differently from their relatives in the Americas. *(Photo by Martin Harvey/Peter Arnold Inc.)*

DID YOU KNOW?

Bats and humans are mammals, but they have developed very different physical structures and lifestyles. One clue that they are related is their pentadactyl (five-fingered) limbs. Humans have evolved hands for grasping, whereas bats have proportionally much longer "fingers" that support their wings for flying.

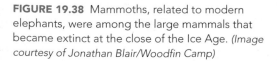

FIGURE 19.38 Mammoths, related to modern elephants, were among the large mammals that became extinct at the close of the Ice Age. *(Image courtesy of Jonathan Blair/Woodfin Camp)*

Large Mammals and Extinction

During the rapid mammal diversification of the Cenozoic era, some groups became very large. For example, by the Oligocene epoch, a hornless rhinoceros evolved that stood nearly 5 meters (16 feet) high. It is the largest land mammal known to have existed. As time passed, many other mammals evolved to larger forms—more, in fact, than now exist. Many of these large forms were common as recently as 11,000 years ago. However, a wave of late Pleistocene extinctions rapidly eliminated these animals from the landscape.

North America experienced the extinction of mastodons and mammoths, both huge relatives of the modern elephant (**FIGURE 19.38**). In addition, saber-toothed cats, giant beavers, large ground sloths, horses, giant bison, and others died out. In Europe, late Pleistocene extinctions included woolly rhinos, large cave bears, and Irish elk. Scientists remain puzzled about the reasons for this recent wave of extinctions that targeted large animals. Having survived several major glacial advances and interglacial periods, it is difficult to ascribe extinctions of these animals to climate change. Some scientists hypothesize that early humans hastened the decline of these mammals by selectively hunting large forms (**FIGURE 19.39**).

CONCEPT CHECK 19.9

❶ What animal group became the dominant land animals of the Cenozoic era?

❷ How did the demise of the large Mesozoic reptiles impact the development of mammals?

❸ Describe one hypothesis that might explain the extinction of large mammals in the late Pleistocene.

FIGURE 19.39 Cave painting of animals that early humans encountered about 17,000 years ago. *(Photo courtesy of Sisse Brimberg/National Geographic Society)*

CHAPTER NINETEEN
Earth's Evolution through Geologic Time
In Review

⊙ The history of Earth began about 13.7 billion years ago when the first elements were created during the *Big Bang*. It was from this material, plus other elements ejected into interstellar space by now-defunct stars, that Earth, along with the rest of the solar system, formed. As material collected, high velocity impacts of chunks of matter called *planetesimals* and the decay of radioactive elements caused the temperature of our planet to steadily increase. Iron and nickel melted and sank to form the metallic core, while rocky material rose to form the mantle and Earth's initial crust.

⊙ Earth's primitive atmosphere, which consisted mostly of water vapor and carbon dioxide, formed by a process called *outgassing*, which resembles the steam eruptions of modern volcanoes. About 3.5 billion years ago, photosynthesizing bacteria began to release oxygen, first into the oceans and then into the atmosphere. This began the evolution of our modern atmosphere. The oceans formed early in Earth's history as water vapor condensed to form clouds, and torrential rains filled low-lying areas. The salinity in seawater came from volcanic outgassing and from elements weathered and eroded from Earth's primitive crust.

⊙ The *Precambrian,* which is divided into the *Archean* and *Proterozoic eons,* spans nearly 90 percent of Earth's history, beginning with the formation of Earth about 4.6 billion years ago and ending approximately 542 million years ago. During this time, much of Earth's stable continental crust was created through a multistage process. First, partial melting of the mantle generated magma that rose to form volcanic island arcs and oceanic plateaus. These thin crustal fragments collided and accreted to form larger crustal provinces, which in turn assembled into larger blocks called *cratons*. Cratons, which form the core of modern continents, were created mainly during the Precambrian.

⊙ Supercontinents are large landmasses that consist of all, or nearly all, existing continents. *Pangaea* was the most recent supercontinent, but other massive continents including an even larger one, *Rodinia,* preceded it. The splitting and reassembling of supercontinents have generated most of Earth's major mountain belts. In addition, the movements of these crustal blocks have profoundly affected Earth's climate and caused sea level to rise and fall.

⊙ The time span following the close of the Precambrian, called the *Phanerozoic eon,* encompasses 542 million years and is divided into three eras: *Paleozoic, Mesozoic,* and *Cenozoic.* The Paleozoic era was dominated by continental collisions as the supercontinent of Pangaea assembled—forming the Caledonian, Appalachian, and Ural Mountains. Early in the Mesozoic, much of the land was above sea level. However, by the middle Mesozoic, seas invaded western North America. As Pangaea began to break up, the westward-moving North American plate began to override the Pacific plate, causing crustal deformation along the entire western margin of North America. Owing to their different relations with plate boundaries, the eastern and western margins of the continent experienced contrasting events. The stable eastern margin was the site of abundant sedimentation as isostatic adjustment raised the modern Appalachians, causing streams to erode with renewed vigor and deposit their sediment along the continental margin. In the West, the *Laramide Orogeny* (responsible for building the Rocky Mountains) was coming to an end, the Basin and Range Province was forming, and volcanic activity was extensive.

⊙ The first known organisms were single-celled bacteria, *prokaryotes,* which lack a nucleus. One group of these organisms, called cyanobacteria, used solar energy to synthesize organic compounds (sugars). For the first time, organisms had the ability to produce their own food. Fossil evidence for the existence of these bacteria includes layered mounds of calcium carbonate called *stromatolites*.

⊙ The beginning of the Paleozoic is marked by the *appearance of the first life-forms with hard parts* such as shells. Therefore, abundant fossils occur, and a far more detailed record of Paleozoic events can be constructed. Life in the early Paleozoic was restricted to the seas and consisted of several invertebrate groups, including trilobites, cephalopods, sponges, and corals. During the Paleozoic, organisms diversified dramatically. Insects and plants moved onto land, and lobe-finned fishes that adapted to land became the first amphibians. By the Pennsylvanian period, large tropical swamps, which became the major coal deposits of today, extended across North America, Europe, and Siberia. At the close of the Paleozoic, a mass extinction destroyed 70 percent of all vertebrate species on land and 90 percent of all marine organisms.

⊙ The Mesozoic era, literally the era of middle life, is often called the "*Age of Reptiles.*" Organisms that survived the extinction at the end of the Paleozoic began to diversify in spectacular ways. *Gymnosperms* (cycads, conifers, and ginkgoes) became the dominant trees of the Mesozoic because they could adapt to the drier climates. Reptiles became the dominant land animals. The most awesome of the Mesozoic reptiles were the *dinosaurs*. At the close of the Mesozoic, many large reptiles, including the dinosaurs, became extinct.

⊙ The Cenozoic is often called the "*Age of Mammals*" because these animals replaced the reptiles as the dominant vertebrate life-forms on land. Two groups of mammals, the marsupials and the placentals, evolved and expanded during this era. One tendency was for some mammal groups to become very large. However, a wave of late *Pleistocene* extinctions rapidly eliminated these animals from the landscape. Some scientists suggest that early humans hastened their decline by selectively hunting the larger animals. The Cenozoic could also be called the "*Age of Flowering Plants.*" As a source of food, flowering plants (angiosperms) strongly influenced the evolution of both birds and herbivorous (plant-eating) mammals throughout the Cenozoic era.

Key Terms

banded iron formations (p. 467)
cratons (p. 470)
eukaryotes (p. 477)

outgassing (p. 467)
planetesimals (p. 465)
prokaryotes (p. 477)

protoplanets (p. 465)
solar nebula (p. 465)
stromatolites (p. 477)

supercontinents (p. 471)
supercontinent cycle (p. 471)
supernova (p. 465)

GIVE IT SOME THOUGHT

❶ Refer to the geologic time scale in Figure 19.3. The Precambrian accounts for nearly 90 percent of geologic time. Why do you think it has fewer divisions than the rest of the time scale?

❷ Referring to Figure 19.4, list the major events that led to the formation of Earth.

❸ What does the existence of iron-rich rocks during the Precambrian indicate about the evolution of Earth's atmosphere?

❹ What are two ways that the appearance of a significant amount of oxygen in the atmosphere relates to life?

❺ Presently, about 71 percent of Earth's surface is ocean covered. This percentage was much higher early in Earth history. Explain.

❻ Figure 19.12 shows the distribution of Earth's ancient rocks. Explain why there are so few rocks older than 3500 million years old on Earth.

❼ Contrast the eastern and western margins of North America during the Cenozoic era in terms of their relationships to plate boundaries.

❽ Why do you think plants moved onto land before animals?

❾ Refer to Figure 19.21. How has biodiversity changed with time? How did mass extinctions contribute to this trend?

❿ Some scientists have proposed that the environments around black smokers may be similar to the extreme conditions that existed early in Earth history. Therefore, they look to the unusual life that exists around black smokers for clues about how earliest life may have survived. Compare and contrast the environment of a black smoker to the environment on Earth approximately 3 to 4 billion years ago. Do you think there are parallels between the two, and if so, do you think black smokers are good examples of the environment that earliest life may have experienced? Explain.

Companion Website

www.mygeoscienceplace.com

The *Essentials of Geology, 11e* companion Website contains numerous multimedia resources accompanied by assessments to aid in your study of the topics in this chapter. The use of this site's learning tools will help improve your understanding of geology. Utilizing the access code that accompanies this text, visit **www.mygeoscienceplace.com** in order to:

- **Review** key chapter concepts.
- **Read** with links to the Pearson eText and to chapter-specific web resources.
- **Visualize** and comprehend challenging topics using the learning activities in *GEODe: Essentials of Geology* and the *Geoscience Animations Library*.
- **Test** yourself with online quizzes.

Global Climate Change

C LIMATE IS MORE THAN JUST AN EXPRESSION OF AVERAGE ATMOSPHERIC conditions. To accurately portray the character of a place or an area, variations and extremes must also be included.

To assist you in learning the important concepts in this chapter, focus on the following questions:

- Why is it important to unravel past climate changes? How are such changes detected?

- What is the composition and structure of Earth's atmosphere? How is the atmosphere heated?

- What are some natural causes of climate change?

- In what ways are humans changing the composition of the atmosphere? How is the atmosphere responding to these changes?

- What are climate feedback mechanisms? What part might they play in global climate change?

- What are some possible consequences of global warming?

Mauna Loa Observatory, 3397 meters (11,140 feet) atop Mauna Loa volcano is an important atmospheric research facility. For more about the observatory, see "Did You Know" on p. 510. (Photo by Forrest M. Mims III)

Climate and Geology Are Linked

Anyone who has the opportunity to travel around the world will find such an incredible variety of climates that it is hard to believe they could all occur on the same planet. Climate strongly influences plant and animal life, the soil, and many of Earth's surface processes (**FIGURE 20.1**). Climate influences people as well.

Climate has a profound impact on many geologic processes. When climate changes, these processes respond. A glance back at the rock cycle in Chapter 1 (p. 19) reminds us about many of the connections. Of course rock weathering has an obvious climate connection, as do processes that operate in arid, tropical, and glacial landscapes. Events such as debris flows and river flooding are often triggered by atmospheric happenings such as periods of extraordinary rainfall. Clearly the atmosphere is a basic link in the hydrologic cycle. Other climate–geology connections involve the impact of internal processes on the atmosphere. For example, the particles and gases emitted by volcanoes can change the composition of the atmosphere, and mountain building can have a significant impact on regional temperature, precipitation, and wind patterns.

Climate not only varies from place to place, but it is also naturally variable over time. Over the great expanse of Earth history, and long before humans were roaming the planet, there were many shifts—from warm to cold and from wet to dry, and back again. The geologic record is a storehouse of data that confirms this fact.

The study of sediments, sedimentary rocks, and fossils clearly demonstrates that, through the ages, practically every place on our planet has experienced wide swings in climate, such as from ice ages to conditions associated with subtropical coal swamps or desert dunes. Chapter 19, *"Earth's Evolution through Geologic Time,"* reinforces this fact. Time scales for climate change vary from decades to millions of years.

What factors have been responsible for climate variations during Earth history? One of the topics treated in this chapter involves natural causes of climate change.

Today, global climate change is more than just a topic of academic interest to a group of scientists curious about Earth history. Rather, the subject is making headlines. Many members of the general

FIGURE 20.1 Washington's Mount Rainier National Park. Glaciers are still sculpting this large volcano. The five major parts of the climate system are represented in this image. (*Photo by Jamie & Judy Wild/DanitaDelimont.com*)

public are not only curious but also concerned about the possibilities. Why is climate change newsworthy? The reason is that research focused on human activities and their impact on the environment has demonstrated that people are inadvertently changing the climate. Unlike changes in the past, which were natural variations, modern climate change is dominated by human influences that are sufficiently large as to exceed the bounds of natural variability. Moreover, these changes are likely to continue for many centuries. The effects of this venture into the unknown with climate could be very disruptive, not only to humans but to many other life forms as well.

How are the details of past climates determined? How is this information useful to us now? In what ways are humans changing global climate? What are the possible consequences?

CONCEPT CHECK 20.1

❶ List at least five examples of connections between climate and geology.

The Climate System

Throughout this book you have been reminded that Earth is a multidimensional system that consists of many interacting parts. A change in any one part can produce changes in any or all of the other parts—often in ways that are neither obvious nor immediately apparent. This fact is certainly true when it comes to the study of climate and climate change.

To understand and appreciate climate, it is important to realize that climate involves more than just the atmosphere (Figure 20.1):

The atmosphere is the central component of the complex, connected, and interactive global environmental system upon which all life depends. Climate may be broadly defined as the long-term behavior of this environmental system. To understand fully and to predict changes in the atmospheric component of the climate system, one must understand the sun, oceans, ice sheets, solid earth, and all forms of life.*

Indeed, we must recognize that there is a **climate system** that includes the atmosphere, hydrosphere, geosphere, biosphere, and cryosphere. (The *cryosphere* refers to the ice and snow that exist at Earth's surface.) The climate system *involves the exchanges of energy and moisture that occur among the five spheres.* These exchanges link the atmosphere to the other spheres so that the whole functions as an extremely complex, interactive unit. Changes to the climate system do not occur in isolation. Rather, when one part of it changes, the other components also react. The major components of the climate system are shown in **FIGURE 20.2**.

CONCEPT CHECK 20.2

❶ List the five parts of the climate system.

*The American Meteorological Society and the University Corporation for Atmospheric Research, "Weather and the Nation's Well-Being," Bulletin of the American Meteorological Society, 73, No. 12 (December 2001), p. 2038.

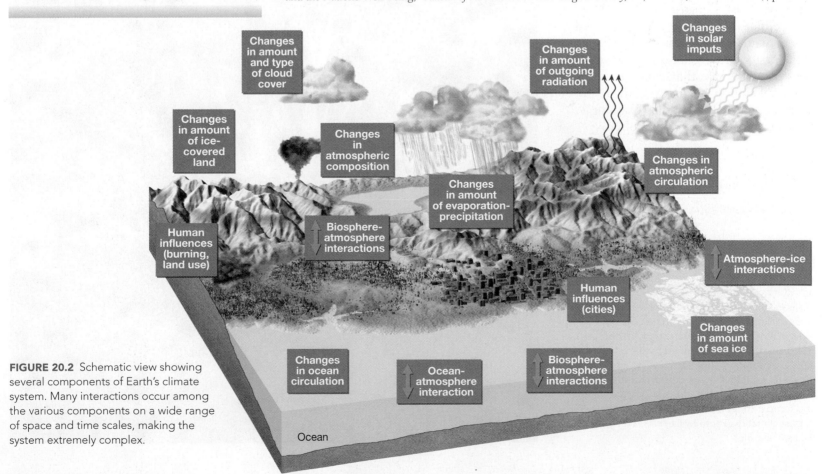

FIGURE 20.2 Schematic view showing several components of Earth's climate system. Many interactions occur among the various components on a wide range of space and time scales, making the system extremely complex.

How Is Climate Change Detected?

High-technology and precision instrumentation are now available to study the composition and dynamics of the atmosphere. Such tools are recent inventions, however, and have been providing data for only a short time span. To understand fully the behavior of the atmosphere and to anticipate future climate change, we must somehow discover how climate has changed over broad expanses of time.

Instrumental records go back only a couple of centuries at best, and the further back we go, the less complete and more unreliable the data become. To overcome this lack of direct measurements, scientists must decipher and reconstruct past climates by using indirect evidence. Such **proxy data** come from natural recorders of climate variability, such as seafloor sediments, glacial ice, fossil pollen, and tree-growth rings (**FIGURE 20.3**). Historical documents can also be useful sources of information about climate variability. Scientists who analyze proxy data and reconstruct past climates are engaged in the study of **paleoclimatology.** The main goal of such work is to understand the climate of the past in order to assess the current and potential future climate in the context of natural climate variability. In the following discussion, we will briefly examine some of the important sources of proxy data.

Seafloor Sediment—A Storehouse of Climate Data

We know that the parts of the Earth system are linked so that a change in one part can produce changes in any or all of the other parts. In this example you will see how changes in atmospheric and oceanic temperatures are reflected in the nature of life in the sea.

Most seafloor sediments contain the remains of organisms that once lived near the sea surface (the ocean–atmosphere interface). When such near-surface organisms die, their shells slowly settle to the floor of the ocean, where they become part of the sedimentary record (**FIGURE 20.4**). These seafloor sediments are useful recorders of worldwide climate change because the number and types of organisms living near the sea surface change with the climate:

We would expect that in any area of the ocean/atmosphere interface the average annual temperature of the surface water of the ocean would approximate that of the contiguous atmosphere. The temperature equilibrium established between surface seawater and the air above it should mean that …changes in climate should be reflected in changes in organisms living near the surface of the deep sea…. When we recall that the seafloor sediments in vast areas of the ocean consist mainly of shells of pelagic foraminifers, and that these animals are sensitive to variations in water temperature, the connection between such sediments and climate change becomes obvious.*

**Richard F. Flint, *Glacial and Quaternary Geology* (New York: John Wiley & Sons, 1971), p. 718.

FIGURE 20.3 Ancient bristlecone pines in California's White Mountains. The study of tree-growth rings is one way scientists reconstruct past climates. Some of these trees are more than 4000 years old. Each year a growing tree produces a layer of new cells beneath the bark. If the tree is felled and the trunk examined (or if a core is taken to avoid cutting the tree), each year's growth can be seen as a ring. Because the amount of growth (thickness of a ring) depends upon precipitation and temperature, tree rings are useful recorders of past climates. *(Photo by Dennis Flaherty/Photo Researchers, Inc.)*

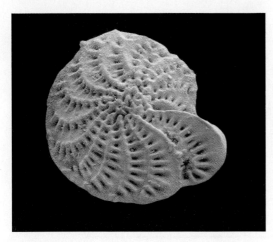

FIGURE 20.4 Image of the shell of a foraminifera using a scanning electron microscope. These tiny, single-celled organisms are sensitive to even small fluctuations in temperature. Seafloor sediments containing fossils such as this are useful recorders of climate change. *(Photo by Andrew Syred/Photo Researchers, Inc.)*

FIGURE 20.5 The *Chikyu* (meaning "Earth" in Japanese), the world's most advanced scientific drilling vessel. It can drill as deep as 7000 meters (nearly 23,000 feet) below the seabed in water as deep as 2500 meters (8200 feet). It is part of the Integrated Ocean Drilling Program (IODP). *(AP Photo/Itsuo Inouye)*

Thus, in seeking to understand climate change as well as other environmental transformations, scientists are tapping the huge reservoir of data in seafloor sediments. The sediment cores gathered by drilling ships and other research vessels have provided invaluable data that have significantly expanded our knowledge and understanding of past climates (**FIGURE 20.5**).

One notable example of the importance of seafloor sediments to our understanding of climate change relates to unraveling the fluctuating atmospheric conditions of the Ice Age. The records of temperature changes contained in cores of sediment from the ocean floor have proven critical to our present understanding of this recent span of Earth history.*

Oxygen Isotope Analysis

Oxygen isotope analysis is based on precise measurement of the ratio between two isotopes of oxygen: ^{16}O, which is the most common, and the heavier ^{18}O. A molecule of H_2O can form from either ^{16}O or ^{18}O. The lighter isotope, ^{16}O, evaporates more readily from the oceans. Because of this, precipitation (and hence the glacial ice that it may form) is enriched in ^{16}O. This leaves a greater concentration of the heavier iso-

*For more information on this topic, see "Causes of Glaciation" in Chapter 11, pages 281–283.

tope, ^{18}O, in the ocean water. Thus, during periods when glaciers are extensive, more of the lighter ^{16}O is tied up in ice, so the concentration of ^{18}O in seawater increases. Conversely, during warmer interglacial periods when the amount of glacial ice decreases dramatically, more ^{16}O is returned to the sea, so the proportion of ^{18}O relative to ^{16}O in ocean water also drops. Now, if we had some ancient recording of the changes of the $^{18}O/^{16}O$ ratio, we could determine when there were glacial periods and, therefore, when the climate grew cooler.

Fortunately, we do have such a recording. As certain microorganisms secrete their shells of calcium carbonate ($CaCO_3$), the prevailing $^{18}O/^{16}O$ ratio is reflected in the composition of these hard parts. When the organisms die, their hard parts settle to the ocean floor, becoming part of the sediment layers there. Consequently, periods of glacial activity can be determined from variations in the oxygen–isotope ratio found in shells of certain microorganisms buried in deep-sea sediments.

The $^{18}O/^{16}O$ ratio also varies with temperature. More ^{18}O is evaporated from the oceans when temperatures are high, and less is evaporated when temperatures are low. Therefore, the heavy isotope is more abundant in the precipitation of warm eras and less abundant during colder periods. Using this principle, scientists studying the layers of ice and snow in glaciers have been able to produce a record of past temperature changes.

Climate Change Recorded in Glacial Ice

Ice cores are an indispensable source of data for reconstructing past climates (see Figure 11.35, p. 283). Research based on vertical cores taken from the Greenland and Antarctic ice sheets has changed our basic understanding of how the climate system works.

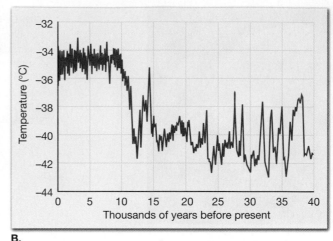

A.

B.

FIGURE 20.6 A. The National Ice Core Laboratory is a physical plant for storing and studying cores of ice taken from glaciers around the world. These cores represent a long-term record of material deposited from the atmosphere. The lab provides scientists the capability to conduct examinations of ice cores, and it preserves the integrity of these samples in a repository for the study of global climate change and past environmental conditions. *(Photo by USGS/National Ice Core Laboratory)* **B.** This graph, showing temperature variations over the past 40,000 years, is derived from oxygen-isotope analysis of ice cores recovered from the Greenland ice sheet. *(After U.S. Geological Survey)*

Scientists collect samples with a drilling rig, like a small version of an oil drill. A hollow shaft follows the drill head into the ice, and an ice core is extracted. In this way, cores that sometimes exceed 2000 meters (6500 feet) in length, and may represent more than 200,000 years of climate history, are acquired for study (**FIGURE 20.6A**).

The ice provides a detailed record of changing air temperatures and snowfall. Air bubbles trapped in the ice record variations in atmospheric composition. Changes in carbon dioxide and methane are linked to fluctuating temperatures. The cores also include atmospheric fallout such as wind-blown dust, volcanic ash, pollen, and modern-day pollution.

Past temperatures are determined by *oxygen isotope analysis*. Using this technique, scientists are able to produce a record of past temperature changes. A portion of such a record is shown in **FIGURE 20.6B**.

Other Types of Proxy Data

Other sources of proxy data that are used to gain insight into past climates include fossil pollen, corals, and historical documents.

FOSSIL POLLEN. Climate is a major factor influencing the distribution of vegetation, so knowing the nature of the plant community occupying an area is a reflection of the climate. Pollen and spores are parts of the life cycles of many plants, and because they have resistant walls, they are often the most abundant, easily identifiable, and best-preserved plant remains in sediments (**FIGURE 20.7**). By analyzing pollen from accurately dated sediments, it is possible to obtain high-resolution records of vegetational changes in an area. From such information, past climates can be reconstructed.

CORALS. Coral reefs consist of colonies of corals that live in warm shallow waters and form atop the hard material left behind by past corals. Corals build their hard skeletons from calcium carbonate ($CaCO_3$) extracted from seawater. The carbonate contains isotopes of oxygen that can be used to determine the temperature of the water in which the coral grew.

The portion of the skeleton that forms in winter has a density that is different from that formed in summer because of variations in growth rates related to temperature and other environmental factors. Thus, corals exhibit seasonal growth bands very much like those observed in trees. The accuracy and reliability of the climate data extracted from corals has been established by comparing recent instrumental

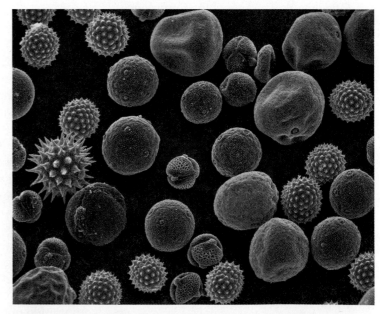

FIGURE 20.7 This false color image from an electron microscope shows an assortment of pollen grains. Note how the size, shape, and surface characteristics differ from one species to another. Analysis of the types and abundance of pollen in lake sediments and peat deposits provides information on how climate has changed over time. *(Photo by David Scharf/Science Photo Library)*

records to coral records for the same period. Oxygen-isotope analysis of coral growth rings can also serve as a proxy measurement for precipitation, particularly in areas where large variations in annual rainfall occur.

Think of coral as a paleothermometer that enables us to answer important questions about climate variability in the world's oceans. The graph in **FIGURE 20.8** is a 350-year sea-surface temperature record based on oxygen-isotope analysis of a core from the Galapagos Islands.

HISTORICAL DATA. Historical documents sometimes contain helpful information. Although it might seem that such records should readily lend themselves to climate analysis, that is not the case. Most manuscripts were written for purposes other than climate description. Furthermore, writers understandably neglected periods of relatively stable atmospheric conditions and mention only droughts, severe storms, memorable blizzards, and other extremes. Nevertheless, records of crops, floods, and the migration of people have furnished useful evidence of the possible influences of changing climate (**FIGURE 20.9**).

CONCEPT CHECK 20.3

❶ What is proxy data and why is it necessary in the study of climate change? List several examples.

❷ Why are seafloor sediments useful in the study of past climates?

Some Atmospheric Basics

In order to better understand climate change, it is helpful to possess some knowledge about the composition and structure of the atmosphere and the process by which it is heated—the *greenhouse effect.*

Composition of the Atmosphere

Air is *not* a unique element or compound. Rather, air is a *mixture* of many discrete gases, each with its own physical

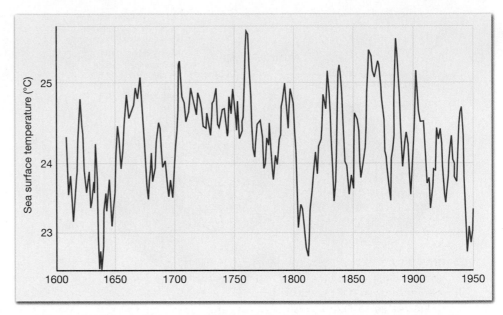

FIGURE 20.8 This graph shows a 350-year record of sea-surface temperatures obtained by oxygen isotope analysis of coral from the Galapagos Islands. (*After National Climatic Data Center/NOAA*)

properties, in which varying quantities of tiny solid and liquid particles are suspended.

CLEAN DRY AIR. As you can see in **FIGURE 20.10**, clean, dry air is composed almost entirely of two gases—78 percent nitrogen and 21 percent oxygen. Although these gases are the most plentiful components of air and are of great significance to life on Earth, they are of little or no importance in affecting weather phenomena. The remaining 1 percent of dry air is mostly the inert gas argon (0.93 percent) plus tiny quantities of a number of other gases. Carbon dioxide, although present in

FIGURE 20.9 Historical records can sometimes be helpful in the analysis of past climates. The date for the beginning of the grape harvest in the fall is an integrated measure of temperature and precipitation during the growing season. These dates have been recorded for centuries in Europe and provide a useful record of year-to-year climate variations. (*Photo by Owen Franken/CORBIS*)

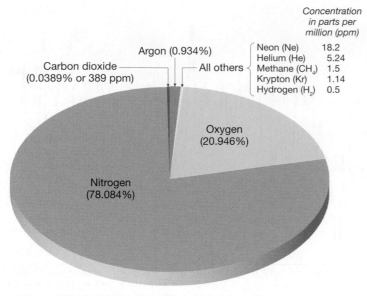

FIGURE 20.10 Proportional volume of gases composing dry air. Nitrogen and oxygen obviously dominate.

only minute amounts (0.0389 percent, or 389 parts per million), is nevertheless an important constituent of air because it has the ability to absorb heat energy radiated by Earth and thus influences the heating of the atmosphere.

Aerosols have both natural and human-made sources. *(AP Photo/Gurindes Osan)*

Air includes many gases and particles that vary significantly from time to time and from place to place. Important examples include water vapor, ozone, and tiny solid and liquid particles.

WATER VAPOR. The amount of water vapor in the air varies considerably, from practically none at all up to about 4 percent by volume. Why is such a small fraction of the atmosphere so significant? Certainly the fact that water vapor is the source of all clouds and precipitation would be enough to explain its importance. However, water vapor has other roles. Like carbon dioxide, it has the ability to absorb heat energy given off by Earth as well as some solar energy. It is, therefore, important when we examine the heating of the atmosphere.

OZONE. Another important component of the atmosphere is *ozone*. It is a form of oxygen that combines three oxygen atoms into each molecule (O_3). Ozone is not the same as the oxygen we breathe, which has two atoms per molecule (O_2). There is very little ozone in the atmosphere, and its distribution is not uniform. It is concentrated well above Earth's surface in a layer called the *stratosphere,* at an altitude of between 10 and 50 kilometers (6 and 31 miles). The presence of the ozone layer in our atmosphere is crucial to those of us

who dwell on Earth. The reason is that ozone absorbs the potentially harmful ultraviolet (UV) radiation from the Sun. If ozone did not filter a great deal of the ultraviolet radiation, and if the Sun's UV rays reached the surface of Earth undiminished, our planet would be uninhabitable for most life as we know it.

AEROSOLS. The movements of the atmosphere are sufficient to keep a large quantity of solid and liquid particles suspended within it. Although visible dust sometimes clouds the sky, these relatively large particles are too heavy to stay in the air for very long. Still, many particles are microscopic and remain suspended for considerable periods of time. They may originate from many sources, both natural and human made, and include sea salts from breaking waves, fine soil blown into the air, smoke and soot from fires, pollen and microorganisms lifted by the wind, ash and dust from volcanic eruptions, and more. Collectively, these tiny solid and liquid particles are called **aerosols.**

From a meteorological standpoint, these tiny, often invisible particles can be significant. First, many act as surfaces on which water vapor can condense, an important function in the formation of clouds and fog. Second, aerosols can absorb or reflect incoming solar radiation. Thus, when an air pollution episode is occurring, or when ash fills the sky following a volcanic eruption, the amount of sunlight reaching Earth's surface can be measurably reduced.

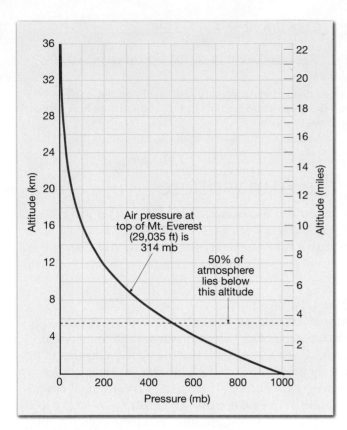

FIGURE 20.11 Atmospheric pressure changes with altitude. Pressure decreases rapidly near Earth's surface and more gradually at greater heights.

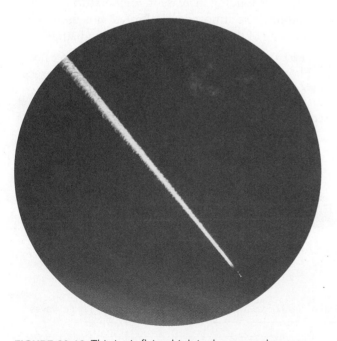

FIGURE 20.12 This jet is flying high in the atmosphere at an altitude of more than 9000 meters (30,000 feet). More than two thirds of the atmosphere is below this height. To someone on the ground, the atmosphere seems to extend for a great distance. However, when compared to the thickness (radius) of the solid Earth, the atmosphere is a very shallow layer. (Photo by Warren Faidley/Weatherstock)

Extent and Structure of the Atmosphere

To say that the atmosphere begins at Earth's surface and extends upward is obvious. However, where does the atmosphere end and outer space begin? There is no sharp boundary; the atmosphere rapidly thins as you travel away from Earth, until there are too few gas molecules to detect.

PRESSURE CHANGES WITH HEIGHT. To understand the vertical extent of the atmosphere, let us examine changes in atmospheric pressure with height. Atmospheric pressure is simply the weight of the air above. At sea level, the average pressure is slightly more than 1000 millibars. This corresponds to a weight of slightly more than 1 kilogram per square centimeter (14.7 pounds per square inch). Obviously, the pressure at higher altitudes is less (**FIGURE 20.11**).

One half of the atmosphere lies below an altitude of 5.6 kilometers (3.5 miles). At about 16 kilometers (10 miles), 90 percent of the atmosphere has been traversed, and above 100 kilometers (62 miles) only 0.00003 percent of all the gases making up the atmosphere remains (**FIGURE 20.12**). Even so, traces of our atmosphere extend far beyond this altitude, gradually merging with the emptiness of space.

TEMPERATURE CHANGES. In addition to vertical changes in air pressure, there are also changes in air temperature as we ascend through the atmosphere. Earth's atmosphere is divided vertically into four layers on the basis of temperature (**FIGURE 20.13**).

The bottom layer in which we live is characterized by a decrease in temperature with an increase in altitude and is called the *troposphere*. The term literally means the region where air "turns over," a reference to the appreciable vertical mixing of air in this lowermost zone. The troposphere is the chief focus of meteorologists, because it is in this layer that essentially all important weather phenomena occur.

The temperature decrease in the troposphere is called the *environmental lapse rate*. Its average value is 6.5 °C per kilometer (3.5 °F per 1000 feet), a figure known as the *normal lapse rate*. It should be emphasized, however, that the environmental lapse rate is not a constant, but rather can be highly variable and must be regularly measured. To determine the actual environmental lapse rate as well as to gather information about vertical changes in pressure, wind, and humidity, radiosondes are used. The *radiosonde* is an instrument package that is attached to a balloon and transmits data by radio as it ascends through the atmosphere.

The thickness of the troposphere is not the same everywhere; it varies with latitude and the season. On average, the temperature drop continues to a height of about 12 kilometers (7.4 miles). The outer boundary of the troposphere is the *tropopause*.

Beyond the tropopause is the *stratosphere*. In the stratosphere, the temperature remains constant to a height of about 20 kilometers (12 miles) and then begins a gradual increase that continues until the *stratopause*, at a height of nearly 50 kilometers (30 miles) above Earth's surface. Below the tropopause, atmospheric properties like temperature and humidity are readily transferred by large-scale turbulence and mixing. Above the tropopause, in the stratosphere, they are not. Temperatures increase in the stratosphere because it is in this layer that the atmosphere's ozone is concentrated. Recall that ozone absorbs ultraviolet radiation from the Sun. As a consequence, the stratosphere is heated.

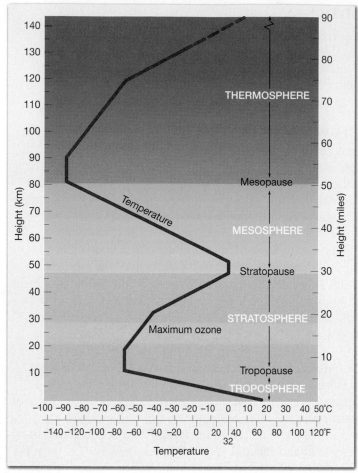

FIGURE 20.13 Thermal structure of the atmosphere.

In the third layer, the *mesosphere,* temperatures again decrease with height until, at the *mesopause,* more than 80 kilometers (50 miles) above the surface, the temperature approaches −90 °C (−130 °F). The coldest temperatures anywhere in the atmosphere occur at the mesopause.

The fourth layer extends outward from the mesopause and has no well-defined upper limit. It is the *thermosphere,* a layer that contains only a tiny fraction of the atmosphere's mass. In the extremely rarefied air of this outermost layer, temperatures again increase, owing to the absorption of very short-wave, high-energy solar radiation by atoms of oxygen and nitrogen.

Temperatures rise to extremely high values of more than 1000 °C in the thermosphere, but such temperatures are not comparable to those experienced near Earth's surface. Temperature is defined in terms of the average speed at which molecules move. Because the gases of the thermosphere are moving at very high speeds, the temperature is very high, but the gases are so sparse that, collectively, they process only an insignificant quantity of heat.

❶ What are the major components of clean, dry air? List two significant *variable* components.

❷ The atmosphere is divided vertically into four layers on the basis of temperature. Name the layers from bottom to top and indicate how temperatures change in each.

Heating the Atmosphere

Nearly all of the energy that drives Earth's variable weather and climate comes from the Sun. Before we can adequately describe how Earth's atmosphere is heated, it is helpful to know something about solar energy and what happens to this energy once it is intercepted by Earth.

Energy from the Sun

From our everyday experience, we know that the Sun emits light and heat as well as the ultraviolet rays that cause suntan. Although these forms of energy comprise a major portion of the total energy that radiates from the Sun, they are only part of a large array of energy called *radiation* or *electromagnetic radiation.* This array, or spectrum, of electromagnetic energy is shown in **FIGURE 20.14**. All radiation—whether X-rays, microwaves, or radio waves—transmits energy through the vacuum of space at 300,000 kilometers (186,000 miles) per second and only slightly slower through our atmosphere. When an object absorbs any form of radiant energy, the result is an increase in molecular motion, which causes a corresponding increase in temperature.

To better understand how the atmosphere is heated, it is useful to have a general understanding of the basic laws governing radiation.

1. *All objects, at whatever temperature, emit radiant energy.* Hence, not only hot objects like the Sun but also Earth, including its polar ice caps, continually emit energy.

2. *Hotter objects radiate more total energy per unit area than do colder objects.*

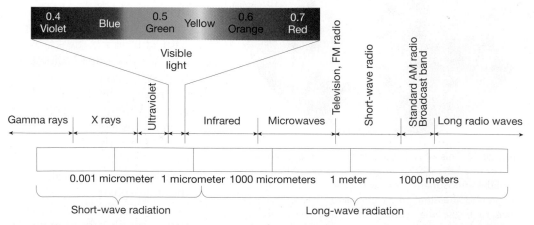

FIGURE 20.14 The electromagnetic spectrum, illustrating the wavelengths and names of various types of radiation.

3. *The hotter the radiating body, the shorter the wavelength of maximum radiation.* The Sun, with a surface temperature of about 5700 °C, radiates maximum energy at 0.5 micrometer, which is in the visible range. The maximum radiation for Earth occurs at a wavelength of 10 micrometers, well within the infrared (heat) range. Because the maximum Earth radiation is roughly 20 times longer than the maximum solar radiation, Earth radiation is often called *longwave radiation,* and solar radiation is called *shortwave radiation.*

4. *Objects that are good absorbers of radiation are good emitters as well.* Earth's surface and the Sun approach being perfect radiators because they absorb and radiate with nearly 100 percent efficiency for their respective temperatures. On the other hand, *gases are selective absorbers and emitters of radiation.* For some wavelengths the atmosphere is nearly transparent (little radiation absorbed). For others, however, it is nearly opaque (a good absorber). Experience tells us that the atmosphere is transparent to visible light; hence, these wavelengths readily reach Earth's surface. This is not the case for the longer wavelength radiation emitted by Earth.

The Fate of Incoming Solar Energy

FIGURE 20.15 shows the fate of incoming solar radiation averaged for the entire globe. Notice that the atmosphere is quite transparent to incoming solar radiation. On average, about 50 percent of the solar energy reaching the top of the atmosphere passes through the atmosphere and is absorbed at Earth's surface. Another 20 percent is absorbed directly by clouds and certain atmospheric gases (including oxygen and ozone) before reaching the surface. The remaining 30 percent is reflected back to space by the atmosphere, clouds, and reflective surfaces such as snow and ice. The fraction of the total radiation that is reflected by a surface is called its *albedo.* Thus, the albedo for Earth as a whole (the *planetary* albedo) is 30 percent.

What determines whether solar radiation will be transmitted to the surface, scattered, or reflected outward? It depends greatly on the wavelength of the energy being transmitted, as well as on the nature of the intervening material.

A rainbow displays the colors that make up the visible portion of the electromagnetic spectrum. *(Photo by Mark Pink/Alamy)*

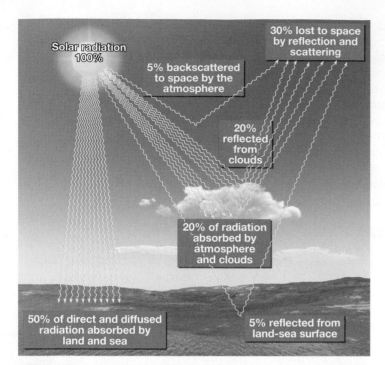

FIGURE 20.15 Average distribution of incoming solar radiation by percentage. More solar energy is absorbed by Earth's surface than by the atmosphere. Consequently, the air is not heated directly by the Sun but is rather heated indirectly from Earth's surface. These percentages can vary. For example, if cloud cover increases or the surface is brighter, the percentage of light reflected increases. This situation leaves less solar energy to take the other two paths.

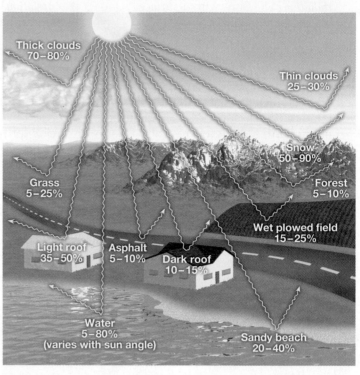

FIGURE 20.16 Albedo (reflectivity) of various surfaces. In general, light-colored surfaces tend to be more reflective than dark-colored surfaces and thus have higher albedos.

The numbers shown in Figure 20.15 represent global averages. The actual percentages can vary greatly. An important reason for much of this variation has to do with changes in the percentage of light reflected and scattered back to space (**FIGURE 20.16**). For example, if the sky is overcast, a higher percentage of light is reflected back to space than when the sky is clear.

The Greenhouse Effect

If Earth had no atmosphere, it would experience an average surface temperature far below freezing. But the atmosphere warms the planet and makes Earth livable. The extremely important role the atmosphere plays in heating Earth's surface has been named the **greenhouse effect.**

As discussed earlier, cloudless air is largely transparent to incoming short-wave solar radiation and, hence, transmits it to Earth's surface. By contrast, a significant fraction of the long-wave radiation emitted by Earth's land–sea surface is absorbed by

water vapor, carbon dioxide, and other trace gases in the atmosphere. This energy heats the air and increases the rate at which it radiates energy, both out to space and back toward Earth's surface. The energy that is emitted back to the surface causes it to heat up more, which then results in greater emissions from the surface. This complicated game of "pass the hot potato" keeps Earth's average temperature much higher than it would otherwise be (**FIGURE 20.17**). Without these absorbant gases in our atmosphere, Earth would not provide a suitable habitat for humans and other life forms.

This natural phenomenon was named the greenhouse effect because it was once thought that greenhouses were heated in a similar manner. The glass in a greenhouse allows short-wave solar radiation to enter and be absorbed by the objects inside. These objects, in turn, radiate energy but at longer wavelengths, to which glass is nearly opaque. The heat, therefore, is "trapped" in the greenhouse. It has been shown, however, that air inside

greenhouses attains higher temperatures than outside air mainly because greenhouses restrict the exchange of air between the inside and outside. Nevertheless, the term "greenhouse effect" remains.

DID YOU KNOW?

If Earth's atmosphere had no greenhouse gases, its average surface temperature would be a frigid –18 °C (–0.4 °F) instead of the relatively comfortable 14.5 °C (58 °F) that it is today.

CONCEPT CHECK 20.5

❶ What are the three paths taken by incoming solar radiation? What might cause the percentage taking each path to vary?

❷ Explain why the atmosphere is heated chiefly by radiation from Earth's surface rather than by direct solar radiation.

❸ Describe or outline the greenhouse effect.

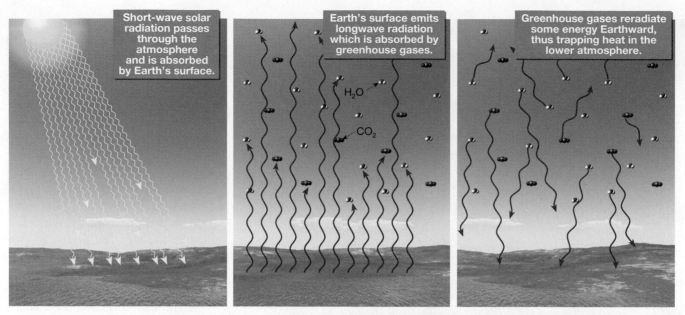

FIGURE 20.17 The heating of the atmosphere. Most of the short-wavelength radiation from the Sun passes through the atmosphere and is absorbed by Earth's land–sea surface. This energy is then emitted from the surface as longer-wavelength radiation, much of which is absorbed by certain gases in the atmosphere. Some of the energy absorbed by the atmosphere will be reradiated Earthward. This so-called greenhouse effect is responsible for keeping Earth's surface much warmer than it would be otherwise.

Natural Causes of Climate Change

A great variety of hypotheses have been proposed to explain climate change. Several have gained wide support, only to lose it and then sometimes regain it again. Some explanations are controversial. This is to be expected, because planetary atmospheric processes are so large-scale and complex that they cannot be reproduced physically in laboratory experiments. Rather, climate and its changes must be simulated mathematically (modeled) using powerful computers.

In Chapter 11, the section on "Causes of Glaciation" (pp. 281–283) described two "natural" mechanisms of climate change. Recall that the movement of lithospheric plates gradually moves Earth's continents closer to or farther from the equator. Although these shifts in latitude are slow, they can have a dramatic impact on climate over spans of millions of years. Moving landmasses can also lead to significant shifts in ocean circulation, which influences heat transport around the globe.*

A second natural mechanism of climate change discussed in Chapter 11 involved variations in Earth's orbit. Changes in the shape of the orbit (*eccentricity*), variations in the angle that Earth's axis makes with the plane of its orbit (*obliquity*), and the wobbling of the axis (*precession*) cause fluctuations in the seasonal and latitudinal distribution of solar radiation. These variations, in turn, contributed to the alternating glacial–interglacial episodes of the Ice Age.

In this section, we describe two additional hypotheses that have earned serious consideration from the scientific community. One involves the role of volcanic activity. How do the gases and particles emitted by volcanoes impact climate? A second natural cause of climate change discussed in this section involves solar variability. Does the Sun vary in its radiation output? Do sunspots affect the output?

After this look at natural factors, we will examine human-made climate changes, including the effects of rising levels of carbon dioxide and other trace gases.

As you read this section, you will find that more than one hypothesis may explain the same climate change. In fact, several mechanisms may interact to shift climate. Also, no single hypothesis can explain climate change on all time scales. A proposal that explains variations over millions of years generally cannot explain fluctuations over hundreds of years. If our atmosphere and its changes ever become fully understood, we will probably see that climate change is caused by many of the mechanisms discussed here, plus new ones yet to be proposed.

Volcanic Activity and Climate Change

The idea that explosive volcanic eruptions might alter Earth's climate was first proposed many years ago. It is still regarded as a plausible explanation for some aspects of climatic variability. Explosive eruptions emit huge quantities of gases and fine-grained debris into the atmosphere (**FIGURE 20.18**). The greatest eruptions are sufficiently powerful to inject material high into the atmosphere, where it spreads around the globe and remains for many months or even years.

*For more on this, see the section titled "Supercontinents and Climate" in Chapter 19, pp. 472–473.

A.

FIGURE 20.18 Mount Etna, a volcano on the island of Sicily, erupting in late October 2002. Mount Etna is Europe's largest and most active volcano. **A.** This photo of Mount Etna looking southeast was taken by a member of the International Space Station. It shows a plume of volcanic ash streaming southeastward from the volcano. **B.** This image from the Atmospheric Infrared Sounder on NASA's *Aqua* satellite shows the sulfur dioxide (SO_2) plume in shades of purple and black. Climate may be affected when large quantities of SO_2 are injected into the atmosphere. *(Images courtesy of NASA)*

in 1991 have given scientists an opportunity to study the atmospheric effects of volcanic eruptions with the aid of more sophisticated technology than had been available in the past. Satellite images and remote-sensing instruments allowed scientists to monitor closely the effects of the clouds of gases and ash that these volcanoes emitted.

MOUNT ST. HELENS. When Mount St. Helens erupted, there was immediate speculation about the possible effects on our climate. Could such an eruption cause our climate to change? There is no doubt that the large quantity of volcanic ash emitted by the explosive eruption had significant local and regional effects for a short period. Still, studies indicated that any longer-term lowering of hemispheric temperatures was negligible. The cooling was so slight, probably less than 0.1 °C (0.2 °F), that it could not be distinguished from other natural temperature fluctuations.

EL CHICHÓN. Two years of monitoring and studies following the 1982 El Chichón eruption indicated that its cooling effect on global mean temperature was greater than that of Mount St. Helens, on the order of 0.3 to 0.5 °C (0.5 to 0.9 °F). The eruption of El Chichón was *less explosive* than the Mount St. Helens blast, so why did it have a greater impact on global temperatures? The reason is that the material emitted by Mount St. Helens was largely fine ash that settled out in a relatively short time. El Chichón, on the other hand, emitted far greater quantities of sulfur dioxide gas (an estimated 40 times more) than Mount St. Helens. This gas combines with water vapor in the stratosphere to produce a dense cloud of tiny sulfuric acid particles (**FIGURE 20.19A**). The particles, called *aerosols,* take several years to settle out completely. They lower the troposphere's mean temperature because they reflect solar radiation back into space (**FIGURE 20.19B**).

We now understand that volcanic clouds that remain in the stratosphere for a year or more are composed largely of sulfuric-acid droplets and not of dust, as was once thought. Thus, the volume of fine debris

THE BASIC PREMISE. The basic premise is that this suspended volcanic material will filter out a portion of the incoming solar radiation, which in turn will lower temperatures in the troposphere. More than 200 years ago Benjamin Franklin used this idea to argue that material from the eruption of a large Icelandic volcano could have reflected sunlight back to space and therefore might have been responsible for the unusually cold winter of 1783–1784.

Perhaps the most notable cool period linked to a volcanic event is the "year without a summer" that followed the 1815 eruption of Mount Tambora in Indonesia. The eruption of Tambora is the largest of modern times. During April 7–12, 1815, this nearly 4000-meter-high (13,000-foot) volcano violently expelled an estimated 100 cubic kilometers (24 cubic miles) of

volcanic debris. The impact of the volcanic aerosols on climate is believed to have been widespread in the Northern Hemisphere. From May through September 1816 an unprecedented series of cold spells affected the northeastern United States and adjacent portions of Canada. There was heavy snow in June and frost in July and August. Abnormal cold was also experienced in much of Western Europe. Similar, although apparently less dramatic, effects were associated with other great explosive volcanoes, including Indonesia's Krakatoa in 1883.

Three major volcanic events have provided considerable data and insight regarding the impact of volcanoes on global temperatures. The eruptions of Washington state's Mount St. Helens in 1980, the Mexican volcano El Chichón in 1982, and the Philippines' Mount Pinatubo

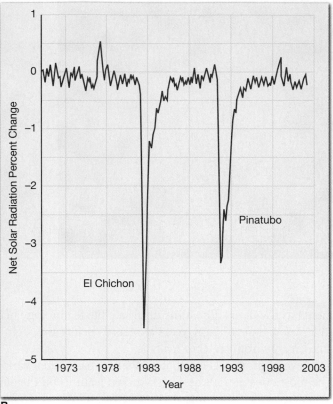

A.

B.

FIGURE 20.19 **A.**This satellite image shows a plume of white haze from Anatahan Volcano, blanketing a portion of the Philippine Sea following a large eruption in April 2005. The haze is *not* volcanic ash. Rather, it consists of tiny droplets of sulfuric acid formed when sulfur dioxide from the volcano combined with water in the atmosphere. The haze is bright and reflects sunlight back to space. *(NASA image)*
B. Net solar radiation at Hawaii's Mauna Loa Observatory relative to 1970 (zero on the graph). The eruptions of El Chichón and Mt. Pinatubo clearly caused a temporary drop in solar radiation reaching the surface. *(After Earth System Research Laboratory/NOAA)*

emitted during an explosive event is not an accurate criterion for predicting the global atmospheric effects of an eruption.

MOUNT PINATUBO. The Philippines' volcano, Mount Pinatubo, erupted explosively in June 1991, injecting 25–30 million tons of sulfur dioxide into the stratosphere. The event provided scientists with an opportunity to study the climatic impact of a major explosive volcanic eruption using NASA's spaceborne Earth Radiation Budget Experiment. During the next year the haze of tiny aerosols increased reflectivity and lowered global temperatures by 0.5 °C (0.9 °F).

The impact on global temperature of eruptions like El Chichón and Mount Pinatubo is relatively minor, but many scientists agree that the cooling produced could alter the general pattern of atmospheric circulation for a limited period.

Such a change, in turn, could influence the weather in some regions. Predicting, or even identifying, specific regional effects still presents a considerable challenge to atmospheric scientists.

The preceding examples illustrate that the impact on climate of a single volcanic eruption, no matter how great, is relatively small and short-lived. The graph in Figure 20.19B reinforces that point. Therefore, if volcanism is to have a pronounced impact over an extended period, many great eruptions, closely spaced in time, need to occur. If this happens, the stratosphere will be loaded with enough gases and volcanic dust to seriously diminish the amount of solar radiation reaching the surface. Because no such period of explosive volcanism is known to have occurred in historic times, it is most often mentioned as a possible contributor to prehistoric climatic shifts.

VOLCANISM AND GLOBAL WARMING. The Cretaceous Period is the last period of the Mesozoic Era, the era of *middle life* that is often called the "age of dinosaurs." It began about 145.5 million years ago and ended about 65.5 million years ago with the extinction of the dinosaurs (and many other life forms as well).*

The Cretaceous climate was among the warmest in Earth's long history. Dinosaurs, which are associated with mild temperatures, ranged north of the Arctic Circle. Tropical forests existed in Greenland and Antarctica, and coral reefs grew as much as 15 degrees latitude closer to the poles than at present. Deposits of peat that would eventually form widespread coal beds accumulated at high latitudes. Sea level was as much as 200 meters (650 feet) higher than it is today, indicating that there were no polar ice sheets.

What was the cause of the unusually warm climates of the Cretaceous Period? Among the significant factors that may have contributed was an enhancement of the greenhouse effect due to an increase in the amount of carbon dioxide in the atmosphere.

Where did the additional CO_2 come from that contributed to the Cretaceous warming? Many geologists suggest that the

*For more on the end of the Cretaceous, see Chapter 19.

FIGURE 20.20 A. Large sunspot group on the solar disk. *(Celestron 8 photo courtesy of Celestron International)* **B.** Sunspots having visible umbra (dark central area) and penumbra (lighter area surrounding umbra). *(Courtesy of National Optical Astronomy Observatories)*

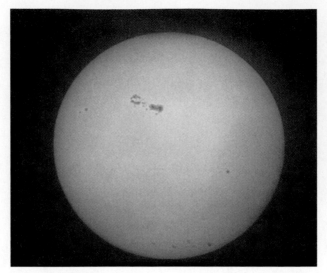

A.

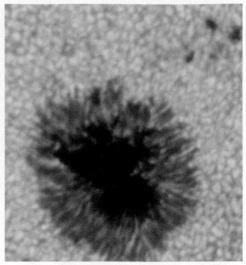

B.

probable source was volcanic activity. Carbon dioxide is one of the gases emitted during volcanism, and there is now considerable geologic evidence that the Middle Cretaceous was a time when there was an unusually high rate of volcanic activity. Several huge oceanic lava plateaus were produced on the floor of the western Pacific during this span. These vast features were associated with hot spots that may have been the product of large mantle plumes. Massive outpourings of lava over millions of years would have been accompanied by the release of huge quantities of CO_2 which in turn would have enhanced the atmospheric greenhouse effect. *Thus, the warmth that characterized the Cretaceous may have had its origins deep in Earth's mantle.*

Solar Variability and Climate

Among the most persistent hypotheses of climate change have been those based on the idea that the Sun is a variable star and that its output of energy varies over time. The effect of such changes would seem direct and easily understood: Increases in solar output would cause the atmosphere to warm, and reductions would result in cooling. This notion is appealing because it can be used to explain climate change of any length or intensity. However, no major *long-term* variations in the total intensity of solar radiation have yet been measured outside the atmosphere. Such measure-

ments were not even possible until satellite technology became available. Now that it is possible, we will need many years of records before we begin to sense how variable (or invariable) energy from the Sun really is.

Several proposals for climate change, based on a variable Sun, relate to sunspot cycles. The most conspicuous and best-known features on the surface of the Sun are the dark blemishes called **sunspots** (**FIGURE 20.20**). Sunspots are huge magnetic storms that extend from the Sun's surface deep into the interior. Moreover, these spots are associated with the Sun's ejection of huge masses of particles that, on reaching Earth's upper atmosphere, interact with gases to produce auroral displays such as the Aurora Borealis, or Northern Lights, in the Northern Hemisphere.

Along with other solar activity, the number of sunspots seems to increase and decrease in a regular way, creating a cycle of about 11 years. The graph in **FIGURE 20.21** shows the annual number of sunspots, beginning in the early 1700s. However, this pattern does not always occur. There have been periods when the Sun was essentially free of sunspots. In addition to the well-known 11-year cycle, there is also a 22-year cycle. This longer cycle is based on the fact that the magnetic polarities of sunspot clusters reverse every successive 11 years.

Interest in possible Sun–climate effects has been sustained by an almost continuous effort to find correlations on time scales ranging from days to tens of thousands of years.

FIGURE 20.21 Mean annual sunspot numbers.

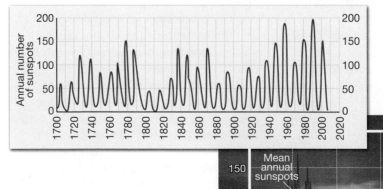

Two widely debated examples are briefly described here.

SUNSPOTS AND TEMPERATURE.
Studies indicate that there have been prolonged periods when sunspots were absent or nearly so. Moreover, these events correspond closely with cold periods in Europe and North America. Conversely, periods characterized by plentiful sunspots have correlated well with warmer times in these regions.

Referring to these matches, some scientists have suggested that such correlations make it appear that changes on the Sun are important to climate change. Other scientists seriously question this notion. In part their hesitation stems from subsequent investigations, using different climate records from around the world, that failed to find a significant correlation between solar activity and climate. Even more troubling is that no testable physical mechanism exists to explain the purported effect.

SUNSPOTS AND DROUGHT. A second possible Sun–climate connection, on a time scale different from the preceding example, relates to variations in precipitation rather than in temperature. An extensive study of tree rings revealed a recurrent 22-year rotation in the pattern of droughts in the western United States. This periodicity coincides with the previously mentioned 22-year magnetic cycle of the Sun.

Commenting on this possible connection, a panel of the National Research Council concluded:

No convincing mechanism that might connect so subtle a feature of the sun to drought patterns in limited

regions has yet appeared. Moreover, the cyclic pattern of droughts found in tree rings is itself a subtle feature that shifts from place to place within the broad region of the study.*

Possible connections between solar variability and climate would be much easier to determine if researchers could identify physical links between the Sun and the lower atmosphere. Despite much research, no connection between solar variations and weather has yet been well established. Apparent correlations have almost always faltered when put to critical statistical examination or when tested with different data sets. As a result, the subject has been characterized by ongoing controversy and debate.

CONCEPT CHECK 20.6

❶ Describe and briefly explain the effect on global temperatures of the eruptions of El Chichón and Mt. Pinatubo.

❷ How might volcanism lead to global warming?

❸ List two examples of climate change linked to solar variability.

Human Impact on Global Climate

So far we have examined four potential natural causes of climate change. In this section we discuss how humans contribute to global climate change. One impact largely results from the addition to the atmosphere of carbon dioxide and other greenhouse gases. A second impact is related to the addition of human-generated aerosols to the atmosphere.

Human influence on regional and global climate did not just begin with the onset of the modern industrial period. There is solid evidence that people have been modifying the environment over extensive areas for thousands of years. The use of fire and the overgrazing of marginal lands by domesticated animals have reduced the abundance and distribution of vegetation. By altering ground cover, humans have modified such important climatological factors as surface

**Solar Variability, Weather, and Climate* (Washington, D.C.: National Academy Press, 1982), 7.

albedo, evaporation rates, and surface winds. Commenting on this aspect of human-induced climate modification, the late astronomer Carl Sagan noted, "In contrast to the prevailing view that only modern humans are able to alter climate, we believe it is more likely that the human species has made a substantial and continuing impact on climate since the invention of fire."**

Subsequently, these ideas were reinforced and expanded upon by a study that used data collected from Antarctic ice cores. This research suggested that humans may have started to have a significant impact on atmospheric composition and global temperatures thousands of years ago.

Humans started slowly ratcheting up the thermostat as early as 8000 years ago, when they began clearing forests for agriculture, and 5000 years ago with the arrival of wet-rice cultivation. The greenhouse gases carbon dioxide and methane given off by these changes would have warmed the world. . . .***

CONCEPT CHECK 20.7

❶ How might people have altered climate thousands of years ago?

Carbon Dioxide, Trace Gases, and Climate Change

Earlier you learned that carbon dioxide (CO_2) represents only about 0.039 percent of the gases that make up clean, dry air. Nevertheless, it is a very significant component meteorologically. Carbon dioxide is influential because it is transparent to incoming short-wavelength solar radiation, but it is not transparent to some of the longer-wavelength outgoing Earth radiation. A portion of the energy leaving the ground is absorbed by atmospheric CO_2.

**Carl Sagan et al., "Anthropogenic Albedo Changes and the Earth's Climate," *Science*, 206, No. 4425 (1980), 367.
***An Early Start for Greenhouse Warming?", Science, Vol. 303, 16 January, 2004. This article is a report on a paper given by paleoclimatologist William Ruddiman at a meeting of the American Geophysical Union in December 2003.

A.

B.

FIGURE 20.22 Clearing the tropical rain forest is a serious environmental issue. In addition to the loss of biodiversity, tropical deforestation is a significant source of carbon dioxide. **A.** This August 2007 satellite image shows deforestation in the Amazon rain forest in western Brazil. Intact forest is dark green, whereas cleared areas are tan (bare ground) or light green (crops and pasture). *(NASA)* **B.** Fires are frequently used to clear the land. This scene is also in Brazil's Amazon Basin. *(Photo by Pete Oxford/Nature Picture Library)*

This energy is subsequently reemitted, part of it back toward the surface, thereby keeping the air near the ground warmer than it would be without CO_2.

Thus, along with water vapor, carbon dioxide is largely responsible for the *greenhouse effect* of the atmosphere. Carbon dioxide is an important heat absorber, and it follows logically that any change in the air's CO_2 content could alter temperatures in the lower atmosphere.

CO_2 Levels Are Rising

Earth's tremendous industrialization of the past two centuries has been fueled—and still is fueled—by burning fossil fuels: coal, natural gas, and petroleum (see Figure 6.24, p. 170). Combustion of these fuels

has added great quantities of carbon dioxide to the atmosphere.

The use of coal and other fuels is the most prominent means by which humans add CO_2 to the atmosphere, but it is not the only way. The clearing of forests also contributes substantially because CO_2 is released as vegetation is burned or decays. Deforestation is particularly pronounced in the tropics, where vast tracts are cleared for ranching and agriculture or subjected to inefficient commercial logging

operations (**FIGURE 20.22**). According to U.N. estimates, nearly 10.2 million hectares (25.1 million acres) of tropical forest were permanently destroyed each year during the decade of the 1990s. Between the years 2000 and 2005 the average figure increased to 10.4 million hectares (25.7 million acres) per year.

Some of the excess CO_2 is taken up by plants or is dissolved in the ocean. It is estimated that 45 to 50 percent remains in the atmosphere. **FIGURE 20.23** is a graphic

DID YOU KNOW?

Mauna Loa Observatory, shown in the chapter-opening photo, has been continuously collecting atmospheric data since the 1950s. It is an ideal place to sample the atmospheric components that influence climate change because of its remote location in the middle of the Pacific Ocean away from major sources of air pollution. In addition, the observatory protrudes through a strong marine temperature inversion layer that acts as a lid to keep local pollutants below it.

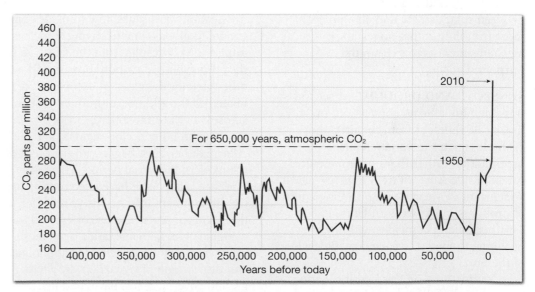

FIGURE 20.23 Carbon dioxide concentrations over the past 400,000 years. Most of the data comes from the analysis of air bubbles trapped in ice cores. The record since 1958 comes from direct measurements of atmospheric CO_2 taken at Mauna Loa Observatory, Hawaii. The rapid increase in CO_2 concentrations since the onset of the industrial revolution is obvious.

Annual CO₂ Contribution of an Average American

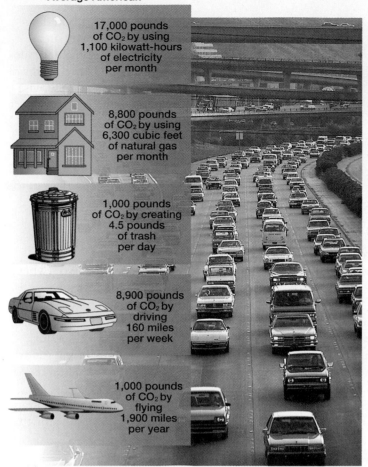

17,000 pounds of CO₂ by using 1,100 kilowatt-hours of electricity per month

8,800 pounds of CO₂ by using 6,300 cubic feet of natural gas per month

1,000 pounds of CO₂ by creating 4.5 pounds of trash per day

8,900 pounds of CO₂ by driving 160 miles per week

1,000 pounds of CO₂ by flying 1,900 miles per year

FIGURE 20.24 Americans are responsible for about 25 percent of the world's greenhouse-gas emissions. That amounts to 24,300 kilograms (54,000 pounds) of carbon dioxide each year for an average American, which is about five times the emissions of the average global citizen. This diagram represents some of the ways that people in the United States contribute. *(Data from various U.S. government agencies; Photo by Jerry Schad/Photo Researchers, Inc.)*

record of changes in atmospheric CO₂ extending back 400,000 years. Over this long span, natural fluctuations varied from about 180 to 300 parts per million (ppm). As a result of human activities, the present CO₂ level is about 30 percent higher than its highest level over at least the last 650,000 years. The rapid increase in CO₂ concentrations since the onset of industrialization is obvious. The annual rate at which atmospheric CO₂ concentrations are growing has been increasing over the last several decades (**FIGURE 20.24**).

The Atmosphere's Response

Given the increase in the atmosphere's carbon dioxide content, have global temperatures actually increased? The answer is yes. According to a 2007 report by the Intergovernmental Panel on Climate Change (IPCC), "Warming of the climate system is unequivocal, as is now evident from observations of increases in global average air and ocean temperatures, widespread melting of snow and ice, and

rising global sea level."* It is *very likely* that most of the observed increase in global average temperatures since the mid-20th century is due to the observed increase in human-generated greenhouse-gas concentrations. (As used by the IPCC, *very likely* indicates a probability of 90 to 99 percent.) Global warming since the mid-1970s is now about 0.6 °C (1 °F), and total warming in the past century is about 0.8 °C (1.4 °F). The upward trend in surface temperatures is shown in **FIGURE 20.25A**.

*IPCC, Summary for Policy Makers. In *Climate Change 2007: The Physical Science Basis.* p. 4. Cambridge University Press, Cambridge, United Kingdom and New York, NY.

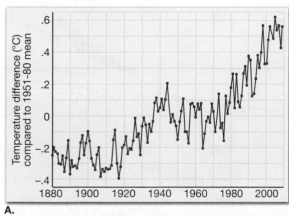

A.

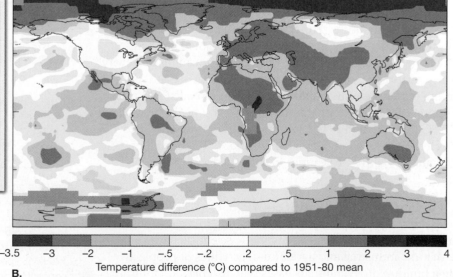

FIGURE 20.25 A. The graph depicts global temperature change in °C since the year 1880. **B.** The world map shows how temperatures in 2009 deviated from the mean for the 1951–80 base period. The high latitudes in the Northern Hemisphere clearly stand out. *(After NASA/Goddard Institute for Space Studies)*

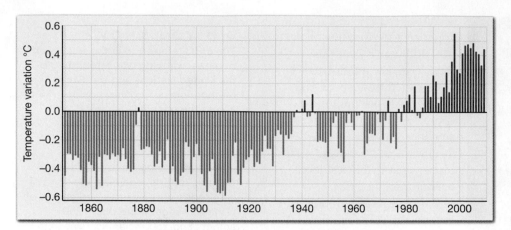

FIGURE 20.26 Annual average global temperature variations for the period 1850 to 2009. The basis for comparison is the average for the 1961–90 period (the 0.0 line on the graph). Each narrow bar on the graph represents the departure of the global mean temperature from the 1961–90 average for a particular year. For example, the global mean temperature for 1862 was more than 0.5 °C below the 1961–90 average, whereas the global mean for 1998 was more than 0.5 °C above. *(After Climate Research Unit/University of East Anglia)*

The world map in **FIGURE 20.25B** compares surface temperatures for 2009 to the base period (1951–1980). You can see that the greatest warming has been in the Arctic and neighboring high latitude regions. Here are some related facts:

- When we consider the time span for which there are instrumental records (since 1850), 14 of the last 15 years (1995–2009) rank among the 15 warmest (**FIGURE 20.26**).
- Global mean temperature is now higher than at any time in at least the past 500 to 1000 years.
- The average temperature of the global ocean has increased to depths of at least 3000 meters (10,000 feet).

Are these temperature trends caused by human activities, or would they have occurred anyway? The scientific consensus of the IPCC is that human activities were *very likely* responsible for most of the temperature increase since 1950.

What about the future? Projections for the years ahead depend in part on the quantities of greenhouse gases that are emitted. **FIGURE 20.27** shows the best estimates of global warming for several different scenarios. The 2007 IPCC report also states that if there is a doubling of the preindustrial level of carbon dioxide (280 ppm) to 560 ppm, the "likely temperature increase will be in the range of 2 to 4.5 °C (3.5 to 8.1 °F). The increase is "very unlikely" (1 to 10 percent probability) to be less than 1.5 °C (2.7 °F) and values higher than 4.5 °C (8.1 °F) cannot be excluded.

The Role of Trace Gases

Carbon dioxide is not the only gas contributing to a global increase in temperature. In recent years atmospheric scientists have come to realize that the industrial and agricultural activities of people are causing a buildup of several trace gases that also play a significant role. The substances are called *trace gases* because their concentra-

tions are so much smaller than that of carbon dioxide. The trace gases that are most important are methane (CH_4), nitrous oxide (N_2O), and chlorofluorocarbons (CFCs). These gases absorb wavelengths of outgoing radiation from Earth that would otherwise escape into space. Although individually their impact is modest, taken together the effects of these trace gases play a significant role in warming the troposphere.

Methane is present in much smaller amounts than CO_2 is, but its significance is greater than its relatively small concentration would indicate (**FIGURE 20.28**). The reason is that methane is about 20 times more effective than CO_2 at absorbing infrared radiation emitted by Earth.

Methane is produced by *anaerobic* bacteria in wet places where oxygen is scarce (anaerobic means "without air," specifically oxygen). Such places include swamps, bogs, wetlands, and the guts of termites and grazing animals like cattle and sheep. Methane is also generated in flooded paddy fields ("artificial swamps") used for growing rice. Mining of coal and drilling for oil and natural gas are other sources because methane is a product of the formation of these fuels (**FIGURE 20.29**).

The concentration of methane in the atmosphere has risen rapidly since 1800, an increase that has been in step with the growth

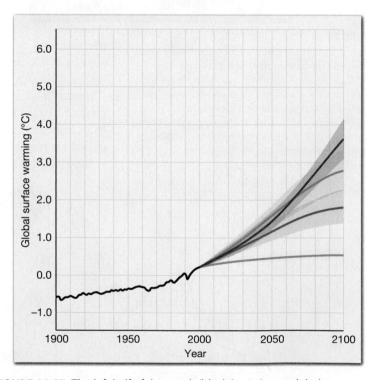

FIGURE 20.27 The left half of the graph (black line) shows global temperature changes for the 20th century. The right half shows projected global warming in different emissions scenarios. The shaded zone adjacent to each colored line shows the uncertainty range for each scenario. The basis for comparison (0.0 on the vertical axis) is the global average for the period 1980–99. The orange line represents the scenario in which carbon dioxide concentrations were held constant at the values for the year 2000. *(After IPCC, 2007)*

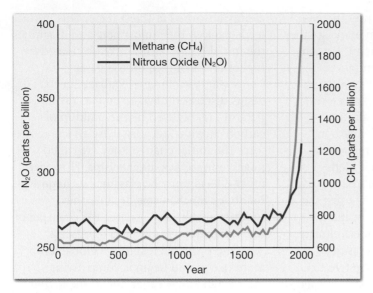

FIGURE 20.28 Increases in the concentrations of two trace gases, methane (CH_4) and nitrous oxide (N_2O), that contribute to global warming. The sharp rise during the industrial era is obvious. *(After U.S. Global Change Research Program)*

in human population. This relationship reflects the close link between methane formation and agriculture. As population has risen, so have the number of cattle and rice paddies.

Nitrous oxide, sometimes called "laughing gas," is also building in the atmosphere, although not as rapidly as methane (see Figure 20.28). The increase results primarily from agricultural activity. When farmers use nitrogen fertilizers to boost crop yields, some of the nitrogen enters the air as nitrous oxide. This gas is also produced by high-temperature combustion of fossil fuels. Although the annual release into the atmosphere is small, the lifetime of a nitrous oxide molecule is about 150 years! If the use of nitrogen fertilizers and fossil fuels grows at projected rates, nitrous oxide may make a contribution to greenhouse warming that approaches half that of methane.

Unlike methane and nitrous oxide, chlorofluorocarbons (CFCs) are not naturally present in the atmosphere. CFCs are manufactured chemicals with many uses; they have gained notoriety because they are responsible for ozone depletion in the stratosphere. The role of CFCs in global warming is less well known. CFCs are very effective greenhouse gases. They were not developed until the 1920s and were not used in great quantities until the 1950s, but they already contribute to the greenhouse effect at a level equal to methane. Although corrective action has been taken, CFC levels will *not* drop rapidly. CFCs remain in the atmosphere for decades, so even if all CFC emissions were to stop immediately,

FIGURE 20.29
A. Coal mining and drilling for oil and natural gas are sources of methane. This large flame is a flare of methane being burned at an oil well. *(Photo on left by Kim Steele/Getty Images)* **B.** Methane is also produced by anaerobic bacteria in wet places, where oxygen is scarce (anaerobic means "without air," specifically oxygen). Such places include swamps, bogs, wetlands, and the guts of termites and grazing animals, like cattle and sheep. Methane is also generated in flooded paddy fields ("artificial swamps") used for growing rice. *(Photo by Moodboard/CORBIS)*

A.

B.

the atmosphere would not be free of them for many years.

Carbon dioxide is clearly the most important single cause for the projected global greenhouse warming. However, it is not the only contributor. When the effects of all human-generated greenhouse gases other than CO_2 are added together and projected into the future, their collective impact significantly increases the impact of CO_2 alone.

Sophisticated computer models show that the warming of the lower atmosphere caused by CO_2 and trace gases will not be the same everywhere. Rather, the temperature response in polar regions could be two to three times greater than the global average. One reason is that the polar troposphere is very stable, which suppresses vertical mixing and thus limits the amount of surface heat that is transferred upward. In addition, the expected reduction in sea ice will also contribute to the greater temperature increase. This topic will be explored more fully in the next section.

CONCEPT CHECK 20.8

1 Why has the CO_2 level of the atmosphere been increasing for more than 150 years?

2 How are temperatures in the lower atmosphere likely to change as CO_2 levels continue to increase?

3 Aside from CO_2, what trace gases are contributing to global temperature change?

Climate-Feedback Mechanisms

Climate is a very complex interactive physical system. Thus, when any component of the climate system is altered, scientists must consider many possible outcomes. These possible outcomes are called **climate-feedback mechanisms.** They complicate climate-modeling efforts and add greater uncertainty to climate predictions.

Types of Feedback Mechanisms

What climate-feedback mechanisms are related to carbon dioxide and other greenhouse gases? One important mechanism is

that warmer surface temperatures increase evaporation rates. This in turn increases the atmosphere's water vapor content. Remember that water vapor is an even more powerful absorber of radiation emitted by Earth than is carbon dioxide. Therefore, with more water vapor in the air, the temperature increase caused by carbon dioxide and the trace gases is reinforced.

Figure 20.25B showed that high-latitude regions warmed more in 2009 than lower-latitude areas. This was not just the case in 2009; it has been true for many years. Scientists who model global climate change indicate that the temperature increase at high latitudes may be two to three times greater than the global average. This assumption is based in part on the likelihood that the area covered by sea ice will decrease as surface temperatures rise. Because ice reflects a much larger percentage of incoming solar radiation than does open water, the melting of the sea ice replaces a highly reflecting surface with a relatively dark surface (**FIGURE 20.30**). The result is a substantial increase in the solar energy absorbed at the surface. This in turn feeds back to the atmosphere and magnifies the initial temperature increase created by higher levels of greenhouse gases.

So far, the climate-feedback mechanisms discussed have magnified the temperature rise caused by the buildup of carbon dioxide. Because these effects reinforce the initial change, they are called **positive-feedback mechanisms.** However, other effects must be classified as **negative-feedback mechanisms** because they produce results that are just the opposite of the initial change and tend to offset it.

One probable result of a global temperature rise would be an accompanying increase in cloud cover due to the higher moisture content of the atmosphere. Most clouds are good reflectors of solar radiation. At the same time, however, they are also good absorbers and emitters of radiation emitted by Earth. Consequently, clouds produce two opposite effects. They are a negative-feedback mechanism because they increase the reflection of solar radiation and thus diminish the amount of solar energy

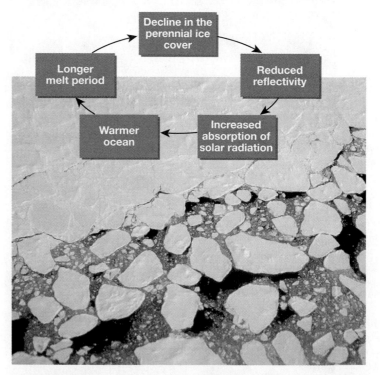

FIGURE 20.30 This satellite image shows the springtime breakup of sea ice near Antarctica. The inset shows a likely feedback loop. A reduction in sea ice acts as a positive-feedback mechanism because surface reflectivity would decrease and the amount of energy absorbed at the surface would increase. *(Reproduced with permission from "Science" (Cover, January 25, 2002). Copyright American Association for the Advancement of Science. Photo: D. N. Thomas.)*

available to heat the atmosphere. On the other hand, clouds act as a positive-feedback mechanism by absorbing and emitting radiation that would otherwise be lost from the troposphere.

Which effect, if either, is stronger? Atmospheric modeling shows that the negative effect of a higher reflectivity is dominant. Therefore, the net result of an increase in cloudiness should be a decrease in air temperature. The magnitude of this negative feedback, however, is not believed to be as great as the positive feedback caused by added moisture and decreased sea ice. Thus, although increases in cloud cover may partly offset a global temperature increase, climate models show that the ultimate effect of the projected increase in CO_2 and trace gases will still be a temperature increase.

The problem of global warming caused by human-induced changes in atmospheric composition continues to be one of the most studied aspects of climate change. Although no models yet incorporate the full range of potential factors and feedbacks, the scientific consensus is that the increasing levels of atmospheric carbon dioxide and trace gases will lead to a warmer planet with a different distribution of climate regimes.

Computer Models of Climate: Important Yet Imperfect Tools

Earth's climate system is amazingly complex. Comprehensive state-of-the-science climate simulation models are among the basic tools used to develop possible climate-change scenarios. They are based on fundamental laws of physics and chemistry and incorporate human and biological interactions. The models simulate many variables, including temperature, rainfall, snow cover, soil moisture, winds, clouds, sea ice, and ocean circulation over the entire globe through the seasons and over spans of decades.

In many other fields of study, hypotheses can be tested by direct experimentation in the laboratory or by observations and measurements in the field. However, this is often not possible in the study of climate. Rather, scientists must construct computer

models of how our planet's climate system works. If we understand the climate system correctly and construct the model appropriately, then the behavior of the model climate system should mimic the behavior of Earth's climate system (**FIGURE 20.31**).

What factors influence the accuracy of climate models? Clearly, mathematical models are *simplified* versions of real Earth and cannot capture its full complexity, especially at smaller geographic scales. Moreover, when computer models are used to simulate future climate change, many assumptions have to be made that significantly influence the outcome. They must consider a wide range of possibilities for future changes in population, economic growth, consumption of fossil fuels, technological development, improvements in energy efficiency, and more.

Despite many obstacles, our ability to use supercomputers to simulate climate continues to improve. Although today's models are far from infallible, they are powerful tools for understanding what Earth's future climate might be like.

CONCEPT CHECK 20.9

❶ Distinguish between positive and negative feedback mechanisms.

❷ Describe at least one example of each type of feedback mechanism.

How Aerosols Influence Climate

Increasing the levels of carbon dioxide and other greenhouse gases in the atmosphere is the most direct human influence on global climate, but it is not the only impact. Global climate is also affected by human activities that contribute to the atmosphere's aerosol content. Recall that *aerosols* are the tiny, often microscopic, liquid and solid particles

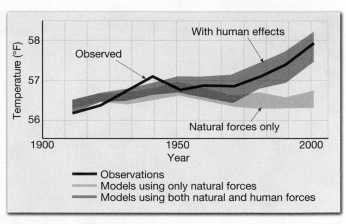

FIGURE 20.31 The blue band shows how global average temperatures would have changed due to natural forces only, as simulated by climate models. The red band shows model projections of the effects of human and natural forces combined. The black line shows actual observed global average temperatures. As the blue band indicates, without human influences, temperature over the past century would actually have first warmed and then cooled slightly over recent decades. Bands of color are used to express the range of uncertainty. *(After U.S. Global Change Research Program)*

DID YOU KNOW?

A *scenario* is an example of what might happen under a particular set of assumptions. Scenarios are a way of examining questions about an uncertain future. For example, future trends in fossil-fuel use and other human activities are uncertain. Therefore, scientists have developed a set of scenarios for how the climate may change based on a wide range of possibilities for these variables.

that are suspended in the air. Unlike cloud droplets, aerosols are present even in relatively dry air. Atmospheric aerosols are composed of many different materials, including soil, smoke, sea salt, and sulfuric acid. Natural sources are numerous and include such phenomena as dust storms and volcanoes.

Presently the human contribution of aerosols to the atmosphere *equals* the quantity emitted by natural sources. Most human-generated aerosols come from the sulfur dioxide emitted during the combustion of fossil fuels and as a consequence of burning vegetation to clear agricultural land. Chemical reactions in the atmosphere convert the sulfur dioxide into sulfate aerosols, the same material that produces acid precipitation.

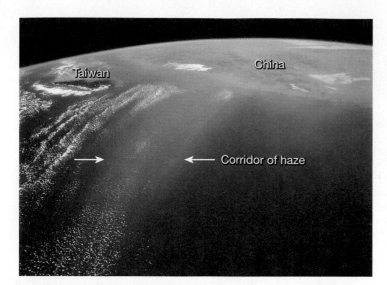

FIGURE 20.32 Human-generated aerosols are concentrated near the areas that produce them. Because aerosols reduce the amount of solar energy available to the climate system, they have a net cooling effect. This satellite image shows a dense blanket of pollution moving away from the coast of China. The plume is about 200 kilometers wide and more than 600 kilometers long. *(NASA Image)*

as carbon dioxide, remain in the atmosphere for many decades. By contrast, aerosols released into the troposphere remain there for only a few days or, at most, a few weeks before they are "washed out" by precipitation. Because of their short lifetime in the troposphere, aerosols are distributed unevenly over the globe. As expected, human-generated aerosols are concentrated near the areas that produce them, namely industrialized regions that burn fossil fuels and land areas where vegetation is burned (**FIGURE 20.32**).

Because their lifetime in the atmosphere is short, the effect of aerosols on today's climate is determined by the amount emitted during the preceding couple of weeks. By contrast, the carbon dioxide and trace gases released into the atmosphere remain for much longer spans and thus influence climate for many decades.

CONCEPT CHECK 20.10

❶ What are the main sources of human-generated aerosols?
❷ What effect do these aerosols have on temperatures in the troposphere?

How do aerosols affect climate? Aerosols act directly by reflecting sunlight back to space and indirectly by making clouds "brighter" reflectors. The second effect relates to the fact that many aerosols (such as those composed of salt or sulfuric acid) attract water and thus are especially effective as cloud condensation nuclei. The large quantity of aerosols produced by human activities (especially industrial emissions) trigger an increase in the number of cloud droplets that form within a cloud. A greater number of small droplets increases the cloud's brightness—that is, more sunlight is reflected back to space.

By reducing the amount of solar energy available to the climate system, aerosols have a net cooling effect. Studies indicate that the cooling effect of human-generated aerosols offsets a portion of the global warming caused by the growing quantities of greenhouse gases in the atmosphere. Unfortunately, the magnitude and extent of the cooling effect of aerosols is highly uncertain. This uncertainty is a significant hurdle in advancing our understanding of how humans alter Earth's climate.

It is important to point out some significant differences between global warming by greenhouse gases and aerosol cooling. After being emitted, greenhouse gases, such

Some Possible Consequences of Global Warming

What consequences can be expected if the carbon dioxide content of the atmosphere reaches a level that is twice what it was early in the 20th century? Because the climate system is so complex, predicting the distribution of particular regional changes is speculative. It is not yet possible to pinpoint specifics, such as where or when it will become drier or wetter. Nevertheless, plausible scenarios can be given for larger scales of space and time.

As noted, the magnitude of the temperature increase will not be the same everywhere. The temperature rise will probably be smallest in the tropics and increase toward the poles. As for precipitation, the models indicate that some regions will experience significantly more precipitation and runoff. However, others will experience a decrease in runoff due to reduced precipitation or greater evaporation caused by higher temperatures.

TABLE 20.1 summarizes some of the more likely effects and their possible consequences. The table also provides the IPCC's estimate of the probability of each effect. Levels of confidence for these projections vary from "*likely*" (67 to 90 percent probability) to "*very likely*" (90 to 99 percent probability) to *virtually certain* (greater than 99 percent probability).

Sea-Level Rise

A significant impact of a human-induced global warming is a rise in sea level (**FIGURE 20.33**). As this occurs, coastal cities, wetlands, and low-lying islands could be threatened with more frequent flooding, increased shoreline erosion, and saltwater encroachment into coastal rivers and aquifers.

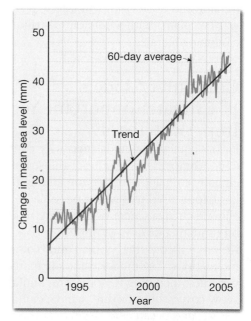

FIGURE 20.33 Using data from satellites and floats, it was determined that sea level rose an average of 3 millimeters (0.1 inch) per year between 1993 and 2005. Researchers attributed about half the rise to melting glacial ice and the other half to thermal expansion. Rising sea level can adversely affect some of the most densely populated areas on Earth. *(NASA/Jet Propulsion Laboratory)*

TABLE 20.1

Projected Changes and Effects of Global Warming in the 21st Century

Projected changes and estimated probability*	Examples of projected impacts
Higher maximum temperatures; more hot days and heat waves over nearly all land areas (*virtually certain*)	Increased incidence of death and serious illness in older age groups and urban poor. Increased heat stress in livestock and wildlife. Shift in tourist destinations. Increased risk of damage to a number of crops. Increased electric cooling demand and reduced energy supply reliability.
Higher minimum temperatures; fewer cold days, frost days, and cold waves over nearly all land areas (*virtually certain*)	Decreased cold-related human morbidity and mortality. Decreased risk of damage to a number of crops, and increased risk to others. Extended range and activity of some pest and disease vectors. Reduced heating energy demand.
Frequency of heavy precipitation events increases over most areas (*very likely*)	Increased flood, landslide, avalanche, and debris-flow damage. Increased soil erosion. Increased flood runoff could increase recharge of some floodplain aquifers. Increased pressure on government and private flood insurance systems and disaster relief.
Area affected by drought increases (*likely*)	Increased damage to building foundations caused by ground shrinkage. Decreased water-resource quantity and quality. Increased risk of forest fire. Decreased crop yields.
Intense tropical cyclone activity increases (*likely*)	Increased risks to human life, risk of Infectious-disease epidemics, and many other risks. Increased coastal erosion and damage to coastal buildings and infrastructure. Increased damage to coastal ecosystems, such as coral reefs and mangroves.

* *Virtually certain* indicates a probability greater than 99 percent, *very likely* indicates a probability of 90–99 percent, and *likely* indicates a probability of 67–90 percent.
Source: IPCC, 2001, 2007

How is a warmer atmosphere related to a global rise in sea level? The most obvious connection, the melting of glaciers, is important, but not the only factor.* An equally important factor is that a warmer atmosphere causes an increase in ocean volume due to thermal expansion. Higher air temperatures warm the adjacent upper layers of the ocean, which in turn causes the water to expand and sea level to rise.

Research indicates that sea level has risen between 10 and 23 centimeters (4 and 8 inches) over the past century and that this trend will continue at an accelerated rate. Some models indicate that the rise may approach or even exceed 50 centimeters (20 inches) by the end of the 21st century. Such a change may seem modest, but scientists realize that any rise in sea level along a *gently* sloping shoreline, such as the Atlantic and Gulf coasts of the United States, will lead to significant erosion and severe, permanent inland flooding (**FIGURE 20.34**). If this happens, many beaches and wetlands will be eliminated

and coastal civilization would be severely disrupted.**

Because rising sea level is a gradual phenomenon, it may be overlooked by coastal residents as an important contributor to shoreline erosion problems. Rather, the blame may be assigned to other forces, especially storm activity. Although a given storm may be the immediate cause, the magnitude of its destruction may result

*For more about the behavior of glaciers, see the sections on "Glaciers in Retreat" and "Greenland's Glacial Budget" in Chapter 11.

**The Rising Sea* by Orrin H. Pilkey and Rob Young (Island Press, 2009) is a highly readable and up-to-date perspective on the impact of sea level rise on the coastal environment by two prominent scientists.

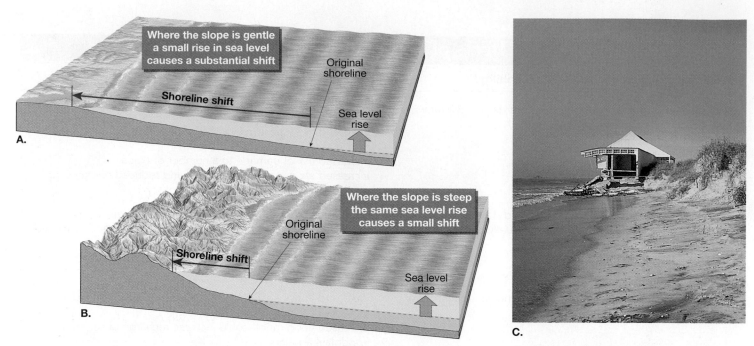

FIGURE 20.34 The slope of a shoreline is critical to determining the degree to which sea-level changes will affect it. **A.** When the slope is gentle, small changes in sea level cause a substantial shift. **B.** The same sea-level rise along a steep coast results in only a small shoreline shift. **C.** As sea level gradually rises, the shoreline retreats, and structures that were once thought to be safe from wave attack are exposed to the force of the sea. *(Photo by Kenneth Hasson)*

from the relatively small sea-level rise that allowed the storm's power to cross a much greater land area.

As mentioned, a warmer climate will cause glaciers to melt. In fact, a portion of the 10- to 25-centimeter (4- to 8-inch) rise in sea level over the past century is attributed to the melting of mountain glaciers. This contribution is projected to continue through the 21st century. Of course, if the Greenland and Antarctic ice sheets were to experience a significant increase in melting, it would lead to a much greater rise in sea level and a major encroachment by the sea in coastal zones. Is this possible? In their 2007 report, the IPCC indicates that such changes may occur over time scales of thousands of years. However, the report goes on to say that a more rapid rise on century time scales *cannot* be excluded.*

The Changing Arctic

A recent study of climate change in the Arctic began with the following statement:

*For nearly 30 years, Arctic sea ice extent and thickness have been falling dramatically. Permafrost temperatures are rising and coverage is decreasing. Mountain glaciers and the Greenland ice sheet are shrinking. Evidence suggests we are witnessing the early stage of an anthropogenically induced global warming superimposed on natural cycles, reinforced by reductions in Arctic ice.***

ARCTIC SEA ICE. Climate models are in general agreement that one of the strongest signals of global warming should be a loss of sea ice in the Arctic. This is indeed occurring. The map in **FIGURE 20.35A** compares the extent of sea ice at the end of the summer melting period in September 2009 to the long-term average for the period 1979–2000. September represents the end of the melt period when the area covered by sea ice is at a minimum. In September 2009, sea ice was 24 percent below the long-term average—third lowest, after

2007 and 2008. The trend is also clear when you examine the graph in **FIGURE 20.35B**. Is it possible that this trend may be part of a natural cycle? Yes, but it is more likely that the sea-ice decline represents a combination of natural variability and human-induced global warming, with the latter becoming increasingly evident in coming decades. As was noted in the section on "Climate Feedback Mechanisms," a reduction in sea ice represents a positive feedback mechanism that reinforces global warming.

PERMAFROST. During the past decade, evidence has mounted to indicate that the extent of permafrost in the Northern Hemisphere has decreased, as would be expected under long-term warming conditions. **FIGURE 20.36** presents one example that such a decline is occurring.

In the Arctic, short summers thaw only the top layer of frozen ground. The permafrost beneath this *active layer* is like the cement bottom of a swimming pool. In summer, water cannot percolate downward, so it saturates the soil above the permafrost and collects on the surface in thousands of lakes. However, as Arctic

*IPCC, Summary for Policymakers of the Synthesis Report of the IPCC Fourth Assessment Report, p. 13.
**J. T. Overpeck, et al. "Arctic System on Trajectory to New, Seasonally Ice-Free States," *EOS, Transactions, American Geophysical Union,* Vol. 86, No. 34, 23 August 2005, p. 309.

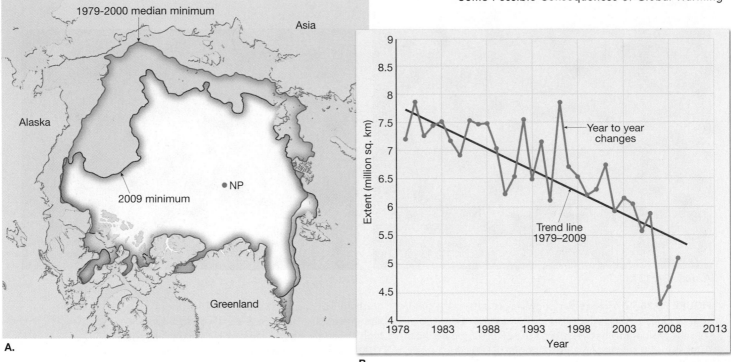

A.

B.

FIGURE 20.35 A. This map shows the extent of sea ice at the end of the summer melting period in 2009 compared to the average for the peiod 1979–2000. In 2009 the minimum occurred on September 12th. In other years this point might be reached on other dates in September. After this date, the extent of sea ice begins its cycle of growth in response to autumn cooling. In 2009 the extent of sea ice was 24 percent below the long-term average for 1979–2000. (*After NASA*) **B.** This graph depicts the decline in Arctic sea ice from 1979–2009. (*National Snow and Ice Data Center*)

temperatures climb, the bottom of the "pool" seems to be "cracking." Satellite imagery shows that over a 20-year span, a significant number of lakes have shrunk or disappeared altogether. As the permafrost thaws, lake water drains deeper into the ground.

Thawing permafrost represents a potentially significant positive feedback mechanism that may reinforce global warming. When vegetation dies in the Arctic, cold temperatures inhibit its total decomposition. As a consequence, over thousands of years a great deal of organic matter has become stored in the permafrost. When the permafrost thaws, organic matter that may have been frozen for millennia comes out of "cold storage" and decomposes. The result is the release of carbon dioxide and methane—greenhouse gases that contribute to global warming.

Most glaciers in mountain valleys are shrinking.
(*Photo by Kurt Werby/Photolibrary*)

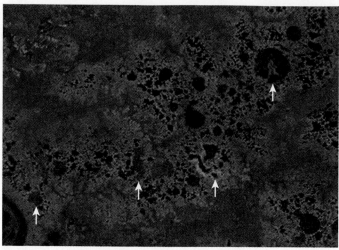

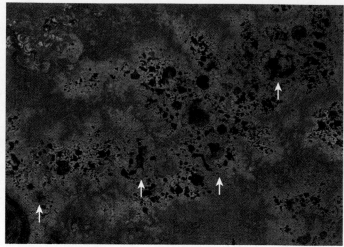

A. June 27, 1973

B. July 2, 2002

FIGURE 20.36 This image pair shows lakes dotting the tundra in northern Siberia in 1973 and 2002. The tundra vegetation is colored a faded red, whereas lakes appear blue or blue-green. Many lakes have clearly disappeared or shrunk considerably between 1973 and 2002. Compare the areas highlighted by the white arrowheads in each image. After studying satellite imagery of about 10,000 large lakes in a 500,000-square-kilometer area in northern Siberia, scientists documented an 11 percent decline in the number of lakes, with at least 125 disappearing completely.

Increasing Ocean Acidity

The human-induced increase in the amount of carbon dioxide in the atmosphere has some serious implications for ocean chemistry and marine life. Recent studies show that about one third of the human-generated carbon dioxide currently ends up in the oceans. As a result, the ocean's pH becomes lower, making seawater more acidic. The pH scale is briefly described in **FIGURE 20.37**.

When atmospheric CO_2 dissolves in seawater (H_2O) it forms carbonic acid (H_2CO_3). This lowers the ocean's pH and changes the balance of certain chemicals found naturally in seawater. In fact, the oceans have already absorbed enough carbon dioxide for surface waters to have experienced a pH decrease of 0.1 pH units since preindustrial times, with an additional pH decrease likely in the future. Moreover, if the current trend in carbon dioxide emissions continues, by the year 2100 the ocean will experience a pH decrease of at least 0.2 pH units, which represents a change in ocean chemistry that has not occurred for millions of years. This shift toward acidity and the changes in ocean chemistry that result make it more difficult for certain marine creatures to build hard parts out of calcium carbonate. The decline in pH thus threatens a variety

of calcite-secreting organisms as diverse as microbes and corals, which concerns marine scientists because of the potential consequences for other sea life that depend on the health and availability of these organisms.

The Potential for "Surprises"

In summary, you have seen that climate in the 21st century, unlike the preceding thousand years, is not expected to be stable. Rather, a constant state of change is very likely. Many of the changes will probably be gradual environmental shifts, imperceptible from year to year. Nevertheless, the effects, accumulated over decades, will have powerful economic, social, and political consequences.

Despite our best efforts to understand future climate shifts, there is also the potential for "surprises." This simply means that, due to the complexity of Earth's climate system, we might experience relatively sudden, unexpected changes or see some aspects of climate shift in an unexpected manner. The report on *Climate Change Impacts on the United States* describes the situation like this:

Surprises challenge humans' ability to adapt, because of how quickly and unexpectedly they occur. For example,

what if the Pacific Ocean warms in such a way that El Niño events become much more extreme? This could reduce the frequency, but perhaps not the strength, of hurricanes along the East Coast, while on the West Coast, more severe winter storms, extreme precipitation events, and damaging winds could become common. What if large quantities of methane, a potent greenhouse gas currently frozen in icy Arctic tundra and sediments, began to be released to the atmosphere by warming, potentially creating an amplifying "feedback loop" that would cause even more warming? We simply do not know how far the climate system or other systems it affects can be pushed before they respond in unexpected ways.

There are many examples of potential surprises, each of which would have large consequences. Most of these potential outcomes are rarely reported, in this study or elsewhere. Even if the chance of any particular surprise happening is small, the chance that at least one such surprise will occur is much greater. In other words, while we can't know which of these events will occur, it is likely that one or more will eventually occur.*

The impact on climate of an increase in atmospheric carbon dioxide and trace gases is obscured by some uncertainties. Yet, climate scientists continue to improve our understanding of the climate system and the potential impacts and effects of global climate change. Policy makers are confronted with responding to the risks posed

*National Assessment Synthesis Team, *Climate Change Impacts on the United States: The Potential Consequences of Climate Variability and Change.* Washington, D.C.: U.S. Global Research Program, 2000, 19.

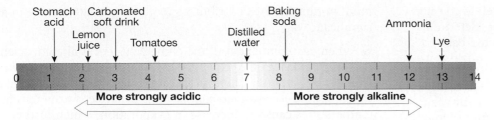

Stomach acid · Lemon juice · Carbonated soft drink · Tomatoes · Distilled water · Baking soda · Ammonia · Lye

| 0 | 1 | 2 | 3 | 4 | 5 | 6 | 7 | 8 | 9 | 10 | 11 | 12 | 13 | 14 |

← **More strongly acidic**   **More strongly alkaline** →

FIGURE 20.37 The *pH scale* is a common measure of the degree of acidity or alkalinity of a solution. The scale ranges from 0 to 14, with a value of 7 denoting a solution that is neutral. Values below 7 indicate greater acidity, whereas numbers above 7 indicate greater alkalinity. The pH values of some familiar substances are shown on the diagram. Although distilled water is neutral (pH 7), rainwater is naturally acidic. It is important to note that the pH scale is logarithmic; that is each whole number increment indicates a tenfold difference. Thus pH 4 is 10 times more acidic than pH 5 and 100 times (10 × 10) more acidic than pH 6.

by emissions of greenhouse gases knowing that our understanding is imperfect. However, they are also faced with the fact that climate-induced environmental changes cannot be reversed quickly, if at all, owing to the lengthy time scales associated with the climate system.

CONCEPT CHECK 20.11

❶ Describe at least four potential consequences of global warming.

CHAPTER TWENTY
Global Climate Change in Review

⊙ The *climate system* includes the atmosphere, hydrosphere, geosphere, biosphere, and cryosphere (the ice and snow that exists at Earth's surface). The system involves the exchanges of energy and moisture that occur among the five spheres.

⊙ Techniques for analyzing Earth's climate history on a scale of hundreds to thousands of years include evidence from *seafloor sediments* and *oxygen isotope analysis*. Seafloor sediments are useful recorders of worldwide climate change because the numbers and types of organic remains included in the sediment are indicative of past sea-surface temperatures. Using oxygen-isotope analysis, scientists can use the $^{18}O/^{16}O$ ratio found in the shells of microorganisms in sediment and layers of ice and snow to detect past temperatures. Other sources of data used for the study of past climates (called *proxy data*) include the growth rings of trees, pollen contained in sediments, coral reefs, and information contained in historical documents.

⊙ Air is a mixture of many discrete gases, and its composition varies from time to time and place to place. After water vapor, dust, and other variable components are removed, two gases, *nitrogen* and *oxygen*, make up 99 percent of the volume of the remaining clean, dry air. *Carbon dioxide*, although present in only minute amounts (0.0389 percent or 389 parts per million), is an efficient absorber of energy emitted by Earth and thus influences the heating of the atmosphere.

⊙ Two important variable components of air are water vapor and aerosols. Like carbon dioxide, water vapor can absorb heat given off by Earth. *Aerosols* (tiny solid and liquid particles) are important because these often invisible particles act as surfaces on which water vapor can condense and are also good absorbers and reflectors (depending on the particles) of incoming solar radiation.

⊙ *Electromagnetic radiation* is energy emitted in the form of rays, or waves, called electromagnetic waves. All radiation is capable of transmitting energy through the vacuum of space. One of the most important differences among electromagnetic waves is their *wavelengths,* which range from very long *radio waves* to very short *gamma rays. Visible light* is the only portion of the electromagnetic spectrum we can see. Some of the basic laws that govern radiation as it heats the atmosphere are (1) all objects emit radiant energy, (2) hotter objects radiate more total energy than do colder objects, (3) the hotter the radiating body, the shorter the wavelengths of maximum radiation, and (4) objects that are good absorbers of radiation are good emitters as well. Gases are selective absorbers, meaning that they absorb and emit certain wavelengths but not others.

⊙ Because the atmosphere gradually thins with increasing altitude, it has no sharp upper boundary but simply blends into outer space. Based on temperature, the atmosphere is divided vertically into four layers. The *troposphere* is the lowermost layer. In the troposphere, temperature usually decreases with increasing altitude. This *environmental lapse rate* is variable, but averages about 6.5 °C per kilometer (3.5 °F per 1000 feet). Essentially all important weather phenomena occur in the troposphere. Above the troposphere is the *stratosphere,* which exhibits warming because of absorption of ultraviolet radiation by ozone. In the mesosphere, temperatures again decrease. Upward from the mesosphere, is the *thermosphere,* a layer with only a tiny fraction of the atmosphere's mass, and no well-defined upper limit.

⊙ Approximately 50 percent of the solar energy that strikes the top of the atmosphere reaches Earth's surface. About 30 percent is reflected back to space. The remaining 20 percent of the energy is absorbed by clouds and the atmosphere's gases. The wavelengths of the energy being transmitted, as well as the size and nature of the absorbing or reflecting substance, determine whether solar radiation will be scattered and reflected back to space or absorbed.

◉ Radiant energy that is absorbed heats Earth and eventually is reradiated skyward. Because Earth has a much lower surface temperature than the Sun, its radiation is in the form of long-wave infrared radiation. Because the atmospheric gases, primarily water vapor and carbon dioxide, are more efficient absorbers of terrestrial (long-wave) radiation, the atmosphere is heated from the ground up. The transmission of short-wave solar radiation by the atmosphere, coupled with the selective absorption of Earth radiation by atmospheric gases, results in the warming of the atmosphere and is referred to as the *greenhouse effect.*

◉ Several explanations have been formulated to explain climate change. Current hypotheses for the "natural" mechanisms (causes unrelated to human activities) of climate change include (1) plate tectonics, rearranging Earth's continents closer to or farther from the equator, (2) variations in Earth's orbit, involving changes in the shape of the orbit (*eccentricity*), angle that Earth's axis makes with the plane of its orbit (*obliquity*), and/or the wobbling of the axis (*precession*), (3) volcanic activity, reducing the solar radiation that reaches the surface, and (4) changes in the Sun's output associated with *sunspots.*

◉ Humans have been modifying the environment for thousands of years. By altering ground cover with the use of fire and the overgrazing of land, people have modified such important climatological factors as surface reflectivity (*albedo*), evaporation rates, and surface winds.

◉ By adding carbon dioxide and other trace gases (methane, nitrous oxide, and chlorofluorocarbons) to the atmosphere, modern humans are contributing to global climate change in a significant way.

◉ When any component of the climate system is altered, scientists must consider the many possible outcomes, called *climate-feedback mechanisms.* Changes that reinforce the initial change are called *positive-feedback mechanisms.* For example, warmer surface temperatures cause an increase in evaporation, which further increases temperature as the additional water vapor absorbs more radiation emitted by Earth. On the other hand, *negative-feedback mechanisms* produce results that are the opposite of the initial change and tend to offset it. An example would be the negative effect that increased cloud cover has on the amount of solar energy available to heat the atmosphere.

◉ Global climate is also affected by human activities that contribute to the atmosphere's *aerosol* (tiny, often microscopic, liquid and solid particles that are suspended in air) content. By reflecting sunlight back to space, aerosols have a net cooling effect. The effect of aerosols on today's climate is determined by the amount emitted during the preceding couple of weeks, while carbon dioxide remains for much longer spans and influences climate for many decades.

◉ Because the climate system is so complex, predicting specific regional changes that may occur as the result of increased levels of carbon dioxide in the atmosphere is speculative. However, some possible consequences of greenhouse warming include (1) altering the distribution of the world's water resources, (2) a probable rise in sea level, (3) a greater intensity of tropical cyclones, and (4) changes in the extent of Arctic sea ice and permafrost.

◉ Due to the complexity of the climate system, not all future shifts can be foreseen. Thus, "surprises" (relatively sudden unexpected changes in climate) are possible.

Key Terms

aerosols (p. 500)
climate-feedback mechanisms
    (p. 514)
climate system (p. 495)

greenhouse effect (p. 504)
negative-feedback mechanism
    (p. 514)
oxygen isotope analysis (p. 497)

paleoclimatology (p. 496)
positive-feedback mechanism
    (p. 514)

proxy data (p. 496)
sunspots (p. 508)

GIVE IT SOME THOUGHT

❶ Refer to Figure 20.2 which illustrates various components of Earth's climate system. Boxes represent interactions or changes that occur in the climate system. Select three boxes and provide an example of an interaction or change associated with each. Explain how these interactions may influence temperature.

❷ Describe one way in which changes in the biosphere can cause changes in the climate system. Next, suggest one way in which the biosphere is affected by changes in some other part of the climate system. Finally, indicate one way in which the biosphere records changes in the climate system.

❸ Recent volcanic events, such as the eruptions of El Chichón and Mt. Pinatubo, were associated with a drop in global temperatures. During the Cretaceous Period, volcanic activity was associated with global warming. Explain this apparent paradox.

❹ Describe the greenhouse effect in your own words. Is the greenhouse effect synonymous with global warming? Explain.

❺ Briefly explain the idea of a feedback mechanism. Next, use your imagination to invent a situation that involves a feedback mechanism—it *does not* have to relate to climate change, geology, or anything scientific. Begin by describing the initial condition or circumstance. Then describe any subsequent changes that take place due to the feedback mechanism. Make sure to indicate whether the feedback mechanism is positive or negative and why.

❻ During a conversation, an acquaintance indicates that he is skeptical about global warming. When you ask him why he feels that way, he says, "The past couple of years in this area have been among the coolest I can remember." While you assure this person that it is useful to question scientific findings, you suggest to him that his reasoning in this case may be flawed. Use your understanding of the definition of climate along with one or more graphs in the chapter to persuade this person to reevaluate his reasoning.

Companion Website

www.mygeoscienceplace.com

The *Essentials of Geology, 11e* companion Website contains numerous multimedia resources accompanied by assessments to aid in your study of the topics in this chapter. The use of this site's learning tools will help improve your understanding of geology. Utilizing the access code that accompanies this text, visit **www.mygeoscienceplace.com** in order to:

- **Review** key chapter concepts.
- **Read** with links to the Pearson eText and to chapter-specific web resources.
- **Visualize** and comprehend challenging topics using the learning activities in *GEODe: Essentials of Geology* and the *Geoscience Animations Library*.
- **Test** yourself with online quizzes.

Metric and English Units Compared

Units

1 kilometer (km) = 1000 meters (m)
1 meter (m) = 100 centimeters (cm)
1 centimeter (cm) = 0.39 inch (in.)
1 mile (mi) = 5280 feet (ft)
1 foot (ft) = 12 inches (in.)
1 inch (in.) = 2.54 centimeters (cm)
1 squre mile (mi^2) = 640 acres (a)
1 kilogram (kg) = 1000 grams (g)
1 pound (lb) = 16 ounces (oz)
1 fathom = 6 feet (ft)

Conversions

When you want to convert:	Multiply by:	To find:
LENGTH		
inches	2.54	centimeters
centimeters	0.39	inches
feet	0.30	meters
meters	3.28	feet
yards	0.91	meters
meters	1.09	yards
miles	1.61	kilometers
kilometers	0.62	miles
AREA		
square inches	6.45	square centimeters
square centimeters	0.15	square inches
square feet	0.09	square meters
square meters	10.76	square feet
square miles	2.59	square kilometers
square kilometers	0.39	square miles
VOLUME		
cubic inches	16.38	cubic centimeters
cubic centimeters	0.06	cubic inches
cubic feet	0.028	cubic meters
cubic meters	35.3	cubic feet
cubic miles	4.17	cubic kilometers
cubic kilometers	0.24	cubic miles
liters	1.06	quarts
liters	0.26	gallons
gallons	3.78	liters

MASSES AND WEIGHTS

ounces	28.33	grams
grams	0.035	ounces
pounds	0.45	kilograms
kilograms	2.205	pounds

TEMPERATURE

When you want to convert degrees Fahrenheit (°F) to degrees Celsius (°C), subtract 32 degrees and divide by 1.8.

When you want to convert degrees Celsius (°C) to degrees Fahrenheit (°F), multiply by 1.8 and add 32 degrees.

When you want to convert degrees Celsius (°C) to kelvins (K), delete the degree symbol and add 273. When you want to convert kelvins (K) to degrees Celsius (°C), add the degree symbol and subtract 273.

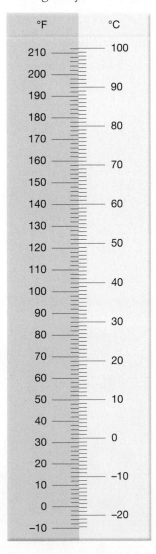

FIGURE A.1 A comparison of Fahrenheit and Celsius temperature scales.

Topographic Maps

A map is a representation on a flat surface of all or a part of Earth's surface drawn to a specific scale. Maps are often the most effective means for showing the locations of both natural and human structures, their sizes, and their relationships to one another. Like photographs, maps readily display information that would be impractical to express in words.

While most maps show only the two horizontal dimensions, geologists, as well as other map users, often require that the third dimension—elevation—be shown on maps. Maps that show the shape of the land are called **topographic maps.** Although various techniques may be used to depict elevations, the most accurate method involves the use of contour lines.

Contour Lines

A **contour line** is a line on a map representing a corresponding imaginary line on the ground that has the same elevation above sea level along its entire length. While many map symbols are pictographs, resembling the objects they represent, a contour line is an abstraction that has no counterpart in nature. It is, however, an accurate and effective device for representing the third dimension on paper.

Some useful facts and rules concerning contour lines are listed as follows. This information should be studied in conjunction with Figure B.1.

1. Contour lines bend upstream or upvalley. The contours form V's that point upstream, and in the upstream direction the successive contours represent higher elevations. For example, if you were standing on a stream bank and wished to get to the point at the same elevation directly opposite you on the other bank without stepping up or down, you would need to walk upstream along the contour at that elevation to where it crosses the stream bed, cross the stream, and then walk back downstream along the same contour.

2. Contours near the upper parts of hills form closures. The top of a hill is higher than the highest closed contour.

3. Hollows (depressions) without outlets are shown by closed, hatched contours. Hatched contours are contours with short lines on the inside pointing downslope.

4. Contours are widely spaced on gentle slopes.

5. Contours are closely spaced on steep slopes.

6. Evenly spaced contours indicate a uniform slope.

7. Contours usually do not cross or intersect each other, except in the rare case of an overhanging cliff.

8. All contours eventually close, either on a map or beyond its margins.

9. A single high contour never occurs between two lower ones, and vice versa. In other words, a change in slope direction is always determined by the repetition of the same elevation either as two different contours of the same value or as the same contour crossed twice.

10. Spot elevations between contours are given at many places, such as road intersections, hill summits, and lake surfaces. Spot elevations differ from control elevation stations, such as benchmarks, in not being permanently established by permanent markers.

Relief

Relief refers to the difference in elevation between any two points. Maximum relief refers to the difference in elevation between the highest and lowest points in the area being considered. Relief determines the **contour interval,** which is the difference in elevation between succeeding contour lines that is used on topographic maps. Where relief is low, a small contour interval, such as 10 or 20 feet, may be used. In flat areas, such as wide river valleys or broad, flat uplands, a contour interval of 5 feet is often used. In rugged mountainous terrain, where relief is many hundreds of feet, contour intervals as large as 50 or 100 feet are used.

Scale

Map **scale** expresses the relationship between distance or area on the map to the true distance or area on Earth's surface. This is generally expressed as a ratio or fraction, such as 1:24,000 or 1/24,000. The numerator, usually 1, represents map distance, and the denominator, a large number, represents ground distance. Thus, 1:24,000 means that a distance of 1 unit on the map represents a distance of 24,000 such units on the surface of Earth. It does not matter what the units are.

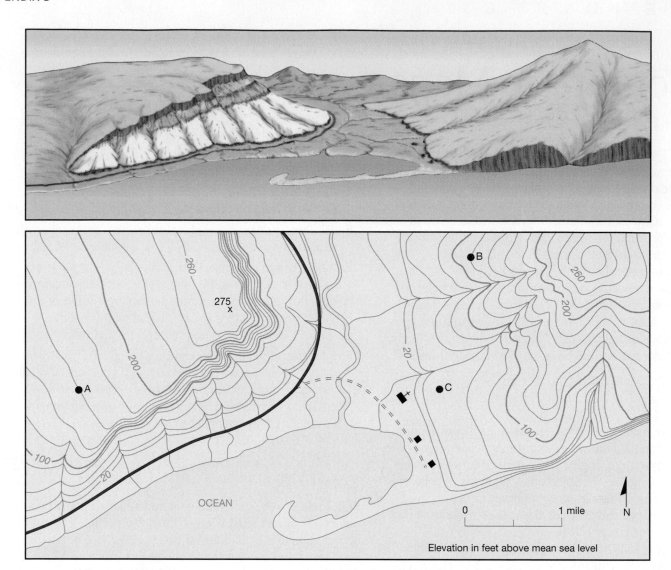

FIGURE B.1 Perspective view of an area and a contour map of the same area. These illustrations show how features are depicted on a topographic map. The upper illustration is a perspective view of a river valley and the adjoining hills. The river flows into a bay, which is partly enclosed by a hooked sandbar. On either side of the valley are terraces through which streams have cut gullies. The hill on the right has a smoothly eroded form and gradual slopes, whereas the one on the left rises abruptly in a sharp precipice, from which it slopes gently and forms an inclined plateau traversed by a few shallow gullies. A road provides access to a church and the two houses situated across the river from a highway that follows the seacoast and curves up the river valley. The lower illustration shows the same features represented by symbols on a topographic map. The contour interval (vertical distance between adjacent contours) is 20 feet. *(After U.S. Geological Survey)*

Often, the graphic or bar scale is more useful than the fractional scale, because it is easier to use for measuring distances between points. The graphic scale (Figure B.2) consists of a bar divided into equal segments, which represent equal distances on the map. One segment on the left side of the bar is usually divided into smaller units to permit more accurate estimates of fractional units.

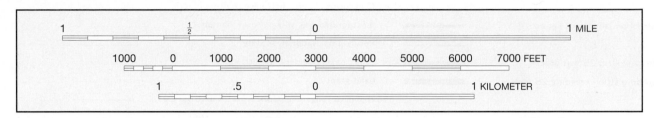

FIGURE B.2 Graphic scale.

Topographic maps, which are also referred to as *quadrangles*, are generally classified according to publication scale. Each series is intended to fulfill a specific type of map need. To select a map with the proper scale for a particular use, remember that large-scale maps show more detail and small-scale maps show less detail. The sizes and scales of topographic maps published by the U.S. Geological Survey are shown in Table B.1.

TABLE B.1
National Topographic Maps

Series	Scale	1 inch Represents	Standard Quadrangle Size (latitude–longitude)	Quadrangle Area (square miles)	Paper Size E–W N–S Width Length (inches)
$7\frac{1}{2}$-minute	1:24,000	2000 feet	$7\frac{1}{2}' \times 7\frac{1}{2}'$	49–70	22×27
Puerto Rico $7\frac{1}{2}$-minute	1:20,000	about 1667 feet	$7\frac{1}{2}' \times 7\frac{1}{2}'$	71	$29\frac{1}{2} \times 32\frac{1}{2}$
15-minute	1:62,500	nearly 1 mile	$15' \times 15'$	197–282	17×21
Alaska 1:63,360	1:63,360	1 mile	$15' \times 20' - 36'$	207–281	18×21
U.S. 1:250,000	1:250,000	nearly 4 miles	$1° \times 2°$	4580–8669	34×22
U.S. 1:1,000,000	1:1,000,000	nearly 16 miles	$4° \times 6°$	73,734–102,759	27×27

Source: U.S. Geological Survey

Color and Symbol

Each color and symbol used on U.S. Geological Survey topographic maps has significance. Common topographic map symbols are shown in Figure B.3. The meaning of each color is as follows:

Blue—water features

Black—works of humans, such as homes, schools, churches, roads, and so forth

Brown—contour lines

Green—woodlands, orchards, and so forth

Red—urban areas, important roads, public land subdivision lines

TOPOGRAPHIC MAP SYMBOLS

VARIATIONS WILL BE FOUND ON OLDER MAPS

Primary highway, hard surface . . . .

Secondary highway, hard surface . . . .

Light-duty road, hard or improved surface . . . .

Unimproved road . . . .

Road under construction, alinement known . . . .

Proposed road . . . .

Dual highway, dividing strip 25 feet or less . . . .

Dual highway, dividing strip exceeding 25 feet . . . .

Trail . . . .

Railroad: single track and multiple track . . . .

Railroads in juxtaposition . . . .

Narrow gage: single track and multiple track . . . .

Railroad in street and carline . . . .

Bridge: road and railroad . . . .

Drawbridge: road and railroad . . . .

Footbridge . . . .

Tunnel: road and railroad . . . .

Overpass and underpass . . . .

Small masonry or concrete dam . . . .

Dam with lock . . . .

Dam with road . . . .

Canal with lock . . . .

Buildings (dwelling, place of employment, etc.) . . . .

School, church, and cemetery . . . .

Buildings (barn, warehouse, etc.) . . . .

Power transmission line with located metal tower . . . . . .

Telephone line, pipeline, etc. (labeled as to type) . . . . . .

Wells other than water (labeled as to type) . . . .

Tanks: oil, water, etc. (labeled only if water) . . . .

Located or landmark object; windmill . . . .

Open pit, mine, or quarry; prospect . . . .

Shaft and tunnel entrance . . . .

Horizontal and vertical control station:

    Tablet, spirit level elevation . . . . BM △ 5653

    Other recoverable mark, spirit level elevation . . . . △ 5455

Horizontal control station: tablet, vertical angle elevation VABM △ 9519

    Any recoverable mark, vertical angle or checked elevation △ 3775

Vertical control station: tablet, spirit level elevation . . . . . BM × 957

    Other recoverable mark, spirit level elevation . . . . × 954

Spot elevation . . . . × 7369   × 7369

Water elevation 670   670

Boundaries: National . . . .

    State . . . .

    County, parish, municipio . . . .

    Civil township, precinct, town, barrio . . . .

    Incorporated city, village, town, hamlet . . . .

    Reservation, National or State . . . .

    Small park, cemetery, airport, etc. . . . .

    Land grant . . . .

Township or range line, United States land survey . . . . .

Township or range line, approximate location . . . .

Section line, United States land survey . . . .

Section line, approximate location . . . .

Township line, not United States land survey . . . .

Section line, not United States land survey . . . .

Found corner: section and closing . . . .

Boundary monument: land grant and other . . . .

Fence or field line . . . .

Index contour . . . .   Intermediate contour . .

Supplementary contour . .   Depression contours . . .

Fill . . . .   Cut . . . .

Levee . . . .   Levee with road . . . . . .

Mine dump . . . .   Wash . . . .

Tailings . . . .   Tailings pond . . . .

Shifting sand or dunes . . .   Intricate surface . . . . .

Sand area . . . .   Gravel beach . . . .

Perennial streams . . . . .   Intermittent streams . . .

Elevated aqueduct . . . .   Aqueduct tunnel . . . . . .

Water well and spring . .   Glacier . . . .

Small rapids . . . .   Small falls . . . .

Large rapids . . . .   Large falls . . . .

Intermittent lake . . . . .   Dry lake bed . . . .

Foreshore flat . . . .   Rock or coral reef . . . .

Sounding, depth curve . . 10   Piling or dolphin . . . . .

Exposed wreck : . . . . . .   Sunken wreck . . . .

Rock, bare or awash; dangerous to navigation . . . .

Marsh (swamp) . . . .   Submerged marsh . . . .

Wooded marsh . . . .   Mangrove . . . .

Woods or brushwood . . .   Orchard . . . .

Vineyard . . . .   Scrub . . . .

Land subject to
controlled inundation . . . .   Urban area . . . .

FIGURE B.3 U.S. Geological Survey topographic map symbols. (Variations will be found on older maps.)

Landforms of the Conterminous United States

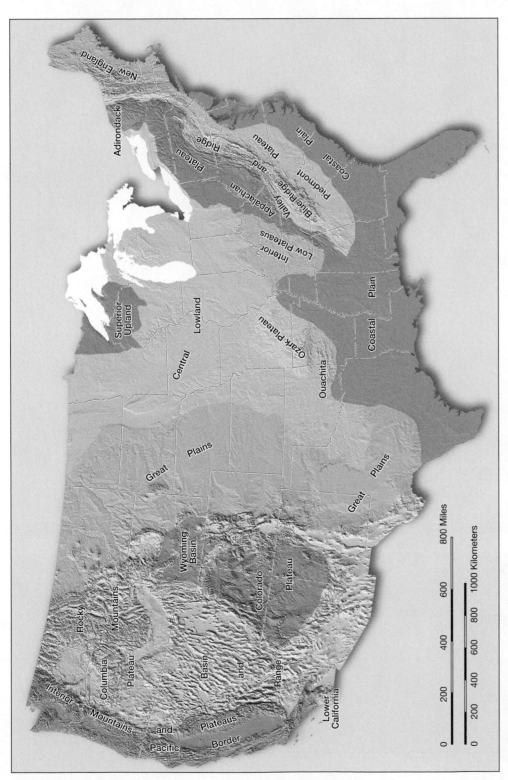

FIGURE C.1 Outline map showing major physiographic provinces of the United States.

Landforms of the Conterminous United States

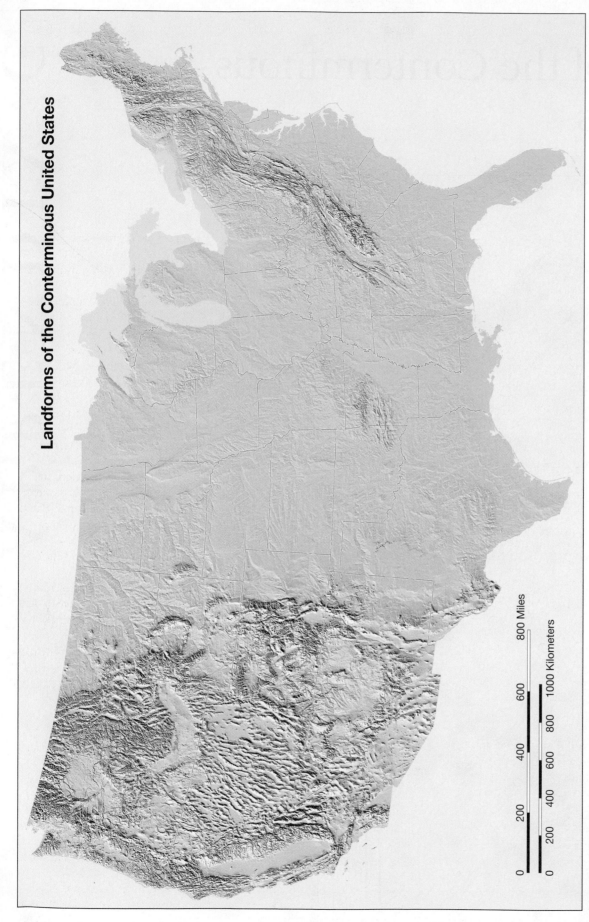

FIGURE C.2 Digital shaded relief landform map of the United States. *(Data provided by the U.S. Geological Survey)*

Glossary

Aa   A type of lava flow that has a jagged, blocky surface.

Ablation   A general term for the loss of ice and snow from a glacier.

Abrasion   The grinding and scraping of a rock surface by the friction and impact of rock particles carried by water, wind, or ice.

Abyssal plain   Very level area of the deep-ocean floor, usually lying at the foot of the continental rise.

Accretionary wedge   A large wedge-shaped mass of sediment that accumulates in subduction zones. Here sediment is scraped from the subducting oceanic plate and accreted to the overriding crustal block.

Active continental margin Usually narrow and consisting of highly deformed sediments, they occur where oceanic lithosphere is being subducted beneath the margin of a continent.

Active layer   The zone above the permafrost that thaws in summer and refreezes in winter.

Aerosols   Tiny solid and liquid particles suspended in the atmosphere.

Aftershock   A smaller earthquake that follows the main earthquake.

Alluvial channel   A stream channel in which the bed and banks are composed largely of unconsolidated sediment (alluvium) that was previously deposited in the valley.

Alluvial fan   A fan-shaped deposit of sediment formed when a stream's slope is abruptly reduced.

Alluvium   Unconsolidated sediment deposited by a stream.

Alpine glacier   See *Valley glacier.*

Andean-type plate margin   A type of convergent boundary in which a slab of oceanic crust is subducting beneath a continental margin.

Andesitic   Igneous rocks having a mineral makeup between that of granite and basalt, after the common volcanic rock andesite. See also *Intermediate.*

Angle of repose   The steepest angle at which loose material remains stationary without sliding downslope.

Angular unconformity   An unconformity in which the older strata dip at an angle different from that of the younger beds.

Anthracite   A hard, metamorphic form of coal that burns clean and hot.

Anticline   A fold in sedimentary strata that resembles an arch.

Aphanitic texture   A texture of igneous rocks in which the crystals are too small for individual minerals to be distinguished with the unaided eye.

Aquifer   Rock or sediment through which groundwater moves easily.

Aquitard   An impermeable bed that hinders or prevents groundwater movement.

Archean eon   The first eon of Precambrian time; the eon preceding the Proterozoic. It extends between 4.5 and 2.5 billion years ago.

Arête   A narrow, knifelike ridge separating two adjacent glaciated valleys.

Arkose   A feldspar-rich sandstone.

Artesian well   A well in which the water rises above the level where it was initially encountered.

Asthenosphere   A subdivision of the mantle situated below the lithosphere. This zone of weak material exists below a depth of about 100 kilometers and in some regions extends as deep as 700 kilometers. The rock within this zone is easily deformed.

Atmosphere   The gaseous portion of a planet; the planet's envelope of air. One of the traditional subdivisions of Earth's physical environment.

Atoll   A continuous or broken ring of coral reef surrounding a central lagoon.

Atom   The smallest particle that exists as an element.

Atomic mass unit   A mass unit equal to exactly one-twelfth the mass of a carbon-12 atom.

Atomic number   The number of protons in the nucleus of an atom.

Atomic weight   The average of the atomic masses of isotopes of a given element.

Aureole   A zone or halo of contact metamorphism found in the host rock surrounding an igneous intrusion.

Back swamp   A poorly drained area on a floodplain, resulting when natural levees are present.

Backshore   The inner portion of the shore, lying landward of the high-tide shoreline. It is usually dry, being affected by waves only during storms.

Bajada   An apron of sediment along a mountain front created by the coalescence of alluvial fans.

Banded iron formations   Finely layered iron- and silica-rich (chert) layers deposited mainly during the Precambrian.

Bar   Common term for sand and gravel deposits in a stream channel.

Barchan dune   A solitary sand dune shaped like a crescent with its tips pointed downwind.

Barchanoid dunes   Dunes forming scalloped rows of sand oriented at right angles to the wind. This form is intermediate between isolated barchans and extensive waves of transverse dunes.

Barrier island   A low, elongated ridge of sand that parallels the coast.

Basal slip   A mechanism of glacial movement in which the ice mass slides over the surface below.

Basalt   An aphanitic igneous rock of mafic composition.

Basaltic   Term used to describe igneous rocks that contain abundant dark (ferromagnesian) minerals and about 50 percent silica.

Base level   The level below which a stream cannot erode.

Basin   A circular downfolded structure.

Batholith   A large mass of igneous rock that formed when magma was emplaced at depth, crystallized, and was subsequently exposed by erosion.

Baymouth bar   A sandbar that completely crosses a bay, sealing it off from the main body of water.

Beach   An accumulation of sediment found along the landward margin of the ocean or a lake.

Beach drift   The transport of sediment in a zigzag pattern along a beach. It is caused by the uprush of water from obliquely breaking waves.

Beach face The wet, sloping surface that extends from the berm to the shoreline.

Beach nourishment Large quantities of sand are added to the beach system to offset losses caused by wave erosion. By building beaches seaward, beach quality and storm protection are both improved.

Bed See *Strata.*

Bed load Sediment rolled along the bottom of a stream by moving water, or particles rubbed along the ground surface by wind.

Bedding plane A nearly flat surface separating two beds of sedimentary rock. Each bedding plane marks the end of one deposit and the beginning of another having different characteristics.

Bedrock A general term for the rock that underlies soil or other unconsolidated surface materials.

Belt of soil moisture A zone in which water is held as a film on the surface of soil particles and may be used by plants or withdrawn by evaporation. The uppermost subdivision of the zone of aeration.

Berm The dry, gently sloping zone on the backshore of a beach at the foot of the coastal cliffs or dunes.

Biochemical Describing a type of chemical sediment that forms when material dissolved in water is precipitated by water-dwelling organisms. Shells are common examples.

Biogenous sediment Seafloor sediments consisting of material of marine-organic origin.

Bituminous coal The most common form of coal, often called soft, black coal.

Block lava Lava having a surface of angular blocks associated with material having andesitic and rhyolitic compositions.

Blowout (deflation hollow) A depression excavated by wind in easily eroded materials.

Body wave A seismic wave that travels through Earth's interior.

Bottomset bed A layer of fine sediment deposited beyond the advancing edge of a delta and then buried by continuous delta growth.

Bowen's reaction series A concept proposed by N. L. Bowen that illustrates the relationship between magma and the minerals crystallizing from it during the formation of igneous rock.

Braided stream A stream consisting of numerous intertwining channels.

Breakwater A structure protecting a nearshore area from breaking waves.

Breccia A sedimentary rock composed of angular fragments that were lithified.

Brittle deformation Deformation that involves the fracturing of rock. Associated with rocks near the surface.

Burial metamorphism Low-grade metamorphism that occurs in the lowest layers of very thick accumulations of sedimentary strata.

Caldera A large depression typically caused by collapse of the summit area of a volcano following a violent eruption.

Caliche A hard layer, rich in calcium carbonate, that forms beneath the *B* horizon in soils of arid regions.

Calving Wastage of a glacier that occurs when large pieces of ice break off into water.

Cap rock A necessary part of an oil trap, the cap rock is impermeable and hence keeps upwardly mobile oil and gas from escaping at the surface.

Capacity The total amount of sediment a stream is able to transport.

Capillary fringe A relatively narrow zone at the base of the zone of aeration. Here water rises from the water table in tiny threadlike openings between grains of soil or sediment.

Catastrophism The concept that Earth was shaped by catastrophic events of a short-term nature.

Cavern A naturally formed underground chamber or series of chambers most commonly produced by solution activity in limestone.

Cementation One way in which sedimentary rocks are lithified. As

material precipitates from water that percolates through the sediment, open spaces are filled and particles are joined into a solid mass.

Cenozoic era A time span on the geologic time scale beginning about 65.5 million years ago following the Mesozoic era.

Chemical bond A strong attractive force that exists between atoms in a substance. It involves the transfer or sharing of electrons that allows each atom to attain a full valence shell.

Chemical compound A substance formed by the chemical combination of two or more elements in definite proportions and usually having properties different from those of its constituent elements.

Chemical sedimentary rock Sedimentary rock consisting of material that was precipitated from water by either inorganic or organic means.

Chemical weathering The processes by which the internal structure of a mineral is altered by the removal and/or addition of elements.

Cinder cone A rather small volcano built primarily of pyroclastics ejected from a single vent.

Cirque An amphitheater-shaped basin at the head of a glaciated valley produced by frost wedging and plucking.

Clastic A sedimentary rock texture consisting of broken fragments of preexisting rock.

Cleavage The tendency of a mineral to break along planes of weak bonding.

Climate feedback mechanism Because the atmosphere is a complex interactive physical system, several different outcomes may result when one of the system's elements is altered. These various possibilities are called *climate-feedback mechanisms.*

Climate system The exchanges of energy and moisture occurring among the atmosphere, hydrosphere, geosphere, biosphere, and cryosphere.

Coarse-grained texture See *Phaneritic texture.*

Coast A strip of land that extends inland from the coastline as far as ocean-related features can be found.

Coastline The coast's seaward edge; the landward limit of the effect of the highest storm waves on the shore.

Col A pass between mountain valleys where the headwalls of two cirques intersect.

Color A phenomenon of light by which otherwise identical objects may be differentiated.

Column A feature found in caves that is formed when a stalactite and stalagmite join.

Columnar joints A pattern of cracks that forms during cooling of molten rock to generate columns.

Compaction A type of lithification in which the weight of overlying material compresses more deeply buried sediment. It is most important in fine-grained sedimentary rocks such as shale.

Competence A measure of the largest particle a stream can transport; a factor dependent on velocity.

Composite cone A volcano composed of both lava flows and pyroclastic material.

Compound See *Chemical compound.*

Compressional mountains Mountains in which great horizontal forces have shortened and thickened the crust. Most major mountain belts are of this type.

Concordant A term used to describe intrusive igneous masses that form parallel to the bedding of the surrounding rock.

Conduit A pipelike opening through which magma moves toward Earth's surface. It terminates at a surface opening called a vent.

Confined aquifer An aquifer that has impermeable layers (aquitards) both above and below.

Cone of depression A cone-shaped depression immediately surrounding a well.

Conformable layers Rock layers that were deposited without interruption.

Conglomerate A sedimentary rock consisting of rounded, gravel-size particles.

Contact metamorphism Changes in rock caused by the heat of a nearby magma body.

Continental drift A hypothesis, credited largely to Alfred Wegener, that suggested all present continents once existed as a single supercontinent. Further, he posited that beginning about 200 million years ago, the supercontinent began breaking into smaller continents, which then drifted to their present positions.

Continental margin See *Active continental margin* and *Passive continental margin*.

Continental rift A linear zone along which continental lithosphere stretches and pulls apart. Its creation may mark the beginning of a new ocean basin.

Continental rise The gently sloping surface at the base of the continental slope.

Continental shelf The gently sloping submerged portion of the continental margin, extending from the shoreline to the continental slope.

Continental slope The steep gradient that leads to the deep-ocean floor and marks the seaward edge of the continental shelf.

Continental volcanic arc Mountains formed in part by igneous activity associated with the subduction of oceanic lithosphere beneath a continent. Examples include the Andes and the Cascades.

Convergent plate boundary A boundary in which two plates move together, resulting in oceanic lithosphere being thrust beneath an overriding plate, eventually to be reabsorbed into the mantle. It can also involve the collision of two continental plates to create a mountain system.

Core Located beneath the mantle, it is Earth's innermost layer. The core is divided into an outer core and an inner core.

Correlation Establishing the equivalence of rocks of similar age in different areas.

Covalent bond A chemical bond produced by the sharing of electrons.

Crater The depression at the summit of a volcano, or that which is produced by a meteorite impact.

Craton That part of the continental crust that has attained stability; that is, it has not been affected by significant tectonic activity during the Phanerozoic eon. It consists of the shield and stable platform.

Creep The slow downhill movement of soil and regolith.

Crevasse A deep crack in the brittle surface of a glacier.

Cross-bedding Structure in which relatively thin layers are inclined at an angle to the main bedding. Formed by currents of wind or water.

Cross-cutting A principle of relative dating. A rock or fault is younger than any rock (or fault) through which it cuts.

Crust The very thin, outermost layer of Earth.

Crystal An orderly arrangement of atoms.

Crystal shape See *Habit*.

Crystalline texture See *Nonclastic*.

Crystallization The formation and growth of a crystalline solid from a liquid or gas.

Curie point The temperature above which a material loses its magnetization.

Cut bank The area of active erosion on the outside of a meander.

Cutoff A short channel segment created when a river erodes through the narrow neck of land between meanders.

Dark silicate Silicate minerals containing ions of iron and/or magnesium in their structure. They are dark in color and have a higher specific gravity than nonferromagnesian silicates.

Daughter product An isotope resulting from radioactive decay.

Debris flow A relatively rapid type of mass wasting that involves a flow of soil and regolith containing a large amount of water; also called *mudflows*.

Debris slide See *Rockslide*.

Decompression melting Melting that occurs as rock ascends due to a drop in confining pressure.

Deep-focus earthquake An earthquake focus at a depth of more than 300 kilometers.

Deep-ocean basin The portion of seafloor that lies between the continental margin and the oceanic ridge system. This region comprises almost 30 percent of Earth's surface.

Deep-ocean trench See *Trench*.

Deflation The lifting and removal of loose material by wind.

Deformation General term for the processes of folding, faulting, shearing, compression, or extension of rocks as the result of various natural forces.

Delta An accumulation of sediment formed where a stream enters a lake or ocean.

Dendritic drainage pattern A stream system that resembles the pattern of a branching tree.

Density The weight per unit volume of a particular material.

Desert One of the two types of dry climate; the drier of the dry climates.

Desert pavement A layer of coarse pebbles and gravel created when wind removes the finer material.

Detachment fault A low-angle fault that represents a major boundary between unfaulted rocks below that exhibit ductile deformation and rocks above that exhibit brittle deformation via faulting.

Detrital sedimentary rocks Rocks that form from the accumulation of materials that originate and are transported as solid particles derived from both mechanical and chemical weathering.

Diagenesis A collective term for all the chemical, physical, and biological changes that take place after sediments are deposited and during and after lithification.

Dike A tabular-shaped intrusive igneous feature that cuts through the surrounding rock.

Dip The angle at which a rock layer is inclined from the horizontal. The direction of dip is at a right angle to the strike.

Dip-slip fault A fault in which the movement is parallel to the dip of the fault.

Discharge The quantity of water in a stream that passes a given point in a given period of time.

Disconformity A type of unconformity in which the beds above and below are parallel.

Discontinuity A sudden change with depth in one or more of the physical properties of the material making up Earth's interior. The boundary between two dissimilar materials in Earth's interior as determined by the behavior of seismic waves.

Discordant A term used to describe plutons that cut across existing rock structures, such as bedding planes.

Disseminated deposit Any economic mineral deposit in which the desired mineral occurs as scattered particles in the rock but in sufficient quantity to make the deposit an ore.

Dissolved load The portion of a stream's load carried in solution.

Distributary A section of a stream that leaves the main flow.

Diurnal tide A tide characterized by a single high and low water height each tidal day.

Divergent plate boundary A boundary in which two plates move apart, resulting in upwelling of material from the mantle to create new seafloor.

Divide An imaginary line that separates the drainage of two streams; often found along a ridge.

Dome A roughly circular, upfolded structure.

Drainage basin The land area that contributes water to a stream.

Drawdown The difference in height between the bottom of a cone of depression and the original height of the water table.

Drift See *Glacial drift*.

Drumlin A streamlined asymmetrical hill composed of glacial till. The steep side of the hill faces the direction from which the ice advanced.

Dry climate A climate in which the yearly precipitation is less than the potential loss of water by evaporation.

Ductile deformation A type of solid-state flow that produces a change in the size and shape of a rock body without fracturing. Occurs at depths where temperatures and confining pressures are high.

Dune   A hill or ridge of wind-deposited sand.

Earthflow   The downslope movement of water-saturated, clay-rich sediment, most characteristic of humid regions.

Earthquake   The vibration of Earth produced by the rapid release of energy.

Echo sounder   An instrument used to determine the depth of water by measuring the time interval between emission of a sound signal and the return of its echo from the bottom.

Elastic deformation Nonpermanent deformation in which rock returns to its original shape when the stress is released.

Elastic rebound   The sudden release of stored strain in rocks that results in movement along a fault.

Electron   A negatively charged subatomic particle that has a negligible mass and is found outside the atom's nucleus.

Element   A substance that cannot be decomposed into simpler substances by ordinary chemical or physical means.

Eluviation   The washing out of fine soil components from the *A* horizon by downward-percolating water.

Emergent coast   A coast where land formerly below the sea level has been exposed by either crustal uplift or a drop in sea level or both.

End moraine   A ridge of till marking a former position of the front of the glacier.

Energy levels   Spherically shaped negatively charged zones that surround the nucleus of an atom.

Eon   The largest time unit on the geologic time scale, next in order of magnitude above *era*.

Epicenter   The location on Earth's surface that lies directly above the focus of an earthquake.

Epoch   A unit of the geological time scale that is a subdivision of a *period*.

Era   A major division on the geologic time scale; eras are divided into shorter units called *periods*.

Erosion   The incorporation and transportation of material by a mobile agent, such as water, wind, or ice.

Eruption column   Buoyant plumes of hot ash-laden gases that can extend thousands of meters into the atmosphere.

Esker   Sinuous ridge composed largely of sand gravel deposited by a stream flowing in a tunnel beneath a glacier near its terminus.

Estuary   A partially enclosed coastal water body that is connected to the ocean. Salinity here is measurably reduced by the freshwater flow of rivers.

Eukaryotes   Organisms whose genetic material is enclosed in a nucleus. Plants, animals, and fungi are *eukaryotes*.

Evaporite   A sedimentary rock formed of material deposited from solution by evaporation of water.

Evapotranspiration   The combined effect of evaporation and transpiration.

Exfoliation dome   Large, dome-shaped structure, usually composed of granite, formed by sheeting.

Exotic stream   A permanent stream that traverses a desert and has its source in well-watered areas outside the desert.

External process   Process such as weathering, mass wasting, or erosion that is powered by the Sun and transforms solid rock into sediment.

Extrusive   Igneous activity that occurs at Earth's surface.

Eye   A zone of scattered clouds and calm averaging about 20 kilometers in diameter at the center of a hurricane.

Eye wall   A doughnut-shaped area of intense cumulonimbus (thunderstorm) development and very strong winds that surrounds the eye of a hurricane.

Fall   A type of movement common to mass-wasting processes that refers to the free falling of detached individual pieces of any size.

Fault   A break in a rock mass along which movement has occurred.

Fault creep   Slow, gradual displacement along a fault that occurs relatively smoothly and with little noticeable seismic activity.

Fault scarp   A cliff created by movement along a fault. It represents the exposed surface of the fault prior to modification by weathering and erosion.

Fault-block mountain   A mountain formed by the displacement of rock along a fault.

Felsic   A term derived from *feld*spar and *sili*ca (quartz). It is a term used to describe granitic igneous rocks.

Ferromagnesian silicate   See *Dark silicate*.

Fetch   The distance that the wind has traveled across the open water.

Fine-grained texture   See *Aphanitic texture*.

Fiord   A steep-sided inlet of the sea formed when a glacial trough was partially submerged.

Fissure   A crack in rock along which there is a distinct separation.

Fissure eruption   An eruption in which lava is extruded from narrow fractures or cracks in the crust.

Flood   The overflow of a stream channel that occurs when discharge exceeds the channel's capacity. The most common and destructive geologic hazard.

Flood basalts   Flows of basaltic lava that issue from numerous cracks or fissures and commonly cover extensive areas to thicknesses of hundreds of meters.

Floodplain   The flat, low-lying portion of a stream valley subject to periodic inundation.

Flow   A type of movement common to mass-wasting processes in which water-saturated material moves downslope as a viscous fluid.

Fluorescence   The absorption of ultraviolet light, which is reemitted as visible light.

Focus (earthquake)   The zone within Earth where rock displacement produces an earthquake.

Fold   A bent layer or series of layers that were originally horizontal and subsequently deformed.

Foliated   A texture of metamorphic rocks that gives the rock a layered appearance.

Foliation   A term for a linear arrangement of textural features often exhibited by metamorphic rocks.

Foreset bed   An inclined bed deposited along the front of a delta.

Foreshocks   Small earthquakes that often precede a major earthquake.

Foreshore   That portion of the shore lying between the normal high- and low-water marks; the intertidal zone.

Fossil   The remains or traces of organisms preserved from the geologic past.

Fossil fuel   General term for any hydrocarbon that may be used as a fuel, including coal, oil, natural gas, bitumen from tar sands, and shale oil.

Fossil magnetism   See *Paleomagnetism*.

Fossil succession   Fossil organisms succeed one another in a definite and determinable order, and any time period can be recognized by its fossil content.

Fractional crystallization   The process that separates magma into components having varied compositions and melting points.

Fracture (mineral)   One of the basic physical properties of minerals. It relates to the breakage of minerals when there are no planes of weakness in the crystalline structure. Examples include conchoidal, irregular, and splintery.

Fracture (rock)   Any break or rupture in rock along which no appreciable movement has taken place.

Fracture zone   Linear zone of irregular topography on the deep-ocean floor; it follows transform faults and their inactive extensions.

Fragmental texture   See *Pyroclastic texture*.

Frost wedging   The mechanical breakup of rock caused by the expansion of freezing water in cracks and crevices.

Fumarole   A vent in a volcanic area from which fumes or gases escape.

Gaining stream   Streams that gain water from the inflow of groundwater through the streambed.

Geology   The science that examines Earth, its form and composition, and the changes it has undergone and is undergoing.

Geosphere   The solid Earth; one of Earth's four basic spheres.

Geothermal energy   Natural steam used for power generation.

Geothermal gradient   The gradual increase in temperature with depth in the crust. The average is 30 °C per kilometer in the upper crust.

Geyser   A fountain of hot water ejected periodically from the ground.

Glacial budget   The balance, or lack of balance, between accumulation at the upper end of a glacier and loss at the other end.

Glacial drift   An all-embracing term for sediments of glacial origin, no matter how, where, or in what shape they were deposited.

Glacial erratic   An ice-transported boulder that was not derived from the bedrock near its present site.

Glacial striations   Scratches and grooves on bedrock caused by glacial abrasion.

Glacial trough   A mountain valley that has been widened, deepened, and straightened by a glacier.

Glacier   A thick mass of ice originating on land from the compaction and recrystallization of snow. The ice shows evidence of past or present flow.

Glass (volcanic)   Natural glass produced when molten lava cools too rapidly to permit crystallization. Volcanic glass is a solid composed of unordered atoms.

Glassy texture   A term used to describe the texture of certain igneous rocks, such as obsidian, that contain no crystals.

Gneissic texture   The texture displayed by the metamorphic rock *gneiss* in which dark and light silicate minerals have separated, giving the rock a banded appearance.

Gondwanaland   The southern portion of Pangaea consisting of South America, Africa, Australia, India, and Antarctica.

Graben   A valley formed by the downward displacement of a fault-bounded block.

Graded bed   A sediment layer characterized by a decrease in sediment size from bottom to top.

Graded stream   A stream that has the correct channel characteristics to maintain the exact velocity required to transport the material supplied to it.

Gradient   The slope of a stream; generally measured in feet per mile.

Granitic   Igneous rocks composed mainly of light-colored silicates (quartz and feldspar) are said to have this composition.

Gravitational collapse   The gradual subsidence of mountains caused by lateral spreading of weak material located deep within these structures.

Greenhouse effect   Carbon dioxide and water vapor in a planet's atmosphere absorb and reradiate infrared wavelengths, effectively trapping solar energy and raising the temperature.

Groin   A short wall built at a right angle to the seashore to trap moving sand.

Ground moraine   An undulating layer of till deposited as the ice front retreats.

Groundmass   The matrix of smaller crystals within an igneous rock that has porphyritic texture.

Groundwater   Water in the zone of saturation.

Guyot   A submerged flat-topped seamount.

Habit   The common or characteristic shape of a crystal or aggregate of crystals.

Hadean eon   An informal term that refers to the earliest interval (eon) of Earth history—before the oldest known rocks.

Half-graben   A tilted fault block in which the higher side is associated with mountainous topography and the lower side is a basin that fills with sediment.

Half-life   The time required for one half of the atoms of a radioactive substance to decay.

Hanging valley   A tributary valley that enters a glacial trough at a considerable height above the floor of the trough.

Hardness   A mineral's resistance to scratching and abrasion.

Head   The vertical distance between the recharge and discharge points of a water table. Also, the source area or beginning of a valley.

Headward erosion   The extension upslope of the head of a valley due to erosion.

Historical geology   A major division of geology that deals with the origin of Earth and its development through time. Usually involves the study of fossils and their sequence in rock beds.

Hogback   A narrow, sharp-crested ridge formed by the upturned edge of a steeply dipping bed of resistant rock.

Horn   A pyramidlike peak formed by glacial action in three or more cirques surrounding a mountain summit.

Horst   An elongated, uplifted block of crust bounded by faults.

Hot spot   A proposed concentration of heat in the mantle capable of introducing magma that in turn extrudes onto Earth's surface. The intraplate volcanism that produced the Hawaiian Islands is one example.

Hot spot track   A chain of volcanic structures produced as a lithospheric plate moves over a mantle plume.

Hot spring   A spring in which the water is 6° to 9 °C (10 °C to 15 °F) warmer than the mean annual air temperature of its locality.

Humus   Organic matter in soil produced by the decomposition of plants and animals.

Hydraulic conductivity   A factor relating to groundwater flow; it is a coefficient that takes into account the permeability of the aquifer and the viscosity of the fluid.

Hydraulic gradient   The slope of the water table. It is determined by finding the height difference between the two points on the water table and dividing by the horizontal distance between the two points.

Hydrogenous sediment   Seafloor sediments consisting of minerals that crystallize from seawater. The principal example is manganese nodules.

Hydrologic cycle   The unending circulation of Earth's water supply. The cycle is powered by energy from the Sun and is characterized by continuous exchanges of water among the oceans, the atmosphere, and the continents.

Hydrolysis   A chemical-weathering process in which minerals are altered by chemically reacting with water and acids.

Hydrosphere   The water portion of our planet; one of the traditional subdivisions of Earth's physical environment.

Hydrothermal metamorphism   Chemical alterations that occur as hot, ion-rich water circulates through fractures in rock.

Hydrothermal solution   The hot, watery solution that escapes from a mass of magma during the latter stages of crystallization. Such solutions may alter the surrounding country rock and are frequently the source of significant ore deposits.

Hypothesis   A tentative explanation that is then tested to determine if it is valid.

Ice cap   A mass of glacial ice covering a high upland or plateau and spreading out radially.

Ice sheet   A very large, thick mass of glacial ice flowing outward in all directions from one or more accumulation centers.

Ice-contact deposit   An accumulation of stratified drift deposited in contact with a supporting mass of ice.

Igneous rock   A rock formed by the crystallization of molten magma.

Immature soil   A soil lacking horizons.

Impact metamorphism   Metamorphism that occurs when meteorites strike Earth's surface.

Incised meander   Meandering channel that flows in a steep, narrow valley. These meanders form either when an area is uplifted or when base level drops.

Inclusion   A piece of one rock unit contained within another; inclusions are used in relative dating. The rock mass adjacent to the one containing the inclusion must have been there first in order to provide the fragment.

Index fossil A fossil that is associated with a particular span of geologic time.

Index mineral A mineral that is a good indicator of the metamorphic environment in which it formed. Used to distinguish different zones of regional metamorphism.

Inertia Objects at rest tend to remain at rest, and objects in motion tend to stay in motion unless acted upon by an outside.

Infiltration The movement of surface water into rock or soil through crack and pore spaces.

Infiltration capacity The maximum rate at which soil can absorb water.

Inner core The solid, innermost layer of Earth, about 1216 kilometers (754 miles) in radius.

Inselberg An isolated mountain remnant characteristic of the late stage of erosion in a mountainous region.

Intensity (earthquake) A measure of the degree of earthquake shaking at a given locale based on the amount of damage.

Interface A common boundary where different parts of a system interact.

Interior drainage A discontinuous pattern of intermittent streams that do not flow to the ocean.

Intermediate Compositional category for igneous rocks found near the middle of Bowen's reaction series, mainly amphibole and the intermediate plagioclase feldspars.

Intermediate focus An earthquake focus at a depth of between 60 and 300 kilometers.

Internal process A process such as mountain building or volcanism that derives its energy from Earth's interior and elevates Earth's surface.

Intraplate volcanism Igneous activity that occurs within a tectonic plate away from plate boundaries.

Intrusive rock Igneous rock that formed below Earth's surface.

Ion An atom or molecule that possesses an electrical charge.

Ionic bond A chemical bond between two oppositely charged ions formed by the transfer of valence electrons from one atom to another.

Island arc See *Volcanic island arc*.

Isostasy The concept that Earth's crust is floating in gravitational balance upon the material of the mantle.

Isotopes Varieties of the same element that have different mass numbers; their nuclei contain the same number of protons but different numbers of neutrons.

Joint A fracture in rock along which there has been no movement.

Kame A steep-sided hill, composed of sand and gravel, originating when sediment collected in openings in stagnant glacial ice.

Kame terrace A narrow, terracelike mass of stratified drift deposited between a glacier and an adjacent valley wall.

Karst A topography consisting of numerous depressions called *sinkholes*.

Kettle holes Depressions created when blocks of ice become lodged in glacial deposits and subsequently melt.

Laccolith A massive, concordant igneous body intruded between preexisting strata.

Lahar Mudflows on the slopes of volcanoes that result when unstable layers of ash and debris become saturated and flow downslope, usually following stream channels.

Laminar flow The movement of water particles in straight-line paths that are parallel to the channel. The water particles move downstream without mixing.

Lateral moraine A ridge of till along the sides of a valley glacier composed primarily of debris that fell to the glacier from the valley walls.

Laterite A red, highly leached soil type found in the tropics and rich in oxides of iron and aluminum.

Laurasia The northern portion of Pangaea consisting of North America and Eurasia.

Lava Magma that reaches Earth's surface.

Lava dome A bulbous mass associated with an old-age volcano, produced when thick lava is slowly squeezed from the vent. Lava domes may act as plugs to deflect subsequent gaseous eruptions.

Lava tube Tunnel in hardened lava that acts as a horizontal conduit for lava flowing from a volcanic vent. Lava tubes allow fluid lavas to advance great distances.

Law of superposition In any undeformed sequence of sedimentary rocks or surface-deposited igneous materials, each layer is older than the one above it and younger than the one below.

Leaching The depletion of soluble materials from the upper soil by downward-percolating water.

Light silicate Silicate minerals that lack iron and/or magnesium. They are generally lighter in color and have lower specific gravities than dark silicates.

Lithification The process, generally by cementation and/or compaction, of converting sediments to solid rock.

Lithosphere The rigid outer layer of Earth, including the crust and upper mantle.

Lithospheric plate A coherent unit of Earth's rigid outer layer that includes the crust and upper mantle.

Local (temporary) base level See *Temporary (local) base level*.

Loess Deposits of windblown silt, lacking visible layers, generally buff colored, and capable of maintaining a nearly vertical cliff.

Longitudinal dunes Long ridges of sand oriented parallel to the prevailing wind; these dunes form where sand supplies are limited.

Longitudinal profile A cross section of a stream channel along its descending course from the head to the mouth.

Longshore current A nearshore current that flows parallel to the shore.

Losing stream Streams that lose water to the groundwater system by outflow through the streambed.

Lower mantle The part of the mantle that extends from a depth of 660 kilometers (410 miles) to the top of the core, at a depth of 2900 kilometers (1800 miles).

Luster The appearance or quality of light reflected from the surface of a mineral.

Mafic Because basaltic rocks contain a high percentage of ferromagnesian minerals, they are also called mafic (from *magnesium* and *ferrum*, the Latin name for iron).

Magma A body of molten rock found at depth, including any dissolved gases and crystals.

Magnetic reversal A change in Earth's magnetic field from normal to reverse or vice versa.

Magnetic time scale The detailed history of Earth's magnetic reversals developed by establishing the magnetic polarity of lava flows of known age.

Magnetometer A sensitive instrument used to measure the intensity of Earth's magnetic field at various points.

Magnitude (earthquake) The total amount of energy released during an earthquake.

Manganese nodules A type of hydrogenous sediment scattered on the ocean floor, consisting mainly of manganese and iron, and usually containing small amounts of copper, nickel, and cobalt.

Mantle The 2885-kilometer (1789-mile) thick layer of Earth located below the crust.

Mantle plume A mass of hotter-than-normal mantle material that ascends toward the surface, where it may lead to igneous activity. These plumes of solid yet mobile material may originate as deep as the core–mantle boundary.

Mass number The sum of the number of neutrons and protons in the nucleus of an atom.

Mass wasting The downslope movement of rock, regolith, and soil under the direct influence of gravity.

Massive An igneous pluton that is not tabular in shape.

Meander A looplike bend in the course of a stream.

Meander scar A floodplain feature created when an oxbow lake becomes filled with sediment.

Mechanical weathering The physical disintegration of rock, resulting in smaller fragments.

Medial moraine A ridge of till formed when lateral moraines from two coalescing valley glaciers join.

Melt The liquid portion of magma excluding the solid crystals.

Mesozoic era   A time span on the geologic time scale between the Paleozoic and Cenozoic eras—from about 251 million to 65.5 million years ago.

Metallic bond   A chemical bond present in all metals that may be characterized as an extreme type of electron sharing in which the electrons move freely from atom to atom.

Metamorphic rock   Rock formed by the alteration of preexisting rock deep within Earth (but still in the solid state) by heat, pressure, and/or chemically active fluids.

Metamorphism   The changes in mineral composition and texture of a rock subjected to high temperature and pressure within Earth.

Metasomatism   A significant change in the chemical composition of a rock, usually by the addition or removal of ions in solution.

Microcontinents   Relatively small fragments of continental crust that may lie above sea level, such as the island of Madagascar, or be submerged, as exemplified by the Campbell Plateau located on the seafloor near New Zealand.

Mid-ocean ridge   See *Oceanic ridge*.

Migmatite   A rock exhibiting both igneous and metamorphic rock characteristics. Such rocks may form when light-colored silicate minerals melt and then crystallize, while the dark silicate minerals remain solid.

Mineral   A naturally occurring, inorganic crystalline material with a unique chemical structure.

Mineral resource   All discovered and undiscovered deposits of a useful mineral that can be extracted now or at some time in the future.

Modified Mercalli intensity scale A 12-point scale developed to evaluate earthquake intensity based on the amount of damage to various structures.

Mohorovičić discontinuity (Moho)   The boundary separating the crust and the mantle, discernible by an increase in seismic velocity.

Mohs scale   A series of 10 minerals used as a standard in determining hardness.

Moment magnitude   A more precise measure of earthquake magnitude than the Richter scale, it is derived from the amount of displacement that occurs along a fault zone.

Monocline   A one-limbed flexure in strata. The strata are usually flat lying or very gently dipping on both sides of the monocline.

Mouth   The point downstream where a river empties into another stream or water body.

Mud crack   A feature in some sedimentary rocks that forms when wet mud dries out, shrinks, and cracks.

Mudflow   See *Debris flow*.

Natural levees   The elevated landforms composed of alluvium that parallel some streams and act to confine their waters, except during floodstage.

Neap tide   The lowest tidal range, occurring near the times of the first and third quarters of the Moon.

Nearshore zone   The zone of a beach that extends from the low-tide shoreline seaward to where waves break at low tide.

Nebular theory   A model for the origin of the solar system that assumes a rotating nebula of dust and gases that contracted to produce the Sun and planets.

Negative feedback mechanism As used in climate change, any effect that is opposite of the initial change and tends to offset it.

Neutron   A subatomic particle found in the nucleus of an atom. The neutron is electrically neutral, with a mass approximately equal to that of a proton.

Nonclastic   A term for the texture of sedimentary rocks in which the minerals form a pattern of interlocking crystals.

Nonconformity   An unconformity in which older metamorphic or intrusive igneous rocks are overlain by younger sedimentary strata.

Nonferromagnesian silicate   See *Light silicate*.

Nonfoliated texture Metamorphic rocks that do not exhibit foliation.

Nonmetallic mineral resource Mineral resource that is not a fuel or processed for the metals it contains.

Nonrenewable resource Resource that forms or accumulates over such long time spans that it must be considered as fixed in total quantity.

Normal fault   A fault in which the rock above the fault plane has moved down relative to the rock below.

Normal polarity   A magnetic field the same as that which presently exists.

Nucleus   The small, heavy core of an atom that contains all of its positive charge and most of its mass.

Nuée ardente   Incandescent volcanic debris that is buoyed up by hot gases and moves downslope in an avalanche fashion.

Numerical date   Date that specifies the actual number of years that have passed since an event occurred.

Oceanic plateau   An extensive region on the ocean floor composed of thick accumulations of pillow basalts and other mafic rocks that in some cases exceed 30 kilometers in thickness.

Oceanic ridge   A continuous mountainous ridge on the floor of all the major ocean basins and varying in width from 500 to 5000 kilometers (300 to 3000 miles). The rifts at the crests of these ridges represent divergent plate boundaries.

Octet rule   Atoms combine in order that each may have the electron arrangement of a noble gas; that is, the outer energy level contains eight electrons.

Offshore zone   The relatively flat submerged zone that extends from the breaker line to the edge of the continental shelf.

Oil trap   A geologic structure that allows for significant amounts of oil and gas to accumulate.

Ore   Usually a useful metallic mineral that can be mined at a profit. The term is also applied to certain nonmetallic minerals such as fluorite and sulfur.

Organic sedimentary rock Sedimentary rock composed of organic carbon from the remains of plants that died and accumulated on the floor of a swamp. Coal is the primary example.

Original horizontality   Layers of sediment are generally deposited in a horizontal or nearly horizontal position.

Orogenesis   The processes that collectively result in the formation of mountains.

Outer core   A layer beneath the mantle about 2270 kilometers (1410 miles) thick that has the properties of a liquid.

Outgassing   The release of gases dissolved in molten rock.

Outlet glacier   A tongue of ice normally flowing rapidly outward from an ice cap or ice sheet, usually through mountainous terrain to the sea.

Outwash plain   A relatively flat, gently sloping plain consisting of materials deposited by meltwater streams in front of the margin of an ice sheet.

Oxbow lake   A curved lake produced when a stream cuts off a meander.

Oxidation   The removal of one or more electrons from an atom or ion. So named because elements commonly combine with oxygen.

Oxygen isotope analysis   A method of deciphering past temperatures based on the precise measurement of the ratio between two isotopes of oxygen, ^{16}O and ^{18}O. Analysis is commonly made of seafloor sediments and cores of glacial ice.

P wave   The fastest earthquake wave; travels by compression and expansion of the medium.

Pahoehoe   A lava flow with a smooth-to-ropy surface.

Paleoclimatology   The study of ancient climates; the study of climate and climate change prior to the period of instrumental records using proxy data.

Paleomagnetism   The natural remnant magnetism in rock bodies. The permanent magnetization acquired by rock that can be used to determine the location of the magnetic poles and the latitude of the rock at the time it became magnetized.

Paleontology   The systematic study of fossils and the history of life on Earth.

Paleozoic era A time span on the geologic time scale between the Proterozoic and Mesozoic eras–from about 542 million to 251 million years ago.

Pangaea The proposed supercontinent that 200 million years ago began to break apart and form the present landmasses.

Parabolic dune A sand dune similar in shape to a barchan dune except that its tips point into the wind. These dunes often form along coasts that have strong onshore winds, abundant sand, and vegetation that partly covers the sand.

Paradigm Theory that is held with a very high degree of confidence and is comprehensive in scope.

Parasitic cone A volcanic cone that forms on the flank of a larger volcano.

Parent material The material upon which a soil develops.

Parent rock The rock from which a metamorphic rock formed.

Partial melting The process by which most igneous rocks melt. Because individual minerals have different melting points, most igneous rocks melt over a temperature range of a few hundred degrees. If the liquid is squeezed out after some melting has occurred, a melt with a higher silica content results.

Passive continental margin A margin that consists of a continental shelf, continental slope, and continental rise. They are *not* associated with plate boundaries and therefore experience little volcanism and few earthquakes.

Pater noster lakes A chain of small lakes in a glacial trough that occupy basins created by glacial erosion.

Pegmatite A very coarse-grained igneous rock (typically granite) commonly found as a dike associated with a large mass of plutonic rock that has smaller crystals. Crystallization in a water-rich environment is believed to be responsible for the very large crystals.

Pegmatitic texture A texture of igneous rocks in which the interlocking crystals are all larger than one centimeter in diameter.

Perched water table A localized zone of saturation above the main water table created by an impermeable layer (aquitard).

Peridotite An igneous rock of ultramafic composition thought to be abundant in the upper mantle.

Period A basic unit of the geologic calendar that is a subdivision of an *era*. Periods may be divided into smaller units called *epochs*.

Periodic table The tabular arrangement of the elements according to atomic number.

Permafrost Any permanently frozen subsoil. Usually found in the subarctic and arctic regions.

Permeability A measure of a material's ability to transmit water.

Phaneritic texture An igneous rock texture in which the crystals are roughly equal in size and large enough so that individual minerals can be identified with the unaided eye.

Phanerozoic eon That part of geologic time represented by rocks containing abundant fossil evidence. The eon extending from the end of the Proterozoic eon (about 540 million years ago) to the present.

Phenocryst Conspicuously large crystals in a porphyry that are imbedded in a matrix of finer-grained crystals (the groundmass).

Physical geology A major division of geology that examines the materials of Earth and seeks to understand the processes and forces acting upon Earth's surface from below.

Piedmont glacier A glacier that forms when one or more valley glaciers emerge from the confining walls of mountain valleys and spread out to create a broad sheet in the lowlands at the base of the mountains.

Pillow lava Basaltic lava that solidifies in an underwater environment and develops a structure that resembles a pile of pillows.

Pipe A vertical conduit through which magmatic materials have passed.

Placer Deposit formed when heavy minerals are mechanically concentrated by currents, most commonly streams and waves. Placers are sources of gold, tin, platinum, diamonds, and other valuable minerals.

Planetesimal A solid celestial body that accumulated during the first stages of planetary formation. Planetesimals aggregated into increasingly larger bodies, ultimately forming the planets.

Plastic deformation Permanent deformation that results in a change in size and shape through folding or flowing.

Plastic flow A type of glacial movement that occurs within the glacier, below a depth of approximately 50 meters, in which the ice is not fractured.

Plate See *Lithospheric plate*.

Plate tectonics The theory that proposes Earth's outer shell consists of individual plates, which interact in various ways and thereby produce earthquakes, volcanoes, mountains, and the crust itself.

Playa The flat central area of an undrained desert basin.

Playa lake A temporary lake in a playa.

Pleistocene epoch An epoch of the Quaternary period beginning about 1.8 million years ago and ending about 10,000 years ago. Best known as a time of extensive continental ice sheets.

Plucking The process by which pieces of bedrock are lifted out of place by a glacier.

Pluton A structure that results from the emplacement and crystallization of magma beneath Earth's surface.

Pluvial lake A lake formed during a period of increased rainfall. For example, this occurred in many nonglaciated areas during periods of ice advance elsewhere.

Point bar A crescent-shaped accumulation of sand and gravel deposited on the inside of a meander.

Polar wandering hypothesis As the result of paleomagnetic studies in the 1950s, researchers proposed that either the magnetic poles migrated greatly through time or the continents gradually shifted their positions.

Polymerization The process of linking the same molecules together to form a chain or three-dimensional structure.

Polymorphs Two or more minerals having the same chemical composition but different crystalline structures, exemplified by the diamond and graphite forms of carbon.

Porosity The volume of open spaces in rock or soil.

Porphyritic texture An igneous rock texture characterized by two distinctively different crystal sizes. The larger crystals are called *phenocrysts*, and the matrix of smaller crystals is termed the *groundmass*.

Porphyroblastic texture A texture of metamorphic rocks in which particularly large grains (porphyroblasts) are surrounded by a fine-grained matrix of other minerals.

Porphyry An igneous rock with a porphyritic texture.

Positive-feedback mechanism As used in climate change, any effect that acts to reinforce the initial change.

Pothole A depression formed in a stream channel by the abrasive action of the water's sediment load.

Precambrian All geologic time prior to the Paleozoic era.

Principal shells See *Energy levels*.

Principle of fossil succession Fossil organisms succeed one another in a definite and determinable order, and any time period can be recognized by its fossil content.

Principle of original horizontality Layers of sediment are generally deposited in a horizontal or nearly horizontal position.

Proterozoic eon The eon following the Archean and preceding the Phanerozoic. It extends between 2.5 billion and 542 million years ago.

Proton A positively charged subatomic particle found in the nucleus of an atom.

Protoplanet A developing planetary body that grows by the accumulation of planetesimals.

Proxy data Data gathered from natural recorders of climate variability such as tree rings, ice cores, pollen, and ocean-floor sediments.

Pyroclastic flow A highly heated mixture, largely of ash and pumice fragments, traveling down the flanks of a volcano or along the surface of the ground.

Pyroclastic material The volcanic rock ejected during an eruption. Pyroclastics include ash, bombs, and blocks.

Pyroclastic texture An igneous rock texture resulting from the consolidation of individual rock fragments that are ejected during a violent eruption.

Quarrying The removal of loosened blocks from the bed of a channel during times of high flow rates.

Radial drainage pattern A system of streams running in all directions away from a central elevated structure, such as a volcano.

Radioactivity The spontaneous decay of certain unstable atomic nuclei.

Radiocarbon (carbon-14) The radioactive isotope of carbon produced continuously in the atmosphere and used in dating events as far back as 75,000 years.

Radiometric dating The procedure of calculating the absolute ages of rocks and minerals containing certain radioactive isotopes.

Rainshadow desert A dry area on the lee side of a mountain range; many middle-latitude deserts are of this type.

Rapids A part of a stream channel in which the water suddenly begins flowing more swiftly and turbulently because of an abrupt steepening of the gradient.

Recessional moraine An end moraine formed as the ice front stagnated during glacial retreat.

Rectangular pattern A drainage pattern that develops on jointed or fractured bedrock and is characterized by numerous right-angle bends.

Refraction A change in direction of waves as they enter shallow water. The portion of the wave in shallow water is slowed, which causes the wave to bend and align with the underwater contours.

Regional metamorphism Metamorphism associated with large-scale mountain building.

Regolith The layer of rock and mineral fragments that nearly everywhere covers Earth's land surface.

Relative dating Rocks are placed in their proper sequence or order; only the chronological order of events is determined.

Renewable resource A resource that is virtually inexhaustible or that can be replenished over relatively short time spans.

Reserve Already identified deposits from which minerals can be extracted profitably.

Reservoir rock The porous, permeable portion of an oil trap that yields oil and gas.

Residual soil Soil developed directly from the weathering of the bedrock below.

Reverse fault A fault in which the material above the fault plane moves up in relation to the material below.

Reverse polarity A magnetic field opposite to that which presently exists.

Richter scale A scale of earthquake magnitude based on the motion of a seismograph.

Ridge push A mechanism that may contribute to plate motion. It involves the oceanic lithosphere sliding down the oceanic ridge under the pull of gravity.

Rift A region of Earth's crust along which divergence (separation) is taking place.

Rip current A strong, narrow surface or near-surface current of short duration and high speed that moves seaward through the breaker zone at nearly a right angle to the shore.

Ripple marks Small waves of sand that develop on the surface of a sediment layer by the action of moving water or air.

River A general term for a stream that carries a substantial amount of water and has numerous tributaries.

Roche moutonnée An asymmetrical knob of bedrock formed when glacial abrasion smoothes the gentle slope facing the advancing ice sheet and plucking steepens the opposite side as the ice overrides the knob.

Rock A consolidated mixture of minerals.

Rock avalanche The very rapid downslope movement of rock and debris. These rapid movements may be aided by a layer of air trapped beneath the debris, and they have been known to reach speeds in excess of 200 kilometers per hour.

Rock cleavage The tendency of rock to split along parallel, closely spaced surfaces. These surfaces are often highly inclined to the bedding planes in the rock.

Rock cycle A model that illustrates the origin of the three basic rock types and the interrelatedness of Earth's materials and processes.

Rock flour Ground-up rock produced by the grinding effect of a glacier.

Rockslide The rapid slide of a mass of rock downslope along planes of weakness.

Runoff Water that flows over the land rather than infiltrating into the ground.

S wave An earthquake wave, slower than a P wave, that travels only in solids.

Salt flat A white crust on the ground produced when water evaporates and leaves its dissolved materials behind.

Saltation Transportation of sediment through a series of leaps or bounces.

Schistosity A type of foliation characteristic of coarser-grained metamorphic rocks. Such rocks have a parallel arrangement of platy minerals such as the micas.

Scoria Hardened lava that has retained the vesicles produced by escaping gases.

Scoria cone See *Cinder cone*.

Sea arch An arch formed by wave erosion when caves on opposite sides of a headland unite.

Sea stack An isolated mass of rock standing just offshore, produced by wave erosion of a headland.

Seafloor spreading The hypothesis first proposed in the 1960s by Harry Hess, suggesting that new oceanic crust is produced at the crests of mid-ocean ridges, which are the sites of divergence.

Seamount An isolated volcanic peak that rises at least 1000 meters (3300 feet) above the deep-ocean floor.

Seawall A barrier constructed to prevent waves from reaching the area behind the wall. Its purpose is to defend property from the force of breaking waves.

Secondary enrichment The concentration of minor amounts of metals that are scattered through unweathered rocks into economically valuable concentrations by weathering processes.

Secondary (S) wave A seismic wave that involves oscillation perpendicular to the direction of propagation.

Sediment Unconsolidated particles created by the weathering and erosion of rock, by chemical precipitations from solution in water, or from the secretions of organisms, and transported by water, wind, or glaciers.

Sedimentary rock Rock formed from the weathered products of preexisting rocks that have been transported, deposited, and lithified.

Seiche The rhythmic sloshing of water in lakes, reservoirs, and other smaller enclosed basins. Some seiches are initiated by earthquake activity.

Seismic gap A segment of an active fault zone that has not experienced a major earthquake over a span when most other segments have. Such segments are probable sites for future major earthquakes.

Seismic sea wave A rapidly moving ocean wave generated by earthquake activity and capable of inflicting heavy damage in coastal regions.

Seismogram The record made by a seismograph.

Seismograph An instrument that records earthquake waves.

Seismology The study of earthquakes and seismic waves.

Settling velocity   The speed at which a particle falls through a still fluid. The size, shape, and specific gravity of particles influence settling velocity.

Shadow zone   The zone between 105 and 140 degrees distance from an earthquake epicenter that direct waves do not penetrate because of refraction by Earth's core.

Shallow-focus earthquake   An earthquake focus at a depth of less than 60 kilometers.

Shear   Stress that causes two adjacent parts of a body to slide past one another.

Sheet flow   Runoff moving in unconfined thin sheets.

Sheeting   A mechanical weathering process characterized by the splitting off of slablike sheets of rock.

Shelf break   The point at which a rapid steepening of the gradient occurs, marking the outer edge of the continental shelf and the beginning of the continental slope.

Shield   A large, relatively flat expanse of ancient metamorphic rock within the stable continental interior.

Shield volcano   A broad, gently sloping volcano built from fluid basaltic lavas.

Shock metamorphism   Metamorphic alteration of rock by high speed impact of an extraterrestrial body such as a meteorite.

Shore   Seaward of the coast, this zone extends from the highest level of wave action during storms to the lowest tide level.

Shoreline   The line that marks the contact between land and sea; it migrates up and down as the tide rises and falls.

Silicate   Any one of numerous minerals that have the silicon-oxygen tetrahedron as their basic structure.

Silicon-oxygen tetrahedron   A structure composed of four oxygen atoms surrounding a silicon atom that constitutes the basic building block of silicate minerals.

Sill   A tabular igneous body that was intruded parallel to the layering of preexisting rock.

Sinkhole   A depression produced in a region where soluble rock has been removed by groundwater.

Slab pull   A mechanism that contributes to plate motion in which cool, dense oceanic crust sinks into the mantle and "pulls" the trailing lithosphere along.

Slaty cleavage   The type of foliation characteristic of slates in which there is a parallel arrangement of fine-grained metamorphic minerals.

Slide   A movement common to mass-wasting processes in which the material moving downslope remains fairly coherent and moves along a well-defined surface.

Slip face   The steep, leeward surface of a sand dune that maintains a slope of about 34 degrees.

Slump   The downward slipping of a mass of rock or unconsolidated material moving as a unit along a curved surface.

Snowfield   An area where snow persists year-round.

Snowline   Lower limit of perennial snow.

Soil   A combination of mineral and organic matter, water, and air; that portion of the regolith that supports plant growth.

Soil horizon   A layer of soil that has identifiable characteristics produced by chemical weathering and other soil-forming processes.

Soil profile   A vertical section through a soil showing its succession of horizons and the underlying parent material.

Soil Taxonomy   A soil classification system consisting of six hierarchical categories based on observable soil characteristics. The system recognizes 12 soil orders.

Solar nebula   The cloud of interstellar gas and/or dust from which the bodies of our solar system formed.

Solifluction   Slow, downslope flow of water-saturated materials common to permafrost areas.

Solum   The O, A, and B horizons in a soil profile; living roots and other plant and animal life are largely confined to this zone.

Solution   The change of matter from the solid or gaseous state into the liquid state by its combination with a liquid.

Sorting   The degree of similarity in particle size in sediment or sedimentary rock.

Specific gravity   The ratio of a substance's weight to the weight of an equal volume of water.

Speleothem   A collective term for the dripstone features found in caverns.

Spheroidal weathering   Any weathering process that tends to produce a spherical shape from an initially blocky shape.

Spit   An elongated ridge of sand that projects from the land into the mouth of an adjacent bay.

Spreading center   See *Divergent plate boundary.*

Spring   A flow of groundwater that emerges naturally at the ground surface.

Spring tide   The highest tidal range; occurs near the times of the new and full moons.

Stable platform   That part of the craton that is mantled by relatively undeformed sedimentary rocks and underlain by a basement complex of igneous and metamorphic rocks.

Stalactite   The iciclelike structure that hangs from the ceiling of a cavern.

Stalagmite   The columnlike form that grows upward from the floor of a cavern.

Star dune   Isolated hill of sand that exhibits a complex form and develops where wind conditions are variable.

Steppe   One of the two types of dry climate. A marginal and more humid variant of the desert that separates the desert from bordering humid climates.

Stock   A pluton similar to but smaller than a batholith.

Strata   Parallel layers of sedimentary rock.

Stratified drift   Sediments deposited by glacial meltwater.

Stratovolcano   See *Composite cone.*

Streak   The color of a mineral in powdered form.

Stream   A general term to denote the flow of water within any natural channel. Thus, a small creek and a large river are both streams.

Stream valley   The channel, valley floor, and the sloping valley walls of a stream.

Stress   The force per unit area acting on any surface within a solid.

Striations   The multitude of fine parallel lines found on some cleavage faces of plagioclase feldspars but not present on orthoclase feldspar.

Striations (glacial)   Scratches or grooves in a bedrock surface caused by the grinding action of a glacier and its load of sediment.

Strike   The compass direction of the line of intersection created by a dipping bed or fault and a horizontal surface. Strike is always perpendicular to the direction of dip.

Strike-slip fault   A fault along which the movement is horizontal.

Stromatolite   Structures that are deposited by algae and that consist of layered mounds or columns of calcium carbonate.

Subduction   The process of thrusting oceanic lithosphere into the mantle along a convergent zone.

Subduction zone   A long, narrow zone where one lithospheric plate descends beneath another.

Subduction zone metamorphism   High-pressure, low-temperature metamorphism that occurs where sediments are carried to great depths by a subducting plate.

Submarine canyon   A seaward extension of a valley that was cut on the continental shelf, slope, and rise by turbidity currents.

Submergent coast   A coast whose form is largely the result of the partial drowning of a former land surface due either to a rise of sea level or subsidence of the crust, or both.

Subsoil   A term applied to the B horizon of a soil profile.

Supercontinent   A huge landmass that consists of all, or nearly all, of the existing continents combined into one.

Supercontinent cycle   The idea that the rifting and dispersal of one supercontinent is followed by a long period during which the fragments gradually reassemble into a new supercontinent.

Supernova   An exploding star that increases its brightness many thousands of times.

Superposition, law of   In any undeformed sequence of sedimentary rocks, each bed is older than the one above and younger than the one below.

Surf   A collective term for breakers; also the wave activity in the area between the shoreline and the outer limit of breakers.

Surface soil   The upper portion of a soil profile consisting of the *O* and *A* horizons.

Surface waves   Seismic waves that travel along the outer layer of Earth.

Surge   A period of rapid glacial advance; surges are typically sporadic and short-lived.

Suspended load   The fine sediment carried within the body of flowing water or air.

Swells   Wind-generated waves that have moved into an area of weaker winds or calm.

Syncline   A linear downfold in sedimentary strata; the opposite of anticline.

System   A group of interacting or interdependent parts that form a complex whole.

Tablemount   See *Guyot.*

Tabular   Describing a feature such as an igneous pluton having two dimensions that are much longer than the third.

Talus   An accumulation of rock debris at the base of a cliff.

Tarn   A small lake in a cirque.

Tectonic plate   See *Lithospheric plate.*

Tectonics   The study of the large-scale processes that collectively deform Earth's crust.

Temporary (local) base level   The level of a lake, resistant rock layer, or any other base level that stands above sea level.

Tenacity   Describes a mineral's toughness or its resistance to breaking or deforming.

Terminal moraine   The end moraine marking the farthest advance of a glacier.

Terrace   A flat, benchlike structure produced by a stream that was left elevated as the stream cut downward.

Terrane   A crustal block bounded by faults whose geologic history is distinct from the histories of adjoining crustal blocks.

Terrigenous sediment   Seafloor sediment derived from terrestrial weathering and erosion.

Texture   The size, shape, and distribution of the particles that collectively constitute a rock.

Theory   A well-tested and widely accepted view that explains certain observable facts.

Thermal metamorphism   See *Contact metamorphism.*

Thrust fault   A low-angle reverse fault.

Tide   Periodic change in the elevation of the ocean's surface.

Till   Unsorted sediment deposited directly by a glacier.

Tillite   A rock formed when glacial till is lithified.

Tombolo   A ridge of sand that connects an island to the mainland or to another island.

Topset bed   An essentially horizontal sedimentary layer deposited on top of a delta during floodstage.

Transform fault   A major strike-slip fault that cuts through the lithosphere and accommodates motion between two plates.

Transform fault boundary   A boundary in which two plates slide past one another without creating or destroying lithosphere.

Transpiration   The release of water vapor to the atmosphere by plants.

Transported soil   Soils that form on unconsolidated deposits.

Transverse dunes   A series of long ridges oriented at right angles to the prevailing wind; these dunes form where vegetation is sparse and sand is very plentiful.

Travertine   A form of limestone that is deposited by hot springs or as a cave deposit.

Trellis drainage   A system of streams in which nearly parallel tributaries occupy valleys cut in folded strata.

Trench   An elongate depression in the seafloor produced by bending of oceanic crust during subduction.

Tsunami   The Japanese word for a seismic sea wave.

Turbidite   Turbidity current deposit characterized by graded bedding.

Turbidity current   A downslope movement of dense, sediment-laden water created when sand and mud on the continental shelf and slope are dislodged and thrown into suspension.

Turbulent flow   The movement of water in an erratic fashion often characterized by swirling, whirlpool-like eddies. Most streamflow is of this type.

Ultimate base level   Sea level; the lowest level to which stream erosion could lower the land.

Ultramafic   Compositional category for igneous rocks made up almost entirely of ferromagnesian minerals (mostly olivine and pyroxene).

Unconformity   A surface that represents a break in the rock record; caused by erosion or nondeposition.

Uniformitarianism   The concept that the processes that have shaped Earth in the geologic past are essentially the same as those operating today.

Unsaturated zone   The area above the water table where openings in soil, sediment, and rock are not saturated with water but filled mostly with air.

Valence electron   The electrons involved in the bonding process; the electrons occupying the highest principal energy level of an atom.

Valley glacier   A glacier confined to a mountain valley, which in most instances had previously been a stream valley.

Valley train   A relatively narrow body of stratified drift deposited on a valley floor by meltwater streams that issue from the terminus of a valley glacier.

Vein deposit   A mineral filling a fracture or fault in a host rock. Such deposits have a sheetlike, or tabular, form.

Ventifact   A cobble or pebble polished and shaped by the sandblasting effect of wind.

Vesicles   Spherical or elongated openings on the outer portion of a lava flow that were created by escaping gases.

Vesicular texture   A term applied to aphanitic igneous rocks that contain many small cavities, called *vesicles.*

Viscosity   A measure of a fluid's resistance to flow.

Volatiles   Gaseous components of magma dissolved in the melt. Volatiles will readily vaporize (form a gas) at surface pressures.

Volcanic   Pertaining to the activities, structures, or rock types of a volcano.

Volcanic bomb   A streamlined pyroclastic fragment ejected from a volcano while the fragment is still molten.

Volcanic island arc   A chain of volcanic islands generally located a few hundred kilometers from a trench where there is active subduction of one oceanic plate beneath another.

Volcanic neck   An isolated, steep-sided, erosional remnant consisting of lava that once occupied the vent of a volcano.

Volcano   A mountain formed from lava and/or pyroclastics.

Wash   A desert stream course that is typically dry except for brief periods immediately following rainfall.

Water gap   A pass through a ridge or mountain in which a stream flows.

Water table   The upper level of the saturated zone of groundwater.

Wave-cut cliff   A seaward-facing cliff along a steep shoreline formed by wave erosion at its base and by mass wasting.

Wave-cut platform   A bench or shelf along a shore at sea level, cut by wave erosion.

Wave height   The vertical distance between the trough and crest of a wave.

Wave length   The horizontal distance separating successive crests or troughs.

Wave period   The time interval between the passage of successive crests at a stationary point.

Weathering   The disintegration and decomposition of rock at or near the surface of the Earth.

Welded tuff   A pyroclastic deposit composed of particles fused together by the combination of heat still contained in the deposit after it has come to rest and the weight of overlying material.

Well   An opening bored into the zone of saturation.

Wetted perimeter   The total distance in a linear cross-section of a stream that is in contact with water.

Wind gap   An abandoned water gap. These gorges typically result from stream piracy.

Xenolith   An inclusion of unmelted country rock in an igneous pluton.

Xerophyte   A plant highly tolerant of drought.

Yardang   A streamlined, wind-sculpted ridge having the appearance of an inverted ship's hull that is oriented parallel to the prevailing wind.

Yazoo tributary   A tributary that flows parallel to the main stream because a natural levee is present.

Zone of accumulation   The part of a glacier characterized by snow accumulation and ice formation.

The outer limit of this zone is the *snowline.*

Zone of aeration   See *Unsaturated zone.*

Zone of fracture   The upper portion of a glacier consisting of brittle ice.

Zone of saturation   Zone where all open spaces in sediment and rock are completely filled with water.

Index

Great Permian Extinction,
481–483
Greenhouse effect, 499, 504
Greenhouse gas, 16
Greenland's glacial budget, 269
Groins, 321
Gros Ventre (rockslide), 209
Ground moraines, 275
Ground subsidence, 348–349
Groundwater, 238–259
aquifers, 245
aquitards, 245
contamination, 251–252
distribution of, 240, 243
environmental problems
associated with, 250–252
flow system, 246
geologic work of, 255–258
See also Caverns
importance of, 240–243
Karst topography, 256–258
movement
factors influencing, 245–249
measuring, 246–247
as nonrenewable resource, 250
occupancy, 241
perched water table, 247
permeability, 245
porosity, 245
springs, 247
storage, factors influencing,
245–249
streams and, interaction
between, 244–245
gaining streams, 244
losing stream
(connected), 244
losing stream
(disconnected), 244
transpiration, 246
unsaturated zone, 243
use, 242
withdrawal, land subsidence
caused by, 250–251
zone of saturation, 243
See also Wells
Gulf coast, 324
Gullies, 142
Guyots, 399–400
Gypsum, 57
uses, 169

H

Hadean, 455
Half-grabens, 422
Half-life, 451–452
Halite, uses, 169
Hanging valley, 271
Hanging wall block, 422
Hard stabilization, shorelines,
320–322
alternatives to, 322

beach nourishment, 322
coarse calcareous sand, 323
muddier calcareous
sand, 323
breakwaters, 321
groins, 321
jetties, 320
seawalls, 321
Hardness, 46
Hawaiian Island–Emperor
Seamount chain, 380
Heat
changes caused by, 178
geothermal gradient, 179
as a metamorphic agent,
178–179
source, 179
Heating the atmosphere,
502–505
electromagnetic spectrum, 502
energy from the Sun, 502–503
fate of incoming solar energy,
503–504
longwave radiation, 502
shortwave radiation, 502
Heavy rains, 328
High-resolution multibeam, 395
Himalayas, 429–430
Historical geology, 2
Histosols, 139
HMS *Beagle*, 401
HMS *Challenger*, 394
Hogbacks, 420
Homo sapiens, 489
Horizons, 137
Horizontal tidal flow, 331
Hornblende, 55
Horns, 272–273
Horsts, 422
Hot spots, 117, 379–380
Hot springs, 249, 252–254
Hubbard Glacier, 268
Hurricanes, 324–329
destruction, 326–327
profile of, 325–326
Hydraulic conductivity, 247
Hydraulic gradient, 246
Hydrologic cycle, 216–217
distribution of Earth's
water, 217
Earth system, 16
glaciers, 262
infiltration, 216
Hydrosphere, 12–13, 216
fresh water of, 240
Hydrothermal (hot-water)
solution, 86
Hydrothermal metamorphism,
189
Hydrothermal processes, 407
disseminated deposits, 86
and metal deposits, 86
and pegmatite, 84
Hypothesis, 8

I

Ice, melting, 264–265
Ice Age Glaciers, 278–280
crustal subsidence, 278
ice dams create proglacial
lakes, 279–280
rebound, 278
rivers before and after,
278–279
sea-level changes, 278
Ice caps, 264
Ice-jam floods, 233–234
Ice sheets, 263
Ice shelves, 263
Ichthyosaurs, 484
Igneous rocks, 18, 20
andesite, 74
andesitic composition, 67
aphanitic, 68
basaltic composition, 67
basic igneous structures, 81
Bowen's reaction series, 78–80
classification of, 71
compositions, 65
crystallization in, 65
diorite, 74
felsic (granitic), 71–72
fragmental, 70
gabbro, 75
glassy, 69–70
granitic composition, 66
intrusive igneous activity, 80–84
mafic (basaltic), 74–75
metallic minerals
occurrences, 85
mineral resources, 84–86
naming of, 71
pegmatitic, 70–71
peridotite, 67
phaneritic, 69
porphyritic, 69
processes, 65
pumice, 74
pyroclastic, 70, 75
rhyolite, 73
silica content in, 67–71
textures, 67
ultramafic, 67
vesicular, 69
See also Magma
Immature soils, 138
Impactiles, 191
Impressions, 447
Inceptisols, 139
Incised meanders, 229–230
Inclusions, 441–442
Index fossils, 448
Index minerals, 192
Inertia, 340
Infiltration, 216
Inland flooding, 327–328
Inner core, 24
Inorganic limestones, 158

Inorganic process, 156
Inselberg, 294
Integrated Ocean Drilling
Program (IODP), 379
Interface, 16, 308
Interior drainage, 293
International Commission on
Stratigraphy (ICS), 456
International Union of
Geological Sciences, 456
Intraplate volcanism, 117
Intrusions, 81
Ionic bond, 42
Ionic compounds, 41
Island arc, 374
Island archipelagos, 116
Isostasy, 432
isostatic adjustment, 432
Isotopes, 44, 450
radioactive decay, 44

J

Jasper, 159
Jetties, 320
JOIDES Resolution, 379
Joints, 127, 424
Jurassic, 474

K

Kames, 277–278
Kaolinite, 54
Karats, 40
Karst topography, 256–258
Kettles, 275–277
Kimberlite, 86
Kutiah Glacier, 266

L

Laccoliths, 83
Lahars
hazards, 118
mass wasting, 209–210
Laki fissure eruption, 111
Laminae, 153
Laminar flow, 219
'Land of fire and ice', 264
Land subsidence groundwater
withdrawal causing,
250–251
Landforms created by glacial
erosion, 270–274
Landslides, 348–349
Blackhawk (California), 207
in Kashmir, 198
at La Conchita (California),
204
as natural disasters, 198
on Oahu (Hawaii), 204
in Pacific Palisades
(California), 198